Thin-Walled Composite Beams

SOLID MECHANICS AND ITS APPLICATIONS
Volume 131

Series Editor: **G.M.L. GLADWELL**
Department of Civil Engineering
University of Waterloo
Waterloo, Ontario, Canada N2L 3GI

Aims and Scope of the Series

The fundamental questions arising in mechanics are: *Why?, How?,* and *How much?*
The aim of this series is to provide lucid accounts written by authoritative researchers
giving vision and insight in answering these questions on the subject of mechanics as it
relates to solids.

The scope of the series covers the entire spectrum of solid mechanics. Thus it includes
the foundation of mechanics; variational formulations; computational mechanics;
statics, kinematics and dynamics of rigid and elastic bodies: vibrations of solids and
structures; dynamical systems and chaos; the theories of elasticity, plasticity and
viscoelasticity; composite materials; rods, beams, shells and membranes; structural
control and stability; soils, rocks and geomechanics; fracture; tribology; experimental
mechanics; biomechanics and machine design.

The median level of presentation is the first year graduate student. Some texts are
monographs defining the current state of the field; others are accessible to final year
undergraduates; but essentially the emphasis is on readability and clarity.

For a list of related mechanics titles, see final pages.

Thin-Walled Composite Beams

Theory and Application

by

LIVIU LIBRESCU

Virginia Polytechnic and State University,
Blacksburg, VA, U.S.A.

and

OHSEOP SONG

Chungnam National University,
Daejeon, Korea

 Springer

A C.I.P. Catalogue record for this book is available from the Library of Congress.

ISBN-10 1-4020-3457-1 (HB)
ISBN-13 978-1-4020-3457-2 (HB)
ISBN-10 1-4020-4203-5 (e-book)
ISBN-13 978-1-4020-4203-4 (e-book)

Published by Springer,
P.O. Box 17, 3300 AA Dordrecht, The Netherlands.

www.springeronline.com

Printed on acid-free paper

Printed in the Netherlands.

Liviu Librescu: To my wife Marilena Librescu, M. D., for her devotion, love, support and encouragement. Her patience and her contributions to technical matters in the book are greatly appreciated.

Ohseop Song: To my wife Bokyong Chun for her endless love and support.

Motto

I will bless the Lord at all times; his praise shall continually
be in my mouth.
David, Psalm 34 verse 2.

To the glory of God who inspired us to believe, and gave
us the ability to think and the wisdom to write this book
for others to read and to use. Blessed be the Lord. It is
with gratitude and happiness that we serve the community.

Thanks be to God.

Liviu Librescu and Ohseop Song

CONTENTS

10. Spinning Thin-Walled Anisotropic Beams   395

PREFACE

There has been a growing interest in the foundation of the theory of thin-walled composite beams and of their incorporation in aeronautical/aerospace, automotive, helicopter and turbomachinery rotor blades, mechanical, civil and naval constructions in the last two decades or so.

The proliferation of the specialized literature, mainly in the form of journal/proceedings papers, and the activity in terms of workshops devoted to this topic attest this interest. A decisive factor that has fueled this growing activity was generated by high diversity and severity of demands and operating conditions imposed on structural elements involved in the advanced technology. In order to be able to survive and fulfill their mission in the extreme environmental conditions in which they operate, new materials and new structural paradigms are required.

The new exotic structures have to provide higher performances, unattainable by the classical structures built of traditional materials. The advent of advanced composite materials, of smart materials and functionally graded materials (FGMs), have constituted the strongest stimuli for such developments. Moreover, their incorporation is likely to expand the use and capabilities of thin-walled beam structures. The new and stringent requirements imposed on aeronautical/aerospace, turbomachinery and shaft structural systems will be best met by such new types of material structures.

However, incorporation of these new material structures in the various areas of advanced technology and the solution of many challenging problems involving their static/dynamic response, stability and control, require a good understanding of the various aspects of their modeling and computational methodologies. While the directionality property of composite materials provides new degrees of freedom to the designer, enabling him to achieve greater structural efficiency, it constitutes an enormous challenge for someone who is not familiar with the capabilities that the implementation of the tailoring technology can provide.

For these structures, a good knowledge is also mandatory when dealing with the use of FGMs, and of smart materials in conjunction with feedback control.

For the authors of this book it was a pleasure to contribute to this most challenging and modern field of structural mechanics with a monograph on composite thin-walled beams, the first in the overall specialized literature.

Issues of the modeling and behavior of such structures composed of fiber composite materials and of FGMs, as well as appropriate feedback control methodologies have been developed and applied to many important problems, to alleviate and contain the oscillations generated by explosive blasts impacting the structure.

The monograph is concerned not only with the foundation and formulation of modern linear and nonlinear theories of composite thin-walled beams developed by the authors with their collaborators, but also provides powerful mathematical tools to address issues of free vibration, dynamic response to external excitation, stability and control of gyroscopic and aeroelastic systems. Special care was exercised to show the power of the tailoring technique in the specific problems treated in the monograph. The effects of transverse shear, warping inhibition, and of various elastic couplings on the behavior of these structures, have been highlighted. The two theories, shearable and unshearable, have been compared from the point of view of their mathematical description and of their static, dynamic and stability predictions, and proper conclusions have been drawn. No effort has been spared to compare our predictions with those available in the literature.

Regarding the foundation and formulation of the 1-D theory of composite thin-walled beams, an effort was made to derive the related field equations from the three-dimensional equations of elasticity theory. This approach has enabled us to present a unified treatment of the general theory of thin-walled beams, and provide a number of theorems with counterparts in the 3-D elasticity theory.

In order to provide a unified formulation and approach to both the linear and nonlinear problems, Chapters 2 through 5 are devoted to the foundation of the theory of thin-walled composite beams. While Chapter 1 has an introductory role, highlighting the importance of the topics treated in this monograph and presenting the basic milestones along the developments of thin-walled beam theory, Chapter 2 deals with the kinematics of open/closed cross-sections, and of uni/multicell TWBs. The issue of free and constrained warping, as well as of those related to the geometrically nonlinear and higher-order kinematics are elaborated.

Chapter 3 is concerned with the derivations of the equations of motion and of the related boundary conditions of open/closed cross-section beams. Since Hamilton's principle was used in their derivation, the associated boundary conditions have been obtained in a consistent manner. This was done for both the shearable/nonshearable and for the linear/nonlinear TWBs theories. The equations associated with the higher-order geometrically linear theory and with dissipative effects have been also accounted for.

Chapter 4 deals with some additional equations associated with the linear theory. The kinetic and strain energies associated with the shearable/unshearable TWBs models are presented. The governing equations are expressed in terms of 1-D displacement quantities for the anisotropic TWBs. It is shown that two decoupling types of the governing equations, each of which involving different elastic couplings, can be obtained through the implementation of two special lay-up configurations; this is presented for both shearable and unshearable TWBs.

Chapter 5 is devoted to the formulation of a few theorems in TWBs having counterparts in the 3-D elasticity theory. Beyond their academic importance, these theorems reveal the affiliation of the derived equations with those in the 3-D elasticity theory, from which they were obtained. Applications emerging from some of these theorems have been used in various parts of the monograph.

Chapter 6 deals with the free vibration of TWBs. Two basic solution methodologies used in the monograph in general, and in free vibration problems in particular, are presented. Related problems for non-uniform cross-section beams, validation of predictions, incorporation of the concept of *smart* TWBs, pertinent results and conclusions are provided.

Chapter 7 is devoted to the dynamic response of TWBs to time-dependent external excitations generated by an explosive blast or by a sonic-boom pulse. The two solution methodologies presented in Chapter 6 are extended to address this problem. In addition, the approach based on the orthogonality property of eigenmodes that was obtained in Chapter 5 by simply applying the reciprocal theorem, is also presented. We discuss the dynamic response of TWBs without/with incorporation of adaptive capabilities in conjunction with the piezoelectric induced strain actuation and of a feedback control methodology, and present results highlighting the beneficial and synergistic effects of tailoring and active feedback control technologies.

Chapter 8 deals with the behavior of TWBs carrying external stores. The problem is treated in the context of the transverse bending only. Issues of the feedback control of smart beams with external stores are addressed and results are reported.

Chapter 9, the largest chapter, deals with rotating TWBs. It addresses issues related to their modeling and incorporation of a number of important effects such as pretwist, presetting, anisotropy of constituent materials, temperature degradation of material properties for turbomachinery blades operating in a high temperature environment, transverse shear, tennis-racket, tension-torsion, geometrical nonlinearities. The model of a rotating blade made up of functionally graded materials (FGMs) is developed, and its performances are supplied.

Models of rotating blades carrying a tip mass, featuring extension-twist elastic coupling that is important for helicopter blades and tilt-rotor aircraft are developed in a general context. Feedback control of smart rotating beams is considered. The validity and complete agreement of the kinematic equations of pretwisted rotating blades as obtained via application of Wagner's multifilamentary concept and through Washizu's most accurate modeling are demonstrated.

Chapter 10 deals with the modeling and behavior of spinning TWBs. It is shown that centrifugal effects generated by the spinning speed produce a bifurcation of eigenfrequencies. Although conservative in nature, in some conditions, such systems can feature instabilities by divergence and flutter that are proper to non-conservative systems. We consider issues related the effects of pretwist, transverse shear, anisotropy, cross-sectional shape, boundary conditions, geometrical nonlinearities on their behavior. We develop the concept of FGMs in this type of structure and emphasize its usefulness in the presence of a high temperature during operating conditions. In this chapter we model the robotic gyroscopic system consisting of a TWB carrying a spinning tip rotor, and discuss its behavior with respect to the instabilities by divergence and flutter. We note the beneficial effects of piezoelectric induced strain actuation and of the tailoring technique.

Chapter 11 is devoted to the vibration and feedback control of spacecraft booms modeled as TWBs exposed to solar radiation. It is well-known that these structural systems are susceptible to flutter instability. This instability can jeopardize the mission of the spacecraft or satellite. An advanced model of spacecraft booms is developed, and issues of vibration, instability, feedback control and thermal compensation are addressed.

Chapter 12 is devoted to the aeroelasticity of aircraft wings modeled as anisotropic thin-walled beams. We develop an advanced model of anisotropic straight/swept wing structures and address issues of static aeroelastic instability and response, and feedback control. In the context of the same structural model, a few results on flutter and dynamic aeroelastic response of aircraft wings in various flight speed regimes are included.

Chapter 13 deals exclusively with the modeling and the static/dynamic behavior of open-section anisotropic beams of the I-profile.

In order to avoid any interruption in the description of the main topics treated in the monograph, the Appendix provides the background for the study of the anisotropy and heterogeneity of TWBs.

Lists of references are included at the end of each chapter.

We trust that this preview will give to the reader an idea on the topics covered by this monograph.

We hope that the monograph will prove useful to many researchers working in industry and academia in applied mechanics and aeronautics, for practicing engineers in the field of aeronautical/aerospace, mechanical, civil, nuclear and naval engineering, and for graduate students also.

All symbols are defined when first introduced. We tried to maintain a complete uniformity in notations. In some instances the same symbol has been used in different contexts. In such cases, the proper indications will avoid any possibility of confusion.

In the preparation of this book, we are indebted to many colleagues, scientists and former doctoral students who, through numerous discussions and research collaborations over many years, have helped us to mature our insight into the subject matter and enrich this work. It is virtually impossible to quantify how much we owe to them.

In particular, Chapters 6 through 8 contain results obtained during a close and fruitful scientific cooperation with Professor Sungsoo Na from Korea University in Seoul, Korea. Chapters 9 and 10 have been greatly influenced by close cooperation with Dr. Sang-Yong Oh, a Research Scientist at Virginia Tech; some issues in Chapter 10 were developed in cooperation with Dr. Nam-Hui Jeong from Chungnam National University in Daejeon, Korea; the parts from Chapters 2 and 3 addressing geometrically non-linear theories of TWBs have been influenced by cooperation with Professor Kanaan Bhaskar from the Indian Institute of Technology, Madras, India, during his working visit at Virginia Tech; the issues of higher-order theory and its applications to rotating blades were developed with Professor Naresh K. Chandiramani and his students, among others Mr. Shete C.D., from the Indian Institute of Technology, Guwahati, India; Chapter 12, in particular, was influenced by joint research with Dr. Zhanming Qin, a multidisciplinary Research Scientist at Virginia Tech who validated predictions generated from our developed structural model against those available in literature, while Chapter 10 benefited from the cooperation with Dr. Hyuck-Dong Kwon, Yonsei University, Seoul, South Korea, now a visting Research Scientist at Virginia Tech.

The close and continuous cooperation with Professor Piergovanni Marzocca from Clarkson University, Potsdam, New York, was also extended in the context of Chapter 12. Special thanks are also due to Dr. I. Yoon and to Mr. C.Y. Lee and Mr. W.K. Park from Chungnam National University in Daejeon, Korea, for their help.

We greatly enjoyed our cooperation with all these talented scientists; their enthusiasm and creativity have been a stimulus in our work.

Many thanks are due to the Series Editor, Professor Graham Gladwell for all his detailed suggestions that have rendered the manuscript more readable, and to Mrs. Nathalie Jacobs, Publishing Editor of Mechanical Engineering at Springer, for her encouragement, kind cooperation and infinite patience.

The authors express their indebtedness to all these people, and last but not least to Mrs. Lisa Smith, *the unknown soldier*, from the Engineering Science and Mechanics Department of Virginia Tech, for preparing the manuscript with such care, skill, efficiency and devotedness. A word of high appreciation is extended also to Mrs. Jolanda Karada (Karada Publishing Services, Nova Gorica, Slovenia) for the skillful work done on the final version of the manuscript.

On a personal note, Liviu Librescu must acknowledge his most influential and most beloved teachers, his mother Mina and his father, the lawyer Isidor Librescu, who in spite of enormous sufferings during two long consecutive terrible dictatorships that ravaged our native country, loved, guided, supported, encouraged, and gave me hope for a better future. Their memory will always be blessed!

Ohseop Song would like to extend his utmost gratitude to his father (Y. S. Song), his late mother (Y. H. Lee) and his loving wife (Bokyong Chun). They have been his warm supporters during this long-time project.

Liviu Librescu
Blacksburg, VA, USA

Ohseop Song
Daejeon, Republic of Korea

Chapter 1

INTRODUCTION

1.1 PRELIMINARY REMARKS. IMPORTANCE OF THE TOPIC

In broad terms, a thin-walled beam is a slender structural element whose distinctive geometric dimensions are all of different orders of magnitude: its thickness is small compared with the cross-sectional dimensions, while its length greatly exceeds the dimensions of its cross-section. Thin-walled beams can be categorized further by geometrical features: uniform or nonuniform cross sections, open or closed cross-sections, straight or curved, etc. Owing to their high efficiency from the point of view of minimum weight for given strength, these structural elements have been used for a long time in civil and mechanical engineering, as well as in ship and offshore constructions, as beams, columns, frame-works, containers, hulls, etc. However, the factor that has greatly contributed to the development of this type of structures, from both theoretical and practical points of view, is related to their wide application in the design of flight vehicle structures. This fact is illustrated by the large number of research works devoted to the modeling and stability of thin-walled metallic structures used in aeronautical industry that have erupted just before and after the World War II. As a result of these works and of the pioneering monographs by Vlasov (1940, 1961) and Umansky (1939), the theory of thin-walled beam structures has emerged as a new chapter in the textbooks of strength of materials, and, most importantly, in every book on strength of airplanes, both in East and West.

The encyclopedic work by Bruhn (1973) that is an essential piece of work in the hands of every engineer involved in the design of aeronautical structures, is devoted in large part to thin-walled metallic beams. Excellent surveys of the state-of-the-art of theory of metallic thin-walled members that include numerous references are Nowinski (1959), Panovko (1960), and the papers included in the volume edited by Chilver (1967).

Further, due to their wide application as long tubular booms installed on satellites, the importance of these structures is still expanding. Since these booms that are exposed to solar radiation experienced thermal flutter instability, an intensive research work beginning in the 60s, was devoted to the study of this phenomenon and to its prevention.

A further stimulus for research on thin-walled beams was generated by the emergence of fiber-reinforced polymeric composite materials and their increased use in structural components of aircraft, spacecraft, robot arms, bridges, and in many other advanced technological systems. Due to the advantages involving their high stiffness-and-strength-to-weigh ratios, high resistance to corrosion, high damping and enhanced fatigue life over conventional metallic materials, as well as due to their anisotropic nature that provides unique opportunities for tailoring the properties according to the design requirements, composite thin-walled beams became very attractive for their use in many secondary and primary load-carrying structures, subjected to complex static and dynamic loading systems.

A surge in the literature devoted to the derivation, extension and generalization of the equations of solid/thin-walled beams occured after the work on *static aeroelastic tailoring* due to Krone (1975), applied to a swept forward wing (SFW) aircraft. In his seminal work he showed that the proper use of the directional property in the fibrous constituent materials, enabling one to create the bending-twist coupling required to counteract the *wash-in effect*, can eliminate the chronic static instability facing this type of aircraft. Krone's theoretical findings have been instrumental in the design of the X29 Grumman SFW aircraft. The concept of aeroelastic tailoring was further expanded, in the sense of the exhaustive use of exotic properties of advanced composite materials to achieve desired design goals, including improved and extended operational and fatigue life. This new technology broadly referred to as *structural tailoring*, enables one to create preferred deformation modes, such as extension-twist or bending-twist, and generate enhanced response characteristics. The pioneering work devoted to the foundation of the theory of composite thin-walled beams by Rehfield (e.g. 1985) was instrumental in the understanding of the elastic coupling mechanism in composite thin-walled beams, and to devise proper methodologies as to capture the most favorable ones.

Thin-walled beams made of advanced composites emerged as primary loading bearing structures in the construction of helicopter, propeller, turbomachinery and wind turbine rotor blades. One of the earliest works devoted to the use of exotic composite materials for aeroelastic tailoring of helicopter rotor blades was conducted by Mansfield and Sobey (1979). Although their study was based on a crude model of composite thin-walled beams, they obtained good insight into the role of elastic couplings on rotor blade aeroelasticity. Moreover, their work has constituted a challenge for the rotorcraft technical community. Motivated by the needs of an accurate theoretical formulation of advanced rotor blades modeled as composite thin-walled beams, a workshop, significantly entitled *Beamology Workshop* was organized by Drs. R. Ormiston and G. L. Anderson, as a result of a suggestion made by Drs. D. Hodges, J. Kosmatka and M. Borri. The Beamology Workshop was held at NASA Ames Research Center, Moffett Field, California, in October 1992.

In addition to its application in rotorcraft structures, structural tailoring has been applied in tilt rotor aircraft to achieve a trade-off between hover and forward-flight regimes. It was found that changes in the blade twist between these flight modes can be developed through the use of extension-twist elastic coupling, as implemented in the XV-15 tilt rotor aircraft. Other examples of elastic tailoring applications can be found in the design of space robot arms required to operate with high precision, at a high speed.

In order to address these issues, an understanding of the implications of the various elastic couplings is needed. In addition, the structural model should incorporate a number of non-classical features, enabling one to get accurate predictions, not only for composite thin-walled beams, but also for moderately thick ones.

However, the ever increasing demands on the structures of aeronautical/ aerospace vehicles, of rotor blades, space robot arms, etc., has brought new challenges. As is well known, the wing of advanced combat aircraft is likely to feature increasing flexibilities and higher maneuverabilities, in spite of the severe environmental conditions in which it should operate. The vibratory motion of the structure, even of a benign type, can result in failure by fatigue, and in the shortening of operational life. The aeroelastic instabilities which can jeopardize their free flight must not appear for any combination of speed and load factor within their flight envelope, and even during aggressive maneuvers which can transgress the conditions imposed by the flight envelope.

The same is valid for reusable space vehicles which require, a prolongation of their life, without impairing upon the security of flight. Robot manipulator arms operating in space are required to be lightweight, strong, and capable of high precision under time-dependent external excitation; helicopter and turbine blades, and spacecraft booms, must satisfy similar conditions.

Structural tailoring is passive in its nature, in the sense that once implemented, the structure cannot respond to the variety of environmental factors under which it is likely to operate. Structural response control via implementation of the adaptive materials technology was suggested. In a structure with adaptive capabilities, the free vibration, the dynamic response characteristics to transient or harmonically oscillating loads and the instabilities (static and dynamic) can be controlled in a known and predictable manner.

In addition, due to the nature of adaptive structures that feature a highly distributed network of sensors and actuators, more encompassing control schemes can be implemented. The adaptive capability can be provided by a system of piezoactuators embedded in or bonded to the external surface of the host structure, that can produce, based on the converse piezoelectric effect, a localized strain field in response to an applied voltage.

The problem of application of both tailoring technique and adaptive capability to control and prevent the occurrence of any instability, and reduce the oscillation amplitude of thin-walled structural beams, has added a new dimension to the issues of thin-walled beams and their behavior.

Another complexity has emerged from the necessity of turbomachinery blades and advanced spinning shafts to operate in a high temperature environment. In order to alleviate the detrimental effects of the temperature on vibration and instability, a new generation of engineered composite materials, known as Functionally Graded Materials have been considered in their construction. This issue has brought another new dimension to the research already carried out on thin-walled beams.

In spite of its great importance, in the present and future advanced technology, and of the tremendous research work accomplished so far, the theory of thin-walled beams and its applications have not yet constituted the focus of any monograph. It is puzzling that in the area of composite shells and plates at least a dozen books have been produced during the last four decades, and not one on composite thin-walled beams. From the perspective of a person who had and still has the chance to be involved in both, (Librescu, 1969, 1975), the only plausible explanation is that the theory of composite thin-walled beams is much more complex than that of their plate/shell counterparts. As a matter of fact, the same state of affairs is valid also for the classical theory of thin-walled metallic beams *vis-à-vis* that of plates and shells. In this sense, prior to the first books by Vlasov (1940) and Umansky (1939), there were already published well established books on plates (Nadai, 1925), and shells (Flügge, 1934).

The modeling of composite thin-walled beams and the treatment of the previously indicated topics constitute the primary focus of this monograph.

1.2 CONTENTS OF THE MONOGRAPH

After Chapters 2 through 5 that are concerned with the foundation of the theory of composite thin-walled beams, applications of the theory, including specific ones related with structures of mechanical and aeronautical/aerospace type are presented.

In the theoretical part devoted to the derivation of the basic equations, an effort was made to provide a treatment of the subject by using the equations of the 3-D elasticity theory. The equations of the linear theory of thin-walled beams are included and fully discussed; these are obtained via the linearization of the those of nonlinear theory. The equations usually obtained for the shearable case are always specialized for the unshearable one. Some results regarding the higher order shear deformable theory of thin-walled beams are also presented.

Non-classical effects such as the primary and secondary warping, transverse shear, and the anisotropy of the constituent materials yielding the coupling of twist-bending (transversal)– bending (lateral)– extension deformation modes are included. In the development of the theory no *ad hoc* assumption beyond those initially stated have been adopted. The issue of the exact decoupling of the governing equation system is addressed. A careful study of the implications of these effects on the static and dynamic response quantities, in a diversity of applications has been carried out. The thermal degradation of the mechanical properties of constituent materials is addressed for both rotating and spinning thin-walled beams. The synergistic implications of the implementations of the tailoring technique and of the feedback control of adaptive structures is addressed.

Functionally graded materials have been included in the context of rotating and spinning blades, and their superior performances as compared to the purely metallic ones have been revealed. A thorough static and dynamic study of composite I beams has been accomplished.

While all the results and structural models presented here are original, in the sense that they have been developed by the authors together with their collaborators, a large number of results included here have not appeared or been discussed previously in the literature.

Although a large number of references on the various aspects of the theory and applications of thin-walled beams, in general, and on composite ones, have been supplied, no attempt has been made to provide an exhaustive list of works on these topics. The survey papers indicated in this work will assist the interested reader to find more information on this matter.

Special care was exercised throughout this work to address and validate the adopted solution methods and supplied results, against those obtained by alternative analytical, numerical or experimental means, that are available in the literature. The additional comparisons and validations carried out by these authors and their collaborators not included in this work, have been indicated in terms of the papers that contain these results.

In order to ease the reading of the text and render it reasonably self-contained, Appendix A devoted to constitutive equations of anisotropic Hookean materials and of piezoelectricity, and to their conversion to a form usable in thin-walled beams, is included at the end of the monograph.

The problems treated in this monograph reveal in a certain sense the increasing complexity of structural systems involving aeronautical and mechanical constructions characterizing the modern technology. New challenges will undoubtedly occur that will generate new developments of the theory of thin-walled beam structures.

It is our strong belief that the developments and results presented in this monograph will be helpful to the research worker involved in the solution of the new forthcoming problems.

REFERENCES

Bruhn, E. F. (Ed.) (1973) *Analysis and Design of Flight Vehicle Structures*, Jacobs Publ. Inc.

Chilver, A. H. (Ed.) (1967) *Thin-Walled Structures*, John Wiley & Sons Inc., New York.

Flügge, W. (1934) *Static and Dynamik der Shalen*, Springer, Berlin.

Krone, N. J. Jr. (1975) "Divergence Elimination with Advanced Composites," *AIAA Paper*, 75-1009, August.

Librescu, L. (1969) *Statics and Dynamics of Elastic Anisotropic and Heterogeneous Structures*, Publishing House of the Romanian Academy of Science, 290 pp. (in Romanian).

Librescu, L. (1975) *Elastostatics and Kinetics of Anisotropic and Heterogeneous Shell-Type Structures*, Noordhoff International Publishing, Leyden, Netherlands, 598 pp.

Mansfield, E. H. and Sobey, A. J. (1979) "The Fibre Composite Helicopter Blade – Part 1: Stiffness Properties – Part 2: Prospect for Aeroelastic Tailoring," *Aeronautical Quarterly*, Vol. 30, pp. 413–449.

Nadai, A. (1925) *Die Elastishen Platten*, Springer, Berlin.

Nowinski, J. (1959) "Theory of Thin-Walled Bars," *Applied Mechanics Reviews*, Vol. 12, No. 4, April, pp. 219–227.

Panovko, Ya. G. (1960) "Thin-Walled Members," in *Structural Mechanics in the U.S.S.R., 1917–1957*, I. M. Rabinovich (Ed.), G. Herrmann, translation ed., Pergamon Press, Oxford, pp. 142–159.

Rehfield, L. W. (1985) "Design Analysis Methodology for Composite Rotor Blades," in *Proceedings of the Seventh DoD/NASA Conference on Fibrous Composites in Structures Design*, AFWAL-TR-85-3094, pp. (V(a)-1)–(V(a)-15).

Umansky, A. A. (1939) *Torsion and Bending of Thin-Walled Aircraft Structures* (in Russian), Oboronghiz, Moscow.

Vlasov, V. Z. (1940) Russian original book; Stroizdat, Moscow [1961, English Translation, National Science Foundation, Washington, DC, Israel, Program for Scientific Translation, Jerusalem, Israel].

Chapter 2

KINEMATICS OF THIN WALLED BEAMS

2.1 GEOMETRICALLY LINEAR THEORY

2.1-1 General Considerations

Modern structural engineering deals with solid bodies in various forms that can be roughly classified as massive bodies; plates and shells; solid cross-section beams; thin/thick walled beams. These bodies are distinguished by their relative physical dimensions.

Massive bodies have all three physical dimensions of comparable magnitude. Plates and shells have one physical dimension, their thickness, small in comparison with their other two. Solid beams have cross-sectional dimensions small compared with the third, the longitudinal dimension. In thin/thick walled beams all three dimensions are of different order of magnitude. For such structures the wall thickness is small compared with any other characteristic dimension of the cross-section, whereas the linear dimensions of the cross-section are small compared with the longitudinal dimension.

One should in principle be able to derive the theory for beams, shells and plates, and massive bodies by using the equations of 3-D continuum theory, and taking advantage of the factors which serve to distinguish each type of structure. In this sense, the theory of plates and shells constitutes a two-dimensional approximation of the three-dimensional elasticity theory, while solid cross-section and thin/thick walled beams are both one-dimensional approximations of three-dimensional continuum theory.

In spite of this commonality, the theory of *thin/thick walled beams* is basically different from that of *solid cross-section beams*.

Due to their wide applications in civil, aeronautical/aerospace and naval engineering, and due to the increased use in their construction of advanced

7

composite material systems, a comprehensive theory of thin/thick walled beams has to be developed: this is one of the aims of this book.

2.1-2 General Definitions. Coordinate Systems

Consider a slender thin/thick walled structure of cylindrical or prismatic uniform cross-section (Fig. 2.1-1). Let h be its wall thickness assumed to be constant along the beam span but variable along the mid-line of the cross-section contour implying that $h \equiv h(s)$; l any characteristic dimension of the cross-section (i.e., its width or height) and L its length. Let h_{max} denote the maximum wall thickness; when

$$h_{\max}/l \le 0.1, \quad l/L \le 0.1, \tag{2.1-1a,b}$$

according to Vlasov (1961), the beam can be categorized as a *slender thin-walled beam*; otherwise it is a *thick-walled beam*.

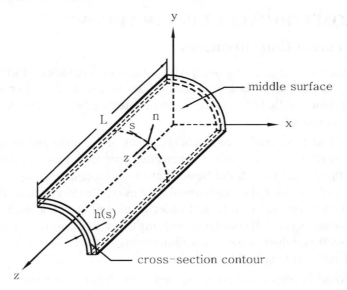

Figure 2.1-1: Geometry of a thin-walled beam

Two classes of thin/thick walled beam cross-sections have to be distinguished; *open* and *closed*.

A thin/thick walled section is said to be closed/open when the locus of points defining the center line of the wall forms a closed/open contour. The area enclosed by the closed contour is called a *cell*.

Closed cross-section thin-walled beams can be of single and multi-cell types. Single and multicellular thin/thick walled type beams can be found in the construction of the aircraft fuselage, wing and tail-surfaces, helicopter blades, spacecraft booms, ship hulls and bridges. Thin-walled beams can be categorized further according to their geometric characteristics straight or curved central lines; uniform or non-uniform cross-sections. In this monograph only straight thin-walled members are considered.

As in the case of plates and shells, in the theory of thin-walled beams an important role is played by the *middle surface*, defined as the locus of points equi-distant from the upper and bottom surfaces of the beam. The middle surface belongs to the class of cylindrical surfaces. The straight lines lying on the middle surface parallel to the beam longitudinal axis are the generators of this surface. The intersection of the middle surface with a plane normal to the generators determines the *mid-line* of the cross-section contour.

We impose no restrictions on the wall thickness. In spite of this, the generic terminology of thin-walled beams will be used.

It is convenient to adopt two coordinate systems for describing the kinematics of the beam. One is the Cartesian coordinate system (x, y, z) shown in Fig. 2.1-1, where the z-axis is along a straight longitudinal reference axis, while x and y are the transverse coordinates of the beam cross-section measured from the reference axis. In order to identify unambiguously the points associated with the mid-surface and those off the mid-surface, the notations (x, y, z) and (X, Y, Z) will be used, respectively (see Fig. 2.1-2). For the same purpose, this convention will be extended to other quantities belonging to the points on and off the mid-surface of the beam.

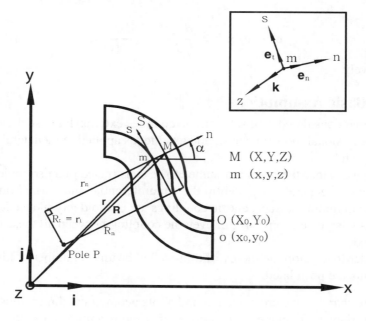

Figure 2.1-2: Coordinates associated with the on and off mid-surface points

The other system is the orthogonal curvilinear coordinate system (n, s, z) shown in Fig. 2.1-2. The coordinate s is measured along the tangent to the middle surface in a counter-clockwise direction with the origin conveniently chosen on the mid-line contour, while $n(-h/2 \leq n \leq h/2)$ is the coordinate perpendicular to the s-coordinate with the origin on the middle line contour.

In order to determine the relationship between the two coordinate systems, one defines the position vector $\mathbf{r}(\equiv \mathbf{r}(s, z))$ from the reference z-axis of the beam to an arbitrary point located on the middle surface as

$$\mathbf{r}(s, z) = x(s)\mathbf{i} + y(s)\mathbf{j} + z\mathbf{k}. \tag{2.1-2}$$

The unit vectors $(\mathbf{i}, \mathbf{j}, \mathbf{k})$ are associated with the Cartesian coordinates (x, y, z), respectively. The position vector $\mathbf{R}$ of an arbitrary point off the mid-surface of the beam can be expressed as

$$\mathbf{R} = \mathbf{r} + n\mathbf{e}_n, \tag{2.1-3}$$

where $\mathbf{e}_n$ denotes the unit vector associated with the n-coordinate.

Equations (2.1-2) and (2.1-3) establish the relationship between the Cartesian (x, y, z) and the curvilinear (n, s, z) coordinate systems. The unit vectors $\mathbf{e}_t$ and $\mathbf{e}_n$ tangent and normal to the mid-line contour are defined as

$$\mathbf{e}_t = \frac{d\mathbf{r}}{ds} = \frac{dx(s)}{ds}\mathbf{i} + \frac{dy(s)}{ds}\mathbf{j}, \tag{2.1-4}$$

and

$$\mathbf{e}_n = \mathbf{e}_t \times \mathbf{k} = \frac{dy(s)}{ds}\mathbf{i} - \frac{dx(s)}{ds}\mathbf{j}, \tag{2.1-5}$$

respectively.

2.1-3 Basic Assumptions

The beam is considered to be subjected to complex external loads such as biaxial bending, torsional moment, shear, and axial loads applied to the lateral surface and at its ends.

Moreover, since the beam is assumed to be constructed of anisotropic materials which, in general, vary both in the circumferential and normal directions to the beam middle surface, elastic couplings are unavoidably induced, such as bending-twist, extension-twist, or even the coupling between all these deformation modes.

In order to develop the theory of thin-walled beams, a number of kinematic statements are postulated:

(a) The shape of the cross-section and all its geometrical dimensions remain invariant in its plane. This implies that the beam cross-sections are assumed rigid in their own planes (i.e. that $\epsilon_{xx} = \epsilon_{yy} = \epsilon_{xy} = 0$), but are allowed to warp out of their original planes. For thin-walled beam structures (such as aircraft wings and fuselage and ship hulls), the original cross-sectional shape is maintained by a system of transverse stiffening members (ribs or bulkheads). These are considered rigid within their plane but perfectly flexible with regard to deformation normal to their own plane, so that the adoption of this assumption leads to a reasonable mathematical model for the actual physical behavior.

(b) The transverse shear strains are uniform over the beam cross-section, and

(c) The ratio of the wall thickness to the radius of curvature at any point of the beam wall is negligibly small compared to unity. It is evident that this assumption is actually exact for prismatic thin-walled beams composed of linear segments, and very accurate for shallow curved surfaces as well.

It is also postulated here that the displacements are infinitesimal. However, at a later stage, the theory of thin-walled beams will be formulated by considering also finite displacements.

2.1-4 Displacement Field

If a section is undeformable in its own plane (Assumption (a)), the only possible motion of the section in its own plane is the rigid body motion. This implies that displacements u and v of any point of a center-line of the beam cross-section in the x and y directions, respectively, can be described in terms of those of an arbitrary point P (x_P, y_P), referred to as the *pole*, and of the angle of rotation $\phi(z, t)$ of the cross-section about P, (positive when counterclockwise).

For small rotation of the section, the displacements u and v are expressed as (see Fig. 2.1-3)

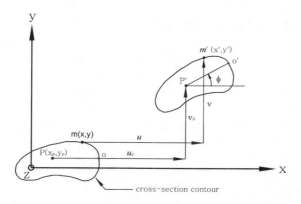

Figure 2.1-3: The displacement of the undistorted beam cross-section

$$u(x, y, z, t) = u_P(z, t) - (y - y_P)\phi(z, t),$$
$$v(x, y, z, t) = v_P(z, t) + (x - x_P)\phi(z, t), \tag{2.1-6a,b}$$

where t denotes the time, u_P and v_P are the x and y direction components of the displacement vector of point P, respectively, and $\phi(z, t)$ is the rotation about point P as given by

$$\phi(z, t) = \frac{1}{2}\left(\frac{\partial v}{\partial x} - \frac{\partial u}{\partial y}\right)_{x=x_P; y=y_P}. \tag{2.1-7}$$

Whereas Eqs. (2.1-6) enable one to represent the displacements of the beam cross-section in its plane in terms of the displacements of the pole and of the rotation about the pole axis (i.e. the axis passing through P and parallel to the z-axis), Eq. (2.1-7) reveals that ϕ is constant across the beam cross-section.

For convenience, two representations of the displacement vector $\mathbf{u}$, of the points of the beam mid-surface in the *global* and *local* coordinate systems are recorded:

$$\mathbf{u}(x, y, z, t) = u\mathbf{i} + v\mathbf{j} + w\mathbf{k}, \tag{2.1-8a}$$

and

$$\mathbf{u}(s, z, t) = u_n\mathbf{e}_n + u_t\mathbf{e}_t + w\mathbf{k}, \tag{2.1-8b}$$

respectively.

In view of the definitions of $\mathbf{e}_n$ and $\mathbf{e}_t$ (Eqs. (2.1-4) and (2.1-5)), the displacement components u_n and u_t are

$$u_n(s, z, t) = \mathbf{u} \cdot \mathbf{e}_n = u\frac{dy}{ds} - v\frac{dx}{ds}, \tag{2.1-9a}$$
$$u_t(s, z, t) = \mathbf{u} \cdot \mathbf{e}_t = u\frac{dx}{ds} + v\frac{dy}{ds}, \tag{2.1-9b}$$

and, in light of Eqs. (2.1-6), these become

$$u_n = u_P\frac{dy}{ds} - v_P\frac{dx}{ds} - r_t\phi, \tag{2.1-10a}$$
$$u_t = u_P\frac{dx}{ds} + v_P\frac{dy}{ds} + r_n\phi. \tag{2.1-10b}$$

In these equations r_n and r_t defined as follows:

$$r_n(s) = (x - x_P)\frac{dy}{ds} - (y - y_P)\frac{dx}{ds},$$
$$r_t(s) = (x - x_P)\frac{dx}{ds} + (y - y_P)\frac{dy}{ds}, \tag{2.1-11a,b}$$

denote the perpendicular distances from point P to the tangent and to the normal of the mid-line beam contour, at the point M, respectively, (see Fig. 2.1-4).

Indeed, upon defining the position vector ρ of a point of the mid-line beam contour measured from point P as $\rho = (x - x_P)\mathbf{i} + (y - y_P)\mathbf{j}$, one can express these quantities as $r_n = \rho \cdot \mathbf{e_n}$ and $r_t = \rho \cdot \mathbf{e_t}$. These relations imply

$$|\rho|^2 \equiv \rho^2 = r_n^2 + r_t^2 = (x - x_P)^2 + (y - y_P)^2, \qquad (2.1\text{-}11c)$$

$$\rho = r_n \mathbf{e_n} + r_t \mathbf{e_t}. \qquad (2.1\text{-}11d)$$

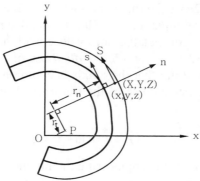

Figure 2.1-4: The pole and local coordinates of the beam cross-section

2.1-5 Free and Constrained Warping

When a thin-walled beam is loaded by equal and opposite torques at the end cross-sections, and the ends are free from any restraint, the beam exhibits **free torsion**. In this case, only shearing stresses on each section of the beam will be produced. The law of distribution of these stresses depends on the form of the cross-section and is the same for all cross-sections. Moreover, the rate of change of twist angle along the beam axis ϕ' ($\equiv d\phi/dz$) is constant.

Another case arises when the cross sections are not free to warp and/or if the twist moment varies along the length of the beam. In this case, warping displacement varies along the beam, and torsion is accompanied by tension or compression of longitudinal fibers. As a result, the rate of change of the angle of twist ϕ varies along the axis of the beam, i.e. ϕ' is no longer constant, but a function of the longitudinal coordinate z, that is $\phi' = \phi'(z, t)$.

Note that the variation of the beam cross-section and/or of twist moment along the beam axis are factors contributing to the non-uniform torsion behavior. As a result, when a partial or total restraint of the warping displacement arises, in addition to the primary system of stresses which equilibrate the applied loads, self-equilibrating systems of normal and tangential stresses are generated in the respective section. When the beam is subjected to twist moments only, in the case of constrained warping, the equilibrium conditions require that the resultant force and bending moments associated with secondary system of stresses be zero.

2.1-6 Open Cross-Section Beams

2.1-6a Warping Displacement and Primary Warping Function

In open cross-section beams, one assumes that the **direct** shear strain component γ_{sz} of the points of the beam mid-surface is zero. Under pure torsion, as shown by Megson (1990), Oden and Ripperger (1981), Wallerstein (2002), the direct shear strain is $\gamma_{sz} = 2n\phi'$. To this expression, the one induced by transverse shear effects should be superposed. This is mandatory for *composite* thin-walled beams and even for metallic ones whose wall thickness does not fulfill the thinness criterion (2.1-1a) for a *thin-walled* beam.

For such a case, recalling that the z-axis is common in both (x, y, z) and (n, s, z) coordinate systems, use of the transformation rule of second order tensors yields

$$\gamma_{sz} = \ell\gamma_{yz} - m\gamma_{xz} + 2n\phi'. \tag{2.1-12}$$

In Eq. (2.1-12) γ_{sz} denotes the engineering membrane shear strain, γ_{yz} and γ_{xz} denote the transverse shear components, while ℓ ($\equiv \cos(n, x)$) and m ($\equiv \cos n, y)$) denote the direction cosines of the outward normal n. These are defined as

$$\ell = \frac{dx}{dn}, \quad m = \frac{dy}{dn}, \tag{2.1-13a,b}$$

and alternatively (see Fig. 2.1-5), as

$$\ell = \frac{dy}{ds} \, (\equiv \cos\alpha), \quad m = -\frac{dx}{ds} \, (\equiv \sin\alpha), \tag{2.1-13c,d}$$

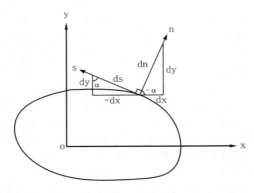

Figure 2.1-5: Geometry for the direction cosines of the normal and tangent to the boundary

where α is the angle between the positive directions of n and x.

In view of Eqs. (2.1-10b) and (2.1-12) considered in conjunction with the definition of γ_{sz}

$$\gamma_{sz} = \frac{\partial u_t}{\partial z} + \frac{\partial w}{\partial s}, \tag{2.1-14}$$

one obtains for the points on the beam mid-surface, that is on the surface corresponding to $n = 0$,

$$\frac{\partial w}{\partial s} = \theta_x(z, t)\frac{dy}{ds} + \theta_y(z, t)\frac{dx}{ds} - r_n(s)\phi'(z, t), \tag{2.1-15}$$

where

$$\theta_x(z, t) = \gamma_{yz} - v'_P, \tag{2.1-16a}$$

$$\theta_y(z, t) = \gamma_{xz} - u'_P, \tag{2.1-16b}$$

represent cross section rotations about the axes x and y at point P, respectively, while the primes indicate differentiation with respect to the z-coordinate. Integration of Eq. (2.1-15) with respect to s from a conveniently chosen contour origin o $(x_o, \ y_o)$ to an arbitrary point m (see Fig. 2.1-2) whose displacement is sought yields

$$w(s, z, t) = w_0(z, t) + y(s)\theta_x(z, t) + x(s)\theta_y(z, t) - \phi'(z, t)\overline{F}(s). \tag{2.1-17}$$

Here

$$w_0(z, t) = \overline{w}(z, t) - y_0\theta_x(z, t) - x_0\theta_y(z, t) \tag{2.1-18}$$

while

$$\overline{F}(s) = \int_o^s r_n(\overline{s})d\overline{s} = 2\Omega_{os}, \tag{2.1-19}$$

represents twice the area Ω_{os} swept by the radius vector r_n with the origin at the pole P when the generic point on the mid-line contour moves from the origin o, defined by $(\overline{s} = 0)$ to the point $m(\overline{s} = s)$. This area is referred to as the sectorial area.

In Eq. (2.1-17) w_0 can be interpreted as the longitudinal displacement of the point o, while $x(s)$ and $y(s)$ are the Cartesian coordinates of point M. Point o is called the sectorial origin, and P the pole of the sectorial areas. $\overline{F}(s)$ referred to as the primary (contour) warping, depends on the chosen positions for pole P and origin o. When the effect of transverse shear strains γ_{yz} and γ_{xz} is neglected, implying that the Euler-Bernoulli hypothesis is adopted, Eq. (2.1-17) reduces to that derived by Vlasov (1961) and Gjelsvik (1981).

Remark 1

It is readily seen that Eq. (2.1-15) providing w, can be obtained in an alternative way. Indeed, by equating the right-hand side member of Eq. (2.1-12) with that obtained when γ_{yz} and γ_{xz} are expressed in terms of displacement quantities as:

$$\gamma_{yz} = \frac{\partial v}{\partial z} + \frac{\partial w}{\partial y}, \ \gamma_{xz} = \frac{\partial u}{\partial z} + \frac{\partial w}{\partial x} \qquad (R1a, b)$$

one gets

$$l\gamma_{yz} - m\gamma_{xz} = l\frac{\partial v}{\partial z} - m\frac{\partial u}{\partial z} + \frac{\partial w}{\partial s} \qquad (R1c)$$

and after using Eqs. (2.1-13), (2.1-6), (2.1-11a) and (2.1-16), one obtains Eq. (2.1-15).

Remark 2

Since the contour origin $o(x_o, y_o)$ and the pole $P(x_P, y_P)$ can be chosen arbitrarily, the warping function $\overline{F}(s)$ may not be unique. It is possible, however to choose the pole and origin as to satisfy the condition

$$\oint \overline{F}(s)ds = 0 \qquad (R2a)$$

where $\oint (\cdot)ds$ denotes the integral along the mid-line contour of the beam cross-section.

Alternatively, for any chosen position of the pole and contour origin, the warping function $\overline{F}(s)$ can be defined as

$$\overline{F}(s) = \int_o^s r_n(\bar{s})d\bar{s} + C \qquad (R2b)$$

where the constant can be determined so as to satisfy condition, Eq. (R2a).

A similar feature associated with the warping function is valid for both open and closed, single and multi-cell cross-section thin-walled beams.

2.1-6b Secondary Warping Function

Based on Eq. (2.1-17) one can determine the distribution of axial strain. This equation implies that these quantities are associated with the mid-line beam contour only. However, when the criterion, Eq. (2.1-1), defining a *thin* walled-beam is violated, a secondary strain/stress system can be developed across the thickness of the beam. In such a case, the cross-sections undergo secondary warping. This secondary state of strain/stress is associated with the axial displacement of the points in the cross-section, off the middle surface.

To quantify this effect, we proceed as Section *2.1-6a*. The tensor transformation law for the shear strain Γ_{nz} associated with the points off the middle surface of the beam, yields

$$\Gamma_{nz} = L\gamma_{xz} + M\gamma_{yz}. \qquad (2.1-20)$$

In Eq. (2.1-20) and in the following, in order to avoid any ambiguity, the quantities associated with the on and off points belonging to the mid-line contour or middle surface counterparts will be denoted by lower and upper case letters, respectively.

In order to obtain the expression of $w(s, z, n, t)$, several preliminary steps have to be carried out. We use the relationships between the coordinates defining the points off and on the mid-line beam contour, (see Fig. 2.1-2),

$$X = \mathbf{R} \cdot \mathbf{i} = x + n\frac{dy}{ds}, \quad Y = \mathbf{R} \cdot \mathbf{j} = y - n\frac{dx}{ds}, \quad Z = z, \quad (2.1\text{-}21a\text{-}c)$$

and one should express Eqs. (2.1-10) for the points off the mid-line contour in a convenient form as:

$$U_n(s, z, n, t) = u_P\frac{dY}{dS} - v_P\frac{dX}{dS} - R_t\phi, \quad (2.1\text{-}22a)$$

$$U_t(s, z, n, t) = u_P\frac{dX}{dS} + v_P\frac{dY}{dS} + R_n\phi. \quad (2.1\text{-}22b)$$

Note that the direction cosines of the normal for the points off the mid-line contour $L \, (\equiv dY/dS)$ and $M (\equiv -dX/dS)$ reduce exactly to $\ell \, (\equiv dy/ds)$ and $m \, (\equiv -dx/ds)$, respectively, that is

$$L = l \text{ and } M = m, \quad (2.1\text{-}23a,b)$$

where

$$dS = \left(1 + \frac{n}{r_n}\right)ds, \quad (2.1\text{-}24)$$

defines the element of arc dS parallel to the mid-line contour.

Remark 3

We can readily verify (2.1-23). Indeed, using (2.1-21) we have

$$\frac{dX}{dS} = \frac{dx}{ds}\frac{ds}{dS} + n\frac{d}{ds}\left(\frac{dy}{ds}\right)\frac{ds}{dS}, \quad (R3.1a)$$

$$\frac{dY}{dS} = \frac{dy}{ds}\frac{ds}{dS} - n\frac{d}{ds}\left(\frac{dx}{ds}\right)\frac{ds}{dS}. \quad (R3.1b)$$

On the other hand, Eqs. (2.1-13) and $r_n d\alpha = ds$, give

$$\frac{d^2x}{ds^2} = -\frac{1}{r_n}\frac{dy}{ds}, \quad \frac{d^2y}{ds^2} = \frac{1}{r_n}\frac{dx}{ds}. \quad (R3.2a, b)$$

Replacement of (R 3.2) in (R 3.1) followed by elementary manipulations results in Eqs. (2.1-23).

Having in view the relationships

$$R_n = r_n + n \quad \text{and} \quad R_t = r_t, \quad (2.1\text{-}25a,b)$$

representing the off mid-line counterparts of Eqs. (2.1-11), and recognizing that

$$\Gamma_{nz} = \frac{\partial U_n}{\partial z} + \frac{\partial W}{\partial n}, \quad (2.1\text{-}26)$$

one obtains

$$\frac{\partial W}{\partial n} = \theta_y(z, t)\frac{dy}{ds} - \theta_x(z, t)\frac{dx}{ds} + r_t\phi'(z, t). \qquad (2.1\text{-}27)$$

In these equations U_n and W represent the off-midline counterparts of u_n and w, respectively, By integrating Eq. (2.1-27) across the wall thickness in the interval $[0, n)$, taking into account Eq. (2.1-17) and returning to the usual notation of w, one obtains

$$w(s, z, n, t) = w_0(z, t) + \theta_y(z, t)\left[x(s) + n\frac{dy}{ds}\right] + \theta_x(z, t)\left[y(s) - n\frac{dx}{ds}\right]$$
$$- \phi'(z, t)F(n, s). \qquad (2.1\text{-}28)$$

Herein

$$F(n, s)\left(\equiv \overline{F}(s) + \overline{\overline{F}}(n, s)\right), \qquad (2.1\text{-}29a)$$

denotes the warping function of the open cross-section beams, where

$$\overline{F}(s) \equiv \int_o^s r_n(\bar{s})d\bar{s} \quad \text{and} \quad \overline{\overline{F}}(n, s) \equiv -nr_t(s) \qquad (2.1\text{-}29b,c)$$

are related to the points on and off the mid-line contour, respectively. For this reason $\overline{F}(s)$ and $\overline{\overline{F}}(n, s)$ are referred to as contour (primary) and thickness (secondary) warping functions, respectively. Equation (2.1-28) represents the transverse shear deformable (referred in the following to as the shearable) counterpart of the warping displacement derived by Gjelsvik [1981] for the non-shearable (referred in the following to as the unshearable) beam theory. In the latter case, one should consider in Eq. (2.1-28), $\theta_x = -v'_P$ and $\theta_y = -u'_P$.

Remark 4
    It can easily be shown that Eq. (2.1-27) is obtainable in an alternative way. Indeed, upon equating the second member of Eq. (2.1-20) with that obtained when γ_{xz} and γ_{yz} are expressed by Eqs. (R1 a,b) and using (2.1-23) (2.1-13) and (2.1-11b) one obtains Eq. (2.1-27).

2.1-7 Single-Cell Closed Beam Cross-Sections

In contrast to the open cross-section beams, the closed ones can carry shear and torsional loads. Incorporating transverse shear effects and assuming that the shear stresses induced in the 3-D medium of the beam exhibit a linear variation in the thickness n-direction, one has for the net shear strain Γ_{sz} of any point of the beam cross-section, see e.g. Nishino et al. (1977),

$$\Gamma_{sz} = L\gamma_{yz} - M\gamma_{xz} + n_{sz}/(hG_{sz}) + nN_{sz}/(hG_{sz}). \qquad (2.1\text{-}30)$$

In this equation $G_{sz}(s)$ denotes the tangential shear stiffness of the laminate; $n_{sz} \left(\equiv (\hat{N}_{sz} + \hat{N}_{sz})/2 \right)$ and $N_{sz} \left(\equiv (\hat{N}_{sz} - \hat{N}_{sz})/h \right)$ denote the average shear flow (referred to as Bredt-Batho shear flow) and the thicknesswise shear flow, respectively, where $\hat{N}_{sz}$ and $\hat{N}_{sz}$ stand for the tangential shear flows in the S-z plane evaluated at $n = h/2$ and $n = -h/2$, respectively.

Expressing the off mid-line counterpart of Eqs. (2.1-14) as

$$\Gamma_{sz} = \frac{\partial W}{\partial S} + \frac{\partial U_t}{\partial z}, \tag{2.1-31}$$

in view of Eqs. (2.1-22b) and (2.1-30) one obtains

$$\frac{\partial W}{\partial S} = \left(\gamma_{xz} - u'_P\right)\frac{dX}{dS} + \left(\gamma_{yz} - v'_P\right)\frac{dY}{dS} + n_{sz}/(hG_{sz}) + nN_{sz}/(hG_{sz}) - R_n\phi'. \tag{2.1-32}$$

Since W must be continuous around the circumference of the closed cross-section contour, one must enforce the condition

$$\oint_S \frac{\partial W}{\partial S}dS = 0, \tag{2.1-33}$$

which, in conjunction with (2.1-32), yields:

$$\oint_S \left[n_{sz}/(hG_{sz}) + nN_{sz}/(hG_{sz}) - R_n\phi'\right]dS = 0. \tag{2.1-34}$$

Assuming further that n_{sz} and N_{sz} are independent on the S-coordinate one obtains

$$n_{sz} + nN_{sz} = \frac{\oint_S R_n dS}{\oint_S \dfrac{dS}{h(S)G_{sz}(S)}}\phi'(z,t). \tag{2.1-35}$$

Invoking the relationships (2.1-25a) and (2.1-24) and keeping in mind that for a structure whose contour is constituted of plane segments, or is shallow, $\oint ds/r_n = 0$, from Eq. (2.1-35) one gets:

$$n_{sz} = \frac{\oint r_n ds}{\oint \dfrac{ds}{h(s)G_{sz}(s)}}\phi'(z,t) \text{ and } N_{sz} = \frac{2\oint ds}{\oint \dfrac{ds}{h(s)G_{sz}(s)}}\phi'(z,t), \tag{2.1-36a,b}$$

where $\oint (\cdot) ds$ and $\oint_S (\cdot) dS$ denote the integral around the mid-line circumference and on a circumference parallel to the mid-line contour, respectively. For
brevity, the following notation will be used:

$$\mathcal{L} = \oint \frac{ds}{h(s) G_{sz}(s)}. \tag{2.1-36c}$$

Substitution of (2.1-36) into (2.1-32), followed by its integration with respect
to S from a conveniently chosen contour origin O (X_o, Y_o) yields

$$W(s, z, n, t) = w_o(z, t) + \theta_y(z, t)X + \theta_x(z, t)Y$$
$$- \phi'(z, t) \int_0^s \left[\left(R_n - \frac{2\Omega}{hG_{sz} \mathcal{L}} - n\frac{2\beta}{hG_{sz}\mathcal{L}} \right) \right] \left(1 + \frac{n}{r_n} \right) ds. \tag{2.1-37}$$

In view of Eqs. (2.1-21) and (2.1-25a), and keeping in mind that $n/r_n \ll 1$,
Eq. (2.1-37) becomes

$$W(s, z, n, t) = w_o(z, t) + \theta_y(z, t) \left(x + n\frac{dy}{ds} \right) + \theta_x(z, t) \left(y - n\frac{dx}{ds} \right)$$
$$- \phi'(z, t) \int_o^s \left[r_n(\bar{s}) + 2n - \frac{2\Omega}{h(\bar{s})G_{sz}(\bar{s})\mathcal{L}} - n\frac{2\beta}{h(\bar{s})G_{sz}(\bar{s})\mathcal{L})} \right] d\bar{s}. \tag{2.1-38}$$

In these equations

$$\oint r_n ds = 2\Omega, \text{ and } \oint ds = \beta, \tag{2.1-39a,b}$$

denote the double area enclosed by the center line of the beam cross-section
and its perimeter, respectively.

Upon retracing these steps for Γ_{nz} as given by Eq. (2.1-20), one obtains

$$\frac{\partial W}{\partial n} = \theta_y \frac{dY}{dS} - \theta_x \frac{dX}{dS} + R_t \phi', \tag{2.1-40}$$

wherefrom, by integrating this with respect to n and keeping in mind Eq.
(2.1-25b), we find:

$$W(s, z, n, t) = w_o(z, t) + w(s, z, t) + n \left[\theta_y \frac{dy}{ds} - \theta_x \frac{dx}{ds} + r_t \phi' \right]. \tag{2.1-41}$$

From Eq. (2.1-38) one can extract the expression of $w(s, z, t)$ which reads

$$w(s, z, t) = \theta_y(z, t)x + \theta_x(z, t)y - \phi'(z, t) \int_o^s \left(r_n(\bar{s}) - \frac{2\Omega}{h(\bar{s})G_{sz}(\bar{s})\mathcal{L}} \right) d\bar{s}. \tag{2.1-42}$$

Substitution of Eq. (2.1-42) into (2.1-41) yields

$$w(s, z, n, t) = w_o(z, t) + \theta_y(z, t)\left(x + n\frac{dy}{ds}\right) + \theta_x(z, t)\left(y - n\frac{dx}{ds}\right)$$
$$- \phi'(z, t)\left\{\overline{F}(s) + \overline{\overline{F}}(n, s) - 2n\int_o^s \left[\underset{\cdots\cdots\cdots\cdots\cdots}{\frac{\beta}{h(\overline{s})G_{sz}(\overline{s})\mathcal{L}} - 1}\right]d\overline{s}\right\},$$

$$(2.1-43)$$

where

$$\overline{F}(s) = \int_o^s (r_n(\overline{s}) - \psi(\overline{s}))\,d\overline{s}, \quad and \quad \overline{\overline{F}}(n, s) = -nr_t. \qquad (2.1\text{-}44a,b)$$

The quantity

$$\psi(s) = \frac{2\Omega}{h(s)G_{sz}(s)\mathcal{L}}, \qquad (2.1\text{-}44c)$$

is referred to as the *torsional function*. When the thickness and membrane shear modulus are uniform along the beam circumference, the torsional function becomes

$$\psi = \oint r_n ds / \oint ds. \qquad (2.1\text{-}44d)$$

Note that for free twist, $\phi(z)$ is a linear function of the z-coordinate (implying that the rate of twist is constant), while for constrained twist $\phi(z)$ can be an arbitrary function of z.

For a beam fulfilling the criterion provided by Eq. (2.1-1), one can assume that N_{sz} is negligible small, and as a result, the term in Eq. (2.1-43) underscored by a dotted line can be neglected. The same conclusion holds valid when h and G_{sz} are uniform along the s-coordinate.
In either case

$$\overline{F}(s) = \int_o^s \left(r_n(\overline{s}) - \frac{2\Omega}{\beta}\right)d\overline{s} = 2\Omega_{os} - \frac{2\Omega}{\beta}s. \qquad (2.1\text{-}45)$$

With the contour parameters

$$\delta = \oint \frac{ds}{h(s)G_{sz}(s)}, \qquad \delta_{os}(s) = \int_o^s \frac{d\overline{s}}{h(\overline{s})G_{sz}(\overline{s})}, \qquad (2.1\text{-}46a,b)$$

Eq. (2.1-45) becomes

$$\overline{F}(s) = -2\Omega\left[\frac{\delta_{os}}{\delta} - \frac{\Omega_{os}}{\Omega}\right]. \qquad (2.1\text{-}47)$$

This primary warping function is identical to that obtained by Smith and Chopra (1990, 1991), and Megson (1974, 1990). This equation shows that under a pure torque, a closed section beam will not warp if, in addition to the conditions that r_n and $G_{sz}h$ are constant, the condition stipulating the absence of secondary warping is fulfilled.

Equation (2.1-47) also shows that for a *circular cross-section beam* whose thickness and shear modulus are circumferentially uniform, $\overline{F}(s)$ becomes identically zero. The same happens when the beam cross-section is in the shape of a regular polygon of constant thickness and uniform shear modulus, or for a rectangular cross-section beam whose thickness of its vertical and horizontal walls, h_w and h_F and their size c_w and c_F, respectively, fulfil the relationship $h_w c_F = h_F c_w$. This thin-walled square cross-section beams of constant thickness do not exhibit primary warping either. In these instances, the only warping stresses induced by twisting are those due to secondary warping. In such cases and/or of thick-walled beam cross-sections, its effect has always to be taken into consideration.

2.1-8 Saint-Venant Free Twist of Closed Cross-Section Beams

Assume that the conditions of free warping are fulfilled. This implies that the shear flow is induced by torque moments applied at the beam ends and that the end cross sections are free to warp. Defining a modulus-weighted thickness as

$$\tilde{h}(s) = \frac{G_{sz}(s)}{\tilde{G}}h(s), \qquad (2.1\text{-}48)$$

where $\tilde{G}$ is a conveniently chosen shear modulus, from Eq. (2.1-36a) one obtains

$$\phi' = \frac{n_{sz}}{2\Omega\tilde{G}}\oint\frac{ds}{\tilde{h}}, \text{ or } n_{sz} = \tilde{G}\frac{\oint r_n ds}{\oint \dfrac{ds}{\tilde{h}}}\phi'. \qquad (2.1\text{-}49a,b)$$

On the other hand, because the total twisting moment developed by the shear flow is

$$\overline{M}_t = \oint r_n n_{sz}ds = n_{sz}\oint r_n ds = 2\Omega n_{sz}, \qquad (2.1\text{-}50)$$

one can express Eq. (2.1-49) in the form

$$\phi' = \frac{\overline{M}_t}{4\Omega^2\tilde{G}}\oint\frac{ds}{\tilde{h}}. \qquad (2.1\text{-}51)$$

Having in view that in the free warping case $\overline{M}_t$ is independent on z, Eq. (2.1-51) implies that the rate of twist is independent of the longitudinal coordinate, as well. This reverts to the conclusion, that in the case of free twist, $\phi(z)$ is a linear function of the longitudinal z-coordinate.

Expressing the twisting moment under the alternative form

$$\overline{M}_t = \tilde{G}\overline{J}\phi', \tag{2.1-52a}$$

where $\tilde{G}\overline{J}$ denotes the stiffness of the beam in free torsion and comparing Eq. (2.1-52a) with Eq. (2.1-51) one obtains

$$\overline{J} = \frac{4\Omega^2}{\oint \dfrac{ds}{\tilde{h}}}. \tag{2.1-52b}$$

Equations (2.1-49), (2.1-51) and (2.1-52), are referred to as the Bredt-Batho equations.

Remark 5

From Eq. (2.1-50) it results that for the same twist moment $\overline{M}_t$, the shear flow n_{sz} is independent of the cross-sectional shape geometry characterized by the same enclosed area Ω.

2.1-9 Unified Form of the Warping Function

A comparison of the expressions for w associated with open and closed beam cross-sections shows that the only difference occurs in the proper definition of the warping function $F(n, s)$:

$$F(n, s) = \begin{cases} \displaystyle\int_0^s r_n(\bar{s})d\bar{s} - nr_t(s) & \text{for open cross-sections} \\[3ex] \displaystyle\int_0^s [r_n(\bar{s})d\bar{s} - nr_t(s)] \\[2ex] \quad -\displaystyle\int_0^s \frac{\displaystyle\oint_{\mathcal{L}} r_n(s)ds}{\mathcal{L}} \frac{d\bar{s}}{h(\bar{s})G_{sz}(\bar{s})} & \text{for closed cross-sections} \\[4ex] \quad -2n\displaystyle\int_0^s \left[\underline{\frac{\oint ds}{h(\bar{s})G_{sz}(\bar{s})\mathcal{L}} - 1} \right] d\bar{s} \end{cases} \tag{2.1-53}$$

The conditions under which the term underscored by the dotted line becomes negligible have been already stipulated in connection with the similar term appearing in Eq. (2.1-43).

When h and G_{sz} are uniform along the beam circumference, Eqs. (2.1-53) can be expressed in a unified form for both open and closed cross-section beams as

$$F(n, s) = \int_o^s r_n(\bar{s})d\bar{s} - nr_t(s) - \delta_c \int_o^s \frac{\oint r_n(s)ds}{\oint ds} d\bar{s}, \qquad (2.1\text{-}54)$$

where the tracer $\delta_c = 1$ and $\delta_c = 0$, for closed and open cross-section beams, respectively. When secondary warping is discarded, Eq. (2.1-54) coincides with that obtained by Nishino and Hasegawa (1979).

2.1-10 The Strain Field

For infinitesimal displacements, consistent with the three-dimensional elasticity theory, the strain - displacement relationships are expressed as

$$\epsilon_{ij} = \frac{1}{2}\left(\frac{\partial V_i}{\partial x_j} + \frac{\partial V_j}{\partial x_i}\right), \qquad (i, j = 1, 2, 3) \qquad (2.1\text{-}55)$$

Assimilation in these equations of (V_1, V_2, V_3) and (x_1, x_2, x_3) with (u, v, w) and (x, y, z), respectively, yields:

$$\epsilon_{xx} = \frac{\partial u}{\partial x}, \quad \epsilon_{yy} = \frac{\partial v}{\partial y}, \quad \epsilon_{xy} = \frac{1}{2}\left(\frac{\partial u}{\partial y} + \frac{\partial v}{\partial x}\right) \qquad (2.1\text{-}56a\text{-}c)$$

$$\epsilon_{zz} = \frac{\partial w}{\partial z}, \quad \epsilon_{xz} = \frac{1}{2}\left(\frac{\partial u}{\partial z} + \frac{\partial w}{\partial x}\right), \quad \epsilon_{yz} = \frac{1}{2}\left(\frac{\partial v}{\partial z} + \frac{\partial w}{\partial y}\right) \qquad (2.1\text{-}56d\text{-}f)$$

In view of Eqs. (2.1-6), the strain components are:

$$\epsilon_{xx} = 0, \quad \epsilon_{yy} = 0, \quad \epsilon_{xy} = 0, \qquad (2.1\text{-}57a\text{-}c)$$

$$\epsilon_{zz}(s, z, n, t) = \epsilon_{zz}^{(0)}(s, z, t) + n\epsilon_{zz}^{(1)}(s, z, t), \qquad (2.1\text{-}57d)$$

where

$$\epsilon_{zz}^{(0)} = w_o'(z, t) + y(s)\theta_x'(z, t) + x(s)\theta_y'(z, t)$$

$$- \phi''(z, t)\left[\int_o^s r_n(\bar{s})d\bar{s} - \delta_c \int_o^s \frac{\oint r_n(s)ds}{\mathcal{L}} \frac{d\bar{s}}{h(\bar{s})G_{sz}(\bar{s})}\right], \qquad (2.1\text{-}58a)$$

and

$$\epsilon_{zz}^{(1)} = \frac{dy}{ds}\theta_y'(z, t) - \frac{dx}{ds}\theta_x'(z, t) + r_t(s)\phi''(z, t). \qquad (2.1\text{-}58b)$$

In Eq. (2.1-58b) it was assumed that the conditions for the underscored term in Eq. (2.1-43) (or (2.1-53)) to be neglected are met.

Whereas Eqs. (2.1-57 a-c) show that the displacement representation given by Eqs. (2.1-6) is consistent with the assumption postulating cross-section non-deformability, Eqs. (2.1-57d) and (2.1-58) show that in an unrestrained beam subjected to pure torque, since w'_o, θ'_x and θ'_y can be shown to be immaterial, $\epsilon_{zz} = 0$. The latter result implies that the generators of the beam surface remain unchanged in length during the free warping displacement.

To obtain the expression of shear strain components $\Gamma_{sz} (\equiv 2\,\epsilon_{sz})$ and $\Gamma_{nz} (\equiv 2\,\epsilon_{nz})$, which should include their variation through the wall thickness, one starts again with Eq. (2.1-31) which expresses the tangential shear strain off the mid-line contour in terms of displacement quantities.

Replacement in Eq. (2.1-31) of Eq. (2.1-22b) and of the expression for W valid for both open and closed cross-section beams

$$W(S, z, n, t) = w_o(z, t) + X(S)\theta_y(z, t) + Y(S)\theta_x(z, t)$$

$$- \phi'(z, t) \int_o^S \left[R_n - 2\delta_0 n - \delta_c \frac{\oint_S R_n dS}{\oint_S dS} \right] dS,$$

(2.1-59)

and invoking Eqs. (2.1-16) yields

$$\Gamma_{sz}(S, z, n, t) = \gamma_{xz}\frac{dX}{dS} + \gamma_{yz}\frac{dY}{dS} + \delta_c \frac{\oint_S R_n dS}{\oint_S dS}\phi'(z, t) + 2\delta_0 n\phi'(z, t).$$

(2.1-60)

Using Eqs. (2.1-21) and (2.1-23) through (2.1-25) we can show that

$$\frac{\oint_S R_n dS}{\oint_S dS} = \frac{\oint r_n ds}{\oint ds} + 2n,$$

(2.1-61)

and consequently

$$\Gamma_{sz}(s, z, n, t) = \gamma_{xz}(z, t)\frac{dx}{ds} + \gamma_{yz}(z, t)\frac{dy}{ds}$$

$$+ \delta_c \left(\frac{\oint r_n ds}{\oint ds} + 2n \right) \phi'(z, t) + 2\delta_0 n\phi'(z, t).$$

(2.1-62)

As concerns Γ_{nz}, employment of (2.1-22a) and (2.1-28) or (2.1-43) in Eq. (2.1-26) yields

$$\Gamma_{nz}(s, z, n, t) = \gamma_{xz}(z, t)\frac{dy}{ds} - \gamma_{yz}(z, t)\frac{dx}{ds}. \tag{2.1-63}$$

Equations (2.1-62) and (2.1-63) show that within the shearable beam theory, Γ_{nz} is uniform through the wall thickness while Γ_{sz} is linear across it. Alternative expressions of Γ_{sz} and Γ_{nz} are obtainable when in Eqs. (2.1-62) and (2.1-63) γ_{xz} and γ_{yz} are replaced by those provided by Eqs. (2.1-16).

In the context of the non-shearable beam theory

$$\Gamma_{sz}(s, z, n, t) = \delta_c \left(\frac{\oint r_n ds}{\oint ds} + 2n \right) \phi'(z, t) + 2\delta_0 n \phi'(z, t), \tag{2.1-64a}$$

and

$$\Gamma_{nz} = 0. \tag{2.1-64b}$$

2.1-11 Open Versus Closed Section Beams

Before addressing this issue, we show (see in this sense Fig. 2.1-6) how open and closed section beams develop shearing stresses as to balance applied twist moments.

For an open circular beam, the stress distribution is linear across its thickness, which leads to a loop of forces as indicated in Fig. 2.1-6a. For closed beams, as shown in Fig. 2.1-6b, there is a totally different shearing stress distribution.

For thin-walled beams we can assume that the shear stresses are uniform across the thickness. However, for thick walled beams this assumption should be regarded with caution.

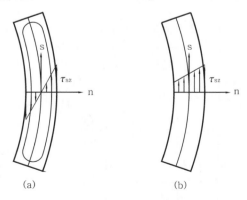

(a)   (b)

Figure 2.1-6: Shearing stresses developed in the open (a), and the closed beams (b)

To emphasize the essential differences between the behavior in free pure torsion of open and closed cross-section non-shearable beams consider the case of a closed circular tube of radius r and of its counterpart which has a slot extending the full length of the tube (so that the cross-section is open).

For the closed beam, from Eq. (2.1-52) one obtains

$$M_t^{\text{closed}} = 2\pi r^3 h G_{sz} \phi'. \tag{2.1-65}$$

For the open beam, from Eq. (2.1-64a) we have $\tau_{sz} = 2nG_{sz}\phi'$ and as a result, use of Eqs. (2.1-24) and (2.1-25) in the expression

$$M_t^{\text{open}} = 2 \int_o^s \int_{-h/2}^{h/2} G_{sz} n\phi'(r_n + n)\left(1 + \frac{n}{r_n}\right) dn ds, \tag{2.1-66a}$$

yields, for the problem at hand.

$$M_t^{\text{open}} = \frac{2\pi}{3} r h^3 G_{sz} \phi'. \tag{2.1-66b}$$

If both tubes are subjected to the same moment, the ratio $\phi'_{\text{(open)}}/\phi'_{\text{closed}} = 3(r/h)^2$. For $r/h = 20$ the closed tube is 1200 times stiffer in twist than the open tube. Short slits in thin cross-sections are unavoidable whenever it is necessary to have access to the interior of tubular structures such as submarine hull or aircraft fuselage. Aircraft wings and helicopter blades are generally closed thin-walled beams, because twisting loads on these structures are commonplace.

2.2 KINEMATICS OF GEOMETRICALLY NONLINEAR THIN-WALLED BEAMS

2.2-1 Preliminaries

The large number of problems of practical importance requiring a geometrically nonlinear formulation, such as the postbuckling behavior as well as the reliable evaluation of load carrying capacity of structures used in aeronautical, aerospace, marine and off shore industries, as well as in mechanical and civil engineering, fully justifies interest in geometrically nonlinear theories. Within the formulation of this theory, the flexural displacements are assumed to be finite, while the twist angle is taken to arbitrarily large so as to account for the inherent flexibility in torsion of an open beam. Most of the available works on nonlinear theories of thin-walled beams encompass metallic structures (see e.g. Ascione and Grimaldi (1983), Attard (1986), Ghobarach and Tso (1971), Meredith and Witmer (1981), Mollmann (1982a,b), Nishino and Hasegawa (1979). For nonlinear modeling of beams, in general, see Grimaldi and Pignataro (1979), Roberts and Azizian (1983), Rosen and Friedmann (1979), Polillo et al. (1998), and the work by Van Erp (1987) where a comprehensive discussion of the state-of-the-art is presented). Additional work addressing this issue, specially in connection with the modeling of rotor blades, was accomplished by Bauchau and Hong (1987, 1988), Borri and Merlini (1986), Fraternali and Feo (2000), Hodges and Dowell (1974), Kvaternik et al. (1978), and by Minguet and Dugundji (1990 a, b). In the same context of solid metallic beams, general modelings have been supplied by Suhubi (1968), Wempner (1981), Iura and Hirashima (1985), and Tsay and Kingsbury (1988), while for composite beams by Pai and Nayfeh (1992, 1994) and Nayfeh and Pai (2004). From the very limited work carried out on nonlinear theory of composite thin-walled beams, the studies by Bhaskar and Librescu (1995 a, b) are indicated.

2.2-2 Coordinate Systems. Assumptions

As before, the straight undeformed beam is referred to the global and local right-handed coordinate systems (x, y, z) and (n, s, z), respectively.

Some of the assumptions under which the nonlinear beam theory will be developed have been already stipulated and used in the construction of the linear theory (see Sect. 2.1-3). Here there are the supplementary assumptions:

(i) The flexural displacements u and v in the x and y direction, respectively, are small but finite, while the twist ϕ of the cross-section can be arbitrarily large.

(ii) The axial displacement w is much smaller than u or v so that products of the derivatives of w can be neglected in the strain-displacement relations.

(iii) Strains are small so that a linear constitutive law can be used to relate the second Piola-Kirchhoff stress tensor to Green-Lagrange's strain tensor.

2.2-3 Displacement Field

By virtue of assumption a), Sect. 2.1-4, the u and v displacements of any point of the beam cross-section can be described in terms of those of a chosen point $P(x_P, y_P)$, that is in terms of $u_P(z, t)$ and $v_P(z, t)$, and the twist $\phi(z, t)$ of the cross section about any longitudinal axis. In a generic form, these expressions are

$$u(x, y, z, t) = u_P(z, t) - (x - x_P)(1 - \cos \phi(z, t)) - (y - y_P) \sin \phi(z, t),$$

$$(2.2\text{-}1a)$$

$$v(x, y, z, t) = v_P(z, t) + (x - x_P) \sin \phi(z, t) - (y - y_P)(1 - \cos \phi(z, t)),$$

$$(2.2\text{-}1b)$$

where ϕ is positive counterclockwise and denotes an average rotational displacement at point P expressed by

$$\sin \phi(z, t) = \frac{1}{2} \left(\frac{\partial v}{\partial x} - \frac{\partial u}{\partial y} \right)_{x=x_P, y=y_P}. \qquad (2.2\text{-}1c)$$

Remark 5

It can readily by proven that based on Eqs. (2.2-1) one obtains, in conjunction with (2.1-9), the nonlinear counterpart of Eqs. (2.1-10) as

$$u_n = u_P \frac{dy}{ds} - v_P \frac{dx}{ds} - r_t \sin \phi - (1 - \cos \phi) r_n, \qquad (R5a)$$

$$u_t = u_P \frac{dx}{ds} + v_P \frac{dy}{ds} + r_n \sin \phi - (1 - \cos \phi) r_t. \qquad (R5b)$$

For moderately small rotations, $\sin \phi \simeq \phi$ and $1 - \cos \phi \simeq \phi^2/2$, while when $\sin \phi \simeq \phi$ and $\cos \phi \simeq 1$, the linearized counterpart of these equations are obtained.

Needless to say, for the points off the mid-line contour, in Eqs. (2.2-1), x and y should be replaced by X and Y, respectively, their connection being given by Eqs. (2.1-21). Starting with the Green-Lagrange strain tensor expressed in terms of the displacement components V_i and in the Cartesian system of coordinates as

$$\epsilon_{ij} = \frac{1}{2} \left(\frac{\partial V_i}{\partial x_j} + \frac{\partial V_j}{\partial x_i} + \frac{\partial V_r}{\partial x_i} \frac{\partial V_r}{\partial x_j} \right), \qquad (2.2\text{-}2a)$$

and assimilating (V_1, V_2, V_3) and (x_1, x_2, x_3) with (u, v, w) and (x, y, z), respectively, one can extract the strain-displacement relations for the three components ϵ_{xx}, ϵ_{xy} and ϵ_{yy} of Lagrange's strain tensor. With due regard to assumption (ii) in *Section 2.2-2*, these are given by

$$\epsilon_{xx} = \frac{\partial u}{\partial x} + \frac{1}{2} \left[\left(\frac{\partial u}{\partial x} \right)^2 + \left(\frac{\partial v}{\partial x} \right)^2 \right], \qquad (2.2\text{-}2b)$$

$$\epsilon_{yy} = \frac{\partial v}{\partial y} + \frac{1}{2} \left[\left(\frac{\partial u}{\partial y} \right)^2 + \left(\frac{\partial v}{\partial y} \right)^2 \right], \qquad (2.2\text{-}2c)$$

$$\epsilon_{xy} = \frac{1}{2} \left[\frac{\partial u}{\partial y} + \frac{\partial v}{\partial x} + \frac{\partial u}{\partial x} \frac{\partial u}{\partial y} + \frac{\partial v}{\partial x} \frac{\partial v}{\partial y} \right]. \qquad (2.2\text{-}2d)$$

It can readily be seen that substitution in these relations of Eqs. (2.2-1) leads to zero values for all the three strains, thus verifying cross-sectional non-deformability. This also implies that assumption (ii) stated in Section 2.2-2 and that of cross-section non-deformability are mutually compatible.

As in the linear case, the warping displacement, w, is obtained by considering the strains Γ_{sz} and Γ_{nz} off the mid-line contour.

Following the procedure used in the linear case (Sect. 2.1-7), upon representing Γ_{sz} by Eq. (2.1-30), one obtains its expression in terms of displacement quantities as:

$$\Gamma_{sz} = L\left(\frac{\partial v}{\partial z} + \frac{\partial w}{\partial y} + \frac{\partial u}{\partial y}\frac{\partial u}{\partial z} + \frac{\partial v}{\partial y}\frac{\partial v}{\partial z}\right) - M\left(\frac{\partial u}{\partial z} + \frac{\partial w}{\partial x} + \frac{\partial u}{\partial x}\frac{\partial u}{\partial z} + \frac{\partial v}{\partial x}\frac{\partial v}{\partial z}\right)$$

$$+ \delta_c n_{sz}/(hG_{sz}) + \delta_c n N_{sz}/(hG_{sz}) + 2\delta_o n\phi'. \tag{2.2-3}$$

Use of the expressions of direction cosines of the normal, L and M, (2.1-13) and (2.1-23) yields:

$$\Gamma_{sz} = L\frac{\partial v}{\partial z} - M\frac{\partial u}{\partial z} + \frac{\partial w}{\partial S} + \frac{\partial u}{\partial S}\frac{\partial u}{\partial z} + \frac{\partial v}{\partial S}\frac{\partial v}{\partial z} + \delta_c n_{sz}/(hG_{sz})$$

$$+ \delta_c n N_{sz}/(hG_{sz}) + 2\delta_o n\phi'. \tag{2.2-4}$$

Substitution of u and v from Eqs. (2.2-1) into Eq. (2.2-4) and then equating its right-hand side with that of Eq. (2.1-30) one obtains

$$\frac{\partial w}{\partial S} = \frac{dY}{dS}\theta_x(z, t) + \frac{\partial X}{\partial S}\theta_y(z, t) + \delta_c n_{sz}/(hG_{sz}),$$

$$+ \delta_c n N_{sz}/(hG_{sz}) - R_n\phi'(z, t) + 2\delta_o n\phi'(z, t), \tag{2.2-5}$$

where the rotation measures assume the form

$$\theta_x(z, t) = \gamma_{yz} + u'_P \sin\phi - v'_P \cos\phi, \tag{2.2-6a}$$

$$\theta_y(z, t) = \gamma_{xz} - v'_P \sin\phi - u'_P \cos\phi. \tag{2.2-6b}$$

Following the procedure used in Sect. 2.1-7, one obtains formally similar expressions for n_{sz}, N_{sz}, and for the warping displacement w. The latter, when compared to its linear counterpart, differs only by terms involving rotation measures θ_x and θ_y.

2.2-4 Strain Field

By virtue of assumption ii)in Sec. 2.2-2, the three non-zero off mid-line contour components, ϵ_{zz}, ϵ_{sz} and ϵ_{nz} of Lagrange's strain tensor are:

$$\epsilon_{zz} = \frac{\partial w}{\partial z} + \frac{1}{2}\left[\left(\frac{\partial u}{\partial z}\right)^2 + \left(\frac{\partial v}{\partial z}\right)^2\right], \tag{2.2-7a}$$

$$\begin{aligned}
\Gamma_{sz}(\equiv 2\,\epsilon_{sz}) = {}& L\left(\frac{\partial v}{\partial z} + \frac{\partial w}{\partial y} + \frac{\partial u}{\partial y}\frac{\partial u}{\partial z} + \frac{\partial v}{\partial y}\frac{\partial v}{\partial z}\right) \\
& - M\left(\frac{\partial u}{\partial z} + \frac{\partial w}{\partial x} + \frac{\partial u}{\partial x}\frac{\partial u}{\partial z} + \frac{\partial v}{\partial x}\frac{\partial v}{\partial z}\right) \\
& + \delta_c n_{sz}/(hG_{sz}) + \delta_c n N_{sz}/(hG_{sz}) + 2\delta_0 n\phi',
\end{aligned} \tag{2.2-7b}$$

$$\begin{aligned}
\Gamma_{nz}(\equiv 2\,\epsilon_{nz}) = {}& L\left(\frac{\partial u}{\partial z} + \frac{\partial w}{\partial x} + \frac{\partial u}{\partial x}\frac{\partial u}{\partial z} + \frac{\partial v}{\partial x}\frac{\partial v}{\partial z}\right) \\
& + M\left(\frac{\partial v}{\partial z} + \frac{\partial w}{\partial y} + \frac{\partial u}{\partial y}\frac{\partial u}{\partial z} + \frac{\partial v}{\partial y}\frac{\partial v}{\partial z}\right).
\end{aligned} \tag{2.2-7c}$$

Substitution in Eq. (2.2-7a) of displacement representations Eq. (2.2-1) and that of w for points off the mid-line contour, and keeping in mind Eqs. (2.1-11) and (2.1-21), one obtains for ϵ_{zz}

$$\epsilon_{zz}(s, z, n, t) = \epsilon_{zz}^{(0)} + n\epsilon_{zz}^{(1)} + n^2\epsilon_{zz}^{(2)}, \tag{2.2-8}$$

where

$$\begin{aligned}
\epsilon_{zz}^{(0)} = {}& w_o' + y\theta_x' + x\theta_y' - \phi''\overline{F}(s) + \frac{1}{2}\left((u_P')^2 + (v_P')^2\right) \\
& + \frac{1}{2}(\phi')^2\left[(x - x_P)^2 + (y - y_P)^2\right] \\
& - \phi'[\{y(u_P'\cos\phi + v_P'\sin\phi) + x(u_P'\sin\phi - v_P'\cos\phi)\} \\
& - \{y_P(u_P'\cos\phi + v_P'\sin\phi) + x_P(u_P'\sin\phi - v_P'\cos\phi)\}],
\end{aligned} \tag{2.2-9a}$$

$$\begin{aligned}
\epsilon_{zz}^{(1)} = {}& \frac{dy}{ds}\theta_y' - \frac{dx}{ds}\theta_x' + r_t\phi'' - \phi'\left\{\frac{dy}{ds}(u_P'\sin\phi - v_P'\cos\phi)\right. \\
& \left. - \frac{dx}{ds}(v_P'\sin\phi + u_P'\cos\phi)\right\} + r_n(\phi')^2,
\end{aligned} \tag{2.2-9b}$$

$$\epsilon_{zz}^{(2)} = \frac{1}{2}(\phi')^2. \tag{2.2-9c}$$

Although Γ_{sz} and Γ_{nz} can be derived from Eqs. (2.2-7b,c) by using the displacement expressions valid for the off mid-line contour, an easier way is to use the expressions

$$\Gamma_{sz} = L\gamma_{yz} - M\gamma_{xz} + \delta_c \left(\frac{\oint r_n ds}{\oint ds} + 2n \right) \phi'(z,t) + 2\delta_0 n\phi'(z,t),$$

$$(2.2\text{-}10\text{a})$$

$$\Gamma_{nz} = L\gamma_{xz} + M\gamma_{yz}. \qquad (2.2\text{-}10\text{b})$$

Employment of Eqs. (2.1-23) and (2.2-6) in Eqs. (2.2-10), yields:

$$\Gamma_{sz}(s,z,n,t) = \gamma_{sz}^{(0)} + n\gamma_{sz}^{(1)}, \qquad (2.2\text{-}11\text{a})$$

$$\Gamma_{nz}(s,z,n,t) = \gamma_{nz}^{(0)}, \qquad (2.2\text{-}11\text{b})$$

where

$$\gamma_{sz}^{(0)} = \underline{\theta_y \frac{dx}{ds} + \theta_x \frac{dy}{ds}} + (u_P' \cos\phi + v_P' \sin\phi) \frac{dx}{ds}$$

$$+ (v_P' \cos\phi - u_P' \sin\phi) \frac{dy}{ds} + \delta_c \frac{\oint r_n ds}{\oint ds} \phi', \qquad (2.2\text{-}12\text{a})$$

$$\gamma_{sz}^{(1)} = 2\phi', \qquad (2.2\text{-}12\text{b})$$

and

$$\gamma_{nz}^{(0)} = (\theta_y + v_P' \sin\phi + u_P' \cos\phi) \frac{dy}{ds}$$

$$- \underline{(\theta_x - u_P' \sin\phi + v_P' \cos\phi)} \frac{dx}{ds}. \qquad (2.2\text{-}13)$$

These expressions include the effect of transverse shear deformations. If transverse shear effect is neglected, implying that Bernoulli-Euler beam model is adopted, γ_{xz} and γ_{yz} in Eqs. (2.2-6) become zero and consequently

$$\theta_x(z,t) = u_P' \sin\phi - v_P' \cos\phi, \qquad (2.2\text{-}14\text{a})$$

$$\theta_y(z,t) = -v_P' \sin\phi - u_P' \cos\phi. \qquad (2.2\text{-}14\text{b})$$

As a result, for the unshearable beam model, in the expressions of γ_{sz} and γ_{nz}, Eqs. (2.2-12a) and (2.2-13), respectively, the terms underscored by a solid line become immaterial, while in Eqs. (2.2-9a) and (2.2-9b), θ_x and θ_y should be expressed in accordance with (2.2-14).

Equations (2.2-11) through (2.2-13) show that the character of the variation of Γ_{sz} and Γ_{nz} across the wall thickness is the same as for the geometrically linear theory counterpart.

2.3 MULTICELL THIN-WALLED BEAMS

2.3-1 Preliminaries

The wing, the fuselage and other structural elements of aeronautical/aero-space constructions and of ship hulls usually consist of multicell thin-walled cross-section beams. Such beams are also used in structural components of civil engineering and in ship hulls.

Due to their importance, there has been a great deal of work related to the modeling of multicell thin-walled beams In spite of this, the development have concerned almost entirely, the metallic thin-walled beams. In this sense, see the books on aeronautical structures, such as the ones by Rivello (1969), Megson (1974, 1990), Petre (1984), Allen and Haisler (1985), Vasiliev (1986), and Donaldson (1993), those on ship structures (Hughes, 1983; Jensen, 2001), as well as to the one by Oden and Ripperger (1981).

For *composite* thin-walled multicell beams, the papers by Rehfield et al. (1988) and Chandra and Chopra (1992b) are among the first ones in this area.

2.3-2 Torsion of Multi-Cell Beams

In this subsection, the kinematics in torsion of multi-cell beams is considered. Consider a thin-walled beam of arbitrary cross-section, see Fig. 2.3-1. Suppose that the multi-cell beam consists of an arbitrary number of cells which are separated by thin webs. Assume that the cross-sectional distortions during twisting are prevented by transverse stiffening members (ribs or frames), which are considered to be rigid within their plane, but perfectly flexible with regard to deformations normal to their planes. We shall consider for the moment that there is no warping restraint at any cross-section and that the beam is loaded by equal and opposite torques at the end cross-sections. Consequently, the warping function obtained from the free torsion model will provide a cross-sectional distribution for an N cell beam which involves N constant shear flows. As a result, additional equations enabling one to determine these shear flows are needed.

To this end, we use Eq. (2.1-49a) specialized for the points on the mid-line beam contour. For the Rth cell of an N cell beam subjected to a pure torque $\overline{M}_t$, one obtains for the rate of twist in the cell

$$\phi'_R = \oint_R \frac{n_{sz}}{2\Omega_R h G_{sz}} ds, \tag{2.3-1}$$

where the line integral extends around the mid-line contour of the Rth cell and Ω_R is the area enclosed by its mid-line contour.

This equation may also be written as

$$\phi'_R = \frac{1}{2\Omega_R \tilde{G}} \oint_R \frac{n_{sz}}{\tilde{h}}, \tag{2.3-2}$$

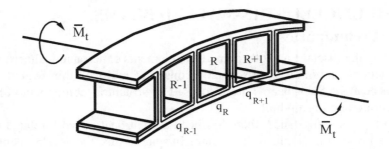

Figure 2.3-1: Multi-cell beam subjected to a twist moment

where $\tilde{h}(s)$ is the modulus-weighted thickness defined by Eq. (2.1-48).

The torque is shared by the N constituent cells, each of them developing in each wall segment a constant shear flow given by Bredt's formula (2.1-49b). For the problem at hand it has the following form:

$$\phi'_R = \frac{1}{2\Omega_R \tilde{G}} \left(q_R \delta_R - q_{R-1} \delta_{R,R-1} - q_{R+1} \delta_{R,R+1}\right). \qquad (2.3\text{-}3)$$

In this equation, for the sake of convenience, the shear flow n_{sz} was denoted as q where q_R, q_{R+1} and q_{R-1} are the shear flows around the Rth, $(R+1)th$ and $(R-1)th$ cells, respectively, while δ_R, $\delta_{R,R-1}$ and $\delta_{R,R+1}$ denote the integral $\int ds/\tilde{h}$ considered for all the walls enclosing the Rth cell, for the web common to the Rth and $(R-1)th$ cells, and for the web common to the Rth and $(R+1)th$ cells, respectively. In an explicit form these are expressed as:

$$\delta_R = \oint_R \frac{ds}{\tilde{h}}, \qquad \delta_{R,R-1} = \int_{R,R-1} \frac{ds}{\tilde{h}}, \qquad \delta_{R,R+1} = \int_{R,R+1} \frac{ds}{\tilde{h}}. \qquad (2.3\text{-}4)$$

Equation (2.3-4) is applicable to multicell beams in which the cells are connected in sequence, in the sense that cell 1 is connected to cell 2, cell 2 to cells 1 and 3, and so on. For the case of a cell R bounded by m cells, Eq. (2.3-3) changes to

$$\phi'_R = \frac{1}{2\Omega_R \tilde{G}} \left(q_R \delta_R - \sum_{r=1}^{m} q_r \delta_{r,R}\right), \qquad (2.3\text{-}5)$$

where

$$\delta_{r,R} = \int_{r,R} \frac{ds}{\tilde{h}}, \qquad (2.3\text{-}6)$$

defines the line integral over the web common to the Rth and rth cells.

Equation (2.3-3) may be applied to each of the Rth cells. As a result, one obtains

$$\boldsymbol{\phi}' = S\mathbf{q}, \qquad (2.3\text{-}7)$$

where, for N constituent cells,

$$\boldsymbol{\phi}' = \{\phi'_1, \phi'_2, \cdots \phi'_{N-1}, \phi'_N\}^T, \qquad (2.3\text{-}8)$$

$$\mathbf{q} = \{q_1, q_2, \cdots q_{N-1}, q_N\}^T. \qquad (2.3\text{-}9)$$

In Eqs. (2.3-7) through (2.3-9) **S** is the flexibility matrix of twist rate, while $\{\ \}^T$ denotes the transpose of the vector $\{\ \}$. For the cross-section beam configuration depicted in Fig. 2.3-2, [S] is given by

$$
\mathbf{S} \equiv
\begin{bmatrix}
S_{11} & S_{12} & 0 & 0 & \cdots 0 & 0 & 0 & 0 & 0 & 0 & 0 \\
S_{21} & S_{22} & S_{23} & 0 & \cdots 0 & 0 & 0 & 0 & 0 & 0 & 0 \\
\cdots \\
0 & 0 & 0 & & \cdots 0 & S_{R,R-1} & S_{R,R} & S_{R,R+1} & 0 & 0 & 0 \\
\cdots \\
\cdots \\
0 & & & & \cdots & & & & 0 & S_{N,N-1} & S_{N,N}
\end{bmatrix}
$$

$$(2.3\text{-}10)$$

where its elements are

$$S_{R,R} = \frac{1}{2\Omega_R \tilde{G}} \delta_R, \quad S_{R,R-1} = \frac{1}{2\Omega_R \tilde{G}} \delta_{R,R-1}, \quad S_{R,R+1} = \frac{1}{2\Omega_R \tilde{G}} \delta_{R,R+1}. \qquad (2.3\text{-}11)$$

Equation (2.3-7) shows that $S_{R,R}$ represents the twist rate of cell R due to a unit shear flow in the cell $R (q_R = 1, q_{R-1} = q_{R+1} = 0)$, while $S_{R,R+1}$ denotes the twist rate of cell R due to a unit shear flow along the web common to the Rth and $(R+1)th$ cells $(q_{R+1} = 1, q_R = q_{R-1} = 0)$.

Imposing the condition of the beam cross-section invariance, implying

$$\phi'_1 = \phi'_2 = \cdots \cdots = \phi'_{N-1} = \phi'_N \equiv \phi', \qquad (2.3\text{-}12)$$

we can write Eq. (2.3-12)

$$\boldsymbol{\phi}' = \mathbf{I}\,\phi', \qquad (2.3\text{-}13)$$

where **I** is the unit vector.

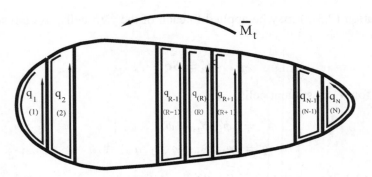

Figure 2.3-2: Shear flows in the multi-cell beam

Inversion of Eq. (2.3-7) with due consideration of (2.3-13) yields

$$\mathbf{q} = \mathbf{H}\phi', \qquad (2.3\text{-}14)$$

where

$$\mathbf{H} = \mathbf{S}^{-1}\mathbf{I}. \qquad (2.3\text{-}15)$$

For a beam with N cells joined together, the total torque is the sum of the individual torques in each cell. Upon using Eq. (2.1-50), one obtains the equation expressing this condition:

$$\overline{M}_t = 2\mathbf{\Omega}^T \mathbf{q}, \qquad (2.3\text{-}16)$$

where

$$\mathbf{\Omega} = \{\Omega_1, \Omega_2, \cdots \Omega_{N-1}, \Omega_N).\}^T$$

Replacement of Eqs. (2.3-14) and (2.3-15) in Eq. (2.3-16) yields

$$\overline{M}_t = 2\mathbf{\Omega}^T \mathbf{S}^{-1}\mathbf{I}\phi'. \qquad (2.3\text{-}17)$$

Equation (2.3-17) considered in conjunction with Eq. (2.1-52a) provides the global torsional stiffness of the multi-cell beam

$$(\tilde{G}\overline{J})_{\text{global}} = \frac{\overline{M}_t}{\phi'} = 2\mathbf{\Omega}^T \mathbf{S}^{-1}\mathbf{I}. \qquad (2.3\text{-}18)$$

In such a way, Eqs. (2.3-14) and (2.3-17) supply (N+1) equations necessary for obtaining the N shear flows and the twist rate ϕ'.

Solution of the set of simultaneous equations yields the shear flows in each cell and, implicitly, in each web. For example, for a wall separating the Rth and the $(R + 1)th$ cells, $q_{R,R+1} = q_R - q_{R+1}$ where $q_{R,R+1}$ is assumed to be positive upward, whereas for a wall separating the $(R-1)th$ and the Rth cells, $q_{R-1,R} = q_{R-1} - q_R$, where $q_{R-1,R}$ is assumed to be positive downward.

Equation (2.3-14) enables one to determine the shear flow in free torsion for each element of the multi-cell beam as

$$\left\{ \begin{array}{c} q_1 \\ q_{12} \\ q_2 \\ q_{32} \\ \vdots \\ q_N \end{array} \right\} = \left\{ \begin{array}{c} q_1 \\ q_1 - q_2 \\ q_2 \\ q_3 - q_2 \\ \vdots \\ q_N \end{array} \right\} = \left\{ \begin{array}{c} H_1 \\ H_2 - H_1 \\ H_2 \\ H_3 - H_2 \\ \vdots \\ H_N \end{array} \right\} \phi'. \qquad (2.3\text{-}19)$$

As a result of Eq. (2.3-19), the distribution of shear strain becomes

$$\left\{ \begin{array}{c} \gamma_{sz}^1 \\ \gamma_{sz}^{12} \\ \gamma_{sz}^2 \\ \gamma_{sz}^{23} \\ \vdots \\ \gamma_{sz}^N \end{array} \right\} = \left\{ \begin{array}{c} H_1/(hG_{sz})_1 \\ (H_2 - H_1)/(hG_{sz})_{12} \\ H_2/(hG_{sz})_2 \\ (H_3 - H_2)/(hG_{sz})_{23} \\ \vdots \\ H_N/(hG_{sz})_N \end{array} \right\} \phi' \qquad (2.3\text{-}20a)$$

or, equivalently as

$$\{\gamma_{sz}\} = \{\lambda\} \phi', \qquad (2.3\text{-}20b)$$

where $\{\gamma_{sz}\}$ and $\{\lambda\}$ are $(2N-1) \times 1$ vectors.

In this way, one can obtain the axial displacement of the beam featuring warping inhibition. Proceeding as in Sect. 2.1-7 and assuming that the axial displacement at a section of a beam exhibits the same transverse distribution as in Saint Venant's theory of torsion, it is possible to determine $w(s, z, t)$. Its expression becomes similar to that of a single-cell beam, the difference consisting of the torsion-related warping function.

2.3-3 Warping Functions

The warping displacement in pure twist can be obtained as the product of the warping function and the rate of twist

$$w(s, z, t) = \overline{F}(s)\phi'(z, t), \qquad (2.3\text{-}21)$$

where

$$\overline{F}(s) = \int_o^s (\lambda - r_n)ds + C, \qquad (2.3\text{-}22)$$

the constant C being determined by satisfying the condition $\oint \overline{F}(s)ds = 0$, thus yielding,

$$C = \oint \overline{F}ds / \oint ds. \qquad (2.3\text{-}23)$$

In (2.3-23) $\oint (\cdot)ds$ denotes the integration around the whole mid-line contour of the multi-cell cross-section.

Note that fulfillment of $\oint \overline{F}(s)ds = 0$, implies

$$\oint_K \frac{\partial w}{\partial s}ds = 0. \qquad K = \overline{(1, N)} \qquad (2.3\text{-}24)$$

and vice-versa.

• *Example 1. Determination of the primary warping function for a single-cell section beam*
In order to underline the implications of the location of the pole and coordinate origin on the determination of the warping function, the case of single-cell thin-walled beams is considered (see Fig. E1a). In this case $\overline{F}(s)$ is expressed by Eq. (2.1-45) and for the problem at hand $\psi = 2\Omega/\beta = 5/3$.

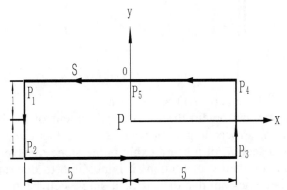

Figure E1a: Single-cell thin-walled beam

Two cases, a) and b) are considered:
Case a) *The pole is selected at the origin of coordinates while the contour origin at (0,1).*
Consequently $\overline{F}(s)$ is

- for the path $0 \to p_1$: $\overline{F}(s) = \int_0^s (1 - \frac{5}{3})ds = -\frac{2}{3}s$ where $s \in [0, 5]$

- for the path $p_1 \to p_2$: $\overline{F}(s) = -\frac{2(5)}{3} + \int_5^s (5 - \frac{5}{3})ds = -\frac{10}{3} + \frac{10}{3}(s - 5)$ where $s \in [5, 7]$

- for the path $p_2 \to p_3$: $\overline{F}(s) = -\frac{10}{3} + \frac{(10)(2)}{3} + \int_7^s (1 - \frac{5}{3})ds = \frac{10}{3} - \frac{2}{3}(s - 7)$ where $s \in [7, 17]$

- for the path $p_3 \to p_4$: $\overline{F}(s) = \frac{10}{3} - \frac{2(10)}{3} + \int_{17}^{s}(5 - \frac{5}{3})ds = -\frac{10}{3} + \frac{10}{3}(s - 17)$
  where $s \in [17, 19]$

- for the path $p_4 \to p_5$: $\overline{F}(s) = -\frac{10}{3} + \frac{10(2)}{3} + \int_{19}^{s}(1 - \frac{5}{3})ds = \frac{10}{3} - \frac{2}{3}(s - 19)$ where $s \in [19, 24]$.

  The graph representing $\overline{F}(s)$ in Fig. E1b, shows that $\oint \overline{F}(s)ds = 0$.

Case b): *The pole is located at the origin of coordinate while the contour origin at (-5, 1).* As a result:

- for the path $p_1 \to p_2$: $\overline{F}(s) = \int_{0}^{s}(5 - \frac{5}{3})ds = \frac{10s}{3}$ where $s \in [0, 2]$

- for the path $p_2 \to p_3$: $\overline{F}(s) = \frac{(10)(2)}{3} + \int_{2}^{s}(1 - \frac{5}{3})ds = \frac{20}{3} - \frac{2}{3}(s - 2)$ where $s \in [2, 12]$

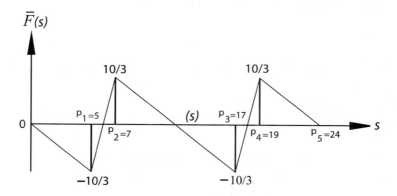

Figure E1b: Distribution of $\overline{F}(s)$

- for the path $s_3 \to s_4$: $\overline{F}(s) = \frac{20}{3} - \frac{(2)(10)}{3} + \int_{12}^{s}(5 - \frac{5}{3})ds = \frac{10}{3}(s - 12)$ where $s \in [12, 14]$

- for the path $s_4 \to s_5$: $\overline{F}(s) = \frac{(10)(2)}{3} + \int_{14}^{s}(1 - \frac{5}{3})ds = \frac{20}{3} - \frac{2}{3}(s - 14)$ where $s \in [14, 24]$

  It results that $\oint \overline{F}(s)ds = \frac{20(2)}{3} + \frac{20(10)}{3} = 80$ and $C = \oint \overline{F}(ds)/\oint ds = \frac{80}{24} = \frac{10}{3}$.

  Shifted by $\frac{10}{3}$ downwards, the graph $\overline{F}(s)$ will yield $\oint \overline{F}ds = 0$.

  The representation of $\overline{F}(s)$ before and after the shift is given in Fig. E1c.

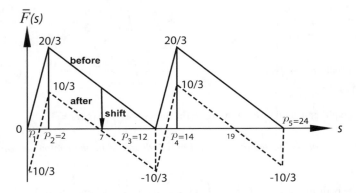

Figure E1c: Representation of $\overline{F}(s)$ before and after the shift

• *Example 2. Determination of the primary warping function for a two-cell section beam*

Consider a two-cell section depicted in Fig. E2a. For the sake of simplicity one assumes that both the wall thickness $\tilde{h}$ and shear modulus $\tilde{G}_{sz}$, properly non-dimensionalized, are constant in all the constituent branches i.e. $\tilde{h} \equiv 1$ and $\tilde{G}_{sz} \equiv G$.

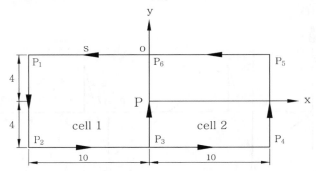

Figure E2a: Two-cell section beam

One assumes that the pole P is chosen to be at the coordinate origin, while the contour origin o at the center of the upper flange. As basic ingredients, δ_R and $\delta_{R,R-1}$ have to be determined.

By virtue of Eq. (2.3-4) we have

$$\delta_1 = \oint_1 \frac{ds}{h} = \frac{10}{1} + \frac{8}{1} + \frac{10}{1} + \frac{8}{1} = 36$$

$$\delta_{1,2} = \int_{1,2} \frac{ds}{h} = \frac{8}{1} = 8 \tag{E2a}$$

$$\delta_2 = \oint_2 \frac{ds}{h} = 36$$

With due consideration of (E1a), Eq. (2.3-3) yields:

$$\phi_1' = \frac{1}{2\Omega_1 G}(\delta_1 q_1 - \delta_{1,2} q_2) = \frac{1}{40G}(9q_1 - 2q_2)$$

$$\phi_2' = \frac{1}{2\Omega_2 G}(\delta_2 q_2 - \delta_{1,2} q_1) = \frac{1}{40G}(-2q_1 + 9q_2). \tag{E2b}$$

The counterpart, for this case, of Eq. (2.3-7) is

$$\left\{ \begin{array}{c} \phi_1' \\ \phi_2' \end{array} \right\} = \frac{1}{40G} \left[\begin{array}{cc} 9 & -2 \\ -2 & 9 \end{array} \right] \left\{ \begin{array}{c} q_1 \\ q_2 \end{array} \right\}, \tag{E2c}$$

and that of Eq. (2.3-14) is

$$\left\{ \begin{array}{c} q_1 \\ q_2 \end{array} \right\} = \left\{ \begin{array}{c} \dfrac{40G}{7} \\[2mm] \dfrac{40G}{7} \end{array} \right\} \phi', \tag{E2d}$$

and Eq. (2.3-19) implies that $H_1 = H_2 = \dfrac{40G}{7}$ and $H_{12} = H_2 - H_1 = 0$.
As a result, in conjunction with Eqs. (2.3-20a,b) we have for elements of $\{\lambda\}$

- for the path $0 \to p_1 \to p_2 \to p_3$ $\qquad\qquad \lambda_1 = \dfrac{H_1}{h_1 G} = \dfrac{40}{7}$

- for the path $p_3 \to p_4 \to p_5 \to p_6$ $\qquad\qquad \lambda_2 = \dfrac{H_2}{h_2 G} = \dfrac{40}{7}$

- for the path $p_3 \to p_6$ $\qquad\qquad\qquad\qquad \lambda_{12} = 0.$

   Now applying Eq. (2.3-22) we have for $\overline{F}(s)$:

- path $0 \to p_1 (r_n = 4)$: $\overline{F}(s) = \displaystyle\int_o^s \left(\frac{40}{7} - 4 \right) ds = \frac{12}{7} s$ where $s \in [0, 10]$

- path $p_1 \to p_2$ $(r_n = 10)$: $\overline{F}(s) = \dfrac{120}{7} + \displaystyle\int_o^s \left(\frac{40}{7} - 10 \right) ds = \frac{120}{7} - \frac{30}{7} s$ where $s \in [0, 8]$

- path $p_2 \to p_3$ $(r_n = 4)$: $\overline{F}(s) = \dfrac{120}{7} - \dfrac{(30)(8)}{7} + \displaystyle\int_o^s \left(\frac{40}{7} - 4 \right) ds = -\frac{120}{7} + \frac{12}{7} s$ where $s \in [0, 10]$

- path $p_3 \to p_4$ $(r_n = 4)$: $\overline{F}(s) = -\dfrac{120}{7} + \dfrac{(12)(10)}{7} + \displaystyle\int_o^s \left(\frac{40}{7} - 4 \right) ds = \frac{12}{7} s$ where $s \in [0, 10]$

- path $p_4 \to p_5$ $(r_n = 10)$: $\overline{F}(s) = \dfrac{(12)(10)}{7} + \displaystyle\int_o^s \left(\frac{40}{7} - 10 \right) ds = \frac{120}{7} - \frac{30}{7} s$ where $s \in [0, 8]$

- path $p_5 \to p_6$ $(r_n = 4)$: $\overline{F}(s) = \dfrac{120}{7} - \dfrac{(30)(8)}{7} + \displaystyle\int_o^s \left(\frac{40}{7} - 4 \right) ds = -\frac{120}{7} + \frac{12}{7} s$ where $s \in [0, 10]$

- path $p_3 \to p_6$ $(r_n = 0)$: $\overline{F}(s) = -\dfrac{120}{7} + \dfrac{(12)(10)}{7} + \displaystyle\int_o^s (0 - 0) ds = 0$

Graphical representations of $\overline{F}(s)$ are displayed in Fig. E2b, which shows that the condition $\displaystyle\oint \overline{F}(s) ds = 0$ is fulfilled, entailing that in Eq. (2.3-22), $C \equiv 0$.

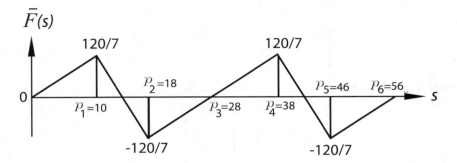

Figure E2b: Distribution of $\overline{F}(s)$ along the beam contour

In this case the contour origin was located on the cross-section symmetry axis. For other locations of the contour origin, it can occur that $\oint \overline{F}(s)ds \neq 0$ and, as a result, the constant C should be determined as to fulfill Eq. (2.3-23).

• *Example 3. The case of a three-cell section beam.*
Consider the three-cell section beam represented in Fig. E3a. As in the *Example 2*, the pole P and contour origin are located at the coordinate origin and at the center of the upper flange, respectively. One assumes that $\tilde{G}_{sz} \equiv G$ for all flanges and webs, while $\tilde{h} = 1$ everywhere excepting the web 2 where $\tilde{h} = 2$. It results that

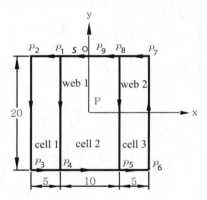

Figure E3a: Three-cell section beam

$$\delta_1 = 50, \quad \delta_{1,2} = 20, \quad \delta_2 = 50, \quad \delta_{2,3} = 10, \quad \delta_3 = 40$$

Based on Eq. (2.3-3) one obtains:

$$\phi_1' = \frac{1}{2\Omega_1 G}(\delta_1 q_1 - \delta_{1,2} q_2) = \frac{1}{2(100)G}(50 q_1 - 20 q_2),$$

$$\phi_2' = \frac{1}{2\Omega_2 G}(-\delta_{1,2} q_1 + \delta_2 q_2 - \delta_{2,3} q_3) = \frac{1}{2(200)G}(-20 q_1 + 50 q_2 - 10 q_3),$$

$$\phi_3' = \frac{1}{2\Omega_3 G}(-\delta_{2,3} q_2 + \delta_3 q_3) = \frac{1}{2(100)G}(-10 q_2 + 40 q_3),$$

or in matrix form as

$$
\left\{\begin{array}{c} \phi_1' \\ \phi_2' \\ \phi_3' \end{array}\right\} = \frac{1}{40G} \left[\begin{array}{ccc} 10 & -4 & 0 \\ -2 & 5 & -1 \\ 0 & -2 & 8 \end{array}\right] \left\{\begin{array}{c} q_1 \\ q_2 \\ q_3 \end{array}\right\}.
$$

Conversely, from the previous result, the counterpart of Eqs. (2.3-14) is obtained, where,

$$
\mathbf{H} = \left\{\frac{740G}{79}, \frac{1060G}{79}, \frac{660G}{79}\right\}^T
$$

As in the previous example, the elements of the vector $\{\lambda\}$ can be obtained as:

$$
\lambda_1 = \frac{H_1}{(hG)_1} = \frac{740}{79} \text{ for the path } p_1 \to p_2 \to p_3 \to p_4
$$

$$
\lambda_{12} = \frac{(H_2 - H_1)}{(hG)_{1,2}} = \frac{320}{79} \text{ for a path } p_1 \to p_4
$$

$$
\lambda_2 = \frac{H_2}{(hG)_2} = \frac{1060}{79} \text{ for a path } 0 \to p_1, p_8 \to p_9, p_4 \to p_5
$$

$$
\lambda_{23} = \frac{H_3 - H_2}{(hG)_{2,3}} = -\frac{400}{79(2)} = -\frac{200}{79} \text{ for a path } p_8 \to p_5
$$

$$
\lambda_3 = \frac{H_3}{(hG)_3} = \frac{660}{79} \text{ for a path } p_5 \to p_6 \to p_7 \to p_8
$$

Using Eq. (2.3-22), $\overline{F}(s)$ is obtained as:

- on the path $0 \to p_1$, $(r_n = 10)$, $\overline{F}(s) = \int_o^s \left(\frac{1060}{79} - 10\right) ds = \frac{270}{79}s$ where $s \in [0, 5]$

- on the path $p_1 \to p_2 \to p_3 \to p_4$, $(r_n = 10)$, $\overline{F}(s) = \frac{(270)(5)}{79} + \int_o^s \left(\frac{740}{79} - 10\right) ds = \frac{1350}{79} - \frac{50}{79}s$, where $s \in [0, 30]$

- on the path $p_4 \to p_5$, $(r_n = 10)$, $\overline{F}(s) = \frac{1350}{79} - \frac{(50)(30)}{79} + \int_o^s \left(\frac{1060}{79} - 10\right) ds = -\frac{150}{79} + \frac{270}{79}s$ where $s \in [0, 10]$

- on the path $p_5 \to p_6 \to p_7 \to p_8$, $(r_n = 10)$, $\overline{F}(s) = -\frac{150}{79} + \frac{270}{79}(10) + \int_o^s \left(\frac{660}{79} - 10\right) ds = \frac{2550}{79} - \frac{130}{79}s$ where $s \in [0, 30]$

- on the path $p_8 \to p_9$, $r_n = 10$, $\overline{F}(s) = \frac{2550}{79} - \frac{(130)(30)}{79} + \int_o^s \left(\frac{1060}{79} - 10\right) ds = -\frac{1350}{79} + \frac{270}{79}s$, where $s \in [0, 5]$

- on the path $p_1 \to p_4$, (web1), $(r_n = 5)$, $\overline{F}(s) = \frac{(270)(5)}{79} + \int_o^s \left(\frac{320}{79} - 5\right) ds = \frac{1350}{79} - \frac{75}{79}s$, where $s \in [0, 20]$

- on the path $p_8 \to p_5$, $(web2)$, $\overline{F}(s) = \frac{2550}{79} - \frac{(130)(30)}{79} + \int_o^s \left(-\frac{200}{79} - (-5) \right) ds =$

$-\dfrac{1350}{79} + \dfrac{195}{79} s$ where $s \in [0, 20]$.

On this path $r_n = x \dfrac{dy}{ds} - y \dfrac{dx}{ds} = 5(-1) - y(0) = -5$

The graphical representation of $\overline{F}(s)$ is displayed in Figs. E3b through E3d. This shows that
$\oint \overline{F}(s) ds = \dfrac{72000}{79} \neq 0$.

As a result, in order to fulfill condition (2.3-23) the whole graph of $\overline{F}(s)$ should be shifted
downward by an amount $\oint \overline{F}(s) ds / \oint ds = \dfrac{72000}{(79)(120)} = \dfrac{600}{79}$.

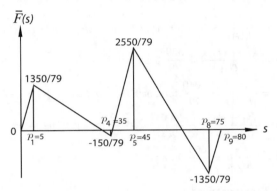

Figure E3b: Distribution of $\overline{F}(s)$ along the beam contour

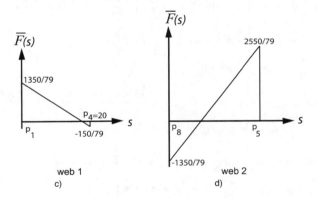

Figure E3c,d: Distribution of $\overline{F}(s)$ along the webs

2.4 HIGHER-ORDER THIN-WALLED BEAM THEORY

2.4-1 Preliminaries

In the previous developments, the concept of the *first order shear deformation theory (FSDT)* was used to derive the kinematical equations. The classical (unshearable) beam counterpart was obtained as a special case of these equations. As was shown in Eqs. (2.1-63) and (2.2.11b), (2.2-13), consistent with the FSDT beam model, transverse shear strains (and hence the transverse shear stresses), are uniform across the beam thickness, in the sense that these are functions of the (s, z) coordinates, and not of the thickness coordinate, n. As a result, within that formulation, the tangential traction conditions on the bounding surfaces $n = \pm h/2$ of the beam are not fulfilled, entailing that the uniform variation of transverse shear strains (and stresses) through the wall thickness violates this physical requirement.

Its fulfilment, together with the need for a more exact approach of *thick-walled* composite structures have prompted the development of higher-order thin-walled beam theories (HSDT).

In striking contrast to plate/shell type structures in which there is an extensive literature addressing problems of modeling and behavior within a higher order theory (see e.g. Librescu, 1975), for thin-walled beams, such a research work, is extremely limited. The reader is referred to the papers by Kim and White (1996, 1997) and Suresh and Nagaraj (1996). Moreover, with the exception of the results due to Dökmeci (1972) for solid beams, and Chandiramani et al. (2001) for geometrically nonlinear thin-walled beams, the literature related to higher order shear deformation theory is quite void of such results.

2.4-2 Assumptions and Basic Equations

To develop the HSDT, a number of assumptions are stipulated: those in *Section* 2.2-2 and assumption a) from Sect. 2.1-3. In addition, consistent with the HSDT, we postulate a quadratic variation through the wall thickness of transverse shear strains

$$\Gamma_{xz}(z, n, t) = \gamma_{xz}^{(0)}(z, t) + n\eta_x(z, t) + n^2\zeta_x(z, t),$$
$$\Gamma_{yz}(z, n, t) = \gamma_{yz}^{(0)}(z, t) + n\eta_y(z, t) + n^2\zeta_y(z, t), \qquad (2.4\text{-}1)$$

where η_x, η_y and ζ_x, ζ_y are for the moment 1-D unknown quantities.

The absence of shear tractions on the external surfaces of the beam requires

$$\Gamma_{nz} = 0 \text{ at } n = \pm h/2 \qquad (2.4\text{-}2)$$

and by virtue of Eq. (2.2-10), Γ_{xz} and Γ_{yz} should vanish at $n = \pm h/2$. This requirement yields the representations

$$\Gamma_{xz}(s, z, n, t) = (1 - 4n^2/h^2)\gamma_{xz}^{(0)}(z, t),$$
$$\Gamma_{yz}(s, z, n, t) = (1 - 4n^2/h^2)\gamma_{yz}^{(0)}(z, t). \qquad (2.4\text{-}3a,b)$$

For simplicity suppose that the pole $P(x_P, y_P)$ is chosen as $P(0, 0)$, Eqs. (2.2-1) reduce to

$$u\,(x, y, z, t)\ = u_P(z, t) - x(1 - \cos\phi) - y\sin\phi,$$
$$v\,(x, t, z, t)\ = v_P(z, t) + x\sin\phi - y(1 - \cos\phi). \tag{2.4-4}$$

On the other hand, Eqs. (2.2-7) specialized for $n = 0$, used in conjunction with (2.1-30), yield

$$\gamma_{sz} = l\,\gamma_{yz} - m\gamma_{xz} + n_{sz}/(hG_{sz}) = l\,\frac{\partial v}{\partial z} - m\frac{\partial u}{\partial z} + \frac{\partial w}{\partial s} + \frac{\partial u}{\partial s}\frac{\partial u}{\partial z}$$
$$+ \frac{\partial v}{\partial s}\frac{\partial v}{\partial z} + n_{sz}/(hG_{sz}), \tag{2.4-5a}$$

$$\gamma_{nz} = l\,\gamma_{xz} + m\,\gamma_{yz} = l\,\frac{\partial u}{\partial z} + m\frac{\partial v}{\partial z} + \frac{\partial w}{\partial n} + \frac{\partial u}{\partial n}\frac{\partial u}{\partial z} + \frac{\partial v}{\partial n}\frac{\partial v}{\partial z}, \tag{2.4-5b}$$

where l and m are the direction cosines defined by Eqs. (2.1-13). Using in (2.4-5a) their expressions, and enforcing the continuity of w over the closed mid-line contour one obtains

$$n_{sz} = \frac{\oint r_n ds}{\oint \dfrac{ds}{h(s)G_{sz}(s)}}\phi'(z, t). \tag{2.4-6}$$

Use of (2.4-6) in (2.4-5a) followed by its integration with respect to s from a conveniently chosen mid-contour origin (x_0, y_0), and of the integration of (2.4-5b) across the wall thickness in the interval $[0, n)$, yields

$$w(s, z, n, t)\ =\ \bar{w}(z, t) - (\bar{F}_w(s) - nr_t)\phi'$$
$$+ \left(y(s) - n\frac{dx}{ds}\right)\left[\delta_t\theta_x + \delta_e(u'_P\sin\phi - v'_P\cos\phi)\right]$$
$$+ \left(x(s) + n\frac{dy}{ds}\right)\left[\delta_t\theta_y - \delta_e(u'_P\cos\phi + v'_P\sin\phi)\right]$$
$$- \delta_h\frac{4n^3}{3h^2}\frac{dy}{ds}[v'_P\sin\phi + u'_P\cos\phi - \theta_y] + \delta_h\frac{4n^3}{3h^2}\frac{dx}{ds}[v'_P\cos\phi$$
$$- u'_P\sin\phi + \theta_x], \tag{2.4-7}$$

where

$$\theta_y(z, t)\ =\ -u'_P\cos\phi - v'_P\sin\phi + \gamma_{xz}^{(0)},$$

$$\theta_x(x, t)\ = u'_P\sin\phi - v'_P\cos\phi + \gamma_{yz}^{(0)}, \tag{2.4-8}$$

$\bar{F}(s)$ being the primary warping function defined by Eq. (2.1-44a), while

$$\bar{w} = w_0(z, t) - y_0\,\theta_x + x_0\theta_y. \tag{2.4-9}$$

The tracers $(\delta_e,\ \delta_t,\ \delta_h)$ appearing in Eq. (2.4-7) take the values $(1, 0,\ 0)$, $(0, 1,\ 0)$ and $(0,\ 1,\ 1)$ according to whether the beam is unshearable, first-order shearable, or higher- order shearable, respectively.

Replacement of (2.4-4), (2.4-7) and (2.4-8) in the expressions of Lagrange's strain components as given by Eqs. (2.2-7), yields

$$\epsilon_{zz} = \sum_{i=0}^{3} n^i \epsilon_{zz}^{(i)}, \quad \Gamma_{sz} = \sum_{i=0}^{3} n^i \gamma_{sz}^{(i)}, \quad \Gamma_{nz} = \left(\delta_t - \frac{4n^2}{h^2}\delta_h\right)\gamma_{nz}^{(0)}.$$

$$\tag{2.4-10a-c}$$

In these expressions the 2-D strain measures assume the form

$$\begin{aligned}
\epsilon_{zz}^{(0)} &= w_0' + \frac{1}{2}\big((u_p'^2) + (v_p'^2)\big) + \frac{1}{2}[(x^2 + y^2)](\phi'^2) - \bar{F}\,\phi'' \\
&\quad - \delta_t\{(x[u_p'\sin\phi - v_p'\cos\phi] + y[u_p'\cos\phi + v_p'\sin\phi])\phi' \\
&\quad - (y\theta_x' + x\theta_y')\} + \delta_e\{y[u_p''\sin\phi - v_p''\cos\phi] \\
&\quad - x[u_p''\cos\phi + v_p''\sin\phi]\}, \tag{2.4-11a}
\end{aligned}$$

$$\begin{aligned}
\epsilon_{zz}^{(1)} &= r_n(s)(\phi'^2) + r_t(s)\phi'' \\
&\quad + \delta_t\{(m\theta_x' + l\theta_y') - [l(u_p'\sin\phi - v_p'\cos\phi) + m(u_p'\cos\phi \\
&\quad + v_p'\sin\phi)]\phi'\} + \delta_e\{m[u_p''\sin\phi - v_p''\cos\phi] \tag{2.4-11b} \\
&\quad - l[u_p''\cos\phi + v_p''\sin\phi]\},
\end{aligned}$$

$$\epsilon_{zz}^{(2)} = (\phi'^2/2), \tag{2.4-11c}$$

$$\begin{aligned}
\epsilon_{zz}^{(3)} &= \frac{4\delta_h}{3h^2}\{l(\theta_y' - v_p'\phi'\cos\phi + u_p'\phi'\sin\phi - v_p''\sin\phi - u_p''\cos\phi) \\
&\quad - m(\theta_x' + v_p'\phi'\sin\phi + u_p'\phi'\cos\phi - v_p''\cos\phi + u_p''\sin\phi)\}, \\
&\tag{2.4-11d}
\end{aligned}$$

$$\begin{aligned}
\gamma_{sz}^{(0)} &= \phi'\psi + \delta_t\{(l[\theta_x - u_p'\sin\phi + v_p'\cos\phi] - m[\theta_y + u_p'\cos\phi \\
&\quad + v_p'\sin\phi]\}, \tag{2.4-11e}
\end{aligned}$$

$$\gamma_{sz}^{(1)} = 2\phi', \quad \gamma_{sz}^{(2)} = \frac{4\delta_h}{h^2}(\phi'\psi - \gamma_{sz}^{(0)}), \tag{2.4-11f,g}$$

$$\gamma_{nz}^{(0)} = \delta_t\{m[\theta_x - u_p'\sin\phi + v_p'\cos\phi] - l[\theta_y + u_p'\cos\phi + v_p'\sin\phi]\}. \tag{2.4-11h}$$

Here ψ is the torsional function as defined by Eq. (2.1-44c). Note that in spite of the presence of higher order terms in the representation of 3-D strains, the same 1-D displacement quantities, namely u_P, v_P, w_0, ϕ, θ_x and θ_y are involved in both FSDT and HSDT theories.

Needless to say, the 2-D strain measures (2.4-11) specialized for the FSDT case, coincide with those supplied in Section 2.2-4, when in the latter ones $x_P = y_P = 0$.

2.5 BIBLIOGRAPHICAL COMMENTS

The specialized literature reveals that there are alternative ways to derive the theory of thin-walled beams. One of them is based on the asymptotic analysis of the equations of the classical shell theory. In this sense, in a pioneering work, by using this methodology, Gol'denveiser (1949) has shown that the assumption of the cross-section non-deformability remains still accurate, even for short thin-walled beams. Models of thin-walled anisotropic beams based on this procedure have been developed by Berdichevsky et al. (1992), Badir et al. (1993). For geometrically nonlinear thin-walled beam theory this idea was followed by Hodges et al. (1992). Using the Kirchhoffean nonlinear shell theory in conjunction with the application of Hamilton's principle, a nonlinear theory of thin-walled beams was also devised by Meredith and Witmer (1981).

• The idea of expressing the warping displacements for the restrained warping case as the superposition of a distribution of the warping displacement for the unrestrained warping and of a number of generalized warping modes was prompted by Capurso (1964). The concept of generalized warping functions was used further by Pedersen (1991), and De Lorenzis and La Tegola (2005). Developments of this concept are also due to Bauchau (1985), Bauchau and Hong (1987a,b, 1988) and Bauchau et al. (1987).

• The problem of the constrained torsion of isotropic closed-section thin-walled beams, was addressed in the pioneering works by Vlasov (1961), von Kármán and Christensen (1944), Fine and Williams (1945). A good discussion on the issue of free/constrained warping on composite beams is provided by Loughlan and Ata (1995,1997), while in the paper by Librescu et al. (2003), new results related with the implications of anisotropy and heterogeneity on statics and dynamics of this problem have been provided.

• A pioneering work devoted to the analysis of thin-walled multi-cell metallic beams was accomplished by Argyris and Dunne (1947, 1949) and later by Benscoter (1954).

REFERENCES

Allen, D. H. and Haisler, W. E. (1985) *Introduction to Aerospace Structural Systems*, John Wiley & Sons, New York, Chichester, Briscone.

Argyris, J. H. and Dunne, P. C. (1947) "The General Theory of Cylindrical and Conical Tubes Under Torsion and Bending Loads," *Journal of Royal Aeronautical Society*, Vol. 51, 199–269, 757–784, 884–930; Vol. 53 (1949) 461–483, 558–620.

Ascione, L. and Grimaldi, A. (1983) "On the Stability and Postbuckling Behavior of Elastic Beams," *Thin-Walled Structures*, (1), pp. 325–351.

Attard, M. M. (1986) "Non-Linear Theory of Non-Uniform Torsion of Thin-Walled Open Beams," *Thin-Walled Structures*, (4), pp. 101–134.

Badir, A. M., Berdichevsky, V. L. and Armanios, E. A. (1993) "Theory of Composite Thin-Walled Opened Cross Section Beams," *34th AIAA/ASME/ASCE/AHS/ASC Structures, Structural Dynamics and Materials Conference*, AIAA/ASME Adaptive Structures Forum, La Jolla, CA, 19–22, April, pp. 2761–2770.

Bauchau, O. A. (1985) "Theory for Anisotropic Materials," *Journal of Applied Mechanics, Trans. ASME*, Vol. 52, pp. 416–422.

Bauchau, O. A., Coffenberry, B. S. and Rehfield, L. W. (1987) "Composite Box Beam Analysis: Theory and Experiments," *Journal of Reinforced Plastics and Composites*, Vol. 6, pp. 25–35.

Bauchau, O. A. and Hong, C. H. (1987a) "Finite Element Approach to Rotor Blade Modeling," *Journal of the American Helicopter Society*, Vol. 32, No. 1, pp. 60–67.

Bauchau, O. A. and Hong, C. H. (1987b) "Large Displacement Analysis of Naturally Curved and Twisted Composite Beams," *AIAA Journal*, Vol. 25, No. 10, pp. 1469–1475.

Bauchau, O. A. and Hong, C. H. (1988) "Nonlinear Composite Beam Theory," *Journal of Applied Mechanics, Trans. ASME*, Vol. 55, No. 1, pp. 156–163.

Benscoter, S. U. (1954) "A Theory of Torsion Bending for Multicell Beams," *Journal of Applied Mechanics*, Vol. 20, pp. 25–34.

Berdichevsky, V., Armanios, E., and Badir, A., (1992) "Theory of Anisotropic Thin-Walled Closed-Cross-Section Beams," *Composites Engineering*, Vol. 2, Nos. 5–7, pp. 411–432.

Bhaskar, K. and Librescu, L. (1995a) "A Geometrically Non-Linear Theory for Laminated Anisotropic Thin-Walled Beams," *International Journal of Engineering Science*, Vol. 33, No. 9, pp. 1331–1344.

Bhaskar, K. and Librescu, L. (1995b) "Buckling under Axial Compression of Thin-Walled Composite Beams Exhibiting Extension Twist Coupling," *Composite Structures*, Vol. 31, No. 3, pp. 203–242.

Borri, M. and Merlini, T. (1986) "A Large Displacement Formulation for Anisotropic Beam Analysis," *Meccanica*, Vol. 21, pp. 30–37.

Capurso, M. (1964) "Sur calcolo delle travi di parete sotlite in presenta de forze e distorsions," *La Ricerca Scientifico*, (**34**), Series (2), Section A, Vol. 6, pp. 213–286, Vol. 7, pp. 5–106.

Chandra, R., Stemple, A. D. and Chopra, I. (1990) "Thin-Walled Composite Beams under Bending, Torsional and Extensional Loads," *Journal of Aircraft*, Vol. 27, No. 7, pp. 619–626.

Chandra, R. and Chopra, I. (1991) "Experimental and Theoretical Analysis of Composite I-Beams with Elastic Couplings," *AIAA Journal*, Vol. 29, No. 12, pp. 2197–2206.

Chandra, R. and Chopra, I. (1992a) "Experimental-Theoretical Investigation of the Vibration Characteristics of Rotating Composite Box Beams," *Journal of Aircraft*, Vol. 29, No. 4, pp. 657–664.

Chandra, R. and Chopra, I. (1992b) "Structural Response of Composite Beams and Blades with Elastic Couplings," *Composites Engineering*, Vol. 2, Nos. 5–7, pp. 347–374.

Chandiramani, N. K., Librescu, L. and Shete, C. D. (2001) "Free-Vibration of Rotating Composite Beams Incorporating Higher-Order Transverse Shear Effects," *International Mechanical Engineering Congress & Exposition*, November 11–16, New York City, NY (on CD-ROM).

Chandiramani, N. K., Librescu, L. and Shete, C. D. (2002) "On the Free-Vibration of Rotating Composite Beams Using a Higher-Order Shear Formulation," *Aerospace Science and Technology*, Vol. 6, pp. 545–561.

De Lorenzis, L. and la Tegola, A. (2005) "Effect of the Actual Distribution of Applied Stresses on Global Buckling of Isotropic and Transversely Isotropic Thin-Walled Members: Theoretical Analysis," *Composite Structures*, Vol. 68, pp. 339–348.

Donaldson, B. K. (1993) *Analysis of Aircraft Structures. An Introduction*, McGraw-Hill Inc., New York, St. Louis.

Dökmeci, M. C. (1972) "A General Theory of Elastic Beams," *International Journal of Solids and Structures*, Vol. 8, pp. 1205–1222.

Fine, M. and Williams, D. (1945) "Effects of End Constraint on Thin-Walled Cylinders Subject to Torque," *Aeronautical Research Council*, Rep. No. 2223.

Fraternali, F. and Feo, L. (2000) "On a Moderate Rotation Theory of Thin-Walled Composite Beams," *Composites: Part B*, Vol. 31, pp. 141–158.

Ghobarach, A. A. and Tso, W. K. (1971) "A Non-Linear Thin-Walled Beam Theory," *International Journal of Mechanical Sciences*, Vol. 13, (12), pp. 1025–1038.

Gjelsvik, A. (1981) *The Theory of Thin Walled Beams*, Wiley, New York, NY.

Gol'denveiser, A. L. (1949) "The General Theory of Thin-Walled Beams" (in Russian), *Prikladnaia Mathematika i Mechanika*, Vol. 13, No. 6, pp. 561–596.

Grimaldi, A. and Pignataro, M. (1979) "Postbuckling Behavior of Thin-Walled Open Cross-Section Compression Members," *Journal of Structural Mechanics*, Vol. 7, No. 2, pp. 143–159.

Hirashima, M. and Yoda, T. (1982) "Finite Displacement Theory of Curved and Twisted Thin-Walled Beams," *Memoirs of the School of Science & Engineering*, Waseda University, No. 46, pp. 277–294.

Hodges, D. H., Atilgan, A. R., Cesnik, C. E. S. and Fulton, M. V. (1992) "On a Simplified Strain Energy Function for Geometrically Nonlinear Behavior of Anisotropic Beams," *Composites Engineering*, Vol. 2, Nos. 5–7, pp. 513–526.

Hodges, D. H. and Dowell, E. H. (1974) "Nonlinear Equations of Motion for the Elastic Bending and Torsion of Twisted Nonuniform Rotor Blades," NASA TN D-7818.

Hong, C. H. and Chopra, I. (1986) "Aeroelastic Stability Analysis of a Composite Bearingless Rotor Blade," *Journal of American Helicopter Society*, Vol. 31, pp. 29–35.

Hughes, O. F. (1983) *Ship Structural Design*, John Wiley & Sons, New York.

Jensen, J. J. (2001) *Load and Global Response of Ships*, Pergamon Press, Oxford.

Jung, S. N., Nagaraj, V. T. and Chopra, I. (1999) "Assessment of Composite Rotor Blade: Modeling Techniques," *Journal of the American Helicopter Society*, Vol. 44, No. 2, pp. 188–205.

Kim, C. and White, S. R. (1996) "Analysis of Thick Hollow Composite Beams Under General Loadings," *Composite Structures*, Vol. 34, pp. 263–277.

Kim, C., and White, S. R. (1997) "Thick-Walled Composite Beam Theory Including 3-D Elastic Effects and Torsional Warping," *International Journal of Solids and Structures*, Vol. 34, Nos. 31–32, pp. 4237–4259.

Kvaternik, R. G., White, S. F., Jr. and Kaza, K. R. V. (1978) "Nonlinear Flap-Lag-Axial Equations of a Rotational Beam with Arbitrary Precone Angle," *AIAA Journal*, AIAA Paper No. 78-491.

Iura, M. and Hirashima, M. (1985) "Geometrically Nonlinear Theory of Naturally Curved and Twisted Rods with Finite Rotations," *Proceedings of the JSCE, Structural Engineering/Earthquake Engineering*, Vol. 2, No. 2, pp. 107–117.

Lee, S. W. and Kim, Y. H. (1987) "Finite-Element Model for Composite Beams with Arbitrary Cross-Sectional Warping," *International Journal of Numerical Method Engineering*, Vol. 24, No. 12, pp. 2327–2341.

Librescu, L. (1975) *Elasto-statics and Kinetics of Anisotropic and Heterogeneous Shell-Type Structures*, Noordhoff International Publishers, Leyden, The Netherlands.

Librescu, L., and Song, O. (1991) "Behavior of Thin-Walled Beams Made of Advanced Composite Materials and Incorporating Non-Classical Effects," *Applied Mechanics Reviews*, Vol. 44, No. 11, Part 2, S174–S180.

Librescu, L. and Song, O. (1992) "On the Static Aeroelastic Tailoring of Composite Aircraft Swept Wings Modelled as Thin-Walled Beam Structures," *Composites Engineering*, Vol. 2, Nos. 5–7 (Special Issue: *Use of Composites in Rotorcraft and Smart Structures*) pp. 497–512.

Librescu, L., Meirovitch, L. and Song, O. (1996) "Refined Structural Modeling for Enhancing Vibrations and Aeroelastic Characteristics of Composite Aircraft Wings," *La Recherche Aérospatiale*, Vol. 1, pp. 23–35.

Librescu, L. Meirovitch, L. and Na, S. S. (1997) "Control of Cantilever Vibration via Structural Tailoring and Adaptive Materials," *AIAA Journal*, Vol. 35, No. 8, pp. 1309–1315.

Librescu, L., Qin, Z. and Ambur, D. R. (2003) "Implications of Warping Restraint on Statics and Dynamics of Elastically Tailored Thin-Walled Composite Beams," *International Journal of Mechanical Sciences*, Vol. 45, No. 8, pp. 1247–1267.

Loughlan, J. and Ata, M. (1995) "The Restrained Torsional Response of Open Section Carbon Fibre Composite Beams," *Composite Structures*, Vol. 32, pp. 13–31.

Loughlan, J. and Ata, M. (1997) "The Behavior of Open and Closed Section Carbon Fibre Composite Beams Subjected to Constrained Torsion," *Composite Structures*, Vol. 38, Nos. 1–4, pp. 631–647.

Mansfield, E. H. and Sobey, A. J. (1979) "The Fibre Composite Helicopter Blade", Part I: Stiffness Properties; Part 2: Prospects for Aeroelastic Tailoring, *Aeronautical Quarterly*, Vol. XXX, pp. 413–499.

Megson, T. H. G. (1974) *Linear Analysis of Thin-Walled Elastic Structures*, John Wiley & Sons, New York.

Megson, T. H. G. (1990) *Aircraft Structures for Engineering Students*, Second Edition, Halstedt Press.

Meredith, D. and Witmer, E. A. (1981), "A Nonlinear Theory of General Thin-Walled Beams," *Computers & Structures*, Vol. 13, pp. 3–9.

Minguet, P. and Dugundji, J. (1990 a,b) "Experiments and Analysis for Composite Blades Under Large Deflections," *AIAA Journal*, Part I, Vol. 28, pp. 1573–1579; Part II: Vol. 28, pp. 1580–1588.

Mollmann, H. (1982a, b) "Finite Displacements of Thin-Walled Beams," Parts 1 and 2, Danish Center for Application Mathematics and Mechanics, Technical University of Denmark, Reports Nos. 252 and 253.

Nayfeh, A. H. and Pai, P. F. (2004) *Linear and Nonlinear Structural Mechanics*, Wiley-Interscience.

Nishino, F., Hasegawa, A. and Natori, E. (1977), "Thin-Walled Rectangular Beams with Shear Deformation and Cross Sectional Distortion," *Mechanics of Engineering*, ASCE-EMD, University of Waterloo.

Nishino, F., and Hasegawa, A. (1979) "Thin-Walled Elastic Members," *Journal of the Faculty of Engineering*, The University of Tokyo CB, Vol. XXXV, No. 2, pp. 109–190.

Oden, J. T. and Ripperger, E. A. (1981) *Mechanics of Elastic Structures*, Second Edition, Hemisphere Publication Corp., Washington.

Pai, P. F. and Nayfeh, A. H. (1992) "A Nonlinear Composite Beam Thoery," *Nonlinear Dynamics*, Vol. 3, pp. 273–303.

Pai, P. F. and Nayfeh, A. H. (1994) "A Fully Nonlinear Theory of Curved and Twisted Composite Rotor Blades Accounting for Warpings and Three-Dimensional Stress Effects," *International Journal of Solids and Structures*, Vol. 3, pp. 1309–1340.

Pedersen, P. T. (1991) "Beam Theories for Torsional-Bending Response of Ship Hulls," *Journal of Ship Research*, Vol. 35, No. 3, pp. 254–265.

Petre, A. (1984) *The Analysis of Aeronautical Structures* (in Romanian), Editura Technica, Bucharest.

Polillo, V. R., Garcia, L. F. T. and Villaca, S. F. (1998) "Discussion about Geometrically Nonlinear Formulations for Combined Flexure and Torsion of Thin-Walled Open Bars," *Journal of the Brazilian Society of Mechanical Sciences*, Vol. XX, No 1, pp. 103–115.

Rehfield, L. W., Atilgan, A. R. and Hodges (1988) "Structural Modeling for Multicell Composite Rotor Blades," *Proceedings 28th AIAA/ASME/ASCE/AHS/ACS, Structures, Structural Dynamics and Materials Conference*, AIAA Paper No. 88-2250.

Rehfield, L. W., Atilgan, A. R. and Hodges, D. H. (1990) "Nonclassical Behavior of Thin-Walled Composite Beams with Closed Cross Sections," *Journal of the American Helicopter Society*, Vol. 35, No. 2, pp. 42–51.

Rehfield, L. W. (1985) "Design Analysis Methodology for Composite Rotor Blades," *7th DoD/NASA Conference on Fibre Composites in Structural Design*, Denver, CO, AFWAL-TR-85-3094, (Va-1)–(Va-15).

Rivello, R. M. (1969) *Theory and Analysis of Flight Structures*, McGraw-Hill Book Company, New York, St. Louis.

Roberts, T. M. and Azizian, Z. G. (1983) "Non Linear Analysis of Thin-Walled Bars of Open Cross Section," *International Journal of Mechanical Sciences*, Vol. 25, No. 8, pp. 565–577.

Rosen, A. and Friedmann, P. P., (1979) "The Non-Linear Behavior of Elastic Slender Straight Beams Undergoing Small Strains and Moderate Rotations," *Journal of Applied Mechanics, Trans. ASME*, Vol. 46, March, pp. 161–168.

Smith, E. C. and Chopra, I. (1991) "Formulation and Evaluation of an Analytical Model for Composite Box-Beams," *Journal of the American Helicopter Society*, Vol. 36, No. 3, pp. 23–35.

Song, O. (1990) "Modeling and Response Analysis of Thin-Walled Beam Structures Constructed of Advanced Composite Materials," Ph.D. Dissertation, Virginia Polytechnic Institute and State University.

Song, O. and Librescu, L. (1993) "Free Vibration of Anisotropic Composite Thin-Walled Beams of Closed Cross-Section Contour," *Journal of Sound and Vibration*, Vol. 167, No. 1, pp. 129–147.

Stemple, A. D. and Lee, S. W. (1988) "Finite-Element Model for Composite Beams with Arbitrary Cross-Sectional Warping," *AIAA Journal*, Vol. 26, No. 12, pp. 1512–1520.

Suhubi, E. S. (1968) "On the Foundations of the Theory of Rods," *International Journal of Engineering Science*, Vol. 6, pp. 169–191.

Suresh, J. K. and Nagaraj, V. T. (1996) "Higher-Order Shear Deformation Theory for Thin-Walled Composite Beams," *Journal of Aircraft*, Vol. 33, No. 5, pp. 978–986.

Tsay, H. S. and Kingsbury, H. B. (1988) "Vibrations of Rods with General Space Curvature," *Journal of Sound and Vibrations*, Vol. 124, No. 2, pp. 539–554.

Van Erp, G. M. (1987) "The Non-linear Flexural Torsional Behavior of Straight Slender Elastic Beams with Arbitrary Cross-Sections," Eindhoven University of Technology, EVT Rept. WRW 87-050, Eindhoven, Netherlands.

Vasiliev, G. V. (1986) *Fundaments of the Analysis of Thin-Walled Aeronautical Structures* (in Romanian), Vol. 1, Editura Academiei R. S. Romania.

Vlasov, V. Z. (1961) *Thin Walled Elastic Beams*, National Science Foundation, Washington, DC, Israel Program for Scientific Translation, Jerusalem, Israel [First edition – Stroizdat (in Russian) Moscow, 1940].

von Kármán, T. and Christensen, N. B. (1944) "Methods of Analysis for Torsion with Variable Twist," *Journal of Aeronautical Science*, Vol. 11, pp. 110–124.

Wallerstein, D. V. (2002) *A Variational Approach to Structural Analysis*, John Wiley & Sons, Inc.

Wempner, G. (1981) *Mechanics of Solids with Applications to Thin Bodies*, Sijthoff & Noordhoff Alphen aan den Rijn, The Netherlands, Rockville, Maryland, USA.

Chapter 3

THE EQUATIONS OF MOTION OF OPEN/ CLOSED CROSS-SECTION BEAMS

3.1 NONLINEAR FORMULATION

3.1-1 Preliminaries

In this chapter the equilibrium/motion equations and the boundary conditions for beams incorporating geometrical nonlinearities will be derived. To this end, the extended Hamilton's Principle of the 3-D elastic medium in conjunction with the concept of the reference state will be used.

Within this concept, and in contrast to the conventional representation of stress measured per unit area of the deformed body, the notion of stress measured per unit area of the undeformed body, taken as reference, is used. In this sense, see e.g. Librescu (1975, 1987). This description makes it possible to properly identify the origin of various additional terms that appear in the equations of motion and boundary conditions and further, in the governing equations of the nonlinear theory of beams. Along this line, the nonlinear approach of thin-walled beams was carried out by Bhaskar and Librescu (1995).

3.1-2 General Background

Consider a continuous body in equilibrium under the action of body forces $_0\mathbf{H}$ and surface tractions $_0\hat{\sigma}$. Let $_0\tau$ be the volume of the undeformed body and $_0\Omega_\sigma$ and $_0\Omega_v$ the two no-overlapping parts of its total undeformed boundary $_0\Omega$, where the tractions $\hat{\sigma}_i$ and the displacements $\underset{\sim}{V}_i$ are prescribed. Let V_i, h_i and $_0H_i$ be, respectively, the components with respect to the base vectors of the undeformed body, of the vectors of displacement, acceleration and body force per unit mass of the undeformed body of density $_0\rho$.

Hamilton's principle stipulates that among all the dynamic paths represented in terms of the displacement components $V_i(x, y, z, t)$, that satisfy the geometric boundary conditions over $_0\Omega_v$ at all times and that start and end with the actual values at two arbitrary instants of time t_0 and t_1, the actual dynamic path (i.e., the one corresponding to the equilibrium configuration), satisfies the equation (see Librescu, 1975)

$$\delta J = \int_{t_0}^{t_1} dt \left[\int_{0\tau} \hat{\sigma}_{ij}\delta\epsilon_{ij} \, d\tau - \delta\mathcal{K} - \int_{0\Omega_\sigma} \tilde{\hat{\sigma}}_i \delta V_i d\Omega - \int_{0\tau} {}_0\rho \, {}_0H_i\delta V_i d\tau \right] = 0. \tag{3.1-1}$$

Here $\hat{\sigma}_{ij}$ are the components of the second Piola-Kirchhoff symmetric stress tensor; $\hat{\sigma}_i$ are the prescribed components of the stress vector per unit area of the deformed body referred to base vectors in the undeformed configuration; ϵ_{ij} is the Green-Lagrange strain tensor; t_0 and t_1 denote two arbitrary instants of time; Latin indices take the values 1, 2, 3 and are summed when repeated, while δ denotes the variation operator. In addition,

$$\mathcal{K} = \frac{1}{2}\int_{0\tau} {}_0\rho(\dot{V}_i \dot{V}_i)d\tau, \tag{3.1-2}$$

denotes the kinetic energy and the superposed dots denote time derivatives.

First, we show how this principle can provide the equations of motion and the boundary conditions of geometrically nonlinear three-dimensional elasticity theory, by identifying the various terms appearing in Eq. (3.1-1). Its first term is the virtual strain energy:

$$\int_{0\tau} \hat{\sigma}_{ij}\delta\epsilon_{ij} \, d\tau = \int_{0\tau} \frac{1}{2}\hat{\sigma}_{ij}(\delta V_{i,j} + \delta V_{j,i} + V_{m,i}\delta V_{m,j} + \delta V_{m,i}V_{m,j})d\tau. \tag{3.1-3}$$

The symmetry of $\hat{\sigma}_{ij}$, the divergence theorem, and the fact that the variations δV_i must vanish over the boundary surface $_0\Omega_v$ where displacements are prescribed, imply

$$\int_{0\tau} \hat{\sigma}_{ij}\delta\epsilon_{ij} = -\int_{0\tau} \left[\hat{\sigma}_{ij}(\delta_{mi} + V_{m,i})\right]_{,j} \delta V_m d\tau + \int_{0\Omega_\sigma} \hat{\sigma}_{ij}(\delta_{mi}+V_{m,i})_0n_j\delta V_m d\Omega, \tag{3.1-4}$$

where δ_{mi} is the Kronecker delta, and $_0n_j$ are the components of the outward unit normal to $_0\Omega$. From (3.1-2) we obtain

$$\delta\mathcal{K} = \int_{0\tau} {}_0\rho\dot{V}_i\delta\dot{V}_i d\tau = \int_{0\tau} {}_0\rho\frac{\partial}{\partial t}(\dot{V}_i\delta V_i)d\tau - \int_{0\tau} {}_0\rho\ddot{V}_i\delta V_i d\tau. \tag{3.1-5}$$

Integration of (3.1-5) over the time interval $[t_0, t_1]$, and keeping in mind that $\delta V_i = 0$ at $t = t_0$ and t_1, yields

$$\int_{t_0}^{t_1} \delta\mathcal{K}dt = -\int_{t_0}^{t_1} dt \int_{0\tau} {}_0\rho\ddot{V}_i\delta V_i d\tau. \tag{3.1-6}$$

In view of (3.1-4) and (3.1-6), Eq. (3.1-1) can be cast as

$$\delta J = \delta(J_1 + J_2) = 0, \tag{3.1-7}$$

where

$$\delta J_1 = \int_{t_0}^{t_1} dt \int_{0\tau} \left\{ \left[\hat{\sigma}_{ij}(\delta_{mi} + V_{m,i}) \right]_{,j} + {}_0\rho\,{}_0H_m - {}_0\rho\,h_m \right\} \delta V_m d\tau = 0, \tag{3.1-8}$$

and

$$\delta J_2 = \int_{t_0}^{t_1} dt \int_{0\Omega_\sigma} \left\{ \hat{\sigma}_{jr}(\delta_{ir} + V_{i,r}){}_0 n_j - \hat{\underline{\sigma}}_i \right\} \delta V_i d\Omega = 0. \tag{3.1-9}$$

Since the variations δV_i are arbitrary in $[t_0, t_1]$ throughout the volume $_0\tau$ and on the boundary $_0\Omega$, Eq. (3.1-7) is fulfilled when their coefficients in the integrands of (3.1-8) and (3.1-9) vanish independently which yields:

$$[\hat{\sigma}_{ij}(\delta_{ri} + V_{r,i})]_{,j} + {}_0\rho\,{}_0H_r = {}_0\rho h_r, \qquad \text{in } _0\tau \tag{3.1-10a}$$
$$[\hat{\sigma}_{ij}(\delta_{ri} + V_{r,i})]_0 n_j - \underline{\sigma}_r = 0, \qquad \text{on } _0\Omega_\sigma \tag{3.1-10b}$$
$$V_i = \underline{V}_i, \qquad \text{on } _0\Omega_v. \tag{3.1-10c}$$

where $h_r \equiv \ddot{V}_r$.

Equations (3.1-10) are the equations of motion and the boundary conditions of the three-dimensional nonlinear elasticity theory in terms of a reference state, as given for instance by Green and Adkins (1960).

For small strains and displacements, the Piola-Kirchhoff $\hat{\sigma}_{ij}$ and Cauchy σ_{ij} stress tensors are identical, and the Lagrangian strain tensor ϵ_{ij} reduces to that given by Eq. (2.1-55).

3.1-3 Application of Hamilton's Principle to Thin-Walled Beams

3.1-3a Strain Energy Functional

To apply Hamilton's principle, we need to identify the various energy quantities appearing in Eq. (3.1-1).

As a first step, consider the strain energy

$$\mathcal{W} = \frac{1}{2} \int_{0\tau} \hat{\sigma}_{ij}\,\epsilon_{ij}\, d\tau. \tag{3.1-11}$$

The element of volume $d\tau$ of the beam is

$$d\tau = dn\,ds\,dz, \tag{3.1-12}$$

and consistent with Eqs. (2.2-8) and (2.2-11), we have

$$
\begin{aligned}
W &= \frac{1}{2} \int_{z_1}^{z_2} \int_C \int_h [\hat{\sigma}_{zz}\, \epsilon_{zz} + \hat{\sigma}_{sz}\Gamma_{sz} + \hat{\sigma}_{nz}\Gamma_{nz}]_{(k)}\, dn\, ds\, dz \\
&= \frac{1}{2} \int_{z_1}^{z_2} \int_C \int_h \{\hat{\sigma}_{zz}^{(k)}[\epsilon_{zz}^{(0)} + n\epsilon_{zz}^{(1)} + n^2\epsilon_{zz}^{(2)}] \\
&\quad + \hat{\sigma}_{sz}^{(k)}[\gamma_{sz}^{(0)} + n\gamma_{sz}^{(1)}] + \hat{\sigma}_{nz}^{(k)}\gamma_{nz}^{(0)}\}\, dn\, ds\, dz.
\end{aligned}
\tag{3.1-13}
$$

Carrying out the integration through the wall thickness and using the expressions of the *shell* stress resultants and stress couples one obtains

$$
\begin{aligned}
W &= \frac{1}{2} \int_{z_1}^{z_2} \int_C \big[N_{zz}\epsilon_{zz}^{(0)} + \delta_n L_{zz}\epsilon_{zz}^{(1)} + \delta_n \Gamma_{zz}\epsilon_{zz}^{(2)} \\
&\quad + N_{sz}\gamma_{sz}^{(0)} + \delta_n L_{sz}\gamma_{sz}^{(1)} + N_{nz}\gamma_{nz}^{(0)} \big]\, ds\, dz.
\end{aligned}
\tag{3.1-14}
$$

This equation involves the two-dimensional stress measures defined by Eqs. (A.56) through (A.58), and the higher-order stress couple Γ_{zz}:

$$
\begin{aligned}
(N_{zz}, L_{zz}, \Gamma_{zz}) &= \int_{-h/2}^{h/2} \hat{\sigma}_{zz}(1, n, n^2)\, dn, \\
(N_{sz}, L_{sz}) &= \int_{-h/2}^{h/2} \hat{\sigma}_{sz}(1, n)\, dn, \\
N_{nz} &= \int_{-h/2}^{h/2} \hat{\sigma}_{nz}\, dn.
\end{aligned}
\tag{3.1-15a-c}
$$

For infinitesimal displacements, according to Eq.(2.2-9c), $\epsilon_{zz}^{(2)}$ vanishes, and as a result, the contribution of Γ_{zz} to the strain energy becomes immaterial. Using the strain measures given by Eqs. (2.2-9), (2.2-12) and (2.2-13), and carrying out the integration around the mid-line contour, we find

$$
\begin{aligned}
W &= \frac{1}{2} \int_{z_1}^{z_2} \Big\{ T_z \Big[w_0' + \frac{1}{2}((u_P')^2 + (v_P')^2) + \phi'[y_P(u_P' \cos\phi + v_P' \sin\phi) \\
&\quad + x_P(u_P' \sin\phi - v_P' \cos\phi)] \Big] \\
&\quad + M_x \Big[\delta_t[\theta_x' - \phi'(u_P' \cos\phi + v_P' \sin\phi)] + \delta_e(u_P'' \sin\phi - v_P'' \cos\phi) \Big] \\
&\quad + M_y \Big[\delta_t[\theta_y' - \phi'(u_P' \sin\phi - v_P' \cos\phi)] - \delta_e(v_P'' \sin\phi + u_P'' \cos\phi) \Big] \\
&\quad + \delta_t Q_x(\theta_y + v_P' \sin\phi + u_P' \cos\phi) + \delta_t Q_y(\theta_x - u_P' \sin\phi + v_P' \cos\phi) \\
&\quad + M_z \phi' - \delta_r B_\omega \phi'' + \frac{1}{2}\Lambda_z (\phi')^2 \Big\}\, dz.
\end{aligned}
\tag{3.1-16}
$$

The one-dimensional stress measures appearing in Eq. (3.1-16), are expressed in terms of *shell* stress resultants and stress-couples as follows:

$$T_z(z, t) = \int_C N_{zz} ds,$$

$$Q_x(z, t) = \int_C \left(N_{sz} \frac{dx}{ds} + \delta_t N_{zn} \frac{dy}{ds} \right) ds,$$

$$Q_y(z, t) = \int_C \left(N_{sz} \frac{dy}{ds} - \delta_t N_{zn} \frac{dx}{ds} \right) ds,$$

$$M_y(z, t) = \int_C \left(x N_{zz} + \delta_n L_{zz} \frac{dy}{ds} \right) ds, \qquad\qquad (3.1\text{-}17\text{a-h})$$

$$M_x(z, t) = \int_C \left(y N_{zz} - \delta_n L_{zz} \frac{dx}{ds} \right) ds,$$

$$M_z(z, t) = \int_C (\delta_c N_{sz} \psi + \delta_n 2 L_{sz}) \, ds,$$

$$B_\omega(z, t) = \int_C [\overline{F}(s) N_{zz} - \delta_n r_t(s) L_{zz}] ds,$$

$$\Lambda_z(z, t) = \int_C [N_{zz}[(x - x_P)^2 + (y - y_P)^2] + 2 L_{zz} r_n + \Gamma_{zz}] ds.$$

The torsional function is expressed as

$$\psi (\equiv \psi(s)) = \frac{\displaystyle\oint r_n(\bar{s}) d\bar{s}}{h(s) G_{sz}(s) \displaystyle\oint \frac{d\bar{s}}{h(\bar{s}) G_{sz}(\bar{s})}}. \qquad\qquad (3.1\text{-}17\text{i})$$

$G_{sz}(s)$ is required only when the laminate properties vary with respect to s (see Smith and Chopra, 1991; Bhaskar and Librescu, 1995); its importance has been emphasized by Jung et al. (1999) and Qin and Librescu (2001, 2002).

If h and G_{sz} are uniform around the beam contour, then

$$\psi = \frac{\displaystyle\oint r_n ds}{\displaystyle\oint ds} \left(\equiv \frac{2\Omega}{\beta} \right). \qquad\qquad (3.1\text{-}17\text{j})$$

Equations (3.1-17) that follow directly from Hamilton's principle, constitute an extension of those obtained e.g. by Rehfield (1985).

For an open beam cross-section $\int_C (\cdot) \, ds$ denotes the curvilinear integral along the mid-line contour between the longitudinal sections $s = s_1$ and $s = s_2$; for a closed beam cross-section, the integral is $\oint (\cdot) ds$, that is around the entire mid-line contour of the cross-section. For brevity the integral $\int_C (\cdot)\, ds$ will be written simply as $\int (\cdot)\, ds$.

In Eqs. (3.1-17) T_z corresponds to the axial force; Q_x and Q_y are associated with the chordwise and flapwise shear forces, respectively; M_x and M_y are the bending moments about the x (flapwise bending moment) and $-y$ (chordwise bending moment) directions, respectively; M_z corresponds to the Saint-Venant twist moment; Γ_z is a higher-order stress-couple contributing to the twist, while B_ω is the bimoment or the warping torque. In its expression, Eq. (3.1-17g), the role of moment arms for the shell stress-resultant N_{zz} and stress-couple L_{zz} are played by the primary and secondary warping functions, respectively. Equation (3.1-17), show that T_z, Q_x and Q_y have the units of force [F]; M_x, M_y and M_z have the units of force times length [F.L], while B_ω and Λ_z have the units of force times length squared $[F.L^2]$.

The terms in the strain energy functional, Eq. (3.1-16), are represented as products of generalized forces and generalized strain measures. In Eqs. (3.1-16) and (3.1-17), the tracer δ_n indicates that the terms accompanying it are generated by the strain components which are different of zero off the mid-line contour, while δ_t and δ_e are tracers indicating the affiliation of the respective quantities to the shearable and unshearable beam models, respectively. For the shearable beam model $\delta_t = 1$ and $\delta_e = 0$; for the Euler-Bernoulli beam model $\delta_t = 0$ and $\delta_e = 1$. We emphasize the generality of the strain energy as given by Eq. (3.1-16), and the possibility of obtaining from it, special cases of practical importance.

Return to Eq. (3.1-14), take virtual variation of the generalized displacements, integrate by parts to relieve the virtual displacements of any differentiation, and obtain the strain energy for the shearable beam model:

$$
\begin{aligned}
\delta \mathcal{W} =& \int_{z_1}^{z_2} \int \Big[N_{zz}\delta\epsilon_{zz}^{(0)} + \delta_n L_{zz}\delta\epsilon_{zz}^{(1)} + \delta_n \Gamma_{zz}\delta\epsilon_{zz}^{(2)} + N_{sz}\delta\gamma_{sz}^{(0)} + \delta_n L_{sz}\delta\gamma_{sz}^{(1)} \\
& + N_{nz}\delta\gamma_{nz}^{(0)} \Big] ds\, dz \\
=& -\int_{z_1}^{z_2} \Big\langle T_z'\delta w_0 + (M_x' - Q_y)\delta\theta_x + (M_y' - Q_x)\delta\theta_y \\
& + \Big[[T_z\{\cos\phi\,(u_P' y_P - v_P' x_P) + \sin\phi\,(v_P' y_P + u_P' x_P)\}]' \\
& - T_z\phi'\{\cos\phi\,(v_P' y_P + u_P' x_P) + \sin\phi\,(v_P' x_P - u_P' y_P)\} \\
& - \{\cos\phi\,(M_x u_P' - M_y v_P') + \sin\phi\,(M_x v_P' + M_y u_P')\}' \\
& + [\cos\phi\,(M_x v_P'\phi' + M_y u_P'\phi' + Q_y u_P' - Q_x v_P')
\end{aligned}
$$

$$+ \sin \phi (-M_x u'_P \phi' + M_y v'_P \phi' + Q_y v'_P + Q_x u'_P)]$$

$$+ \delta_r B''_\omega + (\Lambda_z \phi')' + M'_z \Big] \delta \phi + \Big\{ [T_z \{u'_P + \phi'(y_P \cos \phi + x_P \sin \phi)\}]'$$

$$+ [(Q_x - M_x \phi') \cos \phi - (Q_y + M_y \phi') \sin \phi]' \Big\} \delta u_P$$

$$+ \Big\{ [T_z \{v'_P + \phi'(y_P \sin \phi - x_P \cos \phi)\}]' + [-(Q_x + M_x \phi') \sin \phi$$

$$+ (Q_y + M_y \phi') \cos \phi]' \Big\} \delta v_P \Big) dz \tag{3.1-18}$$

$$+ \Big[T_z \delta w_0 + M_x \delta \theta_x + M_y \delta \theta_y + \Big\{ T_z [(y_P v'_P + x_P u'_P) \sin \phi$$

$$+ (y_P u'_P - x_P v'_P) \cos \phi] - [(M_y u'_P + M_x v'_P) \sin \phi$$

$$+ (M_x u'_P - M_y v'_P) \cos \phi] + \delta_r B'_\omega + \Lambda_z \phi' + M_z \Big\} \delta \phi - \delta_r B_\omega \delta \phi'$$

$$+ \Big\{ T_z [u'_P + \phi'(y_P \cos \phi + x_P \sin \phi)]$$

$$+ [-(Q_y + M_y \phi') \sin \phi + (Q_x - M_x \phi') \cos \phi] \Big\} \delta u_P$$

$$+ \Big\{ T_z [v'_P + \phi'(y_P \sin \phi - x_P \cos \phi)]$$

$$+ (Q_y + M_y \phi') \cos \phi - (Q_x + M_x \phi') \sin \phi \Big\} \delta v_P \Big] \Big|_{z=z_1, z_2}.$$

For the unshearable beam model, replace θ_x and θ_y by their expressions, Eqs. (2.2-6), specialized to this case, and integrate partially to free the virtual displacements of any differentiation.

3.1-3b Kinetic Energy

The kinetic energy of the beam is

$$\mathcal{K} = \frac{1}{2} \int \int \int {}_0 \rho_{(k)} [\dot{u}^2 + \dot{v}^2 + \dot{w}^2] \, dn \, ds \, dz. \tag{3.1-19}$$

In the Hamilton's functional, Eq. (3.1-1), only the quantity $\int_{t_0}^{t_1} \delta \mathcal{K} \, dt$ appears; Eq. (3.1-6) gives

$$\int_{t_0}^{t} \delta \mathcal{K} \, dt = - \int_{t_0}^{t_1} dt \int \int \int {}_0 \rho_{(k)} (\ddot{u} \delta u + \ddot{v} \delta v + \ddot{w} \delta w) \, dn \, ds \, dz. \tag{3.1-20}$$

Using Eqs. (3.1-1), (3.1-5) and (3.1-6), substituting for the virtual displacements, carrying out the integration through the beam wall thickness and around the mid-line of the cross-section contour, and integrating by parts whenever necessary we obtain (3.1-20) as

$$
\int_{t_0}^{t_1} \delta \mathcal{K} dt = - \int_{t_0}^{t_1} dt \int \Big[[K_1 + \delta_e (K_6 \cos\phi - K_8 \sin\phi)'] \delta u_P
$$

$$
+ [K_2 + \delta_e (K_8 \cos\phi + K_6 \sin\phi)'] \delta v_P + K_3 \delta w_0 + \delta_t K_7 \delta\theta_x
$$

$$
+ \delta_t K_5 \delta\theta_y + \{K_4 \cos\phi + K_{10} \sin\phi - K_9' - \delta_e[(K_6 v_P'
$$

$$
- K_8 u_P') \cos\phi - (K_6 u_P' + K_8 v_P') \sin\phi]\} \delta\phi \Big] dz \qquad (3.1\text{-}21)
$$

$$
- \int_{t_0}^{t_1} dt \Big[- \delta_e (K_6 \cos\phi - K_8 \sin\phi) \delta u_P - \delta_e (K_8 \cos\phi
$$

$$
+ K_6 \sin\phi) \delta v_P + K_9 \delta\phi] \Big|_{z=z_1,z_2} .
$$

The terms K_i appearing in Eq. (3.1-21) are displayed in a unified form for both shearable and unshearable beams in Table 3.1-1.

Table 3.1-1: Expression of inertia terms

K_1	$B_1 \ddot{u}_P - B_3 A(\phi) - B_2 B(\phi)$
K_2	$B_1 \ddot{v}_P + B_3 B(\phi) - B_2 A(\phi)$
K_3	$B_1 \ddot{w}_0 + B_{11}(\delta_t \ddot{\theta}_x + \delta_e T) + B_{12}(\delta_t \ddot{\theta}_y + \delta_e S) - B_7 \ddot{\phi}'$
K_4	$B_3 \ddot{v}_P - B_2 \ddot{u}_P + (B_4 + B_5) B(\phi)$
$K_5 = K_6$	$B_{12} \ddot{w}_0 + B_{15}(\delta_t \ddot{\theta}_x + \delta_e T) + B_{14}(\delta_t \ddot{\theta}_y + \delta_e S) - B_9 \ddot{\phi}'$
$K_7 = K_8$	$B_{11} \ddot{w}_0 + B_{13}(\delta_t \ddot{\theta}_x + \delta_e T) + B_{15}(\delta_t \ddot{\theta}_y + \delta_e S) - B_8 \ddot{\phi}'$
K_9	$-[B_7 \ddot{w}_0 + B_8(\delta_t \ddot{\theta}_x + \delta_e T) + B_9(\delta_t \ddot{\theta}_y + \delta_e S) - B_{10} \ddot{\phi}']$
K_{10}	$-[B_3 \ddot{u}_P + \bar{B}_2 \ddot{v}_P] + (B_4 + B_5) A(\phi)$

Here $A(\phi) \equiv \ddot{\phi} \sin\phi + \dot{\phi}^2 \cos\phi$, and $B(\phi) = \ddot{\phi} \cos\phi - \dot{\phi}^2 \sin\phi$, and T and S represent the unshearable beam counterparts of $\ddot{\theta}_x$ and $\ddot{\theta}_y$, respectively; Eqs. (2.2-14) give

$$
T \equiv u_P' \ddot{\phi} \cos\phi - u_P' \dot{\phi}^2 \sin\phi + 2\dot{u}_P' \dot{\phi} \cos\phi
$$

$$
+ \ddot{u}_P' \sin\phi + v_P' \ddot{\phi} \sin\phi + v_P' \dot{\phi}^2 \cos\phi + 2\dot{v}_P' \dot{\phi} \sin\phi - \ddot{v}_P' \cos\phi, \tag{3.1-22a}
$$

$$
S \equiv v_P' \dot{\phi}^2 \sin\phi - v_P' \ddot{\phi} \cos\phi - 2\dot{v}_P' \dot{\phi} \cos\phi - \ddot{v}_P' \sin\phi + u_P' \ddot{\phi} \sin\phi
$$

$$
+ u_P' \dot{\phi}^2 \cos\phi + 2\dot{u}_P' \dot{\phi} \sin\phi - \ddot{u}_P' \cos\phi. \tag{3.1-22b}
$$

The reduced mass terms B_i $(i = \overline{1, 15})$ are defined as:

$$(B_1, B_2, ...B_{15}) = \int\int {}_0\rho\{1, (Y - y_P), (X - x_P), (Y - y_P)^2, (X - x_P)^2,$$

$$(X - x_P)(Y - y_P), F, YF, XF, F^2, Y, X, Y^2, X^2, XY\}dsdn, \quad (3.1\text{-}23)$$

where $F \equiv F(n, s)$ is the warping function considered in its entirety, expressed in a unified form for both open and closed beam cross-section by Eq. (2.1-53), and X and Y are the coordinates of the points off the mid-line cross-section contour as defined by Eqs. (2.1-21).

Considering a symmetrically laminated beam and $x_P = y_P = 0$; with the help of (2.1-21), B_i as defined by (3.1-23) reduce to

$$B_1 = b_1, B_2 = b_2, B_3 = b_3, B_4 = b_4 + \delta_n b_{14}, B_5 = b_5 + \delta_n b_{15}, B_6 = b_6$$
$$- \delta_n b_{13}, B_7 = b_7, B_8 = b_8 + \delta_n b_{16}, B_9 = b_9 - \delta_n b_{17}, B_{10} = b_{10} + \delta_n b_{18},$$
$$B_{11} = B_2, B_{12} = B_3, B_{13} = B_4, B_{14} = B_5, B_{15} = B_6. \quad (3.1\text{-}24a)$$

where

$$(b_1, \ b_2, \ b_3, \ b_4, b_5, \ \ldots, \ b_{10}, \ b_{11}, \ b_{12})$$

$$= \int m_0[1, y, x, y^2, \ x^2, \ xy, \ \bar{F}, \ y\bar{F}, \ x\bar{F}, \ \bar{F}^2, \ dx/ds, dy/ds]ds,$$

$$(b_{13}, \ b_{14}, \ \ldots, \ b_{18}) = \int m_2 \left[\left(\frac{dx}{ds}\right) \left(\frac{dy}{ds}\right), \ \left(\frac{dx}{ds}\right)^2, \right.$$

$$\left. r_t \frac{dx}{ds}, \ r_t \frac{dx}{ds}, \ r_t \frac{dy}{ds}, \ r_t^2 \right] ds. \quad (3.1\text{-}24b)$$

In these equations the reduced mass terms m_0 and m_2 are defined as

$$(m_0, \ m_2) = \sum_{k=1}^{N} \int_{h_{(k-1)}}^{h_{(k)}} {}_0\rho_{(k)} \ (1, n^2)dn. \quad (3.1\text{-}25)$$

3.1-3c Work Performed by the External Loading and Body Forces

Consider the beam subjected to distributed surface loads q_x, q_y, q_z per unit area of the middle surface, conservative end tractions t_x, t_y, t_z per unit area of the beam cross section, body forces H_x, H_y and H_z per unit mass, and for open-beam cross-section, tractions $\hat{t}_x, \hat{t}_y, \hat{t}_z$, per unit area of the longitudinal sections, $s = s_1$ and $s = s_2$. All these quantities are assumed to act on the deformed configuration of the beam, but are specified over the area or volume of the undeformed beam. Corresponding to the last two terms in Eq. (3.1-1), the work performed by these forces is expressible as

$$
\mathcal{J} = \int\int\int \,_o\rho \,(H_x u + H_y v + H_z w)\,ds\,dn\,dz + \int\int (q_x \overline{u} + q_y \overline{v} + q_z \overline{w})\,ds\,dz
$$

$$
+ n_z \int\int (t_x u + t_y v + t_z w)\bigg|_{z=z_1,z_2} ds\,dn + \delta_o n_s \int\int (\hat{t}_x u + \hat{t}_y v + \hat{t}_z w)\bigg|_{s=s_1,s_2} dn\,dz.
$$

$$(3.1\text{-}26)$$

Here, the displacement quantities with a superposed bar are associated with the points of the middle surface, whereas n_z and n_s are defined as:

$$
n_z = \begin{cases} -1 \text{ at } z = z_1 \\ 1 \text{ at } z = z_2 \\ 0 \text{ otherwise} \end{cases}
\qquad
n_s = \begin{cases} -1 \text{ at } s = s_1 \\ 1 \text{ at } s = s_2 \\ 0 \text{ otherwise} \end{cases}
\qquad (3.1\text{-}27)
$$

where $z = z_1$ and $z = z_2$ are the end sections of the beam. The tracer δ_0 takes the values 0 or 1 depending on whether the beam has closed or open cross-section, respectively. Considering Eqs. (2.2-1) in (3.1-26) and taking the variations of u, v and w, carrying out the indicated integrations and integrating by parts whenever necessary, one obtains

$$
\delta\mathcal{J} = \delta\mathcal{J}_1 + \delta\mathcal{J}_2 + \delta\mathcal{J}_3, \qquad (3.1\text{-}28)
$$

where

$$
\delta\mathcal{J}_1 = \int_{z_1}^{z_2} [p_x \delta u_P + p_y \delta v_P + p_z \delta w_0 + m_x \delta\theta_x + m_y \delta\theta_y + (m_z \cos\phi + m'_\omega
$$

$$
+ \lambda_T \sin\phi)\delta\phi]\,dz, \qquad (3.1\text{-}29a)
$$

$$
\delta\mathcal{J}_2 = n_z \{ \underset{\sim}{Q}_x \delta u_P + \underset{\sim}{Q}_y \delta v_P + \underset{\sim}{T}_z \delta w_0 + \underline{M}_x \delta\theta_x + \underline{M}_y \delta\theta_y + (\underline{M}_z \cos\phi
$$

$$
+ \underline{\Gamma}_T \sin\phi - \delta_0 \hat{m}_\omega)\delta\phi - \underline{B}_\omega \delta\phi' \}|_{z=z_1,z_2}, \qquad (3.1\text{-}29b)
$$

and

$$
\delta\mathcal{J}_3 = \delta_0 \int_{z_1}^{z_2} [\hat{p}_x \delta u_P + \hat{p}_y \delta v_P + \hat{p}_z \delta w_0 + \hat{m}_x \delta\theta_x + \hat{m}_y \delta\theta_y + (\hat{m}_z \cos\phi + \hat{m}'_\omega
$$

$$
+ \hat{\lambda}_T \sin\phi)\delta\phi]\,dz. \qquad (3.1\text{-}29c)
$$

Here $p_x(z,t)$, $p_y(z,t)$, $p_z(z,t)$, $m_x(z,t)$, $m_y(z,t)$, $m_z(z,t)$, $\lambda_T(z,t)$, $m_\omega(z,t)$ and their counterparts identified by a circumflex represent equivalent line loads due to the effective external and body forces and couples, measured per unit span of the undeformed beam, and due to the tractions on the longitudinal edges of an open cross-section beam, respectively. These 1-D generalized loadings are expressed as

$$p_x(z, t) = \int q_x ds + \int \int H_x ds dn,$$

$$p_y(z, t) = \int q_y ds + \int \int H_y ds dn,$$

$$p_z(z, t) = \int q_z ds + \int \int H_z ds dn,$$

$$m_x(z, t) = \int q_z y ds + \int \int H_z Y ds dn,$$

$$m_y(z, t) = \int q_z x ds + \int \int H_z X ds dn, \qquad (3.1\text{-}30\text{a-e})$$

$$m_z(z, t) = \int [q_y(x - x_P) - q_x(y - y_P)] ds$$
$$+ \int \int [H_y(X - x_P) - H_x(Y - y_P)] ds dn,$$

$$m_\omega(z, t) = \int q_z \overline{F} ds + \int \int H_z F ds dn, \qquad (3.1\text{-}30\text{f-h})$$

$$\lambda_T(z, t) = - \int [q_x(x - x_P) + q_y(y - y_p)] ds - \int \int [H_x(X - x_P)$$
$$+ H_y(Y - y_P)] ds dn.$$

The edge loads are defined as follows:

$$(\hat{p}_x, \hat{p}_y, \hat{p}_z) = n_s \int [\hat{t}_x, \hat{t}_y, \hat{t}_z]\Big|_{s=s_1, s_2} dn, \qquad (3.1\text{-}30\text{i})$$

$$(\hat{m}_x, \hat{m}_y, \hat{m}_z, \hat{m}_\omega, -\hat{\lambda}_T) = n_s \int [\hat{t}_z Y, \hat{t}_z X, \hat{t}_y(X - x_P)$$
$$- \hat{t}_x(Y - y_P), \hat{t}_z F, \hat{t}_x(X - x_P) + \hat{t}_y(Y - y_P)]\Big|_{s=s_1, s_2} dn, \qquad (3.1\text{-}30\text{j})$$

$$(\underset{\sim}{T}_z, \underset{\approx}{Q}_x, \underset{\approx}{Q}_y) = \int \int (t_z, t_x, t_y) ds dn,$$

$$(\underline{M}_x, \underline{M}_y, \underline{B}_\omega) = \int \int t_z(Y, X, F) ds dn,$$

$$\underline{M}_z = \int \int [t_y(X - x_P) - t_x(Y - y_P)] ds dn, \qquad (3.1\text{-}30\text{k-n})$$

$$\underline{\Gamma}_T = - \int \int [t_x(X - x_P) + t_y(Y - y_P)] ds dn.$$

In these equations $X(\equiv x + ndy/ds)$ and $Y(\equiv y - ndx/ds)$ are the co-ordinates for the points off the beam mid-line contour, while $\overline{F}(\equiv \overline{F}(s))$ and $F(\equiv F(s, n))$ represent the primary and total warping functions, respectively, see Eqs. (2.1-53).

3.1-4 Equations of Motion

Substitute (3.1-18), (3.1-21) and (3.1-29) into the variational Eq. (3.1-1), collect the coefficients of the virtual generalized displacements, and set these coefficients to zero separately: this yields both the equations of motion and the boundary conditions. *The equations of motions are as follows:*

$$\delta u_P : [T_z\{u'_P + \phi'(y_P \cos\phi + x_P \sin\phi)\}]' + [(Q_x - M_x\phi')\cos\phi$$
$$- (Q_y + M_y\phi')\sin\phi]' - K_1 + p_x + \delta_0\hat{p}_x = 0,$$
$$\delta v_P : [T_z\{v'_P + \phi'(y_P \sin\phi - x_P \cos\phi)\}]' + [-(Q_x + M_x\phi')\sin\phi$$
$$+ (Q_y + M_y\phi')\cos\phi]' - K_2 + p_y + \delta_0\hat{p}_y = 0,$$
$$\delta w_0 : T'_z - K_3 + p_z + \delta_0\hat{p}_z = 0,$$
$$\delta\theta_x : M'_x - Q_y - K_7 + m_x + \delta_0\hat{m}_x = 0,$$
$$\delta\theta_y : M'_y - Q_x - K_5 + m_y + \delta_0\hat{m}_y = 0, \qquad (3.1\text{-}31\text{a-f})$$
$$\delta\phi : [T_z\{\cos\phi(u'_P y_P - v'_P x_P) + \sin\phi(v'_P y_P + u'_P x_P)\}]'$$
$$- T_z\phi'[\cos\phi(v'_P y_P + u'_P x_P) + \sin\phi(v'_P x_P - u'_P y_P)]$$
$$- [\cos\phi(M_x u'_P - M_y v'_P) + \sin\phi(M_x v'_P + M_y u'_P)]'$$
$$+ [\cos\phi(M_x v'_P\phi' + M_y u'_P\phi' + Q_y u'_P - Q_x v'_P)$$
$$+ \sin\phi(-M_x u'_P\phi' + M_y v'_P\phi' + Q_y v'_P + Q_x u'_P)]$$
$$+ \delta_r B''_\omega + (\Lambda_z\phi')' + M'_z - K_4 \cos\phi - K_{10}\sin\phi + K'_9 + m_z\cos\phi$$
$$+ \lambda_T \sin\phi + m'_\omega + \delta_0(\hat{m}_z\cos\phi + \hat{\lambda}_T\sin\phi + \hat{m}'_\omega) = 0.$$

In Eqs. (3.1-31) is seen that the traction loads on longitudinal edges of an open-beam are absorbed in the equations of motion, these appearing as statically equivalent line loads. Their inclusion was done in a variationally-consistent manner. For closed cross-section beams, $\delta_0 = 0$, and in Eqs. (3.1-30), $\int(\cdot)ds \to \oint(\cdot)ds$.

3.1-5 The Boundary Conditions

The boundary conditions at $z = z_1$ and z_2 are obtained by setting the coefficients of δu_P, δv_P, δw_0, $\delta\theta_x$, $\delta\theta_y$, $\delta\phi$ and $\delta\phi'$ to zero separately in the non-integral terms to be evaluated at $z = z_1, z_2$, appearing in Eqs. (3.1-18), (3.1-21) and (3.1-29). This yields:

$$u_P = \underline{u}_P, \quad \text{or} \quad T_z\{u'_p + \phi'(y_P \cos \phi + x_P \sin \phi)\} + \{-(Q_y + M_y\phi') \sin \phi$$
$$+ (Q_x - M_x\phi') \cos \phi\} = n_z \underset{\sim}{Q}_x,$$

$$v_P = \underline{v}_P, \quad \text{or} \quad T_z\{v'_p + \phi'(y_P \sin \phi - x_P \cos \phi)\} + \{(Q_y + M_y\phi') \cos \phi$$
$$- (Q_x + M_x\phi') \sin \phi\} = n_z \underset{\sim}{Q}_y,$$

$$\theta_x = \underline{\theta}_x \quad \text{or} \quad M_x = n_z \underset{\sim}{M}_x, \qquad\qquad\qquad (3.1\text{-}32\text{a-g})$$

$$\theta_y = \underline{\theta}_y \quad \text{or} \quad M_y = n_z \underset{\sim}{M}_y,$$

$$w_0 = \underline{w}_0 \quad \text{or} \quad T_z = n_z \underset{\sim}{T}_z,$$

$$\phi = \underline{\phi} \quad \text{or} \quad T_z\{\sin \phi(y_P v'_p + x_P u'_p) + \cos \phi(y_P u'_p - x_P v'_p)\} - \{\sin \phi(M_y u'_p$$
$$+ M_x v'_p) + \cos \phi(M_x u'_p - M_y v'_p)\} + \delta_r B'_\omega + \Lambda_z \phi'$$
$$+ M_z + K_9 + m_\omega = n_z(\underset{\sim}{M}_z \cos \phi + \underset{\sim}{\Gamma}_T \sin \phi - \delta_0 \hat{m}_\omega)$$

$$\phi' = \underline{\phi}' \quad \text{or} \quad B_\omega = n_z \underset{\sim}{B}_\omega.$$

3.1-6 The Equations for Non-Shearable Beam Model

Now neglect the effect of transverse shear; θ_x and θ_y are given by Eqs. (2.2-14). Proceeding as before, we find the following

Equations of motion

$$\delta u_P : [T_z\{u'_p + \phi'(y_P \cos \phi + x_P \sin \phi)\}]' - [(M_x \sin \phi - M_y \cos \phi)''$$
$$+ (m_x \sin \phi - m_y \cos \phi)'] - K_1 - (K_6 \cos \phi - K_8 \sin \phi)'$$
$$+ p_x + \delta_0 \hat{p}_x - \delta_0(\hat{m}_x \sin \phi - \hat{m}_y \cos \phi)' = 0,$$

$$\delta v_P : [T_z\{v'_p + \phi'(y_P \sin \phi - x_P \cos \phi)\}]' + [(M_x \cos \phi + M_y \sin \phi)''$$
$$+ (m_x \cos \phi + m_y \sin \phi)'] - K_2 - (K_8 \cos \phi + K_6 \sin \phi)'$$
$$+ p_y + \delta_0 \hat{p}_y + \delta_0(\hat{m}_x \cos \phi + \hat{m}_y \sin \phi)' = 0,$$

$$\delta w_0 : T'_z - K_3 + p_z + \delta_0 \hat{p}_z = 0, \qquad\qquad (3.1\text{-}33\text{a-d})$$

$$\delta\phi : [T_z\{\cos \phi(u'_p y_P - v'_p x_P) + \sin \phi(v'_p y_P + u'_p x_P)\}]' - T_z \phi'[\cos \phi(v'_p y_P$$
$$+ u'_p x_P) + \sin \phi(v'_p x_P - u'_p y_P)] - [\cos \phi(M_x u''_p - M_y v''_p)$$
$$+ \sin \phi(M_x v''_p + M_y u''_p)] + \delta_r B''_\omega + (\Lambda_z \phi')' + M'_z$$
$$- K_4 \cos \phi - K_{10} \sin \phi + K'_9 + [\cos \phi(K_6 v'_p - K_8 u'_p)$$
$$- \sin \phi(K_6 u'_p + K_8 v'_p)] + m_z \cos \phi + \lambda_T \sin \phi + m'_\omega$$
$$+ [\cos \phi(m_x u'_p - m_y v'_p) + \sin \phi(m_x v'_p + m_y u'_p)]$$
$$+ \delta_0\{\hat{m}_z \cos \phi + \hat{\lambda}_T \sin \phi + \hat{m}'_\omega + \cos \phi(\hat{m}_x u'_p - \hat{m}_y v'_p)$$
$$+ \sin \phi(\hat{m}_x v'_p + \hat{m}_y u'_p)\} = 0.$$

Boundary conditions at $z = z_1, z_2$

$$u_P = \underline{u}_P \text{ or } T_z\{u'_P + \phi'(y_P \cos\phi + x_P \sin\phi)\} - (M_x \sin\phi - M_y \cos\phi)'$$
$$- (K_6 \cos\phi - K_8 \sin\phi) - (m_x \sin\phi - m_y \cos\phi) - \delta_0(\hat{m}_x \sin\phi$$
$$- \hat{m}_y \cos\phi) = n_z \underset{\approx}{Q}_x,$$

$$v_P = \underline{v}_P \text{ or } T_z\{v'_P + \phi'(y_P \sin\phi - x_P \cos\phi)\} + (M_x \cos\phi + M_y \sin\phi)'$$
$$- (K_8 \cos\phi + K_6 \sin\phi) + (m_x \cos\phi + m_y \sin\phi) + \delta_0(\hat{m}_x \cos\phi$$
$$+ \hat{m}_y \sin\phi) = n_z Q_y,$$

$$u'_P = \underline{u}'_P \text{ or } \sin\phi(M_x + n_z \underline{M}_x) - \cos\phi(M_y + n_z \underline{M}_y)] = 0,$$
$$v'_P = \underline{v}'_P \text{ or } \sin\phi(M_y + n_z \underline{M}_y) + \cos\phi(M_x + n_z \underline{M}_x)] = 0,$$
$$w_0 = \underline{w}_0 \text{ or } T_z = n_z \underline{T}_z \tag{3.1-34a-g}$$
$$\phi = \underline{\phi} \text{ or } T_z\{\sin\phi(y_P v'_P + x_P u'_P) + \cos\phi(y_P u'_P - x_P v'_P)\}$$
$$+ \delta_r B'_\omega + \Lambda_z \phi' + M_z + K_9 + m_\omega + n_z\{(\underline{M}_y u'_P$$
$$+ \underline{M}_x v'_P) \sin\phi + \cos\phi(\underline{M}_x u'_P - \underline{M}_y v'_P)\}$$
$$- n_z(\underline{M}_z \cos\phi + \underline{\Gamma}_T \sin\phi - \delta_0 \hat{m}_\omega)] = 0,$$
$$\delta_r \phi' = \delta_r \underline{\phi}' \text{ or } \delta_r B_\omega = n_z \delta_r \underline{B}_\omega.$$

Note that, whether or not transverse shear deformation is included, there are seven boundary conditions at each end of the beam; the order of the governing system should be fourteen. However, for the free warping model (i.e. St. Venant torsional model with $\delta_r = 0$), there are only 6 boundary conditions at each end, and the system is of order 12. For the unshearable beam model, special cases of these results were obtained by Polillo et al. (1992).

3.1-7 Remarks

(i) The equations of motion and the related boundary conditions have been obtained in a consistent way, via Hamilton's principle.

(ii) The geometrically nonlinear theory accounts for end loads, lateral loads and body forces so that the buckling/postbuckling equations corresponding to any combination of such loads can be derived in a straightforward, yet consistent manner. Moreover, because dynamic terms are included, the theory is also useful for dynamic stability analyses and for studying the effect of time-dependent external excitation on dynamic response.

(iii) The equations for shearable and unshearable beam models have been expressed in a unified form, for both closed and open beam cross-sections, by using the tracer δ_0.

(iv) The theory accounts for arbitrarily large twist angles. Hence, approximations for cases where the twist angle is *moderately small, reasonably small* (see Murray and Rajasekaran, 1975), *or small* can be derived from the present theory by expanding $\cos\phi$ and $\sin\phi$ in series: $\sin\phi = \phi - \phi^3/6 + \phi^5/120...$, $\cos\phi = 1 - \phi^2/2 + \phi^4/24...$ and retaining a convenient number of terms.

(v) The use of the tracers δ_t, δ_e, δ_r and δ_n enables reduction of the general theory to special cases, so that various non-classical effects can be identified and their role assessed.

(vi) The present theory employs a reference point P to which the flexural displacements and the twist of the section are referred. By proper choice of P some of the beam stiffness coefficients can be made zero. This has been shown (see e.g. Bauld and Tzeng, 1984), to be possible at least for the case of the Bernoulli-Euler cross-ply beams.

(vii) The previously obtained equations are independent of the constitutive equations; they are equally valid for both elastic and inelastic analyses:

Finally, note that the nonlinear equations of motion and boundary conditions reveal a full coupling between extension, compound bending, twist and transverse shear.

3.2 LINEAR FORMULATION

The linearized equations can be obtained into two steps: (i) approximation of $\cos\phi \approx 1$ and $\sin\phi \approx \phi$, and (ii) linearization of the respective equations after incorporation of this approximation. We will obtain linearized forms of various energy terms, for use in subsequent chapters.

3.2-1 Energy Quantities in Linear Beam Theory

The linearized counterpart of the various energy quantities previously obtained for the nonlinear theory, will be displayed next. These quantities will be identified by the same symbols. Corresponding to both shearable and unshearable thin-walled beam models, the strain energy functional is:

$$W = \frac{1}{2} \int [T_z w_0' + M_x(\delta_t \theta_x' - \delta_e v_P'') + M_y(\delta_t \theta_y' - \delta_e u_P'') + \delta_t Q_x(u_P' + \theta_y)$$
$$+ \delta_t Q_y(v_P' + \theta_x) + M_z \phi' - \delta_r B_\omega \phi''] dz. \tag{3.2-1}$$

For the shearable beam theory, for δW one obtains:

$$
\delta W = -\int \Big\{ T_z' \delta w_0 + (M_y' - Q_x)\delta\theta_y + (M_x' - Q_y)\delta\theta_x + (B_\omega'' + M_z')\delta\phi
$$
$$
+ Q_x' \delta u_P + Q_y' \delta v_P \Big\} dz + [T_z \delta w_0 + M_y \delta\theta_y + M_x \delta\theta_x - B_\omega \delta\phi'
$$
$$
+ (B_\omega' + M_z)\delta\phi + Q_x \delta u_P + Q_y \delta v_P]|_{z=z_1,z_2}. \tag{3.2-2}
$$

The non-shearable version of (3.2-2) is obtained by expressing $\theta_x = -v_P'$ and $\theta_y = -u_P'$, and carrying out an integration by parts whenever necessary. The transformation is straightforward and the result will not be presented here.

The kinetic energy for shearable beams is expressed as:

$$
\mathcal{K} = \frac{1}{2}\int\int\int {}_0\rho_{(k)}\Big[[\dot{u}_P - (Y - y_P)\dot\phi]^2 + [\dot{v}_P + (X - x_P)\dot\phi]^2
$$
$$
+ \Big\{ \dot{w}_0 + x\dot\theta_y + y\dot\theta_x - \bar{F}(s)\dot\phi' + n\Big(\frac{dy}{ds}\dot\theta_y - \frac{dx}{ds}\dot\theta_x + r_t\dot\phi'\Big) \Big\}^2 \Big] dn\,ds\,dz. \tag{3.2-3}
$$

Since the linearized forms $A(\phi)$, $B(\phi)$, T and S appearing in Table 3.1-1 are $A(\phi) = 0$, $B(\phi) = \ddot\phi$, $T = -\ddot{v}_P'$, $S = -\ddot{u}_P'$, Eq. (3.1-21) gives

$$
\int_{t_0}^{t_1}\delta\mathcal{K}dt = -\int_{t_0}^{t_1}dt\int\Big[[K_1 + \delta_e K_6']\delta u_P + [K_2 + \delta_e K_8']\delta v_P
$$
$$
+ K_3\delta w_0 + \delta_t K_7\delta\theta_x + \delta_t K_5\delta\theta_y + [K_4 - K_9'
$$
$$
- \delta_e[(K_6 v_P' - K_8 u_P')]]\delta\phi \Big] dz \tag{3.2-4}
$$
$$
- \int_{t_0}^{t_1}dt\Big\{ -\delta_e K_6\delta u_P - \delta_e K_8\delta v_P + K_9\delta\phi \Big\}\Big|_{z=z_1,z_2}.
$$

The terms K_i appearing in Eq. (3.2-4) are defined in Table 3.2-1.

Table 3.2-1 The linearized counterparts of K_i terms

K_1	$B_1\ddot{u}_P - B_2\ddot\phi$
K_2	$B_1\ddot{v}_P + B_3\ddot\phi$
K_3	$B_1\ddot{w}_0 + B_{11}(\delta_t\ddot\theta_x - \delta_e\ddot{v}_P') + B_{12}(\delta_t\ddot\theta_y - \delta_e\ddot{u}_P') - B_7\ddot\phi'$
K_4	$B_3\ddot{v}_P - B_2\ddot{u}_P + (B_4 + B_5)\ddot\phi$
$K_5 = K_6$	$B_{12}\ddot{w}_0 + B_{15}(\delta_t\ddot\theta_x - \delta_e\ddot{v}_P') + B_{14}(\delta_t\ddot\theta_y - \delta_e\ddot{u}_P') - B_9\ddot\phi\prime$
$K_7 = K_8$	$B_{11}\ddot{w}_0 + B_{13}(\delta_t\ddot\theta_x - \delta_e\ddot{v}_P') + B_{15}(\delta_t\ddot\theta_y - \delta_e\ddot{u}_P') - B_8\ddot\phi\prime$
K_9	$-[B_7\ddot{w}_0 + B_8(\delta_t\ddot\theta_x - \delta_e\ddot{v}_P') + B_9(\delta_t\ddot\theta_y - \delta_e\ddot{u}_P') - B_{10}\ddot\phi\prime]$
K_{10}	$-[B_3\ddot{u}_P + B_2\ddot{v}_P]$

In Table 3.2-1 there are underscored the terms that will be considered in the subsequent developments.

The mass terms B_i are expressed by Eq. (3.1-23) as before. Proceeding as for the nonlinear theory, Hamilton's principle provides the *equations of motion* and the associated boundary conditions expressed in a unified form for both shearable and unshearable thin-walled beams.

3.2-2 Dissipative Effects

To introduce dissipative effects into the equations of motion, we use Rayleigh's dissipation function $\mathcal{R}$:

$$\mathcal{R} = \frac{1}{2} \int_\tau (c_1 \dot{V}_1^2 + c_2 \dot{V}_2^2 + c_3 \dot{V}_3^2) d\tau, \tag{3.2-5}$$

where c_i are the damping coefficients. The effect of dissipative forces can be included into the Hamilton's principle by adding to Eq. (3.1-28), the virtual work done by these forces:

$$\delta \mathcal{J}_4 = - \int_\tau [c_1 \dot{V}_1 \delta V_1 + c_2 \dot{V}_2 \delta V_2 + c_3 \dot{V}_3 \delta V_3] d\tau. \tag{3.2-6}$$

Making the changes

$$V_1 \rightarrow u, \ V_2 \rightarrow v, \ V_3 \rightarrow w, \tag{3.2-7}$$

using the displacement representations, Eqs. (2.1-6) and (2.1-28), the expression of $d\tau$ given by Eq. (3.1-12), and restricting x and y appearing in the expressions of u and v to the points of the mid-line contour, one can convert (3.2-6) to:

$$
\begin{aligned}
\delta \mathcal{J}_4 = - \int \int \int &\left[c_1 \left\{ \dot{u}_P - (y - y_P)\dot{\phi} \right\} \left\{ \delta u_P - (y - y_P)\delta\phi \right\} \right. \\
&+ c_2 \left\{ \dot{v}_P + (x - x_P)\dot{\phi} \right\} \left\{ \delta v_P + (x - x_P)\delta\phi \right\} \\
&+ c_3 \left\{ \dot{w}_0 + x\dot{\theta}_y + y\dot{\theta}_x - \bar{F}(s)\dot{\phi}' \right\} \left\{ \delta w_0 + x\delta\theta_y + y\delta\theta_x - \bar{F}(s)\delta\phi' \right\} \\
&+ c_3 n \left\{ \cdots \cdots \right\} \left\{ \cdots \cdots \right\} + c_3 n^2 \left\{ \frac{dy}{ds}\dot{\theta}_y - \frac{dx}{ds}\dot{\theta}_x + r_t \dot{\phi}' \right\} \\
&\left. \times \left\{ \frac{dy}{ds}\delta\theta_y - \frac{dx}{ds}\delta\theta_x + r_t \delta\phi' \right\} \right] dn\,ds\,dz.
\end{aligned}
\tag{3.2-8}
$$

Since the linear term in the thickness coordinate n becomes immaterial when it is integrated over a symmetric interval (i.e. between - h/2 and h/2), it was omitted in Eq. (3.2-8).

Following the usual procedure, for shearable beams one obtains

$$\delta \mathcal{J}_4 = -\int_{z_1}^{z_2} [F_{D_1}\delta u_P + F_{D_2}\delta v_P + F_{D_3}\delta w_0 + F_{D_4}\delta\phi$$
$$+ F_{D_5}\delta\theta_y + F_{D_6}\delta\theta_x + F_{D_7}\delta\phi']dz, \qquad (3.2\text{-}9)$$

where F_{D_i} are the dissipative forces associated with the generalized displacements of the beam. Equation (3.2-9), gives a unified form of $\delta\mathcal{J}_4$ for both shearable and unshearable beams:

$$\delta\mathcal{J}_4 = [\delta_e F_{D_5}\delta u_P + \delta_e F_{D_6}\delta v_P - F_{D_7}\delta\phi]|_{z=z_1,z_2}$$
$$- \int_{z_1}^{z_2} \left\{ \left(F_{D_1} + \delta_e F'_{D_5}\right)\delta u_P + \left(F_{D_2} + \delta_e F'_{D_6}\right)\delta v_P \right.$$
$$\left. + F_{D_3}\delta w_0 + \left(F_{D_4} - F'_{D_7}\right)\delta\phi + \delta_t F_{D_5}\delta\theta_y + \delta_t F_{D_6}\delta\theta_x \right\} dz. \quad (3.2\text{-}10)$$

The expressions of dissipative forces F_{D_i} are recorded in Table 3.2-2: Here, the generalized damping coefficients are defined as:

$$(d_1; \, d_2; \, d_5) = \int c_1 h \left(1; \, y - y_P; \, (y - y_P)^2\right) ds, \qquad (3.2\text{-}11a)$$

$$(d_3; \, d_4; \, d_6) = \int c_2 h \left(1; \, x - x_P; \, (x - x_P)^2\right) ds, \qquad (3.2\text{-}11b)$$

$(d_7, \, d_8, \, d_9, \, d_{10}, \, d_{11}, \, d_{12}, \, d_{13}, \, d_{14}, \, d_{15}, d_{16}, \, d_{17}, \, d_{18}, \, d_{19}, \, d_{20}, \, d_{21},$
$d_{22}, \, d_{23}, d_{24})$

$$= \int c_3 h \left[1, \, y, \, x, \, xy, \, x^2, \, y^2, \, \frac{h^2}{12}\frac{dx}{ds}, \, \frac{h^2}{12}\frac{dy}{ds}, \, \frac{h^2}{12}\frac{dx}{ds}\frac{dy}{ds}, \, \frac{h^2}{12}\left(\frac{dx}{ds}\right)^2, \right.$$
$$\left. , \frac{h^2}{12}\left(\frac{dy}{ds}\right)^2, \, \bar{F}(s), \, x\bar{F}(s), \, y\bar{F}(s), \, \bar{F}^2(s), \, r_t\frac{h^2}{12}\frac{dx}{ds}, \, r_t\frac{h^2}{12}\frac{dy}{ds}, \, r_t^2\frac{h^2}{12} \right] ds. $$
$$(3.2\text{-}11c)$$

The principal terms in F_{D_i} are underscored by a solid line.

Note that the expressions in Table 3.2-2 can be obtained in an alternative way by starting with Eq. (3.2-5), using Eqs. (3.2-7) and representation of generalized displacements, and recognizing that

Table 3.2-2: Expressions of dissipative forces F_{D_i}

$F_{D_1}(z,t)$	$d_1\ddot{u}_P - d_2\dot{\phi}$
$F_{D_2}(z,t)$	$d_3\dot{v}_P + d_4\dot{\phi}$
$F_{D_3}(z,t)$	$d_7\dot{w}_0 - d_{18}\dot{\phi}' + \delta_t d_8\dot{\theta}_x + \delta_t d_9\dot{\theta}_y - \delta_e d_8\dot{v}'_P - \delta_e d_9\dot{u}'_P$
$F_{D_4}(z,t)$	$d_4\dot{v}_P - d_2\dot{u}_P + (d_5 + d_6)\dot{\phi}$
$F_{D_5}(z,t)$	$d_9\dot{w}_0 + \delta_t(d_{10} - \delta_n d_{15})\dot{\theta}_x + \delta_t(d_{11} + \delta_n d_{17})\dot{\theta}_y$
	$-(d_{19} - \delta_n d_{23})\dot{\phi}' - \delta_e(d_{10} - \delta_n d_{15})\dot{v}'_P - \delta_e(d_{11} + \delta_n d_{17})\dot{u}'_P$
$F_{D_6}(z,t)$	$d_8\dot{w}_0 + \delta_t(d_{10} - \delta_n d_{15})\dot{\theta}_y + \delta_t(d_{12} + \delta_n d_{16})\dot{\theta}_x - (d_{20} + \delta_n d_{22})\dot{\phi}'$
	$- \delta_e(d_{10} - \delta_n d_{15})\dot{u}'_P - \delta_e(d_{12} + \delta_n d_{16})\dot{v}'_P$
$F_{D_7}(z,t)$	$-d_{18}\dot{w}_0 - \delta_t(d_{20} + \delta_n d_{22})\dot{\theta}_x - \delta_t(d_{19} - \delta_n d_{23})\dot{\theta}_y + (d_{21} + \delta_n d_{24})\dot{\phi}'$
	$+\delta_e(d_{19} - \delta_n d_{23})\dot{u}'_P + \delta_e(d_{20} + \delta_n d_{22})\dot{v}'_P$

$$F_{D_1} = -\frac{\partial \mathcal{R}}{\partial \dot{u}_P}, \quad F_{D_2} = -\frac{\partial \mathcal{R}}{\partial \dot{v}_P}, \quad F_{D_3} = -\frac{\partial \mathcal{R}}{\partial \dot{w}_0}, \quad F_{D_4} = -\frac{\partial \mathcal{R}}{\partial \dot{\phi}},$$

$$F_{D_5} = -\frac{\partial \mathcal{R}}{\partial \dot{\theta}_y}, \quad F_{D_6} = -\frac{\partial \mathcal{R}}{\partial \dot{\theta}_x}, \quad F_{D_7} = -\frac{\partial \mathcal{R}}{\partial \dot{\phi}'}. \tag{3.2-12}$$

3.2-3 Equations of Motion and Boundary Conditions

Replacing all the various terms in the variational equations, and considering the variations arbitrary and independent, one obtain the equations of motion:

$$\delta u_P : \; \delta_e M''_y + \delta_t Q'_x - K_1 - \delta_e K'_6 - F_{D_1} - \delta_e F'_{D_5}$$
$$+ p_x + \delta_e m'_y + \delta_0(\hat{p}_x + \delta_e \hat{m}'_y) = 0,$$

$$\delta v_P : \; \delta_e M''_x + \delta_t Q'_y - K_2 - \delta_e K'_8 - F_{D_2} - \delta_e F'_{D_6}$$
$$+ p_y + \delta_e m'_x + \delta_0(\hat{p}_y + \delta_e \hat{m}'_x) = 0,$$

$$\delta w_0 : \; T'_z - K_3 - F_{D_3} + p_z + \delta_0 \hat{p}_z = 0, \tag{3.2-13a-f}$$

$$\delta\phi : \; \delta_r B''_\omega + M'_z - K_4 + K'_9 + \delta_e(K_6 v_P' - K_8 u_P')$$
$$- F_{D_4} + F'_{D_7} + m_z + m'_\omega + \delta_0(\hat{m}_z + \hat{m}'_\omega) = 0,$$

$$\delta\theta_x : \; \delta_t[M'_x - Q_y - K_7 - F_{D_6} + m_x + \delta_0 \hat{m}_x] = 0,$$

$$\delta\theta_y : \; \delta_t[M'_y - Q_x - K_5 - F_{D_5} + m_y + \delta_0 \hat{m}_y] = 0,$$

and the *boundary conditions* at $z = z_1, z_2$:

$$u_P = \underline{u}_P \; \text{or}, \; \delta_e M'_y + \delta_t Q_x - \delta_e K_6 - \delta_e F_{D5} = n_z(\underline{Q}_x - \delta_0 \delta_e \hat{m}_y),$$

$$v_P = \underline{v}_P \; \text{or}, \; \delta_e M'_x + \delta_t Q_y - \delta_e K_8 - \delta_e F_{D6} = n_z(\underline{Q}_y - \delta_0 \delta_e \hat{m}_x),$$

$$w_0 = \underline{w}_0 \; \text{or}, \; T_z = n_z \underline{T}_z,$$

$$\delta_t \theta_y = \delta_t \underline{\theta}_y \; \text{or}, \; M_y = n_z \underline{M}_y,$$

$$\delta_e u'_P = \delta_e \underline{u}'_P \text{ or, } M_y = n_z \underline{M}_y, \qquad\qquad (3.2\text{-}14\text{a-i})$$

$$\delta_t \theta_x = \delta_t \underline{\theta}_x \text{ or, } M_x = n_z \underline{M}_x,$$

$$\delta_e v'_P = \delta_e \underline{v}'_P \text{ or, } M_x = n_z \underline{M}_x,$$

$$\phi = \underline{\phi} \text{ or, } \delta_r B'_\omega + M_z + K_9 + F_{D_7} = n_z (\underline{M}_z - \delta_0 \hat{m}_\omega)$$

$$\delta_r \phi' = \delta_r \underline{\phi}' \text{ or, } \delta_r \bar{B}_\omega = \delta_r n_z \underline{\bar{B}}_\omega.$$

These equations were obtained by Song (1990), Song and Liberscu (1993), Librescu and Song (1991, 1992), Librescu et al. (1996, 1997). In the static case, the equations agree with the those obtained by Rehfield (1985). The unshearable and free warping version of the static equilibrium equations as obtained from Eqs. (3.2-13), agree with those derived by Berdichevsky et al. (1992) and Johnson et al. (2001), respectively. A variationally consistent system of linearized equations of motion and boundary conditions obtained by discarding the secondary warping effect was supplied also by Chen (2000).

3.2-4 Remarks

The linearized equations of motion and boundary conditions displayed here are valid for both open and closed cross-section beams, and include also transverse shear, rotatory inertia and warping restraint effects. To obtain special cases of these equations, proper values (that is 0 or 1) of the tracers identifying the various effects, have to be prescribed.

It is instructive to summarize the generalized stress resultants and the corresponding generalized strains, displacements, body and surface forces. These are displayed in Tables 3.2-3 and 3.2-4.

Table 3.2-3: Generalized static and kinematic variables for thin-walled beams. Nonlinear theory

Generalized 1-D Stress Resultant and Stress Couples Measures	Generalized 1-D Strain Measures
$T_z(z,t)$	$w'_0 + \frac{1}{2}[(u'_P)^2 + (v'_P)^2] + \phi'[y_P(u'_P \cos\phi + v'_P \sin\phi)$ $+ x_P(u'_P \sin\phi - v'_P \cos\phi)] \Rightarrow$ axial strain
$M_x(z,t)$	$\delta_t[\theta'_x - \phi'(u'_P \cos\phi + v'_P \sin\phi)]$ $+ \delta_e(u''_P \sin\phi - v''_P \cos\phi) \Rightarrow$ bending strain
$M_y(z,t)$	$\delta_t[\theta'_y - \phi'(u'_P \sin\phi - v'_P \cos\phi)]$ $- \delta_e(u''_P \cos\phi + v''_P \sin\phi) \Rightarrow$ bending strain
$Q_x(z,t)$	$\delta_t(\theta_y + v'_P \sin\phi + u'_P \cos\phi)(\equiv \gamma_{xz}) \Rightarrow$ transverse shear strain
$Q_y(z,t)$	$\delta_t(\theta_x - u'_P \sin\phi + v'_P \cos\phi)(\equiv \gamma_{yz}) \Rightarrow$ transverse shear strain
$\delta_r B_\omega(z,t)$	$\delta_r \phi'' \Rightarrow$ torsional strain related to warping
$M_z(z,t)$	$\phi' \Rightarrow$ rate of twist
$\Lambda_z(z,t)$	$(\phi')^2 \Rightarrow$ nonlinear rate of twist

Table 3.2-4: Generalized body and surface line loads and corresponding generalized displacement measures[+]

Body and Line Loads	End Loads	Displacements
$(p_x + \delta_0 \hat{p}_x)$ [F/L]	$\underset{\sim}{Q}_x$	u_P
$(p_y + \delta_0 \hat{p}_y)$ [F/L]	$\underset{\sim}{Q}_y$	v_P
$(p_z + \delta_0 \hat{p}_z)$ [F/L]	$\underset{\sim}{T}_z$	w_0
$(m_x + \delta_0 \hat{m}_x)$ [F]	$\underset{\sim}{M}_x$	θ_x
$(m_y + \delta_0 \hat{m}_y)$ [F]	$\underset{\sim}{M}_y$	θ_y
$(m_z + \delta_0 \hat{m}_z)$ [F]	$\underset{\sim}{M}_z$	ϕ
$\delta_r (m_\omega + \delta_0 \hat{m}_\omega)$ [F·L]	$\delta_r \underset{\sim}{B}_\omega$	$\delta_r \phi'$

[+] In square brackets there are indicated the units of the associated quantities

3.3 HIGHER-ORDER THEORY: GEOMETRICALLY LINEAR THIN-WALLED BEAMS

In Sections 3.1 and 3.2, the shearable beams have been approached within the first order transverse shear deformation theory (FSDT). It is interesting to display the equations of the linearized equilibrium and the boundary conditions corresponding to the higher order shear deformation theory, (HSDT).

Following the procedure followed in the case of the FSDT beam model, and using the linearized counterpart of the nonlinear kinematic equations displayed in Section 2.4 one can obtain the equilibrium equations and boundary conditions. For simplicity, the inertia and dissipative terms will be discarded. For thin-walled beams of closed-cross section contour one obtains the following:

The equilibrium equations

$$\delta u_P : \; \delta_t Q'_x - \delta_h M''_{ya} + \delta_e M''_{yb} + p_x + \delta_e m'_y = 0,$$

$$\delta v_P : \; \delta_t Q'_y - \delta_h M''_{xa} + \delta_e M''_{xb} + p_y + \delta_e m'_x = 0,$$

$$\delta \theta_y : \; \delta_t (M'_y - Q_x) + m_y = 0,$$

$$\delta \theta_x : \; \delta_t (M'_x - Q_y) + m_x = 0, \tag{3.3-1a-f}$$

$$\delta w_0 : \; T'_z + p_z = 0,$$

$$\delta \phi : \; \delta_r B''_w + M'_z + m_z + m'_\omega = 0,$$

and the *homogeneous boundary conditions* at $z = z_1, z_2$:

$$u_P = 0 \; \text{or} \; \delta_t Q_x - \delta_h M'_{ya} + \delta_e M'_{yb} = 0,$$

$$v_P = 0 \; \text{or} \; \delta_t Q_y - \delta_h M'_{xa} + \delta_e M'_{xb} = 0,$$

$$\theta_x = 0 \; \text{or} \; \delta_t M_x = 0,$$

$$\theta_y = 0 \; \text{or} \; \delta_t M_y = 0, \tag{3.3-2a-i}$$

$$w_0 = 0 \; \text{or} \; T_z = 0,$$

$$u'_P = 0 \; \text{or} \; \delta_h M_{ya} - \delta_e M_{yb} = 0,$$

$$v'_P = 0 \text{ or } \delta_h M_{xa} - \delta_e M_{xb} = 0,$$

$$\phi = 0 \text{ or } \delta_r B'_w + M_z = 0,$$

$$\delta_r \phi' = 0 \text{ or } \delta_r B_w = 0.$$

The meaning of tracers δ_h, δ_e and δ_t, and the way to obtain various special cases have been presented in Section 2.4. The 1-D stress resultants and stress couples involved in these equations are defined as follows:

$$T_z(z,t) = \oint N_{zz}^{(0)} ds,$$

$$Q_x(z,t) = \oint \left\{ \delta_t \frac{dy}{ds} \left(N_{nz}^{(0)} - \delta_h \frac{4}{h^2} N_{nz}^{(2)} \right) + \frac{dx}{ds} \left(N_{sz}^{(0)} - \delta_h \frac{4}{h^2} N_{sz}^{(2)} \right) \right\} ds,$$

$$Q_y(z,t) = \oint \left\{ \frac{dy}{ds} \left(N_{sz}^{(0)} - \delta_h \frac{4}{h^2} N_{sz}^{(2)} \right) - \delta_t \frac{dx}{ds} \left(N_{nz}^{(0)} - \delta_h \frac{4}{h^2} N_{nz}^{(2)} \right) \right\} ds,$$

$$M_x(z,t) = \oint \left\{ y N_{zz}^{(0)} - \delta_n \frac{dx}{ds} \left(N_{zz}^{(1)} - \delta_h \frac{4}{3h^2} N_{zz}^{(3)} \right) \right\} ds,$$

$$M_y(z,t) = \oint \left\{ x N_{zz}^{(0)} + \delta_n \frac{dy}{ds} \left(N_{zz}^{(1)} - \delta_h \frac{4}{3h^2} N_{zz}^{(3)} \right) \right\} ds,$$

$$M_{xa}(z,t) = \delta_h \frac{4}{3h^2} \oint \frac{dx}{ds} N_{zz}^{(3)} ds, \qquad\qquad\qquad (3.3\text{-}3a\text{-}k)$$

$$M_{ya}(z,t) = \delta_h \frac{4}{3h^2} \oint \frac{dy}{ds} N_{zz}^{(3)} ds,$$

$$M_{xb}(z,t) = \delta_e \oint \left(y N_{zz}^{(0)} - \delta_n \frac{dx}{ds} N_{zz}^{(1)} \right) ds,$$

$$M_{yb}(z,t) = \delta_e \oint \left(x N_{zz}^{(0)} + \delta_n \frac{dy}{ds} N_{zz}^{(3)} \right) ds,$$

$$M_z(z,t) = \oint \left(\psi N_{sz}^{(0)} + \delta_n 2 N_{sz}^{(1)} \right) ds,$$

$$\delta_r B_w(z,t) = \delta_r \oint \left(\bar{F}(s) N_{zz}^{(0)} - \delta_n r_t(s) N_{zz}^{(1)} \right) ds.$$

Herein the stress-resultant and stress couple measures are defined as

$$\left(N_{zz}^{(0)}, N_{zz}^{(1)}, N_{zz}^{(2)}, N_{zz}^{(3)} \right) = \int_{-h/2}^{h/2} \sigma_{zz}(1, n, n^2, n^3) dn,$$

$$\left(N_{sz}^{(0)}, N_{sz}^{(1)}, N_{sz}^{(2)} \right) = \int_{-h/2}^{h/2} \tau_{sz}(1, n, n^2) dn, \qquad (3.3\text{-}4a\text{-}c)$$

$$\left(N_{nz}^{(0)}, N_{nz}^{(2)} \right) = \int_{-h/2}^{h/2} \tau_{nz}(1, n^2) dn.$$

Note that the 2-D stress resultant measures, $N_{zz}^{(0)} (\equiv N_{zz})$, $N_{zz}^{(1)} (\equiv L_{zz})$, $N_{sz}^{(0)} (\equiv N_{sz})$ and $N_{sz}^{(1)} (\equiv L_{sz})$ and $N_{zn}^{(0)} (\equiv N_{zn})$ are the only ones present in the equations of equilibrium and boundary conditions corresponding to the FSDT and the unshearable theories. The remaining ones are higher order stress resultant measures that occur only within the HSDT beam model.

Equation (3.3-2) shows that in the HSDT there are nine boundary conditions at each end of the beam; the order of the system is eighteen, that is, four degrees higher than those of the first order shearable and the unshearable beam models. For the free warping model, as in the FSDT and classical models, the order of the governing equations drops by two, becoming for the HSDT sixteen.

REFERENCES

Bauld, N. R. Jr. and Tzeng, L-S. (1984) "A Vlasov Theory for Fiber-Reinforced Beams with Thin-Walled Open Cross-Sections," *International Journal of Solids and Structures*, Vol. 20, No. 3, pp. 277–297.

Berdichevsky, V. L., Armanios, E. A. and Badir, A. M. (1992) "Theory of Anisotropy Thin-Walled Closed-Section Beams," *Composites Engineering*, Vol. 2, Nos. 5–7, pp. 411–432.

Bhaskar, K. and Librescu, L. (1995) "A Geometrically Non-Linear Theory for Laminated Anisotropic Thin-Walled Beams," *International Journal of Engineering Science*, Vol. 33, No. 9, pp. 1331–1344.

Chandiramani, N. K., Librescu, L. and Shete, C. D., (2002) "On the Free Vibration of Rotating Thin-Walled Composite Beams Using a Higher Order Shear Formulation," *Aerospace Science and Technology*, Vol. 6, pp. 545–561.

Chen, C-N., (2000) "Dynamic Equilibrium Equations of Composite Anisotropic Beams Considering the Effects of Transverse Shear Deformations and Structural Damping," *Composite Structures*, Vol. 48, pp. 287–303.

Green, A. E. and Adkins, J. E. (1960) *Large Elastic Deformations and Non-Linear Continuum Mechanics*, Oxford, Clarendon Press.

Johnson, E. R., Vasiliev, V. V. and Vasiliev, D. V., (2001) "Anisotropic Thin-Walled Beams with Closed Cross-Sectional Contours," *AIAA Journal*, Vol. 39, No. 12, pp. 2389–2393.

Jung, S. N., Nagaraj, V. T. and Chopra, I. (1999) "Assessment of Composite Rotor Blade: Modeling Techniques," *Journal of the American Helicopter Society*, Vol. 44, No. 2, pp. 188–205.

Librescu, L. (1975) *Elastostatics and Kinetics of Anisotropic and Heterogeneous Shell-Type Structures*, Leyden, The Netherlands: Noordhoff International Publishing.

Librescu, L. (1987) "Refined Geometrically Non-Linear Theories of Anisotropic Laminated Shells," *Quarterly of Applied Mathematics*, Vol. 45, No. 1, pp. 1–22.

Librescu, L. and Song, O. (1991) "Behavior of Thin-Walled Beams Made of Advanced Composite Materials and Incorporating Non-Classical Effects," *Applied Mechanics Reviews*, Vol. 44, No. 11, pp. 174–180.

Librescu, L. and Song, O. (1992) "On the Static Aeroelastic Tailoring of Composite Aircraft Swept Wings Modeled as Thin-Walled Beam Structures," *Composites Engineering*, Vol. 2, Nos. 5–7, pp. 497–512.

Librescu, L., Meirovitch, L. and Song, O. (1996) "Refined Structural Modeling for Enhancing Vibration and Aeroelastic Characteristics of Composite Aircraft Wings," *La Recherche Aérospatiale*, Vol. 1, pp. 23–35.

Librescu, L., Meirovitch, L. and Na, S. S. (1997) "Control of Cantilevers Vibration via Structural Tailoring and Adaptive Materials," *AIAA Journal*, Vol. 35, No. 8, pp. 1309–1315.

Murray, D. W. and Rajasekaran, S. (1975) "Technique for Formulating Beam Equations," *Journal of the Engineering Mechanics Division*, EM5, October, pp. 561–573.

Polillo, V. R., Viccaca, S. F. and Garcia, L. F. T. (1992) "Variational Approach for Geometrically Nonlinear Analysis of Combined Flexure and Torsion of Thin-Walled Bars with Open Cross Section Under Dynamic Loading," *Revista Brasileira de Engenharia Estructural*, Vol. 8, No. 2, pp. 33–56.

Qin, Z and Librescu, L., (2001) "Static and Dynamic Validations of a Refined Thin-Walled Composite Beam Model," *AIAA Journal*, Vol. 39, No. 12, pp. 2422–2424.

Qin, Z. and Librescu, L., (2002) "On a Shear-Deformable Theory of Anisotropic Thin-Walled Beams: Further Contribution and Validations," *Composite Structures*, Vol. 56, No. 4, pp. 345–358.

Rehfield, L. W. (1985) "Design Analysis Methodology for Composite Rotor Blades," *Proceedings of the 7th DoD/NASA Conference on Fibrous Composites in Structural Design*, June, Denver, Co.

Smith, E. C. and Chopra, I. (1991) "Formulation and Evaluation of an Analytical Model for Composite Box Beams," *Journal of the American Helicopter Society*, Vol. 36, No. 3, pp. 23–35.

Song, O. (1990) "Modeling and Response Analysis of Thin-Walled Beam Structures Constructed of Advanced Composite Materials," Ph.D. Dissertation, Virginia Polytechnic Institute & State University, VA, USA.

Song, O. and Librescu, L. (1993) "Free Vibration of Anisotropic Composite Thin-Walled Beams of Closed Cross-Section Contour," *Journal of Sound and Vibration*, Vol. 164, No. 1, pp. 129–147.

Chapter 4

ADDITIONAL EQUATIONS OF THE LINEAR BEAM THEORY

In this chapter, we derive in matrix form, the expressions of kinetic, strain and dissipation energies for use in later analysis. In addition, the governing equations of the theory of thin-walled beams will be established.

4.1 KINETIC ENERGY

For shearable beams, the kinetic energy expressed by Eq. (3.2-3) may be written:

$$\mathcal{K} = \int_{z_1}^{z_2} \hat{\mathcal{K}} dz, \qquad (4.1\text{-}1)$$

where $\hat{\mathcal{K}}$ is the kinetic energy density

$$\hat{\mathcal{K}} = \frac{1}{2} \dot{\mathbf{V}}^T \mathbf{M} \dot{\mathbf{V}}, \qquad (4.1\text{-}2a)$$

where

$$\dot{\mathbf{V}}^T \equiv \{\dot{u}_P, \ \dot{v}_P, \ \dot{w}_0, \ \dot{\theta}_x, \ \dot{\theta}_y, \ \dot{\phi}, \ \dot{\phi}'\}, \qquad (4.1\text{-}2b)$$

denotes the vector of generalized velocities. For the case of the pole located at the origin of the coordinate system ($x_P = y_P = 0$), the (7×7) mass matrix $\mathbf{M}$ is

$$\mathbf{M} = \begin{bmatrix} m_{11} & 0 & 0 & 0 & 0 & m_{16} & 0 \\ & m_{22} & 0 & 0 & 0 & m_{26} & 0 \\ & & m_{33} & m_{34} & m_{35} & 0 & m_{37} \\ & & & m_{44} & m_{45} & 0 & m_{47} \\ & & & & m_{55} & 0 & m_{57} \\ & \text{Symm.} & & & & m_{66} & 0 \\ & & & & & & m_{77} \end{bmatrix}. \qquad (4.1\text{-}3a)$$

77

The entries $m_{ij} = m_{ji}$, $(i, j = \overline{1,7})$ of the mass matrix are

$$m_{11} = m_{22} = m_{33} = b_1,$$
$$m_{16} = -m_{34} = -b_2 + \delta_n \hat{b}_2,$$
$$m_{26} = m_{35} = b_3 + \delta_n \hat{b}_3,$$
$$m_{44} = b_4 + \delta_n b_{14} - \delta_n 2\hat{b}_8,$$
$$m_{55} = b_5 + \delta_n b_{15} + 2\delta_n \hat{b}_9,$$
$$m_{66} = b_4 + b_5 + \delta_n(b_{14} + b_{15}) + 2\delta_n(\hat{b}_9 - \hat{b}_8), \qquad (4.1\text{-}3b\text{-}1)$$
$$m_{77} = b_{10} + \delta_n b_{18} - 2\delta_n \hat{b}_7,$$
$$m_{37} = -b_7 + \delta_n \hat{b}_1,$$
$$m_{47} = -b_8 - \delta_n b_{16} + \delta_n \hat{b}_5,$$
$$m_{57} = -b_9 + \delta_n b_{17} - \delta_n \hat{b}_6,$$
$$m_{45} = b_6 - \delta_n b_{13} + \delta_n \hat{b}_4.$$

The b_i obtained from (3.1-24) are

$$(b_1, b_2, b_3, b_4, b_5, b_6, b_7, b_8, b_9, b_{10}, b_{11}, b_{12}) = \int m_0[1, y, x, y^2, x^2, xy,$$

$$\bar{F}, y\bar{F}, x\bar{F}, \bar{F}^2, dx/ds, dy/ds]ds, \qquad (4.1\text{-}4a)$$

$$(b_{13}, b_{14}, b_{15}, b_{16}, b_{17}, b_{18}) = \int m_2[(dx/ds)(dy/ds), (dx/ds)^2, (dy/ds)^2,$$

$$r_t(dx/ds), r_t(dy/ds), r_t^2]ds, (\hat{b}_1, \hat{b}_2, \hat{b}_3, \hat{b}_4, \hat{b}_5, \hat{b}_6, \hat{b}_7, \hat{b}_8, \hat{b}_9) \qquad (4.1\text{-}4b)$$

$$= \int m_1[r_t, dx/ds, dy/ds, ydy/ds - xdx/ds, \bar{F}dx/ds$$

$$+ yr_t, \bar{F}dy/ds - xr_t, \bar{F}r_t, ydx/ds, xdy/ds]ds. \qquad (4.1\text{-}4c)$$

Here, the reduced mass terms m_i are defined as

$$(m_0, m_1, m_2) = \sum_{k=1}^{N} \int_{h(k-1)}^{h(k)} \rho_{(k)}(1, n, n^2)dn. \qquad (4.1\text{-}5)$$

The mass matrix $\mathbf{M}$ is symmetric and positive definite. Its off-diagonal terms introduce dynamic couplings between the various vibration modes. For *a symmetrically laminated thin-walled* beam, the mass terms $\hat{b}_i$ are all zero.

In Eq. (4.1-3) the tracer δ_n takes the value 1 or 0 depending on whether the contributions brought by the terms off the beam mid-surface are or not included, respectively.

For nonshearable beams, implying $\theta_x \to -v'_P$ and $\theta_y \to -u'_P$, and the generalized velocity vector is

$$\dot{\mathbf{V}}^T = \{\dot{u}_P,\ \dot{v}_P,\ \dot{w}_0,\ \dot{v}'_P,\ \dot{u}'_P,\ \dot{\phi},\ \dot{\phi}'\}, \tag{4.1-6}$$

whereas $\mathbf{M}$ is given by Eq. (4.1-3) as before.

4.2 RAYLEIGH'S DISSIPATION FUNCTION

In agreement with Eqs. (3.2-5) and (3.2-7) is possible to express the Rayleigh's dissipation function $\mathcal{R}$ in the form

$$\mathcal{R} = \frac{1}{2}\int_{\tau}(c_1\dot{u}^2 + c_2\dot{v}^2 + c_3\dot{w}^2)d\tau = \int_{z1}^{z2}\hat{\mathcal{R}}dz, \tag{4.2-1}$$

where $\hat{\mathcal{R}}$ is the dissipation density function and c_i are the damping coefficients. Using the expressions Eqs. (2.1-6) and (2.1-28) for the displacements, one can express $\hat{\mathcal{R}}$ as

$$\hat{\mathcal{R}} = \frac{1}{2}\dot{\mathbf{V}}^T\mathbf{C}\dot{\mathbf{V}}, \tag{4.2-2}$$

where $\dot{\mathbf{V}}^T$ is the vector of generalized velocities. For the pole located at the origin ($x_P = y_P = 0$), the symmetric (7×7) damping matrix $\mathbf{C}$ is

$$\mathbf{C} = \begin{bmatrix} c_{11} & 0 & 0 & 0 & 0 & c_{16} & 0 \\ & c_{22} & 0 & 0 & 0 & c_{26} & 0 \\ & & c_{33} & c_{34} & c_{35} & 0 & c_{37} \\ & & & c_{44} & c_{45} & 0 & c_{47} \\ & & & & c_{55} & 0 & c_{57} \\ & \textit{Symm.} & & & & c_{66} & 0 \\ & & & & & & c_{77} \end{bmatrix}, \tag{4.2-3}$$

where its entries $c_{ij} = c_{ji}$ are

$$c_{11} = d_1;\ c_{16} = -d_2;\ c_{22} = d_3;\ c_{26} = d_4,$$
$$c_{33} = d_9;\ c_{34} = d_{10};\ c_{35} = d_{11};\ c_{37} = -d_{20},$$
$$c_{44} = d_{14} + \delta_n d_{18};\ c_{45} = d_{12} - \delta_n d_{17}, \tag{4.2-4}$$
$$c_{47} = -d_{22} + \delta_n d_{24};\ c_{55} = d_{13} + \delta_n d_{19},$$
$$c_{57} = -(d_{21} + \delta_n d_{25});\ c_{66} = d_5 + d_6 + \delta_n(d_7 + d_8);\ c_{77} = d_{23} + \delta_n d_{26}.$$

Here d_{ij} are the generalized damping coefficients defined by Eqs. (3.2-11). For unshearable beams, but the vector of generalized velocities changes according to Eq. (4.1-6), the matrix $\mathbf{C}$ remains formally unchanged.

4.3 STRAIN ENERGY

Before deriving the strain energy for to the linear theory of thin-walled beams, we need several basic relationships: the one-dimensional force and couple measures defined by Eqs.(3.1-17) have to be expressed in terms of one-dimensional generalized displacements. This is done by using the constitutive equations relating the *shell* stress resultants and stress couples to the strain measures in the Eq. (3.1-17). This leads to the matrix equation.

$$\mathbf{F} = \mathbf{AD}, \tag{4.3-1}$$

where $\mathbf{F}$, $\mathbf{D}$ and $\mathbf{A}$ denote the vector of generalized one- dimensional stress measures, the gradient of one-dimensional displacement measures, and the symmetric (7×7) cross-sectional stiffness matrix, respectively:

For shearable beams these are defined as

$$\mathbf{F}^T = \{T_z, \ M_y, \ M_x, \ Q_x, \ Q_y, \ B_\omega, \ M_z\}, \tag{4.3-2a}$$

$$\mathbf{D}^T = \{w'_o, \ \theta'_y, \ \theta'_x, \ u'_P + \theta_y, \ v'_P + \theta_x, -\phi'', \ \phi'\}, \tag{4.3-2b}$$

while

$$\mathbf{A} = \begin{bmatrix} a_{11} & a_{12} & a_{13} & a_{14} & a_{15} & a_{16} & a_{17} \\ & a_{23} & a_{24} & a_{25} & a_{26} & a_{27} \\ & & a_{33} & a_{34} & a_{35} & a_{36} & a_{37} \\ & & & a_{44} & a_{45} & a_{46} & a_{47} \\ & & & & a_{55} & a_{56} & a_{57} \\ & Symm. & & & & a_{66} & a_{67} \\ & & & & & & a_{77} \end{bmatrix}. \tag{4.3-3}$$

The unshearable counterpart of Eqs. (4.3-2) is

$$\mathbf{F}^T = \{T_z, \ M_y, \ M_x, \ B_\omega, \ M_z\}, \tag{4.3-4a}$$

$$\mathbf{D}^T = \{w'_o, \ u''_P, \ v''_P, -\phi'', \ \phi'\}, \tag{4.3-4b}$$

while

$$\mathbf{A} = \begin{bmatrix} a_{11} & -a_{12} & -a_{13} & a_{16} & a_{17} \\ & a_{22} & a_{23} & -a_{26} & -a_{27} \\ & & a_{33} & -a_{36} & -a_{37} \\ & & & a_{66} & a_{67} \\ & Symm. & & & a_{77} \end{bmatrix}. \tag{4.3-5}$$

The quantities $a_{ij} (= a_{ji})$ in the matrices (4.3-3) and (4.3-5) are the stiffnesses characterizing the anisotropic composite thin-walled beams of open/closed cross-sections.

Equation (4.3-1) in conjunction with (4.3-2) and (4.3-4) show that for arbitrary anisotropy of the beam material, there is full coupling between the forces and moments and the various modes of deformation.

The definition, nature, and dimension of stiffness quantities a_{ij} are provided in Table 4.3-1. The first term in the expression of a_{77} accompanied by the tracer δ_c is different of zero for close-closed section beams, only. In this case $\delta_c = 1$; otherwise it is zero.

Replacement of Eqs. (A.92), (A.85), (A.75), (A.54) and (A.37) in the definitions of the stiffness quantities as displayed in Table 4-3.1, provides their expression as a function of the fiber orientation. The flow chart in Fig. 4.3-1 describes the procedure enabling one to accomplish this task: input data should consist of (i) the basic engineering constants of the orthotropic material assigned to each ply, namely E_1, E_2, E_3, G_{12}, G_{13}, G_{23}, ν_{12}, ν_{23}, and ν_{31}, and (ii) the lay-up, stacking sequence architecture and the cross-section geometry of the beam.

The linearized counterpart of Eq. (3.1-16) considered in conjunction with Eq. (4.3-1), gives the strain energy for anisotropic thin-walled beams

$$W = \int_{z_1}^{z_1} \hat{W} dz, \tag{4.3-6}$$

where $\hat{W}$ is the strain energy per unit beam span. In matrix form its expression is

$$\hat{W} = \frac{1}{2} \mathbf{D}^T \mathbf{A} \mathbf{D}, \tag{4.3-7}$$

where matrices $\mathbf{D}$ and $\mathbf{A}$ are expressed either by Eqs.(4.3-2b) and (4.3-3), or by Eqs. (4.3-4) and (4.3-5), depending on whether the beam model is shearable or unshearable. The symmetry of the stiffness and mass matrices $\mathbf{A}$ and $\mathbf{M}$, yields a number of important theorems analogous to the ones proper to the three-dimensional elasticity theory.

The modified shell-stiffness quantities K_{ij} are given by Eqs. (A.85) and (A.92). An account of the *total* and *independent* number of stiffness quantities in the shearable and unshearable models, and the respective number of *coupling* stiffness quantities is supplied in Table 4.3-2. Comparing the data from Table 4.3-2 with their shearable and unshearable anisotropic plates/shell counterparts we should notice that in the latter case the number of independent stiffnesses is 21 and 18, respectively,

4.4 THE GOVERNING SYSTEM

One can represent the governing equation system of the theory of thin-walled beams in terms of various quantities that define their stress-strain state. However, one of the most convenient representations is in terms of displacement measures. This representation will explicitly be carried out for the linear theory of thin-walled beams.

4.4-1 Displacement Formulation

The fundamental unknowns of shearable anisotropic beams are the 1-D displacement measures u_P, v_P, w_0, θ_x, θ_y and ϕ. Determination of the pertinent system of differential equations referred to as the governing system, requires the use of:

Table 4.3-1: Stiffness quantities. Definition and meaning

Stiff-ness	Definition	Nature and/or Coupling Involved	Dimension
a_{11}	$\int K_{11}ds$	extensional	$[F]$
a_{12}	$\int [K_{11}x + K_{14}(dy/ds)]ds$	extension-chordwise bending	$[F.L]$
a_{13}	$\int [K_{11}y - K_{14}(dx/ds)]ds$	extension-flapwise bending	$[F.L]$
a_{14}	$\int K_{12}(dx/ds)ds$	extension-chordwise transverse shear	$[F]$
a_{15}	$\int K_{12}(dy/ds)ds$	extension-flapwise transverse shear	$[F]$
a_{16}	$\int [K_{11}\bar{F} - K_{14}r_t]ds$	extension-warping	$[F.L^2]$
a_{17}	$\int K_{13}ds$	extension-twist	$[F.L]$
a_{22}	$\int [K_{11}x^2 + 2xK_{14}dy/ds + K_{44}(dy/ds)^2]ds$	chordwise bending	$[F.L^2]$
a_{23}	$\int [K_{11}xy - xK_{14}(dx/ds) + yK_{14}dy/ds - K_{44}(dx/ds)(dy/ds)]ds$	chordwise bending-flapwise bending	$[F.L^2]$
a_{24}	$\int [K_{12}x(dx/ds) + K_{24}(dx/ds)(dy/ds)]ds$	chordwise bending -chordwise transverse shear	$[F.L]$
a_{25}	$\int [K_{12}x(dy/ds) + K_{24}(dy/ds)^2 ds]$	chordwise bending - -flapwise transverse shear	$[F.L]$
a_{26}	$\int [K_{11}x\bar{F} - K_{14}xr_t + \bar{F}K_{14}(dy/ds) - K_{44}r_t(dy/ds)]ds$	chordwise bending-warping	$[F.L^3]$
a_{27}	$\int [K_{13}x + K_{43}(dy/ds)]ds$	chordwise bending-twist	$[F.L^2]$

Table 4.3-1: (continued) Stiffness quantities. Definition and meaning

Stiff-ness	Definition	Nature and/or Coupling Involved	Dimension
a_{33}	$\int [K_{11}y^2 - 2yK_{14}(dx/ds) + K_{44}(dx/ds)^2]ds$	flapwise bending	$[F.L^2]$
a_{34}	$\int [K_{12}y(dx/ds) - K_{24}(dx/ds)^2]ds$	flapwise bending - chordwise transverse shear	$[F.L]$
a_{35}	$\int [K_{12}y(dy/ds) - K_{24}(dx/ds)(dy/ds)]ds$	flapwise bending - flapwise transverse shear	$[F.L]$
a_{36}	$\int [K_{11}y\bar{F} - K_{14}yr_t \\ -\bar{F}K_{14}(dx/ds) + K_{44}r_t(dx/ds)]ds$	flapwise bending warping	$[F.L^3]$
a_{37}	$\int [yK_{13} - K_{43}(dx/ds)]ds$	flapwise bending-twist	$[F.L^2]$
a_{44}	$\int [K_{22}(dx/ds)^2 + A_{44}(dy/ds)^2]ds$	chordwise transverse shear	$[F]$
a_{45}	$\int [K_{22}(dx/ds)(dy/ds) \\ -A_{44}(dx/ds)(dy/ds)]ds$	chordwise transverse shear - flapwise transverse shear	$[F]$
a_{46}	$\int [\bar{F}K_{21}(dx/ds) - K_{24}r_t(dx/ds)]ds$	chordwise transverse shear warping	$[F.L^2]$
a_{47}	$\int K_{23}(dx/ds)ds$	chordwise transverse shear twist	$[F.L]$
a_{55}	$\int [K_{22}(dy/ds)^2 + A_{44}(dx/ds)^2]ds$	flapwise transverse shear	$[F]$
a_{56}	$\int [\bar{F}K_{21}(dy/ds) - K_{24}r_t(dy/ds)]ds$	flapwise transverse shear warping	$[F.L^2]$
a_{57}	$\int K_{23}(dy/ds)ds$	flapwise transverse shear twist	$[F.L]$
a_{66}	$\int [K_{11}\bar{F}^2 - 2K_{14}\bar{F}r_t + K_{44}r_t^2)]ds$	warping	$[F.L^4]$
a_{67}	$\int [K_{13}\bar{F} - K_{43}r_t)]ds$	warping-twist	$[F.L^3]$
a_{77}	$\delta_c \int \psi K_{23}ds + 2\int K_{53}ds$	twist	$[F.L^2]$

(a) the equations of equilibrium/motion,

(b) the constitutive equations relating the one-dimensional stress resultants and stress-couples with the one-dimensional strain-measures, and

(c) the kinematic equations relating the strain components to the one-dimensional displacement measures.

Table 4.3-2: An account of the number of stiffness quantities a_{ij}

Theory	Total Number of Stiffness Quantities	Number of Independent Stiffness Quantities	Number of Off-diagonal (coupling) Stiffness Quantities
Shearable	49	28	21
Unshearable	25	15	10

The equations (a) are given by Eqs. (3.2-13); those for (b) and (c) are expressed by Eqs. (4.3-1). Substituting (4.3-1) into the equations of equilibrium/motion associated with the shearable beam theory, Eqs. (3.2-13), considering the beam nonuniformity in its spanwise z-direction, and discarding the dissipative effects, we obtain

$$\delta w_0 : (a_{11}w_0')' + (a_{12}\theta_y')' + (a_{13}\theta_x')' + [a_{14}(u_P' + \theta_y)]' + [a_{15}(v_P' + \theta_x)]'$$
$$- (a_{16}\phi'')' + (a_{17}\phi')' + p_z + \delta_0 \hat{p}_z = K_3,$$

$$\delta \theta_y : (a_{21}w_0')' + (a_{22}\theta_y')' + (a_{23}\theta_x')' + [a_{24}(u_P' + \theta_y)]' + [a_{25}(v_P' + \theta_x)]'$$
$$- (a_{26}\phi'')' + (a_{27}\phi')' - a_{41}w_0' - a_{42}\theta_y' - a_{43}\theta_x' - a_{44}(u_P' + \theta_y)$$
$$- a_{45}(v_P' + \theta_x) + a_{46}\phi'' - a_{47}\phi' + m_y + \delta_0\hat{m}_y = K_5,$$

$$\delta \theta_x : (a_{31}w_0')' + (a_{32}\theta_y')' + (a_{33}\theta_x')' + [a_{34}(u_P' + \theta_y')]' + [a_{35}(v_P' + \theta_x)]'$$
$$- (a_{36}\phi'')' + (a_{37}\phi')' - a_{51}w_o' - a_{52}\theta_y' - a_{53}\theta_x' - a_{54}(u_P' + \theta_y)$$
$$- a_{55}(v_P' + \theta_x) + a_{56}\phi'' - a_{57}\phi' + m_x + \delta_0\hat{m}_x = K_7, \qquad (4.4\text{-}1a\text{-}f)$$

$$\delta u_P : (a_{41}w_0')' + (a_{42}\theta_y')' + (a_{43}\theta_x')' + [a_{44}(u_P' + \theta_y)]' + [a_{45}(v_P' + \theta_x)]'$$
$$- (a_{46}\phi'')' + (a_{47}\phi')' + p_x + \delta_0\hat{p}_x = K_1,$$

$$\delta v_P : (a_{51}w_0')' + (a_{52}\theta_y')' + (a_{53}\theta_x')' + [a_{54}(u_P' + \theta_y)]' + [a_{55}(v_P' + \theta_x)]'$$
$$- (a_{56}\phi'')' + (a_{57}\phi')' + p_y + \delta_0\hat{p}_y = K_2,$$

$$\delta \phi : (a_{61}w_0')'' + (a_{62}\theta_y')'' + (a_{63}\theta_x')'' + [a_{64}(u_P' + \theta_y)]'' + [a_{65}(v_P' + \theta_x)]''$$
$$- (a_{66}\phi'')'' + (a_{67}\phi')'' + (a_{71}w_0')' + (a_{72}\theta_y')' + (a_{73}\theta_x')'$$
$$+ [a_{74}(u_P' + \theta_y)]' + [a_{75}(v_P' + \theta_x)]' - (a_{76}\phi'')' + (a_{77}\phi')'$$
$$+ m_z + m_\omega' + \delta_0(\hat{m}_z + \hat{m}_\omega') = K_4 - K_9'.$$

We express the boundary conditions in terms of the same functions, and find

$$\delta w_0 : a_{11}w_0' + a_{12}\theta_y' + a_{13}\theta_x' + a_{14}(u_P' + \theta_y) + a_{15}(v_P' + \theta_x)$$
$$- a_{16}\phi'' + a_{17}\phi' = n_z \underline{T}_z,$$

$$\delta \theta_y : a_{21}w_0' + a_{22}\theta_y' + a_{23}\theta_x' + a_{24}(u_P' + \theta_y) + a_{25}(v_P' + \theta_x)$$
$$- a_{26}\phi'' + a_{27}\phi' = n_z \underline{M}_y,$$

$$\delta \theta_x : a_{31}w_0' + a_{32}\theta_y' + a_{33}\theta_x' + a_{34}(u_P' + \theta_y) + a_{35}(v_P' + \theta_x)$$
$$- a_{36}\phi'' + a_{37}\phi' = n_z \underline{M}_x,$$

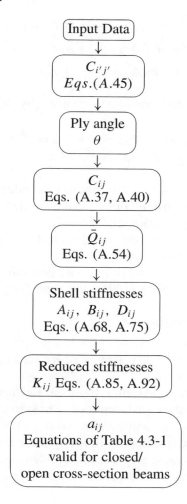

Figure 4.3-1: Flow chart for determination of the variation of a_{ij} as a function of fiber orientation

$$\delta u_P : a_{41}w_o' + a_{42}\theta_y' + a_{43}\theta_x' + a_{44}(u_P' + \theta_y) + a_{45}(v_P' + \theta_x)$$
$$- a_{46}\phi'' + a_{47}\phi' = n_z \underset{\sim}{Q}_x, \qquad (4.4\text{-}2a\text{-}g)$$
$$\delta v_P : a_{51}w_0' + a_{52}\theta_y' + a_{53}\theta_x' + a_{54}(u_P' + \theta_y) + a_{55}(v_P' + \theta_x)$$
$$- a_{56}\phi'' + a_{57}\phi' = n_z \underset{\sim}{Q}_y,$$
$$\delta_r\phi' : a_{61}w_0' + a_{62}\theta_y' + a_{63}\theta_x' + a_{64}(u_P' + \theta_y) + a_{65}(v_P' + \theta_x)$$
$$- a_{66}\phi'' + a_{67}\phi' = \delta_r n_z \underset{\sim}{B}_\omega,$$

$$\delta\phi : (a_{61}w_0')' + (a_{62}\theta_y')' + (a_{63}\theta_x')' + [a_{64}(u_P' + \theta_y)]' + [a_{65}(v_P' + \theta_x)]'$$
$$- (a_{66}\phi'')' + (a_{67}\phi')' + a_{71}w_0' + a_{72}\theta_y' + a_{73}\theta_x' + a_{74}(u_P' + \theta_y)$$
$$+ a_{75}(v_P' + \theta_x) - a_{76}\phi'' + a_{77}\phi' + K_9 = n_z(\underline{M}_z - \delta_0\hat{m}_\omega).$$

The geometrical boundary conditions are given by Eqs. (3.2-14); they are

$$u_P = \underline{u}_P; \quad v_P = \underline{v}_P; \quad w_0 = \underline{w}_0; \quad \theta_x = \underline{\theta}_x;$$
$$\theta_y = \underline{\theta}_y; \quad \phi = \underline{\phi}; \quad \delta_r\phi' = \delta_r\underline{\phi}'. \tag{4.4-3a-g}$$

The governing system (4.4-1) and the boundary conditions (4.4-2) and (4.4-3) belong to the theory of anisotropic composite thin-walled beams that incorporate transverse shear, warping restraint and spanwise cross-section nonuniformity. The equations include exotic elastic couplings that can be exploited to enhance the response behavior of the structure.

Equations (4.4-1) through (4.4-3) represent a generalization of those given by Song (1990), Librescu and Song (1991, 1992) and Song and Librescu (1993).

4.4-2 A Few Comments

The system of governing equations, (4.4-1) through (4.4-3), as well as the specialized counterparts to be displayed next, are analogues of Lamé equations of the 3-D elasticity theory:

The order of the governing system is fourteen, and there are, seven boundary conditions to be prescribed at each beam edge. The system applies to both closed and open cross-section beams, and governs: (i) static boundary-value problems for a beam under static external loads, (ii) the dynamic boundary-value problem consisting e.g. of the dynamic response to time-dependent external excitation, and (iii) solution of various (conservative and non-conservative) eigenvalue problems.

The *dynamic response solution* obtained from Eqs. (4.4-1) through (4.4-3) consists of two parts: the homogeneous solution and the particular solution. The homogeneous solution corresponds to zero time-dependent external excitation, and hence, it can be interpreted as the response to *initial conditions*. For uniqueness of the solution, the following initial conditions have to be prescribed:

$$u_P(z, t = 0) = \hat{u}_P; \quad \dot{u}_P(z, t = 0) = \hat{\dot{u}}_P,$$
$$v_P(z, t = 0) = \hat{v}_P; \quad \dot{v}_P(z, t = 0) = \hat{\dot{v}}_P,$$
$$\theta_x(z, t = 0) = \hat{\theta}_x; \quad \dot{\theta}_x(z, t = 0) = \hat{\dot{\theta}}_x,$$
$$\theta_y(z, t = 0) = \hat{\theta}_y; \quad \dot{\theta}_y(z, t = 0) = \hat{\dot{\theta}}_y, \tag{4.4-4a-n}$$
$$w_0(z, t = 0) = \hat{w}_0; \quad \dot{w}_0(z, t = 0) = \hat{\dot{w}}_0,$$

$$\phi(z, t = 0) = \hat{\phi}; \quad \dot{\phi}(z, t = 0) = \dot{\hat{\phi}},$$

$$\phi'(z, t = 0) = \hat{\psi}; \quad \dot{\phi}'(z, t = 0) = \dot{\hat{\psi}}.$$

On the other hand, the particular solution is the response to the external excitations alone, that is the response to the external excitations, with zero initial conditions. Various special cases will be considered in the next sections.

4.4-3 Unshearable Thin-Walled Beams

Within this theory, transverse shear effects are discarded. Extracting $a_{44}(u'_P + \theta_y)$ and $a_{55}(v'_P + \theta_x)$ from Eqs. (4.4-1b) and (4.4-1c), respectively, replacing these expressions in the remaining equations of motion and boundary conditions, and letting $\theta_y \rightarrow -u'_P$ and $\theta_x \rightarrow -v'_P$, we find the Bernoulli-Euler counterpart of Eqs. (4.4-1) through (4.4-3):

Governing Equations:

$$\delta u_P : \quad (a_{22}u''_P)'' + (a_{23}v''_P)'' - (a_{21}w'_0)'' + (a_{26}\phi'')'' - (a_{27}\phi')''$$
$$= p_x + \delta_0 \hat{p}_x + m'_y + \delta_0 \hat{m}'_y - (K_1 + K'_5),$$

$$\delta v_P : \quad (a_{33}v''_P)'' + (a_{32}u''_P)'' - (a_{31}w'_0)'' + (a_{36}\phi'')'' - (a_{37}\phi')''$$
$$= p_y + \delta_0 \hat{p}_y + m'_x + \delta_0 \hat{m}'_x - (K_2 + K'_7),$$

$$\delta w_0 : \quad (a_{11}w'_0)' - (a_{12}u''_P)' - (a_{13}v''_P)' - (a_{16}\phi'')' + (a_{17}\phi')'$$
$$= -(p_z + \delta_0 \hat{p}_z) + K_3, \quad\quad\quad (4.4\text{-}5a\text{-}d)$$

$$\delta\phi : \quad - (a_{66}\phi'')'' + (a_{67}\phi')'' - (a_{76}\phi'')' + (a_{77}\phi')' - (a_{62}u''_P)''$$
$$- (a_{63}v''_P)'' - (a_{72}u''_P)' - (a_{73}v''_P)' + (a_{61}w'_0)''$$
$$+ (a_{71}w'_0)' = -(m_z + m'_\omega) - \delta_0(\hat{m}_z + \delta_0 \hat{m}'_\omega) + (K_4 - K'_9).$$

Boundary Conditions:

$$(a_{22}u''_P)' + (a_{23}v''_P)' + (a_{26}\phi'')' - (a_{27}\phi')' - (a_{21}w'_0)' - (m_y + \delta_0 \hat{m}_y)$$
$$+ K_5 = -n_z \underset{\sim}{Q}_x \text{ or } \delta u_P = 0,$$

$$(a_{32}u''_P)' + (a_{33}v''_P)' + (a_{36}\phi'')' - (a_{37}\phi')' - (a_{31}w'_0)' - (m_x + \delta_0 \hat{m}_x)$$
$$+ K_7 = -n_z \underset{\sim}{Q}_y \text{ or } \delta v_P = 0, \quad\quad\quad (4.4\text{-}6a\text{-}g)$$

$$a_{22}u''_P + a_{23}v''_P + a_{26}\phi'' - a_{27}\phi' - a_{21}w'_0 = -n_z \underset{\sim}{M}_y \text{ or } \delta u'_P = 0,$$

$$a_{33}v''_P + a_{32}u''_P + a_{36}\phi'' - a_{37}\phi' - a_{31}w'_0 = -n_z \underset{\sim}{M}_x \text{ or } \delta v'_P = 0,$$

$$a_{11}w'_0 - a_{12}u''_P - a_{13}v''_P - a_{16}\phi'' + a_{17}\phi' = n_z \underset{\sim}{T}_z \text{ or } \delta w_0 = 0$$

$$-(a_{66}\phi'')' + (a_{67}\phi')' - a_{76}\phi'' + a_{77}\phi' - (a_{62}u''_P)' - (a_{63}v''_P)' - a_{72}u''_P$$
$$- a_{73}v''_P + (a_{61}w'_0)' + a_{71}w'_0 = -K_9 + n_z(\underset{\sim}{M}_z - \delta_0 \hat{m}_\omega)$$
$$\text{or } \delta\phi = 0,$$

$$-a_{66}\phi'' + a_{67}\phi' - a_{62}u_P'' - a_{63}v_P'' + a_{61}w_0' = \delta_r n_z \underline{B}_\omega \text{ or } \delta_r \delta\phi' = 0.$$

A few remarks are in order: (i) even through thirteen terms associated with transverse shear stiffness vanish, (that is stiffnesses a_{14}, a_{15}, a_{24}, a_{25}, a_{34}, a_{35}, a_{44}, a_{55}, a_{45}, a_{46}, a_{47}, a_{56} and a_{57}), for an anisotropic material, the system of equations, Eqs. (4.4-5) through (4.4-6) exhibits a full coupling between flapwise bending-chordwise bending-extension and twist motions, (ii) the order of the governing system and the associated number of boundary conditions remain the same as for the shearable beam model.

Another special case corresponds to the free warping model. The order of the governing system is reduced from fourteen to twelve, and there are six boundary conditions at each end.

At this point, a word of caution is in order. As Table 4.3-1 shows, a_{46} and a_{56} are transverse shear-warping coupling stiffnesses; when either of these effects is discarded, both stiffnesses should be taken to be zero.

The governing systems, for both shearable and unshearable beams, are applicable to both closed and open beam cross-sections. However, for each of these cases, the stiffness quantities a_{ij}, as well as the load terms, should be properly defined.

4.4-4 Two Structural Coupling Configurations

In Section A.8.11 we described two structural configurations that induce special elastic coupling mechanisms. These two structural couplings are achievable by skewing the ply-angles with respect to the z-axis. Some of their implications in the various equations will now be explored.

4.4-4a Circumferentially Uniform Stiffness Configuration (CUS)

This structural configuration, referred to in the followings by the acronym CUS, can be generated by skewing the ply-angles in the top and bottom walls (flanges) and in the lateral ones (webs) according to the law

$$\theta(y) = \theta(-y), \quad \theta(x) = \theta(-x). \tag{4.4-7a,b}$$

Now, the extensional, extensional-flexural, and flexural local stiffness quantities A_{ij}, B_{ij}, and D_{ij}, respectively, are uniform along the cross-section contour, and as a result, the modified local stiffness quantities at any point of the top and bottom walls, satisfy

$$K_{ij}(y) = K_{ij}(-y); \quad K_{ij}(x) = K_{ij}(-x). \tag{4.4-8a,b}$$

These equations and the symmetry conditions of a closed cross-section contour, for which make a number of global stiffnesses a_{ij} vanish: $a_{12} = a_{13} = a_{14} = a_{15} = a_{16} = a_{23} = a_{24} = a_{26} = a_{27} = a_{35} = a_{36} = a_{37} = a_{45} = a_{46} =$

$a_{47} = a_{56} = a_{57} = a_{67} = 0.$

$$\oint x[1, y, (dx/ds), \bar{F}, r_t]ds = 0, \qquad \oint y[1, x, (dy/ds), \bar{F}, r_t]ds = 0,$$

$$\oint (dx/ds)[1, (dy/ds), \bar{F}, r_t]ds = 0,$$

$$\oint (dy/ds)[1, (dx/ds), \bar{F}, r_t]ds = 0, \qquad \oint (\bar{F}, r_t)ds = 0, \qquad (4.4\text{-}9)$$

Consequently, the direct stiffness quantities $a_{11}, a_{22}, a_{33}, a_{44}, a_{55}, a_{66}, a_{77}$, as well as the cross-coupling stiffnesses a_{17}, a_{25}, and a_{34}, remain, within the CUS, different from zero.

By virtue of Eqs. (4.4-9), it also results that a number of mass terms vanish; the ones surviving are $b_1, b_4, b_5, b_{10}, b_{14}, b_{15}$ and b_{18}. As a result, the vectors **F** and **D** can be represented in a decoupled form:

$$\mathbf{F}_1 = \mathbf{A}_1 \mathbf{D}_1, \qquad (4.4\text{-}10)$$

and

$$\mathbf{F}_2 = \mathbf{A}_2 \mathbf{D}_2, \qquad (4.4\text{-}11)$$

where

$$\mathbf{F}_1^T = \{T_z, M_z, B_\omega\}, \qquad (4.4\text{-}12a)$$
$$\mathbf{F}_2^T = \{M_y, M_x, Q_x, Q_y\}, \qquad (4.4\text{-}12b)$$
$$\mathbf{D}_1^T = \{w_0', \phi', -\phi''\}, \qquad (4.4\text{-}12c)$$
$$\mathbf{D}_2^T = \{\theta_y', \theta_x', u_P' + \theta_y, v_P' + \theta_x\}, \qquad (4.4\text{-}12d)$$

while

$$\mathbf{A}_1 = \begin{bmatrix} a_{11} & a_{17} & 0 \\ & a_{77} & 0 \\ Symm & & a_{66} \end{bmatrix}, \qquad (4.4\text{-}13a)$$

and

$$\mathbf{A}_2 = \begin{bmatrix} a_{22} & 0 & 0 & a_{25} \\ & a_{33} & a_{34} & 0 \\ & & a_{44} & 0 \\ Symm. & & & a_{55} \end{bmatrix}. \qquad (4.4\text{-}13b)$$

Similarly, the decoupled counterparts of the mass matrix, Eq. (4.1-3a), are

$$\mathbf{M}_1 = \begin{bmatrix} m_{33} & 0 & 0 \\ & m_{66} & 0 \\ Symm. & & m_{77} \end{bmatrix}, \qquad (4.4\text{-}13c)$$

and

$$\mathbf{M}_2 = \begin{bmatrix} m_{11} & 0 & 0 & 0 \\ & m_{22} & 0 & 0 \\ Symm. & & m_{44} & 0 \\ & & & m_{55} \end{bmatrix}. \qquad (4.4\text{-}13d)$$

This ply-angle distribution, renders the governing system and the associated boundary conditions to split exactly and entirely into two independent groups.

Within the former group, the coupled extension - twist motion is involved. The governing system is

$$\delta w_0: \ (a_{11}w_0')' + (a_{17}\phi')' + p_z + \delta_0\hat{p}_z = b_1\ddot{w}_0,$$

$$\delta\phi: \ -\underline{(a_{66}\phi'')''} + (a_{71}w_0')' + (a_{77}\phi')' + m_z + m_\omega' + \delta_0(\hat{m}_z + \hat{m}_\omega')$$

$$= [(b_4+b_5) + \delta_n(b_{14}+b_{15})]\ddot{\phi} - \underline{\underline{[(b_{10}+\delta_nb_{18})\ddot{\phi}']'}} \ . \qquad (4.4\text{-}14a,b)$$

The associated *homogeneous* boundary conditions are:

the *geometric ones*:

$$w_0 = \phi = \underline{\phi'} = 0, \qquad (4.4\text{-}15a\text{-}c)$$

and *the static ones*:

$$\delta w_0: \ a_{11}w_0' + a_{17}\phi' = 0,$$

$$\delta\phi: \ -\underline{(a_{66}\phi'')'} + a_{71}w_0' + a_{77}\phi' + \underline{\underline{(b_{10}+\delta_nb_{18})\ddot{\phi}'}} = 0,$$

$$\delta\phi': \ \underline{a_{66}\phi''} = 0. \qquad (4.4\text{-}16a\text{-}c)$$

In Eqs. (4.4-14) and related boundary conditions (4.4-15) and (4.4-16), the terms underscored by one and two dashed lines are associated with the warping inhibition and the dynamic warping, respectively.

The order of the governing system is six. However, when the effect of warping restraint is discarded, the order of the associated governing system reduces to four, and, correspondingly, the number of boundary conditions to be fulfilled at each edge reduces to two. In the absence of the warping restraint, Eqs. (4.4-14) coincide with those obtained by Rehfield and Atilgan (1989).

The latter group of governing equations belonging to the CUS, involve the coupling between vertical bending (referred herein also to as flapwise bending or simply flapping), lateral bending (referred to as chordwise bending or lagging) and flapwise and chordwise transverse shear:

$$\delta u_P : \ (a_{43}\theta_x')' + [a_{44}(u_P' + \theta_y)]' + p_x + \delta_0 \hat{p}_x = b_1 \ddot{u}_P,$$

$$\delta v_P : \ (a_{52}\theta_y')' + [a_{55}(v_P' + \theta_x)]' + p_y + \delta_0 \hat{p}_y = b_1 \ddot{v}_P,$$

$$\delta\theta_y : \ (a_{22}\theta_y')' + [a_{25}(v_P' + \theta_x)]' - a_{43}\theta_x' - a_{44}(u_P' + \theta_y) + m_y + \delta_0 \hat{m}_y$$

$$= \ \underbrace{(b_5 + \delta_n b_{15})}\, \ddot{\theta}_y$$

$$\delta\theta_x : \ (a_{33}\theta_x')' + [a_{34}(u_P' + \theta_y)]' - a_{52}\theta_y' - a_{55}(v_P' + \theta_x) + m_x + \delta_0 \hat{m}_x$$

$$= \ \underbrace{(b_4 + \delta_n b_{14})}\, \ddot{\theta}_x \ . \qquad (4.4\text{-}17\text{a-d})$$

The associated homogeneous boundary conditions are
The geometric ones:

$$u_P = v_P = \theta_y = \theta_x = 0, \qquad (4.4\text{-}18\text{a-d})$$

whereas *the static boundary conditions* in terms of displacement measures are expressed as

$$\delta u_P : \ a_{43}\theta_x' + a_{44}(u_P' + \theta_y) = 0,$$

$$\delta u_P : \ a_{52}\theta_y' + a_{55}(v_P' + \theta_x) = 0,$$

$$\delta\theta_y : \ a_{22}\theta_y' + a_{25}(v_P' + \theta_x) = 0, \qquad (4.4\text{-}19\text{a-d})$$

$$\delta\theta_x : \ a_{33}\theta_x' + a_{34}(u_P' + \theta_y) = 0.$$

The system is of the eighth-order, and there are, four boundary conditions to be prescribed at each beam end.

In Eqs. (4.4-17), the terms underscored by a wavy line are associated with the rotatory inertia. The equations of motion (4.4-17) are identical to those obtained by Atilgan and Rehfield (1990).

4.4-4b Circumferentially Asymmetric Stiffness Configuration (CAS)

As mentioned in Section A.8.11, this structural configuration is achievable when in the top and bottom walls (flanges) and in the lateral ones, (webs), the ply-angle distributions follow the law

$$\theta(y) = -\theta(-y), \ \theta(x) = -\theta(-x). \qquad (4.4\text{-}20\text{a,b})$$

By virtue of relationships (A.95), we have also

$$\oint (K_{12}, K_{13}, K_{14}, K_{24}, K_{43})ds = 0. \qquad (4.4\text{-}21)$$

Equations (4.4-9) in conjunction with (4.4-21), considered in the definitions of a_{ij}, yields as in the case of CUS configuration, the same non-vanishing *direct* stiffness quantities, i.e. a_{11}, a_{22}, a_{33}, a_{44}, a_{55}, a_{66} and a_{77}, and in addition, the non-vanishing coupling stiffnesses a_{14}, a_{37}, and a_{56}. The non-vanishing mass terms are b_1, b_4, b_5, b_{10}, b_{14} and b_{15}.

As a result, a different decoupling of Eqs. (4.4-1) and (4.4-2) occurs:

$$\mathbf{F}_1 = \mathbf{A}_1 \mathbf{D}_1 \text{ and } \mathbf{F}_2 = \mathbf{A}_2 \mathbf{D}_2 \qquad (4.4\text{-}22a,b)$$

where

$$\mathbf{F}_1^T = \{M_x, Q_y, B_\omega, M_z\}, \quad \mathbf{F}_2^T = \{T_z, M_y, Q_x\}, \qquad (4.4\text{-}23a,b)$$

$$\mathbf{D}_1^T = \{\theta_x', v_P' + \theta_x, -\phi'', \phi'\}, \quad \mathbf{D}_2^T = \{w_0', \theta_y', u_P' + \theta_y\}, \qquad (4.4\text{-}24a,b)$$

$$\mathbf{A}_1 = \begin{bmatrix} a_{33} & 0 & 0 & a_{37} \\ & a_{55} & a_{56} & 0 \\ & & a_{66} & 0 \\ Symm. & 0 & 0 & a_{77} \end{bmatrix}, \quad \mathbf{A}_2 = \begin{bmatrix} a_{11} & 0 & a_{14} \\ & a_{22} & 0 \\ Symm. & & a_{44} \end{bmatrix}$$

$$(4.4\text{-}25a,b)$$

and the associated mass matrices

$$\mathbf{M}_1 = \begin{bmatrix} m_{44} & 0 & 0 & 0 \\ & m_{22} & 0 & 0 \\ & & m_{66} & 0 \\ Symm. & & & m_{77} \end{bmatrix}, \qquad (4.4\text{-}25c)$$

$$\mathbf{M}_2 = \begin{bmatrix} m_{33} & 0 & 0 \\ & m_{11} & 0 \\ Symm. & & m_{55} \end{bmatrix}. \qquad (4.4\text{-}25d)$$

Now the fully coupled governing system and boundary conditions split exactly into two sets: one of these involves the coupling between twist, flapwise bending and flapwise shear, and the other, involves the coupling between extension, chordwise bending and chordwise shear.

The governing system featuring the twist-flapwise bending-flapwise transverse shear cross-coupling is

$$\delta v_P : \ [a_{55}(v'_P + \theta_x)]' - (a_{56}\phi'')' + p_y + \delta_0 \hat{p}_y = b_1 \ddot{v}_P,$$

$$\delta \theta_x : \ (a_{33}\theta'_x)' + (a_{37}\phi')' - a_{55}(v'_P + \theta_x) + a_{56}\phi'' + m_x + \delta_0 \hat{m}_x$$

$$= (b_4 + \delta_n b_{14}) \ddot{\theta}_x,$$

$$\delta \phi : \ - (a_{66}\phi'')'' + (a_{77}\phi')' + [a_{65}(v'_P + \theta_x)]'' + (a_{73}\theta'_x)' + m_z \quad (4.4\text{-}26\text{a-c})$$

$$+ m'_\omega + \delta_0 (\hat{m}_z + \hat{m}'_\omega) = [(b_4 + b_5) + \delta_n(b_{14} + b_{15})]\ddot{\phi}$$

$$- [(b_{10} + \delta_n b_{18})\ddot{\phi}']'.$$

The associated homogeneous boundary conditions are:

The geometric ones:

$$v_P = 0, \ \theta_x = 0, \ \phi = 0, \ \phi' = 0, \quad\quad (4.4\text{-}27\text{a-d})$$

and the static ones:

$$\delta v_P : \ a_{55}(v'_P + \theta_x) - a_{56}\phi'' = 0,$$

$$\delta \theta_x : \ a_{33}\theta'_x + a_{37}\phi' = 0,$$

$$\delta \phi : \ - (a_{66}\phi'')' + a_{77}\phi' + [a_{65}(v'_P + \theta_x)]' + a_{73}\theta'_x$$

$$+ (b_{10} + \delta_n b_{18})\ddot{\phi}' = 0, \quad\quad (4.4\text{-}28\text{a-d})$$

$$\delta \phi' : \ a_{65}(v'_P + \theta_x) - a_{66}\phi'' = 0.$$

In these equations, the terms induced by warping - transverse shear cross-coupling are underscored by a dotted line $(- \bullet - \bullet - \bullet - \bullet -)$. These terms are different from zero when *both* transverse shear and warping restraint effects are accounted for. Otherwise, these terms are zero.

The system of governing equations incorporating warping restraint effect is of the eighth-order, and there are four boundary conditions to be prescribed at each edge. For the free-warping model, the terms underscored by the interrupted and dotted lines become immaterial, and as such, the order of governing equations reduces to six, and the number of boundary conditions reduces to three.

The second group of governing equations belonging to CAS, involves the coupling between the extension, chordwise bending and transverse shear. It is given by

$$\delta w_0 : \ (a_{11}w_0')' + [a_{14}(u_P' + \theta_y)]' + p_z + \delta_0 \hat{p}_z = b_1 \ddot{w}_0,$$

$$\delta u_P : \ (a_{41}w_0')' + [a_{44}(u_P' + \theta_y)]' + p_x + \delta_0 \hat{p}_x = b_1 \ddot{u}_P, \qquad (4.4\text{-}29a\text{-}c)$$

$$\delta \theta_y : \ -a_{41}w_0' - a_{44}(u_P' + \theta_y) + (a_{22}\theta_y')' + m_y + \delta_0 \hat{m}_y = \underwave{(b_5 + \delta_n b_{15})\, \ddot{\theta}_y} \ .$$

The associated homogeneous boundary conditions are:

The geometric ones:

$$w_o = u_P = \theta_y = 0, \qquad (4.4\text{-}30a\text{-}c)$$

and the static ones, expressed in terms of displacements:

$$\delta w_0 : a_{11}w_0' + a_{14}(u_P' + \theta_y) = 0,$$

$$\delta u_P : a_{41}w_0' + a_{44}(u_P' + \theta_y) = 0,$$

$$\delta \theta_y : a_{22}\theta_y' = 0. \qquad (4.4\text{-}31a\text{-}c)$$

The governing system is of the sixth order, and there are three boundary conditions at each beam end. The specialized counterparts of the two sets of dynamic governing equations belonging to the CAS configuration coincide with those supplied by Atilgan and Rehfield (1990).

4.4-5 Other Special Cases

4.4-5a Transversely Isotropic Laminate

For beams symmetrically made up of laminae of transversely-isotropic materials whose plane of isotropy is parallel at each point to the beam mid-surface, CUS and CAS are identical.   In this case, since the modified stiffness quantities $K_{14}, K_{12}, K_{13}, K_{24}, K_{52}, K_{43}, K_{51},$ K_{54} vanish, a reduction of the number of global stiffness quantities will occur. Assuming also that the loads are distributed along the x, y and z-axes, the entire system of equations, Eqs. (4.4-1) through (4.4-3), splits into four independent systems governing the extension expressed in terms of w_o; the flapwise bending, (in terms of v_P and θ_x); the chordwise bending, (in terms of u_P and θ_y), and the twist, (in terms of ϕ):

• *Extension*

$$\delta w_0 : \ (a_{11}w_0')' + p_z + \delta_0 \hat{p}_z = b_1 \ddot{w}_0, \qquad (4.4\text{-}32)$$

with the homogeneous boundary conditions:

$$w_0' = 0 \quad \text{or} \quad w_0 = 0. \tag{4.4-33a,b}$$

• *Flapwise bending - transverse shear*

$$\delta v_P : \quad [a_{55}(v_P' + \theta_x)]' + p_y + \delta_0 \hat{p}_y = b_1 \ddot{v}_P, \tag{4.4-34a,b}$$

$$\delta \theta_x : \quad (a_{33}\theta')_x' - a_{55}(v_P' + \theta_x) + m_x + \delta_0 \hat{m}_x = \underset{\sim}{(b_4 + \delta_n b_{14})\,\ddot{\theta}_x} \,,$$

and the boundary conditions:

$$v_P = 0, \qquad\qquad \theta_x = 0, \tag{4.4-35a,b}$$

$$\text{or} \qquad a_{55}(v_P' + \theta_x) = 0, \quad a_{33}\theta_x' = 0. \tag{4.4-35c,d}$$

• *Chordwise bending - transverse shear*

$$\delta u_P : \quad [a_{44}(u_P' + \theta_y)]' + p_x + \delta_0 \hat{p}_x = b_1 \ddot{u}_P, \tag{4.4-36a,b}$$

$$\delta \theta_y : \quad -a_{44}(u_P' + \theta_y) + (a_{22}\theta_y')' + m_y + \delta_0 \hat{m}_y = \underset{\sim}{(b_5 + \delta_n b_{15})\,\ddot{\theta}_y} \,,$$

and the boundary conditions:

$$u_P = \theta_y = 0, \tag{4.4-37a,b}$$

$$\text{or} \qquad a_{44}(u_P' + \theta_y) = 0, \qquad a_{22}\theta_y' = 0. \tag{4.4-37c,d}$$

• *Pure Twist*

$$- \underline{(a_{66}\phi'')''} + (a_{77}\phi')' + m_z + \underline{m_\omega'} + \delta_0(\hat{m}_z + \underline{\hat{m}_\omega'})$$

$$= [(b_4 + b_5) + \delta_n(b_{14} + b_{15})]\ddot{\phi} - \underline{\underline{[(b_{10} + \delta_n b_{18})\ddot{\phi}']'}}. \tag{4.4-38}$$

and the boundary conditions

$$\phi = \underline{\phi'} = 0, \tag{4.4-39a,b}$$

or

$$- \underline{(a_{66}\phi'')'} + a_{77}\phi' = - \underline{\underline{(b_{10} + \delta_n b_{18})\ddot{\phi}'}},$$

$$\underline{a_{66}\phi''} = 0. \tag{4.4-39c,d}$$

A special case of these equations can be found in the paper by Senjanovic and Fan (1991). Their static counterparts associated with the restrained twist were obtained for the first time by Vlasov (1940, 1961).

4.4-5b Remarks on Related Stiffness Quantities

For thin-walled beams made up from transversely isotropic materials, the plane of isotropy being parallel to the plane $x_1 - x_2$, (see Eqs. (A.85) and (A.92)), the expressions of non-vanishing stiffness quantities K_{ij} are

$$K_{11} = A_{11} - \frac{A_{12}^2}{A_{11}}, \quad K_{22} = A_{66},$$

$$K_{23} = 2A_{66}\frac{\Omega}{\beta}, \quad K_{44} = D_{11} \equiv D, \, K_{53} = 2D_{66} \qquad (4.4\text{-}40\text{a-e})$$

As a result, the stiffness quantities a_{ij} proper to this case are

$$a_{11} = \int K_{11}ds,$$

$$a_{22} = \int [K_{11}x^2 + D(dy/ds)^2]ds,$$

$$a_{33} = \int [K_{11}y^2 + D(dx/ds)^2]ds,$$

$$a_{44} = \int [A_{66}(dx/ds)^2 + A_{44}(dy/ds)^2]ds, \qquad (4.4\text{-}41\text{a-g})$$

$$a_{55} = \int [A_{66}(dy/ds)^2 + A_{44}(dx/ds)^2]ds,$$

$$a_{66} = \int [K_{11}\overline{F}^2 + Dr_t^2]ds,$$

$$a_{77} = \delta_c \int 4\frac{\Omega^2}{\beta^2}A_{66}ds + \int 4D_{66}ds.$$

4.4-6 Unshearable CAS and CUS Beam Configurations

Now transverse shear effects are discarded; this is possible when the slenderness requirement Eq. (2.1-1) is fulfilled, and/or when transverse shear stiffness of constituent materials is rather high.

For arbitrary thin-walled cross-section beams designed for the CUS and CAS configurations, the pertinent governing equations are provided in the following sections.

4.4-6a Unshearable CUS Beam Configuration

For this case Eqs. (4.4-10), (4.4-12a) and (4.4-13a) remain unchanged, whereas $\mathbf{A}_2$, $\mathbf{F}_2$ and $\mathbf{D}_2$ become

$$\mathbf{A}_2 = \begin{bmatrix} a_{22} & 0 \\ 0 & a_{33} \end{bmatrix}, \quad \mathbf{F}_2^T = \{M_y, \, M_x\}, \quad \mathbf{D}_2^T = \{-u_P'', \, -v_P''\} \qquad (4.4\text{-}42\text{a-c})$$

Within the context of CUS, the governing equations (4.4-14) and the associated boundary conditions (4.4-15) and (4.4-16) coupling the extension-twist motion do not involve transverse shear effect. In contrast to this, Eqs. (4.4-17), and boundary conditions (4.4-18) and (4.4-19) strongly involve transverse shear effects. Their unshearable counterpart is obtained by eliminating $(u'_P + \theta_y)$ from Eqs. (4.4-17a,c), and $(v'_P + \theta_x)$ from Eqs. (4.4-17b,d), and then letting $\theta_x \to -v'_P$ and $\theta_y \to -u'_P$.

The equations are

$$\delta u_P : \quad (a_{22}u''_P)'' = p_x + m'_y + \delta_0(\hat{p}_x + \hat{m}'_y)$$

$$+ \; [(b_5 + \delta_n b_{15})\,\ddot{u}'_P]' \; - b_1 \ddot{u}_P,$$

$$\delta v_P : \quad (a_{33}v''_P)'' = p_y + m'_x + \delta_0(\hat{p}_y + \hat{m}'_x)$$

$$+ \; [(b_4 + \delta_n b_{14})\,\ddot{v}'_P]' \; - b_1 \ddot{v}_P, \qquad (4.4\text{-}43a,b)$$

whereas the boundary conditions become

The static BCs:

$$\delta u_P : \quad (a_{22}u''_P)' - \; (b_5 + \delta_n b_{15})\,\ddot{u}'_P \; = 0,$$

$$\delta u'_P : \quad a_{22}u''_P = 0, \qquad\qquad (4.4\text{-}44a\text{-}d)$$

$$\delta v_P : \quad (a_{33}v''_P)' - \; (b_4 + \delta_n b_{14})\,\ddot{v}'_P \; = 0,$$

$$\delta v'_P : \quad a_{33}v''_P = 0,$$

and the *geometric* BCs:

$$u_P = 0, \quad u'_P = 0, \quad v_P = 0, \quad v'_P = 0. \qquad (4.4\text{-}45a\text{-}d)$$

4.4-6b Unshearable CAS Beam Configuration

The counterparts of vector $\mathbf{F}_1$, $\mathbf{F}_2$, $\mathbf{D}_1$, $\mathbf{D}_2$ and matrices $\mathbf{A}_1$ and $\mathbf{A}_2$ become:

$$\mathbf{F}_1^T = \{T_z, \; M_y\}, \quad \mathbf{F}_2^T = \{M_x, \; B_\omega, \; M_z\}, \qquad (4.4\text{-}46a,b)$$

$$\mathbf{D}_1^T = \{w'_0, \; u''_P\}, \quad \mathbf{D}_2^T = \{v''_P, \; -\phi'', \; \phi'\}, \qquad (4.4\text{-}47a,b)$$

and consequently

$$\mathbf{A}_1 = \begin{bmatrix} a_{11} & 0 \\ 0 & a_{22} \end{bmatrix}, \quad \mathbf{A}_2 = \begin{bmatrix} a_{33} & 0 & -a_{37} \\ 0 & a_{66} & 0 \\ -a_{73} & 0 & a_{77} \end{bmatrix}. \qquad (4.4\text{-}48a,b)$$

Proceeding as before we can express the unshearable counterpart of Eqs. (4.4-26) as:

$$\delta v_P: \ (a_{33}v_P'')'' - (a_{37}\phi')'' = p_y + m_x' + \delta_0(\hat{p}_y + \hat{m}_x') - b_1\ddot{v}_P$$

$$+ \ [(b_4 + \delta_n b_{14})\,\ddot{v}_P']' \ ,$$

$$\delta\phi: \ - \ (a_{66}\phi'')'' \ + (a_{77}\phi')' - (a_{73}v_P'')' = -(m_z + \hat{m}_\omega') - \delta_0(\hat{m}_z + \hat{m}_\omega')$$

$$+ \ [(b_4 + b_5) + \delta_n(b_{14} + b_{15})]\ddot{\phi} - \ [(b_{10} + \delta_n b_{18})\ddot{\phi}']'. \qquad (4.4\text{-}49a,b)$$

The boundary conditions become

The static ones:

$$\delta v_P: \ (a_{33}v_P'')' - (a_{37}\phi')' = \ (b_4 + \delta_n b_{14})\,\ddot{v}_P' \ , \qquad (4.4\text{-}50a\text{-}d)$$

$$\delta v_P': \ a_{37}\phi' - a_{33}v_P'' = 0,$$

$$\delta\phi: - \ (a_{66}\phi'')' \ + a_{77}\phi' - a_{73}v_P'' = - \ (b_{10} + \delta_n b_{18})\ddot{\phi}',$$

$$\delta\phi': \ a_{66}\phi'' \ = 0,$$

and the *geometrical BCs*:

$$v_P = v_P' = \phi = \ \phi' \ = 0. \qquad (4.4\text{-}51a\text{-}d)$$

The unshearable counterpart of the system (4.4-29) governing the extension-chordwise bending becomes

$$\delta u_P: \ (a_{22}u_P'')'' = p_x + m_y' + \delta_0(\hat{p}_x + \hat{m}_y') - b_1\ddot{u}_P + \ [(b_5 + \delta_n b_{15})\,\ddot{u}_P']',$$

$$\delta w_0: \ (a_{11}w_0')' + p_z + \delta_0\hat{p}_z = b_1\ddot{w}_0, \qquad (4.4\text{-}52a,b)$$

whereas the associated boundary conditions are

$$u_P = w_0 = u_P' = 0 \qquad (4.4\text{-}53a\text{-}c)$$

or

$$\delta u_P: \ (a_{22}u_P'')' - \ (b_5 + \delta_n b_{15})\,\ddot{u}_P' \ = 0,$$

$$\delta u_P': \ a_{22}u_P'' = 0, \qquad (4.4\text{-}53d\text{-}f)$$

$$\delta w_0: \ a_{11}w_0' = 0.$$

For the unshearable case, there is a complete decoupling between the chord-wise bending and extension.

4.4-7 Cross-Ply Beam Configuration

4.4-7a Shearable Case

When each constituent ply is made up of orthotropic materials, a number of cross-coupling stiffness quantities vanish: a_{14}, a_{17}, a_{24}, a_{27}, a_{34}, a_{25}, a_{35}, a_{46}, a_{56}, a_{67}, a_{37}. There is a full decoupling of the governing equations and boundary conditions:

- *Pure extension*

$$\delta w_0: \quad (a_{11}w_0')' + p_z + \delta_0 \hat{p}_z = b_1 \ddot{w}_0, \tag{4.4-54}$$

with the boundary conditions

$$w_0 = 0 \text{ or } w_0' = 0. \tag{4.4-55a,b}$$

- *Pure twist*:

$$\delta\phi: \quad - \underline{(a_{66}\phi'')''} + (a_{77}\phi')' + m_z + \underline{m_\omega'} + \delta_0(\hat{m}_z + \hat{m}_\omega') = [(b_4 + b_5)$$

$$+ \delta_n(b_{14} + b_{15})]\ddot{\phi} - \underline{\underline{[(b_{10} + \delta_n b_{18}) \ddot{\phi}']'}} , \tag{4.4-56}$$

with the boundary conditions:

$$\phi = \underline{\phi'} = 0, \tag{4.4-57a,b}$$

or

$$\delta\phi: \quad - \underline{(a_{66}\phi'')'} + a_{77}\phi' = -\underline{\underline{(b_{10} + \delta_n b_{18}) \ddot{\phi}'}} ,$$

$$\delta\phi': \quad \underline{a_{66}\phi''} = 0. \tag{4.4-58a,b}$$

- *Flapwise bending-flapwise transverse shear*:

$$\delta v_P: \quad [a_{55}(v_P + \theta_x)]' + p_y + \delta_0 \hat{p}_y = b_1 \ddot{v}_P,$$

$$\delta\theta_x: \quad (a_{33}\theta_x')' - a_{55}(v_P' + \theta_x) + m_x + \delta_0 \hat{m}_x \tag{4.4-59a,b}$$

$$= \underwave{(b_4 + \delta_n b_{14}) \ddot{\theta}_x} ,$$

with the associated boundary conditions:

$$v_P = \theta_x = 0, \qquad\qquad\qquad (4.4\text{-}60a,b)$$

or

$$\delta v_P : \ a_{55}(v'_P + \theta_x) = 0,$$
$$\delta \theta_x : \ a_{33}\theta'_x = 0. \qquad\qquad (4.4\text{-}61a,b)$$

• *Chordwise bending - chordwise transverse shear*

$$\delta u_P : [a_{44}(u'_P + \theta_y)]' + p_x + \delta_0 \hat{p}_x = b_1 \ddot{u}_P,$$

$$\delta \theta_y : (a_{22}\theta')'_y - a_{44}(u'_P + \theta_y) + m_y + \delta_0 \hat{m}_y = \underwavy{(b_5 + \delta_n b_{15})\, \ddot{\theta}_y} ,$$

$$\qquad\qquad\qquad\qquad\qquad\qquad\qquad (4.4\text{-}62a,b)$$

with the associated boundary conditions:

$$u_P = \theta_y = 0, \qquad\qquad\qquad (4.4\text{-}63a,b)$$

or

$$\delta u_P : \ a_{44}(u'_P + \theta_y) = 0,$$
$$\delta \theta_y : \ a_{22}\theta'_y = 0. \qquad\qquad (4.4\text{-}64a,b)$$

The equations in Section 4.4-7a formally coincide with those in Section 4.4-5a: the differences involve the explicit expressions for stiffness quantities.

4.4-7b Unshearable Case

For unshearable cross-ply laminated beams, there is a complete decoupling: flapwise bending, chordwise bending, extension, and twist become independent each other. The equations of flapwise bending are (4.4-43b), (4.4-44c,d) and (4.4-45c,d); those for chordwise bending are (4.4-43a), (4.4-44a,b) and (4.4-45a,b), the equations of extension are given by (4.4-32) and (4.4-33), the equations for twist being given by (4.4-38) and (4.4-39). Moreover, if we discard the terms underscored once or twice in the equations governing twist, we obtain a second order ordinary differential equation governing the Saint-Venant twist. For this case there is a single boundary condition at each edge of the beam. Within various contexts and approaches, for metallic beams, solutions of the equation of pure torsion including the warping inhibition have been obtained for the static case e.g. by Karyakin (1960), Petre and Atanasiu (1960), Petre et al. (1961), Kollbrunner and Basler (1969), Pilkey (2002), Murray (1984), Musat and Epureanu (1996), while for the dynamic case by Senjanovic and Fan (1991).

4.4-8 Spanwise Beam Uniformity

For a beam with elastic and geometric uniformity along its span, both the stiffness and mass quantities are independent upon the z-coordinate, so that $a'_{ij} = 0$, $b'_i = 0$. In an explicit form the specialized versions of the governing equation will be considered in the context of the next applications.

Remark

Within the unshearable theory of open/closed cross-section uniform beams, there is an interesting similarity between the equations governing the restrained twist, and the ones associated with the flapwise/chordwise bending, involving simultaneously transversal/lateral forces and moments and a compressive axial load. To this end Eqs. (4.4-43) considered in conjunction with (3.1-33a,b), where $T_z \rightarrow -N$, N being a constant compressive axial load yield:

Flapwise Bending
Governing Equation

$$v_P'''' - \frac{N}{a_{33}} v_P'' = \frac{p_y}{a_{33}} + \frac{m_x'}{a_{33}} + \delta_0 \Big(\frac{\hat{p}_y + \hat{m}_x'}{a_{33}} \Big) + \frac{1}{a_{33}} (b_4 + \delta_n b_{14}) \ddot{v}_P'' - \frac{b_1}{a_{33}} \ddot{v}_P,$$

and the boundary conditions

$$v_P = 0 \text{ and } v_P' = 0 \quad \text{or} \quad v_P''' - \frac{N}{a_{33}} v_P' - \frac{1}{a_{33}} (b_4 + \delta_n b_{14}) \ddot{v}_P' = 0, \text{ and } v_P'' = 0.$$

Chordwise Bending
Governing Equation:

$$u_P'''' - \frac{N}{a_{22}} u_P'' = \frac{p_x}{a_{22}} + \frac{m_y'}{a_{22}} + \delta_0 \frac{1}{a_{22}} (\hat{p}_x + \hat{m}_y') + \frac{1}{a_{22}} (b_5 + \delta_n b_{15}) \ddot{u}_P'' - \frac{b_1}{a_{22}} \ddot{u}_P,$$

and the boundary conditions:

$$u_P = 0 \text{ and } u_P' = 0 \quad \text{or} \quad u_P''' - \frac{N}{a_{22}} u_P' - \frac{1}{a_{22}} (b_5 + \delta_n b_{15}) \ddot{u}_P' = 0 \text{ and } u_P'' = 0.$$

Restrained Twist
Governing Equation:

$$\phi'''' - \frac{a_{77}}{a_{66}} \phi'' = \frac{m_z}{a_{66}} + \frac{m_\omega'}{a_{66}} + \delta_0 \frac{1}{a_{66}} (\hat{m}_z + \hat{m}_\omega')$$

$$- \frac{1}{a_{66}} [(b_4 + b_5) + \delta_n (b_{14} + b_{15})] \ddot{\phi} + \frac{1}{a_{66}} (b_{10} + \delta_n b_{18}) \ddot{\phi}'',$$

and boundary conditions:

$$\phi = 0 \text{ and } \phi' = 0 \quad \text{or} \quad \phi''' - \frac{a_{77}}{a_{66}} \phi' - \frac{1}{a_{66}} (b_{10} + \delta_n b_{18}) \ddot{\phi}' = 0 \text{ and } \phi'' = 0.$$

These equations show that the three equations and the associated boundary conditions have an identical structural. The correspondence is displayed as in Table R1.

Table R1: Analogous quantities in bending (flapwise and chordwise) and twist

Flapwise bending	Chordwise bending	Twist
v_P	u_P	ϕ
$\dfrac{N}{a_{33}}$	$\dfrac{N}{a_{22}}$	$\dfrac{a_{77}}{a_{66}}$
$\dfrac{p_y}{a_{33}}, \left(\dfrac{\hat{p}_y}{a_{33}}\right)$	$\dfrac{p_x}{a_{22}}, \left(\dfrac{\hat{p}_x}{a_{22}}\right)$	$\dfrac{m_z}{a_{66}}, \left(\dfrac{\hat{m}_z}{a_{66}}\right)$
$\dfrac{m'_x}{a_{33}}, \left(\dfrac{\hat{m}'_x}{a_{33}}\right)$	$\dfrac{m'_y}{a_{22}} \left(\dfrac{\hat{m}'_y}{a_{22}}\right)$	$\dfrac{m'_\omega}{a_{66}}, \left(\dfrac{\hat{m}'_\omega}{a_{66}}\right)$
$\dfrac{m'_x}{a_{33}}, \left(\dfrac{\hat{m}'_x}{a_{33}}\right)$	$\dfrac{m'_y}{a_{22}} \left(\dfrac{\hat{m}'_y}{a_{22}}\right)$	$\dfrac{m'_\omega}{a_{66}}, \left(\dfrac{\hat{m}'_\omega}{a_{66}}\right)$
$\dfrac{1}{a_{33}}(b_4 + \delta_n b_{14})$	$\dfrac{1}{a_{22}}(b_5 + \delta_n b_{15})$	$\dfrac{1}{a_{66}}(b_{10} + \delta_n b_{18})$
$\dfrac{b_1}{a_{33}}$	$\dfrac{b_1}{a_{22}}$	$\dfrac{1}{a_{66}}[(b_4 + b_5) + \delta_n(b_{14} + b_{15})]$

These similarities can help conversion of anyone of these three groups of equations (or solution) to the other ones. In specialized contexts, this similarity was pointed out by Vlasov (1940, 1961) Djanelidzhe and Panovko (1948) and Karyakin (1960).

4.5 DISCUSSION

We have derived the governing equations of anisotropic thin-walled beams featuring complex elastic cross-couplings. These can play an essential role in enhancing, without weight penalties, the response behavior of beam structures. In various contexts, the directionality property of fibrous composite materials inducing such couplings has played an essential role toward eliminating a chronic aeroelastic instability featured by forward wing aircraft, of enhancing the aeroelastic response in the subcritical speed range and flutter instability of modern aircraft wings, is used in helicopter and turbine rotor blades, as well as in tilt-rotor aircraft in order to improve their twisting motion at different rotor speeds. The dramatic implications of cross-couplings induced by the directional property of composite materials will be emphasized throughout this work. When appropriately specialized, the displayed equations can be applied to open and closed cross-section beams.

Under various specialized contexts, the governing equations supplied here have been used toward the study of free vibration, Song and Librescu (1993), Librescu et al. (1996a), Na and Librescu (1999); dynamic response to time-dependent external excitation, Song and Librescu (1995), Librescu and Na (1998a,b), Song et al. (1998, 2001), Librescu et al. (2000); vibration feed-back control, Librescu et al. (1993, 1996b, 1997), Librescu and Na (1998c,d, 2001), Na and Librescu (1998, 2000, 2001, 2002), and aeroelastic response and instability of thin-walled beams, Song et al. (1992), Librescu et al. (1996a), Qin and Librescu (2002a), Qin et al. (2002). Finally the theory of thin-walled beams presented here was validated by comparing its static/dynamic predictions with the experimental and analytical findings spread throughout the specialized literature, see Qin and Librescu (2001, 2002b).

REFERENCES

Atilgan, A. R. and Rehfield, L. W. (1990) "Vibrations of Composite Thin-Walled Beams with Designing in Elastic Couplings," in *Achievements in Composites in Japan and the United States*, A. Kobayashi (Ed.), Tokyo, Japan, pp. 687–694.

Djanelidzhe, G. Iu. and Panovko, Ja. G. (1948) *Statics of Elastic Thin-Walled Beams* (in Russian), Gostechizdat, Moscow.

Karyakin, N. I. (1960) *Basic Research on Thin-Walled Beams* (in Russian), Publishing House, Vishaia Schola, Moscow.

Kollbrunner, C. and Basler, K. (1969) *Torsion in Structures*, Springer-Verlag, Berlin.

Librescu, L. and Song, O. (1991) "Behavior of Thin-Walled Beams Made of Advanced Composite Materials and Incorporating Non-Classical Effects," *Applied Mechanics Reviews*, Vol. 44, No. 11, Part 2, November, pp. 174–180.

Librescu, L and Song, O. (1992) "On the Static Aeroelastic Tailoring of Composite Aircraft Swept Wings Modelled as Thin-Walled Beam Structures," *Composites Engineering*, Vol. 2, Nos. 5–7 (Special Issue: Use of *Composites in Rotorcraft and Smart Structures*), pp. 497–512.

Librescu, L., Song, O. and Rogers, C. A. (1993) "Adaptive Vibrational Behavior of Cantilevered Structures Modeled as Composite Thin-Walled Beams," *International Journal of Engineering Science*, Vol. 31, No. 5, pp. 775–792.

Librescu, L, Meirovitch, L. and Song, O. (1996a) "Refined Structural Modeling for Enhancing Vibrational and Aeroelastic Characteristics of Composite Aircraft Wings," *La Recherche Aerospatiale*, 1, pp. 23–35.

Librescu, L., Meirovitch, L. and Song, O. (1996b) "Integrated Structural Tailoring and Control Using Adaptive Materials for Advanced Aircraft Wings," *Journal of Aircraft*, Vol. 33, No. 1, pp. 203–213.

Librescu, L., Meirovitch, L. and Na, S. S. (1997) "Control of Cantilevers Vibration via Structural Tailoring and Adaptive Materials," *AIAA Journal*, Vol. 35, No. 8, pp. 1309–1315.

Librescu, L. and Na, S. S. (1998a) "Dynamic Response Control of Thin-Walled Beams to Blast Pulses Using Structural Tailoring and Piezoelectric Actuation," *Journal of Applied Mechanics, Trans. ASME*, Vol. 65, No. 2, pp. 497–504.

Librescu, L. and Na, S. S. (1998b) "Dynamic Response of Cantilevered Thin-Walled Beams to Blast and Sonic-Boom Loadings," *Shock and Vibration*, Vol. 5, pp. 23–33.

Librescu, L. and Na, S. S. (1998c) "Bending Vibration Control of Cantilevers Via Boundary Moment and Combined Feedback Control Law," *Journal of Vibration and Controls*, Vol. 4, No. 6, pp. 733–746.

Librescu, L. and Na, S. S. (1998d) "Boundary Control of Free and Forced Oscillation of Shearable Thin-Walled Beam Cantilevers," *European Journal of Mechanics/A Solids*, Vol. 17, No. 4, pp. 687–700.

Librescu, L., Na, S. S. and Park, S. (2000) "Dynamics and Active Feedback Control of Wing-Type Structures Modeled as Thin-Walled Beams: Implication of Beam Cross-Section Nonuniformity," *Journal of the Chinese Society of Mechanical Engineers*, Vol. 21, No. 1, pp. 87–96.

Librescu, L. and Na, S. S. (2001) "Active Vibration Control of Thin-Walled Tapered Beams Using Piezoelectric Strain Actuation," *Journal of Thin-Walled Structures*, Vol. 39, No. 1, pp. 65–82.

Murray, N. W. (1984) *Introduction to the Theory of Thin-Walled Structures*, Clarendon Press, Oxford.

Musat, S. D. and Epureanu, B. I. (1996) "Study of Warping Torsion of Thin-Walled Beams with Closed Cross-Section Using Micro-Elements," *Communications in Numerical Methods in Engineering*, Vol. 12, pp. 873–884.

Na, S. S. and Librescu, L. (1998) "Oscillation Control of Cantilevers via Smart Materials Technology and Optimal Feedback Control: Actuator Location and Power Consumption Issues," *Smart Materials and Structures*, Vol. 7, pp. 833–842.

Na, S. S. and Librescu, L. (1999) "Dynamic Behavior of Aircraft Wings Modeled as Doubly-Tapered Composite Thin-Walled Beams," P.V.P., Recent Advances in Solids and Structures, Vol. 38, Y. W. Kwon and H. H. Chung (Eds.), 1999 ASME International Mechanical Engineering Congress and Exposition, Nashville, TN November 14–19, pp. 59–68.

Na, S. S. and Librescu, L. (2000) "Optimal Vibration Control of Thin-Walled Anisotropic Cantilevers Exposed to Blast Loadings," *Journal of Guidance, Control and Dynamics*, Vol. 23, No. 3, pp. 491–500.

Na, S. S. and Librescu, L. (2001) "Dynamic Response of Elastically Tailored Adaptive Cantilevers of Nonuniform Cross Section Exposed to Blast Pressure Pulses," *International Journal of Impact Engineering*, Vol. 25, Issue 9, pp. 1847–1867.

Na, S. S. and Librescu, L. (2002) "Optimal Dynamic Response Control of Elastically Tailored Nonuniform Thin-Walled Adaptive Beams," *Journal of Aircraft*, Vol. 39, No. 3, May–June, pp. 469–479.

Petre, A., Stanescu, C. and Librescu, L. (1961) "Aeroelastic Divergence of Multicell Wings Taking Their Fixing Restraint into Account," *Revue de Mécanique Appliquée*, Vol. 6, No. 6, pp. 689–698.

Petre, A. and Atanasiu, M. (1960) *Thin-Walled Beams* (in Romanian), Editura Tehnica, Bucharest, Romania.

Pilkey, W. D. (2002) *Analysis and Design of Elastic Beams: Computational Methods*, Wiley, New York.

Qin, Z. and Librescu, L. (2001) "Static and Dynamic Validations of a Refined Thin-Walled Composite Beam Model," *AIAA Journal*, Vol. 39, No. 12, pp. 2422–2424.

Qin, Z. and Librescu, L. (2002a) "Dynamic Aeroelastic Response of Aircraft Wings Modeled as Anisotropic Thin-Walled Beams," *Journal of Aircraft*.

Qin, Z. and Librescu, L. (2002b) "On a Shear-Deformable Theory of Anisotropic Thin-Walled Beams: Further Contribution and Validation," *Journal of Composite Structures*, Vol. 56, No. 4, pp. 345–358.

Qin, Z., Marzocca, P. and Librescu, L. (2002) "Aeroelastic Instability and Response of Advanced Aircraft Wings at Subsonic Flight Speeds," *Aerospace Science and Technology*, Vol. 6, No. 3, pp. 195–208.

Rehfield, L. W. and Atilgan, A. R. (1989) "Toward Understanding the Tailoring Mechanisms for Thin Walled Composite Tubular Beams," in *Proceedings of the First U.S.S.R.–U.S.A. Symposium on Mechanics of Composite Materials*, Riga, Latvia, pp. 23–26.

Rehfield, R. L., Atilgan, A. R. and Hodges, D. H. (1990) "Dynamic Characteristics of Thin-Walled Composite Beams," Presented at the *AHS National Specialists' Meeting on Rotorcraft Dynamics*, November 13–14, Arlington, Texas.

Senjanovic, I. and Fan, Y. (1991) "On Torsional and Warping Stiffness of Thin-Walled Girders," *Thin-Walled Structures*, Vol. 11, No. 3, pp. 233–276.

Song, O. (1990) "Modeling and Response Analysis of Thin-Walled Beams Structures Constructed of Advanced Composite Materials," Ph.D. Dissertation, Virginia Polytechnic Institute and State University, Blacksburg, VA.

Song, O., Librescu, L. and Rogers, C. A. (1992) "Application of Adaptive Technology to Static Aeroelastic Control of Wing Structures," *AIAA Journal*, Vol. 30, No. 12, pp. 2882–2889.

Song, O. and Librescu, L., (1993) "Free Vibration of Anisotropic Composite Thin-Walled Beams of Closed Cross-Section Contour," *Journal of Sound and Vibration*, Vol. 167, No. 1, pp. 129–147.

Song, O. and Librescu, L. (1995) "Bending Vibration of Cantilevered Thin-Walled Beams Subjected to Time-Dependent External Excitations," *Journal of the Acoustical Society of America*, June, Vol. 98, No. 1, pp. 313–319.

Song, O., Ju, J. S. and Librescu, L. (1998) "Dynamic Response of Anisotropic Thin-Walled Beams to Blast and Harmonically Oscillating Loads," *International Journal of Impact Engineering*, Vol. 21, No. 8, pp. 663–682.

Song, O., Kim, J-B. and Librescu, L. (2001) "Synergistic Implications of Tailoring and Adaptive Materials Technology on Vibration Control of Anisotropic Thin-Walled Beams," *International Journal of Engineering Science*, Vol. 39, No. 1, December, pp. 71–94.

Vlasov, V. Z. (1961) *Thin Walled Elastic Beams*, National Science Foundation, Washington, DC, Israel Program for Scientific Translations, Jerusalem, Israel. (The Russian version of this book entitled "Tonkostennye Uprugie Sterzhni" has appeared in Stroizdat, Moscow in 1940.)

Chapter 5

SEVERAL THEOREMS IN LINEAR THIN-WALLED BEAM THEORY

A number of general theorems of the three-dimensional linear theory of elasto-statics/kinetics (see e.g. Gurtin, 1972; Hetnarski and Ignaczak, 2003), have ana-logues in the theory of open and closed cross-section thin-walled beams. Prior to establishing such theorems, a few basic results from the three-dimensional elastokinetics are supplied.

5.1 THEOREM OF POWER AND ENERGY

Start from the linearized version of Eqs. (3.1-10)

$$\sigma_{rj,j} + \rho H_r = \rho \ddot{U}_r \quad \text{on } \tau, \tag{5.1-1a}$$

$$\sigma_{rj} n_j = \underset{\sim}{\sigma}_r \quad \text{on } \Omega_\sigma, \tag{5.1-1b}$$

$$V_r = \underset{\sim}{V}_r \quad \text{on } \Omega_v. \tag{5.1-1c}$$

Multiplying Eq. (5.1-1a) by $U_r (\equiv \dot{V}_r)$ and integrating it over the whole volume yields

$$\int_\tau (\sigma_{rj,j} + \rho H_r) U_r d\tau = \int_\tau \rho \ddot{U}_r U_r d\tau. \tag{5.1-2}$$

Transforming the first integral on the left-hand side of (5.1-2)

$$\int_\tau \sigma_{rj,j} U_r d\tau = \int_\tau (U_r \sigma_{rj})_{,j} d\tau - \int_\tau U_{r,j} \sigma_{rj} d\tau, \tag{5.1-3}$$

in accordance with Gauss' theorem and use of Eq. (5.1-1b) we have

$$\int_\tau (U_r \sigma_{rj})_{,j} d\tau = \int_\Omega \underset{\sim}{\sigma}_r U_r d\Omega. \tag{5.1-4}$$

We cast $U_{r,j}$ into symmetric and skew-symmetric parts:

$$U_{r,j} = \dot{\epsilon}_{rj} + \dot{\omega}_{rj}, \tag{5.1-5a}$$

where

$$\dot{\epsilon}_{rj}(\equiv (1/2)(U_{r,j} + U_{j,r})) \text{ and } \dot{\omega}_{rj}(\equiv (1/2)(U_{r,j} - U_{j,r})), \tag{5.1-5b,c}$$

denote the rate of strain and vorticity tensors, respectively. Since σ_{rj} is symmetric and ω_{rj} is skew-symmetric, we have $\sigma_{rj}\dot{\omega}_{rj} = 0$. Now, from (5.1-2) considered in conjunction with (5.1-3) and (5.1-4) one obtains

$$\int_\tau \rho H_r U_r d\tau + \int_\Omega \underline{\sigma}_r U_r d\Omega = \int_\tau \rho \dot{U}_r U_r d\tau + \int_\tau \sigma_{rj}\dot{\epsilon}_{rj} d\tau. \tag{5.1-6}$$

The right hand side of Eq. (5.1-6) represents the rate of change of the total energy, (i.e., of the kinetic and strain energies). Thus Eq. (5.1-6) gives

$$\frac{d\mathcal{U}}{dt} = \int_\tau \rho H_r U_r d\tau + \int_\Omega \underline{\sigma}_r U_r d\Omega, \tag{5.1-7a}$$

where

$$\mathcal{U} = \mathcal{K} + \mathcal{W}. \tag{5.1-7b}$$

This theorem, asserting that the time rate of change of the total mechanical energy equals the rate of work done by the surface, edge, and body forces, has an analogue in thin-walled beam theory: for this case $\mathcal{K}$ and $\mathcal{W}$ are supplied by (4.1-1) and (4.3-6), respectively, and $\mathcal{U} = \mathcal{K} + \mathcal{W}$. In the absence of body forces, the right hand side of (5.1-7a) for thin-walled beam theory is $\mathcal{T}_d + \mathcal{T}_e$. Here

$$\mathcal{T}_d = \int_{z_1}^{z_2} \left[(p_x + \delta_0 \hat{p}_x)\dot{u}_P + (p_y + \delta_0 \hat{p}_y)\dot{v}_P + (p_z + \delta_0 \hat{p}_z)\dot{w}_0 + (m_x \right.$$
$$\left. + \delta_0 \hat{m}_x)\dot{\theta}_x + (m_y + \delta_0 \hat{m}_y)\dot{\theta}_y + [m_z + m'_\omega + \delta_0(\hat{m}_z + \hat{m}'_\omega)]\dot{\phi} \right] dz, \tag{5.1-8a}$$

and

$$\mathcal{T}_e = n_z\{\underline{Q}_x \dot{u}_P + \underline{Q}_y \dot{v}_P + \underline{T}_z \dot{w}_0 + \underline{M}_x \dot{\theta}_x + \underline{M}_y \dot{\theta}_y$$
$$+ \underline{M}_z \dot{\phi} - \underline{B}_\omega \dot{\phi}'\}_{z=z_1,z_2}, \tag{5.1-8b}$$

denote the rate of work done by the distributed and end loads, respectively. Thus, for thin-walled beams, the theorem of power and energy is expressed, as

$$\frac{d\mathcal{U}}{dt} = \mathcal{T}_d + \mathcal{T}_e. \tag{5.1-9}$$

Integrating (5.1-9) with respect to time over the interval $[0, t)$, one obtains

$$\mathcal{U}(t) - \mathcal{U}(0) = \int_0^t (\mathcal{T}_d + \mathcal{T}_e) dt. \qquad (5.1\text{-}10)$$

This equation provides a basis for establishing the uniqueness of the solution.

From (5.1-10), assuming that the initial state is an unstrained state of rest, implying that $\mathcal{U}(0) = 0$, and the total strain energy stored in the structure is positive semi-definite (i.e. $\mathcal{W} \geq 0$), since the kinetic energy is always positive definite, it results that

$$\mathcal{U}(t) = \int_0^t (\mathcal{T}_d + \mathcal{T}_e) d\tau \geq 0. \qquad (5.1\text{-}11)$$

This states that the work performed in the time interval $[0, t)$ starting from an unstrained rest state is always nonnegative, and vanishes only when the displacement field corresponds to a rigid motion. This result is the analogue of the *theorem of positive work* of 3-D elasticity theory.

5.2 UNIQUENESS OF SOLUTION

In the preceding chapters, we have derived the equations of the linear theory of anisotropic thin-walled beams. The complete system of equations consists of the kinematic equations; the equations of equilibrium/motion; the constitutive equations; and the boundary and the initial conditions. In addition, the governing system for the most general case of anisotropy of the constituent materials of the beam, as well as its specialized counterparts, have also been supplied.

The solution of the governing system of equations, provide in conjunction with the boundary conditions and the initial conditions, the displacement quantities. From the displacements the entire state of stress and strain in space and time can be determined. It is natural to wonder if the functions characterizing the solution of the problem are determined without ambiguity or, in other words, if the solution of the problem is unique. To establish conditions yielding the uniqueness for the initial mixed boundary-value problems of the dynamical theory of thin-walled beams, we proceed as in the three-dimensional linear elastodynamics.

Suppose that there are two distinct solutions generically denoted as $(\overset{1}{\cdot})$ and $(\overset{2}{\cdot})$. Because of the linearity of the boundary value problem, the difference solution identified by the superposed circumflex namely $(\hat{\cdot}) = (\overset{1}{\cdot}) - (\overset{2}{\cdot})$, satisfies the constitutive equations, the kinematic equations, the *homogeneous* counterparts of equations of motion (3.2-13), as well as the *homogeneous* counterparts of boundary and initial conditions (4.4-2), (4.4-3) and (4.4-4), respectively.

For the difference solution the external and body forces are zero, so that (5.1-7) gives

$$\frac{d}{dt}(\hat{\mathcal{K}} + \hat{\mathcal{W}}) = 0. \tag{5.2-1}$$

Integration of (5.2-1) with respect to time over the interval $[0, t)$, yields

$$\hat{\mathcal{K}} + \hat{\mathcal{W}} = Const. \tag{5.2-2}$$

But since the initial velocities and displacements for the $(\hat{\cdot})$ state vanish, we have $\hat{\mathcal{K}} + \hat{\mathcal{W}} = 0$. Having in view that both the kinetic and strain energies are positive definite quantities, from (5.2-2) it results that $\hat{\mathcal{K}} = \hat{\mathcal{W}} = 0$ at any time t. $\hat{\mathcal{K}} = 0$ implies that $\hat{\dot{u}}_P, \hat{\dot{v}}_P, \hat{\dot{w}}_0, \hat{\dot{\theta}}_x, \hat{\dot{\theta}}_y, \hat{\dot{\phi}}$ and $\hat{\dot{\phi}}'$ are zero for $t \geq t_0$, or equivalently, that $(\overset{1}{\dot{u}}_P) = (\overset{2}{\dot{u}}_P), \ldots (\overset{1}{\dot{\phi}'}) = (\overset{2}{\dot{\phi}'})$. $\hat{\mathcal{W}} = 0$ implies that the strains for the difference solution are zero, and that the displacements are unique to within a rigid displacement. Note that this uniqueness theorem as given above remains valid also in the case of the thermoelastic theory of beams, when, it is reduced to an equivalent elastic one with thermal effects appearing as non-homogeneous terms in the static boundary conditions.

5.3 VIRTUAL WORK PRINCIPLE

As in the three-dimensional elasticity theory, the virtual work principle can be stated as

$$\delta \mathcal{W} = \delta \mathcal{J}, \tag{5.3-1}$$

where

$$\delta \mathcal{J} = \int_{z_1}^{z_2} \{(p_x + \delta_0 \hat{p}_x)\delta u_P + (p_y + \delta_0 \hat{p}_y)\delta v_P + (p_z + \delta_0 \hat{p}_z)\delta w_0$$

$$+ [m_z + m'_\omega + \delta_0(\hat{m}_z + \hat{m}'_\omega)]\delta\phi + (m_x + \delta_0 \hat{m}_x)\delta\theta_x$$

$$+ (m_y + \delta_0 \hat{m}_y)\delta\theta_y\}dz + n_z[\underline{T}_z \delta w_0 + \underline{M}_y \delta\theta_y + \underline{M}_x \delta\theta_x$$

$$- \underline{B}_\omega \delta\phi' + (\underline{B}'_\omega + \underline{M}_z)\delta\phi + \underline{Q}_x \delta u_P + \underline{Q}_y \delta v_P]|_{z=z_1, z_2}, \tag{5.3-2}$$

represents the mechanical work performed by the external, body forces and edge loads through the virtual displacements, whereas $\mathcal{W}$ is the strain energy, $\delta \mathcal{W}$ being supplied by Eq. (3.2-2).

In 3-D elasticity theory, the validity of this principle requires that the stresses satisfy the equations of equilibrium and *boundary conditions*. For the present case this condition is clearly fulfilled, but the role of 3-D stresses is played by the generalized 1-D stress measures as defined by (3.1-17).

One can formulate within the theory of thin-walled beams analogues of all the variational theorems of the classical 3-D elasticity, including the minimum potential and complementary energy theorems. The formulation of these latter theorems within the theory of shells (see Librescu, 1975a,b, 1976), can be helpful in their conversion to the present 1-D case.

5.4 CLAPEYRON'S THEOREM (THEOREM OF WORK AND ENERGY)

In three-dimensional elasticity theory this theorem states that if a body is in equilibrium under a given system of surface and edge loading, then the work done by the external forces of the equilibrium state, acting through the displacements of the strained equilibrium state, is equal to twice the strain energy of the body. Within the theory of thin-walled beams, considering in Eq. (5.3-1) instead of the virtual displacements δu_p, δv_p, δw_0, $\delta \phi$, $\delta \theta_x$ and $\delta \theta_y$, those corresponding to the strained equilibrium state, one obtains:

$$
\begin{aligned}
\int_{z_1}^{z_2} & [(p_x + \delta_0 \hat{p}_x) u_P + (p_y + \delta_0 \hat{p}_y) v_P + (p_z + \delta_0 \hat{p}_z) w_0 + (m_x + \delta_0 \hat{m}_x) \theta_x \\
& + (m_y + \delta_0 \hat{m}_y) \theta_y + [m_z + m'_\omega + \delta_0 (\hat{m}_z + \hat{m}'_\omega)] \phi] dz \\
& + n_z \{ \underset{\sim}{Q}_x u_P + \underset{\sim}{Q}_y v_P + \underset{\sim}{T}_z w_0 + \underset{\sim}{M}_x \theta_x + \underset{\sim}{M}_y \theta_y + \underset{\sim}{M}_z \phi - \underset{\sim}{B}_\omega \phi' \} \big|_{z=z_1, z_2} \\
& = 2\mathcal{W}.
\end{aligned}
\tag{5.4-1}
$$

The quantity on the left-hand side of this equation represents the work performed by the external and end forces, while $\mathcal{W}$ is the strain-energy stored in the structure.

Equation (5.4-1) represents the 1-D analogue of Clapeyron's theorem of the 3-D elasticity theory. From this theorem it clearly appears that in the case of the rigid displacements of the beam, the strains and also the strain energy vanish, and as a result, the work done by the external force systems is zero.

In the next equations of this chapter, for the sake of brevity, the terms associated with the tracer δ_0 will be no longer explicitly displayed, but absorbed in the similar terms without the tracer. In spite of this, there is no doubt that the theorems are valid for both the closed and open cross-section beams.

5.5 BILINEAR FORM ASSOCIATED WITH THE STRAIN ENERGY $\mathcal{W}$

The quadratic form featured by the strain energy, Eq. (4.3-7), enables one to associate to it a symmetric bilinear form in terms of two distinct displacement vectors, $\tilde{\mathbf{D}}$ and $\tilde{\tilde{\mathbf{D}}}$ defined at the same point of the beam. By virtue of the symmetry of the stiffness matrix $\mathbf{A}$ (see Eqs. (4.3-3) and (4.3-5)), the bilinear form $\mathcal{W}(\tilde{\mathbf{D}}, \tilde{\tilde{\mathbf{D}}})$ satisfies

$$\mathcal{W}(\tilde{\mathbf{D}}, \tilde{\tilde{\mathbf{D}}}) = \mathcal{W}(\tilde{\tilde{D}}, \tilde{D}) \tag{5.5-1}$$

where

$$\mathcal{W}(\tilde{\mathbf{D}}, \tilde{\tilde{\mathbf{D}}}) = \int_{z_1}^{z_2} \hat{W}(\tilde{D}, \tilde{\tilde{D}}) dz. \tag{5.5-2}$$

When $\tilde{\mathbf{D}} = \tilde{\tilde{\mathbf{D}}} \equiv \mathbf{D}$, Eq. (5.5-2) reduces to (4.3-6); when $\mathbf{D} \Rightarrow \tilde{\mathbf{D}} + \tilde{\tilde{\mathbf{D}}}$, one obtains

$$\mathcal{W}(\tilde{\mathbf{D}} + \tilde{\tilde{\mathbf{D}}}, \tilde{\mathbf{D}} + \tilde{\tilde{\mathbf{D}}}) = \mathcal{W}(\tilde{\mathbf{D}}, \tilde{\mathbf{D}}) + \mathcal{W}(\tilde{\tilde{\mathbf{D}}}, \tilde{\tilde{\mathbf{D}}}) + 2\mathcal{W}(\tilde{\mathbf{D}}, \tilde{\tilde{\mathbf{D}}}). \tag{5.5-3}$$

As for any quadratic form, the strain energy associated with the sum of the two states is not equal to the sum of the strain energies associated with each of the states.

5.6 BETTI'S RECIPROCAL THEOREM

This theorem was proven to be of a great practical and theoretical importance in a number of problems involving among others, the three-dimensional elasticity theory (see e.g. Achenbach, 2000; de Hoop, 1995). According to this theorem, for the one and the same elastic system subjected to two external force systems and featuring, correspondingly, two elastic states, the work done by the first system of external forces over the displacements of the second elastic state, equals the work done by the second system of external forces over the displacements of the first elastic state.

Identifying the two force systems, and correspondingly, the two elastic states by one and two superposed tildes, Betti's theorem obtained from Eq. (5.4-1) considered in conjunction with (5.5-1) is expressed as:

$$\int_{z_1}^{z_2} [\tilde{\tilde{p}}_x \tilde{u}_P + \tilde{\tilde{p}}_y \tilde{v}_P + \tilde{\tilde{p}}_z \tilde{w}_0 + \tilde{\tilde{m}}_x \tilde{\theta}_x + \tilde{\tilde{m}}_y \tilde{\theta}_y + (\tilde{\tilde{m}}_z + \tilde{\tilde{m}}'_\omega) \tilde{\phi}] dz$$

$$- \int_{z_1}^{z_2} [\tilde{\tilde{K}}_1 \tilde{u}_P + \tilde{\tilde{K}}_2 \tilde{v}_P + \tilde{\tilde{K}}_3 \tilde{w}_0 + (\tilde{\tilde{K}}_4 - \tilde{\tilde{K}}'_9) \tilde{\phi} + \tilde{\tilde{K}}_7 \tilde{\theta}_x + \tilde{\tilde{K}}_5 \tilde{\theta}_y] dz$$

$$+ n_z \{\tilde{\tilde{Q}}_x \tilde{u}_P + \tilde{\tilde{Q}}_y \tilde{v}_P + \tilde{\tilde{T}}_z \tilde{w}_0 + \tilde{\tilde{M}}_x \tilde{\theta}_x + \tilde{\tilde{M}}_y \tilde{\theta}_y + \tilde{\tilde{M}}_z \tilde{\phi} - \tilde{\tilde{B}}_\omega \tilde{\phi}'\}|_{z=z_1, z_2}$$

$$= \int_{z_1}^{z_2} [\tilde{p}_x \tilde{\tilde{u}}_P + \tilde{p}_y \tilde{\tilde{v}}_P + \tilde{p}_z \tilde{\tilde{w}}_0 + \tilde{m}_x \tilde{\tilde{\theta}}_x + \tilde{m}_y \tilde{\tilde{\theta}}_y + (\tilde{m}_z + \tilde{m}'_\omega) \tilde{\tilde{\phi}}] dz$$

$$\tag{5.6-1}$$

$$- \int_{z_1}^{z_2} [\tilde{K}_1 \tilde{\tilde{u}}_P + \tilde{K}_2 \tilde{\tilde{v}}_P + \tilde{K}_3 \tilde{\tilde{w}}_0 + (\tilde{K}_4 - \tilde{K}'_9) \tilde{\tilde{\phi}} + \tilde{K}_7 \tilde{\tilde{\theta}}_x + \tilde{K}_5 \tilde{\tilde{\theta}}_y] dz$$

$$+ n_z \{\tilde{Q}_x \tilde{\tilde{u}}_P + \tilde{Q}_y \tilde{\tilde{v}}_P + \tilde{T}_z \tilde{\tilde{w}}_0 + \tilde{M}_x \tilde{\tilde{\theta}}_x + \tilde{M}_y \tilde{\tilde{\theta}}_y + \tilde{M}_z \tilde{\tilde{\phi}} - \tilde{B}_\omega \tilde{\tilde{\phi}}'\}|_{z=z_1, z_2}.$$

Here, the static quantities in the braces { } are prescribed. It should also be mentioned that in the next developments, the inertia terms K_i are evaluated in conjunction with the underscored terms in Table 3.2-1 and the expression of mass terms in Eq. (3.1-24). It becomes evident that at the basis of Eq. (5.6-1) we have the equality

$$\int_{z_1}^{z_2} [\tilde{T}_z \, \tilde{\tilde{w}}_0' + \tilde{M}_x(\delta_t \tilde{\tilde{\theta}}_x' - \delta_e \tilde{\tilde{v}}_P'') + \tilde{M}_y(\delta_t \tilde{\tilde{\theta}}_y' - \delta_e \tilde{\tilde{u}}_P'')$$

$$+ \delta_t \tilde{Q}_x(\tilde{\tilde{\theta}}_x + \tilde{\tilde{v}}_P) + \delta_t \tilde{Q}_y(\tilde{\tilde{\theta}}_y + \tilde{\tilde{u}}_P) + \tilde{M}_z \tilde{\tilde{\phi}}' - \delta_r \tilde{B}_\omega \tilde{\tilde{\phi}}'']dz$$

$$= \int_{z_1}^{z_2} [\tilde{\tilde{T}}_z \tilde{w}_0' + \tilde{\tilde{M}}_x(\delta_t \tilde{\theta}_x' - \delta_e \tilde{v}_P'') + \tilde{\tilde{M}}_y(\delta_t \tilde{\theta}_y' - \delta_e \tilde{u}_P'')$$

$$+ \delta_t \tilde{\tilde{Q}}_x(\tilde{\theta}_x + \tilde{v}_P') + \delta_t \tilde{\tilde{Q}}_y(\tilde{\theta}_y + \tilde{u}_P') + \tilde{\tilde{M}}_z \tilde{\phi}' - \delta_r \tilde{\tilde{B}}_w \tilde{\phi}'']dz. \quad (5.6\text{-}2)$$

This states, (see Eq. (3.2-1)), that the strain energy associated with the state of stress "$\approx$"acting on the strains of state "$\sim$" is equal to the strain energy associated with the state of stress " $\sim$" acting on the strains of state "$\approx$".

It should be noticed that fulfillment of Betti's reciprocal theorem yields the condition that the underlying system of field equations is self-adjoint. Note that a necessary condition, for the validity of Betti's theorem is that the stiffness matrix **A** should be symmetric. As is clearly seen, (see Eqs. (4.3-3) and (4.3-5)), this condition is fulfilled.

5.7 SELF-ADJOINTNESS OF THE BOUNDARY-VALUE PROBLEM

The spectral theory of self-adjoint operators shows that their eigenvalues and eigenfunctions enjoy special features, namely that the eigenvalues of a self-adjoint operator form an infinite denumerable sequence of real values, and the eigenfunctions corresponding to two distinct eigenvalues are orthogonal. If the boundary value problem is not self-adjoint, the eigenvalues can be complex. We will now show that the system of equations in the theory of thin-walled beams is self-adjoint. To do this, we will make use of the governing equation system (4.4-1) expressed in terms of displacement measures u_P, v_P, w_0, θ_x, θ_y and ϕ.

Write this system as $\mathcal{L}_i \mathbf{v} = \mathbf{K}_i$ where $\mathbf{K}_i$ and $\mathbf{v}$ denote the vectors of the external loads and of displacements, respectively.

Consider two vectors $\tilde{\mathbf{u}}$ and $\tilde{\mathbf{v}}$ satisfying the boundary conditions (4.4-2) and (4.4-3) at $z = z_1$ and $z = z_2$. The boundary value problem is self-adjoint if, by virtue of the boundary conditions

$$\mathcal{E} = \int_{z_1}^{z_2} [\tilde{u}_i \mathcal{L}_i(\tilde{v}) - \tilde{v}_i \mathcal{L}_i(\tilde{u})]dz = 0, \quad (5.7\text{-}1)$$

where u_i and v_i are the components of **u** and **v**, respectively. Considering the integral $\int_{z_1}^{z_2} \tilde{u}_i \mathcal{L}_i(\tilde{v})dz$, integrating by parts and collecting terms, one obtains

$$\int_{z_1}^{z_2} \tilde{u}_i \mathcal{L}_i(\tilde{v})dz = 2\int_{z_1}^{z_2} \mathcal{W}(\tilde{u}, \tilde{v})dz - \int_{z_1}^{z_2} \tilde{v}\mathbf{P}(\tilde{u})dz - n_z[\tilde{v}\mathbf{S}(\tilde{u})]|_{z=z_1,z_2}.$$

(5.7-2)

Here, $\mathcal{W}(\tilde{u}, \tilde{v})$ is the bilinear symmetric form as expressed by (5.5-1), while the remaining two terms in Eq. (5.7-2) correspond, in the same succession, to the work done by the distributed and the edge loads, respectively. These expressions are provided, for example, by the terms belonging to the first and third lines of Eq. (5.6-1), respectively, where "$\approx$" should be replaced by "$\sim$". Proceeding similarly with the integral $\int_{z_1}^{z_2} \tilde{v}_i \mathcal{L}_i(\tilde{u})dz$, from (5.7-1) one finally obtains

$$\int_{z_1}^{z_2} [\tilde{v}_i \mathcal{L}_i(\tilde{u}) - \tilde{u}_i \mathcal{L}_i(\tilde{v})]dz$$

$$= \int_{z_1}^{z_2} \tilde{v}\mathbf{P}(\tilde{u})dz - \int_{z_1}^{z_2} \tilde{u}\mathbf{P}(\tilde{v})dz \qquad (5.7\text{-}3)$$

$$+ n_z[\tilde{v}\mathbf{S}(\tilde{u})]|_{z=z_1,z_2} - n_z[\tilde{u}\mathbf{S}(\tilde{v})]|_{z=z_1,z_2}.$$

By virtue of Betti's reciprocal theorem, whenever $\tilde{u}$ and $\tilde{v}$ correspond to equilibrium states, one obtains

$$\mathcal{E} = 0, \qquad (5.7\text{-}4)$$

and consequently, the boundary value problem is self-adjoint. It is clearly seen that the self-adjointness follows from the reciprocal theorem.

One should remark that in the above proof, for the sake of generality, the loadings were assumed to be functions of displacements.

5.8 ORTHOGONALITY OF MODES OF FREE VIBRATION

Let establish the relationship between the eigenfunctions corresponding to the distinct eigenfrequencies ω_r and ω_s of a thin-walled beam. First, we specialize the reciprocal theorem (5.6-1) for free vibration in the absence of surface and edge loads. Consider the displacement components associated with the *state* identified by the index r expressed as

$$\{u_P(z,t),\ v_P(z,t),\ w_0(z,t),\ \theta_x(z,t), \theta_y(z,t),\ \phi(z,t)\}=$$

$$[u_r(z),\ v_r(z),\ w_r(z),\ \theta_{xr}(z),\ \theta_{yr}(z),\ \phi_r(z)]q_r(t) \qquad (5.8\text{-}1)$$

where $u_r(z), v_r(z), w_r(z), \theta_{xr}(z), \theta_{yr}(z), \phi_r(z)$ play the role of vibration modes, while

$$q_r(t) = \exp(i\omega_r t), \quad i = \sqrt{-1} \qquad (5.8\text{-}2)$$

are the generalized coordinates associated with the *r-th* natural frequency ω_r. Corresponding to the state involving the natural frequency ω_s, in the representation of the displacement vector, Eq. (5.8-1), and of generalized coordinates, Eq. (5.8-2), the index r should be replaced by s. Identifying as indicated in the reciprocal theorem the two states, one obtains

$$(\omega_r^2 - \omega_s^2) \int_{z_1}^{z_2} [b_1 u_r u_s + b_1 v_r v_s + b_1 w_r w_s + (b_5 + \delta_n b_{15})\theta_{yr}\theta_{ys}$$

(5.8-3)

$$+ (b_4 + \delta_n b_{14})\theta_{xr}\theta_{xs} + (b_4 + b_5)\phi_r\phi_s + \underline{(b_{10} + \delta_n b_{18})\ \phi_r'\phi_s'}]dz\ = 0$$

Equation (5.8-3) implies that for real and distinct eigenfrequencies, the integral multiplying $(\omega_r^2 - \omega_s^2)$ must vanish. In the case when $r = s$, the integral does not vanish and is denoted by N_r.

As a result, the orthogonality property of the modes of free vibration is expressed as

$$\int_{z_1}^{z_2} [b_1 u_r u_s + b_1 v_r v_s + b_1 w_r w_s + (b_5 + \delta_n b_{15})\theta_{yr}\theta_{ys}$$

$$+ (b_4 + \delta_n b_{14})\theta_{xr}\theta_{xs} + [(b_4 + b_5) + \delta_n (b_{14} + b_{15})]\phi_r\phi_s$$

$$+ \underline{(b_{10} + \delta_n b_{18})\phi_r'\phi_s'}]dz = \begin{cases} 0 \text{ when } r \neq s \\ N_r \text{ when } r = s \end{cases}$$

(5.8-4)

where

$$N_r = \int_{z_1}^{z_2} [b_1 u_r^2 + b_1 v_r^2 + b_1 w_r^2 + (b_5 + \delta_n b_{15})\theta_{yr}^2 + (b_4 + \delta_n b_{14})\theta_{xr}^2$$

$$+ [(b_4 + b_5) + \delta_n (b_{14} + b_{15})]\phi_r^2 + \underline{(b_{10} + \delta_n b_{18})\phi_r'^2}]dz, \quad (5.8\text{-}5)$$

is the generalized mass of the structure (also referred to as *the norm*).

For the structural coupling configurations, CAS and CUS, the orthogonality conditions split into two parts:

For CAS:

$$\int_{z_1}^{z_2} \{b_1 v_r v_s + (b_4 + \delta_n b_{14})\theta_{xr}\theta_{xs} + [(b_4 + b_5) + \delta_n (b_{14} + b_{15})]\phi_r\phi_s$$

$$+ \underline{(b_{10} + \delta_n b_{18})\ \phi_r'\phi_s'}\}dz = \begin{cases} 0 \text{ for } r \neq s \\ N_r \text{ for } r = s \end{cases}$$

(5.8-6)

and

$$\int_{z_1}^{z_2} [b_1 u_r u_s + b_1 w_r w_s + (b_5 + \delta_n b_{15})\theta_{yr}\theta_{ys}]dz = \begin{cases} 0 \text{ for } r \neq s \\ N_r \text{ for } r = s \end{cases}$$

(5.8-7)

while for CUS:

$$\int_{z_1}^{z_2} \{b_1 w_r w_s + [(b_4 + b_5) + \delta_n(b_{14} + b_{15})]\phi_r \phi_s$$

$$+ \ \underline{\underline{(b_{10} + \delta_n b_{18})\ \phi_r' \phi_s'}}\ \}dz = \begin{cases} 0 \text{ for } r \neq s \\ N_r \text{ for } r = s \end{cases} \tag{5.8-8}$$

and

$$\int_{z_1}^{z_2} [b_1 u_r u_s + b_1 v_r v_s + (b_4 + \delta_n b_{14})\theta_{xr}\theta_{xs}$$

$$+ (b_5 + \delta_n b_{15})\theta_{yr}\theta_{ys}]dz = \begin{cases} 0 \text{ for } r \neq s \\ N_r \text{ for } r = s \end{cases} \tag{5.8-9}$$

Needless to say, for the above displayed cases, the proper norm is obtainable from (5.8-5) by discarding the coupling terms that do not belong to the specific considered case.

5.9 DYNAMIC RESPONSE

As an application of the Betti's reciprocal theorem, we consider the dynamic response of thin-walled beams to time-dependent external excitations or edge loads.

Consider the two states involved in the reciprocal theorem as corresponding to the free vibration and forced vibration problems. Whereas in the free undamped vibration problem the representation of displacement quantities is under the form (5.8-1) considered in conjunction with (5.8-2), for the dynamic response, it is under the form of Eqs. (5.8-1). This yields readily the equation governing the generalized coordinate $q_r(t)$ namely

$$\ddot{q}_r + \omega_r^2 q_r = \frac{Q_r(t)}{N_r}, \tag{5.9-1}$$

where Q_r is the generalized load expressed as

$$Q_r(t) = \int_{z_1}^{z_2} \{p_x(z,t)u_r + p_y(z,t)v_r + p_z(z,t)w_r + m_x(z,t)\theta_{xr}$$

$$+ m_y(z,t)\theta_{yr} + [m_z(z,t) + m_\omega'(z,t)]\phi_r\}dz + n_z\{\underset{\sim}{Q}_x u_r + \underset{\sim}{Q}_y v_r$$

$$+ \underset{\sim}{T}_z w_r + \underset{\sim}{M}_x \theta_{xr} + \underset{\sim}{M}_y \theta_{yr} + \underset{\sim}{M}_z \phi_r - \underset{\sim}{B}_\omega \phi_r'\}|_{z=z_1,z_2} \tag{5.9-2}$$

N_r being given by Eq. (5.8-5). Being uncoupled, Eq. (5.9-1) can be solved independently for each r-th mode to obtain

$$q_r(t) = q_r(0)\cos\omega_r t + \frac{\dot{q}_r(0)}{\omega_r}\sin\omega_r t$$

$$+ \frac{1}{\omega_r N_r}\int_0^t Q_r(\tau)\sin\omega_r(t-\tau)d\tau. \tag{5.9-3}$$

The initial conditions yield

$$u_P(z, t = 0) \equiv \hat{u}_P = \sum q_r(0)u_r(z),$$

$$\dot{u}_P(z, t = 0) \equiv \dot{\hat{u}}_P = \sum \dot{q}_r(0)u_r(z),$$

. . . .

$$\phi(z, t = 0) \equiv \hat{\phi} = \sum q_r(0)\phi_r(z),$$

$$\dot{\phi}(z, t = 0) \equiv \dot{\hat{\phi}} = \sum \dot{q}_r(0)\phi_r(z),$$

$$\phi'(z, t = 0) \equiv \hat{\psi} = \sum q_r(0)\phi_r'(z),$$

$$\dot{\phi}'(z, t = 0) \equiv \dot{\hat{\psi}} = \sum \dot{q}_r(0)\phi_r'(z), \tag{5.9-4}$$

and using the orthogonality conditions, one obtains

$$q_r(0) = \frac{1}{N_r} \int_{z_1}^{z_2} \{b_1\hat{u}_P u_r + b_1\hat{v}_P v_r + b_1\hat{w}w_r$$

$$+ (b_5 + \delta_n b_{15})\hat{\theta}_y\theta_{yr} + (b_4 + \delta_n b_{14})\hat{\theta}_x\theta_{xr}$$

$$+ [(b_4 + b_5) + \delta_n(b_{14} + b_{15})]\hat{\phi}\phi_r + \underline{\underline{(b_{10} + \delta_n b_{18})\,\hat{\psi}\phi_r'}}\}dz \tag{5.9-5}$$

$$\dot{q}_r(0) = \frac{1}{N_r} \int_{z_1}^{z_2} \{b_1\dot{\hat{u}}_P u_r + b_1\dot{\hat{v}}_P v_r + b_1\dot{\hat{w}}w_r$$

$$+ (b_5 + \delta_n b_{15})\dot{\hat{\theta}}_y\theta_{yr} + (b_4 + \delta_n b_{14})\dot{\hat{\theta}}_x\theta_{xr}$$

$$+ [(b_4 + b_5) + \delta_n(b_{14} + b_{15})]\dot{\hat{\phi}}\phi_r + \underline{\underline{(b_{10} + \delta_n b_{18})\dot{\hat{\psi}}\phi_r'}}\}dz. \tag{5.9-6}$$

The complete solution of the dynamic response problem of thin walled beams under arbitrary time-dependent excitations is obtained by inserting (5.9-5) and (5.9-6) in (5.9-3). As indicated by Pilkey (1967), this approach of the dynamic response can accommodate also the presence of thermal forces and arbitrary changes in material and geometric parameters. Equations (5.9-2) through (5.9-7) were obtained via an alternative and more elaborated procedure by Song et al. (1998).

For the dynamic response featuring non-homogeneous time-dependent boundary conditions, following Mindlin and Goodman's (1950), the original dynamic response problem is converted to a counterpart one characterized by homogeneous boundary conditions and a modified time- dependent external load. Again the orthogonality of modes of free vibration should be used. For a recent description of this procedure restricted to 1-D systems, see Meirovitch (1997). As it will be seen later the Extended Galerkin Method can also be used to address this problem.

It should be recalled that δ_t, δ_e δ_o and δ_n intervening in the equations of this chapter are tracing quantities that take the values 1 or 0, depending on whether the effect identified by each of these is accounted for or discarded, respectively.

BIBLIOGRAPHICAL COMMENTS

- The general theorems of the 3-D elasticity theory were shown to be applicable also to the classical, (Librescu (1975a), and shearable, Librescu (1975b, 1976), shell theories.

- Betti's reciprocal theorem obtained, for the static case can easily be extended to the dynamic one, to give Graffi's theorem (Graffi, 1963). This can be obtained by following the procedure described for the 3-D elasticity theory by Fung (1965). For thin shell theory, the dynamic reciprocal theorem was formulated by Greif (1964), while for the dynamic thermoelasticity theory, by Ionescu-Cazimir (1964) and Nowacki (1976). An interesting application of Graffi's theorem in the formulation of a dynamic shape control methodology of structures was done by Irschik and Pichler (2001).

REFERENCES

Achenbach, J. D. (2000) "Calculation of Surface Wave Motion Due to a Subsurface Point Force: An Application of Elastodynamic Reciprocity," *Journal of Acoustical Society of America*, Vol. 107, No. 4, pp. 1892–1897.

Fung, Y. C. (1965) *Foundations of Solid Mechanics*, Prentice-Hall, Inc., Englewood Cliffs, New Jersey.

de Hoop, A. T. (1995) *Handbook of Radiation and Scattering of Waves*, Academic Press, London.

Graffi, D. (1963) "Sui Teoremi di Reciprocita nei Fenomeni Non Stazionari," *Atti Accad. Sci. Bologna*, Vol. 11, No. 10, pp. 33–40.

Greif, D. (1964) "A Dynamic Reciprocal Theorem for Thin Shells," *Journal of Applied Mechanics*, (Trans. ASME), December, pp. 724–726.

Gurtin, M. E. (1972) "The Linear Theory of Elasticity," in *Hanbuck der Physik*, C. Truesdell (Ed.), Springer Verlag, Berlin, pp. 1–273.

Hetnarski, R. B. and Ignaczak, J. (2003) *Mathematical Theory of Elasticity*, Taylor & Francis.

Ionescu-Cazimir, V. (1964) "Problem of Linear Coupled Thermoelasticity Reciprocal Theorems for the Dynamic Problem of Thermoelasticity," *Bull. Acad. Polon. Sci., Série Sci. Techn.*, Vol. 12, No. 9, pp. 473–480.

Irschik, H. and Pichler, U. (2001) "Dynamic Shape Conrol of Solids and Structures by Thermal Expansion Strain," *Journal of Thermal Stresses*, Vol. 24, pp. 565–576.

Librescu, L. (1975a) *Elastostatics and Kinetics of Anisotropic and Heterogeneous Shell-Type Structures*, Noordhoff International Publishing, Leyden, Netherlands, pp. 560–598.

Librescu, L., (1975b) "Some Results Concerning the Refined Theory of Elastic Multilayered Shells, Part II, "The Linearized Field Equations of Anisotropic Laminated Shells," *Revue Roumaine des Science Techniques -Mécanique Appliquée*, Vol. 20, No. 4, pp. 573–583.

Librescu, L. (1976) "On The Theory of Multilayered Anisotropic Shells," Part II – *Mechanika Polimerov*, Vol. 12, No. 1, pp. 100–109 (in Russian) [English translation in *Polymer Mechanics*, Plenum Press, January 1977, pp. 82–90].

Meirovitch, L. (1997) *Principles and Techniques of Vibration*, Prentice Hall, New Jersey.

Mindlin, R. D. and Goodman, L. E. (1950) "Beam Vibrations with Time-Dependent Boundary Conditions," *Journal of Applied Mechanics*, (Trans. ASME), Vol. 72, pp. 377–380.

Nowacki, W. (1976) *Dynamic Problems of Thermoelasticity*, Noordhoff International Publisher.

Pilkey, W. (1967) "Dynamic Response of Elastic Bodies Using the Reciprocal Theorem," *Journal of Applied Mechanics*, (Trans. ASME), September, pp. 774–775.

Song, O., Ju, J. S. and Librescu, L. (1998) "Dynamic Response of Anisotropic Thin-Walled Beams to Blast and Harmonically Oscillating Loads," *International Journal of Impact Engineering*, Vol. 21, No. 8, pp. 663–682.

Chapter 6

FREE VIBRATION

6.1 INTRODUCTION

With their high strength-to-weight and stiffness-to-weight ratios, thick and thin-walled beam structures made of fiber reinforced composite materials are likely to play an important role in the design of aircraft and space vehicles, helicopter blades and of other advanced structural systems.

In addition, the various elastic couplings that are the result of the inherent directional properties of fibrous composite materials, can be successfully used to enhance their static and dynamic response behavior.

The goal of this chapter is the study of the free vibration of anisotropic thick/thin walled beams of closed cross-section contour.

The anisotropy induced by the directional nature of fibrous composites is shown to be able to enhance the free vibration characteristics.

Despite the practical importance of the problem, the research work in this field is scarce. Some results on the dynamic behavior of metallic thin-walled beams of closed cross-section contour are contained in the works by Budiansky and Kruszewski (1952), Gay and Boudet (1980) and Bishop et al. (1983), and to composite ones by Bank and Kao (1989), Rehfield et al. (1990), Atilgan and Rehfield (1990), Laudiero and Savoia (1991), Hodges et al. (1991), Armanios and Badir (1995), Song and Librescu (1991, 1993, 1995, 1996, 1997), Librescu et al. (1993, 1996), Librescu and Na (1997, 1998, 2001), Na and Librescu (1999), Qin and Librescu (2001, 2002), Song et al. (2001, 2002).

Note that, due to the incorporation of transverse shear, the beam theory is applicable to both thin and thick-walled beams, in the sense of $h_{\max}/l \lessgtr 0.1$, respectively, where $h_{\max}$ denotes the maximum thickness of the wall and l a typical cross-sectional dimension, but the standard terminology of thin-walled beams is used throughout. As is well-known (see e.g. Ambartsumian, 1964;

121

Librescu, 1975), the necessity of incorporating transverse shear effects arises not only because composite structures tend to be thicker than their standard metallic counterparts, but also because advanced fiber composites exhibit large flexibilities in transverse shear.

In most cases, the *composite* aircraft wings as well helicopter and turbomachinery blades were modeled as *solid* beams (see e.g. Weisshaar and Foist, 1985; Karpouzian and Librescu, 1994, 1996; Gern and Librescu, 1998). The composite thin-walled beam model permits not only incorporation of effects not accounted for previously, but also provides a much more advanced structural model.

In the last sections of this chapter, we will consider active controllable thin-walled structures, with inherent self-sensing and control capabilities; the introductory elements in Section A.9.2 will be used.

6.2 BASIC ASSUMPTIONS AND GOVERNING EQUATIONS

Consider a thin walled single-cell beam of arbitrary closed-section. The points of the beam structure are referred to a 3-D coordinate system (x, y, z), where the spanwise z coordinate coincides with a straight, unspecified reference axis, while x and y are the cross-section coordinates with the origin on the reference axis (see Fig. 2.1-2).

We recall the assumptions formulated in Chapter 2: (i) beam cross-sections do not deform in their own planes, (ii) transverse shear effects are significant, and as a result are included in the model, (iii) the twist varies along the beam span, i.e., the rate of twist $d\phi/dz$ is no longer assumed to be constant (as in the Saint-Venant torsional model), but a function of the spanwise coordinate, where $\phi' \equiv d\phi/dz$ constitutes a measure of the torsion-related nonuniform warping, (iv) primary and secondary warping effects are included, (v) in the absence of an internal pressure field, the hoop stress resultant is negligible compared to the remaining stresses, (vi) the beam is clamped at the root, $z = 0$, and free at the tip, $z = L$; and (vii) $x_P = y_P = 0$, implying that the pole P lies on the reference beam axis. To be consistent with this convention, the index P will be replaced by the index zero and u_P and v_P will be denoted as u_0 and v_0, respectively.

The governing equations follow from the analysis of Chapter 4; in the context of free vibration there are no external loads. For the *Circumferentially Uniform Stiffness Configuration* (CUS), for cantilevered beams, Eqs. (4.4-14) through (4.4-19) give

- The equations governing the extension-twist motion are:

$$\delta w_0 : \quad a_{11} w_0'' + a_{17} \phi'' = b_1 \ddot{w}_0,$$

$$\delta\phi: \quad \underline{-a_{66}\phi''''} \quad + a_{77}\phi'' + a_{71}w_0'' = [(b_4+b_5)+\delta_n(b_{14}+b_{15})]\ddot{\phi}$$

$$- \underline{\underline{(b_{10}+\delta_n b_{18})\ddot{\phi}''}}, \qquad (6.2\text{-}1\text{a,b})$$

with the associated boundary conditions at $z = L$,

$$\delta w_0: \quad a_{11}w_0' + a_{17}\phi' = 0,$$

$$\delta\phi: \quad \underline{-a_{66}\phi'''} \quad + a_{77}\phi' + a_{71}w_0' = - \underline{\underline{(b_{10}+\delta_n b_{18})}}\,\ddot{\phi}', \quad (6.2\text{-}2\text{a-c})$$

$$\delta\phi': \quad \underline{a_{66}\,\phi''} = 0,$$

and at $z = 0$,

$$w_0 = \phi = \underline{\phi'} = 0, \qquad (6.2\text{-}3\text{a-c})$$

• The equations governing the bending-bending-transverse shear motion

$$\delta u_0: \quad a_{43}\theta_x'' + a_{44}(u_0'' + \theta_y') = b_1\ddot{u}_0,$$

$$\delta v_0: \quad a_{52}\theta_y'' + a_{55}(v_0'' + \theta_x') = b_1\ddot{v}_0, \qquad (6.2\text{-}4\text{a-d})$$

$$\delta\theta_y: \quad a_{22}\theta_y'' + a_{25}(v_0'' + \theta_x') - a_{44}(u_0' + \theta_y) - a_{43}\theta_x' = \underline{(b_5 + \delta_n b_{15})\ddot{\theta}_y},$$

$$\delta\theta_x: \quad a_{33}\theta_x'' + a_{34}(u_0'' + \theta_y') - a_{55}(v_0' + \theta_x) - a_{52}\theta_y' = \underline{(b_4 + \delta_n b_{14})\ddot{\theta}_x}\;.$$

with the associated boundary conditions at $z = L$,

$$\delta u_0: \quad a_{43}\theta_x' + a_{44}(u_0' + \theta_y) = 0,$$
$$\delta v_0: \quad a_{52}\theta_y' + a_{55}(v_0' + \theta_x) = 0, \qquad (6.2\text{-}5\text{a-d})$$
$$\delta\theta_y: \quad a_{22}\theta_y' + a_{25}(v_0' + \theta_x) = 0,$$
$$\delta\theta_x: \quad a_{33}\theta_x' + a_{34}(u_0' + \theta_y) = 0.$$

and at $z = 0$,

$$u_0 = v_0 = \theta_x = \theta_y = 0. \qquad (6.2\text{-}6\text{a-d})$$

The equations pertinent for the *Circumferentially Asymmetric Stiffness Configuration* (CAS), are obtained from Eqs. (4.4-26) through (4.4-31) and are:

- The system of equations (of the sixth order) governing extension – lateral bending-lateral transverse shear:

$$\delta w_0 : \quad a_{11}w_0'' + a_{14}(u_0'' + \theta_y') = b_1\ddot{w}_0,$$
$$\delta u_0 : \quad a_{41}w_0'' + a_{44}(u_0'' + \theta_y') = b_1\ddot{u}_0, \qquad (6.2\text{-}7\text{a-c})$$

$$\delta\theta_y : \quad -a_{41}w_0' - a_{44}(u_0' + \theta_y) + a_{22}\theta_y'' = \underbrace{(b_5 + \delta_n b_{15})\ddot{\theta}_y}.$$

with the associated boundary conditions at $z = L$,

$$\delta w_0 : \quad a_{11}w_0' + a_{14}(u_0' + \theta_y) = 0,$$
$$\delta u_0 : \quad a_{41}w_0' + a_{44}(u_0' + \theta_y) = 0, \qquad (6.2\text{-}8\text{a-c})$$
$$\delta\theta_y : \quad a_{22}\theta_y' = 0,$$

and at $z = 0$,

$$w_0 = u_0 = \theta_y = 0. \qquad (6.2\text{-}9\text{a-c})$$

- The system of equations (of eighth order) governing the twist – flapwise bending – flapwise transverse shear:

$$\delta\phi : \quad \underline{-a_{66}\phi''''} + a_{77}\phi'' + a_{73}\theta_x'' + \underline{a_{65}(v_0' + \theta_x)''}$$

$$= [(b_4 + b_5) + \delta_n(b_{14} + b_{15})]\ddot{\phi} \; \underline{\underline{-(b_{10} + \delta_n b_{18})\ddot{\phi}''}},$$

$$\delta\theta_x : \quad a_{33}\theta_x'' + a_{37}\phi'' - a_{55}(v_0' + \theta_x) + \underline{a_{56}\phi''} = \underline{\underline{(b_4 + \delta_n b_{14})\ddot{\theta}_x}},$$

$$\delta v_0 : \quad a_{55}(v_0' + \theta_x)' - \underline{a_{56}\phi'''} = b_1\ddot{v}_0, \qquad (6.2\text{-}10\text{a-c})$$

with the associated boundary conditions at $z = L$,

$$\delta\phi : \quad \underline{-a_{66}\phi'''} + a_{77}\phi' + a_{73}\theta_x' + \underline{a_{65}(v_0' + \theta_x)'}$$

$$= - \underline{\underline{(b_{10} + \delta_n b_{18})\ddot{\phi}'}},$$

$$\underline{\delta\phi'} : \quad -a_{66}\phi'' + \underline{a_{65}(v_0' + \theta_x)} = 0, \qquad (6.2\text{-}11\text{a-d})$$
$$\delta\theta_x : \quad a_{33}\theta_x' + a_{37}\phi' = 0,$$
$$\delta v_0 : \quad a_{55}(v_0' + \theta_x) - \underline{a_{56}\phi''} = 0,$$

and at $z = 0$,

$$\phi = v_0 = \theta_x = \underline{\phi'} = 0. \qquad (6.2\text{-}12\text{a-d})$$

For cross-ply laminated composite beams, $a_{17}(\equiv a_{71}) = 0$; $a_{34}(\equiv a_{43}) = 0$; $a_{25}(\equiv a_{52}) = 0$, $a_{14}(\equiv a_{41}) = 0$; $a_{37}(\equiv a_{73}) = 0$, and $a_{56}(\equiv a_{65}) = 0$. As a result, Eqs. (6.2-1) through (6.2-3) split into two independent sub-systems governing pure extension and pure twist; Eqs. (6.2-4) through (6.2-6) decouple into two sub-systems governing the chordwise (lateral) bending-chordwise transverse shear and the flapwise bending-flapwise transverse shear; Eqs. (6.2-7) through (6.2-9) decouple into two sub-systems governing the chordwise bending-chordwise transverse shear, and the pure extension motion, whereas Eqs. (6.2-10) through (6.2-12) split into the two sub-systems governing the pure twist and the flapwise bending - flapwise shear motion.

For cross-ply laminated beams, the difference between the CUS and CAS configurations becomes immaterial, in the sense that within both of them the same types of elementary motions occur.

These equations correspond to the shearable beam model; the equations for the unshearable beams can be obtained from Eqs. (4.4-43) through (4.4-53), and will not be displayed here.

6.3 THE EIGENVALUE PROBLEM

6.3-1 Preliminary Considerations

Due to the intricacy of the governing equations, and the fact that the boundary conditions contain also the eigenvalue, the determination of a closed-form solution of the eigenvalue problem for anisotropic beams is precluded; we consider numerical solutions. For a discussion on some issues involving this type of problems, see Belinskiy and Dauer (1998).

To reduce the undamped free vibration problem, to an eigenvalue problem, the unknown functions $\big($i.e. $u_0(z,t)$, $v_0(z,t)$, $w_0(z,t)$, $\theta_x(z,t)$, $\theta_y(z,t)$, and $\phi(z,t)\big)$, denoted generically as $\mathcal{F}(z,t)$, are written

$$\mathcal{F}(z,t) = \overline{\mathcal{F}}(z)\exp(i\omega t), i = \sqrt{-1}. \qquad (6.3\text{-}1)$$

The natural frequency ω plays the role of eigenvalue. Two different solutions are used. The first, based on the *Laplace Transform Method* applied in the space domain, was devised by Librescu and Thangjitham (1991), Song and Librescu (1993) and Karpouzian and Librescu (1994, 1996), for nonconservative eigenvalue problems, and by Thangjitham and Librescu (1991) and Song and Librescu (1995), for dynamic response problems to harmonic and arbitrary time dependent external loads. Its application does not require the discretization of the eigenvalue problem; being exact, it provides a benchmark for approximate method.

The other, the *Extended Galerkin Method*, requires the expansion of $\overline{\mathcal{F}}(z)$ in series of admissible functions, that exactly satisfy only the *geometric boundary conditions*.

The terms arising as a result of the non-fulfillment of static boundary conditions remain as residual terms in the energy functional itself, which are then minimized in the Galerkin sense, thus yielding excellent accuracy and rapid convergence. This method was successfully used in various contexts, including determination of aeroelastic instability and aeroelastic response by Librescu et al. (1996), Gern and Librescu (1998a,b), Qin and Librescu (2001, 2002) in the dynamic response to time- dependent external pulses and other problems of thin-walled and solid beams (see Librescu and Song, 1992; Librescu et al., 1996, 1997, 2000, 2003; Song and Librescu, 1993, 1996, 1997a,b; Librescu and Na, 1997, 2001; Na and Librescu, 1998, 2000, 2001a; Song et al., 1992, 1998, 2001, 2002; Oh et al., 2003; Chandiramani et al., 2002).

6.3-2 The Laplace Transform Method (LTM)

We illustrate the eigenfrequency for the free warping counterpart of beams featuring the twist-transversal (flapwise) bending-tranverse shear coupling as obtained from Eqs. (6.2-10) through (6.2-12). Now, the dashed terms and those underscored by dotted lines have, to be discarded. Before applying Laplace transformation, we write the independent field variables in the form

$$\{v_0(z, t)/L, \ \theta_x(z, t), \phi(z, t)\} = \{V(z), \ X(z), \ \Phi(z)\} \exp(i\omega t), \quad (6.3\text{-}2)$$

and cast the governing equation (6.2-10) and boundary conditions in dimensionless form. We use the dimensionless spanwise coordinate $\eta \equiv z/L$, $(0 \le \eta \le 1)$, and the dimensionless parameters $f_i(i = \overline{1,6})$ defined in Table 6.3-1:

Table 6.3-1: Expression of parameters f_i

f_i	Their expression	f_i	Their expression
f_1	$(b_1/a_{55})\,L^2$	f_4	$[(b_4 + \delta_n b_{14})/a_{33}]\,L^2$
f_2	a_{37}/a_{33}	f_5	(a_{37}/a_{77})
f_3	$(a_{55}/a_{33})\,L^2$	f_6	$\{[(b_4 + b_5) + \delta_n(b_{14} + b_{15})]/a_{77}\}\,L^2$

The modified counterpart of governing equations (6.2-10) becomes

$$\Phi_{,\eta\eta} + f_5 X_{,\eta\eta} + \omega^2\, f_6 \Phi = 0,$$
$$X_{,\eta\eta} + f_2 \Phi_{,\eta\eta} - f_3(V_{,\eta} + X) + \omega^2\, f_4 X = 0, \qquad (6.3\text{-}3\text{a-c})$$
$$V_{,\eta\eta} + X_{,\eta} + \omega^2\, f_1 V = 0,$$

while the boundary conditions become

at $\eta = 0$:

$$V = 0, \quad X = 0, \quad \Phi = 0, \tag{6.3-4a-c}$$

and at $\eta = 1$:

$$\Phi_{,\eta} + f_5 X_{,\eta} = 0, \tag{6.3-5a-c}$$
$$X_{,\eta} + f_2 \Phi_{,\eta} = 0,$$
$$V_{,\eta} + X = 0.$$

Here $(\cdot)_{,\eta} \equiv d(\cdot)/d\eta$. Laplace transformation will be carried out with respect to the spanwise coordinate η. Note that the unilateral Laplace transform of the nth-order derivative of a generic function $F \equiv F(\eta)$ is

$$\mathcal{L}[F^{(n)}(\eta)] = s^n \overline{F}(s) - s^{n-1}\hat{F} - s^{n-2}\hat{F}^{(1)} + \cdots - \hat{F}^{(n-1)}, \tag{6.3-6}$$

where

$$\overline{F}(s) \equiv \mathcal{L}[F(\eta)]; \quad F^{(n)}(\eta) \equiv dF^n/d\eta^n,$$
$$\hat{F} \equiv F]_{\eta=0}; \quad \hat{F}^{(n-1)} = dF^{(n-1)}/d\eta^{n-1}]_{\eta=0}. \tag{6.3-7a-d}$$

$\mathcal{L}$ is the Laplace transform operator

$$\mathcal{L}(\cdot) = \int_0^\infty e^{-s\eta}(\cdot)d\eta, \tag{6.3-8}$$

while s is Laplace transform variable. As an example, for $n = 2$, Eq. (6.3-6) reduces to

$$\mathcal{L}[d^2 F/d\eta^2] = s^2 \overline{F}(s) - sF|_{\eta=0} - (dF/d\eta)|_{\eta=0}. \tag{6.3-9}$$

The Laplace transform of the equation system (6.3-3), considered in conjunction with the boundary conditions at $\eta = 0$, Eqs. (6.3-4), yields

$$s^2 f_5 \overline{X} + (s^2 + \omega^2 f_6)\overline{\Phi} = f_5 \hat{X}_{,\eta} + \hat{\Phi}_{,\eta},$$
$$(s^2 - f_3 + \omega^2 f_4)\overline{X} - f_3 s\overline{V} + s^2 f_2 \overline{\Phi} = \hat{X}_{,\eta} + f_2 \hat{\Phi}_{,\eta}, \tag{6.3-10a-c}$$
$$s\overline{X} + (s^2 + \omega^2 f_1)\overline{V} = \hat{V}_{,\eta}.$$

In matrix form, Eqs. (6.3-10) are written as

$$\begin{bmatrix} g_{11} & g_{12} & g_{13} \\ g_{21} & g_{22} & g_{23} \\ g_{31} & g_{32} & g_{33} \end{bmatrix} \begin{Bmatrix} \overline{V} \\ \overline{X} \\ \overline{\Phi} \end{Bmatrix} = \begin{Bmatrix} f_5\hat{X}_{,\eta} + \hat{\Phi}_{,\eta} \\ \hat{X}_{,\eta} + f_2\hat{\Phi}, \eta \\ \hat{V}_{,\eta} \end{Bmatrix}. \tag{6.3-11}$$

The entries g_{ij} are functions of the Laplace variable s, of the eigenfrequency and of physical and geometrical characteristics of the beam. Their expressions result simply by comparing Eqs. (6.3-10) and (6.3-11).

As a next step, we solve for $\overline{V}(s)$, $\overline{X}(s)$, and $\overline{\Phi}(s)$ in Eqs. (6.3-11) and apply the inverse Laplace transform via the use of the partial fraction expansion technique to obtain their counterparts in the physical space. The expressions for $V(\eta)$, $X(\eta)$ and $\Phi(\eta)$ contain as unknown quantities, $\hat{V}_{,\eta}$, $\hat{X}_{,\eta}$ and $\hat{\Phi}_{,\eta}$.

Replacement of $V(\eta)$, $X(\eta)$ and $\Phi(\eta)$ in the boundary conditions at the beam tip, Eqs. (6.3-5), gives a system of three algebraic equations for $\hat{V}_{,\eta}$, $\hat{X}_{,\eta}$ and $\hat{\Phi}_{,\eta}$ evaluated at the beam root.

$$
\begin{bmatrix} B_{11} & B_{12} & B_{13} \\ B_{21} & B_{22} & B_{23} \\ B_{31} & B_{32} & B_{33} \end{bmatrix} \begin{Bmatrix} \hat{V}_{,\eta} \\ \hat{X}_{,\eta} \\ \hat{\Phi}_{,\eta} \end{Bmatrix} = \begin{Bmatrix} 0 \\ 0 \\ 0 \end{Bmatrix}. \tag{6.3-12}
$$

The elements B_{ij} contain ω^2 and also involve, in terms of stiffness quantities, the overall characteristics of the beam. The vector $\hat{\mathbf{V}}^T = \{\hat{V}_{,\eta}, \hat{X}_{,\eta}, \hat{\Phi}_{,\eta}\}$ plays the role of the eigenvector. The condition of nontriviality of $\hat{\mathbf{V}}^T$ requires that $det\ (B_{ij}) = 0$, wherefrom the eigenfrequencies are obtained.

In the context of the restrained twist model of beams given by the original equations (6.2-10) through (6.2-12), a study of the statics and dynamics of thin-walled beams via the use of the Laplace Transform Method was accomplished by Librescu et al. (2003).

6.3-3 Extended Galerkin Method (EGM)

Now the discretization is accomplished directly in the Hamilton's principle functional. Hamilton's principle, Eq. (3.1-1), can be expressed as

$$
\int_{t_0}^{t_1} (\delta \mathcal{K} + \delta \mathcal{J} - \delta \mathcal{W}) dt = 0,
$$

$$
\delta v_0 = 0, \ \delta \theta_x = 0, \ \delta \phi = 0, \ \text{at } t = t_0, \ t_1 \tag{6.3-13}
$$

where $\mathcal{K}$ is the kinetic energy, $\mathcal{W}$ the strain energy, $\mathcal{J}$ the work performed by the external loads, while t_0 and t_1 are two instants of time. $\mathcal{W}$ and $\mathcal{K}$ are provided by Eqs. (3.2-1) and (3.2-3), while for the free vibration problem $\delta \mathcal{J}$ is discarded.

The eigenvalue problem is discretized by representing v_0, θ_x and ϕ in the form:

$$
\begin{aligned}
v_0(z, t) &= \mathbf{V}_0^T(z)\mathbf{q}_v(t), \\
\theta_x(z, t) &= \mathbf{X}_0^T(z)\mathbf{q}_x(t), \\
\phi(z, t) &= \mathbf{\Phi}^T(z)\mathbf{q}_\phi(t),
\end{aligned} \tag{6.3-14a-c}
$$

where $\mathbf{V}_0$, $\mathbf{X}_0$ and Φ are vectors of suitable trial functions, while $\mathbf{q}_v$, $\mathbf{q}_x$ and $\mathbf{q}_\phi$ are vectors of generalized coordinates. Now, the kinetic and strain energies in discrete form are

$$\mathcal{K} = \frac{1}{2}\dot{\mathbf{q}}^T \mathbf{M}\,\dot{\mathbf{q}} \ \ \text{and} \ \ \mathcal{W} = \frac{1}{2}\mathbf{q}^T \mathbf{K}\,\mathbf{q}. \tag{6.3-15a,b}$$

Here

$$\mathbf{q}(t) = \{\mathbf{q}_v^T, \ \mathbf{q}_x^T, \ \mathbf{q}_\phi^T\}^T, \tag{6.3-16}$$

is the $3N \times 1$ overall generalized coordinate vector, while $\mathbf{M}$ and $\mathbf{K}$ denote the mass and the stiffness matrix, respectively.

For the present case these are

$$\mathbf{M} =$$

$$\int_0^L \begin{bmatrix} b_1 \mathbf{V}_0\, \mathbf{V}_0^T & \mathbf{0} & \mathbf{0} \\ & (b_4 + \delta_n b_{14})\mathbf{X}_0\, \mathbf{X}_0^T & \mathbf{0} \\ Symm. & & [(b_4 + b_5) + \delta_n(b_{14} + b_{15})]\Phi\Phi^T \\ & & + (b_{10} + \delta_n b_{18})\Phi'(\Phi')^T \end{bmatrix} dz,$$

$$(6.3\text{-}17a,b)$$

$$\mathbf{K} =$$

$$\int_0^L \begin{bmatrix} a_{55}\mathbf{V}_0'\,(\mathbf{V}_0')^T & a_{55}\mathbf{V}_0'\mathbf{X}_0^T & -a_{56}\mathbf{V}_0'(\phi'')^T \\ & a_{55}\mathbf{X}_0\mathbf{X}_0^T + a_{33}\mathbf{X}_0'(\mathbf{X}_0')^T & (a_{37} + a_{56})\mathbf{X}_0'(\Phi')^T \\ & & -a_{56}\mathbf{X}_0(\Phi')^T|_{z=L} \\ Symm. & & a_{77}\Phi'(\Phi')^T \\ & & + a_{66}\Phi''(\Phi'')^T \end{bmatrix} dz.$$

Inserting Eqs. (6.3-15) considered in conjunction with Eqs. (6.3-17) and the proper representation of trial functions in Eq. (6.3-14), integrating with respect to time and recognizing that $\delta\mathbf{q} = \mathbf{0}$ at $t = t_0$, t_1, one obtains the discretized equations of motion

$$\mathbf{M}\ddot{\mathbf{q}}(t) + \mathbf{K}\mathbf{q}(t) = 0. \tag{6.3-18}$$

Expressing

$$\mathbf{q}(t) = \mathbf{x}e^{i\omega t}, \tag{6.3-19}$$

we get from (6.3-18) the eigenvalue problem

$$\mathbf{A}\mathbf{x} - \lambda\mathbf{x} = 0, \tag{6.3-20}$$

where

$$A = M^{-1}K, \tag{6.3-21}$$

and $\lambda = \omega^2$. x and λ denote the eigenvector and the eigenvalues, respectively, while ω is the natural frequency.

Hamilton's principal provides both the governing equations of motion and the boundary condition: the trial functions have to satisfy only the kinematic boundary conditions (i.e. the ones at the beam root). An alternative way of addressing this solution is to express the generalized displacement quantities in the form of Eq. (6.3-1); the variational equation (6.3-13), where $\delta \mathcal{J}$ was discarded, can be cast as

$$\int_0^L [(6.2\text{-}10a)\delta\overline{\Phi} + (6.2\text{-}10b)\delta\overline{\theta}_x + (6.2\text{-}10c)\delta\overline{v}_0]dz$$
$$+ [(6.2\text{-}11a)\delta\overline{\Phi} + (6.2\text{-}11b)\delta\overline{\Phi}' + (6.2\text{-}11c)\delta\overline{v}_0 \tag{6.3-22}$$
$$+ (6.2\text{-}11d)\delta\overline{\theta}_x]|_{z=L} = 0.$$

In this equation the displacement quantities, affected by an overbar are functions of the spanwise beam coordinate z only, and are defined by Eq. (6.3-1). They are selected to satisfy the kinematic boundary conditions i.e. those at the beam root. These *admissible functions* (see Meirovitch, 1997) are taken as polynomials in the z-coordinate. The numbers in brackets in (6.3-22) identify the left-hand side expressions of the equations of motion and boundary conditions at $z = L$, modified by transferring to the left-hand side the right-hand side terms, using Eq. (6.3-1), and suppressing the common factor $exp(2i\omega t)$.

Due to the non-fulfilment of the boundary conditions at the beam tip ($z = L$), residual terms, instead of zero-valued ones appear within the functional of (6.3-13) (or more explicitly in Eq. (6.3-22)).

Expressing the displacement quantities as

$$[\overline{\Phi}(z), \; \overline{\theta}_x(z), \; \overline{v}_0(z)] = \sum_{j=1}^{N}[a_j F_j(z), \; b_j X_j(z), \; c_j V_j(z)], \tag{6.3-23}$$

$F_j(z)$, $X_j(z)$ and $V_j(z)$ being suitable trial functions and a_j, b_j and c_j arbitrary coefficients, by virtue of (6.2-23) we also have

$$[\delta\overline{\Phi}, \; \delta\overline{\theta}_x, \; \delta\overline{v}_0] = \sum_{j=1}^{N}[F_j(z)\delta a_j, \; X_j(z)\,\delta b_j, \; V_j(z)\delta c_j]. \tag{6.3-24}$$

Substituting Eqs. (6.3-23) and (6.3-24), in Hamilton's principle, or equivalently in (6.3-22), carrying out the required spanwise integration and applying the fundamental lemma of the calculus of variation, lead to the following matrix equation

$$\mathcal{L}\mathbf{A} = 0. \qquad (6.3\text{-}25)$$

Here $\mathcal{L}$ is a $3N \times 3N$ matrix whose elements contain ω^2, while, $\mathbf{A} = \{a_1, a_2 \ldots a_N, b_1, b_2 \ldots \ldots b_N, c_1, c_2, \ldots c_N\}^T$ is a $3N \times 1$ arbitrary vector that plays the role of an eigenvector.

The condition of nontriviality of $\mathbf{A}$ requires that $det(\mathcal{L}) = 0$. This condition provides the natural frequencies. Note that EGM was largely used by these authors; see also Na and Librescu (1998, 2000, 2001).

6.4 RESULTS

6.4-1 General Considerations

The goal of the numerical simulations is to assess the various non-classical effects on the natural frequencies of thin-walled beams. Two structural systems will be considered: i) a cantilevered thin-walled beam of rectangular cross-section (box-beam) with the CUS and ii) a cantilevered beam of a bi-convex cross-section with the CAS configuration. The former ply-angle configuration is preferred in the design of helicopter blades and tilt rotor aircraft (see e.g. Lake (1990), Lake et al. (1992, 1993), Kosmatka and Lake (1996); the latter one is most important in the design of aircraft wings. In fact, due to the bending-twist elastic coupling generated via the incorporation of fibrous composite materials it was possible to eliminate, without weight penalties, the divergence instability featured by sweptforward aircraft wings.

The two structural systems are represented in Figs. 6.4-1 and 6.4-2, and their geometrical characteristics are provided in Table 6.4-1. The material properties of the graphite-epoxy composite, are displayed in Table 6.4-2.

Table 6.4-1: Dimensions of thin-walled beam structures as considered in the next numerical simulations

L [m]	c [m]	b [m]	$\mathcal{R}(\equiv b/c)$	h [m]
2.032	0.254	0.0508(CUS)	0.2 (CUS)	0.01
		0.068 (CAS)	0.268 (CAS)	

Table 6.4-2: Material properties

E_1 [GPa]	E_2 [GPa]	E_3 [GPa]
206.75	5.17	5.17
G_{12} [GPa]	G_{13} [GPa]	G_{23} [GPa]
3.10	2.55	2.55
$\nu_{21} = \nu_{31}$	ν_{32}	ρ [kg/m³]
0.00625	0.25	1528.15

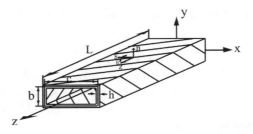

Figure 6.4-1: Geometry of a thin-walled box-beam featuring CUS configuration

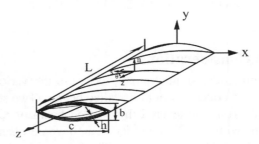

Figure 6.4-2: Geometry of a thin-walled bi-convex cross-sections beam featuring CAS configuration

6.4-2 Results for the CUS Beam Configuration

The beneficial effect mentioned above is due to the fact that the increase of the ply-angle induces an increase of the bending stiffnesses a_{33} and a_{22}. Figures 6.4-3 and 6.4-4 show that at $\theta = 90$ deg., the bending stiffnesses reach maximum values.

Figure 6.4-5 displays the effect of ply-angle on the fundamental eigenfrequencies of the lateral and transversal bending vibration modes. The results incorporate both the warping (secondary) and transverse shear effects.

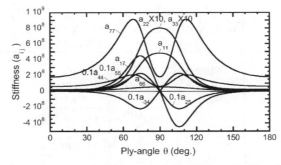

Figure 6.4-3: Variation of a number of stiffness quantities of thin-walled box-beam vs. ply-angle, CUS configuration

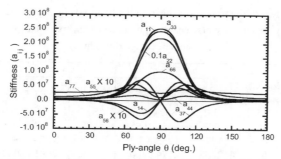

Figure 6.4-4: Variation of a number of stiffness quantities of thin-walled beam of a biconvex cross-section vs. ply-angle, CAS configuration

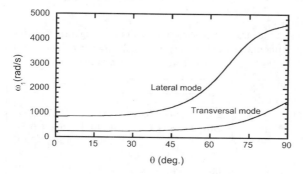

Figure 6.4-5: Variation of lateral and transverse fundamental eigenfrequencies vs. ply-angle

The results show that, for larger ply-angles, higher bending eigenfrequencies are obtained. This fact is justifiable by comparing the magnitudes of the bending stiffnesses for various ply-angles. Figure 6.4-5 also reveals that, corresponding to the same ply-angle, the eigenfrequencies of the lateral bending mode are higher than those of the transversal bending vibrational mode. These trends can be predicted by comparing the bending stiffnesses for each bending vibration mode and those corresponding to various ply-angles (see Fig. 6.4-3). For the specific geometry of the cross-section characterized by $\mathcal{R}(\equiv b/c) = 0.2$, the lateral bending stiffness is much larger than the transverse one.

Tables 6.4-3 and 6.4-4 present results highlighting the influence of transverse shear on the first two eigenfrequencies associated with the lateral and transverse bending vibrational modes, respectively.

Table 6.4-3: Effect of the transverse shear on the first two eigenfrequencies of the lateral bending mode: I, without transverse shear; II, with the transverse shear (values in parentheses denote percentage difference)

Ply-angle	ω_1 (rad/s)		ω_2 (rad/s)	
θ (deg)	I	II	I	II
0	847	843 (-0.5)	5226	4193 (-19.8)
15	861	857 (-0.5)	5317	4256 (-20.0)
30	943	935 (-0.85)	5823	4563 (-21.6)
45	1276	1235 (-3.2)	7875	54 24 (-31.1)
60	2388	2174 (-9.0)	14742	7491 (-49.2)
75	4593	3908 (-14.9)	28351	12273 (-56.7)
90	5355	4580 (-14.5)	33054	18057 (-45.4)

Table 6.4-4: Effect of the transverse shear on the first two eigenfrequencies for the transverse bending mode: I, without transverse shear; II, with the transverse shear (values in parentheses denote percentage difference)

Ply-angle	ω_1 (rad/s)		ω_2 (rad/s)	
θ (deg)	I	II	I	II
0	242	241 (-0.4)	1512	1507(-0.3)
15	246	245 (-0.4)	1540	1530 (-0.6)
30	272	263 (-3.3)	1704	1642 (-3.6)
45	372	314 (-15.6)	2330	1962 (-15.8)
60	689	440 (-36.1)	4313	2764 (-35.9)
75	1312	761 (-42.0)	8210	4913 (-40.2)
90	1527	1499 (-1.8)	9561	8503 (-11.1)

These results show the significant effect of transverse shear decreasing natural frequencies. This implies that discarding transverse shear for a beam whose material is weak in transverse shear (that is equivalent to adopting the classical unshearable beam model), leads to an overprediction of eigenfrequencies.

With the increase of θ, the effect of transverse shear becomes more prominent: there are larger differences between predictions (I) and (II). This trend, remarked also in another context (see Chandiramani et al., 2002), is due to the increase of bending stiffnesses that is associated with the increase of θ, consequently, of the relative increase of transverse shear flexibility when comparing to that at $\theta = 0$.

At this point it should be remarked that the first two natural frequencies of the lateral and transversal bending modes as obtained by Song and Librescu (1993) and reproduced in Tables 6.4-3 and 6.4-4 were compared with the ones obtained via the FEM by Ko and Kim (2003), and a good agreement was reached. Table 6.4-5 compares the first two eigenfrequencies of the torsional mode obtained within the free warping (FW) and warping restraint (WR) beam models, in the case of the ply-angle $\theta = 90$ deg..

Table 6.4-5: Eigenfrequencies of the torsional mode obtained within FW and WR beam models ($\theta = 90$ deg.) (values in parentheses denote percentage differences)

	ω_1 (rad/s)	ω_2 (rad/s)
FW	4307	12918
WR	4726 (9.7)	15090 (16.8)

The results reveal that the warping restraint yields an increase of eigenfrequencies in the torsional vibration mode and that its effect is stronger in the higher modes than in the lower ones. Table 6.4-6 displays the variation of the first three eigenfrequencies in the axial vibration mode for two selected ply-angles. The results reveal large variations of eigenfrequencies as a function of the ply-angle.

Table 6.4-6: Eigenfrequencies of the axial vibration mode for two selected ply-angles (values in parentheses denote percentage differences)

Ply-angle θ (deg)	ω_1 (rad/s)	ω_2 (rad/s)	ω_3 (rad/s)
0	11376	34127	56879
90	71947 (532.0)	215841 (532.0)	359735 (532.0)

Figure 6.4-6 shows the coupled effects of the warping restraint and ply-angle on the fundamental extension-torsion coupled eigenfrequency.

For the free warping beam model, the frequency features a symmetric variation with respect to $\theta = 45$ deg. In contrast to this trend, when warping is included, the peak value of the frequency reached at $\theta = 45$ deg. is not followed by an abrupt decay as for the free-warping case, but by a more moderate one. This trend is due to the warping stiffness whose effect becomes more prominent in that ply-angle range.

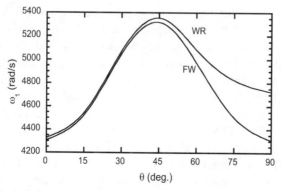

Figure 6.4-6: Effect of the warping restraint on the fundamental extension-torsion coupled eigenfrequency

Table 6.4-7: Variation of the extension-twist coupled eigenfrequencies vs. ply-angle by including (II), and discarding (I) the secondary warping effect (SWE)

Ply-angle	ω_1 (rad/s)		ω_2 (rad/s)	
θ (deg)	I	II	I	II
0	4328	4328	11376	11376
15	4543	4544	11580	11580
30	5070	5072	12773	12773
45	5351	5355	16097	16114
60	5079	5085	15405	15437
75	4820	4827	15117	15155
90	4719	4726	15051	15090

For the box-beam structure featuring the CUS configuration, Table 6.4-7 exhibits the coupled influence played by both the material ply-angle and the secondary warping on the fundamental and second extension-twist coupled eigenfrequencies in bending.

Whereas the increase of the ply-angle in the range $\theta \leq 45$ deg., yields a beneficial effect reflected in the increase of both eigenfrequencies, the inclusion of the secondary warping (implying consideration in the relevant equations of $\delta_n = 1$), results only in a slight increase of natural frequencies as compared to the case when this effect is discarded. However in the ply-angle range 45 deg. $< \theta \leq 90$ deg., a decrease of the eigenfrequencies is experienced, whereas the increase due to the SWE is marginal. In spite of this, it is believed that for special cross-sections for which the primary warping becomes immaterial, larger contributions of the secondary warping are likely to occur. Figures 6.4-3 and 6.4-4 help to explain the variation of eigenfrequencies as a function of the ply-angle.

6.4-3 Results for the CAS Beam Configuration

Figure 6.4-7a displays the first three eigenfrequencies as a function of the ply angle θ with the bending-twist coupling included, ($a_{37} \neq 0$), and discarded, ($a_{37} = 0$). In both cases, the frequency plots versus θ feature symmetry about $\theta = 90$ deg., and experience as many peaks as the eigenfrequency mode number. When $a_{37} = 0$, the second eigenfrequency has a local maximum near $\theta = 75$ deg., and $\theta = 105$ deg., and at these ply angles it is close to the third eigenfrequency. On the other hand, when the effect of bending-twist coupling is included, ($a_{37} \neq 0$), Fig. 6.4-7a reveals that frequency near merging is precluded. This phenomenon was also observed for a solid beam model (Weisshaar and Foist, 1985). The frequency of the fundamental mode has a maximum at $\theta = 90$ deg. At $\theta = 0$ (and 180 deg.), where the bending and twist become decoupled, the fundamental mode is a pure bending mode (denoted by B).

The frequency associated with the second mode first increases for $0 < \theta \leq 80$ deg. and then decreases for 80 deg. $\leq \theta \leq 90$ deg., the variation being symmetric about $\theta = 90$ deg. Another notable trend occurs when the cross-coupling stiffness is discarded; the second eigenfrequency is overestimated as compared to the real case that incorporates the cross-coupling stiffness. At $\theta = 0$ (or $\theta = 180$ deg.) and $\theta = 90$ deg., where the decoupling occurs, the second mode can be identified as the second pure bending mode and the first pure torsional mode, respectively, where the latter is denoted by T.

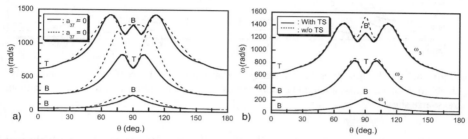

Figure 6.4-7: The first three bending-twist eigenfrequencies versus ply-angle. a) With and without transverse bending-twist elastic coupling a_{37}, b) With and without transverse shear

For the third eigenfrequency, a more complex variation with the ply angle is observed. Indeed, the variation about $\theta = 90$ deg. is drastically attenuated when the cross coupling is ignored. At $\theta = 0$ (or $\theta = 180$ deg.) this mode can be identified as the first pure torsional mode and second pure bending mode, respectively.

Figure 6.4-7b displays the first three transverse bending-twist coupled eigenfrequencies as functions of the ply-angle for the cases where transverse shear effects are incorporated and ignored. Clearly, for ply-angles for which bending is dominant, omission of the transverse shear causes an overestimation of natural frequencies.

Figure 6.4-8a contains plots of the first three lateral bending-extension eigenfrequencies versus ply-angle θ; the lateral bending-extension cross-coupling parameter a_{14} plays a similar role to the bending-twist stiffness a_{37}. However, these three eigenfrequencies are well separated within the entire range of ply-angles θ and, in addition, removal of a_{14} does not cause the second and third eigenfrequencies to approach each other, regardless of the value of θ. The plots are symmetric about $\theta = 90$ deg. and, as expected, the lateral eigenfrequencies are much larger than their transverse counterparts. At $\theta = 90$ deg. the first three modes are identified as the first three pure lateral bending modes, denoted by B, and at $\theta = 0$ and $\theta = 180$ deg. as the first two pure lateral bending modes and the first pure axial mode, where the latter is denoted by A.

Figure 6.4-8b portrays the first three coupled lateral bending-extension eigenfrequencies versus ply-angle θ when transverse shear effect is included and discarded. The comparison of Figs. 6.4-7b and 6.4-8b reveals that the overestimation of eigenfrequencies when the transverse shear is discarded is more pronounced in the lateral than in the transverse bending vibration mode.

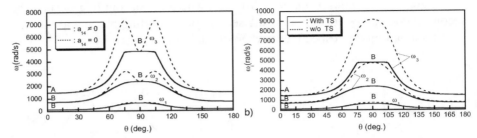

Figure 6.4-8: The first three lateral bending-extension coupled eigenfrequencies versus ply-angle. a) With and without lateral bending-extension elastic coupling a_{14}, b) With and without transverse shear

   Table 6.4-8 shows the effect of rotatory inertia effect coupled with that of the ply-angle on the first three bending-twist coupled eigenfrequencies; while the fundamental eigenfrequency is not affected at all, the higher mode eigenfrequencies are marginally decreased.

   Notice that here the frequencies of the first branch of the frequency spectrum have been considered. However, as is well-known (see Mindlin, 1951; Librescu, 1975), while the neglect of rotatory inertia effects is justifiable in the case of the first branch of the frequency spectrum, for determination of the higher frequency branches their preservation is essential.

Table 6.4-8: The first three mode eigenfrequencies evaluated with and without rotatory inertia (RI) effect, for selected ply-angles. The numbers in brackets indicate the percentage difference when rotatory inertias are neglected

Fiber orientation θ (deg)	ω_1 (rad/s)		ω_2 (rad/s)		ω_3 (rad/s)	
	with RI	without RI	with RI	without RI	with RI	without RI
0	40.3	40.3 (0%)	250.6	251.2 (.2%)	625.6	625.6 (0%)
30	43.5	43.5 (0%)	270.1	270.7 (.2%)	746.7	750.9 (.6%)
45	52.4	52.4 (0%)	324.4	325.2 (.2%)	892.0	896.8 (.5%)
60	77.5	77.5 (0%)	472.7	473.9 (.3%)	1270	1275 (.4%)
75	136	136 (0%)	773	776 (.4%)	1333	1333 (0%)
90	239	239 (0%)	640	640 (0%)	1280	1282 (.2%)

   Figure 6.4-9 presents the bending and twist in the three lowest modes as a function of the normalized position $\eta = z/L$, for $\theta = 45$ deg. The eigenmodes are normalized so that the value at the tip is equal to unity. For other ply-angles and for lateral bending, similar plots were obtained, but are not displayed here. Note that the position of the nodal points changes with θ.

This trend, coupled with large variations of these modes with θ, is likely to have a significant effect on the dynamic behavior of these structures.

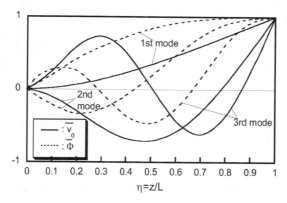

Figure 6.4-9: The first three normalized transverse bending and twist eigenmodes, $\theta = 45$ deg.

6.5 FREE VIBRATION OF NON-UNIFORM CROSS-SECTION BEAMS

6.5-1 Preliminaries

We have studied non-classical effects on uniform beams. An important effect that was not yet been included is related to the nonuniformity of the beam cross-section. Even though this feature constitutes a basic characteristic encountered in many advanced structural systems, such as aircraft wings, rotor blades, satellite antennas, etc. it appears that the specialized literature is quite void of investigations related to *nonuniform thin-walled beams.*

There are a few results for non-uniform composite beams Karpouzian and Librescu (1994, 1996). Results associated with the vibration of nonuniform solid beams can be found e.g. in the papers by Housner and Keightley (1962), Downs (1977) and the monograph by Rao (1991). For non-uniform thin-walled beams, the only results related to the *free vibration problem* are those by Librescu et al. (1993a) and Na and Librescu (1999), Na (2001b).

6.5-2 Basic Assumptions

We consider single-cell thin-walled beams with geometrical characteristics varying linearly along the beam span. In addition to the cross-section nonuniformity, the beam model incorporates transverse shear, warping inhibition, anisotropy of constituent materials. As in Section 6.4-1, a biconvex profile of the beam cross-section will be considered in the numerical simulations. A sketch of the beam is presented in Fig. 6.5-1.

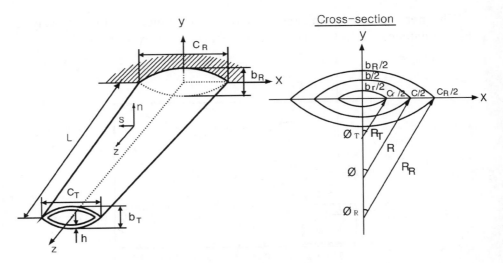

Figure 6.5-1: Geometry of a doubly-tapered thin-walled beam.

Consider (Librescu et al., 1993a), the following linear variation along the beam span of the chord $c(z)$ and height $b(z)$ of the mid-line cross-section profiles:

$$\left\{ \begin{array}{c} c(\eta) \\ b(\eta) \end{array} \right\} = [1 - \eta\,(1 - \sigma)] \left\{ \begin{array}{c} c_R \\ b_R \end{array} \right\}. \tag{6.5-1}$$

Herein $\sigma \equiv c_T/c_R$ (= b_T/b_R) defines the taper ratio, $0 \le \sigma \le 1$; $\sigma = 1$ and $\sigma = 0$, correspond to the two extreme taper ratios, that is to the uniform and triangular beams, respectively; $\eta \equiv z/L$ is the dimensionless spanwise coordinate, and superscripts R and T refer to the beam-root and tip cross-section, respectively. The radius of curvature of the circular arc associated with the midline contour at section η along the beam span is expressed as

$$R(\eta) = [1 - \eta(1 - \sigma)]R_R. \tag{6.5-2}$$

If $\sigma > 1$ the taper is toward the beam root.

Due to the geometrical similarity of the cross-sections, the angles ϕ between the axis y and the radii of curvature of the arcs at points $c_R/2$, $c/2$ and $c_T/2$ are equal, i.e. $\phi_R = \phi_T = \phi \equiv \phi_0$, (see Fig. 6.5-1). As before, the points on the beam cross-sections are identified by the *global* coordinates x, y, z, where z is the spanwise coordinate, and by the local ones s, n, z. As a result of the cross-section nonuniformity, the primary warping function $\overline{F}$ (see Eq. 2.1-44a), becomes a function of both s and η:

$$\overline{F}(s, \eta) = \int_0^s [r_n(\overline{s}, \eta) - \psi(\overline{s}, \eta]d\overline{s}, \tag{6.5-3a}$$

where

$$\psi(s, \eta) = \frac{\oint r_n(\hat{s}, \eta)d\hat{s}}{h(s)G_{sz}(s)\oint d\hat{s}/(h(\hat{s})G_{sz}(\hat{s}))}, \tag{6.5-3b}$$

denotes the torsional function.

By virtue of (6.5-3), even for a beam subjected to pure torques acting at its extreme ends, free warping is precluded.

Straightforward geometrical considerations show that

$$\Omega(\eta) = R^2(\eta)(2\phi_0 - \sin 2\phi_0), \tag{6.5-4}$$

and

$$\beta(\eta) = 4R(\eta)\phi_0, \tag{6.5-5}$$

where ϕ_0 is a prescribed angle, we take $\phi_0 = \pi/6$. Note that since $x = x(s, \eta)$ and $y = y(s, \eta)$, also r_n and r_t will depend on s and η.

In the context of the associated governing equations, all stiffness quantities a_{ij} and inertia coefficients b_i depend on η, i.e. $a_{ij} = a_{ij}(\eta)$, and $b_i = b_i(\eta)$. In spite of this, their general expressions remain similar to those displayed in Table 4.1. With this in mind, the governing equations for the two main ply-angle configurations, CUS and CAS, are provided by Eqs. (4.4-14) through (4.4-19), and by Eqs. (4.4-26) through (4.4-31), respectively.

The spanwise dependence of stiffness quantities, forces us to use the Extended Galerkin Method.

6.5-3 A Few Numerical Results

To show the effect of taper ratio coupled with the transverse shear on eigenfrequencies, we will restrict the analysis to a biconvex cross-section beam profile. In addition to the assumptions stated in Section 6.2, we use those on the beam nonuniformity in the sense of Eq. (6.5-1). The material of the beam is assumed to feature transverse-isotropy, the surface of isotropy being parallel to the beam mid-surface.

Table 6.5-1 (Librescu et al., 2000), record the variation of the first eight natural frequencies in pure transverse bending as a function of the taper parameter σ, for selected values of the transverse shear flexibility ratio E/G'. The geometrical beam characteristics correspond to $L = 80$in. (2m), and $h = 0.4$in. (0.01m). The taper ratio affects the frequencies of each mode somewhat differently, with the first mode behavior in sharp contrast to the other modes.

Table 6.5-1: Variation of the first eight natural frequencies ω_i (rad/s) against the taper ratio σ and of transverse shear flexibility E/G'

E/G'	0			10			100		
σ	0.5	0.8	1.0	0.5	0.8	1.0	0.5	0.8	1.0
ω_1	269	253	246	268	253	246	267	252	245
ω_2	1290	1443	1543	1275	1422	1512	1247	1382	1462
ω_3	3292	3907	4323	3229	3777	4110	3081	3553	3825
ω_4	6324	7505	8135	6091	7089	7708	5607	6427	6919
ω_5	10913	13627	14473	10069	11830	12691	8930	10123	10743
ω_6	19986	21952	24124	16234	18187	19517	13305	14751	15436
ω_7	51978	70379	80949	37225	41032	42065	26751	27847	28119
ω_8	124322	127534	129385	55691	63530	67095	40205	45384	47633

The special case $E/G' = 0$, corresponds to the Bernoulli-Euler unshearable beam model. Similar trends have been obtained for solid beams by Downs (1977). The results also reveal that for a fixed taper ratio, the unshearable beam model overestimates the frequencies (see Na and Librescu, 1999).

6.5-4 Validation of the Accuracy of the Extended Galerkin Method

First to assess the convergence and accuracy of this method, we apply it to the cantilevered uniform thin-walled beam discussed by Hashemi and Richard (2000). In Table 6.5-2, comparisons of the first six eigenfrequencies are carried out. The exact solution (indicated as Exact), as well as results predicted by finite element method (FEM), dynamic finite element method (DFE), and by the present one, (EGM), are provided. The mode shapes obtained by Qin and Librescu (2002) via EGM agree well with those in Hashemi and Richard (using the DFE method). These are not presented here. Only simple polynomials (z^i, $i = 1, 2, 3, \ldots, 7$) are used in the EGM as admissible functions.

Table 6.5-2: Accuracy of the Extended Galerkin's Method (EGM) eigenfrequencies ω_i [rad/s]

Mode No.	"Exact"[a]	DFE[b]	Error[c]	FEM[d]	Error[c]	Present[e]	Error[c]
1	49.62	49.62	0.00(%)	49.62	0.00(%)	49.61	0.02(%)
2	97.04	97.05	0.01(%)	97.04	0.00(%)	97.04	0.00(%)
3	248.87	249.00	0.05(%)	248.88	0.04(%)	248.87	0.00(%)
4	355.59	357.54	0.55(%)	355.59	0.00(%)	355.59	0.00(%)
5	451.46	452.57	0.25 (%)	451.48	0.004(%)	451.53	0.02(%)
6	610.32	610.63	0.05(%)	610.39	0.01(%)	613.48	0.52(%)

[a] These data are provided in paper by H.&R. [Hashemi and Richard (2000)] and obtained via the DSM as developed by Banerjee (1989)

[b] Six dynamic elements are used in the paper by H. & R.

[c] Relative error, $\dfrac{[\text{approximate}] - [\text{exact}]}{[\text{exact}]} \times 100\%$.

[d] 200 static beam elements are used (H. & R.) in conjunction with FEM (Hallauer and Liu, 1982;

[e] Qin and Librescu (2002)

Clearly, the convergence and accuracy are excellent. Also, to provide accurate and robust numerical solutions, Cholesky decomposition technique (see Meirovitch, 1997) for matrix inversion was adopted.

Further validations of dynamic characteristics of thin walled beams are based on data provided in the papers by Chandra and Chopra (1992) and Armanios and Badir (1995), that are not repeated here. The results predicted by the present model are listed in Table 6.5-3; they show a reasonable agreement with the experimental data. Notice that lower natural frequencies for the CAS layups are obtained by Armanios and Badir (1995), even though transverse shear flexibility was not considered in their model.

Table 6.5-3: Dynamic validation: comparison of theoretical and experimental natural frequencies (unit: Hz)

Lay-up	Mode No.	Exp^a. (C & C)	$Analytical^b$ (A & B)	$Error^c$	$Present^d$	$Error^c$
$[30]_6$ CAS	1	20.96	19.92	-4.96%	21.8	4.00%
	2	128.36	124.73	-2.83%	123.28	-3.96%
$[45]_6$ CAS	1	16.67	14.69	-11.88%	15.04	-9.78%
	2	96.15	92.02	-4.30%	92.39	-3.91%
$[15]_6$ CUS	1	28.66	28.67	0.03%	30.06	4.88%
$[0/30]_3$ CUS	1	30.66	34.23	11.7%	34.58	12.79%
$[0/45]_3$ CUS	1	30.0	32.75	9.1%	32.64	8.80%

[a] The experimental results are extracted from the paper by C. & C. (Chandra and Chopra, 1992).
[b] Results listed in the paper by A. & B. (Armanios and Badir, 1995).
[c] Relative error, $\frac{[analytical]-[experimental]}{[experimental]} \times 100\%$.
[d] Qin and Librescu (2002).

In the context of the aeroelastic flutter of aircraft wings modeled as solid beams a comparison of predictions via Laplace Transform and EGM, was carried out and an excellent agreement of results was obtained (see Karpouzian and Librescu, 1994b). For similar results obtained in the context of the dynamic response, see Song et al. (1998).

Having in view that the predictions by LTM are exact, comparisons of natural frequencies obtained via EGM and LTM (see Librescu et al., 2003) are provided in Table 6.5-4.

Table 6.5-4: Comparisons of dimensionless natural frequency predictions $\hat{\omega}_i \, (= \omega_i/\overline{\omega})^+$ via EGM and LTM of a beam featuring the CAS configuration

Mode No.	EGM	LTM	Error of EGM (%)
1	3.465	3.465	0.0
2	7.482	7.481	0.01
3	20.063	20.047	0.08
4	23.071	23.065	0.03
5	40.433	30.418	0.04
6	50.625	50.548	0.15
7	60.560	60.477	0.14

[+] The normalizing $\overline{\omega} = [\hat{a}_{33}(\theta = 90 \text{ deg.})/b_1 L^4]^{1/2}$

The results displayed are for a bi-convex cantilevered beam whose layers disposed in the sequence [90₆] are of a graphite-epoxy material. The geometrical characteristics are: $b = 0.997 \times 10^{-1}$ m, $c = 0.75$m, $t_{(k)} = 0.34 \times 10^{-2}$m being the wall ply thickness. The frequency predictions are based on Eqs. (6.2-10) through (6.2-12), the warping restraint effect being included.

The results in Table 6.5-4 reveal an excellent agreement of the EGM predictions with the exact ones.

6.6 ADAPTIVE/SMART THIN-WALLED BEAMS

6.6-1 Introduction

Adaptive/smart structure technology is an emerging engineering field, that deals with structures that are characterized by self-adaptability, self- sensing and control capabilities.

Such structural systems represent a dramatic evolution of traditional passive elastic systems to actively controllable ones. This field has evolved over the past decades and is likely to become one of the key technologies of our century.

The new high performance piezoelectric polymeric and piezoelectric ceramic composites, are likely to play a major role in the design of advanced aeronautical and aerospace structural systems, helicopter rotor blades, robot manipulator arms, etc. The new generation of structural systems have to be designed to operate safely in complex environmental conditions.

Much research into this technology has been carried out in the last decade; see the survey papers by Crawley (1994), Loewy (1997), Sunar and Rao (1999) and Chopra (2002), as well as the monograph by Tzou (1993).

In contrast to the standard case where the structures are characterized by fixed natural frequencies and mode shapes, in an adaptive structure these dynamic characteristics can be controlled in a known and predictable manner. This control enables one to avoid structural resonance and of any other dynamic instability, and improve response to time-dependent external excitation. The free vibration behavior of adaptive cantilevered structures, featuring the capability referred to as the *induced strain actuation* will be considered next.

6.6-2 Piezoelectrically Induced Bending Moments. Piezoactuator Location

In order to express the 1-D piezoelectrically induced bending moments, the expressions of $\tilde{N}_{zz}$ and $\tilde{L}_{zz}$ provided by Eqs. (A.103a,b) will be used; replacement of (A.103a,b) in M_x and M_y as given by Eqs. (3.1-17d,e), provides the piezoelectrically induced counterpart of these quantities. These are denoted as

$\tilde{M}_x$ and $\tilde{M}_y$:

$$\tilde{M}_x = \int \mathcal{E}_3(n_+ - n_-)\bar{e}_{31} R(s, z)\left[y(1 - \frac{A_{12}^*}{A_{11}^*}) + \frac{dx}{ds}\frac{B_{12}^*}{A_{11}^*}\right]ds$$
$$- \frac{1}{2}\int\left[\frac{dx}{ds}\mathcal{E}_3(n_+^2 - n_-^2)\bar{e}_{31} R(s, z)\right]ds,$$

and

(6.6-1a,b)

$$\tilde{M}_y = \int \mathcal{E}_3(n_+ - n_-)\bar{e}_{31} R(s, z)\left[x(1 - \frac{A_{12}^*}{A_{11}^*}) + \frac{dy}{ds}\frac{B_{12}^*}{A_{11}^*}\right]ds$$
$$+ \frac{1}{2}\int\left[\frac{dy}{ds}\mathcal{E}_3(n_+^2 - n_-^2)\bar{e}_{31} R(s, z)\right]ds,$$

where n_+ and n_- denote the upper and lower faces of a constituent layer, respectively. A_{ij}^* and B_{ij}^* also denote the local stiffness quantities associated with the piezoactuators.

Equations (6.6-1) reveal that the piezoelectrically induced bending moments are proportional to the applied electric voltage $\mathcal{E}_3$.

For future use, Eqs. (6.6-1a,b) are expressed in a condensed form as

$$\{\tilde{M}_x, \ \tilde{M}_y\} = \mathcal{E}_3 \{\mathcal{C}_{Mx}, \ \mathcal{C}_{My}\},$$

(6.6-1c,d)

The global bending moments M_x and M_y can be cast as

$$M_x = \hat{M}_x - \tilde{M}_x, \quad M_y = \hat{M}_y - \tilde{M}_y,$$

(6.6-2a,b)

where the overcaret (^) and overtilde (~) identify the pure mechanical and piezoactuation contributions. In the sequel, the over caret marks will be omitted

For piezoactuators location, two options, Case (a) and Case (b), will be considered. In Case (a) the piezoactuator is a single patch, distributed somewhere along the beam span, and within Case (b) the piezoactuator is spread over the entire beam span. In the latter case, one assumes that the electrical current $\mathcal{E}_3$ is uniform in space, but dependent on time t. In both instances, the actuators are modeled as symmetric pairs mounted on the top and bottom and on the lateral beam surfaces.

In Case (a) the piezoelectrically induced bending moments $\tilde{M}_x$ and $\tilde{M}_y$ appear in the equations of motion and implicitly, in the governing equations, while in Case (b) they appear solely in the boundary conditions at $z = L$, their contribution in the equations of motion being immaterial (see in this context Librescu and Na, 1997, 1998, 2001; Na and Librescu, 1998, 2000, 2001a).

6.6-3 Governing Equations of Smart Thin-Walled Beams

We consider two cases, (A) and (B) related to the type of the elastic coupling. Case (A) corresponds to beams featuring lateral-transverse bending cross-coupling, while Case (B) to those with transverse bending-twist cross-coupling.

The governing equations can be expressed in a convenient form, as to include both Cases (a) and (b) of piezoactuator location.

For Case (A), the governing system is

$$\delta u_0 : (a_{43}\theta'_x)' + [a_{44}(u'_0 + \theta_y)]' - b_1\ddot{u}_0 + p_x(z,t) = 0, \tag{6.6-3a}$$

$$\delta v_0 : (a_{52}\theta'_y)' + [a_{55}(v'_0 + \theta_x)]' - b_1\ddot{v}_0 + p_y(z,t) = 0, \tag{6.6-3b}$$

$$\delta\theta_y : (a_{22}\theta'_y)' + [a_{25}(v'_0 + \theta_x)]' - a_{43}\theta'_x - a_{44}(u'_0 + \theta_y)$$

$$- \underline{(b_5 + \delta_n b_{15})\ddot{\theta}_y} - \delta_P \tilde{M}'_y + m_y(z,t) = 0, \tag{6.6-3c}$$

$$\delta\theta_x : (a_{33}\theta'_x)' + [a_{34}(u'_0 + \theta_y)]' - a_{52}\theta'_y - a_{55}(v'_0 + \theta_x)$$

$$- \underline{(b_4 + \delta_n b_{14})\ddot{\theta}_x} - \delta_P \tilde{M}'_x + m_x(z,t) = 0. \tag{6.6-3d}$$

The homogeneous boundary conditions, for cantilevered beams are

at $z = 0$:

$$u_0 = 0, \quad v_0 = 0, \quad \theta_y = 0, \quad \theta_x = 0 \tag{6.6-4a-d}$$

and at $z = L$:

$$\delta u_0 : a_{43}\theta'_x + a_{44}(u'_0 + \theta_y) = 0, \tag{6.6-5a}$$

$$\delta v_0 : a_{52}\theta'_y + a_{55}(v'_0 + \theta_x) = 0, \tag{6.6-5b}$$

$$\delta\theta_y : a_{22}\theta'_y + a_{25}(v'_0 + \theta_x) - \delta_S\tilde{M}_y = 0, \tag{6.6-5c}$$

$$\delta\theta_x : a_{33}\theta'_x + a_{34}(u'_0 + \theta_y) - \delta_S\tilde{M}_x = 0. \tag{6.6-5d}$$

For Case (B), the governing system is

$$\delta v_0 : [a_{55}(v'_0 + \theta_x)]' - \underline{(a_{56}\phi'')'} + p_y(z,t) = b_1\ddot{v}_0, \tag{6.6-6a}$$

$$\delta\theta_x : (a_{33}\theta'_x)' + (a_{37}\phi')' - a_{55}(v'_0 + \theta_x) + \underline{a_{56}\phi''} + m_x(z,t)$$

$$- \delta_P\tilde{M}'_x = \underline{(b_4 + \delta_n b_{14})\ddot{\theta}_x}, \tag{6.6-6b}$$

$$\delta\phi : - \underline{(a_{66}\phi'')''} + (a_{73}\theta'_x)' + (a_{77}\phi')' + \underline{[a_{65}(v'_0 + \theta_x)]''}$$

$$= [(b_4 + b_5) + \delta_n(b_{14} + b_{15})]\ddot{\phi} - \underline{\underline{[(b_{10} + \delta_n b_{18})\ddot{\phi}']'}}. \tag{6.6-6c}$$

The associated boundary conditions for a cantilevered beam are:

At $z = 0$:

$$\phi = 0, \quad v_0 = 0, \quad \theta_x = 0, \quad \underline{\phi'} = 0, \qquad (6.6\text{-}7a\text{-}d)$$

and at $z = L$:

$$\delta v_0 : \ a_{55}(v_0' + \theta_x) - \underline{a_{56}\phi''} = 0, \qquad (6.6\text{-}8a)$$

$$\delta \theta_x : \ a_{33}\theta_x' + a_{37}\phi' = \delta_S \tilde{M}_x, \qquad (6.6\text{-}8b)$$

$$\delta \phi : \ - \underline{(a_{66}\phi'')'} + a_{73}\theta_x' + a_{77}\phi' + [a_{65}(v_0' + \theta_x)]'$$

$$= - \underline{\underline{(b_{10} + \delta_n b_{18})\ddot{\phi}'}} \ , \qquad (6.6\text{-}8c)$$

$$\delta \phi' : \ \underline{-a_{66}\phi''} + \underline{a_{65}(v_0' + \theta_x)} = 0. \qquad (6.6\text{-}8d)$$

The unshearable counterpart of this governing system is

$$\delta v_0 : \ (a_{33}v_0'')'' - (a_{37}\phi')'' + \delta_P \tilde{M}_x'' = p_y(z, t) - b_1 \ddot{v}_0$$
$$+ [(b_4 + \delta_n b_{14})\ddot{v}_0']', \qquad (6.6\text{-}9a)$$

$$\delta \phi : \ - \underline{(a_{66}\phi'')''} + (a_{77}\phi')' - (a_{73}v_0'')' + m_z(z, t)$$
$$= [(b_4 + b_5) + \delta_n(b_{14} + b_{15})]\ddot{\phi} - [(b_{10} + \delta_n b_{18})\ddot{\phi}']'. \qquad (6.6\text{-}9b)$$

with the BCs:

at $z = 0$:

$$\phi = 0, \quad \underline{\phi'} = 0, \quad v_0 = 0, \quad v_0' = 0, \qquad (6.6\text{-}10a\text{-}d)$$

and at $z = L$:

$$\delta v_0 : (a_{33}v_0'')' - (a_{37}\phi')' = (b_4 + \delta_n b_{14})\ddot{v}_0', \qquad (6.6\text{-}11a)$$

$$\delta v_0' : a_{37}\phi' - a_{33}v_0'' = \delta_S \tilde{M}_x, \qquad (6.6\text{-}11b)$$

$$\delta \phi : \ - \underline{(a_{66}\phi'')'} + a_{77}\phi' - a_{73}v_0'' = 0, \qquad (6.6\text{-}11c)$$

$$\delta \phi' : \ -a_{66}\phi'' = 0. \qquad (6.6\text{-}11d)$$

For a beam made of an orthotropic material whose axes of orthotropy coincide with the geometrical axes, the bending-twist cross-coupling stiffnesses a_{37} and a_{56} become immaterial. In this case, both the system of equations, (6.6-9) and the boundary conditions (6.6-10) and (6.6-11) split into two independent systems belonging to pure transverse bending and twist; the control moments appear only in the equations for transverse bending. It is clear from the above that, for the *coupled* system, the twist motion is controlled in an indirect way only, namely via the bending-twist elastic coupling. It should be mentioned that alternative procedures enabling one to piezoelectrically generate a twist moment have been devised; see Park and Chopra (1996) and the most recent survey-paper by Chopra (2002) where pertinent information about this issue can be found.

For further use, the equations associated with the decoupled system in pure transverse bending will be recorded here. Its shearable counterpart is

$$\delta v_0 : a_{55}(v_0' + \theta_x)' + p_y(z, t) = b_1 \ddot{v}_0,$$

$$\delta \theta_x : (a_{33}\theta_x')' - a_{55}(v_0' + \theta_x) + m_x(z, t) - \delta_P \tilde{M}_x'$$
$$= (b_4 + \delta_n b_{14})\ddot{\theta}_x. \qquad (6.6\text{-}12\text{a,b})$$

and the boundary conditions:

$$v_0 = 0; \quad \theta_x = 0, \text{ at } z = 0, \qquad (6.6\text{-}12\text{c,d})$$

and

$$v_0' + \theta_x = 0; \quad a_{33}\theta_x' = \delta_S \tilde{M}_x, \text{ at } z = L. \qquad (6.6\text{-}12\text{e,f})$$

The classical equations in transverse bending of smart thin-walled beams can be obtained from Eqs. (6.6-12) by using the procedure already explained. Alternatively, these can be obtained from Eqs. (6.6-9a) and (6.6-11a,b) by letting $a_{37} \to 0$.

This yields the governing equation

$$\delta v_0 : (a_{33}v_0'')'' + \delta_P \tilde{M}_x'' = p_y(z, t) - b_1 \ddot{v}_0 + \left[(b_4 + \delta_n b_{14}) \ddot{v}_0' \right]', \qquad (6.6\text{-}13)$$

and the boundary conditions

$$v_0 = v_0' = 0, \text{ at } z = 0, \qquad (6.6\text{-}14\text{a,b})$$

and

$$\delta v_0 : (a_{33}v_0'')' = (b_4 + \delta_n b_{14})\ddot{v}_0',$$

$$\delta v_0' : a_{33}v_0'' = -\delta_S \tilde{M}_x, \text{ at } z = L. \qquad (6.6\text{-}14\text{c,d})$$

Equations (6.6-3) through (6.6-14) show that in the Case (a) of piezoactuator location, the tracers have to be taken as $\delta_P = 1, \delta_S = 0$, while in Case (b), as $\delta_S = 1$ and $\delta_P = 0$. In the latter case, the control is achieved via the piezoelectrically induced boundary bending moments. This type of control has been adopted in many works (see e.g. Bailey and Hubbard, 1985; Tzou and Zhong, 1992; Tzou, 1993; Baz, 1997; Librescu et al., 1993; Librescu and Na, 1998; Song et al., 1992, 1996, 2001). For a mathematical substantiation of the boundary control method, see Lagnese and Lions (1988) and Lagnese (1989).

6.6-4 Sensing and Actuation. Feedback Control

Within the feedback control the voltage output from the sensor(s) is amplified and fed back to the actuator(s). This is achieved by collecting electrical charge from sensor(s) and then redistributing it via controller as applied voltage to the actuator after assigning proper weights.

For the closed-loop control, the output voltage of the sensor which in turn is fed back to the actuator(s) should be known.

One assumes that the piezoelectric elements can be employed concurrently for sensing and actuation. For sensing operation, $\mathcal{E}_3 = 0$, the electric displacement is given by

$$D_3 = \bar{e}_{31}\varepsilon_{zz}, \qquad (6.6\text{-}15a)$$

while the sensor output voltage and the electric charge are expressed respectively by

$$V_s(t) = \frac{q_s(t)}{C_p}, \qquad (6.6\text{-}15b)$$

$$q_s(t) = \int_{A_s} D_3 dA_s = \int_{A_s} \bar{e}_{31}\varepsilon_{zz} dA_s. \qquad (6.6\text{-}15c)$$

In (6.6-15b,c), C_p and A_s denote the sensor's capacitance and area over which the electric displacement has to be integrated as to get the net electric charge.

Assuming that the sensor patches are located symmetrically on the opposite walls, i.e., on $(y = \pm b/2)$ and $(x = \pm c/2)$, based on the previous equations one can represent $V_S^x(t)$ and $V_S^y(t)$ as

$$V_S^x(t) = C_x^S \theta_x(L, t), \quad V_S^y = C_y^S \theta_y(L, t), \qquad (6.6\text{-}16a,b)$$

where the expressions of C_y^S and C_x^S that are not displayed here, are obtained by considering Eqs. (6.6-15b) and (6.6-15c) in conjunction with the expression of ε_{zz}.

Two feedback control laws will be used next: (i) the proportional feedback control law, and (ii) the velocity control law. For proportional feedback the actuating voltage is proportional to the sensor output voltage:

$$\mathcal{E}_3^x(t) = k_p V_S^x(t)/h_a, \quad \mathcal{E}_3^y = k_p V_S^y(t)/h_a, \quad (6.6\text{-}17\text{a,b})$$

whereas for velocity feedback

$$\mathcal{E}_3^x(t) = k_v \frac{dV_S^x(t)/dt}{h_a}, \quad \mathcal{E}_3^y = k_v \frac{dV_S^y(t)/dt}{h_a}, \quad (6.6\text{-}18\text{a,b})$$

where h_a is the thickness of the piezopatch. In Eqs. (6.6-17) and (6.6-18), k_p and k_v stand for the proportional and the velocity feedback gain, respectively.

For the proportional control Eqs. (6.6-16), and (6.6-17) and (6.6-1), give the piezoelectrically induced bending moments

$$\tilde{M}_x = -\frac{k_p C_{M_x^a}}{h_a}[C_x^S \theta_x(L,t)] \equiv -k_p C_{11}\theta_x(L,t), \quad (6.6\text{-}19\text{a})$$

$$\tilde{M}_y = -\frac{k_p C_{M_y^a}}{h_a}[C_y^S \theta_y(L,t)] \equiv -k_p C_{22}\theta_y(L,t). \quad (6.6\text{-}19\text{b})$$

For velocity feedback control, following the same procedure as above, but in connection with (6.6-18) yields

$$\tilde{M}_x = -\frac{k_v C_{M_x^a}}{h_a}[C_x^S \dot{\theta}_x(L,t)] \equiv -k_v C_{11}\dot{\theta}_x(L,t), \quad (6.6\text{-}20\text{a})$$

$$\tilde{M}_y = -\frac{k_v C_{M_y^a}}{h_a}[C_y^S \dot{\theta}_y(L,t)] \equiv -k_v C_{22}\dot{\theta}_y(L,t). \quad (6.6\text{-}20\text{b})$$

It is worth remarking that as per the two previously considered control laws, the associated feedback gains include all the information regarding piezo- sensor and actuator.

6.6-5 The Discretized Governing Equations of Adaptive Beams

6.6-5a Case (A) (Bending-Bending Cross-Coupling)

For computational reasons, it is necessary to discretize the boundary value problem, which amounts to representing u_0, v_0, θ_x, and θ_y as series of space-dependent trial functions satisfying at least the geometric boundary conditions, multiplied by time-dependent generalized coordinates,

$$u_0(z,t) = \mathbf{U}_0^T(z)\mathbf{q}_u(t); \quad v_0(z,t) = \mathbf{V}_0^T(z)\mathbf{q}_v(t),$$
$$\theta_x(z,t) = \mathbf{X}_0^T(z)\mathbf{q}_x(t); \quad \theta_y(z,t) = \mathbf{Y}_0^T(z)\mathbf{q}_y(t), \quad (6.6\text{-}21\text{a-d})$$

where

$$\mathbf{U}_0(z) = \{u_1, u_2, \ldots u_N\}^T, \quad \mathbf{V}_0(z) = \{v_1, v_2, \ldots v_N\}^T,$$
$$\mathbf{X}_0(z) = \{x_1, x_2, \ldots x_N\}^T, \quad \mathbf{Y}_0(z) = \{y_1, y_2, \ldots y_N\}^T, \quad (6.6\text{-}22\text{a-d})$$

are vectors of suitable trial functions, and

$$\mathbf{q}_u(t) = \{q_1^u, q_2^u, \ldots q_N^u\}^T, \quad \mathbf{q}_v(t) = \{q_1^v, q_2^v, \ldots q_N^v\}^T,$$
$$\mathbf{q}_x(t) = \{q_1^x, q_2^x, \ldots q_N^x\}^T, \quad \mathbf{q}_y(t) = \{q_1^y, q_2^y, \ldots q_N^y\}^T, \qquad \text{(6.6-22e-h)}$$

are vectors of time-dependent generalized coordinates.

Based on previous relations one can express the discretized kinetic and strain energies as

$$\mathcal{K} = \frac{1}{2}\dot{\mathbf{q}}^T \mathbf{M}\dot{\mathbf{q}}, \quad \mathcal{W} = \frac{1}{2}\mathbf{q}^T \mathbf{K}\,\mathbf{q}, \qquad \text{(6.6-23a,b)}$$

where

$$\mathbf{M} =$$
$$\int_0^L \begin{bmatrix} b_1 \mathbf{U}_0 \mathbf{U}_0^T & 0 & 0 & 0 \\ & b_1 \mathbf{V}_0 \mathbf{V}_0^T & 0 & 0 \\ & & (b_5 + \delta_n b_{15})\mathbf{Y}_0 \mathbf{Y}_0^T & 0 \\ \text{Symm.} & & & (b_4 + \delta_n b_{14})\mathbf{X}_0 \mathbf{X}_0^T \end{bmatrix} dz,$$
$$\text{(6.6-24a,b)}$$

$$\mathbf{K} =$$
$$\int_0^L \begin{bmatrix} a_{44}\mathbf{U}_0' \mathbf{U}_0'^T & 0 & a_{44}\mathbf{U}_0' \mathbf{Y}_0^T & a_{34}\mathbf{U}_0' \mathbf{X}_0'^T \\ & a_{55}\mathbf{V}_0' \mathbf{V}_0'^T & a_{25}\mathbf{V}_0' \mathbf{Y}_0'^T & a_{55}\mathbf{V}_0' \mathbf{X}_0^T \\ & & a_{44}\mathbf{Y}_0 \mathbf{Y}_0^T & a_{34}\mathbf{Y}_0 \mathbf{X}_0'^T \\ & & +a_{22}\mathbf{Y}_0' \mathbf{Y}_0'^T & +a_{25}\mathbf{Y}_0' \mathbf{X}_0^T \\ \text{Symm.} & & & a_{55}\mathbf{X}_0 \mathbf{X}_0^T \\ & & & +a_{33}\mathbf{X}_0' \mathbf{X}_0'^T \end{bmatrix} dz,$$

while

$$\mathbf{q}(t) = \left\{ \mathbf{q}_u^T(t), \; \mathbf{q}_v^T(t), \; \mathbf{q}_y^T(t), \; \mathbf{q}_x^T(t) \right\}^T, \qquad \text{(6.6-24c)}$$

is the $4N \times 1$ overall generalized coordinate vector.

The virtual work associated with the distributed, concentrated and the edge loads is

$$\delta\mathcal{J} = \tilde{M}_x(L, t)\delta\theta_x(L, t) + \tilde{M}_y(L, t)\delta\theta_y(L, t) + \int_0^L [p_x^d(z, t)\delta u_0(z, t)$$
$$+ p_y^d(z, t)\delta v_0(z, t) + m_x^d(z, t)\delta\theta_x(z, t) + m_y^d(z, t)\delta\theta_y(z, t)]dz$$
$$+ \sum_i [p_x^c(z_i, t)\delta u_0(z_i, t) + p_y^c(z_i, t)\delta v_0(z_i, t) \qquad \text{(6.6-25a)}$$
$$+ m_x^c(z_i, t)\delta\theta_x(z_i, t) + m_y^c(z_i, t)\delta\theta_y(z_i, t)],$$

where superscripts "*d*" and "*c*" identify the forces and moments, distributed and concentrated (at $z = z_i$, $0 < z_i < L$), respectively. $\tilde{M}_x$ and $\tilde{M}_y$ denote the piezoelectrically induced bending moments as defined by Eqs. (6.6-19), or (6.6-20).

Consistent with (6.6-25a), considered with Eqs. (6.6-19) and (6.6-20), the discretized virtual work for the smart beam is

$$\delta \mathcal{J} = \mathbf{Q}^T \delta\mathbf{q} - \delta_p \mathbf{q}^T \mathbf{P}_p \delta\mathbf{q} - \delta_v \dot{\mathbf{q}}^T \mathbf{P}_v \delta\mathbf{q}, \qquad (6.6\text{-}25\mathrm{b})$$

where

$$(\mathbf{P}_p, \mathbf{P}_v) = \\ - (k_p, k_v) \begin{bmatrix} 0 & 0 & 0 & 0 \\ 0 & 0 & 0 & 0 \\ 0 & 0 & C_{22}\mathbf{Y}_0(L)\mathbf{Y}_0^T(L) & 0 \\ 0 & 0 & 0 & C_{11}\mathbf{X}_0(L)\mathbf{X}_0(L)^T \end{bmatrix}, \qquad (6.6\text{-}26)$$

and

$$\mathbf{Q}(t) = \left\{ \begin{array}{c} \int_0^L p_x^d(z, t)\mathbf{U}_0^T dz \\ \int_0^L p_y^d(z, t)\mathbf{V}_0^T dz \\ \int_0^L m_y^d(z, t)\mathbf{Y}_0^T dz \\ \int_0^L m_x^d(z, t)\mathbf{X}_0^T dz \end{array} \right\} + \sum_i \left\{ \begin{array}{c} p_x^c(z_i, t)\mathbf{U}_0^T(z_i, t) \\ p_y^c(z_i, t)\mathbf{V}_0^T(z_i, t) \\ m_y^c(z_i, t)\mathbf{Y}_0^T(z_i, t) \\ m_x^c(z_i, t)\mathbf{X}_0^T(z_i, t) \end{array} \right\}. \qquad (6.6\text{-}27)$$

In Eq. (6.6-25) as well as the next ones, δ_p and δ_v are two tracers identifying the effects induced by the application of the proportional or of the velocity feedback control, respectively. These take the value 0 or 1, depending on whether the respective control is or is not applied, respectively.

Replacing Eqs. (6.6-23) and (6.6-25) in Eq. (6.3-13), integrating with respect to the spanwise z-coordinate and with respect to time, and using Hamilton's condition, $\delta\mathbf{q} = \mathbf{0}$ at t_0, t_1, we obtain the closed-loop discretized system

$$\mathbf{M}\,\ddot{\mathbf{q}}(t) + \delta_v \mathbf{P}_v \dot{\mathbf{q}}(t) + \mathbf{K}\,\mathbf{q}(t) + \delta_p \mathbf{P}_p \mathbf{q}(t) = \mathbf{Q}(t). \qquad (6.6\text{-}28)$$

Here $\mathbf{M}$ and $\mathbf{K}$ should be considered for the structure in its entirety (i.e. for the host and piezoactuators as a whole). Equation (6.6-28) is similar to that supplied in generic form by Inman (2000).

A solution of (6.6-28) can be obtained by casting it first in state-space form; we define the state-space vector $\mathbf{x}(t) = \{\mathbf{q}^T(t)\,,\ \dot{\mathbf{q}}^T(t)\}^T$ where $\mathbf{q}$ is the overall generalized coordinate vector, and adjoin the identity $\dot{\mathbf{q}} \equiv \dot{\mathbf{q}}$.

As a result, (6.6-28) cast in state-space form is given by

$$\dot{\mathbf{x}}(t) = \mathbf{A}\mathbf{x}(t) + \mathbf{W_F}(t), \tag{6.6-29}$$

where

$$\mathbf{A} = \begin{bmatrix} \mathbf{0} & \mathbf{I} \\ -\mathbf{M}^{-1}\hat{\mathbf{K}} & -\delta_v\mathbf{M}^{-1}\mathbf{P}_v \end{bmatrix}, \tag{6.6-30a}$$

and

$$\mathbf{W_F}(t) = \left\{ \begin{array}{c} \mathbf{0} \\ \mathbf{M}^{-1}\mathbf{Q}(t) \end{array} \right\}, \tag{6.6-30b}$$

$$\hat{\mathbf{K}} = \mathbf{K} - \delta_p\mathbf{P}_p, \tag{6.6-30c}$$

$\mathbf{I}$ and $\mathbf{0}$ being the identity and zero matrices of appropriate dimensions. For the present case since the vector $\mathbf{q}$ is $4N \times 1$, $\mathbf{x}$ is $8N \times 1$ and state matrix $\mathbf{A}$ is $8N \times 8N$, where N is the number of generalized coordinates associated to each of the four displacement quantities. Equation (6.6-30a) shows that $\mathbf{A}$ corresponds to the closed-loop system. The homogeneous solution of Eq. (6.6-29) has the exponential form

$$\mathbf{x}(t) = \mathbf{X}e^{\lambda t}, \tag{6.6-31}$$

where $\mathbf{X}$ is a constant vector and λ a constant scalar, both generally complex. Equation (6.6-29) in conjunction with (6.6-31) and $\mathbf{Q} = 0$ yields the standard eigenvalue problem

$$\mathbf{A}\mathbf{X} = \lambda\mathbf{X}, \tag{6.6-32}$$

that can be solved for the eigenvalues λ_r and the eigenvectors $\mathbf{x}_r$.

For the case $k_v \neq 0$, the closed-loop eigenvalues are complex valued quantities

$$(\lambda_r, \bar{\lambda}_r) = \eta_r \pm i\omega_{dr}, \quad (r = 1, 2, \ldots 8N) \tag{6.6-33}$$

which depend on the feedback control gains, and the system parameters.

Here η_r is a measure of the damping in the rth mode and ω_{dr} is the rth frequency of damped oscillation.

From Eq. (6.6-33) one obtains the damping factor in the rth mode

$$\zeta_r = -\eta_r/(\eta_r^2 + \omega_{dr}^2)^{1/2}. \tag{6.6-34}$$

The obtained results reveal that even in the absence of the structural damping, the velocity feedback control generates damping.

6.6-5b Case (B) (Transverse Bending-Twist Cross-Coupling)

Express v_0, θ_x and ϕ in the form of Eqs. (6.3-14), form $\mathcal{K}$ and $\mathcal{W}$ from Eqs. (6.3-15), and **M** and **K** as in Eqs. (6.3-17).

The virtual work of the closed-loop system can be expressed as in Eq. (6.6-25b), where

$$
(\mathbf{P}_p,\ \mathbf{P}_v) = -(k_p, k_v)
\begin{bmatrix}
0 & 0 & 0 \\
0 & C_{11}\mathbf{X}_0(L)\mathbf{X}_0^T(L) & 0 \\
0 & 0 & 0
\end{bmatrix},
\qquad (6.6\text{-}35)
$$

and

$$
\mathbf{Q}(t) =
\left\{
\begin{array}{c}
\int_0^L p_y^d(z,t)\mathbf{V}_0^T dz \\
\int_0^L m_x^d(z,t)\mathbf{X}_0^T dz \\
\int_0^L [m_z^d(z,t) + m_\omega'(z,t)]\mathbf{\Phi}^T dz
\end{array}
\right\}
+ \sum_i
\left\{
\begin{array}{c}
p_y^c(z_i,t)\mathbf{V}_0^T(z_i) \\
m_x^c(z_i,t)\mathbf{X}_0^T(z_i) \\
m_z^c(z_i,t)\mathbf{\Phi}^T(z_i)
\end{array}
\right\}.
$$

$$(6.6\text{-}36)$$

Again one obtains the closed-loop discretized system and its state-space counterpart, in a form similar to that in Eq. (6.6-28), and Eq. (6.6-29), respectively.

6.6-6 Numerical Simulations

The numerical illustrations will involve both the case of thin-walled beams featuring a rectangular ($b \times c$) cross-section profile, Case (A), and that of a bi-convex cross-section, Case (B). The actuators are symmetrically embedded in the host structure as indicated in Figs. 6.6-1a,b.

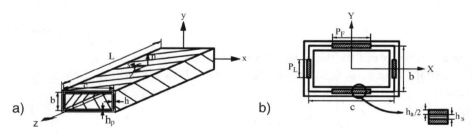

Figure 6.6-1: Smart box-beam (Case (A)). (a) Generic ply-angle distribution in the material of a box-beam inducing bending-bending-transverse shear coupling. The piezoactuators/sensors are also indicated. (b) Piezoactuator/sensor location. The opposite piezopatches P_L and P_F control the lagging and flapping motions

The overall fixed dimensions of the box beam as considered in the numerical simulations are $c_R = 10$ in. (0.25m) and $L = 80$ in. (2.032m). The host structure is made from graphite-epoxy composite whose elastic properties are given in Table 6.4-2 and the piezopatches are manufactured of PZT-4 piezoceramic.

As concerns its properties, those extracted from Berlincourt et al. (1964) (see also Librescu et al., 1993), are supplied in Table 6.6-1.

Table 6.6-1: Properties of the PZT-4 piezoceramic

$C_{11} = 2.016 \times 10^7$ psi, (139GPa)	$C_{12} = 1.128 \times 10^7$ psi, (77.773GPa)
$C_{13} = 1.0776 \times 10^7$ psi, (74.298GPa)	$C_{33} = 1.6679 \times 10^7$ psi, (115GPa)
$C_{44} = 3.7128 \times 10^6$ psi(25.59GPa)	$\rho = 7.0135 \times 10^{-4}$ lb.s^2/in.^{4},
	$(7.498 \times 10^3$ kg/m^3)
$e_{31} = -0.0297$ lb/in. V, $(-5.2014$N/V.m)	$e_{33} = 0.08623$ lb/in.V, (15.101N/V.m)

6.6-6a Case (A) Coupled with Case (b)

In Eqs. (6.6-3) through (6.6-5), $\delta_P = 0$ and $\delta_S = 1$. Figure 6.6-2 show the effect of proportional feedback on the first eigenfrequency. Figure 6.6-2a displays the combined effects of tailoring and adaptive control; Figure 6.6-2b shows the influence of the beam cross-section parameter $\mathcal{R}(\equiv b/c)$, combined with the active control.

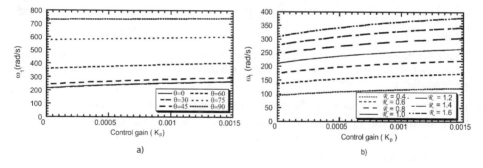

Figure 6.6-2: Effects of ply-angle θ (deg.) and cross-section parameter $\mathcal{R}$ on closed-loop fundamental eigenfrequency. (a) Variation of the fundamental natural frequency ω_1 (rad/s) versus dimensionless proportional feedback gain $K_p (\equiv k_p L^2/a_{33})$, for selected values of the ply-angle θ (deg.). (a_{33} in K_p corresponds to $\theta = 0$ and $\mathcal{R} = 1$), (b) Variation of ω_1 (rad/s) versus K_p for selected values of cross-section parameter, $\theta = 0$

These figures reveal the high sensitivity of the fundamental frequency to the variation of the ply-angle and cross-sectional parameter $\mathcal{R}$, and the efficiency of proportional feedback control in enhancing the dynamic behavior of the beam. Figure 6.6-3 shows the effect played by the beam span on the fundamental closed-loop eigenfrequency under proportional feedback control; with increase in L, the efficiency of the control, together with a general decrease of eigenfrequency is experienced.

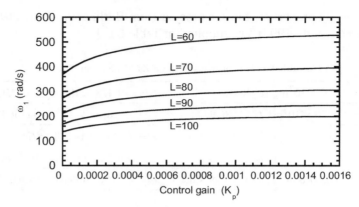

Figure 6.6-3: Variation of the fundamental natural frequency ω_1 versus K_p for selected values of the beam length L (in.), $\mathcal{R} = 1$

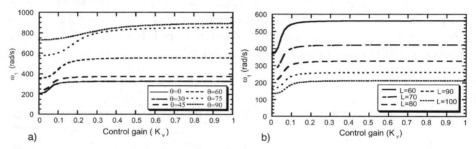

Figure 6.6-4: (a) Counterpart of Fig. 6.6-2a for the case of the velocity feedback control. (b) Counterpart of Fig. 6.6-3. $K_v (\equiv k_v L^2 / a_{33})$ is the dimensionless velocity feedback gain, a_{33} corresponding to $\theta = 0$ and $\mathcal{R} = 1$

Figures 6.6-4a,b that are the counterparts of Figs. 6.6-2a and 6.6-3, respectively, display the effect of the velocity feedback control on the fundamental eigenfrequency; for any ply-angle considered in this analysis, the increase of the velocity feedback gain until a fixed value of K_v increases the fundamental natural frequency. Beyond that fixed value that is specific to each θ, the fundamental frequency varies slightly. Figure 6.6-4 shows that the increase of the ply-angle, yielding an increase of bending stiffness, makes it more difficult to increase of the fundamental natural frequency via piezoelectric actuation.

We conclude that velocity feedback control is better than proportional feedback control. Figure 6.6-5 shows that the control induced modal damping increases with the increase of K_v until a certain value, beyond which, ζ_1 drops sharply. This trend is consistent with that reported by Tzou et al. (1999), where the experimental results agree well with the numerical simulations. Corresponding to various ply-angles, the limiting values of K_v at which the maxima for ζ_1 occur are consistent to those at which ω_1 does not exhibit any further increase with the increase of K_v. Moreover, although at $\theta = 90$ deg. maximum values of the transverse and lateral bending stiffnesses, as well as a maximum fundamental frequency are obtained, the damping does not reach a maximum as compared to the counterpart results obtained for the other ply-angles considered in this analysis. One can say that the increase of the structural stiffness makes it more difficult the task to induce damping via control. However, with the increase of θ, implying the increase of bending stiffness, the sharp drop of ζ_1 tends to disappear, being replaced with the increase of K_v, by a smoother decay of ζ_1.

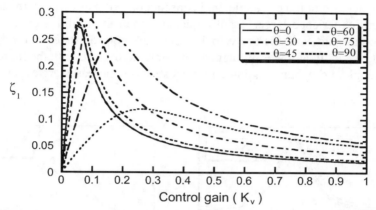

Figure 6.6-5: Piezoelectrically induced damping ζ_1 versus K_v for selected values of θ (deg.), $\mathcal{R} = 1$

6.6-6b Case (B) Coupled with Case (b)

In Eqs. (6.6-6) through (6.6-8), $\delta_P = 0$, $\delta_S = 1$.

Figures 6.6-6a-c, give a good idea of the effects of beam taper-ratio coupled with the ply-angle on the first three open/closed loop eigenfrequencies. These figures display the variation of the first three eigenfrequencies vs. the ply-angle, for both the unactivated ($K_v = 0$) and activated beam ($K_v = 0.007$), for three selected values of the beam taper-ratio, namely $\sigma = 1$; 0.7 and 0.5. Here $K_v = k_v L \bar{\omega}/a^\circ_{33}$, where $\bar{\omega} = 320$ rad/s is a reference natural frequency corresponding to a thin-walled beam characterized by $\sigma = 0.1$, $\theta = 90$ deg., L = 80 in. (2.03 m), while a°_{33} is the normalizing bending stiffness at the beam root (= 1.44117×10^7 lb. in.^{2} ($4.136 \times 10^4 N.m^2$)).

Here we consider the unshearable model of a cantilevered beam. We can draw some conclusions from these figures: i) as a result of the symmetrical variations of the bending and twisting stiffnesses with respect to $\theta = 90$ deg., the graphs are symmetrical with respect to that ply-angle. Increase in ply-angle leads to increase in stiffness, and hence of eigenfrequencies. This increase is experienced by both the open/closed loop eigenfrequencies, ii) the result that the taper ratio affects the frequencies of each mode somewhat differently, with the first mode behavior in sharp contrast to the other modes. In this sense, as Fig. 6.6-6a shows, for uniform cross-section beams the fundamental eigenfrequency in transverse bending is smaller than its doubly-tapered beam counterpart, a trend that remains valid for any ply-angle θ. This result was pointed out by Na and Librescu (1999), Na (2001b) and Librescu et al. (2000), and for solid beams by Downs (1977). Figures. 6.6-6b,c shows that until a certain value of the ply-angle, the trends of variation of the second and third eigenfrequencies is different from that of the fundamental frequency, in the sense that the uniform beams feature larger eigenfrequencies than their tapered beam counterparts. However, with the increase of the ply-angle, and consequently of the stiffness, the tapered beam will experience larger eigenfrequencies than the uniform cross-section beam counterpart, and iii) the velocity feedback control appears to be rather efficient, specially at lower ply-angles and starts to decline with the increase of the ply-angle, i.e. when bending and twist stiffness increases are experienced.

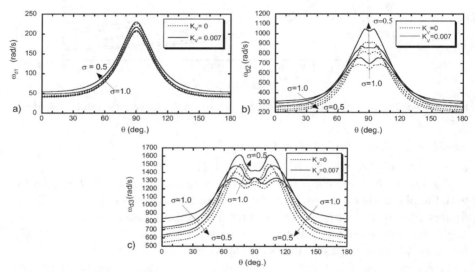

Figure 6.6-6: Variation of the first three open/closed loop damped eigenfrequencies with that of the ply-angle, for three values of the taper ratio σ and two values of K_v. a) fundamental eigenfrequency, b) second eigenfrequency, c) third eigenfrequency

Figures 6.6-7a-c highlight the effects played by the transverse shear flexibility considered in conjunction with the taper ratio on the first three piezoelectrically induced damping factors ζ_i.

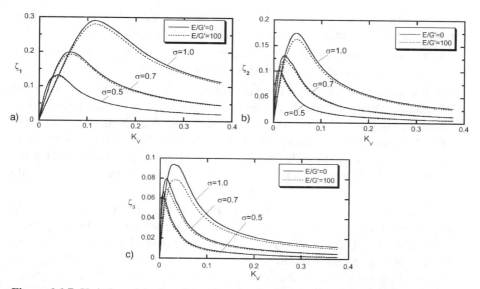

Figure 6.6-7: Variation of the first three piezoelectrically induced damping factors ζ_i with that of K_v for selected values of σ and transverse shear flexibility parameter E/G'. a) ζ_1 vs. K_v, b) ζ_2 vs. K_v, c) ζ_3 vs. K_v

Here the purely bending case is considered. In the plots, the control induced damping in a non-shearable beam ($E/G' = 0$) is compared to the one in a highly shearable ($E/G' = 100$) beam counterpart. The results reveal (see Librescu and Na, 2001), the marginal role played by transverse shear flexibility and the significant role of the taper-ratio.

6.7 CLOSING REMARKS

In this chapter, the free vibration problem of uniform and doubly-tapered thin-walled beams of closed cross-section contour was addressed. The implications of a number of effects related to the transverse shear, warping restraint, taper ratio, rotatory inertia, directionality property of composite materials, and of various elastic couplings that are an inherent result of the anisotropy of material systems, have been highlighted and pertinent conclusions have been outlined.

As it was shown, the proper use of the directionality property of composite materials can results in the dramatic enhancement of the free vibration behavior, and of the generation of most beneficial elastic couplings. In fact, the related technique, referred to as *structural tailoring* constitutes one of the main capabilities offered by fibrous composite materials; as a result, it is likely to have an important role on the design of weight sensitive structural systems such as space vehicles, helicopter blades, robot arms working in space, spacecraft booms, etc. At the same time, the displayed results suggest that the active vibration control via the piezoelectric strain actuation technology can constitute a most promising avenue toward the vibrational control.

In this context, the synergistic interaction of the piezoelectric induced actuation with the structural tailoring was also highlighted.2

At this point a similarity, in a certain sense, between structural tailoring and feedback control can be remarked; the tailored structure can be viewed as a plant, wherein the control law is pre-programmed and embedded in the structure in the form of the material constitutive law. In this analogy, the tailored structure acts both as sensor and actuator. This *kind of control* is however passive in its nature, in the sense that the control law cannot be changed during the operational life of the system.

As it will be seen in the next chapter, the dynamic response of thin-walled beams to time-dependent external excitation is governed not only by the properties of the input, but also by the system eigenvalues characterizing the autonomous (unforced) associated problem. Moreover, the eigenvalues of the free vibration system are also essential toward determination of the flutter instability boundary of flight vehicle structural systems.

Thus, the search of the free vibration eigenvalue problem constitutes a basic pre-requisite toward investigating more complex dynamic problems such as the forced response, dynamic instability in general and, of the flutter instability, in particular.

REFERENCES

Ambartsumian, S. A. (1964) *Theory of Anisotropic Shells*, *NASA Report*, TTF-118.

Armanios E. A. and Badir, A. M. (1995) "Free Vibration Analysis of Anisotropic Thin-Walled Closed-Section Beams," *AIAA Journal*, Vol. 33, No. 10, pp. 1905–1910.

Atilgan, A. R. and Rehfield, L. W. (1990) "Vibrations of Composite Thin-Walled Beams with Design in Elastic Couplings," in *Achievements in Composites in Japan and United States, Proceedings of the Fifth Japan–U.S. Conference on Composite Materials*, June 24–27, Tokyo, Japan, pp. 687–694.

Bailey, T. and Hubbard, J. E., Jr. (1985) "Distributed Piezoelectric-Polymer Active Vibration Control of a Cantilever Beam," *Journal of Guidance, Control and Dynamics*, Vol. 8, No. 5, Sept.–Oct., pp. 605–611.

Bank, L. C. and Kao, C. K. (1989) "The Influence of Geometric and Material Design Variables on the Free Vibration of Thin-Walled Composite Material Beams," *Journal of Vibration, Acoustics, Stress and Reliability in Design*, Vol. 111, pp. 290–297.

Banerjee, J. R. (1989) "Coupled Bending-Torsional Dynamic Stiffness Matrix for Beam Elements," *International Journal of Numer. Meth. Eng.*, Vol. 28, pp. 1283–1298.

Baz, A. (1997) "Dynamic Boundary Control Beams Using Active Constrained Layer Damping," *Journal of Vibration and Acoustics*, Trans. ASME, Vol. 119, April, pp. 266–272.

Belinskiy, B. P. and Dauer, J. P. (1998) "Eigenoscillations of Mechanical Systems with Boundary Conditions Containing the Frequency," *Quarterly of Applied Mathematics*, Vol. LVI, No. 3, pp. 521–541.

Berlincourt, D. A., Curran, Dr. R. and Jaffe, H. (1964) "Piezoelectric and Piezomagnetic Materials and Their Function in Transducers," in *Physical Acoustics – Principles and Methods*, E. P. Mason (Ed.), Vol. 1, Part A, Academic Press, New York, pp. 202–204.

Bishop, R. E. D., Price, W. G. and Chang, Z. X. (1983) "On The Structural Dynamics of a Vlasov Beam," *Proceedings of the Royal Society London*, Vol. A388, pp. 49–83.

Budiansky, B. and Kruszewski, T. (1952) "Transverse Vibrations of Hollow Thin-Walled Cylindrical Beams," *NACA TN 2682*.

Chopra, I. (2002) "Review of State of Art of Smart Structures and Integrated Systems," *AIAA Journal*, Vol. 40, No. 11, pp. 2145–2187.

Crawley, E. F. (1994) "Intelligent Structures for Aerospace; A Technology Overview and Assessment," *AIAA Journal*, Vol. 31, No. 8, pp. 1689–1699.

Chandiramani, N. K., Librescu, L. and Shete, C. D. (2002) "On the Free-Vibration of Rotating Composite Beams Using a Higher-Order Transverse Shear Formulation," *Aerospace Science and Technology*, Vol. 6, pp. 545–561.

Chandra, R. and Chopra, I. (1992a) "Experimental-Theoretical Investigation of the Vibration Characteristics of Rotating Composite Box Beams," *Journal of Aircraft*, Vol. 29, No. 4, pp. 657–664.

Chandra, R. and Chopra, I. (1992b) "Structural Response of Composite Beams and Blades with Elastic Couplings," *Composites Engineering*, Vol. 2, No. 5–7, pp. 437–374.

Downs, B. (1977) "Transverse Vibrations of Cantilever Beams Having Unequal Breadth and Depth," *Journal of Applied Mechanics*, Trans. ASME, Vol. 44, pp. 737–742.

Gay, D. and Boudet, R. (1980) "A Technical Theory of Dynamical Torsion for Beams of Any Cross-Section Shapes," *ASME Paper No. 79-DET-59*, pp. 1–6.

Gern, F. H. and Librescu, L. (1998a) "Static and Dynamic Aeroelasticity of Advanced Aircraft Wings Carrying External Stores," *AIAA Journal*, Vol. 36, No. 7, July, pp. 1121–1129.

Gern, F. H. and Librescu (1998b) "Effects of Externally Mounted Stores on Flutter Characteristics of Advanced Swept Cantilevered Aircraft Wings," *Aerospace Science and Technology*, Vol. 2, No. 5, pp. 321–333.

Hallauer, W. L., Liu, R. Y. L. (1982) "Beam Bending-Torsion Dynamic Stiffness Method for Calculation of Exact Vibration Modes," *Journal of Sound and Vibration*, Vol. 85, pp. 105–113.

Hashemi, S. M. and Richard, J. M. (2000) "A Dynamic Finite Element (DFE) Method for Free Vibrations of Bending-Torsion Coupled Beams," *Aerospace Science and Technology*, Vol. 4, No. 1, pp. 41–55.

Hodges, D. H., Atilgan, A. R., Fulton, M. V. and Rehfield, L. W., (1991) "Free Vibration Analysis of Composite Beams," *Journal of the American Helicopter Society*, Vol. 36, No. 3, pp. 36–47.

Housner, G. W. and Keightley, W. O. (1962) "Vibration of Linearly Tapered Cantilever Beams," *Journal of the Engineering Mechanics Division*, Trans. ASME, April EM2, pp. 95–122.

Inman, D. J. (2000) "Active Modal Control for Smart Structures," *Phil. Trans. R. Soc. London, A*, **359**, pp. 205–219.

Karpouzian, G. and Librescu, L. (1994a) "Comprehensive Model of Anisotropic Composite Aircraft Wings Suitable for Aeroelastic Analyses," *Journal of Aircraft*, Vol. 31, No. 3, May–June, pp. 703–712.

Karpouzian, G. and Librescu, L. (1994b) "Three-Dimensional Flutter Solution of Aircraft Wings Composed of Advanced Composite Materials," in *Proceedings of the AIAA/ASME/ASCE/AHS/ASC 35th Structures, Structural Dynamics and Materials Conference*, Washington, DC, AIAA Paper 94-1490, pp. 2851–2856.

Karpouzian, G. and Librescu, L. (1996) "Non-Classical Effects on Divergence and Flutter of Anisotropic Swept Aircraft Wings," *AIAA Journal*, Vol. 34, No. 4, April, pp. 786–794.

Ko, K-E and Kim, J-H. (2003) "Thermally Induced Vibrations of Spinning Thin-Walled Composite Beams," *AIAA Journal*, Vol. 41, No. 2, pp. 296–303.

Kosmatka, J. B. and Lakes, R. C. (1996) "Extension-Twist Behavior of Initially Twisted Composite Spars for Till-Rotor Applications," *AIAA-96-1565-CP*.

Lagnese, J. E. and Lions, J. L. (1988) *Boundary Stabilization of Thin Plates*, Collection RMA, Masson, Paris.

Lagnese, J. E. (1989) *Boundary Stabilization of Thin Plates*, SIAM Studies in Applied Mechanics, SIAM, Philadelphia, PA.

Lake, R. C. (1990) "Investigation of the Use of Extension-Twist Coupling in Composite Rotor Blades for Application to Tiltrotor Aircraft," *The Third Workshop on Dynamic and Aeroelastic Stability Modeling of Rotorcraft Systems*, Duke University, Durham, North Carolina, March 12–14.

Lake, R. C., Nixon, M. W., Wilbur, M. L., Singleton, J. D. and Mirick, P. H. (1992) "A Demonstration of Passive Blade Twist Control Using Extension Twisting Coupling," *NASA TM-107642 AVSCOMTR-92-B-010*.

Lake, R. C., Izadpanah, A. P. and Bancom, R. B. (1993) "Experimental and Analytical Investigation of Dynamic Characteristics of Extension-Twist-Coupled Composite Tubular Spars," *NASA TP-3225, ARL-TR-30*.

Laudiero, F. and Savoia, M. (1991) "The Shear Strain Influence on the Dynamic of Thin-Walled Beams," *Thin-Walled Structures*, Vol. 11, pp. 375–407.

Librescu, L. (1975) *Elastostatics and Kinetics of Anisotropic and Heterogeneous Shell-Type Structures*, Noordhoff, Leyden, The Netherlands.

Librescu, L. and Thangjitham, S. (1991) "Analytical Studies on Static Aeroelastic Behavior of Forward Swept Composite Wings Structures," *Journal of Aircraft*, Vol. 28, No. 2, pp. 151–157.

Librescu, L. and Song, O. (1992) "On the Static Aeroelastic Tailoring of Composite Aircraft Swept Wings Modelled as Thin-Walled Beam Structures," *Composites Engineering*, Vol. 2, Nos. 5–7 (special issue: *Use of Composites in Rotorcraft and Smart Structures*), pp. 497–512.

Librescu, L., Meirovitch, L. and Song, O. (1993a) "A Refined Structural Model of Composite Aircraft Wings for the Enhancement of Vibrational and Aeroelastic Response Characteristics," in *Proceedings of the 34th AIAA/ASME/ASCE/AHS/ASC Structures, Structural Dynamics and Materials Conference*, LaJolla, CA, April 19–22, AIAA Paper 93-1536CP.

Librescu, L., Song, O. and Rogers, C. A. (1993b) "Adaptive Vibrational Behavior of Cantilevered Structures Modelled as Composite Thin-Walled Beams," *International Journal of Engineering Science*, Vol. 31, No. 5, pp. 775–792.

Librescu, L., Meirovitch, L. and Song, O. (1996a) "Integrated Structural Tailoring and Control Using Adaptive Materials for Advanced Aircraft Wings," *Journal of Aircraft*, Vol. 33, No. 1, Jan.–Feb., pp. 203–213.

Librescu, L., Meirovitch, L. and Song, O. (1996b) "Refined Structural Modeling for Enhancing Vibrational and Aeroelastic Characteristics of Composite Aircraft Wings," *La Recherche Aerospatiale*, Vol. 1, pp. 23–35.

Librescu, L., Meirovitch, L. and Na, S. S. (1997) "Control of Cantilever Vibration Via Structural Tailoring and Adaptive Materials," *AIAA Journal*, Vol. 35, No. 8, August, pp. 1309–1315.

Librescu, L. and Na, S. S. (1997) "Vibration and Dynamic Response Control of Cantilevers Carrying Externally Mounted Stores," *Journal of the Acoustical Society of America*, Vol. 102, No. 6, December, pp. 3516–3522.

Librescu, L. and Na, S. S. (1998a) "Bending Vibration Control of Cantilevers Via Boundary Moment and Combined Feedback Control Law," *Journal of Vibration and Controls*, Vol. 4, No. 6, pp. 733–746.

Librescu, L. and Na, S. S. (1998b) "Boundary Control of Free and Forced Oscillation of Shearable Thin-Walled Beam Cantilevers," *European Journal of Mechanics/A Solids*, Vol. 17, 4, pp. 687–700.

Librescu, L., Na, S. S. and Park, S. (2000) "Dynamics and Active Feedback Control of Wing-Type Structures Modeled as Thin-Walled Beams: Implications of Beam Cross-Section Nonuniformity," *Journal of the Chinese Society of Mechanical Engineers*, Vol. 21, No. 1, pp. 87–96.

Librescu, L. and Na, S. S. (2001) "Active Vibration Control of Doubly Tapered Thin-Walled Beams Using Piezoelectric Actuation," *Thin-Walled Structures*, Vol. 39, No. 1, pp. 65–82.

Librescu, L., Qin, Z. and Ambur, D. R. (2003) "Implications of Warping Restraint on Statics and Dynamics of Elastically Tailored Thin-Walled Composite Beams," *International Journal of Mechanical Sciences*, Vol. 45, No. 8, pp. 1247–1267.

Loewy, R. G. (1997) "Recent Developments in Smart Structures with Aeronautical Applications," *Smart Materials and Structures*, Vol. 5, Oct. pp. 11–42.

Meirovitch, L. (1997) *Principles and Techniques of Vibrations*, Prentice Hall, Englewood Cliffs, NJ.

Mindlin, R. D. (1951) "Influence of Rotatory Inertia and Shear Flexural Motions of Isotropic, Elastic Plates," *Journal of Applied Mechanics*, Trans. ASME, Vol. 18, No. 1, pp. 31–38.

Na, S. S. and Librescu, L. (1998) "Oscillation Control of Cantilevers via Smart Materials Technology and Optimal Feedback Control: Actuator Location and Power Consumption Issues," *Smart Materials and Structures*, Vol. 7, No. 6, pp. 833–842.

Na, S. S. and Librescu, L. (1999) "Dynamic Behavior of Aircraft Wings Modeled as Doubly-Tapered Composite Thin-Walled Beams," PVP, Vol. 398, Recent Advances in Solids and Structures, 1999, Y. W. Kwon and H. H. Chung (Eds.), 1999 ASME International Mechanical Engineering Congress Exposition, Nashville, TN, November 14–19, pp. 59–68.

Na, S. S. and Librescu, L. (2000a) "Dynamic Response of Adaptive Cantilevers Carrying External Stores and Subjected to Blast Loading," *Journal of Sound and Vibration*, Vol. 231, No. 4, pp. 1039–1055.

Na, S. S. and Librescu, L. (2000b) "Optimal Vibration Control of Thin-Walled Anisotropic Cantilevers Exposed to Blast Loading," *Journal of Guidance, Control, and Dynamics*, Vol. 23, No. 3, pp. 491–500.

Na, S. S. and Librescu, L. (2001a) "Dynamic Response of Elastically Tailored Adaptive Cantilevers of Nonuniform Cross Section Exposed to Blast Pressure Pulses," *International Journal of Impact Engineering*, Vol. 25, No. 9, pp. 847–867.

Na, S. S. (2001b) Personal Communication.

Oh, S-Y., Song, O. and Librescu, L. (2003) "Effects of Pretwist and Presetting on Coupled Bending Vibrations of Rotating Composite Beams," *International Journal of Solids and Structures*, Vol. 40, pp. 1203–1224.

Park, C. and Chopra, I. (1996) "Modeling Piezoceramic Actuation of Beams in Torsion," *AIAA Journal*, Vol. 34, No. 12, pp. 2582–2589.

Qin, Z. and Librescu, L. (2001) "Static and Dynamic Validations of a Refined Thin-Walled Composite Beam Model," *AIAA Journal*, Vol. 39, No. 12, pp. 2422–2424.

Qin, Z. and Librescu, L. (2002a) "Static/Dynamic Solutions and Validation of a Refined Anisotropic Thin-Walled Beam Model," AIAA-2002-1394, in *Proceedings of 43rd AIAA/ASME/ASCE/AHS/ASC Structures, Structural Dynamics and Materials Conference, 10th AIAA/ASME/AHS Adaptive Structures Conference*, April 22–25, Denver, Co.

Qin, Z. and Librescu, L. (2002b) "On a Shear-Deformable Theory of Anisotropic Thin-Walled Beams: Further Contribution and Validations," *Composite Structures*, Vol. 56, No. 4, June, pp. 435–358.

Rao, J. S. (1991) *Turbomachine Blade Vibration*, John Wiley & Sons, New York, Chester, Toronto.

Rehfield, L. W., Atilgan, A. R. and Hodges, D. H. (1990) "Nonclassical Behavior of Thin-Walled Composite Beams with Closed Cross Sections," *Journal of the American Helicopter Society*, Vol. 35, pp. 42–50.

Song, O. and Librescu, L. (1991) "Free Vibration and Aeroelastic Divergence of Aircraft Wings Modelled as Composite Thin-Walled Beams," in *Proceedings of the 32nd Structures, Structural Dynamics and Material Conference*, Baltimore, Maryland, Paper AIAA 91-1187-CP.

Song, O., Librescu, L. and Rogers, C. A. (1992) "Application of Adaptive Technology to Static Aeroelastic Control of Wing Structures," *AIAA Journal*, Vol. 30, No. 12, pp. 2882–2889, December.

Song, O. and Librescu, L. (1993) "Free Vibration of Anisotropic Composite Thin-Walled Beams of Closed Cross-Section Contour," *Journal of Sound and Vibration*, Vol. 167, No. 1, pp. 129–147.

Song, O. and Librescu, L. (1995) "Bending Vibration of Cantilevered Thin-Walled Beams Subjected to Time-Dependent External Excitations," *Journal of the Acoustical Society of America*, Vol. 98, No. 1, June, pp. 313–319.

Song, O. and Librescu, L. (1996) "Bending Vibrations of Adaptive Cantilevers with External Stores," *International Journal of Mechanical Sciences*, Vol. 38, No. 5, pp. 483–498.

Song, O. and Librescu, L. (1997a) "Structural Modeling and Free Vibration Analysis of Rotating Composite Thin-Walled Beams," *Journal of the American Helicopter Society*, Vol. 42, No. 4, October, pp. 358–369.

Song, O. and Librescu, L. (1997b) "Anisotropy and Structural Coupling on Vibration and Instability of Spinning Thin-Walled Beams," *Journal of Sound and Vibration*, Vol. 204, No. 3, pp. 477–494.

Song, O., Ju, J. S. and Librescu L. (1998) "Dynamic Response of Anisotropic Thin-Walled Beams to Blast and Harmonically Oscillating Loads," *International Journal of Impact Engineering*, Vol. 21, No. 8, pp. 663–682.

Song, O., Kim, J-B. and Librescu, L. (2001) "Synergistic Implications of Tailoring and Adaptive Materials Technology on Vibration Control of Anisotropic Thin-Walled Beams," *International Journal of Engineering Science*, Vol. 39, No. 1, December, pp. 71–94.

Song, O., Oh, S-Y. and Librescu, L. (2002) "Dynamic Behavior of Elastically Tailored Rotating Blades Modeled as Pretwisted Thin-Walled Beams and Incorporating Adaptive Capabilities," *International Journal of Rotating Machinery*, Vol. 8, No. 1, pp. 13–25.

Sunar, M. and Rao, S. S. (1999) "Recent Advances in Sensing and Control of Flexible Structures via Piezoelectric Materials Technology," *Applied Mechanics Reviews*, Vol. 52, No. 1, pp. 1–16.

Thangjitham, S. and Librescu, L. (1991) "Free Vibration Characteristics of Anisotropic Composite Wing Structures," in *Proceedings of the 32nd SDM Structures, Structural Dynamics and Material Conference*, Paper AIAA-01-1185, Baltimore, Maryland, April.

Tzou, H. S. and Zhong, J. P. (1992) "Adaptive Piezoelectric Structures: Theory and Experiment," *Active Materials and Adaptive Structures, Materials and Structures Series*, G. J. Knowles (Ed.), Institute of Physics Publ., pp. 219–224.

Tzou, H. S. (1993) *Piezoelectric Shells, Distributed Sensing and Control of Continua*, Kluwer Academic Publication, Dordrecht.

Tzou, H. S., Johnson, D. D. and Liu, K. J. (1999) "Damping Behavior of Cantilevered Structronic Systems with Boundary Control," *Journal of Vibration and Acoustics*, Trans. ASME, Vol. 126, July, pp. 402–407.

Weisshaar, T. A. and Foist, B. L. (1985) "Vibration Tailoring of Advanced Composite Lifting Surfaces," *Journal of Aircraft*, Vol. 25, No. 4, pp. 364–371.

Chapter 7

DYNAMIC RESPONSE TO TIME-DEPENDENT EXTERNAL EXCITATION

7.1 INTRODUCTION

Most research on dynamic response of composite structures relates to *plates and shells* (see e.g. Birman and Bert, 1987; Cederbaum et al., 1988, 1989; Librescu and Nosier, 1990; Nosier et al., 1990; Reddy, 1997; Vinson, 1999), and solid beams (Thangjitham and Librescu, 1991; Icardi et al., 2000). Little work has been done on dynamic response of thin-walled beams impacted by blast pulses: see e.g. Bank and Kao (1988), Song and Librescu, (1995), Song et al. (1998, 2001), Librescu and Na (1998a-c), Na and Librescu, (1998, 2000, 2001, 2002). The absence of such results is more intriguing as this structural model is basic when dealing with a number of important constructions such as airplane wing and fuselage, helicopter and turbine blades, tilt rotor aircraft, marine constructions, as well as many other structural systems widely used in mechanical engineering.

During their operational life, the structure of aeronautical and space vehicles as well as of marine and terrestrial ones are exposed to a variety of time-dependent loads arising from blast, sonic boom, shock wave, fuel explosion, thermal impact, etc. The next generation of aeronautical and space vehicles is likely to feature increasing structural flexibilities and operate in more severe environmental conditions. It is thus vital to understand their response to such blast pulses.

The new composite material structures exhibit distinguishing features as compared to their metallic counterparts. For this reason, in order to accurately predict the response behavior of such structures, refined structural models including anisotropy, transverse shear as well as other non-classical effects must be considered.

The aim of this chapter is threefold: (i) determine the response of thin-walled cantilevers to time-dependent excitation, (ii) discuss the use of composite tailoring, and (iii) analyse the dynamic response of *smart* thin-walled structures.

7.2 GOVERNING EQUATIONS

7.2-1 Shearable Model

As before, we consider a single-cell thin-walled beam of arbitrary closed- cross section. Both the global coordinate system (x, y, z) and the local one (n, s, z) are used. The basic assumptions are as follows:

(i) beam cross-sections do not deform in their own planes,

(ii) transverse shear effects are included in the structural model,

(iii) the twist varies along the beam span, i.e., the rate of twist $d\phi/dz$ is no longer assumed to be constant (as in the Saint-Venant torsional model), but a function of the spanwise coordinate, where $\phi' \equiv d\phi/dz$ constitutes a measure of the torsion-related nonuniform warping,

(iv) primary and secondary warping effects are included (the first being related to the warping displacement of the points on the midline cross-section, and the second to points off the midline contour),

(v) the hoop stress resultant is negligibly small compared to the remaining stresses, and

(vi) the constituent materials of the beam are anisotropic.

The numerical simulations will be carried out for a clamped-free thin-walled beam. In order to induce the transverse bending-twist elastic coupling that was proven to be beneficial in aircraft wing structures and helicopter blades, *the circumferentially asymmetric stiffness* configuration (CAS) will be used. Consistent with this ply-angle configuration, the governing equations associated with the transversal bending-twist coupled motion of thin-walled beams of nonuniform cross-section are:

- The system of equations (of eighth order) governing the twist-transversal bending-transverse shear:

$$\delta\phi: \quad -(\underline{a_{66}\phi''}\,)'' + (a_{73}\theta_x)' + [a_{65}(\underline{v_0' + \theta_x})\,]'' + (a_{77}\phi')'$$

$$+ m_z(z,t) = [(b_4 + b_5) + \delta_n(b_{14} + b_{15})]\ddot{\phi} - [(\underline{\underline{b_{10} + \delta_n b_{18}}})\ddot{\phi}']' \ ,$$

$$\delta\theta_x : \quad \left(a_{33}\theta_x'\right)' + \left(a_{37}\phi'\right)' - a_{55}(v_0' + \theta_x) + \underline{a_{56}\phi''} = \underline{\underline{(b_4 + \delta_n b_{14})\ddot{\theta}_x}},$$

$$\delta v_0 : \quad \left[a_{55}(v_0' + \theta_x)\right]' - \underline{(a_{56}\phi'')'} + p_y(z, t) = b_1 \ddot{v}_0, \tag{7.2-1a-c}$$

with the associated clamped-free boundary conditions

at $z = L$:

$$\delta\phi : \quad -(\underline{a_{66}\phi''})' + a_{73}\theta_x' + a_{77}\phi' + [\underline{a_{65}(v_0' + \theta_x)}]' = -\underline{\underline{(b_{10} + \delta_n b_{18})\ddot{\phi}'}},$$

$$\delta\phi' : \quad -\underline{a_{66}\phi''} + \underline{a_{65}(v_0' + \theta_x)} = 0, \tag{7.2-2a-d}$$

$$\delta v_0 : \quad a_{55}(v_0' + \theta_x) - \underline{a_{56}\phi''} = 0,$$

$$\delta\theta_x : \quad a_{33}\theta_x' + a_{37}\phi' = 0,$$

and at $z = 0$:

$$\phi = v_0 = \theta_x = \underline{\phi'} = 0. \tag{7.2-3a-d}$$

As in the previous chapter, the terms underscored by single and double dot-ted lines are associated with the warping inhibition and the warping inertia, respectively, whereas the term underscored by a undulated line is associated with the rotatory inertia.

In these equations $p_y(z, t)$ and $m_z(z, t)$ denote the distributed force per unit beam span and the distributed twist moment about the z axis, respectively.

7.2-2 Purely Bending Shearable Beam Model

Assuming that the beam has a large torsional stiffness, which yields $\phi \to 0$, the system of governing equations (7.2-1) reduces to

$$[a_{55}(v_0' + \theta_x)]' + p_y(z, t) = b_1 \ddot{v}_0,$$

$$(a_{33}\theta_x')' - a_{55}(v_0' + \theta_x) = \underline{\underline{(b_4 + \delta_n b_{14})\ddot{\theta}_x}}. \tag{7.2-4a,b}$$

The boundary conditions associated to the clamped-free beams are:

at the beam root ($z = 0$):

$$v_0 = \theta_x = 0, \tag{7.2-5a,b}$$

and at the beam tip $(z = L)$:

$$v_0' + \theta_x = 0, \quad \theta_x' = 0. \qquad (7.2\text{-}6a,b)$$

For this case, since the bending-stretching and in-plane stretching-shear coupling stiffness quantities, B_{ij} and A_{16} are zero, the 1-D stiffnesses a_{33} and a_{55} reduce to:

$$a_{33} = \oint \left[K_{11} y^2 + K_{44} \left(\frac{dx}{ds} \right)^2 \right] ds, \qquad (7.2\text{-}7a,b)$$

$$a_{55} = \oint \left[A_{66} \left(\frac{dy}{ds} \right)^2 + A_{44} \left(\frac{dx}{ds} \right)^2 \right] ds.$$

For uniform cross-section beams, the specialized counterpart of Eqs. (7.2-4) and of the associated boundary conditions (7.2-5) and (7.2-6) can be expressed in terms of $v_0(z, t)$ alone as we now show. To this end

a. Extract θ_x' from Eq. (7.2-4a),

b. differentiate θ_x' twice to give θ_x''',

c. differentiate Eq. (7.2-4b) once with respect to the z-coordinate and insert θ_x' and θ_x''' from steps a) and b).

This yields the governing equation in terms of the transverse displacement alone:

$$a_{33}\, v_0'''' - \frac{b_1}{a_{55}} \left[a_{33} \ddot{v}_0'' - (b_4 + \delta_n b_{14}) \ddot{v}_0 \right] + b_1 \ddot{v}_0 - (b_4 + \delta_n b_{14}) \ddot{v}_0''$$

$$+ \frac{1}{a_{55}} \left[a_{33}\, p_y''(z, t) - (b_4 + \delta_n b_{14}) \ddot{p}_y(z, t) \right] - p_y(z, t) = 0. \quad (7.2\text{-}8)$$

Similar, treatment of the boundary conditions yields their expressions solely in terms of v_0:

at $z = 0$:

$$v_0 = 0; \quad v_0' + \frac{a_{33}}{a_{55}} \left(v_0''' - b_1 \frac{1}{a_{55}} \ddot{v}_0' + \frac{1}{a_{55}} p_y' \right) = 0, \qquad (7.2\text{-}9a,b)$$

and at $z = L$:

$$v_0'' - \frac{b_1}{a_{55}} \ddot{v}_0 + \frac{p_y}{a_{55}} = 0,$$

$$a_{33} v_0''' - (b_4 + \delta_n b_{14}) \ddot{v}_0' - b_1 \frac{a_{33}}{a_{55}} \ddot{v}_0' + \frac{a_{33}}{a_{55}} p_y' = 0. \qquad (7.2\text{-}10a,b)$$

Clearly, neglect of rotatory inertia terms simplifies these equations. Specialized versions of Eq. (7.2-8) have been used, among others, by Stephen (1981) to study the eigenfrequency characteristics of shearable isotropic solid beams and of their Euler-Bernoulli counterpart (obtained by letting $a_{55} \to \infty$ in Eqs. (7.2-8) through (7.2-10)), and by Senjanovic et al. (2001) to study pontoon transient vibration due to slamming. In the same context, the static counterpart of above equations coincides with those derived by Özbeck and Suhubi (1967).

7.2-3 Unshearable Beam Model

For the beam infinitely rigid in transverse shear (i.e. for the classical Bernoulli-Euler beam theory), Eqs. (7.2-1) through (7.2-3) reduce to

$$\delta v_0 : \quad \left(a_{33} v_0''\right)'' - \left(a_{37} \phi'\right)'' - p_y(z,t) = -b_1 \ddot{v}_0 + [\,\underline{(b_4 + \delta_n b_{14}) \ddot{v}_0'}\,]',$$

$$\delta \phi : \quad -(\,\underline{a_{66} \phi''}\,)'' + \left(a_{77}\phi'\right)' - \left(a_{73} v_0''\right)' + m_z(z,t)$$

$$= [(b_4 + b_5) + \delta_b (b_{14} + b_{15})] \ddot{\phi} \qquad (7.2\text{-}11a,b)$$

$$- [\,\underline{\underline{(b_{10} + \delta_n b_{18}) \ddot{\phi}'}}\,]'.$$

The boundary conditions are:

at $z = 0$:

$$\phi = \underline{\phi'} = v_0 = v_0' = 0, \qquad (7.2\text{-}12a\text{-}d)$$

and at $z = L$:

$$\delta v_0 : \quad (a_{33} v_0'')' - (a_{37}\phi')' = \underline{(b_4 + \delta_n b_{14}) \ddot{v}_0'} \,,$$

$$\delta v_0' : \quad -a_{33} v_0'' + a_{37}\phi' = 0,$$

$$\delta \phi : \quad -(\,\underline{a_{66}\phi''}\,)' + a_{77}\phi' - a_{73} v_0'' = -\underline{\underline{(b_{10} + \delta_n b_{18})}}\, \ddot{\phi}' \,, \qquad (7.2\text{-}13a\text{-}d)$$

$$\delta \phi' : \quad \underline{a_{66}\phi''} = 0.$$

For this case the transverse shear stiffness a_{55} and transverse shear -warping coupling stiffness a_{56} vanish.

If the torsional beam stiffness is very large, one can reduce Eqs. (7.2-11) through (7.2-13) to the purely bending case. The corresponding governing equation is:

$$\delta v_0 : (a_{33} v_0'')'' - p_y(z, t) = -b_1 \ddot{v}_0 + [(b_4 + \delta_n b_{14}) \ddot{v}_0']' , \qquad (7.2\text{-}14)$$

with the associated boundary conditions:

at $z = 0$:

$$v_0 = v_0' = 0, \qquad\qquad (7.2\text{-}15\text{a,b})$$

and at $z = L$:

$$\delta v_0 : (a_{33} v_0'')' = (b_4 + \delta_n b_{14}) \ddot{v}_0',$$

$$\delta v_0' : v_0'' = 0. \qquad\qquad (7.2\text{-}16\text{a,b})$$

The same equations can be obtained by letting $a_{55} \to \infty$ in Eqs. (7.2-8) through (7.2-10), that is, by considering the transverse shear stiffness infinite.

These equations show that for both shear deformable and their infinitely rigid in transverse shear beam counterparts, the governing equations have the same order, namely eight for beams featuring transverse bending-twist coupling, and fourth order for beams featuring purely transverse bending. Consistent with the equation order, four and two boundary conditions have to be prescribed at each edge.

7.3 TIME-DEPENDENT EXTERNAL EXCITATIONS

7.3-1 Explosive Pressure Pulses

For blast loadings, various analytical expressions have been proposed (see e.g. Crocker, 1967; Crocker and Hudson, 1969; Cheng and Benveniste, 1968; Houlston et al., 1985; Gupta et al., 1985). These have been used in varous contexts by Nosier et al. (1990), Librescu and Nosier (1990), Song et al. (1998), Librescu and Na, (1998), Na and Librescu, (2000, 2001, 2002, 2004), Marzocca et al. (2001), Chandiramani et al. (2004), Librescu et al. (2004), and Hause and Librescu (2005). The blast wave reaches its peak in such a short time that the structure can be assumed to be loaded instantly. Experimental evidence suggests that the pressure generated by an explosive blast or a sonic boom is uniformly distributed over the beam span. The overpressure associated with the blast pulses can be described in terms of the modified Friedlander exponential decay equation:

$$p_y(s, z, t)(\equiv p_y(t)) = P_m \left(1 - \frac{t}{t_p}\right) e^{-a't/t_p}, \qquad (7.3\text{-}1)$$

where the negative phase of the blast is included. In Eq. (7.3-1), P_m denotes the peak reflected pressure in excess to the ambient one; t_p denotes the positive phase duration of the pulse measured from the time of impact of the structure, and a' denotes a decay parameter that has to be adjusted to approximate the overpressure signature from the blast tests. A typical depiction of the ratio p_y/P_m vs. time for various values of the ratio a'/t_p and a fixed value of $t_p (= 0.1s.)$ is displayed in Fig. 7.3-1. The triangular shock pulse may be viewed as a limiting case of Eq. (7.3-1), when $a'/t_p \to 0$.

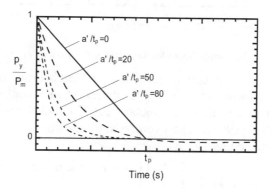

Figure 7.3-1: Explosive blast loading time-history

In such a case, the overpressure signature can be described by:

$$p_y(s, z, t)(\equiv p_y(t)) = \begin{cases} P_m \left(1 - t/t_p\right) \text{ for } 0 < t < t_p \\ \\ 0 \text{ for } 0 > t > t_p \end{cases} \qquad (7.3\text{-}2)$$

where, as in the previous cases, t_p denotes the duration of the shock pulse measured from the impact of the structure. In a few studies related to blast pressure signature, for the exponentially decaying pressure pulse simulating a high explosive blast, the pressure time-history is described by

$$p_y(s, z, t)(\equiv p_y(t)) = P_m e^{-kt}, \qquad (7.3\text{-}3)$$

where k denotes a decay parameter. Mäkinen (1999), see also Librescu et al. (2004), have provided analytical expressions for both the maximum pressure P_m and decay constant k, that are functions of the weight of the explosive charge, its type, and of the stand-off distance.

A sonic-boom, can be modeled as an N-shaped pressure pulse arriving at a normal incidence, see Fig. 7.3-2.

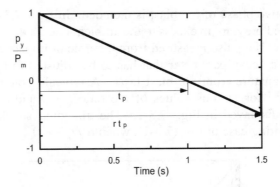

Figure 7.3-2: Sonic-boom loading time-history ($t_p = 1s, r = 1.5$)

Such a pulse corresponds to an idealized far-field overpressure produced by an aircraft flying supersonically in the Earth's atmosphere or by any supersonic projectile rocket or missile (see Crocker, 1967; Crocker and Hudson, 1969; Gottlieb and Ritzel, 1988). The overpressure signature of the N-wave shock pulse can be described by

$$p_y(s, z, t)(\equiv p_y(t)) = \begin{cases} P_m \left(1 - t/t_p\right) \text{ for } 0 < t < rt_p \\ \\ 0 \text{ for } t < 0 \text{ and } t > rt_p \end{cases} \quad (7.3\text{-}4)$$

where r denotes the shock pulse length factor, and P_m and t_p maintain the same meaning as for blast pulses. Note that: (i) for $r = 1$, the N- shaped pulse degenerates into a triangular pulse; (ii) for $r = 2$ a symmetric N-shaped pressure pulse is obtained; while (iii) for $1 < r < 2$ the N-shaped pulse becomes an asymmetric one.

Another special case of blast and sonic-boom pulses corresponds to a step pulse. This case is obtained from either Eq. (7.3-1) or (7.3-2) by considering $t_p \to \infty$.

The half-sine and rectangular pressure pulses described as:

$$p_y(s, z, t) \equiv p_y(t) = \begin{cases} \begin{array}{l} \left.\begin{array}{ll} P_m \sin \pi t/t_p & 0 \leq t \leq t_p \\ 0 & t > t_p \end{array}\right\} \text{half-sine pulse} \\ \\ \left.\begin{array}{ll} P_m & 0 \leq t \leq t_p \\ 0 & t > t_p \end{array}\right\} \text{rectangular pulse} \end{array} \end{cases}$$
$$(7.3\text{-}5)$$

will also be considered in the study of the dynamic response.

7.3-2 Time-Dependent Harmonic Excitation

The dynamic response to a harmonically time-dependent point load will also be considered. In addition to its intrinsic importance, this information can be used to obtain, the response to any time-dependent excitation. For a concentrated arbitrarily located load along the beam span

$$p_y(z, t) = P_0 \delta (z - z_0) \exp(i\omega_f t), \tag{7.3-6}$$

where $\delta(\cdot)$, z_0, P_0, and ω_f denote Dirac's distribution, spanwise location of the load, its amplitude and the excitation frequency, respectively. When p_y as represented by (7.3-6) is located off the longitudinal axis of symmetry by the offset x_0, then it gives rise to the concentrated moment

$$m_z(z, t) = P_0 \delta(z - z_0) x_0 \exp(i\omega_f t). \tag{7.3-7}$$

7.4 SOLUTION METHODOLOGY
7.4-1 Preliminaries

We can extend the analysis developed in Section 6.3 or we can use the modal approach to determine the dynamic response.

7.4-2 Laplace Transform Method
7.4-2a General Considerations

Considering again the free warping counterpart of Eqs. (6.2-10) through (6.2-12) and (6.3-2), and expressions f_i of Table 6.3-1, applying Laplace Transform with respect to the dimensionless spanwise coordinate η in conjunction with Eq. (6.3-6), and following the same steps indicated there, we find that Eqs. (6.3-11) can be written in martrix form as

$$
\begin{bmatrix}
g_{11} & g_{12} & g_{13} \\
g_{21} & g_{22} & g_{23} \\
g_{31} & g_{32} & g_{33}
\end{bmatrix}
\left\{
\begin{array}{c}
\overline{V} \\
\overline{X} \\
\overline{\Phi}
\end{array}
\right\}
=
\left\{
\begin{array}{c}
f_5 \hat{X}_{,\eta} + \hat{\Phi}_{,\eta} + e^{-s\eta_0} f_8 \\
\hat{X}_{,\eta} + f_2 \hat{\Phi}_{,\eta} \\
\hat{V}_{,\eta} + e^{-s\eta_0} f_7
\end{array}
\right\},
\tag{7.4-1}
$$

where $f_7 = -P_0 L/a_{55}$ and $f_8 = -P_0 \eta_0 L^2/a_{77}$ are dimensionless coefficients, additional to the ones supplied in Table 6.3-1.

In contrast to the free vibration problem, for the present case, the counterpart of the system of equations (6.3-12) is nonhomogeneous; the right-hand side member terms are related with the external loading. The vector $\hat{V}^T = \{\hat{V}_{,\eta}, \hat{X}_{,\eta}, \hat{\Phi}_{,\eta}\}$ can be determined, and, from Eq. (7.4-1) one can obtain $\overline{V}$, $\overline{X}$ and $\overline{\Phi}$, and via a Laplace inversion transform, one determine $v_0(\eta, \omega)$, $\theta(\eta, \omega)$ and $\phi(\eta, \omega)$.

From the frequency response functions $\overline{v}_0(z, \omega)$, $\overline{\theta}_x(z, \omega)$ and $\overline{\phi}(z, \omega)$ corresponding to a given excitation frequency ω, we can obtain their counterparts in the time domain and one can use this information as to determine the dynamic response to arbitrary time-dependent external excitation $F(z, t)$.

This conversion is accomplished as:

$$\left\{ \begin{array}{c} v_0(z,t) \\ \theta_x(z,t) \\ \phi(z,t) \end{array} \right\} = \int_0^z \int_0^t \left\{ \begin{array}{c} \bar{v}_0(\bar{z},\bar{t}) \\ \bar{\theta}_x(\bar{z},\bar{t}) \\ \bar{\phi}(\bar{z},\bar{t}) \end{array} \right\} F(z-\bar{z}, t-\bar{t}) d\bar{z} d\bar{t}, \qquad (7.4\text{-}2)$$

where

$$\left\{ \begin{array}{c} \bar{v}_0(\bar{z},\bar{t}) \\ \bar{\theta}_x(\bar{z},\bar{t}) \\ \bar{\phi}(\bar{z},\bar{t}) \end{array} \right\} = \mathcal{F}^{-1} \left\{ \begin{array}{c} v_0(\bar{z},\omega) \\ \theta_x(\bar{z},\omega) \\ \phi(\bar{z},\omega) \end{array} \right\}$$

$$= \frac{1}{2\pi} \int_{-\infty}^{+\infty} \left\{ \begin{array}{c} v_0(\bar{z},\omega) \\ \theta_x(\bar{z},\omega) \\ \phi(\bar{z},\omega) \end{array} \right\} \exp(i\omega\bar{t}) d\omega, \qquad (7.4\text{-}3)$$

$\mathcal{F}^{-1}$ denoting the inverse Fourier transform, while $\bar{z}$ and $\bar{t}$ denote dummy spanwise and time variables. In addition to its role in determining the response to arbitrary time-dependent loads, the frequency response function has a direct role in determining, for example, the dynamic response of an aircraft wing equipped with an engine located along the wing span.

7.4-2b Purely Transverse Bending Case

Laplace Transform Methodology will be applied to address the problem of the dynamic response to a harmonically time-dependent point load. In another context, this method was advanced by Thangjitham and Librescu (1991). For this purpose, consider a uniform thin-walled beam in pure transverse bending. First convert the governing system consisting of equations (7.2-4), the transverse load expressed by (7.3-6), and the boundary conditions, Eqs. (7.2-5) and (7.2-6), to a dimensionless form; write

$$\tilde{v}_0 \equiv v_0/L; \ \eta \equiv z/L; \ \eta_0 \equiv z_0/L; \ \tilde{t} \equiv \omega_0 t; \ \tilde{\omega}_f \equiv \omega_f/\omega_0; \ \tilde{F}_0 \equiv P_0/a_{55}, \qquad (7.4\text{-}4)$$

where $\omega_0 \left(\equiv \beta_0^2 (a_{33}/b_1 L^4)^{1/2}, \ \beta_0 = 1.8751 \right)$ is a reference frequency. For a harmonic time-dependent input, due to the linear character of the governing equations, the response is also harmonic, of the same frequency.

Upon replacing in Eqs. (7.2-4) and boundary conditions (7.2-5) and (7.2-6) $\tilde{v}_0$ and θ_x as given by:

$$\left\{ \begin{array}{c} \tilde{v}_0(\eta; \tilde{\omega}_f) \\ \theta_x(\eta; \tilde{\omega}_f) \end{array} \right\} = \left\{ \begin{array}{c} V(\eta) \\ S(\eta) \end{array} \right\} \exp(i\tilde{\omega}_f \tilde{t}), \qquad (7.4\text{-}5)$$

in conjunction with (7.4-4), the governing system becomes:

$$V,_{\eta\eta} + S,_{\eta} + \tilde{F}_0\delta(\eta - \eta_0) + \frac{a_{33}\beta_0^4\tilde{\omega}_f^2}{a_{55}L^2} V = 0,$$

$$S,_{\eta\eta} - \frac{a_{55}}{a_{33}} L^2(V,_{\eta} + S) + \frac{(b_4 + \delta_n b_{14})}{b_1 L^2} \beta_0^4\tilde{\omega}_f^2 S = 0, \qquad \text{(7.4-6a,b)}$$

and the boundary conditions change to

at $\eta = 0$,

$$V = S = 0, \qquad \text{(7.4-7a,b)}$$

and

at $\eta = 1$,

$$V,_{\eta} + S = S,_{\eta} = 0. \qquad \text{(7.4-8a,b)}$$

In Eqs. (7.4-6) through (7.4-8), $(\cdot),_{\eta} \equiv d(\cdot)/d\eta$. An exact solution to the problem can be obtained via the Laplace Transform technique applied in the space domain.

Applying the Laplace transform to Eqs. (7.4-6) and (7.4-8) and using the transformed counterpart of the boundary conditions at $\eta = 0$, (Eqs. (7.4-7)), we obtain a systems of algebraic equations

$$\begin{bmatrix} \tilde{g}_{11} & \tilde{g}_{12} \\ \tilde{g}_{21} & \tilde{g}_{22} \end{bmatrix} \begin{Bmatrix} \overline{V} \\ \overline{S} \end{Bmatrix} = \begin{Bmatrix} \mu_1 \\ \mu_2 \end{Bmatrix} + \begin{Bmatrix} \hat{V},_{\eta} \\ \hat{S},_{\eta} \end{Bmatrix}. \qquad \text{(7.4-9)}$$

Here $\overline{V}\{\equiv \overline{V}(s) = \mathcal{L}(V(\eta))\}$; $\overline{S}\{\equiv \overline{S}(s) = \mathcal{L}(S(\eta))\}$ where s denotes Laplace transform variable; $\hat{V},_{\eta} \equiv dV/d\eta|_{\eta=0}$, while the elements $\tilde{g}_{ij}$ and μ_i are given by:

$$\tilde{g}_{11} = s^2 + \frac{a_{33}\beta_0^4\tilde{\omega}_f^2}{a_{55}L^2}; \quad \tilde{g}_{12} = s,$$

$$\tilde{g}_{21} = -\frac{a_{55}L^2}{a_{33}}s; \quad \tilde{g}_{22} = s^2 + \frac{(b_4 + \delta_n b_{14})\beta_0^4\tilde{\omega}_f^2}{b_1 L^2} - \frac{a_{55}L^2}{a_{33}}. \qquad \text{(7.4-10a-d)}$$

Upon solving Eqs. (7.4-9) for $\overline{V}(s)$ and $\overline{S}(s)$ one obtains:

$$\begin{Bmatrix} \overline{V}(s) \\ \overline{S}(s) \end{Bmatrix} = \frac{1}{\mathcal{G}} \begin{bmatrix} g_{11} & g_{12} \\ g_{21} & g_{22} \end{bmatrix} \begin{Bmatrix} -\tilde{F}_0 e^{-s\eta_0} + \hat{V},_{\eta} \\ \hat{S},_{\eta} \end{Bmatrix}, \qquad \text{(7.4-11)}$$

where

$$\mathcal{G} = \tilde{g}_{11}\tilde{g}_{22} - \tilde{g}_{12}\tilde{g}_{21}; \quad g_{11} = \tilde{g}_{22},$$
$$g_{12} = -\tilde{g}_{12}; \quad g_{21} = -\tilde{g}_{21}; \quad g_{22} = \tilde{g}_{11}. \tag{7.4-12a-e}$$

The inverse Laplace transform of Eq. (7.4-11) yields:

$$\left\{ \begin{array}{c} V(\eta) \\ S(\eta) \end{array} \right\} = \left[\begin{array}{cc} G_{11} & G_{12} \\ G_{21} & G_{22} \end{array} \right] \left\{ \begin{array}{c} -\tilde{F}_0 Y(\eta - \eta_0) \\ 0 \end{array} \right\} \tag{7.4-13}$$

$$+ \left[\begin{array}{cc} F_{11} & F_{12} \\ F_{21} & F_{22} \end{array} \right] \left\{ \begin{array}{c} \hat{V}_{,\eta} \\ \hat{S}_{,\eta} \end{array} \right\},$$

where

$$G_{ij} = \sum_{m=1}^{M} \frac{g_{ij}(s)}{d\mathcal{G}/ds} \bigg]_{s=s_m} e^{s_m(\eta-\eta_0)},$$

and

$$F_{ij} = \sum_{m=1}^{M} \frac{g_{ij}(s)}{d\mathcal{G}/ds} \bigg]_{s=s_m} e^{s_m \eta}. \tag{7.4-14a,b}$$

Here s_m denotes the mth root of the polynomial $\mathcal{G}$; $M(=4)$ denotes the number of roots of $\mathcal{G}$, while $Y(\cdot)$ denotes Heaviside's distribution.

Enforcement in conjunction with (7.4-13) of boundary conditions at $\eta = 1$, Eqs. (7.4-8), yields a system of equations for $\hat{V}_{,\eta}$ and $\hat{S}_{,\eta}$ given by:

$$\left[\begin{array}{cc} d_{11} & d_{12} \\ d_{21} & d_{22} \end{array} \right] \left\{ \begin{array}{c} \hat{V}_{,\eta} \\ \hat{S}_{,\eta} \end{array} \right\} = \left\{ \begin{array}{c} f_1 \\ f_2 \end{array} \right\}, \tag{7.4-15}$$

where d_{ij} and $f_i(i, j = 1, 2)$ are given by

$$d_{11} = (F_{11,\eta} + F_{21})|_{\eta=1}; \quad d_{12} = (F_{12,\eta} + F_{22})|_{\eta=1};$$
$$d_{21} = F_{21,\eta}|_{\eta=1}; \quad d_{22} = F_{22,\eta}|_{\eta=1}, \tag{7.4-16a-d}$$

$$f_1 = \tilde{F}_0 \left[G_{11,\eta} Y(\eta - \eta_0) + G_{11}\delta(\eta - \eta_0) + G_{21}Y(\eta - \eta_0) \right]_{\eta=1};$$
$$f_2 = \tilde{F}_0 \left[G_{21,\eta} Y(\eta - \eta_0) + G_{21}\delta(\eta - \eta_0) \right]_{\eta=1}. \tag{7.4-17a,b}$$

The response functions $\tilde{v}_0(\eta; \tilde{\omega})$ and $\theta_x(\eta; \tilde{\omega})$ corresponding to a given excitation frequency (referred to as frequency response functions), are determined by replacing the solution of Eq. (7.4-15) i.e., $\hat{V}_{,\eta}$ and $\hat{S}_{,\eta}$ into Eq. (7.4-13), considered in conjunction with Eq. (7.4-5). The natural frequencies of the composite thin-walled beam are obtained as the roots of the determinantal equation, $det(d_{ij}) = 0$.

Knowing the frequency response functions of the system, $\tilde{v}_0(\eta; \tilde{\omega})$ and $\theta_x(\eta; \tilde{\omega})$, we can find the response to any arbitrary time-dependent excitation $F(\eta; \tilde{t})$; response quantities $\tilde{v}_0(\eta; \tilde{t})$ and $\theta_x(\eta; \tilde{t})$ can be determined in terms of a convolution integral given by Eqs. (7.4-2) and (7.4-3).

7.4-3 Extended Galerkin Method

To address the dynamic response of thin-walled beams featuring transverse bending-transverse shear-twist cross-coupling to both arbitrary and harmonic time-dependent excitation via EGM, one should use Hamilton's principle in the form given by Eq. (6.3-13). If there is both a transverse load and moment, p_y and m_z, then

$$\delta \mathcal{J} = \int_0^L [p_y(z, t)\delta v_0(z, t) + m_z(z, t)\delta\phi(z, t)]dz. \tag{7.4-18}$$

Use in Eq. (7.4-18) of representations (6.3-14), yields

$$\delta \mathcal{J} = \mathbf{Q}^T \delta\mathbf{q}, \tag{7.4-19}$$

where

$$\mathbf{Q} = \int_0^L \{p_y \mathbf{V}_0^T, \ \mathbf{0}^T, \ m_z\Phi^T\}^T dz, \tag{7.4-20}$$

$\mathbf{q}$ being the overall generalized coordinate vector, Eq. (6.3-16).

Replacement of Eqs. (7.4-19) considered in conjunction with (7.4-20), (6.3-15) and (6.3-16) in Eq. (6.3-13), integrating with respect to time and recognizing that $\delta\mathbf{q} = \mathbf{0}$ at $t = t_0$, t_1, gives

$$\mathbf{M}\ddot{\mathbf{q}}(t) + \mathbf{K}\mathbf{q}(t) = \mathbf{Q}(t). \tag{7.4-21}$$

This equation constitutes a set of coupled differential equations. Here $\mathbf{q}$ is the overall generalized coordinate vector, Eq. (6.3-16), $\mathbf{M}$ and $\mathbf{K}$ are $3N \times 3N$ square matrices, the latter containing the corrective boundary terms, while $\mathbf{Q}(\equiv \mathbf{Q}(t))$ is the input $3N \times 1$ column matrix.

This equation can provide the vector $\mathbf{q}(t)$; Eqs. (6.3-14) provide the dynamic response.

To this end, multiplying Eq. (7.4-21) by $\mathbf{M}^{-1}$, defining the generalized velocities $\dot{\mathbf{q}}$ as auxiliary variables by means of the matrix identity $\dot{\mathbf{q}} - \dot{\mathbf{q}} = \mathbf{0}$, and considering the $6N \times 1$ state-space vector

$$\mathbf{x} = \left\{ \begin{array}{c} \mathbf{q} \\ \dot{\mathbf{q}} \end{array} \right\}, \tag{7.4-22}$$

one can express (7.4-21) in state space form as:

$$\dot{\mathbf{x}} = \mathbf{A}\mathbf{x} + \mathbf{B}\mathbf{Q}, \tag{7.4-23}$$

where

$$A = \begin{bmatrix} 0 & I \\ -M^{-1}K & 0 \end{bmatrix} \text{ and } B = \begin{Bmatrix} 0 \\ M^{-1} \end{Bmatrix}. \tag{7.4-24a,b}$$

The homogeneous solution of Eq. (7.4-23) has the exponential form

$$x(t) = Xe^{\lambda t}, \tag{7.4-25}$$

in which X is a constant vector and λ a constant scalar, both generally complex. Substituting Eq. (7.4-25) into Eq. (7.4-23) with $Q = 0$ and dividing through by $e^{\lambda t}$, one obtains the eigenvalue problem

$$AX = \lambda X, \tag{7.4-26}$$

which can be solved for the eigenvalues λ_r and eigenvectors $X_r(r = 1, 2, \ldots 6N)$.

To derive the particular solution, we first substitute Eqs. (7.3-6) and (7.3-7) into Eq. (7.4-20) and obtain the generalized force vector

$$Q(t) = \int_0^L [P_0\delta(z - z_0)V_0^T(z), \ 0^T, \ P_0x_0\delta(z - z_0)\Phi^T(z)]dze^{i\omega_f t}$$

$$\equiv Q_0 e^{i\omega_f t}, \tag{7.4-27}$$

where ω_f is the excitation frequency, and

$$Q_0 = \{P_0V_0^T(z_0), \ 0^T, \ P_0x_0\Phi^T(z_0)\}^T, \tag{7.4-28}$$

is a constant vector. Then, letting the particular solution of Eq. (7.4-23) under the form

$$x(t) = X_p e^{i\omega_f t}, \tag{7.4-29}$$

we conclude that

$$X_p = [i\omega_f I - A]^{-1}BQ_0, \tag{7.4-30}$$

in which I is the identity matrix. Equation (7.4-29) can be used in conjunction with Eqs. (6.3-14) to obtain the frequency response behavior corresponding to v_0, θ_x and ϕ.

7.4-4 Normal Mode Approach

In Section 5.9 it was shown that the normal mode approach can be used to address the problem of the dynamic response to arbitrary time-dependent loads, by expressing the displacements in the form,

$$\{v_0(z, t), \theta_x(z, t), \phi(z, t)\} = \sum_{j=1}^{\infty} [v_{or}(z), \theta_{xr}(z), \phi_r(z)] q_r(t), \tag{7.4-31}$$

where $v_{or}(z)$, $\theta_{xr}(z)$ and $\phi_r(z)$ play the role of eigenfunctions obtained from the free vibration analysis, while $q_r(t)$ are the generalized coordinates.

Using the orthogonality condition for the eigenfunctions corresponding to the distinct eigenfrequencies ω_m and ω_n expressed in Chapter 5 as

$$\int_0^L [b_1 v_{om} v_{on} + (b_4 + \delta_n b_{14})\theta_{xm}\theta_{xn} + [(b_4 + b_5) + \delta_n(b_{14} + b_{15})]\phi_m\phi_n$$

$$+ (b_{10} + \delta_n b_{18})\phi_m'\phi_n'] \, dz = 0, \quad \text{when } m \neq n \qquad (7.4\text{-}32)$$

and defining the norm N_m

$$N_m = \int_0^L [b_1 v_{om}^2 + (b_4 + \delta_n b_{14})\theta_{xm}^2$$

$$+ [(b_4 + b_5) + \delta_n(b_{14} + b_{15})]\phi_m^2 + (b_{10} + \delta_n b_{18})(\phi_m')^2] \, dz, \quad (7.4\text{-}33)$$

one can obtain the dynamic response

$$q_r(t) = q_r(0)\cos\omega_r t + \frac{\dot{q}_r(0)}{\omega_r}\sin\omega_r t + \frac{1}{\omega_r N_r}\int_0^t Q_r(\tau)\sin\omega_r(t-\tau)d\tau.$$

$$(7.4\text{-}34)$$

Here $q_r(0)$ and $\dot{q}_r(0)$ are the initial conditions

$$q_r(0) = \frac{1}{N_r}\int_0^L \left[b_1\hat{v}_0 v_{or} + \underline{(b_4 + \delta_n b_{14})\hat{\theta}_x\theta_{xr}} \right.$$

$$+ [(b_4 + b_5) + \delta_n(b_{14} + b_{15})]\hat{\phi}\phi_r + \underline{\underline{(b_{10} + \delta_n b_{18})\hat{\phi}'\phi_r'}} \left. \right] dz,$$

$$\dot{q}_r(0) = \frac{1}{N_r}\int_0^L \left[b_1\dot{\hat{v}}_0 v_{or} + \underline{(b_4 + \delta_n b_{14})\dot{\hat{\theta}}_x\theta_{xr}} \right. \qquad (7.4\text{-}35\text{a,b})$$

$$+ [(b_4 + b_5) + \delta_n(b_{14} + b_{15})]\dot{\hat{\phi}}\phi_r + \underline{\underline{(b_{10} + \delta_n b_{18})\dot{\hat{\phi}}'\phi_r'}} \left. \right] dz,$$

where $v_0(z, t = 0) \equiv \hat{v}_0(z, 0)$; $\dot{v}_0(z, t = 0) \equiv \dot{\hat{v}}_0(z, 0)$; $\theta_x(z, t = 0) \equiv \hat{\theta}_x(z, 0)$; $\dot{\theta}_x(z, t = 0) \equiv \dot{\hat{\theta}}_x(z, 0)$; $\phi(z, t = 0) \equiv \hat{\phi}(z, 0)$; $\dot{\phi}(z, t = 0) \equiv \dot{\hat{\phi}}(z, 0)$.

$Q_r(t)$ denotes the generalized load

$$Q_r(t) = \frac{\int_0^L [p_y(z, t)v_{or}(z) + m_z(z, t)\phi_r(z)] \, dz}{N_r}. \qquad (7.4\text{-}36)$$

With Eqs. (7.4-34), (7.4-36) and (7.4-31), the full solution of the dynamic response of thin-walled beams to arbitrary time-dependent excitations can be obtained.

7.5 NUMERICAL SIMULATIONS

7.5-1 Preliminaries

To get an understanding of the implications of a number of effects, we will analyse different types of structures and time-dependent loads.

7.5-2 Box-Beam in Pure Transverse Bending

Consider the frequency response of a cantilevered box-beam in pure transverse motion (see Fig. 7.5-1). The beam features transversely-isotropic properties, the plane of isotropy being parallel at each point to the mid-surface of the beam walls.

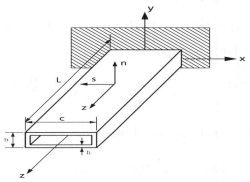

Figure 7.5-1: Geometry of the cantilevered box-beam

Suppose that the concentrated load is located along the centroidal axis of the beam so that it can induce pure bending only. The geometrical and material properties are displayed in Tables 6.4-1 and 7.5-1, respectively. Here E and G' denote the in-plane Young's modulus and transverse shear modulus, respectively.

Table 7.5-1: Material properties

E [GPa]	E/G'	$\nu = \nu'$	$\rho[\text{kg/m}^3]$
206.84	5	0.25	7495

Figures 7.5-2 and 7.5-3 display the distributions of $V(\eta)$ and $S(\eta)$ determined for three selected locations of $p_y(\eta, t)$ along the beam span and for a prescribed normalized excitation frequency $\tilde{\omega}_f \equiv \omega_f/\omega_0 = 0.2$.

The results show that, irrespective of the location of the concentrated load, both transverse deflection and rotation amplitudes increase toward the beam tip. When the location of the point load is shifted toward the beam tip, the deflection and rotation increase. Figure 7.5-3 reveals that beyond the loading point location, the rotation angles of beam cross-sections feature a uniform variation.

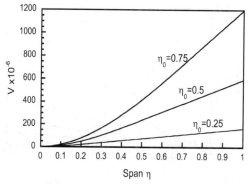

Figure 7.5-2: Spanwise distribution of $V \times 10^{-6}$ for three locations of the loading and for $\tilde{\omega}_f = 0.2$

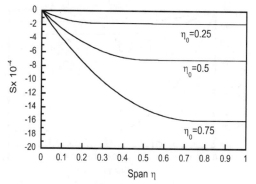

Figure 7.5-3: Spanwise distribution of $S \times 10^{-4}$ for three locations of the loading and for $\tilde{\omega}_f = 0.2$

Figures 7.5-4 and 7.5-5 depict the spanwise distributions of $V(\eta)$ and $S(\eta)$ for three selected values of the amplitude of the point load $p_y(\eta, t)$, located at $\eta_0 = 0.5$ and for $\tilde{\omega}_f = 0.2$.

From Figs. 7.5-4 and 7.5-5 one can get the obvious conclusion that the increase of the load amplitude yields an increase of both the deflection and rotation in each cross-section of the beam.

Figure 7.5-6 display the spanwise distributions of $V(\eta)$ for selected values of the dimensionless excitation frequency of the concentrated load of amplitude $\tilde{F}_0 = 5 \times 10^{-6}$ applied at $\eta_0 = 0.75$.

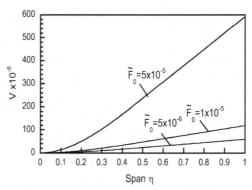

Figure 7.5-4: Spanwise distribution of $V \times 10^{-5}$ for three amplitudes of the loading applied at $\eta_0 = 0.5$ and for $\tilde{\omega}_f = 0.2$

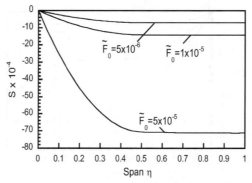

Figure 7.5-5: Spanwise distribution of $S \times 10^{-4}$ for three amplitudes of the loading applied at $\eta_0 = 0.5$ and for $\tilde{\omega}_f = 0.2$

The first three natural frequencies of the structure are $\tilde{\omega}(\equiv \omega/\omega_0) = 1; 6.16$ and 16.9; as Figs. 7.5-7 reveal, for excitation frequencies near these eigen-frequencies, the deflection approach infinity. A similar trend is experienced by the elastic rotation, as well. These plots show that due to the absence of the structural damping, for values of the excitation frequencies near the natural frequencies, the displacement amplitudes become infinite, indicating the occurrence of the resonance phenomenon. As it will be shown later, the occurrence of this phenomenon can be postponed by using in a proper way the directionality property of composite material structures, and even eliminated altogether by implementing the velocity feedback control.

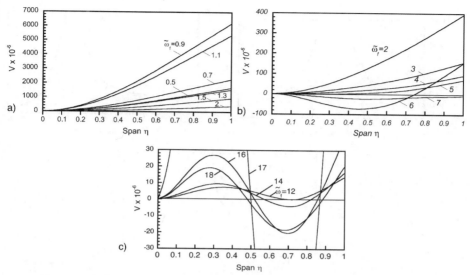

Figure 7.5-6: Spanwise distribution of $V \times 10^{-6}$ for selected excitation frequencies of the loading of amplitude $\tilde{F}_0 = 5 \times 10^{-6}$ applied at $\eta_0 = 0.75$. a) $0.5 \le \tilde{\omega}_f \le 2$, b) $2 \le \tilde{\omega}_f \le 7$, c) $12 \le \tilde{\omega}_f \le 18$

7.5-3 Bi-Convex Symmetric Cross-Section Beams Featuring Bending-Twist Cross-Coupling

7.5-3a Steady-State Response

The elastic and geometrical characteristics of the beam are displayed in Tables 6.4-1 and 6.4-2. Suppose a harmonically oscillating concentrated load located along the centroidal axis of the beam. In order to validate the accuracy of predictions provided by the Extended Galerkin Method (EGM) versus the ones supplied by the exact Laplace Transform Method (LTM), in Table 7.5-2 we present the steady-state dimensionless deflection amplitudes at the beam tip, as supplied by both methods. We consider a load of amplitude $\tilde{F}_0 = 5 \times 10^{-9}$ at the mid-span ($\eta_0 = 0.5$).

Table 7.5-2: Comparison of steady-state dimensionless deflection amplitude at the beam tip, versus excitation frequency as predicted by *EGM* and *LTM*

Ply-angle (θ deg.)	Excitation frequency ω_f (rad/s)	Extended Galerkin Method	Laplace Transform Method
0	50	0.00278918	0.00278918
	100	0.000375133	0.000375139
90	50	0.0000413359	0.0000413359
	100	0.0000483027	0.0000483027

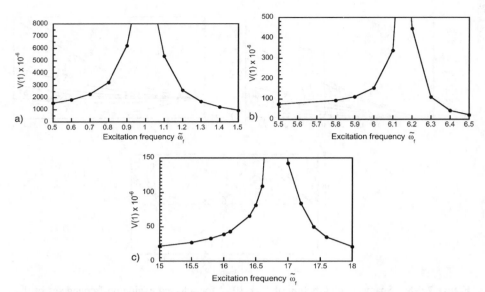

Figure 7.5-7: Deflection amplitude $V \times 10^{-6}$ at the beam tip vs. excitation frequency for a loading of amplitude $\tilde{F}_0 = 5 \times 10^{-6}$ applied at $\eta_0 = 0.75$, a) $0.5 \le \tilde{\omega}_f \le 1.5$, b) $5.5 \le \tilde{\omega}_f \le 6.5$, c) $15 \le \tilde{\omega}_f \le 18$

Table 7.5-3 displays the first three natural frequencies that correspond to two selected ply-angles and predicted by the *EGM* and *LTM*. In addition to the excellent agreement of predictions, these tables show that the increase in the ply-angle, that is accompanied by an increase of the bending stiffness a_{33}, increases the natural frequencies and reduces the steady-state deflection amplitude.

Table 7.5-3: The first three natural frequencies ω_i (rad/s), as predicted by *EGM* and *LTM*, and obtained for two selected ply-angles

Ply-angle	Extended Galerkin Method			Laplace Transform Method		
(θ deg.)	ω_1	ω_2	ω_3	ω_1	ω_2	ω_3
0	40.3	250.6	693.9	40.3	250.6	693.8
90	239.3	1280	3014	239.3	1280	3014

Figures 7.5-8 and 7.5-9 record the normalized steady-state beam tip deflection amplitudes versus the excitation frequency, for selected values of the ply-angle; with the increase of the ply-angle θ, and hence of the flexural stiffness a_{33}, the domains of resonance are shifted toward larger excitation frequencies.

At the same time, increase of θ narrows the ranges of excitation frequencies at which the resonance occurs. These figures show that this trend remains valid throughout the range of excitation frequencies.

As concerns the implications of the secondary warping (SW) on the steady-state response, results not displayed here reveal that due to its stiffening effect in bending, with the increase of the wall-thickness, its influence becomes stronger, this inducing a shift of the resonance domains toward larger excitation frequencies.

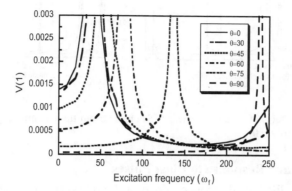

Figure 7.5-8: Effects of the ply-angle, θ (deg.), on steady-state deflection amplitude at the beam tip as a function of ω_f, for an oscillating load concentrated at the tip. The excitation frequency is in the range $\omega_f = 0–250$ rad/s

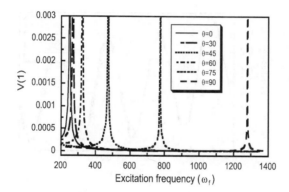

Figure 7.5-9: The counterpart of Fig. 7.5-8 for excitation frequencies in the range $\omega_f = 200–1400$ rad/s

Figures 7.5-10a,b highlight the effects of transverse shear (TS) on the frequency response. Figure 7.5-10a shows that at the flexurally stiffest ply-orientation $\theta = 90$ deg. transverse shear has a significant influence, especially at high or moderate excitation frequencies, by shifting the domains of resonant toward lower excitation frequencies. The results show that at high excitation frequencies, the classical theory discarding transverse shear effects, grossly overestimates the resonant frequency. However, for $\theta = 0$, at which the lowest flexural bending stiffness is experienced, the effects of transverse shear are significantly attenuated. In contrast to the significant effect of transverse shear, that of rotatory inertia on steady-state response is negligible.

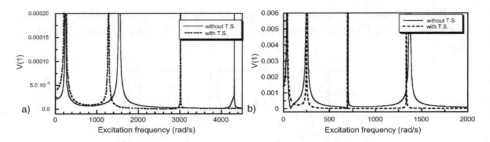

Figure 7.5-10: Effects of transverse shear (TS) on the beam tip dynamic response in bending. (a) $\theta = 90$ deg., (b) $\theta = 0$

7.5-3b Dynamic Response to Blast Pulses

Figure 7.5-11 displays the convergence of the transverse deflection solution at the beam tip, $\tilde{v}_0 (\equiv v_0(L, t)/L)$, achieved via the Extended Galerkin Method, with the number of terms in the polynomial representation.

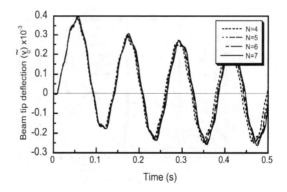

Figure 7.5-11: Convergence of the solution as per the *EGM* as *N* varies, for the dimensionless beam tip deflection time-history for an exponentially decaying load concentrated at the beam tip, $\theta = 45$ deg., $k = 10$, $\eta_0 = 0.5$

The load is concentrated at $\eta_0 = 0.5$ along the centroidal axis, and involves an exponentially decaying time-variation, i.e. $p_y(\eta, t) = P_0 \delta(\eta - \eta_0) \exp(-kt)$ where $\bar{F}_0 (\equiv P_0 L/a_{55}) = 5 \times 10^{-9}$. This graph shows, the extremely fast convergence of the numerical procedure and that four terms are sufficient.

Figures 7.5-12 and 7.5-13 display plots of the dimensionless deflection of the beam tip, $\tilde{v}_0$, and bending slope time-history of the beam tip, $\theta_x(L, t)$, when exposed to an explosive blast as described by Eq. (7.3-3); selected values of the ply-angle θ and a fixed value $k = 10$ have been considered.

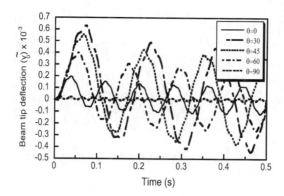

Figure 7.5-12: Influence of the ply-angle, θ (deg.) on the beam tip deflection time-history to exponentially decaying explosive pulse, ($k = 10$)

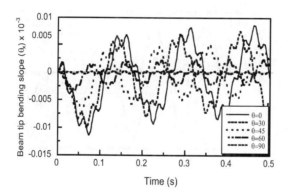

Figure 7.5-13: Influence of the ply-angle, θ (deg.) on beam tip bending slope time-history to an exponentially decaying explosive pulse, ($k = 10$)

The results reveal the strong role played by the directionality property of the composite materials; large ply-angles yielding the increase of the bending stiffness a_{33}, give smaller transversal deflections and rotations.

Figure 7.5-14 highlights the influence of the blast decay parameter on deflection time-history of the beam tip for a ply-angle, $\theta = 45$ deg; increasing the decay parameter results in the decrease of the amplitude of transverse displacement.

Figure 7.5-15 shows the influence of the time-duration of the triangular pulse on the beam tip deflection time-history; the increase of the duration of the pulse yields an increase of the displacement amplitude.

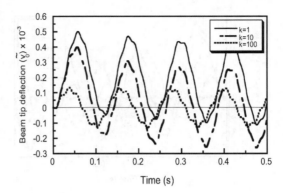

Figure 7.5-14: Influence of the blast decaying parameter on beam tip deflection time-history, (θ = 45 deg.)

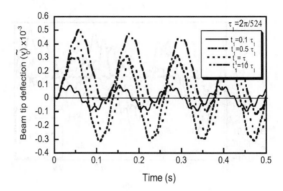

Figure 7.5-15: Influence of the time duration of the triangular pressure pulse on beam tip deflection time-history, (θ = 45 deg.)

7.6 CLOSED-LOOP DYNAMIC RESPONSE

7.6-1 Preliminaries

In Chapter 6, we considered the modeling of smart thin-walled beams, and addressed the closed-loop free vibration problem. The response of a composite structural system to time varying loads can be controlled either passively, by structural tailoring, or actively, by the use of piezoelectric actuation; we will consider both.

A control law relating the applied electrical field with the kinematical response of the host structure according to a prescribed functional relationship, leads to a dynamic boundary-value problem. Its solution yields the *closed-loop dynamic response* characteristics.

The pressure pulses considered in the numerical simulations feature small to moderate intensities; the goal of the control here is to contain and even suppress the induced vibrations, and not to prevent the damage inflicted by large loading intensities.

7.6-2 Closed-Loop Response to Harmonic Loads

Return to Section 6.6-4 and especially to the equations in Section 6.6-5a: we will consider Case (A) and Case (B) involving Eqs. (6.6-3) through (6.6-5) and Eqs. (6.6-6) through (6.6-8), respectively. For harmonically time-dependent loads concentrated at $z = z_0$, and for those distributed along the beam span, we use the load representations

$$\left\{ p_x^c(z_0, t), \, p_y^c(z_0, t), \, m_y^c(z_0, t), \, m_x^c(z_0, t) \right\}$$
$$= \left\{ f_1^c(z), \, f_2^c(z), \, f_3^c(z), \, f_4^c(z) \right\} \delta(z - z_0) \exp(i\omega_f t), \qquad (7.6\text{-}1\text{a})$$

and

$$\left\{ p_x^d(z, t), \, p_y^d(z, t) m_y^d(z, t), \, m_x^d(z, t) \right\}$$
$$= \left\{ f_1^d(z), \, f_2^d(z) f_3^d(z), \, f_4^d(z) \right\} \exp(i\omega_f t), \qquad (7.6\text{-}1\text{b})$$

where indexes c and d identify the concentrated and distributed loads, respectively.

The specialized counterpart of Eq. (7.4-27) becomes

$$\mathbf{Q}(t) = \left\{ \begin{array}{c} \int_0^L f_1^d(z) \mathbf{U}_0^T dz \\ \int_0^L f_2^d(z) \mathbf{V}_0^T dz \\ \int_0^L f_3^d(z) \mathbf{Y}_0^T dz \\ \int_0^L f_4^d(z) \mathbf{X}_0^T dz \end{array} \right\} e^{i\omega_f t} \qquad (7.6\text{-}2)$$

$$+ \left\{ \begin{array}{c} f_1^c(z_0) \mathbf{U}_0^T(z_0) \\ f_2^c(z_0) \mathbf{V}_0^T(z_0) \\ f_3^c(z_0) \mathbf{Y}_0^T(z_0) \\ f_4^c(z_0) \mathbf{X}_0^T(z_0) \end{array} \right\} e^{i\omega_f t} \equiv (\mathbf{F}_E^d + \mathbf{F}_E^c) e^{i\omega_f t},$$

and $\mathbf{BQ}(t) \equiv \mathbf{W}_f(t)$ can be expressed as

$$\mathbf{W}_f(t) = \left\{ \begin{array}{c} 0 \\ \mathbf{M}^{-1}(\mathbf{F}_E^d + \mathbf{F}_E^c) \end{array} \right\} e^{i\omega_f t}. \qquad (7.6\text{-}3)$$

The particular solution of Eq. (7.4-23) considered in conjunction with (7.6-2), that is of Eq. (7.4-30) can be used to address the problem of the frequency response. The same procedure is applicable to Case (B) as well.

7.6-3 Dynamic Response Control

7.6-3a Non-Optimal Control

For the non-optimal *dynamic response control* to arbitrary time-dependent loadings, Eq. (6.6-29) cast in state space form is used. Applying Laplace transform with respect to the time variable to each side of Eq. (6.6-29) and using the relation $\overline{\mathbf{W}}_f(s) = \mathbf{B}\bar{\mathbf{Q}}(s)$ we find

$$\bar{\mathbf{x}}(s) = (s\mathbf{I} - \mathbf{A})^{-1}\mathbf{x}(0) + (s\mathbf{I} - \mathbf{A})^{-1}\mathbf{B}\bar{\mathbf{Q}}(s), \qquad (7.6\text{-}4)$$

where the overbars denote Laplace transform of the unbarred original quantities, $\mathbf{x}(0)$ denotes the initial state vector, while s denotes Laplace transform variable.

The time response $\mathbf{x}(t)$ is obtained as the inverse Laplace transform of Eq. (7.6-4):

$$\mathbf{x}(t) = \mathcal{L}^{-1}\{\mathbf{T}(s)\mathbf{x}(0)\} + \mathcal{L}^{-1}[\mathbf{T}(s)\mathbf{B}\bar{\mathbf{Q}}(s)], \qquad (7.6\text{-}5a)$$

where

$$\mathbf{T}(s) \equiv (s\mathbf{I} - \mathbf{A})^{-1}, \qquad (7.6\text{-}5b)$$

and $\mathcal{L}^{-1}$ denotes the inverse Laplace transform operation.

In this form the solution for $\mathbf{x}(t)$ contains the contributions of both initial conditions, forcing function and of feedback control.

7.6-3b Optimal Control. Linear Quadratic Regulator

Optimal control based on the use of a full state feedback scheme such as linear quadratic regulator concept (LQR) will be considered. Equation (6.6-29) in conjunction with (6.6-30) can be cast in state-space form

$$\dot{\mathbf{x}}(t) = \hat{\mathbf{A}}\mathbf{x}(t) + \mathbf{B}\mathbf{Q}(t) + \mathbf{W}\mathbf{u}(t), \qquad (7.6\text{-}6)$$

where

$$\hat{\mathbf{A}} = \begin{bmatrix} \mathbf{0} & \mathbf{I} \\ -\mathbf{M}^{-1}\mathbf{K} & \mathbf{0} \end{bmatrix}, \quad \mathbf{B} = \left\{ \begin{array}{c} \mathbf{0} \\ \mathbf{M}^{-1} \end{array} \right\}, \quad \mathbf{W} = \left\{ \begin{array}{c} \mathbf{0} \\ -\mathbf{M}^{-1}\mathbf{F} \end{array} \right\}, \qquad (7.6\text{-}7a\text{-}c)$$

$\mathbf{F}$ being the piezoactuator influence vector.

Within the linear quadratic regulator (LQR) control algorithm, both the response of the closed-loop system and the control effort should be minimized simultaneously; we minimize the cost functional

$$J = \frac{1}{2}\int_{t_0}^{t_f} (\mathbf{x}^T\mathbf{Z}\mathbf{x} + \mathbf{u}^T\mathbf{R}\mathbf{u})dt, \qquad (7.6\text{-}8)$$

subject to Eq. (7.6-6), where the state weighting matrix and the control weighting matrix $\mathbf{Z}$ and $\mathbf{R}$, are positive semidefinite and positive definite symmetric, respectively, while t_0 and t_f denote the present and the final time, respectively. The weighting matrices $\mathbf{Z}$ and $\mathbf{R}$ are conveniently chosen to balance the rather conflicting requirements, namely of minimizing the response with that of minimizing the control effort; the optimal feedback control law is given by

$$\mathbf{u}(t) = -\mathbf{G}\mathbf{x}(t), \tag{7.6-9a}$$

where

$$\mathbf{G} = \mathbf{R}^{-1}\mathbf{W}^T\mathbf{P}, \tag{7.6-9b}$$

is the optimal gain matrix, while $\mathbf{P}$ is the positive-definite solution to the steady-state Riccati equation

$$\mathbf{Z} + \mathbf{P}\hat{\mathbf{A}} + \hat{\mathbf{A}}^T\mathbf{P} - \mathbf{P}\mathbf{W}\mathbf{R}^{-1}\mathbf{W}^T\mathbf{P} = 0. \tag{7.6-10}$$

Following Belvin and Park (1990), we choose the weighting matrices $\mathbf{Z}$ and $\mathbf{R}$ proper to a trade off between control effectiveness and control energy consumption by taking

$$\mathbf{Z} = \begin{bmatrix} \alpha\mathbf{K} & \mathbf{0} \\ \mathbf{0} & \beta\mathbf{M} \end{bmatrix}, \quad \mathbf{R} = \eta\mathbf{F}^T\mathbf{K}^{-1}\mathbf{F}, \tag{7.6-11a,b}$$

where α and β are weighting coefficients, ($\alpha\beta \geq 0$ and $(\alpha + \beta) > 0$), while η a scale factor. Equation (7.6-11a) shows that $\mathbf{Z}$ represents the sum of the system kinetic and potential energies in the sense of

$$\frac{1}{2}\int_{t_0}^{t_f} \mathbf{x}^T\mathbf{Z}\mathbf{x}dt = \frac{1}{2}\int_{t_0}^{t_f} [\dot{\mathbf{q}}^T\beta\mathbf{M}\dot{\mathbf{q}} + \mathbf{q}^T\alpha\mathbf{K}\mathbf{q}]dt. \tag{7.6-12}$$

Solution of the Riccati equation yields the control law which depends only on α, β and on the structural mass and stiffness matrices.

At this point one should remark the assumption, tacitly adopted, that the full state vector $\mathbf{x}$ is available for measurement. In this light one should see Eq. (7.6-9a) relating the current control vector with the current state vector.

If some of the states are not available for feedback, a state observer can be incorporated into the design to estimate the unknown states, provided the states are observable through the system output.

For more details on these issues that are not pursued here, the reader is referred to the monographs by Meirovitch (1990), Preumont (1997), and to the paper by Na et al. (2005).

7.6-4 Numerical Simulations and Discussion

7.6-4a Active Control to Arbitrary Time Dependent Loads

The numerical simulations are carried out for Case (A), Section 6.6-5a, that is for a thin-walled beam with a rectangular ($b \times c$) cross-section profile. The cantilevered box beam has sensors and actuators symmetrically embedded in the host structure, as indicated in Fig. 6.6-1.

Figures 7.6-1 and 7.6-2 depict the transverse tip deflection time-history $v_0(L, t)$, to a Dirac delta-pulse $\delta(t)$; Fig. 7.6-3 to a rectangular pressure pulse, whereas Figs. 7.6-4a,b, depict these pressure-pulses.

These plots compare the closed-loop responses when optimal control, and the proportional and the velocity feedback control methodologies are applied. The results reveal that in the free vibration range, the optimal control provides the best way of damping out the oscillations.

Figures 7.6-5 and 7.6-6, highlight the implications of the simultaneous use of the tailoring and velocity feedback control as to damp out the oscillations induced by a Dirac delta-pulse $\delta(t)$ and a rectangular pressure pulse, respectively. The results reveal the remarkable efficiency achieved via the simultaneous application of the tailoring and velocity feedback control.

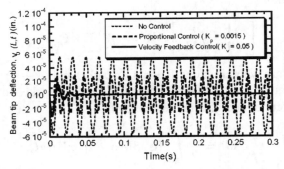

Figure 7.6-1: Beam tip deflection time-history, $v_0(L, t)$ (in.), to a Dirac impulse $\delta(t)$ applied at the beam tip. Open and closed-loop dynamic response obtained within proportional and velocity feedback controls, $(\theta = 0, \mathcal{R} = 1)$

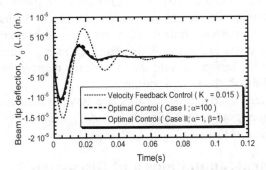

Figure 7.6-2: Beam tip deflection (in.) time-history to a Dirac impulse $\delta(t)$ applied at the beam tip. LQR and velocity feedback controls, $(\theta = 0, \mathcal{R} = 1)$

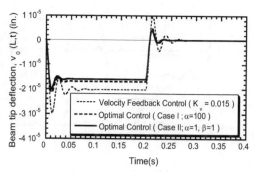

Figure 7.6-3: Beam tip deflection time-history to a rectangular pressure pulse applied at the beam tip. LQR and velocity feedback controls, $(\theta = 0, \mathcal{R} = 1)$

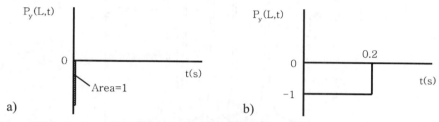

Figure 7.6-4: Pressure pulses involved in Figs. 7.6-1 through 7.6-3 and 7.6-5 and 7.6-6. (a) Dirac-delta pulse; (b) rectangular pressure pulse

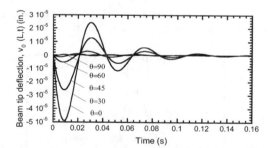

Figure 7.6-5: Implication of the ply-angle, θ (deg.), on beam tip deflection time-history induced by a Dirac pulse applied at the beam tip, velocity feedback control, $(K_v = 0.05)$

7.6-4b Active Control to Harmonically Time-Dependent Loads

Return to the issues addressed in Section 7.5-3a in the absence of the control. Figures 7.6-7 and 7.6-8 display plots of the beam tip deflection amplitude $V(1)$, versus the excitation frequency ω_f (rad/s), for both the controlled/uncontrolled systems. The increase of the proportional feedback gain shifts the resonance frequencies towards larger excitation frequencies, (see Fig. 7.6-7). In the case of the use of the velocity feedback control, the increase of the corresponding feedback gain yields a dramatic decrease of the deflection amplitude and even the suppression of the resonance.

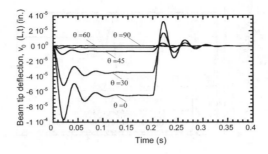

Figure 7.6-6: Counterpart of Fig. 7.6-5 for the case of a rectangular pulse

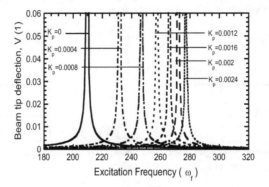

Figure 7.6-7: Steady-state open/closed-loop dimensionless beam tip deflection V(1) versus ω_f for selected values of the feedback gain K_p, ($\mathcal{R} = 1, \theta = 0$)

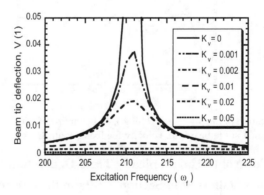

Figure 7.6-8: The counterpart of Fig. 7.6-7 for the case of the velocity feedback control

This trend is consistent with that reported by Song et al. (2001) and Na and Librescu (1998a). A combined feedback control will have more encompassing results then each separate one (see also Na and Librescu, 1998a, and Na, 1997).

Note the result reported by Librescu et al. (1997) showing that the neglect of the mass and stiffness of actuators yields, in a most detrimental case, a resonance frequency about 10% smaller and response amplitudes slightly larger than the actual ones, occurring when actuator mass and stiffness are considered.

7.7 INSTANTANEOUS OPTIMAL FEEDBACK CONTROL

7.7-1 Preliminaries

In this section, we consider a more encompassing control law representing an extension of the standard LQR. It will be applied to Case (B), for a thin-walled beam featuring transversal deflection-twist-transverse shear elastic cross-coupling. The governing equations are Eqs. (7.2-1) through (7.2-3) and their unshearable bending counterparts, Eqs. (7.2-11) through (7.2-13).

7.7-2 Control Law

When the piezoactuator is in the form of a patch located along the beam span, the piezoelectric bending moment $\tilde{M}_x$ appears solely in the governing equations. If it is spread over the entire beam span it intervenes only in BCs at $z = L$. If the electric field intensity is constant across the piezoactuator thickness, then one may approximate it as

$$\mathcal{E}_3(t) = V(t)/h, \qquad (7.7\text{-}1)$$

then from (6.6-1a) one can express $\hat{M}_x$ as

$$\tilde{M}_x(z, t) = \mathcal{C}V(t)[Y(z - z_1) - Y(z - z_2)], \qquad (7.7\text{-}2)$$

where $\mathcal{C}$ is a constant dependent upon the mechanical and geometrical properties of piezoactuator and of the main structure, while the Heaviside distribution $Y(\cdot)$ locates the distributed moment between the points z_1 and z_2 along the beam span.

To discretize the eigenvalue problem we use the Extended Galerkin Method. The displacements $v_0(z, t), \theta_x(z, t)$ and $\phi(z, t)$ will be expressed via Eqs. (6.6-21) and (6.6-22). We use the shifting property of the derivatives of Dirac's distribution

$$\int_{z_0}^{z_1} \delta^j(z - \zeta)\, g(z)dz = (-1)^j\, g^j(\zeta), \qquad (7.7\text{-}3)$$

which is valid when $\zeta \in [z_0, z_1]$, and the fact that $Y''(z) = \delta'(z)$ and $Y'(z) = \delta(z)$, where $\delta(\cdot)$ is the spatial Dirac's distribution, $g(z)$ is an arbitrary function dependent on z, while $(g^j, \delta^j) \equiv d^j(g, \delta)/dz^j$.

Applying the procedure described in Section 6.6 one obtains the discrete equations of motion that are similar to (6.6-28), namely

$$\mathbf{M}\ddot{\mathbf{q}} + \mathbf{K}\,\mathbf{q} = \mathbf{Q} - \mathbf{Fu}. \qquad (7.7\text{-}4)$$

Here $\mathbf{M}$ and $\mathbf{K}$ are the mass and stiffness matrices for the entire structure of host and piezoactuators.

In this context, one can express the stiffness a_{ij} and the mass terms for a smart beam:

$$a_{ij} = \bar{a}_{ij} + \delta_P \hat{a}_{ij} + \delta_S \hat{\hat{a}}_{ij},$$

$$b_i = \bar{b}_i + \delta_P \hat{b}_i + \delta_S \hat{\hat{b}}_i, \qquad\qquad (7.7\text{-}5a,b)$$

where the terms affected by an overbar and the overcarets are associated with the host structure and the piezoactuators, respectively; the tracers δ_P and δ_S were defined in Section 6.6-3.

In Eq. (7.7-4) $\mathbf{u}$ is the vector of the control input, while

$$\mathbf{Q} = \int_0^L p_y(z,t)\mathbf{V}_0^T dz,$$

and $\qquad\qquad (7.7\text{-}6a,b)$

$$\mathbf{F} = [\mathbf{X}_0(z_2) - \mathbf{X}_0(z_1)],$$

are the generalized vectors of the time-dependent external excitation, and piezo-electrically induced bending moment, respectively. As in Section 7.6-3, the equation can be cast in state-space form as

$$\dot{\mathbf{x}}(t) = \mathbf{A}\mathbf{x}(t) + \mathbf{B}\mathbf{Q}(t) + \mathbf{W}\mathbf{u}(t), \qquad\qquad (7.7\text{-}7)$$

where

$$\mathbf{A} = \begin{bmatrix} \mathbf{0} & \mathbf{I} \\ -\mathbf{M}^{-1}\mathbf{K} & \mathbf{0} \end{bmatrix}, \ \mathbf{B} = \left\{ \begin{array}{c} \mathbf{0} \\ \mathbf{M}^{-1} \end{array} \right\} \ \text{and} \ \mathbf{W} = \left\{ \begin{array}{c} \mathbf{0} \\ -\mathbf{M}^{-1}\mathbf{F} \end{array} \right\},$$
$$(7.7\text{-}8a\text{-}c)$$

while $\mathbf{x}(t) = \{\mathbf{q}^T(t), \dot{\mathbf{q}}^T(t)\}^T$ denotes the full state vector.

Now, instead of the functional, Eq. (7.6-8), to be minimized, a more complete one, see e.g. Soong (1990), that incorporates the time-dependent load will be used. Incorporation of this quantity leads to a time-dependent performance index which has to be minimized at any time interval; we call this modified LQR the *instantaneous optimal control (IOC)*.

Following the approach by Chang (1990), Chang and Librescu (1999), and Na and Librescu (1998, 2000), who used this procedure, the augmented performance index including the constraint equation is represented as:

$$J_a = \frac{1}{2}\mathbf{x}^T(t_f)\mathbf{S}\mathbf{x}(t_f) + \int_{t_0}^{t_f} \left\{ \frac{1}{2}(\mathbf{x}^T\mathbf{Z}\mathbf{x} + \mathbf{u}^T\mathbf{R}\mathbf{u}) \right.$$

$$\left. + \mathbf{k}^T[\mathbf{A}\mathbf{x}(t) + \mathbf{W}\mathbf{u}(t) + \mathbf{B}\mathbf{Q}(t) - \dot{\mathbf{x}}] \right\} dt, \qquad (7.7\text{-}9)$$

in which $\mathbf{k}(t)$ is the vector of Lagrangian multipliers (referred to as the costates); $\mathbf{Z}$ and $\mathbf{R}$ are, respectively, positive semidefinite and positive-definite weighting matrices, $\mathbf{S}$ is the non-negative definite weighting matrix associated with the error in the terminal state at $t = t_f$; $\mathbf{u}^T(t) = \{u_1 \ldots u_m\}$ is the control input vector that drives the initial state $\mathbf{x}(t_0)$ to the final one $\mathbf{x}(t_f)$, where m is the number of piezoactuator patches. For large weighting matrix $\mathbf{Z}$ the response in terms of the components of the state vector is small and the required active control $\mathbf{u}(t)$ will be large, while big values of $\mathbf{R}$ require the control not to be excessive. As a result, suitable $\mathbf{Z}$ and $\mathbf{R}$ that provide a desired trade off between the state variable responses and control efforts, while satisfying performance requirements and constraints should be determined.

Expressing Eq. (7.7-9) as

$$J_a = \frac{1}{2}\mathbf{x}^T(t_f)\mathbf{S}\mathbf{x}(t_f) + \int_{t_0}^{t_f} g(\mathbf{x}, \dot{\mathbf{x}}, \mathbf{u}, \mathbf{k}; t)dt, \qquad (7.7\text{-}10a)$$

where

$$g(\mathbf{x}, \dot{\mathbf{x}}, \mathbf{u}, \mathbf{k}; t) = 1/2(\mathbf{x}^T\mathbf{Z}\mathbf{x} + \mathbf{u}^T\mathbf{R}\mathbf{u})$$
$$+\mathbf{k}^T[\mathbf{A}\mathbf{x}(t) + \mathbf{W}\mathbf{u}(t) + \mathbf{B}\mathbf{Q}(t) - \dot{\mathbf{x}}]; \qquad (7.7\text{-}10b)$$

a necessary condition for the optimality is that the first variation of Eq. (7.7-10a) is zero.

Following Na and Librescu (1998, 2000), and Chang and Librescu (1999) one obtains the optimal control law expressed as

$$\mathbf{u}(t) = -\mathbf{R}^{-1}\mathbf{W}^T\mathbf{P}(t)\mathbf{x}(t) - \mathbf{R}^{-1}\mathbf{W}^T\mathbf{d}(t), \qquad (7.7\text{-}11)$$

where $\mathbf{P}(t)$ and $\mathbf{d}(t)$ in (7.7-11) are

$$\dot{\mathbf{P}}(t) = -\mathbf{Z} - \mathbf{A}^T\mathbf{P}(t) - \mathbf{P}(t)\mathbf{A} + \mathbf{P}(t)\mathbf{W}\mathbf{R}^{-1}\mathbf{W}^T\mathbf{P}(t), \qquad (7.7\text{-}12a)$$
$$\dot{\mathbf{d}}(t) = -(\mathbf{A}^T - \mathbf{P}(t)\mathbf{W}\mathbf{R}^{-1}\mathbf{W}^T)\mathbf{d}(t) - \mathbf{P}(t)\mathbf{B}\mathbf{Q}(t). \qquad (7.7\text{-}12b)$$

In addition, one also obtains the condition

$$\mathbf{S}\mathbf{x}(t_f) = \mathbf{P}(t_f)\mathbf{x}(t_f) + \mathbf{d}(t_f), \qquad (7.7\text{-}13)$$

by virtue of which the terminal conditions of Eqs. (7.7-12a,b) result as

$$\mathbf{P}(t_f) = \mathbf{S}, \quad \mathbf{d}(t_f) = \mathbf{0}. \qquad (7.7\text{-}14a,b)$$

The $\mathbf{P}(t)$ and $\mathbf{d}(t)$ in the optimal control law, Eq. (7.7-11), can be determined from Eqs. (7.7-12a,b) considered in conjunction with (7.7-14a,b), by integrating these equations backward in time from the terminal time t_f until the initial time t_0. Within the classical approach of the optimal control, the external excitation is discarded in the performance index which implies $\mathbf{BQ} = 0$, and, $\mathbf{d} = 0$. In this case, t_0 is taken to be zero, and $\mathbf{S} = 0$. If in addition, the terminal time t_f approaches infinity, the Riccati gain matrix $\mathbf{P}(t)$ becomes a constant matrix $\mathbf{P}_c$, the solution to the nonlinear algebraic Riccati equation

$$\mathbf{A}^T\mathbf{P}_c + \mathbf{P}_c\mathbf{A} - \mathbf{P}_c\mathbf{WR}^{-1}\mathbf{W}^T\mathbf{P}_c + \mathbf{Z} = \mathbf{0}, \qquad (7.7\text{-}15)$$

obtained as a special case of Eq. (7.7-12). In this case, the corresponding steady-state linear optimal control law is given by

$$\mathbf{u}(t) = -\mathbf{G}\mathbf{x}(t), \qquad (7.7\text{-}16)$$

where

$$\mathbf{G} = \mathbf{R}^{-1}\mathbf{W}^T\mathbf{P}_c \qquad (7.7\text{-}17)$$

is the optimum gain matrix.

In spite of the fact that within this control law, $t_f \to \infty$, the objective of control can be reached, generally, within a finite period of time. Under a number of conditions, the algebraic Riccati equation has a unique, positive definite solution $\mathbf{P}_c$ which minimizes the performance index

$$J = \frac{1}{2}\int_0^\infty [\mathbf{x}^T\mathbf{Z}\mathbf{x} + \mathbf{u}^T\mathbf{R}\mathbf{u}]dt, \qquad (7.7\text{-}18)$$

when the control law, Eq. (7.7-16), in conjunction with (7.7-17) is used. Note that the state weighting matrix $\mathbf{Z}$ was chosen so that the first term in the cost functional is represented in a form consistent with Eq. (7.6-12), while the weighting matrix of control inputs $\mathbf{R}$, in accordance to (7.6-11b).

This enables one to synthesize the controllers to achieve a proper trade-off between control effectiveness and control energy consumption by varying the relative magnitudes of $\mathbf{Z}$ and $\mathbf{R}$.

For piezoactuators spread over the entire beam span, the piezoelectrically induced bending moment $\tilde{M}_x$ appears in the boundary condition at $z = L$, only;

$$\tilde{M}_x = \mathbf{F}u, \text{ and } \mathbf{F} = \mathbf{X}_0'(L), \text{ while } u = C\,V(t). \qquad (7.7\text{-}19\text{a,b})$$

In addition to this modification, the mass and the stiffness are obtained by setting in Eqs. (7.7-5a,b), $\delta_P = 0$ and $\delta_S = 1$.

7.7-3 Saturation Constraint

In some situations, the required control moment can exceed the moment output capability provided by the actuator. This is due to the limited voltage which can be applied to the piezoactuator. Beyond this limited value, the piezoactuator is saturated. Large amplitude control voltage is ineffective outside this

range. There are several ways to handle optimal control subjected to saturation constraint of the controller. We follow Helgeson and Szustak (1995), and
Khot et al. (1996) for optimal control in both unsaturated and saturated regions
of the actuator. Rather than reducing the feedback gains to accommodate the
constraint, the actuator limitation is explicitly addressed in the optimal control
algorithm: the control voltage must satisfy the condition

$$-u_{max} \leq u(t) \leq u_{max} \quad \text{for } t \in [t_0, t_\infty] \tag{7.7-20}$$

For unsaturated controller, the optimal control is obtained as

$$\mathbf{u}(t) = -\mathbf{R}^{-1}\mathbf{W}^T\mathbf{P}_c(t)\mathbf{x}(t). \tag{7.7-21}$$

In light of Eq. (7.7-20), this is valid for

$$|-\mathbf{R}^{-1}\mathbf{W}^T\mathbf{P}_c(t)\mathbf{x}(t)| \leq u_{max}. \tag{7.7-22}$$

To obtain the bounded optimal control we used the trial and error procedure
outlined by Khot et al. (1996) and followed by Na and Librescu (1998, 2000) to
fulfil the constraint condition, Eq. (7.7-22), and obtain a convergent solution.

7.7-4 Sensor Output Equation

Assume that the piezoelectric elements can be employed concurrently for sensing and actuation. For sensing operation, $\mathcal{E}_3 = \mathbf{0}$ gives electric displacement

$$D_3 = \bar{e}_{31}\varepsilon_{zz}^{(0)}, \tag{7.7-23}$$

and hence

$$D_3 = \bar{e}_{31}y(s)\theta_x'. \tag{7.7-24}$$

The charge developed on the sensor can be found through integration of the
electric displacement over the sensor area, so that

$$q_p(t) = \int_A D_3 dA = \int\int \bar{e}_{31}y(s)\theta_x' ds dz. \tag{7.7-25}$$

The limits of integration in Eq. (7.7-25) depend on the sensor patch configuration.

The voltage across the piezoelectric sensor can be obtained by dividing the
charge developed in the sensor by its capacitance, C_p (see Bailey and Hubbard,
1985),

$$V_p(t) = \frac{q_p(t)}{C_p}, \tag{7.7-26}$$

where C_p depends on the patch area A_p, thickness h^a, and the permittivity of the piezoelectric material ζ_{33}^p, according to (see Dosch et al., 1992)

$$C_p = \frac{\xi_{33}^p A_p}{h^a}. \tag{7.7-27}$$

For a single sensor patch, the sensor electric current output is

$$I_p = \frac{dq_p(t)}{dt} = \int \int \bar{e}_{31} y(s) \dot{\theta}_x' ds dz$$

$$= \int \bar{e}_{31} y(s) [\dot{\theta}_x(z_2) - \dot{\theta}_x(z_1)] ds. \tag{7.7-28}$$

Assume that the actuator voltage required for vibration suppression is proportional to the sensor output voltage (see e.g. Song et al., 2001, and Dosh et al., 1992). The electric power consumption during vibration control can be obtained as the product of the sensor output voltage and current as $P = V_p I_p$.

Next, the matrices $\mathbf{M}$ and $\mathbf{K}$ for a number of cases and types of anisotropy of the constituent materials will be provided; the structure has the CAS ply-angle configuration.

7.8 EXPRESSIONS OF MATRICES M AND K

7.8-1 Consider a transversely isotropic beam, the plane of isotropy coinciding with the beam mid-surface. The bending and twist become decoupled. Here only the bending-transverse shear is considered:

$$\mathbf{M} = \int_0^L \left[\begin{array}{cc} (\bar{b}_1 + \delta_S \hat{b}_1) \mathbf{V}_0 \mathbf{V}_0^T & 0 \\ 0 & (\bar{B}_4 + \delta_S \hat{B}_4) \mathbf{X}_0 \mathbf{X}_0^T \end{array} \right] dz + \delta_P \int_{z_1}^{z_2} \left[\begin{array}{cc} \hat{b}_1 \mathbf{V}_0 \mathbf{V}_0^T & 0 \\ 0 & \hat{B}_4 \mathbf{X}_0 \mathbf{X}_0^T \end{array} \right] dz, \tag{7.8-1a}$$

$$\mathbf{K} = \int_0^L \left[\begin{array}{cc} (\bar{a}_{55} + \delta_S \hat{a}_{55}) \mathbf{V}_0' \mathbf{V}_0'^T & (\bar{a}_{55} + \delta_S \hat{a}_{55}) \mathbf{V}_0' \mathbf{X}_0^T \\ (\bar{a}_{55} + \delta_S \hat{a}_{55}) \mathbf{X}_0 \mathbf{V}_0'^T & (a_{55} + \delta_S \hat{a}_{55}) \mathbf{X}_0 \mathbf{X}_0^T + (a_{33} + \delta_S \hat{a}_{33}) \mathbf{X}_0' \mathbf{X}_0'^T \end{array} \right] dz$$

$$+ \delta_P \int_{z_1}^{z_2} \left[\begin{array}{cc} \hat{a}_{55} \mathbf{V}_0' \mathbf{V}_0'^T & \hat{a}_{55} \mathbf{V}_0' \mathbf{X}_0^T \\ \hat{a}_{55} \mathbf{X}_0 \mathbf{V}_0'^T & \hat{a}_{55} \mathbf{X}_0 \mathbf{X}_0^T + \hat{a}_{33} \mathbf{X}_0' \mathbf{X}_0'^T \end{array} \right] dz, \tag{7.8-1b}$$

where, for the sake of brevity, (see Eq. 3.1-24a), the following equations were used

$$B_4 = b_4 + \delta_n b_{14}, \quad B_5 = b_5 + \delta_n b_{15}, \quad B_{10} = b_{10} + \delta_n b_{18}. \tag{7.8-1c-e}$$

7.8-2 Anisotropic unshearable host structure incorporating warping inhibition:

$$\mathbf{M} = \int_0^L \begin{bmatrix} (\overline{b}_1 + \delta_S \hat{b}_1)\mathbf{V}_0\mathbf{V}_0^T + (\overline{B}_4 + \delta_S \hat{B}_4)\mathbf{V}_0'\mathbf{V}_0'^T & 0 \\ 0 & [\overline{B}_4 + \overline{B}_5 + \delta_S(\hat{B}_4 + \hat{B}_5)]\Phi\Phi^T \\ & + (\overline{B}_{10} + \delta_S \hat{B}_{10})\Phi'\Phi'^T \end{bmatrix} dz$$

$$+ \delta_P \int_{z_1}^{z_2} \begin{bmatrix} \hat{b}_1 \mathbf{V}_0\mathbf{V}_0^T + \hat{B}_4 \mathbf{V}_0'\mathbf{V}_0'^T & 0 \\ 0 & (\hat{B}_4 + \hat{B}_5)\Phi\Phi^T + \hat{B}_{10}\Phi'\Phi'^T \end{bmatrix} dz, \qquad (7.8\text{-}2a)$$

$$\mathbf{K} = \int_0^L \begin{bmatrix} (\overline{a}_{33} + \delta_S \hat{a}_{33})\mathbf{V}_0''\mathbf{V}_0''^T & -(\overline{a}_{37} + \delta_S \hat{a}_{37})\mathbf{V}_0''\Phi'^T \\ -(\overline{a}_{37} + \delta_S \hat{a}_{37})\Phi'\mathbf{V}_0''^T & (\overline{a}_{66} + \delta_S \hat{a}_{66})\Phi''\Phi''^T + (\overline{a}_{77} + \delta_S \hat{a}_{77})\Phi'\Phi'^T \end{bmatrix} dz$$

$$+ \delta_P \int_{z_1}^{z_2} \begin{bmatrix} \hat{a}_{33}\mathbf{V}_0''\mathbf{V}_0''^T & -\hat{a}_{37}\mathbf{V}_0''\Phi'^T \\ -\hat{a}_{37}\Phi'\mathbf{V}_0''^T & \hat{a}_{66}\Phi''\Phi''^T + \hat{a}_{77}\Phi'\Phi'^T \end{bmatrix} dz. \qquad (7.8\text{-}2b)$$

7.8-3 An anisotropic and shearable host structure with bending-transverse shear-free warping:

$$\mathbf{M} = \int_0^L \begin{bmatrix} (\overline{b}_1 + \delta_S \hat{b}_1)\mathbf{V}_0\mathbf{V}_0^T & 0 & 0 \\ 0 & (\overline{B}_4 + \delta_S \hat{B}_4)\mathbf{X}_0\mathbf{X}_0^T & 0 \\ 0 & 0 & [\overline{B}_4 + \overline{B}_5 + \delta_S(\hat{B}_4 + \hat{B}_5)]\Phi\Phi^T \end{bmatrix} dz$$

$$+ \delta_P \int_{z_1}^{z_2} \begin{bmatrix} \hat{b}_1 \mathbf{V}_0\mathbf{V}_0^T & 0 & 0 \\ 0 & \hat{B}_4 \mathbf{X}_0\mathbf{X}_0^T & 0 \\ 0 & 0 & (\hat{B}_4 + \hat{B}_5)\Phi\Phi^T \end{bmatrix} dz, \qquad (7.8\text{-}3a)$$

$$\mathbf{K} = \int_0^L \begin{bmatrix} (\overline{a}_{55} + \delta_S \hat{a}_{55})\mathbf{V}_0'\mathbf{V}_0'^T & (\overline{a}_{55} + \delta_S \hat{a}_{55})\mathbf{V}_0'\mathbf{X}_0^T & 0 \\ (\overline{a}_{55} + \delta_S \hat{a}_{55})\mathbf{X}_0\mathbf{V}_0'^T & (\overline{a}_{55} + \delta_S \hat{a}_{55})\mathbf{X}_0\mathbf{X}_0^T & (\overline{a}_{37} + \delta_S \hat{a}_{37})\mathbf{X}_0'\Phi'^T \\ & + (\overline{a}_{33} + \delta_S \hat{a}_{33})\mathbf{X}_0'\mathbf{X}_0'^T & \\ 0 & (\overline{a}_{37} + \delta_S \hat{a}_{37})\Phi'\mathbf{X}_0'^T & (\overline{a}_{77} + \delta_S \hat{a}_{77})\Phi'\Phi'^T \end{bmatrix} dz$$

$$+ \delta_P \int_{z_1}^{z_2} \begin{bmatrix} \hat{a}_{55}\mathbf{V}_0'\mathbf{V}_0'^T & \hat{a}_{55}\mathbf{V}_0'\mathbf{X}_0^T & 0 \\ \hat{a}_{55}\mathbf{X}_0\mathbf{V}_0'^T & \hat{a}_{55}\mathbf{X}_0\mathbf{X}_0^T + \hat{a}_{33}\mathbf{X}_0'\mathbf{X}_0'^T & \hat{a}_{37}\mathbf{X}_0'\Phi'^T \\ 0 & \hat{a}_{37}\Phi'\mathbf{X}_0'^T & \hat{a}_{77}\Phi'\Phi'^T \end{bmatrix} dz. \qquad (7.8\text{-}3b)$$

7.8-4 An anisotropic shearable host structure including the warping inhibition:

$$\mathbf{M} = \int_0^L \begin{bmatrix} (\overline{b}_1 + \delta_S \hat{b}_1)\mathbf{V}_0\mathbf{V}_0^T & 0 & 0 \\ 0 & (\overline{B}_4 + \delta_S \hat{B}_4)\mathbf{X}_0\mathbf{X}_0^T & 0 \\ 0 & 0 & [\overline{B}_4 + \overline{B}_5 + \delta_S(\hat{B}_4 + \hat{B}_5)]\Phi\Phi^T \\ & & + (\overline{B}_{10} + \delta_S \hat{B}_{10})\Phi'\Phi'^T \end{bmatrix} dz$$

$$+ \delta_P \int_{z_1}^{z_2} \begin{bmatrix} \hat{b}_1 \mathbf{V}_0 \mathbf{V}_0^T & 0 & 0 \\ 0 & \hat{B}_4 \mathbf{X}_0 \mathbf{X}_0^T & 0 \\ 0 & 0 & (\hat{B}_4 + \hat{B}_5)\Phi\Phi^T + \hat{B}_{10}\Phi'\Phi'^T \end{bmatrix} dz, \qquad (7.8\text{-}4a)$$

$$\mathbf{K} = \int_0^L \begin{bmatrix} (\bar{a}_{55} + \delta_S \hat{a}_{55})\mathbf{V}_0'\mathbf{V}_0'^T & (\bar{a}_{55} + \delta_S \hat{a}_{55})\mathbf{V}_0'\mathbf{X}_0^T & -(\bar{a}_{56} + \delta_S \hat{a}_{56})\mathbf{V}_0'\Phi''^T \\ & (\bar{a}_{55} + \delta_S \hat{a}_{55})\mathbf{X}_0\mathbf{X}_0^T & (\bar{a}_{37} + \delta_S \hat{a}_{37})\mathbf{X}_0'\Phi'^T \\ (\bar{a}_{55} + \delta_S \hat{a}_{55})\mathbf{X}_0\mathbf{V}_0'^T & +(\bar{a}_{33} + \delta_S \hat{a}_{33})\mathbf{X}_0'\mathbf{X}_0^T & +(\bar{a}_{56} + \delta_S \hat{a}_{56})\mathbf{X}_0'\Phi'^T \\ & (\bar{a}_{37} + \delta_S \hat{a}_{37})\Phi'\mathbf{X}_0'^T & (\bar{a}_{77} + \delta_S \hat{a}_{77})\Phi'\Phi'^T \\ -(\bar{a}_{56} + \delta_S \hat{a}_{56})\Phi''\mathbf{V}_0'^T & +(\bar{a}_{56} + \delta_S \hat{a}_{56})\Phi''\mathbf{X}_0'^T & +(\bar{a}_{66} + \delta_S \hat{a}_{66})\Phi''\Phi''^T \end{bmatrix} dz$$

$$+ \delta_P \int_{z_1}^{z_2} \begin{bmatrix} \hat{a}_{55}\mathbf{V}_0'\mathbf{V}_0'^T & \hat{a}_{55}\mathbf{V}_0'\mathbf{X}_0^T & \hat{a}_{56}\mathbf{V}_0'\Phi''^T \\ \hat{a}_{55}\mathbf{X}_0\mathbf{V}_0'^T & \hat{a}_{55}\mathbf{X}_0\mathbf{X}_0^T + \hat{a}_{33}\mathbf{X}_0'\mathbf{X}_0'^T & (\hat{a}_{37} + \hat{a}_{56})\mathbf{X}_0'\Phi'^T \\ \hat{a}_{56}\Phi''\mathbf{V}_0'^T & (\hat{a}_{37} + \hat{a}_{56})\Phi'\mathbf{X}_0'^T & \hat{a}_{77}\Phi'\Phi'^T + \hat{a}_{66}\Phi''\Phi''^T \end{bmatrix} dz. \quad (7.8\text{-}4b)$$

7.8-5 An anisotropic unshearable host structure featuring transverse bending-free twist coupling:

$$\mathbf{M} = \int_0^L \begin{bmatrix} (\bar{b}_1 + \delta_S \hat{b}_1)\mathbf{V}_0\mathbf{V}_0^T + (\bar{B}_4 + \delta_S \hat{B}_4)\mathbf{V}_0'\mathbf{V}_0'^T & 0 \\ 0 & [\bar{B}_4 + \bar{B}_5 + \delta_S(\hat{B}_4 + \hat{B}_5)]\Phi\Phi^T \end{bmatrix} dz$$

$$+ \delta_P \int_{z_1}^{z_2} \begin{bmatrix} \hat{b}_1 \mathbf{V}_0\mathbf{V}_0^T + \hat{B}_4\mathbf{V}_0'\mathbf{V}_0'^T & 0 \\ 0 & (\hat{B}_4 + \hat{B}_5)\Phi\Phi^T \end{bmatrix} dz, \qquad (7.8\text{-}5a)$$

$$\mathbf{K} = \int_0^L \begin{bmatrix} (\bar{a}_{33} + \delta_S \hat{a}_{33})\mathbf{V}_0''\mathbf{V}_0''^T & -(\bar{a}_{37} + \delta_S \hat{a}_{37})\mathbf{V}_0''\Phi^T \\ -(\bar{a}_{37} + \delta_S \hat{a}_{37})\Phi'\mathbf{V}_0''^T & (\bar{a}_{77} + \delta_S \hat{a}_{77})\Phi'\Phi'^T \end{bmatrix} dz$$

$$+ \delta_P \int_{z_1}^{z_2} \begin{bmatrix} \hat{a}_{33}\mathbf{V}_0''\mathbf{V}_0''^T & -\hat{a}_{37}\mathbf{V}_0''\Phi^T \\ -\hat{a}_{37}\Phi'\mathbf{V}_0''^T & \hat{a}_{77}\Phi'\Phi'^T \end{bmatrix} dz. \qquad (7.8\text{-}5b)$$

In these equations $\mathbf{V}_0$, $\mathbf{X}_0$ and Φ are vectors of trial functions associated with displacements v_0, θ_x and ϕ, respectively, (see Eqs. (6.3-14)).

7.9 RESULTS

7.9-1 Preliminaries

The results are intended to highlight the effects induced by the various features of the beam model, pressure pulse and feedback control. We consider Case (B) characterized by the *CAS* ply-angle configuration applied to beams of bi-convex cross-section and associated with Cases (a) and (b) of piezoactuator location as defined in Section 6.6-3.

7.9-2 Case (B) Coupled with Case (a)

For numerical simulations, the host structure $L = 80\text{in.}(2.03$ m$)$ and $h = 0.4\text{in.}$ (0.01 m), is constructed of a graphite-epoxy composite and the piezoactuators of *PZT*-4 piezoceramic. The amplitude of the time-dependent load, $P_m = 50\text{lb/L}$ (222 N/L).

We use, the dimensionless velocity feedback gain $K_V = k_v L \bar{\omega}/a_{33}^\circ$, where $\bar{\omega} = 320$ rad/s is a reference natural frequency corresponding to a thin-walled beam characterized by $\sigma = 0.1$, $\theta = 90$ deg., $L = 80$in. (2.03 m), while a_{33}° is the normalizing bending stiffness at the beam root cross-section (= 1.44117×10^7 lb. in.$^2 = 41{,}366$ $N.m^2$).

Figure 7.9-1 displays the effect of beam taper-ratio on the normalized restoring bending moment $\tilde{M}(\equiv ML/a_{33}^\circ)$ time-history at the beam root when the pulse impacting the beam is due to a sonic-boom (shown in the inset). For the forced motion regime, the open-loop amplitudes of the restoring moment are independent of the taper-ratio, whereas for free-motion regime the amplitudes of $\tilde{M}$ for the uniform beam are smaller than for the tapered beam counterpart, and their maxima are shifted towards larger times.

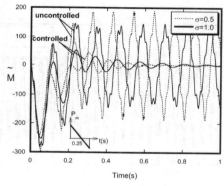

Figure 7.9-1: Open/closed dimensionless restoring bending moment time-history at the beam root for uniform/tapered beam exposed to a sonic boom pulse. $K_V = 0.0018$, $\theta = 45$ deg., $r = 1.2$, $t_p = 0.25$s, unshearable and free warping beam model

This is due to the fact that the taper beam ($\sigma = 0.5$) is more flexible than its uniform beam counterpart, ($\sigma = 1$). For the activated beam, because the uniform beam is stiffer than its non-uniform counterpart, the control is less efficient, see also Librescu et al. (2000) and Librescu and Na (2001).

Figures 7.9-2 and 7.9-3, show the implications of the ply-angle and taper ratio, respectively, on the open/closed-loop dimensionless acceleration $\tilde{a}(\equiv \ddot{v}_0(L, t)/g)$ time-history of the beam tip, when exposed to a blast load.

With the increase of the ply-angle, and hence of stiffness, the beam tip acceleration decreases (see Fig. 7.9-2). On the other hand, for the open-loop acceleration time-history, the uniform beam with its larger bending stiffness than the tapered one, experiences a lower acceleration (see Fig. 7.9-3).

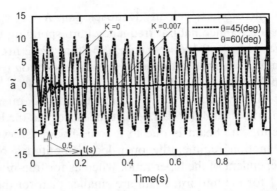

Figure 7.9-2: Open/closed loop, ($K_V = 0.007$), dimensionless acceleration time-history of the beam tip, as influenced by the ply-angle, under a blast pulse, $\sigma = 0.5$. Unshearable and free warping beam model

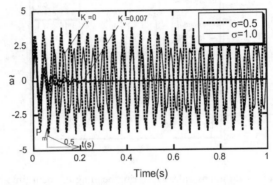

Figure 7.9-3: Open/closed loop, ($K_V = 0.007$), dimensionless acceleration time-history of the beam tip, as influenced by the taper-ratio, under a blast pulse, $\theta = 60$ deg. Unshearable and free warping beam model

For the actuated beam, however, except for a short time during the forced motion regime where slight oscillations subsist, the motion is completely damped out.

7.9-3 Case (B) Coupled with Case (b)

Figure 7.9-4 shows the distribution of the piezopatch along the beam span.

Figures 7.9-5 and 7.9-6 show the time-history of the dimensionless deflection $\tilde{v}_0 (\equiv v_0(L, t)/L)$ of the beam tip when classical LQR and the IOC are applied; the effects of the ply-angle and taper-ratio are also shown.

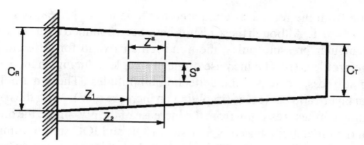

Figure 7.9-4: Distribution of the piezopatch actuator S^a = 3.5in. (0.089m); t^a = 0.00787in. (0.0002m)

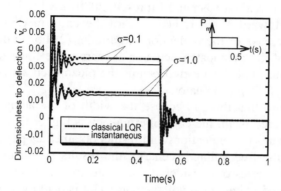

Figure 7.9-5: Dimensionless beam tip transverse deflection time-history, $v_0(L,t)/L$, subjected to a rectangular pulse (see the inset), for two values of σ and for classical and instantaneous optimal control: L = 30in., θ = 45 deg., z^a = (0.2-0.1)L, nonshearable and free warping beam model

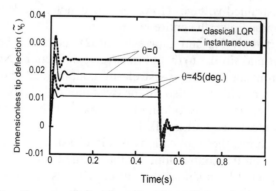

Figure 7.9-6: Implications of ply-angle and of two optimal control methodologies on dimensionless tranverse deflection time-history of the beam tip subjected to a rectangular pressure pulse, (see inset Fig. 7.9-5): L = 30in., σ = 1, nonshearable and free warping beam model

We draw the following conclusions: (i) The classical LQR optimal control provides an upper-bound of the deflection amplitude as compared to that resulting from the IOC. The relative percentage overestimation of the deflection amplitude from the former as compared to the latter, is 12.5% for $\sigma = 1$ and 11% for $\sigma = 0.1$. Although these results have been obtained for a beam exposed to a rectangular pressure pulse, the trend remains valid for different pressure signature profiles. (ii) The increase of the ply-angle inducing an increase of the bending stiffness, decreases the deflection amplitudes. The same holds when the beam taper-ratio decreases (implying an increase of σ). With the increase of the bending stiffness resulting from the increase of the ply-angle, the differences between the predictions based on classical LQR and IOC start to diminish.

Related to the effects of the piezoactuator size and beam taper-ratio on beam tip transverse deflection time-history when subjected to a blast, the results not displayed here reveal (see Na and Librescu, 2000, 2001b, 2002) that: (1) The increase of σ toward $\sigma = 1$, i.e. toward the uniform cross-section beam, results in a decrease of the deflection amplitude, (2) For uniform beams, the effectiveness of the piezopatch actuator appears to be lower than that of the full span actuator, but, for tapered beams the piezopatch actuator becomes as effective as the full span actuator.

We conjecture that this is because the width of the actuators is constant, and for the tapered beam the control efficiency is higher than for the uniform beam counterpart. In connection with results obtained as per the voltage constraint versus the free voltage scenario, the results not displayed here show that: (a) In both cases of considered piezoactuators, in contrast to the case of the unconstrained voltage, the voltage constraint is paid by a slight increase of the deflection amplitude at which the motion is contained, (b) During the first instances of motion and irrespective of the beam taper- ratio, with voltage constraint, the discrete piezoactuator is less effective toward containing the motion than with unconstrained voltage. However, this conclusion does not hold when the actuator is spread over the entire beam span.

Results reported by Na and Librescu (2002) related to beam twist time-history reflect a trend similar to that of transverse deflection. Note that the twist motion arises from the bending-twist elastic coupling. Although the present control appears to be effective also for the twist motion, piezoactuator elements skewed with respect to the beam longitudinal axis can control torsional modes better (in this context, see Park et al., 1996).

Figure 7.9-7 shows the implications of the classical LQR, IOC and of the beam taper-ratio, on the control input voltage time-history. One assumes that the beam is impacted by a sinusoidal pulse. The results show the following: (1) For forced motion regime, the uniform beam ($\sigma = 1$), that is stiffer than the tapered beam, needs a higher voltage to contain the vibrations. However, in free motion regime, (i.e. for $t > 0.5s$), as time unfolds, the effect of the taper-ratio starts to become marginal. (2) For the forced motion regime, for tapered beams, ($\sigma = 0.1$), IOC yields larger voltages than

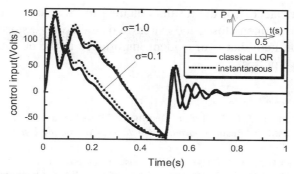

Figure 7.9-7: Effects of the optimal type control and of the beam taper-ratio on the control input voltage time-history, $z^a = (0.2\text{-}0.1)L$, $\theta = 60$ deg.

those in the classical LQR control. In contrast, for uniform cross-section beams the voltages predicted by the two optimal control methodologies are close. However, for free motion (i.e. for $(t > 0.5s)$, the voltage input predictions from the two control methodologies, irrespective of the beam taper ratio, are the same.

Note that rotatory inertia, warping restraint, and the transverse shear flexibility have been included in the results. Special cases have been discussed, e.g. by Oyibo (1989), Song and Librescu (1993), Song et al. (1998, 2001), Librescu et al. (1996), and Na and Librescu (2000), and also within Chapters 6 and 7. Whereas the rotatory and warping inertia terms have little impact on the open and closed-loop response characteristics, transverse shear effect can be important, especially for structures constructed of materials featuring large transverse shear flexibilities and/or high thickness ratios.

Na and Librescu (2000) showed that neglecting transverse shear flexibility, as in the classical Bernoulli-Euler beam model, can severely underestimate the open-loop response amplitudes, but has little effect on the closed-loop response.

The taper-ratio has an appreciable effect on the open-loop deflection time-history, especially for small taper ratios, but little effect on the closed-loop response. Similarly, the warping restraint significantly reduces the twist angle and deflection amplitude, even for high aspect ratio composite beams; it has little effect on closed-loop response. Na and Librescu (2001a) and Na et al. (2004) developed a modified bang-bang control methodology (in the sense of Bruch et al., 1999), for an optimal control design in the conditions of a maximum input voltage constraint, while in Librescu and Na (2005), it was presented a comparative study on the efficiency of LQR, fuzzy logic control (FLC) and on modified bang-bang (MBB), to control and suppress the oscillations of thin-walled anisotropic cantilevers impacted by a pressure pulse. In the same context, one should indicate the control methodology developed by Irsckik et al. (2003), referred to as the *dynamic shape control*.

It should be added that the results obtained in this chapter can be used to determine also the open/closed-loop stress and strain time-histories at any point of the beam, an item that is of evident importance in the design of structures exposed to blast pulses.

REFERENCES

Bailey, T. and Hubbard, J. E. Jr. (1985) "Distributed Piezoelectric-Polymer Active Vibration Control of a Cantilever Beam," *Journal of Guidance, Control and Dynamics*, Vol. 8, No. 5, pp. 605–611.

Bank, L. C. and Kao, C. H. (1988) "Dynamic Response of Composite Beams," in *Recent Advances in Macro and Micro-Mechanics*, D. Hui and J. R. Vinson (Eds.), ASME, pp. 13–19.

Belvin, W. K. and Park, K. C. (1990) "Structural Tailoring and Feedback Control Synthesis: An Interdisciplinary Approach," *Journal of Guidance, Control and Dynamics*, Vol. 13, No. 13, pp. 424–429.

Birman, V. and Bert, C. W. (1987) "Behavior of Laminated Plates Subjected to Conventional Blast," *International Journal of Impact Engineering*, Vol. 6, No. 3, pp. 145–155.

Bruch, J. C., Jr., Sloss, J. M., Adali, S. and Sadek I. S. (1999) "Modified Bang-Bang Piezoelectric Control of Vibrating Beams," *Smart Materials and Structures*, Vol. 8, pp. 647–653.

Cederbaum, G., Librescu, L. and Elishakoff, I. (1988) "Dynamic Response of Flat Panels Made of Advanced Composite Materials Subjected to Random Excitation," *Journal of the Acoustical Society of America*, Vol. 84, No. 2, pp. 660–666.

Cederbaum, G., Librescu, L. and Elishakoff, I. (1989) "Response of Laminated Plates to Non-Stationary Random Excitation," *Structural Safety*, Vol. 6, pp. 99–113.

Chandiramani, N. K., Librescu, L., Saxena, V. and Kumar, A. (2004) "Optimal Vibration Control of a Rotating Composite Beam with Distributed Piezoelectric Sensing and Actuation," *Smart Materials and Structures*, Vol. 13, pp. 433–442.

Chang, M-Y. (1990) "Active Vibration Control of Composite Structures," Ph.D. Thesis, Virginia Polytechnic Institute and State University, Blacksburg, Virginia.

Chang, M-Y. and Librescu, L. (1999) "Vibration Control of Shear Deformable Laminated Plates – Modeling Implications," *Journal of Sound and Vibration*, Vol. 228, No. 4, pp. 731–760.

Cheng, D. H. and Benveniste, J. E. (1968) "Sonic Boom Effects on Structures - A Simplified Approach," *Trans. N. Y. Acad. Sci.*, Series II, No. 30, pp. 457–478.

Crocker, M. J. (1967) "Multimode Response of Panels to Normal and to Travelling Sonic Boom," *Journal of the Acoustical Society of America*, Vol. 42, 1070.

Crocker, M. J. and Hudson, R. R. (1969) "Structural Response to Sonic Booms," *Journal of Sound and Vibrations*, Vol. 9, No. 3, pp. 454–468.

Dosch, J. J., Inman, D. J. and Garcia, E. (1992) "A Self-Sensing Piezoelectric Actuator for Collocated Control," *Journal of Intelligent Material Systems and Structures*, Vol. 3, pp. 166–184.

Gottlieb, J. J. and Ritzel, D. V. (1988) "Analytical Study of Sonic Boom from Supersonic Projectiles," *Progress in Aerospace Sciences*, Vol. 25, pp. 131–188.

Gupta, A. D., Gregory, F. H. and Bitting, R. L. (1985) "Dynamic Response of a Simply Supported Rectangular Plate to an Explosive Blast," *Proc. XIII Southeastern Conference on Theoretical and Appl. Mech.*, Vol. 1, pp. 385–390.

Hause, T. and Librescu,, L. (2005) "Dynamic Response of Anisotropic Sandwich Flat Panels to Explosive Pressure Pulses," *Interational Journal of Impact Engineering*, Vol. 31, No. 5, pp. 607–628.

Helgeson, R. J. and Szustak, P. W. (1995) "Saturation Constrained LQR," in *Tenth Symposium on Structural Dynamics and Control*, May 8–10, L. Meirovitch (Ed.), Blacksburg, VA., pp. 25–36.

Houlston, R., Slater, J. E., Pegg, N. and Des Rochers, C. G. (1985) "On the Analysis of Structural Response of Ship Panels Subjected to Air Blast Loading," *Computer and Structures*, Vol, 21, pp. 273–289.

Icardi, U., Di Sciuva, M. and Librescu, L. (2000) "Dynamic Response of Adaptive Cross-Ply Cantilevers Featuring Interlaminar Bonding Imperfections," *AIAA Journal*, Vol. 38, No. 2, pp. 499–506.

Irschik, H., Krommer, M. and Pickler, U. (2003) "Dynamic Shape Control of Beam-Type Structures by Piezoelectric Actuation and Sensing," *International Journal of Applied Electromagnetics and Mechanics*, 17, pp. 251–258.

Khot, N. S., Veley, D. E. and Bolonkin, A. A. (1996) "Design of Smart Structures with Bounded Controls," in *Smart Structures and Materials; Mathematics and Control in Smart Structures, Proc. SPIE 2715*, V. V. Varadan and J. Chandra (Eds.), pp. 173–183.

Librescu, L., Meirovitch, L. and Song, O. (1996) "Integrated Structural Tailoring and Adaptive Materials Control for Advanced Aircraft Wings," *Journal of Aircraft*, Vol. 33, No. 1, pp. 203–213.

Librescu, L., Meirovitch, L. and Na. S. S. (1997) "Control of Cantilever Vibration via Structural Tailoring and Adaptive Materials," *AIAA Journal*, Vol. 35, No. 8, pp. 1309–1315.

Librescu, L. and Na, S. S. (1998a) "Dynamic Response of Cantilevered Thin-Walled Beams to Blast and Sonic Boom Loadings," *Journal of Shock and Vibration*, Vol. 5, pp. 23–33.

Librescu, L. and Na, S. S. (1998b) "Dynamic Response Control of Thin-Walled Beams to Blast and Sonic-Boom Using Structural Tailoring and Piezoelectrically Induced Boundary Moment," *Journal of Applied Mechanics*, Trans. ASME, Vol. 65, June, pp. 497–504.

Librescu, L. and Na. S. S. (1998c) "Boundary Control of Free and Forced Oscillation of Shearable Thin-Walled Beam Cantilevers," *European Journal of Mechanics/A Solids*, Vol. 17, No. 4, pp. 687–700.

Librescu, L. and Na, S. S. (2001) "Active Vibration Control of Thin-Walled Tapered Beams Using Piezoelectric Strain Actuation," *Journal of Thin-Walled Structures*, Vol. 39, No. 1, pp. 65–82.

Librescu, L. and Na, S. S. (2005) "Comparative Study on Vibration Control Methodologies Applied to Adaptive Thin-Walled Anisotropic Cantilevers," *European Journal of Mechanics/Solids A* (in press).

Librescu, L. and Nosier, A. (1990) "Response of Shear Deformable Elastic Laminated Composite Flat Panels to Sonic Boom and Explosive Blast Loadings," *AIAA Journal*, Vol. 28, No. 2, February, pp. 345–352.

Librescu, L., Na, S. S. and Park, S. (2000) "Dynamics and Active Feedback Control of Wing-Type Structures Modeled as Thin-Walled Beams: Implications of Beam Cross-Section Nonuniformity," *Journal of the Chinese Society of Mechanical Engineers*, Vol. 21, No. 1, pp. 87–96.

Librescu, L., Oh, S. Y. and Hohe, J. (2004) "Linear and Non-Linear Dynamic Response of Sandwich Panels to Blast Loading," *Composites B*, Special Issue on Marine Composites, Y. Rajapakse and D. Hui (Eds.), Vol. 35, Nos. 6–8, pp. 673–683.

Mäkinen, K. (1999), "The Transverse Response of Sandwich Panels to an Underwater Shock Wave," *Journal of Fluids and Structures*, Vol. 13, pp. 631–646.

Marzocca, P., Librescu, L. and Chiocchia, G. (2001) "Aeroelastic Response of 2-D Lifting Surfaces to Gust and Arbitrary Explosive Loading Signature," *International Journal of Impact Engineering*, Vol. 25, No. 1, pp. 67–85.

Meirovitch, L. (1990) *Dynamic and Control of Structures*, John Wiley & Sons.

Na, S. S. (1997) "Control of Dynamic Response of Thin-Walled Composite Beams Using Structural Tailoring and Piezoelectric Actuation," Virginia Polytechnic Institute and State University, Blacksburg, VA, USA.

Na, S. S. and Librescu, L. (1998a) "Bending Vibration Control of Cantilevers via Boundary Moment and Combined Feedback Control Laws," *Journal of Vibration and Controls*, Vol. 4, No. 6, pp. 733–746.

Na, S. S. and Librescu, L. (1998b) "Oscillation Control of Cantilevers via Smart Materials Technology and Optimal Feedback Control: Actuator Location and Power Consumption Issues," *Journal of Smart Materials and Structures*, Vol. 7, No. 6, pp. 833–842.

Na, S. S. and Librescu, L. (2000) "Optimal Vibration Control of Thin-Walled Anisotropic Cantilevers Exposed to Blast Loadings," *Journal of Guidance, Control and Dynamics*, Vol. 23, No. 3, pp. 491–500.

Na, S. S. and Librescu, L. (2001a) "Modified Bang-Bang Vibration Control Strategy Applied to Adaptive Nonuniform Cantilevered Beams," *Proceedings of the 42nd AIAA/ASME/ASCE/AHS/ASC Structures, Structural Dynamics, and Material Conference and Exhibit*, Vol. 3, AIAA, Reston, VA, pp. 1888–1898.

Na, S. S. and Librescu, L. (2001b) "Dynamic Response of Elastically Tailored Adaptive Cantilevers of Nonuniform Cross-Section Exposed to Blast Pressure Pulses," *International Journal of Impact Engineering*, Vol. 25, Issue 9, pp. 1847–1867.

Na, S. S. and Librescu, L. (2002) "Optimal Dynamic Response Control of Elastically Tailored Nonuniform Thin-Walled Adaptive Beams," *Journal of Aircraft*, Vol. 39, No. 3, May–June, pp. 469–479.

Na, S. S., Librescu, L. and Shim, J. H. (2004) "Modified Bang-Bang Vibration Control Applied to Adaptive Thin-Walled Beams Cantilevers," *AIAA Journal*, Vol. 42, No. 8, pp. 1717–1721.

Na, S. S., Librescu, L., Kim, M. H., Leong, I. J. and Marzocca, P. (2005) "Aeroelastic Response of Flapping Wing Systems Using Robust Estimation Control Methodology," *Journal of Guidance, Control and Dynamics*, Vol. 28, in press.

Nosier, A., Librescu, L. and Frederick, D. (1990) "The Effect of Time-Dependent Excitation on the Oscillatory Motion of Viscously Damped Laminated Composite Flat Panels," in *Advances in the Theory of Plates and Shells*, G. Z. Voyiadjis and D. Karamanlides (Eds.), Elsevier Science Publishers, Amsterdam, pp. 249–268.

Oyibo, G. A. (1989) "Some Implications of Warping Restraint on the Behavior of Composite Anisotropic Beams," *Journal of Aircraft*, Vol. 26, No. 2, February, pp. 187–189.

Özbeck, T. and Suhubi, E. S. (1967) "On the Theory of Bending of Thick Straight Bars," *Bulletin of the Technical University of Istanbul*, Vol. 20, pp. 1–17.

Park, C., Waltz, C. and Chopra, I. (1996) "Bending and Torsion Models of Beams with Induced-Strain Actuators," *Smart Materials and Structures*, Vol. 5, Feb., pp. 98–113.

Preumont, A. (1997) *Vibration Control of Active Structures – An Introduction*, Kluwer Acad. Publishers (2nd edition, 2002).

Reddy, J. N. (1997) *Mechanics of Laminated Composite Plates. Theory and Application*, CRC Press, Boca Raton, FL.

Senjanovic, I., Tomasevic, S. and Parunov, J. (2001) "Analytical Solution of Pontoon Transient Vibration Related to Investigation of Ship Whipping Due to Slamming," *International Shipbuilding Progress*, Vol. 48, No. 4, pp. 305–331.

Song, O. and Librescu. L. (1993) "Free Vibration of Anisotropic Composite Thin-Walled Beams of Closed Cross-Section Contour," *Journal of Sound and Vibration*, Vol. 167, No. 1, pp. 129–147.

Song, O. and Librescu, L. (1995) "Bending Vibration of Cantilevered Thin-Walled Beams Subjected to Time-Dependent External Excitations," *Journal of the Acoustical Society of America*, Vol. 98, No. 1, June pp. 313–319.

Song, O., Ju, J. S. and Librescu, L. (1998) "Dynamic Response of Anisotropic Thin-Walled Beams to Blast and Harmonically Oscillating Loads," *International Journal of Impact Engineering*, Vol. 25, No. 8, pp. 663–682.

Song, O., Kim, J-B. and Librescu, L. (2001) "Synergistic Implications of Tailoring and Adaptive Materials Technology on Vibration Control of Anisotropic Thin-Walled Beams," *International Journal of Engineering Science*, Vol. 39, No. 1, December, pp. 71–94.

Soong, T. T. (1990) *Active Structural Control: Theory and Practice*, Longman Scientific & Technical.

Stephen, N. G. (1981) "Considerations on Second Order Beam Theories," *International Journal of Solids and Structures*, Vol. 17, pp. 325–333.

Thangjitham, S. and Librescu, L. (1991) "Free Vibration Characteristics of Anisotropic Composite Wing Structures," in *Proceedings of the 32nd AIAA/ASME/ASCE/AHS/ASC Structures, Structural Dynamics and Materials Conference*, Paper AIAA-91-1185, Baltimore, Maryland.

Vinson, J. R. (1999) *The Behavior of Sandwich Structures of Isotropic and Composite Materials*, Kluwer Academic Publishers, Dordrecht.

Chapter 8

THIN-WALLED BEAMS CARRYING STORES

8.1 PRELIMINARIES

Thin-walled beams carrying heavy concentrated masses along their span can serve as a basic model for a number of structures used in the aeronautical, aerospace, and other areas of the modern technology. Flexible robot arms, space booms or antennas operating in space must be lightweight, strong and precise under the action of heavy masses located along their span.

Military airplanes with lifting surfaces designed to carry heavy mounted stores along their span and at their tip, constitute another application of this structural system. During their complex missions and escape maneuvers, they experience high load factors giving rise to excessive deflections of the wing tip and large bending moments at the wing root. Moreover, depending on store magnitude and location, drastic reductions of the natural frequencies of the structure and modification of their eigenmodes are experienced (see e.g. Bruch and Mitchell, 1987; Abramovich and Hamburger, 1992).

For all these reasons, it is important to obtain the governing system of thin-walled beams carrying an arbitrary number of stores. These equations are useful not only for determining the influence of stores on statics and dynamics of thin-walled beams, but, as it will be shown later, also for the approach of the free vibration and dynamic response feedback control.

8.2 TRANSVERSE BENDING OF BEAMS CARRYING DISTRIBUTED STORES

8.2-1 Basic Assumptions

Consider a cantilevered thin-walled beam of arbitrary cross-section; the beam walls are symmetrically composed of transversely-isotropic layers, the surface of isotropy being parallel to the reference surface of the beam.

213

In addition to the concentrated stores distributed along the beam span, the beam carries a tip mass, see Fig 8.2-1; the distributed stores, located at $z = z_j$, $(j = \overline{1, J})$, are of mass m_j and mass moment of inertia I_j about their centroid, while the tip store is of mass M and moment of inertia I_M; for an airplane wing/robot arm, the tip mass can simulate the presence of a tank/payload; J is the total number of mounted stores distributed along the beam span.

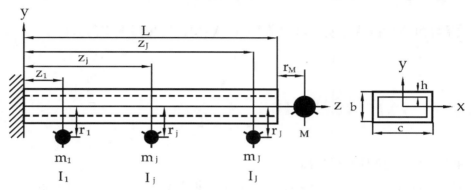

Figure 8.2-1: Cantilevered beam with external stores. Front view and cross section

We consider the following assumptions (see Song et al., 1994; Song and Librescu, 1996; Librescu and Song, 1997): (i) the cross-sections of the beam do not deform in their own planes; (ii) the effects of transverse shear are included; (iii) the hoop stress resultant N_{ss} is negligibly small with respect to the remaining ones, and iv) the centroids of the attached masses lie on the global centroidal axis of the system. In this case the transverse bending decouples from the other displacement modes. Due to its importance, the study will be confined to only transverse bending. In the context of solid beams with stores, Gern and Librescu (1998) developed a general model that included the bending-twist coupling. For the present case, the 3-D displacement vector components expressed in terms of 1-D bending displacement quantities (see Eqs. (2.1-6) and (2.1-38)) are

$$u(x, y, z, t) = 0; \quad v(x, y, z, t) = v_0(z, t),$$

$$w(x, y, z, t) = y(s)\theta_x(z, t) - n\frac{dx}{ds}\theta_x(z, t), \tag{8.2-1a-c}$$

and consequently, the associated strain components become

$$\epsilon_{zz}^{(0)} = y(s)\theta_x'(z, t); \quad \epsilon_{zz}^{(1)} = -\frac{dx}{ds}\theta_x'(z, t),$$

$$\gamma_{sz}^{(0)} = (v_0' + \theta_x)\frac{dy}{ds}; \quad \gamma_{zn} = -(v_0' + \theta_x)\frac{dx}{ds}. \tag{8.2-2a-d}$$

The equations of motion and boundary conditions are obtained from Hamilton's principle.

8.2-2 Equations of Motion and Boundary Conditions. Shearable Beams

The Hamilton's principle of the 3-D elasticity theory is

$$\delta J = 0, \quad \text{where,} \quad J \equiv \int_{t_0}^{t_1} (\mathcal{W} - \mathcal{K} - \mathcal{J}) dt, \quad (8.2\text{-}3a,b)$$

$\mathcal{W}$, $\mathcal{K}$ and $\mathcal{J}$ denoting, respectively, the strain energy, the kinetic energy and the work performed by surface and dead loads; t_0 and t_1 denote the two instants of time, while δ is the variation operator.

For the present case

$$\mathcal{K} = \int_0^L \oint \sum_{k=1}^N \int_{h_{k-1}}^{h_k} \rho_{(k)} \left[\dot{v}_0^2 + \left(y\dot{\theta}_x - n\frac{dx}{ds}\dot{\theta}_x \right)^2 \right] dndsdz$$

$$+ \frac{1}{2} \sum_{j=1}^J \int_0^L (m_j \dot{v}_0^2 + I_j \dot{\theta}_x^2) \delta_D(z - z_j) dz + \frac{1}{2}[M\dot{v}_0^2 + I_M\dot{\theta}_x^2]_{z=L},$$

$$(8.2\text{-}4a)$$

$$\mathcal{W} = \frac{1}{2} \int_0^L \oint \sum_{k=1}^N \int_{h_{k-1}}^{h_k} [\sigma_{zz}^{(k)}(\varepsilon_{zz}^{(0)} + n\varepsilon_{zz}^{(1)}) + \sigma_{sz}\gamma_{sz}^{(0)} + \sigma_{zn}\gamma_{zn}] dndsdz$$

$$= \frac{1}{2} \int_0^L [M_x\theta_x' + Q_y(v_0' + \theta_x)] dz, \quad (8.2\text{-}4b)$$

and

$$\mathcal{J} = -\sum_{j=1}^J \left[\int_0^L m_j g \delta_D(z - z_j) v_0 dz \right] + \int_0^L (-b_1 g + p_y) v_0 dz$$

$$- [M(g - r_M\ddot{\theta}_x)v_0 + M(\ddot{v}_0 + g)r_M\theta_x]\big|_{z=L}. \quad (8.2\text{-}4c)$$

In these equations $\oint (\cdot) ds$ is the integral around the circumference of the mid-line cross-section contour of the beam, g denotes the acceleration of gravity; L is the beam span, r_M is the distance between the beam tip and the centroid of the tip mass, $\delta_D(\cdot)$ is Dirac's distribution, ρ denotes the mass density of constituent materials, the overdots denote time derivatives, while N is the total number of layers of the composite beam.

Performing the integrations in Eqs. (8.2-4a,b) across the beam thickness and around the circumference, carrying out the variations in the expressions involved in Eqs. (8.2-4), replacing these ones in (8.2-3), collecting the coefficients of each of the virtual displacements (δv_0, $\delta\theta_x$) and having in view that these are independent and arbitrary within $z \in [0, L]$, and $t \in [t_0, t_1]$, one obtain the equations of motion associated with the transverse bending:

$$\delta v_0: \quad Q'_y - \left[b_1 \ddot{v}_0 + \sum_{j=1}^{J} m_j \ddot{v}_0 \, \delta_D(z - z_j) \right] - \sum_{j=1}^{J} m_j g \delta_D(z - z_j)$$

$$- b_1 g + p_y(z, t) = 0,$$

$$\delta\theta_x: \quad M'_x - Q_y - \left[\underbrace{(b_4 + \delta_n b_{14})\ddot{\theta}_x} + \sum_{j=1}^{J} \underbrace{I_j \ddot{\theta}_x \delta_D(z - z_j)} \right] = 0.$$

$$(8.2\text{-}5a,b)$$

and the boundary conditions. For cantilevered beams these are:

At the root ($z = 0$):

$$v_0 = \theta_x = 0, \qquad\qquad\qquad (8.2\text{-}6a,b)$$

and at the tip ($z = L$):

$$\delta v_0: \quad Q_y + M\ddot{v}_0 + M(g - \underbrace{r_M\ddot{\theta}_x}) = 0,$$

$$\delta\theta_x: \quad M_x + I_M\ddot{\theta}_x - Mr_M(\ddot{v}_0 + g) = 0. \qquad (8.2\text{-}6c,d)$$

In addition, as it results from Eqs. (4.3-2)

$$\begin{Bmatrix} M_x \\ Q_y \end{Bmatrix} = \begin{bmatrix} a_{33} & 0 \\ 0 & a_{55} \end{bmatrix} \begin{Bmatrix} \theta'_x \\ v'_0 + \theta_x \end{Bmatrix}, \qquad (8.2\text{-}7)$$

where a_{33} and a_{55} denote the bending and transverse shear stiffnesses given by Eqs. (4.4-41).

To determine the motion about the equilibrium position, we represent v_0, θ_x and p_y in Eqs. (8.2-5) through (8.2-7) as

$$v_0(z; t) = \bar{v}_0(z) + \hat{v}_0(z, t),$$
$$\theta_x(z; t) = \bar{\theta}_x(z) + \hat{\theta}_x(z, t), \qquad (8.2\text{-}8a\text{-}c)$$
$$p_y(z; t) = \bar{p}_y(z) + \hat{p}_y(z, t).$$

where $\bar{v}_0(z)$, $\bar{\theta}_x(z)$ and $\bar{p}_y(z)$ are associated with the equilibrium position, while $\hat{v}_0(z, t)$, $\hat{\theta}_x(z, t)$ and $\hat{p}_y(z, t)$ denote the variations about the equilibrium position.

Equation (8.2-8), the equations of motion and the boundary conditions yield the *static equations of bending* and those governing the *transverse motion* about the equilibrium position.

The *static* governing equations are

$$a_{55}(\overline{v}_0'' + \overline{\theta}_x') - \sum_{j=1}^{J} m_j g \delta_D(z - z_j) - b_1 g + \overline{p}_y = 0,$$

$$a_{33}\overline{\theta}_x'' - a_{55}(\overline{v}_0' + \overline{\theta}_x) = 0. \qquad (8.2\text{-}9a,b)$$

and the associated boundary conditions:

at $z = 0$:

$$\overline{v}_0 = \overline{\theta}_x = 0, \qquad (8.2\text{-}10a,b)$$

and at $z = L$:

$$a_{55}(\overline{v}_0' + \overline{\theta}_x) + Mg = 0,$$

$$a_{33}\overline{\theta}_x' - Mgr_M = 0. \qquad (8.2\text{-}11a,b)$$

The *dynamic* equations are

$$\delta \hat{v}_0 : a_{55}(\hat{v}_0'' + \hat{\theta}_x') - \left[b_1 \ddot{\hat{v}}_0 + \sum_{j=1}^{J} m_j \ddot{\hat{v}}_0 \delta_D(z - z_j) \right] + \hat{p}_y(z,t) = 0,$$

$$(8.2\text{-}12a,b)$$

$$\delta \hat{\theta}_x : a_{33}\hat{\theta}_x'' - a_{55}(\hat{v}_0' + \hat{\theta}_x) - \left[\underline{(b_4 + \delta_n b_{14})\,\ddot{\hat{\theta}}_x} + \sum_{j=1}^{J} \underwave{I_j \ddot{\hat{\theta}}_x \delta_D(z - z_j)} \right] = 0,$$

with the associated boundary conditions

at $z = 0$:

$$\hat{v}_0 = \hat{\theta}_x = 0, \qquad (8.2\text{-}13a,b)$$

and at $z = L$:

$$\delta \hat{v}_0 : a_{55}(\hat{v}_0' + \hat{\theta}_x) + M\ddot{\hat{v}}_0 - \underwave{Mr_M \ddot{\hat{\theta}}_x} = 0,$$

$$\delta \hat{\theta}_x : a_{33}\hat{\theta}_x' + I_M\,\ddot{\hat{\theta}}_x - Mr_M\,\ddot{\hat{v}}_0 = 0. \qquad (8.2\text{-}14a,b)$$

8.2-3 Unshearable Beams

The unshearable counterpart of the previously static and dynamic governing equations will be displayed next. We proceed as in Section 4.4-3 and obtain:

The governing equation

$$a_{33}\overline{v}_0'''' + \sum_{j=1}^{J} m_j g \delta_D(z - z_j) + b_1 g = \overline{p}_y(z), \qquad (8.2\text{-}15)$$

and *the boundary conditions*

at $z = 0$:

$$\overline{v}_0 = \overline{v}_0' = 0, \qquad (8.2\text{-}16\text{a,b})$$

and at $z = L$:

$$a_{33}\overline{v}_0''' = Mg, \quad a_{33}\overline{v}_0'' = -Mgr_M. \qquad (8.2\text{-}17\text{a,b})$$

As for the unshearable version of dynamic equations, these are

The governing equation:

$$a_{33}\hat{v}_0'''' - \left[(b_4 + \delta_n b_{14})\ddot{\hat{v}}_0'' + \sum_{j=1}^{J} I_j \ddot{\hat{v}}_0'' \delta_D(z - z_j) \right.$$

$$\left. + \left[b_1 \ddot{\hat{v}}_0 + \sum_{j=1}^{J} m_j \ddot{\hat{v}}_0 \delta_D(z - z_j) \right] = \hat{p}_y(z, t), \qquad (8.2\text{-}18)$$

with the boundary conditions:

at $z = 0$:

$$\hat{v}_0 = \hat{v}_0' = 0 \qquad (8.2\text{-}19\text{a,b})$$

and at $z = L$:

$$a_{33}\hat{v}_0''' = M\ddot{\hat{v}}_0' + Mr_M\ddot{\hat{v}}_0,$$

$$a_{33}\hat{v}_0'' = -I_M\ddot{\hat{v}}_0' - Mr_M\ddot{\hat{v}}_0. \qquad (8.2\text{-}20\text{a,b})$$

Note that for both transverse shear deformable and classical models, the fourth order governing equations require prescription of two boundary conditions at each edge.

8.3 ADAPTIVE BEAMS CARRYING EXTERNAL STORES

8.3-1 Preliminaries

The presence of heavy concentrated masses located along the span of a cantilevered beam, may cause excessive deflections at its tip accompanied by large bending moments at the beam root. Moreover, the stores can drastically reduce the natural frequencies and change the eigenmodes. These changes can worsen the dynamic response to time-dependent external excitation.

To prevent such undesirable structural responses we could reinforce the structure; the accompanying weight increase is unacceptable. Instead, we consider the use of smart materials technology (see Librescu et al., 1993; Song et al., 1994; Song and Librescu, 1996; Librescu and Song, 1997; Na and Librescu, 2000a,b; Na, 1997).

8.3-2 Governing Equations

We approach the statics and dynamics of smart structures separately.

For shearable beams:

- the static equations are

$$\delta \overline{v}_0: \quad a_{55}(\overline{v}_0'' + \overline{\theta}_x') - \sum_{j=1}^{J} m_j \delta_D(z - z_j) - b_1 g + \overline{P}_y = 0,$$

$$\delta \overline{\theta}_x: \quad a_{33}\overline{\theta}_x'' - a_{55}(\overline{v}_0' + \overline{\theta}_x) - \delta_P \tilde{M}_x'(z) = 0, \qquad (8.3\text{-}1a,b)$$

and the associated boundary conditions:

at $z = 0$:

$$\overline{v}_0 = \overline{\theta}_x = 0, \qquad (8.3\text{-}2a,b)$$

and at $z = L$:

$$a_{55}(\overline{v}_0' + \overline{\theta}_x) + Mg = 0, \quad a_{33}\overline{\theta}_x' - Mgr_M = \delta_S \tilde{M}_x. \qquad (8.3\text{-}3a,b)$$

- The equations governing the transverse motion about equilibrium position are

$$\delta \hat{v}_0 : a_{55}(\hat{v}_0'' + \hat{\theta}_x') - \left[b_1 \ddot{\hat{v}}_0 + \sum_{j=1}^{J} m_j \ddot{\hat{v}}_0 \delta_D(z - z_j) \right] + \hat{p}_y(z, t) = 0,$$

$$\delta \hat{\theta}_x : a_{33}\hat{\theta}_x'' - a_{55}(\hat{v}_0' + \hat{\theta}_x) - \delta_P \tilde{M}_x'(z, t)$$

$$- \left[\underline{(b_4 + \delta_n b_{14})\ddot{\hat{\theta}}_x} + \sum_{j=1}^{J} \underwave{I_j \ddot{\hat{\theta}}_x \delta_D(z - z_j)} \right] = 0, \qquad (8.3\text{-}4a,b)$$

with the associated boundary conditions

at $z = 0$:

$$\hat{v}_0 = \hat{\theta}_x = 0, \tag{8.3-5a,b}$$

and at $z = L$:

$$\delta \hat{v}_0 : a_{55}(\hat{v}_0' + \hat{\theta}_x) + M \ddot{\hat{v}}_0 - \underline{Mr_M \ddot{\hat{\theta}}_x} = 0,$$

$$\delta \hat{\theta}_x : a_{33} \hat{\theta}_x' + I_M \ddot{\hat{\theta}}_x - Mr_M \ddot{\hat{v}}_0 = \delta_S \tilde{M}_x. \tag{8.3-6a,b}$$

For unshearable beams counterpart the equations are as follows:

- The static governing equation:

$$\delta \overline{v}_0 : a_{33} \overline{v}_0'''' + \delta_P \tilde{M}_x''(z) + \sum_{j=1}^{J} m_j g \delta_D(z - z_j) + b_1 g = \overline{P}_y(z), \tag{8.3-7}$$

and the boundary conditions:

at $z = 0$:

$$\overline{v}_0 = \overline{v}_0' = 0, \tag{8.3-8a,b}$$

and at $z = L$:

$$\delta \overline{u}_0 : a_{33} \overline{v}_0''' = Mg,$$

$$\delta \overline{v}_0' : a_{33} \overline{v}_0'' + \delta_S \tilde{M}_x = -Mgr_M, . \tag{8.3-9a,b}$$

- The dynamic governing equation:

$$\delta \hat{v}_0 : a_{33} \hat{v}_0'''' + \delta_P \tilde{M}_x''(z, t) - \left[(b_4 + \delta_n b_{14}) \ddot{\hat{v}}_0'' + \sum_{j=1}^{J} \underline{I_j \ddot{\hat{v}}_0'' \delta_D(z - z_j)} \right]$$

$$+ \left[b_1 \ddot{\hat{v}}_0 + \sum_{j=1}^{J} m_j \ddot{\hat{v}}_0 \delta_D(z - z_j) \right] = \hat{p}_y(z, t), \tag{8.3-10}$$

with the boundary conditions:

at $z = 0$:

$$\hat{v}_0 = \hat{v}_0' = 0, \tag{8.3-11a,b}$$

and at $z = L$:

$$\delta \hat{v}_0 : a_{33} \hat{v}_0''' = M \ddot{\hat{v}}_0' + \underline{M r_M \ddot{\hat{v}}_0'},$$

$$\delta \hat{v}_0' : a_{33} \hat{v}_0'' + \delta_S \tilde{M}_x = -\underline{I_M \ddot{\hat{v}}_0'} - M r_M \ddot{\hat{v}}_0. \qquad (8.3\text{-}12\text{a,b})$$

In the previously displayed equations $\tilde{M}_x$ denotes the piezoelectrically induced bending moment, while the tracers δ_S and δ_P are as in Section 6.6-3. White and Heppler (1993) showed that the terms involving r_M have little influence. However, Librescu and Na (1997), showed that in the condition of the increase of the mass ratio M/m_b as is common place in space applications (see Kumar, 1974), where m_b is the mass of the clean beam, r_M can appreciably decrease the eigenfrequencies. In the numerical applications however, we will neglect r_M.

8.3-3 Static and Dynamic Feedback Control

Two smart beam configurations will be considered next. These are labeled as SBA and SBB.

8.3-3a Smart Beam Configuration A (SBA)

Suppose that the adaptive beam consists of a symmetric box-beam, with piezoactuators (selected to be of PZT-4 piezoceramic) bonded on the upper and bottom surfaces of the master structure and spread over the entire beam span.

For the non-optimal feedback control, the electric field $\mathcal{E}_3$ on which the induced bending moment depends, should have a prescribed functional relationship to the mechanical quantities characterizing the wing response.

Within SBA, the following control law is adopted (see Tzou, 1993; Tzou and Zhong, 1992):

$$\tilde{M}_x(L, t) = k_p \hat{v}_0(L, t), \qquad (8.3\text{-}13)$$

where k_p is the proportional feedback gain: the applied electrical field $\mathcal{E}_3$ inducing the boundary bending moment $\tilde{M}_x$ (see Eq. (6.6-1a)), is proportional to the vertical deflection at the beam tip, $\hat{v}_0(L, t)$.

8.3-3b Smart Beam Configuration B (SBB)

Consider a beam of a bi-convex cross-section, with actuators either spread over the entire beam span, $(SBB)_1$, or are in the form of a patch, $(SBB)_2$. In both cases the piezoactuators are symmetrically located over the top and bottom surfaces of the beam.

For $(SBB)_1$, a velocity feedback control law is used: the piezoelectrically induced boundary moment is

$$\tilde{M}_x(L, t) = -k_v \dot{\hat{\theta}}_x(L, t), \tag{8.3-14}$$

where k_v is the velocity feedback control. As a result, the expression of the external work changes to

$$\overline{\mathcal{J}} = \mathcal{J} + \tilde{M}_x(L, t)\hat{\theta}_x(L, t), \tag{8.3-15}$$

where $\mathcal{J}$ is given by Eq. (8.2-4c).

Proceeding as in Section 6.6-5a, replacing the representations of displacements $\hat{v}_0(z, t)$ and $\hat{\theta}_x(z, t)$ from Eqs. (6.6-21b,c) in the Hamilton's functional, one obtains the discretized governing equation

$$\mathbf{M}\ddot{\mathbf{q}}(t) + \mathbf{H}\dot{\mathbf{q}}(t) + \mathbf{K}\mathbf{q}(t) = \mathbf{Q}(\mathbf{t}). \tag{8.3-16}$$

Here $\mathbf{q}(t)$ is the $2N \times 1$ generalized co-ordinate column vector, $\mathbf{M}$, $\mathbf{H}$ and $\mathbf{K}$ are $2N \times 2N$ square matrices corresponding to mass, damping and stiffness, respectively, while $\mathbf{Q}(t)$ is the generalized input force vector:

$$\mathbf{q}(t) = \left\{ \mathbf{q}_v^T(t), \ \mathbf{q}_x^T(t) \right\}^T,$$

$$\mathbf{M} = \int_0^L \begin{bmatrix} b_1 \mathbf{V}_0 \mathbf{V}_0^T & \mathbf{0} \\ \mathbf{0} & (b_4 + \delta_n b_{14})\mathbf{X}_0 \mathbf{X}_0^T \end{bmatrix} dz$$

$$+ \begin{bmatrix} \displaystyle\sum_{j=1}^{J} m_j(z_j)\mathbf{V}_0(z_j)\mathbf{V}_0^T(z_j) & \mathbf{0} \\ \mathbf{0} & \displaystyle\sum_{j=1}^{J} I_j \mathbf{X}_0(z_j)\mathbf{X}_0^T(z_j) \end{bmatrix},$$

$$\mathbf{H} = \begin{bmatrix} \mathbf{0} & \mathbf{0} \\ \mathbf{0} & k_v \mathbf{X}_0(L)\mathbf{X}_0^T(L) \end{bmatrix},$$

$$\mathbf{K} = \int_0^L \begin{bmatrix} a_{55}\mathbf{V}_0'\mathbf{V}_0'^T & a_{55}\mathbf{V}_0'\mathbf{X}_0^T \\ Symm. & a_{55}\mathbf{X}_0\mathbf{X}_0^T + a_{33}\mathbf{X}_0'\mathbf{X}_0'^T \end{bmatrix} dz, \tag{8.3-17a-e}$$

$$\mathbf{Q} = \int p_y \mathbf{V}_0^T dz.$$

As in Section 6.6-5, we cast Eq. (8.3-16) in state-space form: with the state vector $\mathbf{x} = \{\mathbf{q}^T, \dot{\mathbf{q}}^T\}^T$ and the identity $\dot{\mathbf{q}} \equiv \dot{\mathbf{q}}$, Eq. (8.3-16) reduces to

$$\dot{\mathbf{x}}(t) = \mathbf{A}\mathbf{x}(t) + \mathbf{B}\mathbf{Q}(t), \tag{8.3-18}$$

where

$$A = \begin{bmatrix} 0 & I \\ -M^{-1}K & -M^{-1}H \end{bmatrix} \text{ and } B = \{0, \ M^{-1}\}^T. \qquad (8.3\text{-}19a,b)$$

For free vibration, the homogeneous solution of equation (8.3-18) has the form $x(t) = Xe^{\lambda t}$, where X is a constant vector and λ is a constant scalar, both generally complex; we obtain a standard eigenvalue problem

$$AX = \lambda X, \qquad (8.3\text{-}20)$$

The solution of the algebraic eigenvalue problem yields the closed-loop eigenvalues

$$(\lambda_r, \ \bar{\lambda}_r) = \sigma_r \pm i\omega_{dr}, \qquad (8.3\text{-}21)$$

where σ_r is a measure of the piezoelectrically induced damping in the rth mode and ω_{dr} is the rth frequency of damped oscillation. Consideration of $k_v = 0$, yields $\sigma_r = 0$: the free vibration is undamped.

In $(SBB)_2$, proceeding as in Section 7.7, one obtains the discrete equation of motion

$$M\ddot{q} + Kq = Q - Fu, \qquad (8.3\text{-}22)$$

that can be cast in state-space form

$$\dot{x}(t) = Ax(t) + BQ(t) + Wu(t), \qquad (8.3\text{-}23)$$

where

$$A = \begin{bmatrix} 0 & I \\ -M^{-1}K & 0 \end{bmatrix}, \ B = \left\{ \begin{matrix} 0 \\ M^{-1} \end{matrix} \right\} \text{ and } W = \left\{ \begin{matrix} 0 \\ -M^{-1}F \end{matrix} \right\}.$$
$$(8.3\text{-}24a\text{-}c)$$

In these expressions (see Na and Librescu, 2000a,b), the mass and stiffness matrices, M and K, are

$$M = \int_0^L \begin{bmatrix} (\bar{b}_1 + \delta_s \hat{b}_1)V_0V_0^T & 0 \\ 0 & [\bar{B}_4 + \delta_s \hat{B}_4]X_0X_0^T \end{bmatrix} dz$$

$$+ \delta_P \int_{z_1}^{z_2} \begin{bmatrix} \hat{b}_1 V_0V_0^T & 0 \\ 0 & \hat{B}_4 X_0 X_0^T \end{bmatrix} dz$$

$$+ \begin{bmatrix} \sum_{j=1}^{J} m_j V_0(z_j)V_0^T(z_j) & 0 \\ \\ 0 & \sum_{j=1}^{J} I_J X_0(z_j)X_0^T(z_j) \end{bmatrix}, \qquad (8.3\text{-}25a,b)$$

$$\mathbf{K} = \int_0^L \begin{bmatrix} (\bar{a}_{55} + \delta_S \hat{\bar{a}}_{55}) \mathbf{V}_0' \mathbf{V}_0'^T & (\bar{a}_{55} + \delta_S \hat{\bar{a}}_{55}) \mathbf{V}_0' \mathbf{X}_0^T \\ (\bar{a}_{55} + \delta_S \hat{\bar{a}}_{55}) \mathbf{X}_0 \mathbf{V}_0'^T & (a_{55} + \delta_S \hat{\bar{a}}_{55}) \mathbf{X}_0 \mathbf{X}_0^T + (a_{33} + \delta_S \hat{\bar{a}}_{33}) \mathbf{X}_0' \mathbf{X}_0'^T \end{bmatrix} dz$$

$$+ \delta_P \int_{z_1}^{z_2} \begin{bmatrix} \hat{a}_{55} \mathbf{V}_0' \mathbf{V}_0'^T & \hat{a}_{55} \mathbf{V}_0' \mathbf{X}_0^T \\ \hat{a}_{55} \mathbf{X}_0 \mathbf{V}_0'^T & \hat{a}_{55} \mathbf{X}_0 \mathbf{X}_0^T + \hat{a}_{33} \mathbf{X}_0' \mathbf{X}_0'^T \end{bmatrix} dz.$$

As indicated, $\mathbf{V}_0$ and $\mathbf{X}_0$ are vectors of trial functions associated with displacements v_0 and θ_x, respectively; B_4 was defined by Eq. (7.8-1c), while the quantities with an overbar, a single or double overcarets are associated with the host beam, the piezopatch or with the piezoactuator spread over the entire beam span, respectively.

8.3-4 Results

In Case SBA, we carry out numerical simulations restricted to the control of the static response and of free vibration. The two distributions of concentrated masses in Fig. 8.3-1 are identified as Case I and Case II.

For Case I, Figs. 8.3-2 give the distribution of the dimensionless static tip deflection $\tilde{v}_0(L) (\equiv \bar{v}_0(L)/L)$ versus the location $\eta_1 (\equiv z_1/L)$ of the mass m_1 along the wing-span, for the non-activated and the activated systems. For the activated system, the response is represented for selected values of the dimensionless feedback gain $K_p (\equiv k_p L^2 / a_{33})$.

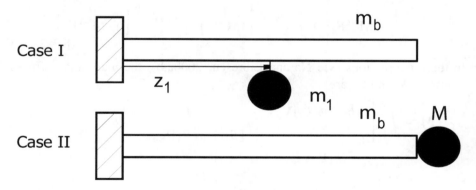

Figure 8.3-1: Selected cases of external store configuration labelled as Case I and Case II

Figures 8.3-2 show that for small values of $\alpha_1 (\equiv m_1/m_b)$, that is $\alpha_1 = 0.1$, the distribution of the deflection is slightly sensitive to the location of the concentrated mass along the beam span, especially for the non-activated beam. However, as Fig. 8.3-2b shows, with the increase of α_1, such as for $\alpha_1 = 0.5$, the deflection distribution is strongly affected by the location of the concentrated mass.

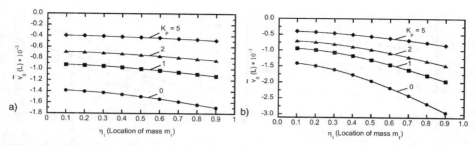

Figure 8.3-2: (a) Distribution of the normalized beam tip deflection against dimensionless span location of store of mass m_1, for the non-activated ($K_p = 0$) and activated ($K_p \neq 0$) beam, $\alpha_1 = 0.1$. (b) Counterpart of Fig. 8.3-2a for $\alpha_1 = 0.5$

Note that for a relatively large feedback gain, the deflection distribution becomes almost insensitive to the mass location. Both graphs emphasize the efficiency of the adaptive technology in reducing the beam tip deflection.

For the store configuration, Case II, the distribution of the normalized tip deflection as a function of $\alpha_2 (\equiv M/m_b)$, for the controlled and uncontrolled beam is depicted in Fig. 8.3-3.

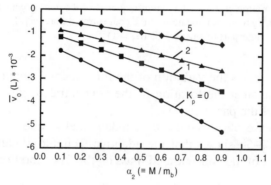

Figure 8.3-3: Distribution of the normalized beam tip deflection against mass ratio α_2 for the controlled and uncontrolled beam

This graph, shows that for the uncontrolled beam, and in contrast to its controlled counterpart, there is a strong variation of the tip deflection with α_2. The great efficiency of the application of the adaptive technology, especially in the regions of large mass ratios α_2, becomes apparent.

Figures 8.3-4a and 8.3-4b depict for store configuration, Case I, the variation of the dimensionless bending moment at the beam root as a function of the mass location. The control reduces the bending root moment $\mathcal{M}_x(0)(\equiv M_x(0)L/a_{33})$. With the increase of the feedback gain, the bending root moment becomes less and less sensitive to the location of the mass. However, this trend is strongly dependent upon the value of the mass ratio α_1.

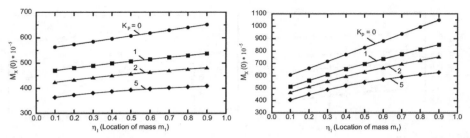

Figure 8.3-4: (a) Variation of the dimensionless bending root moment $\mathcal{M}_x(0)$ with span location of the mass m_1 ($\alpha_1 = 0.1$); store configuration Case I. (b) Counterpart of Fig. 8.3-4a for $\alpha_1 = 0.5$

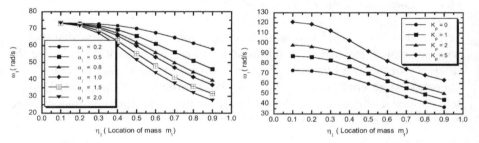

Figure 8.3-5: (a) Variation of the fundamental frequency of the non-activated beam, as per store configuration Case I, against span location of the concentrated mass m_1, for various mass ratios α_1. (b) Controlled counterpart of Fig. 8.3-5a ($\alpha_1 = 1$)

Figure 8.3-5a shows the variation of the fundamental frequency of the non-activated beam against the location of the concentrated mass for Case I, for selected values of the parameter α_1.

With the increase of η_1 and α_1, the fundamental frequency decays dramatically. Figure 8.3-5b shows that the adaptive technology can counteract the detrimental effect of the concentrated mass on the free vibration characteristics of the beam.

It is interesting also to see how the second natural frequency varies with the span location of the store, and how this variation is affected by the control. Figure 8.3-6a depicts the second open-loop natural frequency as a function of the span location of the store, for selected values of the mass ratio α_1; Fig. 8.3-6b shows the closed-loop counterpart.

As in Fig. 8.3-6a, the second natural frequency decreases when the store moves toward the beam tip. However, moving the store towards the node line of the second bending mode, increases the vibration frequency of this mode.

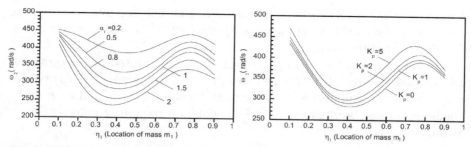

Figure 8.3-6: (a) Variation of the open-loop second natural frequency as a function of the location of the mass m_1, for selected values of α_1. (b) Variation of the closed-loop second natural frequency as a function of the location of the mass m_1, for selected values of feedback gain ($\alpha_1 = 1$)

This trend is consistent with that obtained experimentally by Runyan and Sewall (1947), and also analytically, for a solid beam by Gern and Librescu (1998). It is found that the node line is shifted outward by shifting the store location toward the beam tip. The control works well increasing also the second natural frequency.

For higher mode frequencies, this control becomes less and less efficient. For $(SBB)_1$, some results by Na and Librescu (2000a,b) are related to the store configuration, Case I, and concern the dynamic response to external excitations.

Figure 8.3-7 shows the effects played by the velocity feedback control in conjunction with transverse shear flexibility on dynamic response of a beam impacted by a non-symmetric sonic boom.

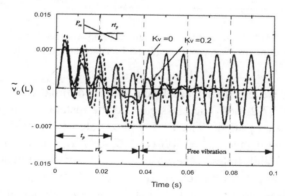

Figure 8.3-7: Dynamic response control of a thin-walled beam impacted by sonic-boom characterized by $r = 1.5$; $t_p = 0.025s$. The beam is of $AR = 3$, and the unistore of $m_1/m_b = 0.1$ is located at $\eta_1 = 0.9$. The material of the host structure is characterized by $E/G' = 0$ (———) and $E/G' = 100$ (– – – –). Here $K_v = k_v \bar{\omega} L/a_{33}$, where $\bar{\omega} = 128.16$ rad/s is a open-loop reference natural frequency corresponding to a beam characterized by $E/G' = 50$ and $AR = 4$

This plot shows that the unshearable beam model underestimates the deflection. However, for the activated beam in the free motion regime (i.e., when

$t > rt_p = 0.0375s$), the differences between the closed-loop dynamic response amplitudes of shearable and unshearable beams are small. This graph indicates the efficiency of this control.

Recall that within the velocity feedback control, the closed-loop eigenvalues are complex (see Eq. (8.3-21)), and there is generated damping. This explains why, within the application of the velocity feedback control and even in the conditions of the absence of structural damping, the response amplitude decays as time unfolds.

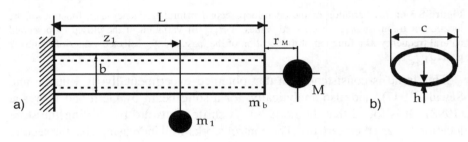

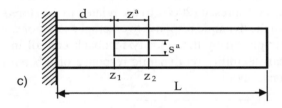

Figure 8.3-8: (a) Cantilevered thin-walled beam carrying a store and a tip mass, $(SBB)_2$, (b) Beam cross-section, (c) Piezopatch location. Piezoactuator thickness, $t^a = 0.00787$ in. (0.199 mm)

For $(SBB)_2$, we consider Case I, and use an optimal control. As in Sections 7.6-3 and 7.7-2, both the classical and the instantaneous LQR controls methodologies are applied. The PZT-4 piezopatch is located as indicated in Fig. 8.3-8. We consider $P_m = 50lb(= 222.4N)/L$, the beam aspect ratio is $AR(\equiv L/c) = 3$ and, unless otherwise stated, it carries a unistore, of relative mass $\alpha_1 = 0.1$. The pressure pulses are indicated in the figure insets. It is also considered, unless otherwise stated, that the piezopatch actuator is of a finite size $z^a = 0.1L$, and $d = 0.1L$.

Figure 8.3-9 shows the effect played by the size and location of a piezopatch on the closed-loop deflection time-history of an unshearable beam carrying a one store located at $\eta_1 = 0.1$ and exposed to a triangular blast pulse. The location and size of the piezopatch are indicated by the two coordinates $\hat{\eta}_1(\equiv z_1/L)$ and $\hat{\eta}_2(\equiv z_2/L)$ of its free edges, appearing in the brackets associated with z^a, as $z^a = (\hat{\eta}_2 - \hat{\eta}_1)L$. For $z^a = (1 - 0)L$ the piezoactuator is spread over the entire beam span. The results show that, with the exception of a short time during forced vibration range (corresponding to the time when the blast load acts on the structure i.e. for $t < 0.02s$), within the free vibration regime, i.e. for $t > 0.02s$, the controls achieved via a piezopatch and via a tip moment (i.e. when the piezoactuator is spread over the entire span) yield the same results, consisting in the total extinction of any oscillation.

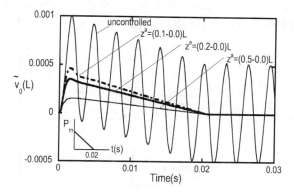

Figure 8.3-9: Influence of the piezopatch location and size on beam tip deflection time-history; triangular blast. The case of the unactivated beam is also displayed

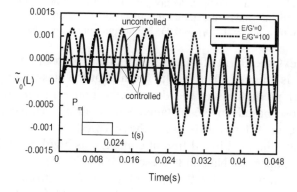

Figure 8.3-10: Influence of transverse shear flexibility of the material of the host structure on beam tip deflection time-history. Rectangular pulse ($\alpha_1 = 0.1$; $\eta_1 = 0.1$)

Figure 8.3-10 shows the effect of transverse shear flexibility of the material of the host structure (measured in terms of the ratio E/G') on the vibration control of the beam carrying a unistore of mass ratio $\alpha_1 = 0.1$, located at $\eta = 0.1$ and exposed to a rectangular pressure pulse: the classical model ($E/G' = 0$), underestimates both the amplitude of oscillation and the time at which their maxima occur; this underestimation is worse in the free-vibration range (i.e. for $t \geq 0.024s$).

For the activated beam, it is seen that: a) in the forced vibration range the *classical beam model* underestimates the amplitude of the vibration response, and b) in the free motion the differences in the closed-loop predictions provided by the unshearable and shearable beam models are marginal.

Figure 8.3-11 shows the implications on response played by incorporation/ discard from the performance index of external excitation. In the former case the *instantaneous optimal control* (IOC) is carried out, whereas in the latter one the *standard optimal control* (LQR) is applied.

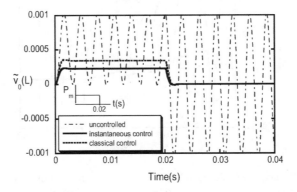

Figure 8.3-11: Influence on beam tip deflection time-history of the inclusion (——) and discard (- - -) in the performance index of external time-dependent excitation. ($-- \cdot --$) denotes unactivated beam, ($m_1/m_b = 0.1$; $\eta_1 = 0.1$). Rectangular pressure pulse

In this context, the Bernoulli-Euler beam counterpart of the case considered in Fig. 8.3-10 is adopted. The results reveal that the classical optimal control provides larger predictions of the dynamic response in the forced-vibration range than those based on the instantaneous control. However, in the free motion regime, the differences between the two predictions become immaterial. A similar result, obtained in a different context was highlighted in Section 7.9 and also in the papers by Na and Librescu (2000c, 2002).

Additional results related to the optimal feedback control of beams carrying external stores may be found in Na and Librescu (2000b), Librescu and Na (1997), and Na (1997). A more general beam model that would include the anisotropy of the material of the host structure would involve more complex couplings than that considered here. For clean beams, see Librescu et al. (1996, 1997), Librescu and Na (2001), Na and Librescu (1998, 2000, 2002), and Song et al. (2001).

REFERENCES

Abramovich, H. and Hamburger, O. (1992) " Vibration of a Uniform Cantilever Timoshenko Beam with Translational and Rotational Springs and with a Tip Mass," *Journal of Sound and Vibration*, Vol. 148, No. 1, pp. 162–170.

Bruch, Jr., J. C. and Mitchell, T. P. (1987) "Vibrations of a Mass-Loaded Clamped-Free Timoshenko Beam," *Journal of Sound and Vibration*, Vol. 114, No. 2, pp. 341–345.

Gern, H. F. and Librescu, L. (1998a) "Static and Dynamic Aeroelasticity of Advanced Aircraft Wings Carrying External Stores," *AIAA Journal*, Vol. 36, No. 7, pp. 1121–1129.

Gern, H. F. and Librescu, L. (1998b) "Effect of Externally Mounted Stores on Flutter Characteristics of Advanced Swept Cantilevered Aircraft Wings," *Aerospace Science and Technology*, Vol. 2, No. 5 pp. 321–333.

Kumar, R. (1974) "Vibration of Space Booms under Centrifugal Force Field," *Canadian Aeronaut. Space Inst. (CASI) Trans.*, Vol. 7, pp. 1–6.

Librescu, L., Song, O. and Rogers, C. A. (1993) "Adaptive Vibrational Behavior of Cantilevered Structures Modeled as Composite Thin-Walled Beams," *International Journal of Engineering Science*, Vol. 31, pp. 775–792.

Librescu, L. Meirovitch, L. and Song, O. (1996) "Integrated Structural Tailoring and Adaptive Materials Control for Advanced Aircraft Wings," *Journal of Aircraft*, Vol. 33, No. 1, pp. 203–213.

Librescu, L. and Song, O. (1997) "Static and Dynamic Behavior of Adaptive Aircraft Wings Carrying Externally Mounted Stores," *Structronic Systems: Smart Structures, Devices and Systems*, Vol. 1, H. S. Tzou (Ed.), World Scientific, Singapore, pp. 113–138.

Librescu, L. and Na, S. S. (1997) "Vibration and Dynamic Response Control of Cantilevers Carrying Externally Mounted Stores," *Journal of the Acoustical Society of America*, Vol. 102, No. 6, pp. 3516–3522.

Librescu, L. Meirovitch, L. and Na, S. S. (1997) "Control of Cantilever Vibration via Structural Tailoring and Adaptive Materials," *AIAA Journal*, Vol. 35, No. 8, pp. 1309–1315.

Librescu, L. and Na, S. S. (2001) "Active Vibration Control of Thin-Walled Tapered Beams Using Piezoelectric Strain Actuation," *Journal of Thin-Walled Structures*, Vol. 39, No. 1, pp. 65–82.

Na, S. S. (1997) "Control of Dynamic Response of Thin-Walled Composite Beams Using Structural Tailoring and Piezoelectric Actuation," *Ph.D. Dissertation*, Virginia Polytechnic Institute and State University, Blacksburg, Virginia.

Na, S. S. and Librescu, L. (1998) "Oscillation Control of Cantilevers via Smart Materials Technology and Optimal Feedback Control: Actuator Location and Power Consumption Issues," *Journal of Smart Materials and Structures*, Vol. 7, pp. 833–842.

Na, S. S. and Librescu, L. (2000a) "Dynamic Response of Adaptive Cantilevers Carrying External Stores and Subjected to Blast Loading," *Journal of Sound and Vibration*, Vol. 231, No. 4, pp. 1039–1055.

Na, S. S. and Librescu, L. (2000b) "Optimal Dynamic Response Control of Adaptive Thin-Walled Cantilevers Carrying Heavy Stores and Exposed to Blast Pulses," *Journal of Intelligent Materials and Structures*, Vol. 11, No. 9, pp. 703–712.

Na, S. S. and Librescu, L. (2000c) "Optimal Vibration Control of Thin-Walled Anisotropic Cantilevers Exposed to Blast Loadings," *Journal of Guidance, Control and Dynamics*, Vol. 23, No. 3, pp. 491–500.

Na, S. S. and Librescu, L. (2002) "Optimal Dynamic Response Control of Elastically Tailored Nonuniform Thin-Walled Adaptive Beams," *Journal of Aircraft*, Vol. 39, No. 3, pp. 469–479.

Runyan, H. L. and Sewall, J. L. (1947) "Experimental Investigation of the Effects of Concentrated Weights on Flutter Characteristics of a Straight Cantilever Wing," *NACA, TN*, 1594.

Song, O. and Librescu L. and Rogers, C. A. (1994) "Adaptive Response Control of Cantilevered Thin-Walled Beams Carrying Heavy Concentrated Masses," *Journal of Intelligent Material Systems and Structures* Vol. 5, pp. 42–48.

Song, O. and Librescu, L. (1996) "Bending Vibrations of Adaptive Cantilevers with External Stores," *International Journal of Mechanical Sciences*, Vol. 38, No. 5, pp. 483–498.

Song, O., Kim, J-B. and Librescu, L. (2001) "Synergistic Implications of Tailoring and Adaptive Material Technology on Vibration Control of Anisotropic Thin-Walled Beams," *International Journal of Engineering Science*, Vol. 39, No. 1, pp. 71–94.

Tzou, H. S. and Zhong, J. P. (1992) "Adaptive Piezoelectric Structures: Theory and Experiment," in *Active Materials and Adaptive Structures, Materials and Structures Series*, G. J. Knowles (Ed.), Institute of Physics Publ., pp. 219–224.

Tzou, H. S. (1993) *Piezoelectric Shells, Distributed Sensing and Control of Continua*, Kluwer Academic Publ., Dordrecht.

White, M. W. D. and Heppler, G. R. (1993) "Dynamics of Timoshenko Beams with Attached Masses," in *Dynamics and Control of Structures in Space II*, Cranfield Inst. of Technology, U.K., C. L. Kirk and R. C. Hughes (Eds.), Computational Mechanics Publ., Southampton/Boston, pp. 209–224.

Chapter 9

ROTATING THIN-WALLED ANISOTROPIC BEAMS

9.1 UNTWISTED BEAMS

9.1-1 Introduction

Rotor blades are vital and critical components of jet engines, tilt rotor aircraft, helicopters and turbomachinery. A thorough understanding of their dynami behavior is essential for their reliable design and the avoidance of catastrophic failures. The determination of dynamic characteristics is also essential for the accurate prediction of their forced response, resonant behavior and flutter instability characteristics.

The design of advanced rotor blades is becoming significantly influenced by composite materials technology. Such composite designs offer considerable advantages with respect to strength and weight criteria, in addition to providing adequate means of efficiently controlling static and dynamic response via implementation of structural tailoring; the pioneering works by Mansfield and Sobey (1979), Bauchau (1985), and Rehfield (1985) should be indicated.

In spite of the large amount of work devoted to the modeling and behavior of rotor blades (see in the sense the surveys by Hodges, 1990; Rosen, 1991; Kunz, 1994; Rand, 1991, 1996; Jung et al., 2001), the literature devoted to the dynamics of *rotating composite blades modeled as thin-walled beams* appears to be rather scarce.

Among the contributing factors, the following deserve to be invoked:

(a) The analysis of *rotating* blade structures is more complex than that of their nonrotating counterparts. In the former case, in addition to the acceler ations resulting from elastic structural deformations, the centrifugal and Coriolis accelerations have to be included in the modeling, and

(b) The anisotropy of constitute materials and, consequently, the induced elastic couplings, render this problem more intricate since it is compounded with the complexities arising from point (a) above.

With this in mind, the objectives of this chapter are: (a) to provide a refined dynamic theory for anisotropic rotating cantilevered blades modeled as thin-walled beams of arbitrary closed cross-section contour, (b) to adopt a ply-angle configuration as to induce elastic couplings of beneficial influence on vibration of rotating blades, and (c) to use a mathematical method to determine their vibration eigencharacteristics accurately.

9.1-2 Preliminaries

9.1-2a Coordinate Systems

Consider a straight untwisted flexible blade experiencing constant rotational motion. The origin of the beam coordinate system (x, y, z) is located at the blade root at an offset R_0 from the rotation axis fixed in space. R_0 denotes the radius of the hub (considered to be rigid) in which the blade is mounted and which rotates about its polar axis through the origin O.

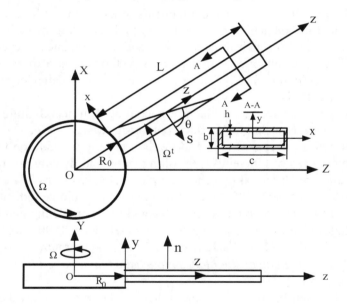

Figure 9.1-1: A schematic description of the blade and of a typical cross-section

The inertial reference system (X, Y, Z) is attached to the center of the hub O and (i, j, k) and (I, J, K) denote the unit vectors associated with the rotating and absolute coordinate systems (x, y, z) and (X, Y, Z), respectively. In addition, a local coordinate system (s, z, n) associated with the blade is also considered. A typical cross-section is presented in Fig. 9.1-1 along with the associated coordinate systems.

Here, one assumes that the precone, presetting and sweep angles of the blade are zero, and that the rotation takes place in the (X, Z) plane with constant angular velocity $\mathbf{\Omega}(\equiv \Omega \boldsymbol{J} = \Omega \boldsymbol{j})$. This representation of $\mathbf{\Omega}$ reflects the assumption that the location and orientation of the spin axis remains fixed in inertial space.

9.1-2b Basic Assumptions

Consider a single-cell thin-walled beam of uniform, but arbitrary cross-sectional shape.

We use the following assumptions: (i) the original cross-section of the beam is preserved, implying that the projection of the cross-section on the plane normal to the z-axis does not distort during deformation, (ii) the concept of the Saint-Venant torsional model is discarded in favor of the non-uniform torsional one, (iii) in addition to the primary warping, the secondary warping effect associated with points off the mid-line contour is also incorporated, (iv) transverse shear effects are accounted for. In this connection, it is assumed that transverse shear strains γ_{xz} and γ_{yz} are uniform in each cross-section, (v) the circumferential stress resultant N_{ss} (i.e., the hoop stress resultant) is assumed negligibly small when compared to the other ones, (vi) the effects of pretwist, presetting and precone angles, as well as the cross-section beam non-uniformity are not included in the present modeling, and (vii) one considers $x_P = y_P = 0$, implying that the pole P lies on the reference beam axis. As a result, u_P and v_P will be denoted u_0 and v_0, respectively.

9.1-3 Kinematic Equations

By virtue of assumptions i) through vii), in order to reduce the 3-D elasticity problem to an equivalent 1-D one, the components of the displacement vector are expressed by Eqs. (2.1-6) and (2.1-43).

Consistent with these representations, one obtains the following expressions of beam strains in terms of 1-D displacement measures:

$$
\begin{aligned}
\varepsilon_{zz}^{(0)}(s, z, t) &= w_0' + x(s)\,\theta_y' + y(s)\,\theta_x' - F(s)\phi'' \\
&\quad + \frac{1}{2}[(u_0' - y\phi')^2 + (v_0' + x\phi')^2], \\
\varepsilon_{zz}^{(1)}(s, z, t) &= \theta_y'\frac{dy}{ds} - \theta_x'\frac{dx}{ds} - r_t(s)\phi'', \\
\gamma_{sz}^{(0)}(s, z, t) &= [u_0' + \theta_y]\frac{dx}{ds} + [v_0' + \theta_x]\frac{dy}{ds} + \psi(s)\phi', \\
\gamma_{sz}^{(1)}(s, z, t) &= 2\phi',
\end{aligned}
\qquad (9.1\text{-}1a\text{-}e)
$$

$$\gamma_{zn}^{(0)}(s, z, t) = [u_0' + \theta_y]\frac{dy}{ds} - [v_0' + \theta_x]\frac{dx}{ds}.$$

In Eq. (9.1-1c), consistent with (2.1-44) and (2.1-36c), the torsional function ψ is

$$\psi(s) = \frac{\oint r_n(\bar{s})d\bar{s}}{h(s)G_{sz}(s)\oint \dfrac{d\bar{s}}{h(\bar{s})G_{sz}(\bar{s})}}, \qquad (9.1\text{-}2)$$

$\bar{s}$ is the dummy circumferential coordinate while $G_{sz}(s)$ denotes the effective membrane shear stiffness of the laminate.

Note that the primary axial strain $\varepsilon_{zz}^{(0)}$, Eq. (9.1-1a), includes nonlinear terms associated with the flapwise and chordwise displacements. As it will be shown later, these nonlinearities enable one to capture in the equations of motion the effect of centrifugal forces which, for rotating blades, are essential. The disregard in Eq. (9.1-1a) of the nonlinearity related with the extensional displacement, w_0, is based on order of magnitude considerations. For a full derivation of this equation, see Chapter 2, Section 2.2-4.

It is also useful to express the position vector $\boldsymbol{R}$ of a point of the deformed beam, measured from a fixed origin O (coinciding with the center of the hub), and its time derivatives. In this sense we have

$$\boldsymbol{R} = \boldsymbol{R}_0 + \boldsymbol{r} + \boldsymbol{\Delta}, \qquad (9.1\text{-}3)$$

where

$$\boldsymbol{R}_0 = R_0\boldsymbol{k}, \text{ and } \boldsymbol{r} = x\boldsymbol{i} + y\boldsymbol{j} + z\boldsymbol{k}, \qquad (9.1\text{-}4a,b)$$

define the position vector of the points of the hub periphery and the undeformed position vector of a point measured in the beam coordinate system, respectively, while

$$\boldsymbol{\Delta} = u\boldsymbol{i} + v\boldsymbol{j} + w\boldsymbol{k}, \qquad (9.1\text{-}5)$$

denotes the displacement vector whose components are defined in Eqs. (2.1-6) and (2.1-43).

From Eqs. (R9), keeping in mind that the rotation takes place solely in the XZ plane, it results

$$\dot{\boldsymbol{i}} = -\Omega\boldsymbol{k}; \quad \dot{\boldsymbol{j}} = 0; \quad \dot{\boldsymbol{k}} = \Omega\boldsymbol{i}, \qquad (9.1\text{-}6a\text{-}c)$$

and making use of (9.1-3) through (9.1-6) one can obtain $\dot{\boldsymbol{R}}$ and $\ddot{\boldsymbol{R}}$. Their expressions are:

$$\dot{\boldsymbol{R}} = [\dot{u} + \Omega(R_0 + z + w)]\boldsymbol{i} + \dot{v}\boldsymbol{j} + [\dot{w} - \Omega(x + u)]\boldsymbol{k}, \qquad (9.1\text{-}7a,b)$$

$$\ddot{\boldsymbol{R}} = \left[\ddot{u} + 2\dot{w}\Omega - \underline{(x + u)\Omega^2}\right]\boldsymbol{i} + \ddot{v}\boldsymbol{j} + \left[\ddot{w} - 2\dot{u}\Omega - \underline{(R_0 + z + w)\Omega^2}\right]\boldsymbol{k}$$

The terms in Eq. (9.1-7b) underscored by one and two superposed solid lines are associated with Coriolis and centrifugal acceleration effects, respectively.

Remark 9.1

For some developments in this chapter and of the next one, it is useful to present a brief derivation of the time derivatives of the unit vectors $\mathbf{i}$, $\mathbf{j}$, $\mathbf{k}$ associated with the rotating Cartesian coordinate system (x, y, z).

For the sake of convenience, here the unit vectors are denoted as $\mathbf{i}_j$, the coordinates as x_j, and the components of the angular velocity vector $\mathbf{\Omega}$ as Ω_j, i.e.

$$\mathbf{\Omega} = \Omega_r \mathbf{i}_r, \tag{R1}$$

where the convention of summation on the repeated indexes that have the range 1, 2 and 3 should be applied.

Having in view the evident relationship

$$\mathbf{i}_s \cdot \mathbf{i}_r = \delta_{sr}, \tag{R2}$$

where δ_{sr} is the Kronecker delta, and appling the operator d/dt, to (R2) one obtains

$$\dot{\mathbf{i}}_s \cdot \mathbf{i}_r = -\mathbf{i}_r \cdot \dot{\mathbf{i}}_s. \tag{R3}$$

Equation (R3) suggests the representation

$$\dot{\mathbf{i}}_s = -\Omega_{sp} \mathbf{i}_p. \tag{R4}$$

Replacement of (R4) in (R3) used in conjunction with (R2) yields

$$\Omega_{sr} = -\Omega_{rs}, \tag{R5}$$

implying that the matrix Ω_{rs} is skew-symmetric. One the other hand, one can associate to vector $\mathbf{\Omega}$ the 3×3 skew-symmetric matrix

$$\tilde{\mathbf{\Omega}} \equiv \Omega_{ij} = \begin{vmatrix} 0 & -\Omega_3 & \Omega_2 \\ \Omega_3 & 0 & -\Omega_1 \\ -\Omega_2 & \Omega_1 & 0 \end{vmatrix}, \tag{R6}$$

where the components Ω_i are related with those of the matrix Ω_{ij} as

$$\Omega_i = \frac{1}{2}\epsilon_{ijk}\Omega_{kj}, \tag{R7a}$$

and conversely

$$\Omega_{mn} = \epsilon_{nmr}\Omega_r, \tag{R7b}$$

ϵ_{ijk} being the Levi-Civita permutation symbol. (R4) in conjunction with (R6) gives

$$\dot{\mathbf{i}}_s = \mathbf{\Omega} \times \mathbf{i}_s, \tag{R8}$$

or in an equivalent form

$$\dot{\mathbf{i}}_s = \Omega_i \mathbf{i}_i \times \mathbf{i}_s = \epsilon_{spi}\Omega_i \mathbf{i}_p. \tag{R9}$$

Coming back to the previously used notation i.e., having in view that $i_1, i_2, i_3 \Rightarrow i, j, k$, and that for a no-preconed rotating blade $\Omega_1 = \Omega_3 = 0$ and $\Omega_2 \equiv \Omega$, by using (R9) one get Eqs. (9.1-6), whereas for the case of the beam spinning about the z-axis, we have $\Omega_1 = \Omega_2 = 0$ and $\Omega_3 \equiv \Omega$ and in conjunction with (R9) one obtain Eqs. (10.1-2) and (10.1-3).

Remark 9.2

A more inclusive derivation of Eqs. (9.1-7) and (10.1-2) and (10.1-3), that was considered by Chelu and Librescu (2005) is based on the concept of the absolute derivative of a vector, in particular that of $\mathbf{R}$ (defined by Eqs. (9.1-3) in conjuction with (9.1-4) and (9.1-5)).

In this case

$$\frac{d\mathbf{R}}{dt} = \dot{\mathbf{R}} + \widetilde{\boldsymbol{\Omega}}\mathbf{R}, \tag{R10a}$$

and

$$\frac{d^2\mathbf{R}}{dt^2} = \ddot{\mathbf{R}} + 2\widetilde{\boldsymbol{\Omega}}\dot{\mathbf{R}} + \widetilde{\boldsymbol{\Omega}}^2\mathbf{R}. \tag{R10b}$$

Herein $d\mathbf{R}/dt$ and $\dot{\mathbf{R}}$ denote the absolute and relative derivatives, respectively, while $\widetilde{\boldsymbol{\Omega}}$ is provided by Eq. (R6).

Not only that Eqs. (R10) can be specialized for the special cases of the beam experiencing pure rotation and pure spinning, but they can provide pertinent results when both motions are simultaneously considered.

9.1-4 The Equations of Motion and Boundary Conditions

In order to derive the equations of motion and the associated boundary conditions, Hamilton's principle, Eq. (3.1-1), is used. As a necessary prerequisite, the various energies that are involved have to be rendered explicitly.

For this case, we have:

$$\int_{t_0}^{t_1} \delta\mathcal{K}dt = \int_{t_0}^{t_1}\int_{\tau} \rho(\dot{\mathbf{R}}\cdot\delta\dot{\mathbf{R}})d\tau dt = -\int_{t_0}^{t_1} dt \int_{\tau} \rho\ddot{\mathbf{R}}\cdot\delta\mathbf{R}d\tau, \tag{9.1-8}$$

where the last expression was obtained by enforcing Hamilton's condition, $\delta\mathbf{R} = 0$, for $t = t_0, t_1$.

Using (9.1-3), (9.1-7) and (9.1-8), and performing the integrations in the circumferential and thickness directions, one can express Eq. (9.1-8) as:

$$\int_{t_0}^{t_1} \delta\mathcal{K}dt = \int_{t_0}^{t_1}\left\{\int_0^L -\{b_1(\ddot{u}_0 + 2\dot{w}_0\Omega - u_0\Omega^2)\delta u_0\right.$$
$$+ b_1\ddot{v}_0\delta v_0 + b_1[\ddot{w}_0 - 2\dot{u}_0\Omega - (R_0 + z + w_0)\Omega^2]\delta w_0 + (b_5 + \delta_n b_{15})$$

$$(\ddot{\theta}_y - \Omega^2\theta_y)\ \delta\theta_y + [(b_4 + \delta_n b_{14})(\ddot{\theta}_x - \Omega^2\theta_x) + 2b_4\Omega\dot{\phi}]\delta\theta_x$$

$$+ [(b_4 + b_5)\ddot{\phi} - 2b_4\Omega\dot{\theta}_x - (b_4 - b_5)\Omega^2\phi - (b_{10} + \delta_n b_{18})(\ddot{\phi}''$$

$$- \Omega^2\phi'')]\delta\phi\}dz - [(b_{10} + \delta_n b_{18})(\ddot{\phi}' - \Omega^2\phi')]\delta\phi]_0^L \Big\}dt.$$

$$(9.1\text{-}9)$$

For the variation of the strain energy functional we have

$$\int_{t_0}^{t_0} \delta\mathcal{W}dt = \int_{t_0}^{t_1} dt \int_{\tau} \sigma_{ij}\delta\varepsilon_{ij}d\tau$$

$$= \int_{t_0}^{t_1} dt \int_{\tau} [\tau_{sz}\delta\gamma_{sz} + \sigma_{zz}\delta\varepsilon_{zz} + \tau_{nz}\delta\gamma_{nz}]d\tau. \qquad (9.1\text{-}10)$$

Having in view Eqs. (9.1-1), performing the integrations over the n and s coordinates and integrating by parts wherever necessary as to relieve the virtual displacements (δu_0, δv_0, δw_0, $\delta\theta_x$, $\delta\theta_y$, $\delta\phi$ and $\delta\phi'$) of any differentiation, and using the definitions of 1-D stress-resultant and stress couple measures, Eqs. (3.1-17a-h), one obtains

$$\int_{t_0}^{t_1} \int_{\tau} \sigma_{ij}\delta\varepsilon_{ij}d\tau dt = \int_{t_0}^{t_1} -\{T_z'\delta w_0 + (M_y' - Q_x)\delta\theta_y + (M_x' - Q_y)\delta\theta_x$$

$$+ [\ B_\omega'' + M_z' + (T_r\phi')']\delta\phi + [Q_x' + (T_z u_0')']\delta u_0 \qquad (9.1\text{-}11)$$

$$+ [Q_y' + (T_z v_0')']\delta v_0\}dz dt + \int_{t_0}^{t_1} \Big[T_z\delta w_0 + M_y\delta\theta_y + M_x\delta\theta_x$$

$$- B_\omega\delta\phi' + (\ B_\omega' + M_z + T_r\phi')\delta\phi$$

$$+ (Q_x + T_z u_0')\delta u_0 + (Q_y + T_z v_0')\delta v_0 \Big]_0^L dt.$$

As is readily seen in this equation, in addition to the already known 1-D stress-resultants and stress-couples, an extra stress-resultant $T_r(z, t)$ appears, that is defined by

$$T_r(z, t) = \oint N_{zz}(x^2 + y^2)ds. \qquad (9.1\text{-}12)$$

Considering also the virtual work of external loads, consistent with Eqs. (3.1-29), one obtains:

$$\delta \mathcal{J} = \int_0^L \left[p_x \delta u_0 + p_y \delta v_0 + p_z \delta w_0 + m_x \delta \theta_x + m_y \delta \theta_y + (m_z + m'_\omega) \delta \phi \right] dz$$

$$+ \left[\underset{\sim}{Q}_x \delta u_0 + \underset{\sim}{Q}_y \delta v_0 + \underset{\sim}{T}_z \delta w_0 + \underset{\sim}{M}_x \delta \theta_x + \underset{\sim}{M}_y \delta \theta_y + \underset{\sim}{M}_z \delta \phi - \underset{\sim}{B}_\omega \delta \phi' \right] \Bigg|_0^L.$$

$$(9.1-13)$$

Using (9.1-9), (9.1-11) and (9.1-13) in the Hamilton's principle, collecting the terms associated with the same variations, invoking the stationarity of the functional within the time interval $[t_0, \ t_1]$ and the fact that the variations (δu_0, δv_0, δw_0, $\delta \theta_x$, $\delta \theta_y$, $\delta \phi$ and $\delta \phi'$) are independent and arbitrary, their coefficients in the two integrands must vanish independently. This yields the equations of motion and the boundary conditions, that are provided next.

● *Equations of Motion*:

$$\delta u_0 : Q'_x + (T_z u'_0)' - b_1 (\ddot{u}_0 + \underline{2 \dot{w}_0 \Omega - \Omega^2 u_0}) + p_x = 0,$$

$$\delta v_0 : Q'_y + (T_z v'_0)' - b_1 \ddot{v}_0 + p_y = 0,$$

$$\delta w_0 : T'_z - b_1 (\ddot{w}_0 - \underline{2 \dot{u}_0 \Omega}) + b_1 \Omega^2 (R_0 + z + w_0) + p_z = 0,$$

$$\delta \theta_y : M'_y - Q_x - (b_5 + \delta_n b_{15})(\underset{\cdots}{\ddot{\theta}_y} - \underwiggle{\Omega^2 \theta_y}) \ + m_y = 0, \qquad (9.1-14\text{a-f})$$

$$\delta \theta_x : M'_x - Q_y - (b_4 + \delta_n b_{14})(\underset{\cdots}{\ddot{\theta}_x} \ - \ \underwiggle{\Omega^2 \theta_x}) - \underline{2 b_4 \Omega \dot{\phi}} + m_x = 0,$$

$$\delta \phi : \ \underline{B''_\omega} \ + (T_r \phi')' + M'_z - (b_4 + b_5) \ddot{\phi} + \underwiggle{(b_4 - b_5) \Omega^2 \phi}$$

$$+ \underline{2 b_4 \Omega \dot{\theta}_x} + (b_{10} + \delta_n b_{18})(\underdashed{\ddot{\phi}''} \ - \ \underdoubleline{\Omega^2 \phi''}) \ + m_z + \underline{m'_\omega} \ = 0.$$

● *Boundary Conditions*:
Helicopter rotors with cantilever blades are commonly termed hingeless rotors. For cantilevered beams the boundary conditions at the root are entirely kinematic, while at the tip entirely static. As a result, we have:

At $z = 0$:

$$u_0 = \underset{\sim}{u}_0, \; v_0 = \underset{\sim}{v}_0, \; w_0 = \underset{\sim}{w}_0, \; \theta_y = \underset{\sim}{\theta}_y, \; \theta_x = \underset{\sim}{\theta}_x, \; \phi = \underset{\sim}{\phi}, \; \phi' = \underset{\sim}{\phi}'$$

$$(9.1\text{-}15\text{a-g})$$

and at $z = L$:

$$\delta u_0 : \; Q_x + \underset{\cdots\cdots}{T_z u_0'} = \underset{\sim}{Q}_x,$$

$$\delta v_0 : \; Q_y + \underset{\cdots\cdots}{T_z v_0'} = \underset{\sim}{Q}_y,$$

$$\delta w_0 : \; T_z = \underset{\sim}{T}_z,$$

$$\delta \theta_y : \; M_y = \underset{\sim}{M}_y,$$

$$\delta \theta_x : \; M_x = \underset{\sim}{M}_x,$$

$$(9.1\text{-}16\text{a-g})$$

$$\delta \phi : \; \underline{B_\omega'} + \underset{\cdots\cdots}{T_r \phi'} + M_z + (b_{10} + \delta_n b_{18}) \, (\underline{\ddot{\phi}'} - \underline{\underline{\underline{\Omega^2 \phi'}}}) = \underset{\sim}{M}_z,$$

$$\delta \phi' : \; \underline{B_\omega} = \underline{\underset{\sim}{B}}_\omega.$$

In Eqs. (9.1-15) and (9.1-16), the quantities underscored by a tilde represent prescribed quantities. The expressions of prescribed static quantities are supplied in Eqs. (3.1-30). In the previous equations, in addition to the convention adopted in Eqs. (9.1-7), the terms associated with the inertial warping are underscored by a dashed line superposed on a solid line ($-\,-\,-$); the centrifugal warping terms by two superposed dashed lines ($\equiv\equiv\equiv$); rotatory inertia terms by a dotted line ($\ldots$), and centrifugal-rotatory effect by a solid line superposed on a wavy line ($\overline{\sim}$).

In addition, the terms appearing in Eq. (9.1-14f) underscored by a wavy line ($\sim\!\!\sim\!\!\sim$) are associated with the tennis-racket effect, and as usual, the terms associated with warping restraint, by a dashed line ($-\,-\,-\,-$). In connection with the derivation of the second term belonging to tennis-racket effect (i.e. $-b_5 \Omega^2 \phi$), a word of caution is in order. As shown in the paper by Song and Librescu (1997) and by Vorob'ev (1988), this term cannot be captured without including in the representations of $u(x, y, z, t)$ and $v(x, y, z, t)$, (Eqs. (2.2-1)), a quadratic term in ϕ coming from the development in series of $\cos\phi$, i.e., $-x\phi^2/2$ and $-y\phi^2/2$, respectively. Inclusion of these terms followed by the application of the required variations in Eq. (9.1-8), results in the centrifugal term $-b_5 \Omega^2 \phi$, which is of the same order of magnitude as $b_4 \Omega^2 \phi$. Excepting the previously mentioned generated term, the higher order terms in u and v do not yield other linear contributions, and consequently, are disregarded.

As concerns the structural and mass cross-sectional blade characteristics appearing in Eqs. (9.1-14) and (9.1-16f), their expressions are displayed in Table 9.1-1. Under this form, Eqs. (9.1-9) and (9.1-10) were obtained for the first time by Song and Librescu (1997).

Table 9.1-1: Cross-sectional blade characteristics

Mass cross-sectional characteristics	*Definition*
b_1, b_4, b_5, b_{10}, b_{14}, b_{15}, b_{18}	$\int \oint \rho \left\{ 1, y^2,\ x^2,\ \bar{F}^2, n^2 \left(\dfrac{dx}{ds} \right)^2, n^2 \left(\dfrac{dy}{ds} \right)^2,\ n^2 r_t^2 \right\} ds\, dn$
I_p	$b_4 + b_5$
$\tilde{I}_p$	$(b_4 + b_5)/m_0$
$\hat{I}_p$	$\tilde{I}_p / \beta$

9.1-5 Several Remarks on the Equations of Motion and Boundary Conditions

In connection with these equations several comments are in order:

(i) The quantities $(T_z u_0')'$, $(T_z v_0')'$ and $(T_r \phi')'$ occurring in the equations of motion (9.1-14a,b, and f), respectively, and $T_z u_0'$, $T_z v_0'$ and $T_r \phi'$ that are present in the natural boundary conditions, Eqs. (9.1-16a,b and f), respectively, are clearly nonlinear quantities.

(ii) The equations of motion feature couplings generated by gyroscopic forces of the Coriolis type. In this sense, Eqs. (9.1-14a,c) reveal the existence of a lag-extension, $(u_0 \longleftrightarrow w_0)$ coupling, while Eqs. (9.1-14e,f), of the flapwise bending-twist $(\theta_x \longleftrightarrow \phi)$ coupling.

Upon discarding the Coriolis effect that, according to Leissa and Co (1984), Ewins and Henry (1992), and others is negligibly small, these couplings become immaterial. Invoking in addition the physically evident fact that the blade is much stiffer in the longitudinal direction than in the flapping and lagging ones, and implicitly that the effect of the axial inertia is much smaller than the others, this one will be discarded.

Consequently, Eqs. (9.1-14a,c) associated with extension and lagging motion, respectively, are no longer coupled. Hence, the direct integration of (9.1-14c) in conjunction with the boundary condition at the free end,

(Eq. 9.1-16c), stipulating zero axial force yields

$$T_z(z, t) = b_1 \Omega^2 R(z) + \int_z^L p_z(\bar{z}, t) d\bar{z}, \tag{9.1-17a}$$

where $\bar{z}$ denotes the dummy spanwise coordinate, while

$$R(z) = R_0(L - z) + \frac{1}{2} (L^2 - z^2), \tag{9.1-17b}$$

(iii) Based on the expression of T_r supplied by Eq. (9.1-12), of T_z, and of $\tilde{I}_p$ and $\hat{I}_p$ provided in Table 9.1-1, one obtains

$$T_r = T_z \hat{I}_p. \tag{9.1-18}$$

In view of (9.1-14) and (9.1-17), it appears that T_r plays the role of a torsional stiffness induced by the centrifugal force field.

By virtue of Eqs. (9.1-17) and (9.1-18), Eqs. (9.1-14a,b,f) become linear and incorporate the centrifugal stiffening effect. Finally,

(iv) For homogeneous boundary conditions at $z = L$, the terms in Eqs. (9.1-16a, b and f) underscored by two superposed dotted lines become immaterial.

9.1-6 The Governing System

9.1-6a Circumferentially Uniform Stiffness Configuration

The governing dynamic system of rotating elastic blades will be expressed in terms of displacements. This can be accomplished by representing the 1-D stress resultants and stress couples measures in terms of 1-D displacement variables. The Coriolis effect will be discarded in these equations.

Using the procedure detailed in the previous chapters, one can obtain, for the case of general anisotropy of constituent materials and ply- stacking sequence, the governing system and the associated boundary conditions expressed in terms of 1-D displacements. In such a case, a complete coupling between the various modes, that is warping (primary and secondary), flapping and lagging, twist and transverse shearing is experienced. The assessment of the influence of these couplings and their proper exploitation should constitute an important task in a rational design of blade structures, and of the proper use of the exotic material characteristics (i.e., anisotropy and ply-stacking sequence) generating these couplings.

In the present analysis, however, a special case of ply-angle distribution inducing special elastic couplings will be considered. This consists of the lamination scheme

$$\theta(x) = \theta(-x), \quad \theta(y) = \theta(-y), \tag{9.1-19}$$

where θ denotes the ply angle orientation considered to be positive when measured from the positive s-axis towards the positive z-axis (see Fig. A.4).

As was previously shown, this ply-angle configuration, referred to as *Circumferentially Uniform Stiffness* configuration (see Fig. A.4), achievable via the usual filament winding technology, leads to an exact decoupling between the compound flapping-lagging bending-transverse shear, and the extension-twist coupled motions, see also Rehfield and Atilgan (1989).

As a result, from the associated equations, a system of four coupled equations of the eighth order that governs the (flap and lag) bending-transverse shear motion, and another one, of the sixth order, governing the twist-extension motion are obtained. Within the former system of equations the effect of centrifugal force field was absorbed by representing T_z via Eq. (9.1-17).

On the other hand, Eqs. (9.1-14c,f) and the associated boundary conditions (Eqs.9.1-16c,f and g) considered in conjunction with Eq. (9.1-18) constitute the system of equations governing the extension-twist coupled motion. The equations associated with both bending-transverse shear and extension-twist motion as obtained by Song and Librescu (1997) are displayed next.

- *The equations governing the compound flap-lag bending-transverse shear motion:*

$$\delta u_0 : a_{44}(u_0'' + \theta_y') + a_{43}\theta_x'' + b_1\Omega^2[R(z)u_0']' + b_1\Omega^2 u_0 - b_1\ddot{u}_0 = 0,$$

$$\delta v_0 : a_{55}(v_0'' + \theta_x') + a_{52}\theta_y'' + b_1\Omega^2[R(z)v_0']' - b_1\ddot{v}_0 = 0,$$

$$\delta\theta_y : a_{22}\theta_y'' + a_{25}(v_0'' + \theta_x') - a_{43}\theta_x' - a_{44}(u_0' + \theta_y)$$

$$- (b_5 + \delta_n b_{15})(\ddot{\theta}_y - \Omega^2\theta_y) = 0, \qquad (9.1\text{-}20\text{a-d})$$

$$\delta\theta_x : a_{33}\theta_x'' + a_{34}(u_0'' + \theta_y') - a_{52}\theta_y' - a_{55}(v_0' + \theta_x)$$

$$- (b_4 + \delta_n b_{14})(\ddot{\theta}_x - \Omega^2\theta_x) = 0.$$

The *associated homogeneous boundary conditions of rotating beams clamped at $z = 0$ and free at $z = L$ are:*

At $z = 0$:

$$u_0 = v_0 = \theta_y = \theta_x = 0, \qquad (9.1\text{-}21\text{a-d})$$

and at $z = L$:

$$\delta u_0 : \quad a_{44}(u_0' + \theta_y) + a_{43}\theta_x' = 0,$$
$$\delta v_0 : \quad a_{55}(v_0' + \theta_x) + a_{52}\theta_y' = 0,$$
$$\delta \theta_y : \quad a_{22}\theta_y' + a_{25}(v_0' + \theta_x) = 0, \qquad \text{(9.1-22a-d)}$$
$$\delta \theta_x : \quad a_{33}\theta_x' + a_{34}(u_0' + \theta_y) = 0.$$

- *The equations governing extension-twist motion:*

$$\delta w_0 : \quad a_{11}w_0'' + a_{17}\phi'' - b_1\ddot{w}_0 + \underline{b_1\Omega^2(R_0 + z + w_0)} = 0,$$

$$\delta \phi : \quad \underline{-a_{66}\phi''''} + a_{71}w_0'' + a_{77}\phi'' + \underline{\underline{(b_{10} + \delta_n b_{18})[\ddot{\phi}'' - \Omega^2\phi'']}}$$

$$\quad - (b_4 + b_5)\ddot{\phi} + \underline{b_1\Omega^2\hat{I}_p[R(z)\phi']'} + \underwave{\Omega^2(b_4 - b_5)\phi} = 0.$$

$$\text{(9.1-23a,b)}$$

The associated BCs:

At $z = 0$:

$$w_0 = \phi = \underline{\phi'} = 0, \qquad \text{(9.1-24a-c)}$$

and at $z = L$:

$$\delta w_0 : \quad a_{11}w_0' + a_{17}\phi' = 0,$$

$$\delta \phi : \quad - \underline{a_{66}\phi'''} + a_{71}w_0' + a_{77}\phi' + (b_{10} + \delta_n b_{18})\,(\ddot{\phi}' - \underline{\underline{\Omega^2\phi'}}) = 0,$$

$$\delta \phi' : \quad \underline{a_{66}\phi''} = 0. \qquad \text{(9.1-25a-c)}$$

In these equations, a_{ij} denote global stiffness quantities whose expressions are displayed in Table 4.3-1. It should be remarked that for ply orientations $\theta = 0$ and 90 deg., the coupling stiffness quantities $a_{17}(\equiv a_{71})$, $a_{25}(\equiv a_{52})$ and $a_{34}(\equiv a_{43})$ vanish, and hence Eqs. (9.1-20) through (9.1-25) decouple into four equations, individually describing lag and flap, extension and twist motions.

9.1-6b Special Cases

(a) If the beam is infinitely rigid in transverse shear, from Eqs. (9.1-20) through (9.1-22), two uncoupled bending motions can be obtained. These correspond to the pure flapping and lagging ones. In order to obtain the equation in the flapping motion, $a_{55}(v_0' + \theta_x)$ derived from Eq. (9.1-20d) is replaced in Eqs. (9.1-20b) and (9.1-22b), where $\theta_x = -v_0'$ is considered. In addition, using the representation $v_0(z, t) = \bar{v}_0(z) \exp(i\omega t)$, one gets the flapping eigenvalue problem in the form of the differential equation:

$$a_{33}\bar{v}_0'''' - b_1\Omega^2[R(z)\bar{v}_0']' + (b_4+\delta_n b_{14})\,(\omega^2 + \underset{\cdots}{\Omega^2}\,)\,\bar{v}_0'' - b_1\omega^2\bar{v}_0 = 0,$$

$$(9.1\text{-}26)$$

where $\bar{v}_0$ is subject to the boundary conditions:

At $z = 0$:

$$\bar{v}_0 = \bar{v}_0' = 0, \qquad\qquad (9.1\text{-}27\text{a,b})$$

and at $z = L$:

$$\bar{v}_0'' = 0, \text{ and } \bar{v}_0''' = -\frac{(b_4+\delta_n b_{14})}{a_{33}}\,(\omega^2 + \underset{\cdots}{\Omega^2}\,)\bar{v}_0'. \qquad (9.1\text{-}28\text{a,b})$$

A similar procedure applied to Eqs. (9.1-20a,c) as well as to (9.1-22a,c) leads to the eigenvalue problem for the lagging motion:

$$a_{22}\bar{u}_0'''' - b_1\Omega^2[R(z)\bar{u}_0']' + (b_5 + \delta_n b_{15})(\omega^2 + \underset{\cdots}{\Omega^2}\,)\bar{u}_0''$$

$$- b_1(\Omega^2 + \omega^2)\bar{u}_0 = 0, \qquad\qquad (9.1\text{-}29)$$

with the associated boundary conditions:

At $z = 0$:

$$\bar{u}_0 = \bar{u}_0' = 0, \qquad\qquad (9.1\text{-}30\text{a,b})$$

and at $z = L$:

$$\bar{u}_0'' = 0, \text{ and } \bar{u}_0''' = -\frac{(b_5 + \delta_n b_{15})}{a_{22}}(\omega^2 + \underset{\cdots}{\Omega^2}\,)\bar{u}_0'. \qquad (9.1\text{-}31\text{a,b})$$

(b) Assuming in Eqs. (9.1-23) and the associated boundary conditions the case of an orthotropic material whose principal axes coincide with the geometrical axes of the blade, this implying $a_{17} = a_{71} = 0$, the system of two equations decouples into two independent equations, one governing pure extension and the other pure twist motion.

In the following we will be interested in the latter equation. For $\phi(z, t) = \bar{\phi}(z) \exp(i\omega t)$, the eigenvalue problem in twist motion of the rotating blade reads:

$$-a_{66}\underline{\overline{\phi}''''} + a_{77}\overline{\phi}'' + I_p\omega^2\overline{\phi} - (b_{10} + \delta_n b_{18})[\ \underline{\omega^2}\ +\ \underline{\underline{\Omega^2}}]\overline{\phi}''$$

$$+\ \underline{\underline{b_1\Omega^2\hat{I}_p[R(z)\overline{\phi}']'}}\ +\ \underaccent{\sim}{\Omega^2(b_4 - b_5)\overline{\phi}} = 0, \tag{9.1-32}$$

where $\overline{\phi}(z)$ should fulfill the boundary conditions:

At $z = 0$:

$$\overline{\phi} = \overline{\phi}' = 0, \tag{9.1-33a,b}$$

and at $z = L$:

$$\overline{\phi}'' = 0; \quad \underline{a_{66}\overline{\phi}'''} - a_{77}\overline{\phi}' + (b_{10} + \delta_n b_{18})\ (\underline{\omega^2}\ +\ \underline{\underline{\Omega^2}}\ )\overline{\phi}' = 0.$$
$$\tag{9.1-34a,b}$$

9.1-7 Decoupled Natural Frequencies of Rotating Beams

For the pure flapping and lagging of unshearable rotating beams it is possible to determine the associated natural frequencies via Rayleigh's quotient method. The procedure will be applied in detail for the flapping case, while for lagging and twist, only the final results will be displayed.

Consider the eigenvalue problem consisting of Eq. (9.1-26) that must be satisfied over the domain $0 < z < L$, where $\overline{v}_0$ is subject to the boundary conditions (9.1-27) and (9.1-28).

Multiplying Eq. (9.1-26) by $\overline{v}_0$, integrating over the domain $0 < z < L$, and integrating whenever possible by parts and considering the boundary conditions, one obtain for the various constituent parts of the equation, the following expressions:

$$\int_0^L a_{33}\overline{v}_0''''\overline{v}_0 dz = a_{33}\overline{v}_0(L)\overline{v}_0'''(L) + \int_0^L a_{33}(\overline{v}_0'')^2 dz,$$

$$\Omega^2\int_0^L b_1[R(z)\overline{v}_0']'\overline{v}_0 dz = -\Omega^2\int_0^L b_1 R(z)(\overline{v}_0')^2 dz, \tag{9.1-35a-c}$$

$$(b_4 + \delta_n b_{14})\ (\omega^2 + \Omega^2)\int_0^L \overline{v}_0''\,\overline{v}_0 dz$$

$$= (b_4 + \delta_n b_{14})(\omega^2 + \Omega^2)\left[\overline{v}_0(L)\overline{v}_0'(L) - \int_0^L (\overline{v}_0')^2 dz\right].$$

Finally, the last term in Eq. (9.1-26) becomes $\omega^2 \int_0^L b_1 \bar{v}_0^2 dz$.

Using the boundary condition (9.1-28b) to express $\bar{v}_0'''(L)$, replacing (9.1-35) in (9.1-26), and solving for $\omega^2 = \omega_{Flap}^2$ one gets

$$\omega_{Flap}^2 = \frac{\int_0^L \left[a_{33}(\bar{v}_0'')^2 + b_1 \Omega^2 R(z)(\bar{v}_0')^2 - \Omega^2(b_4 + \delta_n b_{14})(\bar{v}_0')^2 \right] dz}{\int_0^L \left[b_1 \bar{v}_0^2 + (b_4 + \delta_n b_{14})(\bar{v}_0')^2 \right] dz}.$$

$$(9.1-36)$$

Using the same procedure applied to the lagging eigenvalue problem consisting of the Eq. (9.1-29) and the associated boundary conditions (9.1-30,31), one obtains

$$\omega_{Lag}^2 = \frac{M}{N}, \qquad (9.1-37a)$$

where

$$M = \int_0^L \left[a_{22}(\bar{u}_0'')^2 + b_1 \Omega^2 R(z)(\bar{u}_0')^2 - \Omega^2(b_5 + \delta_n b_{15})(\bar{u}_0')^2 - b_1 \Omega^2 \bar{u}_0^2 \right] dz$$

and

$$N = \int_0^L \left[b_1 \bar{u}_0^2 + (b_5 + \delta_n b_{15})(\bar{u}_0')^2 \right] dz. \qquad (9.1-37b,c)$$

In the Southwell form, Eqs. (9.1-36) and (9.1-37) can be written as

$$\omega_{Flap;\ Lag}^2 = (K_1)_{Flap;\ Lag} + \Omega^2 (K_2)_{Flap;\ Lag} \qquad (9.1-38)$$

where $(K_1)_{Flap,\ Lag}$, and $(K_2)_{Flap,\ Lag}$ are, respectively, the nonrotating blade natural frequencies (associated with structural blade characteristics), and rotating contributions accounting for the centrifugal stiffening in the flapping and lagging motions. K_2 is also referred to as the rise factor or Southwell coefficient (see e.g. Bielawa, 1992).

Note that the eigenvalue problem in flapping appears to be similar *in form* to that of lagging motion, except for a centrifugal force term appearing in the lagging equation. The same trend is reflected in the expressions of natural frequencies, Eqs. (9.1-36) and (9.1-37).

Rayleigh's quotient procedure applied to the eigenvalue problem in twist, Eqs. (9.1-32) through (9.1-34), provides

$$\omega_{Twist}^2 = \frac{A}{B}, \qquad (9.1-39)$$

where

$$A = \int_0^L \{ \underline{a_{66}(\overline{\phi}'')^2} + a_{77}(\overline{\phi}')^2 + b_1 \hat{I}_p \Omega^2 R(z)(\overline{\phi}')^2$$

$$- \Omega^2 [\underwave{(b_4 - b_5)\overline{\phi}^2} + \underline{\underline{(b_{10} + \delta_n b_{18})(\overline{\phi}')^2}}]\} dz,$$

and

(9.1-40a,b)

$$B = \int_0^L [I_p \overline{\phi}^2 + \underline{(b_{10} + \delta_n b_{18})(\overline{\phi}')^2}] dz.$$

As in the case of pure flapping and lagging motions, the natural frequencies of the twist motion can also be expressed in Southwell form.

Equations (9.1-36), (9.1-37) and (9.1-39) represent the generalized counterparts of those associated with isotropic rotating solid beams (see Isakson and Eisley, 1960; Young, 1973; Johnson, 1980; Peters, 1995).

If $\overline{u}_0(z)$, $\overline{v}_0(z)$ and $\overline{\phi}(z)$ are selected to be the eigenfunctions of the associated lagging, flapping and twist eigenvalues, then these equations provide the respective eigenfrequencies.

From Eqs. (9.1-36), (9.1-37) and (9.1-39), a number of conclusions can be drawn:

(i) the rotatory inertia and inertial warping terms tend to diminish the eigenfrequencies,

(ii) the centrifugal stiffening effect is stronger for flapping frequencies than for lagging ones, especially at high rotational speeds,

(iii) Equation (9.1-39), (9.1-40) show that the centrifugal tension-torsion coupling contributes to the increase of the torsional eigenfrequency, whereas the tennis-racket and the centrifugal warping terms constitute centrifugal softening effects.

In short, the procedure for determining the decoupled rotating eigenfrequencies based on Rayleigh's quotient provides an expedient and reliable way to estimate them.

Moreover, based on the closed form solution of rotating eigenfrequencies as provided by this method, a better understanding of the implications of various involved effects can be reached.

9.1-8 Comparisons with Available Predictions

As mentioned previously, the equations derived for rotating *thin-walled beams* are similar in form to those corresponding to a *solid beam* model. The difference occurs only in the *proper* expression of cross-sectional stiffness quantities and mass terms. For this reason, use of dimensionless parameters in which these quantities are absorbed, enables one to obtain universal results, valid for both solid and thin-walled beams. In order to validate both the solution methodology and the structural model developed in this chapter, comparisons with a number of results scattered throughout the specialized literature will be presented. The two methods already used in different contexts of the present work, namely the exact one based on the Laplace Transform Technique (see Chapter 6.3-2 as well as Librescu and Thangjitham, 1991; Karpouzian and Librescu, 1994, 1996; and Song and Librescu, 1995, where this solution methodology was used to address aeroelastic stability and dynamic response problems), and the approximate one, referred to as the Extended Galerkin Method (see Chapter 6.3-3, as well as Palazotto and Linnemann, 1991; Librescu and Song, 1991, 1992; Song and Librescu, 1993, 1996, 1997, 1999; Song et al., 1994; Librescu et al., 1993, 1996, 1997, 1999; Gern and Librescu, 1998, 1999, 2000, 2001, where the *EGM* was used in different contexts), are applied to obtain solutions of respective eigenvalue problems. Since the former method is more laborious than the latter, and since the agreement between the predictions supplied by the two procedures was excellent (see e.g. Karpouzian and Librescu, 1995), the Extended Galerkin Method will be used for numerical illustrations.

Tables 9.1-2a through 9.1-2c compare the results of flapping frequencies obtained in the context of the unshearable (i.e. Euler-Bernoulli) beam model, for two extreme values of the hub ratio $\overline{R}_0$ $(\equiv R_0/L)$. In these tables the normalized angular velocity $\overline{\Omega}$ and eigenfrequencies $\overline{\omega}_i$ are defined as $(\overline{\Omega},\ \overline{\omega}_i) \equiv (\Omega,\ \omega_i)\,(b_1 L^4/a_{33})^{1/2}$. The displayed frequencies have been obtained by Hodges (1981) and Peters (1973) via the asymptotic method, those by Hodges and Rutkowski (1981), Yokoyama (1988), and Yokoyama and Markiewicz (1993) via the FEM, and those by Wright et al. (1982) and Du et al. (1994) via the Frobenius and power series technique, respectively. Tables 9.1-3a,b compare the flapping frequency predictions obtained also for a rotating unshearable beam. In these tables the present predictions are compared with the exact results (corresponding to a power series solution derived upon consideration of 200 terms in the expansion of $\overline{v}_0(z)$), with those obtained with a singular perturbation technique by Eick and Mignolet (1996), as well as that based on Southwell's sum approximation by Flax (1996). Comparisons of flapping frequencies for a shearable rotating beam are displayed in Table 9.1-4. The predictions obtained by Lee and Lin (1994) based on a superposition procedure, of those by Wang et al. (1976) whose approach is based on Legendre polynomials considered in conjunction with the Extended Galerkin Method, and of those by Yokoyama (1988) and Yokoyama and Markiewicz (1993) using the FEM, are compared with those obtained in the context of this study.

Table 9.1-2a: Dimensionless fundamental frequency $\bar{\omega}_1$ versus angular speed $\bar{\Omega}$

$\bar{\Omega}$	$\bar{R}_o$	H.	Y.	Y.& M.	P.	(Exact) H. & R; W.; D.	Present
0	0	3.51602	3.516	3.5160	3.5160	3.51602	3.51602
	1	3.51602	3.516	3.5160		3.51602	3.51602
1	0	3.68163				3.68165	3.68165
	1	3.88874				3.8882	3.88883
2	0	4.13710	4.137			4.13732	4.13732
	1	4.83279	4.834			4.83369	4.83366
3	0	4.79644		4.7973	4.7964	4.79728	4.79727
	1	6.07891		6.0818		6.08175	6.08174
4	0	5.58316	5.585			5.58500	5.58501
	1	7.46955	7.475			7.47505	7.47496
5	0	6.4450				6.44954	6.44958
	1	8.93210				8.94036	8.94050
6	0	7.35614	7.36	7.364	7.3561	7.36037	7.36038
	1	10.4331	10.444	10.4441		10.4439	10.4437
7	0	8.29440				8.29964	8.29964
	1	11.9563				11.9691	11.9695
8	0	9.25085	9.257			9.25684	9.2568
	1	13.4931	13.507			13.5074	13.5072
9	0	10.2192		10.2258		10.2257	10.2257
	1	15.0388		15.0549		15.0541	15.0546
10	0	11.1956	11.203			11.2023	11.2026
	1	16.5905	16.606			16.6064	10.6066
11	0	12.1775				12.1843	12.1841
	1	18.1464				18.1626	18.1621
12	0	13.1634		13.1706	13.1634	13.1702	13.1701
	1	19.7055		19.7237		19.7215	19.7224

H. → Hodges (1981); Y. → Yokoyama (1988); H.& R. → Hodges and Rutkowski (1981); W. → Wright et al. (1982); D. → Du et al. (1994); Y. & M. → Yokoyama and Markiewicz (1993); P → Peters (1973)

Finally, Table 9.1-5 compares the rotating/unrotating eigenfrequencies of an orthotropic blade in pure twist motion obtained in the context of the present approach with those derived by Nagaraj and Sahu (1982) via the FEM. For the sake of comparison, the present model is specialized for a free warping model (implying $a_{66} = 0$).

Table 9.1-2b: Dimensionless frequency $\bar\omega_2$ versus angular speed $\bar\Omega$

$\bar\Omega$	$\bar{R}_o$	Y.	Y. & H.	P.	(Exact) H. & R; W.; D.	Present
0	0	22.036	22.0352	22.0345	22.0345	22.0344
	1	22.036	22.0352		22.0345	22.0344
1	0				22.1810	22.1810
	1				22.3750	22.3749
2	0	22.617			22.6149	22.6150
	1	23.368			23.3660	23.3661
3	0		23.3210	23.3188	23.3203	23.3203
	1		24.9285		24.9277	24.9278
4	0	24.275			24.2734	24.2733
	1	26.959			26.9573	26.9574
5	0				25.4461	25.4461
	1				29.3528	29.3527
6	0	26.811	26.8098	26.7933	26.8091	26.8091
	1	32.030	32.0285		32.0272	32.0275
7	0				28.3341	28.3341
	1				34.9116	34.9114
8	0	29.998			29.9954	29.9954
	1	37.959			37.9538	37.9542
9	0		31.7716		31.7705	31.7705
	1		41.1187		41.1154	41.1159
10	0	33.643			33.6404	33.6402
	1	44.378			44.3682	44.3696
11	0				35.5890	35.5982
	1				47.6916	47.6945
12	0		37.6050	37.5155	37.6031	37.6033
	1		51.0776		51.0701	51.0742

In Table 9.1-5 the dimensionless rotational velocity $(\bar\Omega)$ and frequencies $(\bar\omega_i)$ are represented as $(\bar\Omega^2, \bar\omega_i^2) = (\Omega^2, \omega_i^2)[b_1 L^4/a_{77}, L^2(b_4 + b_5)/a_{77}]$.

Throughout these comparisons, an excellent agreement, even with the exact solution can be remarked.

9.1-9 Numerical Simulations-Extended Galerkin Method Applied to Rotating Beams

As was previously mentioned, the extended Galerkin Method is applied to solve the two decoupled eigenvalue problems associated with the rotating blades that undergo, on the one hand, flap-lag-transverse shear motion, and on the other hand, extension-twist coupled motion. In order to induce such elastic couplings, the circumferentially uniform stiffness (CUS) configuration achievable through the ply-angle scheme expressed by Eq. (9.1-19) was implemented. As

Table 9.1-2c: Dimensionless frequencies $\bar{\omega}_3$ and $\bar{\omega}_4$ versus angular speed $\bar{\Omega}$

$\bar{\Omega}$	$\bar{R}_0$	$\bar{\omega}_3$			$\bar{\omega}_4$		
		P.	H. & R. W. & D.	*Present*	P.	H. & R. W. & D.	*Present*
0	0	61.6973	61.6972	61.7148	120.901	120.902	121.120
	1		61.6972	61.7148		120.902	121.120
1	0		61.8418	61.8595		121.051	121.266
	1		62.0431	62.0604		121.263	121.481
2	0		62.2732	62.2904	122.194	121.497	121.710
	1		63.0676	63.0845		122.340	122.554
3	0	62.9720	62.9850	63.0013		122.236	122.447
	1		64.7338	64.7494		124.109	124.324
4	0		63.9668	63.9819		123.362	123.470
	1		69.9868	67.0006		126.537	126.758
5	0		65.2050	65.2189		124.566	124.772
	1		69.7607	69.7726		129.580	129.818
6	0	66.5421	66.6839	66.6962	125.759	126.141	126.345
	1		72.9863	72.9960		133.187	133.453
7	0		68.3860	68.3967		127.972	128.176
	1		76.5965	76.6042		137.305	137.618
8	0		70.2930	70.3022		130.049	130.258
	1		80.5295	80.5352		141.878	142.254
9	0		72.3867	72.3945		132.358	132.573
	1		84.7315	84.7354		134.884	135.110
	1		89.1563	89.1584		152.183	152.745
11	0		77.0638	77.0686		137.614	137.854
	1		93.7654	93.7664		157.820	158.503
12	0	78.7188	79.6145	79.6180	138.322	140.534	140.795
	1		98.5268	98.5280		163.724	164.543

was shown in Chapter 6, towards solving the related eigenvalue problem, the discretization is accomplished directly via the use of Hamilton's principle (see also Librescu, 1975) given in this case by

$$\int_{t_0}^{t_1} (\delta \mathcal{K} - \delta \mathcal{W}) dt = 0, \qquad (9.1\text{-}41)$$

where $\mathcal{K}$ and $\mathcal{W}$ denote the kinetic and strain energies. For these two independent eigenvalue problems, the 1-D displacement measures are represented by means of series of space-dependent trial-functions multiplied by time-dependent generalized coordinates as

$$u_0(z, t) = \mathbf{U}_0^T(z)\mathbf{q}_u(t), \qquad v_0(z, t) = \mathbf{V}_0^T(z)\mathbf{q}_v(t),$$
$$\theta_x(z, t) = \mathbf{X}^T(z)\mathbf{q}_x(t). \qquad \theta_y(z, t) = \mathbf{Y}^T(z)\mathbf{q}_y(t). \qquad (9.1\text{-}42)$$

Table 9.1-3a: First dimensionless frequency squared, $\bar{\omega}_1^2 (\equiv \omega_1^2/\Omega^2)$ versus ϵ

ϵ ($\equiv a_{33}/b_1\Omega^2 L^4$)	*(Exact)* E. & M.	*(Southwell Sum) F.*	*Present*
0.001	1.07	1.01	1.01
0.004	1.15	1.05	1.14
0.010	1.25	1.12	1.15
0.040	1.66	1.49	1.66
0.100	2.42	2.24	2.42
0.400	6.14	5.94	6.14
1.000	13.55	13.36	13.55
4.00	50.64	49.94	50.64

E. & M. $\rightarrow$ Eick and Mignolet (1996); F. $\rightarrow$ Flax (1996)

Table 9.1-3b: Second dimensionless frequency squared, $\bar{\omega}_2^2 (\equiv \omega_2^2/\Omega^2)$ versus ϵ

ϵ ($\equiv a_{33}/b_1\Omega^2 L^4$)	*(Exact)* E. & M.	*(Southwell Sum) F.*	*Present*
0.001	6.78	6.48	6.98
0.004	8.36	7.94	8.35
0.010	11.3	10.9	11.3
0.040	25.9	25.4	25.9
0.100	55.0	55.0	55.0
0.400	200.7	200.2	200.7
1.000	492.0	491.5	492.0
4.000	1948.6	1947.9	1948.4

Table 9.1-4: Dimensionless flapping frequencies $\bar{\omega}_i$ versus angular speed $\bar{\Omega}$, for a shear-deformable beam ($\bar{R}_0 = 3$)

$\eta(\equiv I_{xx}/b_1 L^2)$	$\mu(\equiv a_{33}/L^2 a_{55})$	$\bar{\Omega}$	$\bar{\omega}_i$	L. & L.	Y.	W.	*Present*
		0	i=1	-	3.435	3.444	3.436
			i=2	-	19.148	19.174	19.139
0.05	0.007647		i=3	-	46.890	46.828	46.758
			i=4	-	79.795	79.367	80.172
		10	1	23.491	23.524	23.514	23.510
			2	55.984	56.105	56.072	56.062
			3	96.913	97.188	97.011	97.002
			4	143.171	144.490	143.815	144.194
		0	1	-	3.231	3.232	3.231
			2	-	14.542	14.535	14.531
0.1	0.030588		3	-	31.836	31.679	31.673
			4	-	48.724	48.237	48.419
		10	1	22.938	23.050	23.037	23.036
			2	44.781	45.598	45.428	45.429
			3	66.287	67.716	66.854	66.866
			4	71.967	73.076	72.313	72.315

L. & L. $\rightarrow$ Lee and Lin (1994); Y. $\rightarrow$ Yokoyama (1988); W. $\rightarrow$ Wang et al. (1976)

Table 9.1-5: Dimensionless twist frequencies $\bar{\omega}_i$ versus angular speed $\bar{\Omega}^2$
$(\hat{I}_p/L^2 = 0.241 \times 10^{-3}; \; (b_4 - b_5)/L^3 = 0.581 \times 10^{-2})$

$\bar{\Omega}^2$		$\bar{\omega}_1$	$\bar{\omega}_2$	$\bar{\omega}_3$	$\bar{\omega}_4$
0	N. & S.	1.57080	4.71253	7.85705	11.01642
	Present	1.57080	4.71239	7.85409	11.0043
100	N. & S.	1.75335	4.79303	7.92561	11.08667
	Present	1.75335	4.79290	7.92263	11.07470
200	N. & S.	1.91861	4.87217	7.99352	11.15638
	Present	1.91861	4.87204	7.99060	11.14410

N. & S. $\to$ Nagaraj and Sahu (1982)

- For the lagging-flapping-transverse shear motion, labeled as (Case I)

- For the extension-twist problem, labeled as (Case II), the representation of associated displacement quantities is

$$w_0(z, t) = \mathbf{W}_0^T(z)\mathbf{q}_w(t); \quad \phi(z, t) = \mathbf{\Phi}^T(z)\mathbf{q}_\phi(t). \qquad (9.1\text{-}43\text{a,b})$$

Associated with these two cases, the kinetic and strain energies in discrete form are:

$$\mathcal{K}_I = \frac{1}{2}\int_0^L [b_1(\dot{u}_0)^2 + b_1\Omega^2 u_0^2 + (b_4 + \delta_n b_{14})(\dot{\theta}_x)^2 + (b_5 + \delta_n b_{15})(\dot{\theta}_y)^2$$
$$+ \Omega^2(b_4 + \delta_n b_{14})(\theta_x)^2 + \Omega^2(b_5 + \delta_n b_{15})(\theta_y)^2 + b_1(\dot{v}_0)^2]dz,$$
$$\qquad (9.1\text{-}44\text{a,b})$$

$$\mathcal{K}_{II} = \frac{1}{2}\int_0^L [b_1\Omega^2(R_0 + z + w_0)^2 + b_1(\dot{w}_0)^2 + (b_4 + b_5)(\dot{\phi})^2$$
$$+ (b_{10} + \delta_n b_{18})(\dot{\phi}')^2 + \Omega^2(b_{10} + \delta_n b_{18})(\phi')^2 + \Omega^2(b_4 - b_5)\phi^2]dz,$$

$$\mathcal{W}_I = \frac{1}{2}\int_0^L [a_{22}(\theta_y')^2 + a_{33}(\theta_x')^2 + a_{44}(u_0' + \theta_y)^2 + a_{55}(v_0' + \theta_x)^2$$
$$+ 2a_{25}\theta_y'(v_0' + \theta_x) + 2a_{34}\theta_x'(u_0' + \theta_y) + T_z((u_0')^2 + (v_0')^2)]dz,$$
$$\qquad (9.1\text{-}45\text{a,b})$$

$$\mathcal{W}_{II} = \frac{1}{2}\int_0^L [a_{11}(w_0')^2 + a_{66}(\phi'')^2 + a_{77}(\phi')^2 + 2a_{17}w_0'\phi' + T_r(\phi')^2]dz.$$

Inserting Eqs. (9.1-42), (9.1-43) in (9.1-44) and (9.1-45), we can express the kinetic and strain energies in discrete form, for Cases I and II:

$$\mathcal{K}_I = \frac{1}{2}\dot{\mathbf{q}}_I^T \mathbf{M}_I \dot{\mathbf{q}}_I + \frac{1}{2}\mathbf{q}_I^T \overline{\mathbf{M}}_I \mathbf{q}_I,$$

$$\mathcal{K}_{II} = \frac{1}{2}\dot{\mathbf{q}}_{II}^T \mathbf{M}_{II} \dot{\mathbf{q}}_{II} + \frac{1}{2}\mathbf{q}_{II}^T \overline{\mathbf{M}}_{II} \mathbf{q}_{II} + \int_0^L b_1\Omega^2(R_0+z)^2 dz \quad (9.1\text{-}46a,b)$$

$$+ 2b_1\Omega^2 \int_0^L (R_0+z)w_0 dz,$$

$$\mathcal{W}_I = \frac{1}{2}\mathbf{q}_I^T \kappa_I \mathbf{q}_I, \quad \mathcal{W}_{II} = \frac{1}{2}\mathbf{q}_{II}^T \kappa_{II} \mathbf{q}_{II}. \tag{9.1-47a,b}$$

Substituting Eqs. (9.1-46) and (9.1-47) in conjunction with the proper representation of trial functions in Eq. (9.1-41), integrating with respect to time and recognizing that $\delta\mathbf{q} = 0$ at $t = t_0, t_1$, one obtains the equations of motion for the two decoupled problems as:

$$\mathbf{M}_I \ddot{\mathbf{q}}_I + \mathbf{K}_I \mathbf{q}_I = 0,$$
$$\mathbf{M}_{II} \ddot{\mathbf{q}}_{II} + \mathbf{K}_{II} \mathbf{q}_{II} - \mathbf{C} = 0 \tag{9.1-48a,b}$$

In these equations:

$$\mathbf{C} = [\, 2b_1\Omega^2 \int_0^L (R_0+z)\mathbf{W}_0^T dz, \ \ \mathbf{0}\,]^T, \tag{9.1-49}$$

$$\mathbf{q}_I = \{\mathbf{q}_u^T, \mathbf{q}_v^T, \mathbf{q}_x^T, \mathbf{q}_y^T\}^T, \quad \mathbf{q}_{II} = \{\mathbf{q}_w^T, \mathbf{q}_\phi^T\}^T, \tag{9.1-50a,b}$$

$$\mathbf{M}_I = \int_o^L \begin{vmatrix} b_1\mathbf{U}_0\mathbf{U}_0^T & & & \mathbf{0} \\ & b_1\mathbf{V}_0\mathbf{V}_0^T & & \\ & & (b_4+\delta_n b_{14})\mathbf{X}\mathbf{X}^T & \\ & & \cdots\cdots\cdots\cdots & \\ \mathbf{0} & & & (b_5+\delta_n b_{15})\mathbf{Y}\mathbf{Y}^T \\ & & & \cdots\cdots\cdots\cdots \end{vmatrix} dz,$$

$$\mathbf{M}_{II} = \int_0^L \begin{vmatrix} b_1\mathbf{W}_0\mathbf{W}_0^T & \mathbf{0} \\ \mathbf{0} & (b_4+b_5)\mathbf{\Phi}\mathbf{\Phi}^T + (\,\underline{b_{10}+\delta_n b_{18}}\,)\mathbf{\Phi}'(\mathbf{\Phi}')^T \end{vmatrix} dz,$$

$$\tag{9.1-51a-d}$$

$$\overline{\mathbf{M}}_I = \Omega^2 \int_o^L \begin{vmatrix} b_1\mathbf{U}_0\mathbf{U}_0^T & & & \mathbf{0} \\ & \mathbf{0} & & \\ & & (b_4 + \delta_n b_{14})\mathbf{X}\mathbf{X}^T & \\ & & \text{------} & \\ \mathbf{0} & & & (b_5 + \delta_n b_{15})\mathbf{Y}\mathbf{Y}^T \\ & & & \text{------} \end{vmatrix} dz,$$

$$\overline{\mathbf{M}}_{II} = \Omega^2 \int_0^L \begin{vmatrix} b_1\mathbf{W}_0\mathbf{W}_0^T & \mathbf{0} \\ \mathbf{0} & (b_4 - b_5)\mathbf{\Phi}\mathbf{\Phi}^T + (b_{10} + \delta_n b_{18})\mathbf{\Phi}'(\mathbf{\Phi}')^T \end{vmatrix} dz$$

$\mathbf{K}_{(I);(II)} = \kappa_{(I);(II)} - \overline{\mathbf{M}}_{(I);(II)}$, where $\kappa_{(I);(II)}$, $\mathbf{M}_{(I);(II)}$ and $\overline{\mathbf{M}}_{(I);(II)}$ are the matrices appearing in Eqs. (9.1-47) and (9.1-46), associated with the systems (I) and (II). For the stiffness matrices $\mathbf{K}_I = |K_{ij}|$ $(i, j = \overline{1, 4})$ and $\mathbf{K}_{II} = |K_{ij}|$ $(i, j = 1, 2)$, their elements are supplied in Tables 9.1-6.

Table 9.1-6a: The entries of $\mathbf{K}_I$, $(K_{ij} = K_{ji})$

K_{ij}	Their expression
K_{11}	$\int_0^L [(T_z + a_{44})\mathbf{U}_0'(\mathbf{U}_0')^T - b_1\Omega^2\mathbf{U}_0\mathbf{U}_0^T]dz$
K_{12}	0
K_{13}	$\int_0^L [a_{34}\mathbf{U}_0'(\mathbf{X}')^T]dz$
K_{14}	$\int_0^L [a_{44}\mathbf{U}_0'(\mathbf{Y}')^T]dz$
K_{22}	$\int_0^L [(T_z + a_{55})\mathbf{V}_0'(\mathbf{V}_0')^T]dz$
K_{23}	$\int_0^L [a_{55}\mathbf{V}_0'\mathbf{X}^T]dz$
K_{24}	$\int_0^L [a_{25}\mathbf{V}_0'(\mathbf{Y}')^T]dz$
K_{33}	$\int_0^L [a_{55}\mathbf{X}\mathbf{X}^T + a_{33}\mathbf{X}(\mathbf{X}')^T - \Omega^2(b_4 + \delta_n b_{14})\mathbf{X}\mathbf{X}^T]dz$
K_{34}	$\int_0^L [a_{34}\mathbf{Y}(\mathbf{X}')^T + a_{25}\mathbf{X}(\mathbf{Y}')^T]dz$
K_{44}	$\int_0^L [a_{44}\mathbf{Y}\mathbf{Y}^T + a_{22}\mathbf{Y}'(\mathbf{Y}')^T + \Omega^2(b_5 + \delta_n b_{15})\mathbf{Y}\mathbf{Y}^T]dz$

Table 9.1-6b: The entries of $\mathbf{K}_{II}$, $(K_{ij} = K_{ji})$

K_{ij}	Their expression
K_{11}	$\int_0^L [a_{11}\mathbf{W}_0'(\mathbf{W}_0')^T - \Omega^2 b_1 \mathbf{W}_0 \mathbf{W}_0^T]dz$
K_{12}	$\int_0^L [a_{17}\mathbf{W}_0'(\mathbf{\Phi}')^T]dz$
K_{22}	$\int_0^L (T_r + a_{77})\mathbf{\Phi}'(\mathbf{\Phi}')^T + a_{66}\mathbf{\Phi}''(\mathbf{\Phi}'')^T dz$
	$- \int_0^L [\Omega^2(b_4 - b_5)\mathbf{\Phi}'(\mathbf{\Phi}')^T + \Omega^2(b_{10} + \delta_n b_{18})\mathbf{\Phi}'(\mathbf{\Phi}')^T]dz$

In order to reduce Eq. (9.1-48) to an eigenvalue problem, we define (see Chandiramani et al., 2002, 2003):

$$\mathbf{C} = \mathbf{K}_{(II)}\mathbf{q}_{st}, \qquad (9.1\text{-}52)$$

and

$$\bar{\mathbf{q}}_{II} = \mathbf{q}_{II} - (\mathbf{q}_{st})_{II}. \qquad (9.1\text{-}53)$$

Upon replacement of (9.1-52) into (9.1-48b) one obtains

$$\mathbf{M}_{II}\ddot{\bar{\mathbf{q}}}_{II} + K_{II}\bar{\mathbf{q}}_{II} = \mathbf{0}, \qquad (9.1\text{-}54)$$

which is similar to (9.1-48a). Following the procedure indicated in Chapter 6, one can get the eigenvector and eigenvalues of the system.

9.1-10 Numerical Simulations and Behavior

With the exception of a few relatively simple special cases for which it was possible to express, in closed form, the decoupled natural frequencies (see Eqs. (9.1-36), (9.1-37) and (9.1-39)), for the more general cases, this is no longer feasible.

Consequently, for the solution of the intricate eigenvalue problem, the Extended Galerkin Method is applied. In the context of the following numerical simulations, the effects of the rotational speed and ply angle on eigenvibration characteristics of rotating blades will be highlighted. To this end, the blade is modelled as a single-layer thin-walled beam of *rectangular uniform cross-section* whose geometric characteristics are identical to those supplied in Table 6.4-1. In addition to those characteristics we also consider that $R_0/L = 0.1$.

As concerns the material used, its properties corresponding to the graphite-epoxy material have been listed in Table 6.4-2. Throughout the numerical simulations, the circumferentially uniform stiffness beam configuration was considered.

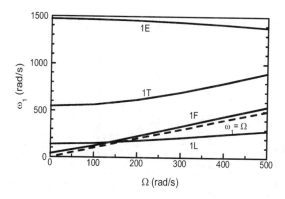

Figure 9.1-2: First decoupled natural frequencies vs. blade rotational speed Ω for ply-angle $\theta = 0$ (L: lagging; F: flapping; E: extensional; T: torsional modes; on the graph also the $\omega_1 = \Omega$ line is displayed)

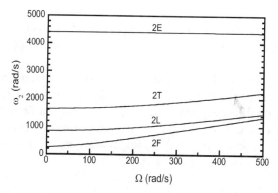

Figure 9.1-3: Second decoupled natural frequencies vs. Ω for the ply-angle $\theta = 0$

Figures 9.1-2 and 9.1-3 highlight the effects of rotor speed Ω and ply-angle ($\theta = 0$, 90 deg.) on the in-vacuo decoupled natural frequencies.

For the ply angle $\theta = 0$, the results emerging from Figs. 9.1-2 and 9.1-3 appear to be similar to those reported by Stafford and Giurgiutiu (1975), Giurgiutiu and Stafford (1977) and Ewins and Henry (1992) for metallic blades. These results (see Fig. 9.1-2) show that at moderate rotational speeds the first flapwise mode exhibits the lowest frequency. However, due to the centrifugal stiffening effect (which is much more significant in the flapping modes), beyond a certain angular speed the lagging frequency becomes the lower of the two. Figure 9.1-2 reveals that in the vicinity of $\Omega \doteq 120$ rad/s, a frequency crossing of lagging and flapping modes occurs.

As Fig. 9.1-3 reveals, this trend remains valid also for higher mode frequencies (i.e., in the present case for the second one). However in these cases, a shift of that specific rotational speed at which the frequency crossing occurs, towards higher values of Ω, is observed.

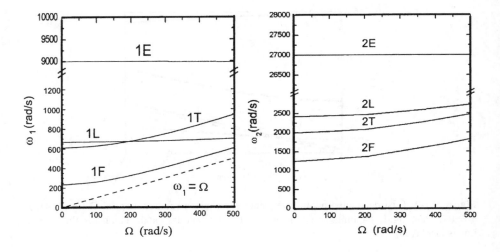

Figure 9.1-4: First decoupled natural frequencies vs. Ω for ply-angle $\theta = 90$ deg

Figure 9.1-5: Second decoupled natural frequencies vs. Ω for ply-angle $\theta = 90$ deg

For the ply angle $\theta = 90$ deg. and for the first mode frequency (Fig. 9.1-4), a similar trend occurs. In this case, however, the frequency crossing takes place at much higher rotational speeds (i.e. approximately at $\Omega = 600$ rad/s). For higher modes (see Fig. 9.1-5) the frequencies remain separated in the usual range of rotational speeds, in the sense that no frequency crossing occurs in that range.

The results displayed in Figs. 9.1-6 through 9.1-8 reveal the sensitivity of modal frequencies to ply-angle orientation. This characteristic of composite materials provides a powerful tool in the effort to structurally tailor helicopter and tilt rotor aircraft blades for improved dynamic responses.

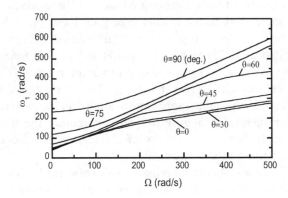

Figure 9.1-6: First coupled flap-lag bending frequency vs. Ω for different ply-angles

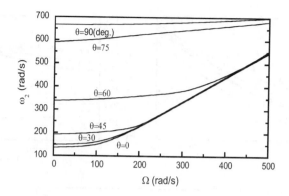

Figure 9.1-7: Second coupled flap-lag bending frequency vs. Ω for different ply-angles

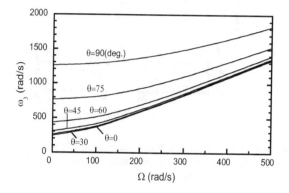

Figure 9.1-8: Third coupled flap-lag bending natural frequency vs. Ω for different ply-angles

As opposed to the centrifugal stiffening effect on the flapping, twist and lagging frequencies, the extension frequency exhibits a slight decrease with increasing rotational speed (see Figs. 9.1-2 through 9.1-5). The specific value of Ω for which the extension frequency becomes zero corresponds to the instability by divergence. For the blade configuration considered here, this instability occurs beyond the range of rotational speeds of practical importance.

This *centrifugal destiffening* phenomenon was emphasized for the first time by Bhuta and Jones (1965). At this point, we recall the apparent controversy raised in the literature (see e.g. Anderson, 1975; Hodges, 1977) related to the character of the influence of the rotor angular velocity on the frequency of axial vibration. A clarification of this controversy was supplied by Kvaternik et al. (1978) and Venkatesan and Nagaraj (1981) and, as concerns our approach, it is consistent with the point of view presented in these references.

Figures 9.1-9a-c highlight the effects played by rotational speed and ply-angle orientation on the natural frequencies of the coupled extensional-twist motion.

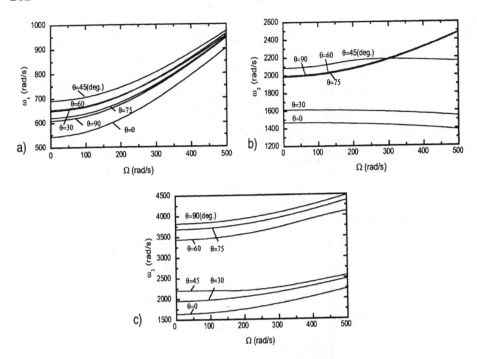

Figure 9.1-9: The first three natural coupled extension-twist natural frequencies vs. Ω for selected ply-angles, θ (deg.). a) First coupled natural frequency; b) Second coupled natural frequency; c) Third coupled natural frequency

Figure 9.1-9a shows that the first coupled extension-twist (E-T) frequency has a maximum for any Ω around the ply angle $\theta = 45$ deg. For this coupled frequency the effect of the extension is rather weak. However, for the second coupled E-T frequency (see Fig. 9.1-9b), for $\theta < 45$ deg. the extension effect is rather dominant and decays for $\theta > 45$ deg., where the usual trend related to the influence of the rotational speed is emerging. As concerns the third E-T coupled frequency (see Fig. 9.1-9c), in contrast to the second frequency, in the present case the twist effect is dominant within the entire domain $0 < \theta < 90$ deg. and this explains the trend of variation of the frequency vs. θ for various rotational speeds. Moreover, with the increase of the frequency mode, the ply-angle θ at which the maximum occurs is shifted towards larger ply-angles.

For a rotating thin-walled beam whose material is characterized by the ply-angle $\theta = 0$, the variation of the first three uncoupled mode shapes for various rotational speeds is portrayed in Figs. 9.1-10a-c.

The mode shapes are normalized to a value of unity at the blade tip. As a general remark which is valid for the (lag and flap) bending and twist modes, with the increase of the rotational speed the blade has the tendency to straighten itself. In connection with the flapping eigenmodes, a similar trend for rotating metallic blades was reported by Bhat (1986). As concerns the extensional mode shapes, the results not displayed here reveal that they are insensitive to the variation of the rotational speed.

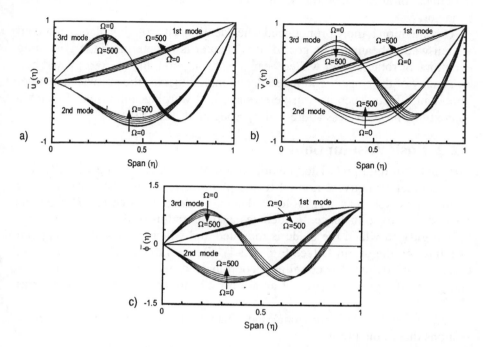

Figure 9.1-10: The first three normalized eigenmodes for ply-angle $\theta = 0$ and for $\Omega = (0$ to $500)$ rad/s. a) the eigenmode in lagging; b) eigenmode in flapping; c) eigenmode in torsion

Although in the computational process the previously mentioned nonclassical effects have been considered in their entirety, a number of them, such as the inertial and centrifugal warpings appear to be negligibly small. Related to the former effect, a similar conclusions was outlined by Kaza and Kielb (1984) for rotating solid blades, and by Stemple and Lee (1989) for their thin- walled counterparts. As concerns the other effects such as warping inhibition and transverse shear, their influence appears to be significant. Just for this reason these effects have to be considered altogether.

9.2 PRETWISTED ROTATING BEAMS
9.2-1 Preliminaries

In order to ensure performance advantages, rotor blades of turbomachinery, tilt rotor aircraft and helicopters are usually twisted. In spite of this feature, very few research works have addressed the problem of the modeling and free vibration of composite pretwisted rotor blades modeled as thin-walled beams.

The available survey-papers reviewing the state-of-the art on this topic, such as the ones by Rosen (1991) and Jung et al. (1999), as well as the monograph by Rao (1991), reveal that even for *solid beams*, there are few studies that include the effects of anisotropy coupled to that of the pretwist on their free vibration behavior.

This situation is more pronounced when dealing with rotating blades modeled as thin-walled beams. The goal of this chapter is to determine the governing equations of rotating thin-walled beams featuring both the anisotropy and heterogeneity of the constituent material, and including also the pretwist and other important effects, such as transverse shear and warping inhibition. In such a context this problem was addressed by Song et al. (2001a)

9.2-2 Basic Assumptions

Consider a straight pretwisted flexible beam of length L rotating with the constant angular velocity Ω normal to the plane of rotation. The origin of the rotating axis system (x, y, z) is located at the blade root at an offset R_o from the rotation axis fixed in space. R_o denotes also the radius of the hub (considered to be rigid), in which the blade is mounted and which rotates about its polar axis through the origin O (see Fig. 9.2-1). Besides the coordinates (x, y, z), there are also defined the cervilinear coordinate system (x^P, y^P, z^P), where x^P and y^P are the *principal axes* of an arbitrary beam cross-section (see Song et al., 2001).

Axes x^P and y^P are orthogonal and rotate around the z-axis with the cross-sections due to the pretwist.

The two coordinate systems are related by the following transformation formulae:

$$x(s, z) = x^P(s) \cos\beta(z) - y^P(s) \sin \beta(z),$$
$$y(s, z) = x^P(s) \sin \beta(z) + y^P(s) \cos \beta(z), \qquad (9.2\text{-}1a\text{-}c)$$
$$z = z^P,$$

where $\beta(z) = \beta_o z/L$, denotes the pretwist angle of a current section, whereas β_o denotes the pretwist at the beam tip. It is assumed that the pretwist exists prior any deformation of the blade.

The inertial reference system (X, Y, Z) is attached to the center of the hub O. By $(\mathbf{i}, \mathbf{j}, \mathbf{k})$ and $(\mathbf{I}, \mathbf{J}, \mathbf{K})$ we define the unit vectors associated with the frame coordinates (x, y, z) and (X, Y, Z), respectively. In addition, a local (surface) coordinate system (s, z, n) associated with the beam is considered. The beam's geometric configuration and the typical cross-section along with the associated system of coordinates are presented in Fig. 9.2-1. Within the present work, it is further assumed that the rotation takes place in the plane (X, Z) with the constant angular velocity $\Omega(\equiv \Omega\mathbf{J} = \Omega\mathbf{j})$, the spin axis coincides with the

Y-axis. The structural model corresponds to a thin/ thick walled beam. In this context, the case of a single-cell thin-walled beam of uniform closed-section is considered, where the spanwise, z-coordinate axis, coincides with a straight unspecified reference axis. Herein, the precone and pre-setting angles of the blade are assumed to be zero.

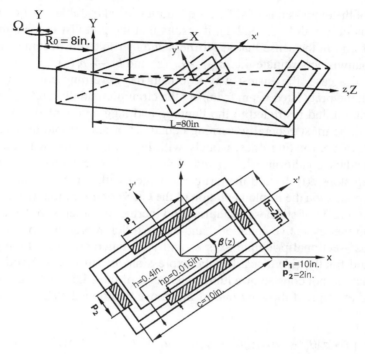

Figure 9.2-1: a) Geometry of the pretwisted beam, and b), its cross-section with the embedded piezoactuators (enlarged view)

In its modeling, the following assumptions are adopted: (i) the original cross-section shape of the beam is preserved, (ii) both primary and secondary (thickness) warping effects are included, (iii) transverse shear, rotatory inertia and centrifugal acceleration effects are included, and finally, (iv) the circumferentially uniform stiffness configuration inducing flapping-lagging coupling, on the one hand, and twist- extension coupling, on the other hand is implemented.

COMMENT

Two approaches, namely those by Wagner and Washizu are used to deal with the problem of pretwisted blades. While the former approach adopted in these developments is based on the widely accepted *helical multifilament concept* as discussed by Kaza and Kielb (1984) and used, among others, by Houbolt and Brooks (1958), the latter one prompted by Washizu (1964) and applied by Hodges (1980) and Rosen (1978, 1980) uses the tensorial relationships for transforming the quantities evaluated in the curvilinear coordinate system (η, ζ, z) to the orthogonal local one (s, n, z).

More details about Washizu's approach and the comparison of its predictions with those based on Wagner's concept will be presented later in this section.

9.2-3 The Governing System

9.2-3a Shearable Beam Model

In light of the expressions of stiffness quantities listed in Table 4.3-1 considered in conjunction with Eqs. (9.2-1), it results that the pretwist renders the beam non-uniform, in the sense that its stiffness and inertia terms become a function of the spanwise coordinate, i.e. $a_{ij} \rightarrow a_{ij}(z)$ and $b_i \rightarrow b_i(z)$.

In such a case, the easiest way to obtain the governing equations associated with the pretwisted beam featuring the circumferentially uniform stiffness configuration and restricted to the flapping and lagging coupled motion, is to start with the most general governing equations of anisotropic beams, for the moment, of nonrotating ones, namely with Eqs. (4.4-1) and with the associated boundary conditions (4.4-2) and (4.4-3). In these equations we apply the following steps: (i) discard the terms associated with extension and twist that should appear, via the implementation of the CUS configuration, (ii) transformation via (9.2-1) of stiffness quantities and inertia coefficients in the terms that subsist in the respective equations, and their explicit representation in terms of their cross-sectional principal axes x_p and y_p counterparts and of the pretwist angle, and finally, (iii) discard in the expressions of stiffness quantities of the terms that are not consistent with the character of the CUS configuration.

As an example, following one of the above mentioned steps, Eq. (4.4-1d) becomes:

$$[a_{42}(z)\theta'_y + a_{43}(z)\theta'_x + a_{44}(z)(u'_0 + \theta_y) + a_{45}(z)(v'_0 + \theta_x)]'$$
$$+ p_x(z, t) = K_1(z, t). \tag{9.2-2}$$

Now, the stiffness quantities have to be expressed in terms of their cross-sectional principal axes counterparts and the pretwist angle. Considering as an example the case of the stiffness quantity $a_{42}(z)$, by using expressions from Table 4.3-1 considered in conjunction with the transformation relationships, Eqs. (9.2-1), one obtains:

$$a_{42}(z) = \oint [K_{12}x\frac{dx}{ds} + K_{24}\frac{dx}{ds}\frac{dy}{ds}]ds$$
$$= \oint \{K_{12}(x^p\frac{dx^p}{ds}\cos^2\beta - x^p\frac{dy^p}{ds}\cos\beta\sin\beta \tag{9.2-3}$$
$$- y^p\frac{dx^p}{ds}\cos\beta\sin\beta + y^p\frac{dy^p}{ds}\sin^2\beta) + K_{24}[(\frac{dx^p}{ds})^2\cos\beta\sin\beta$$
$$+ \frac{dx^p}{ds}\frac{dy^p}{ds}\cos^2\beta - \frac{dx^p}{ds}\frac{dy^p}{ds}\sin^2\beta - (\frac{dy^p}{ds})^2\cos\beta\sin\beta]\}ds.$$

In conjunction with the definitions of stiffness quantities from Table 4.3-1, Eq. (9.2-3) becomes:

$$a_{42}(z) = a_{42}^P \cos^2 \beta + a_{35}^P \sin^2 \beta - (a_{25}^P + a_{34}^P) \cos \beta \sin \beta. \qquad (9.2\text{-}4)$$

For CUS, a_{42} and a_{35} are among the stiffness quantities to become zero, so that Eq. (9.2-4) reduces to $a_{42}(z) = a_{24}(z) = -(a_{25}^P + a_{34}^P) \cos \beta \sin \beta$. Furthermore, for doubly symmetric CUS box beams, since $a_{25}^P = -a_{34}^P$, $a_{42}(z) = a_{24}(z)$ become zero quantities. The same process concerns also the inertia terms.

For example, the inertia terms associated with Eqs. (4.4-1c) are contained in K_7 whose expression was supplied in Table 3.1-1. By discarding the terms that are not proper to this ply-angle configuration (see Sect. 4.4-4a), we have

$$K_7 = B_{13}(z)\ddot{\theta}_x + B_{15}(z)\ddot{\theta}_y, \qquad (9.2\text{-}5)$$

where according to Eqs. (3.1-23)

$$(B_{13}(z); B_{15}(z)) = \int \oint \rho \left\{ \left[\left(y - n\frac{dx}{ds} \right)^2 ; \left(x + n\frac{dy}{ds} \right) \left(y - n\frac{dx}{ds} \right) \right] \right\} ds\, dn. \qquad (9.2\text{-}6)$$

Using the transformation formulae (9.2-1), and the integration in (9.2-6) across the beam thickness over a symmetric interval, (that implies a symmetric laminate), one obtains that

$$\begin{aligned}
\{B_{13}(z); B_{15}(z)\} &= \left\{ \left[b_4^P \cos^2 \beta + b_5^P \sin^2 \beta + 2b_6^P \sin \beta \cos \beta \right] \right. \\
&\quad + \delta_n \left[b_{14}^P \cos^2 \beta + b_{15}^P \sin^2 \beta - 2b_{13}^P \sin \beta \cos \beta \right] \\
&\quad ; \left[(b_5^P - b_4^P) - \delta_n (b_{14}^P - b_{15}^P) \right] \sin \beta \cos \beta \\
&\quad \left. + b_6^P (\cos^2 \beta - \sin^2 \beta) \right\}. \qquad (9.2\text{-}7)
\end{aligned}$$

However, since for CUS, b_6^P and b_{13}^P are vanishing terms, these inertia coefficients should be discarded.

The process of re-expressing the dynamic terms of rotating and pretwisted beams has to be done however, in the expression of the kinetic energy of the straight blade, Eq. (3.1-21).

In such a way, the governing equations of pretwisted rotating blades are expressed as (see Song et al., 2001a,b)

$$\begin{aligned}
\delta u_0 : &\left[a_{42}(z)\theta_y' + a_{43}(z)\theta_x' + a_{44}(z)(u_0' + \theta_y) + a_{45}(z)(v_0' + \theta_x) \right]' \\
&- b_1 \ddot{u}_0 + \underline{\underline{b_1 u_0 \Omega^2}} + \underline{\underline{b_1 \Omega^2 [R(z)u_0']'}} = 0, \\
\delta v_0 : &\left[a_{52}(z)\theta_y' + a_{53}(z)\theta_x' + a_{55}(z)(v_0' + \theta_x) + a_{54}(z)(u_0' + \theta_y) \right]' \\
&- b_1 \ddot{v}_0 + \underline{\underline{b_1 \Omega^2 [R(z)v_0']'}} = 0,
\end{aligned}$$

$$\delta\theta_y \; : \; \left[a_{22}(z)\theta_y' + a_{25}(z)(v_0' + \theta_x) + a_{24}(z)(u_0' + \theta_y) + a_{23}(z)\theta_x'\right]'$$
$$-a_{44}(z)(u_0' + \theta_y) - a_{43}(z)\theta_x' - a_{45}(z)(v_0' + \theta_x) - a_{42}(z)\theta_y'$$

$$- \left(b_5(z) + \delta_n b_{15}(z)\right)(\; \underset{\cdots}{\ddot{\theta}_y} - \underset{\wwave}{\Omega^2\theta_y})$$

$$- \left(b_6(z) - \delta_n b_{13}(z)\right)(\; \underset{\cdots}{\ddot{\theta}_x} \; - \; \underset{\wwave}{\Omega^2\theta_x}) \; = 0, \qquad (9.2\text{-}8\text{a-d})$$

$$\delta\theta_x \; : \; \left[a_{33}(z)\theta_x' + a_{32}(z)\theta_y' + a_{34}(z)(u_0' + \theta_y) + a_{35}(z)(v_0' + \theta_x)\right]'$$

$$- a_{55}(z)(v_0' + \theta_x) - a_{52}(z)\theta_y' - a_{54}(z)(u_0' + \theta_y) - a_{53}(z)\theta_x'$$

$$- \left(b_4(z) + \delta_n b_{14}(z)\right)(\; \underset{\cdots}{\ddot{\theta}_x} \; - \; \underset{\wwave}{\Omega^2\theta_x})$$

$$- \left(b_6(z) - \delta_n b_{13}(z)\right)(\; \underset{\cdots}{\ddot{\theta}_y} \; - \; \underset{\wwave}{\Omega^2\theta_y}) \; = 0.$$

If the blade is clamped at $z = 0$ and free at $z = L$, the boundary conditions are as follows:

At $z = 0$:

$$u_0 = 0, \;\; v_0 = 0, \;\; \theta_x = 0, \;\; \theta_y = 0, \qquad (9.2\text{-}9\text{a-d})$$

and at $z = L$:

$$\delta u_0 : a_{42}(L)\theta_y' + a_{43}(L)\theta_x' + a_{44}(L)(u_0' + \theta_y) + a_{45}(L)(v_0' + \theta_x) = 0,$$
$$\delta v_0 : a_{52}(L)\theta_y' + a_{53}(L)\theta_x' + a_{55}(L)(v_0' + \theta_x) + a_{54}(L)(u_0' + \theta_y) = 0,$$
$$\delta\theta_y : a_{22}(L)\theta_y' + a_{25}(L)(v_0' + \theta_x) + a_{24}(L)(u_0' + \theta_y) + a_{23}(L)\theta_x' = 0,$$
$$\delta\theta_x : a_{33}(L)\theta_x' + a_{34}(L)(u_0' + \theta_y) + a_{35}(L)(v_0' + \theta_x) + a_{32}(L)\theta_y' = 0.$$
$$(9.2\text{-}10\text{a-d})$$

In Eqs. (9.2-8) and the following ones

$$R(z) \equiv [R_0(L - z) + \frac{1}{2}(L^2 - z^2)], \qquad (9.2\text{-}11)$$

whereas the coefficients $a_{ij}(z) = a_{ji}(z)$ and $b_i(z)$ denote stiffness and reduced mass terms, respectively.

9.2-3b Nonshearable Rotating Pretwisted Beams

Extracting from Eqs. (9.2-8c) and (9.2-8d), the expressions $a_{44}(u_0' + \theta_y) + a_{45}(v_0' + \theta_x)$ and $a_{55}(v_0' + \theta_x) + a_{54}(u_0' + \theta_y)$, respectively, and their corresponding replacement in Eqs. (9.2-8a), and (9.2-8b), followed by consideration of $\theta_x \rightarrow -v_0'$ and $\theta_y = -u_0'$, yields the Bernoulli-Euler counterpart of the shearable beam model.

As a result, the governing equations read as:

$$\delta u_0 : [a_{22}(z)u_0'' + a_{23}(z)v_0'']'' - [(b_5(z) + \delta_n b_{15}(z))(\ddot{u}_0' - \Omega^2 u_0')$$

$$+ (b_6(z) - \delta_n b_{13}(z))(\ddot{v}_0' - \Omega^2 v_0')]' + b_1 \ddot{u}_0$$

$$- b_1 \Omega^2 u_0 - b_1 \Omega^2 [R(z)u_0']' = 0, \qquad (9.2\text{-}12\text{a,b})$$

$$\delta v_0 : [a_{33}(z)v_0'' + a_{32}(z)u_0'']'' - [(b_4(z) + \delta_n b_{14}(z))(\ddot{v}_0' - \Omega^2 v_0')$$

$$+ (b_6(z) - \delta_n b_{13}(z))(\ddot{u}_0' - \Omega^2 u_0')]'$$

$$+ b_1 \ddot{v}_0 - b_1 \Omega^2 [R(z)v_0']' = 0.$$

The associated boundary conditions are:

At $z = 0$:

$$u_0 = v_0 = u_0' = v_0' = 0, \qquad (9.2\text{-}13\text{a-d})$$

and at $z = L$:

$$\delta u_0 : [a_{22}(L)u_0'' + a_{23}(L)v_0'']' - (b_5(L) + \delta_n b_{15}(L))(\ddot{u}_0' - \Omega^2 u_0')$$

$$- (b_6(L) - \delta_n b_{13}(L))(\ddot{v}_0' - \Omega^2 v_0') = 0,$$

$$\delta v_0 : [a_{33}(L)v_0'' + a_{32}(L)u_0'']' - (b_4(L) + \delta_n b_{14}(L))(\ddot{v}_0' - \Omega^2 v_0')$$

$$- (b_6(L) - \delta_n b_{13}(L))(\ddot{u}_0' - \Omega^2 u_0') = 0. \qquad (9.2\text{-}14\text{a-d})$$

$$\delta u_0' : a_{22}(L)u_0'' + a_{23}(L)v_0'' = 0,$$
$$\delta v_0' : a_{33}(L)v_0'' + a_{32}(L)u_0'' = 0.$$

The expressions of stiffnesses $a_{ij}(z)$ and reduced mass terms are provided in Table 9.2-1 and 9.2-2, respectively.

The quantities affected by superscript p, indicate their affiliation to the beam cross-sections referred to the principal axes (x^p, y^p).

Table 9.2-1: Expressions of stiffness quantities $(a_{ij}(z) = a_{ji}(z))$ in terms of their cross-sectional principal axes x^p and y^p counterparts, $a_{ij}^p (= a_{ji}^p)$, and pretwist angle

$a_{ij}(z)$	Their expression
a_{22}	$a_{22}^p \cos^2 \beta + a_{33}^p \sin^2 \beta - 2a_{23}^p \sin \beta \cos \beta$
a_{33}	$a_{33}^p \cos^2 \beta + a_{22}^p \sin^2 \beta + 2a_{23}^p \sin \beta \cos \beta$
a_{44}	$a_{44}^p \cos^2 \beta + a_{55}^p \sin^2 \beta - 2a_{45}^p \sin \beta \cos \beta$
a_{55}	$a_{55}^p \cos^2 \beta + a_{44}^p \sin^2 \beta + 2a_{54}^p \sin \beta \cos \beta$
a_{23}	$(a_{22}^p - a_{33}^p) \sin \beta \cos \beta$
$a_{24} = -a_{35}$	$-(a_{25}^p + a_{34}^p) \sin \beta \cos \beta$
a_{25}	$a_{25}^p \cos^2 \beta - a_{34}^p \sin^2 \beta + (a_{24}^p - a_{35}^p) \cos \beta \sin \beta$
a_{34}	$a_{34}^p \cos^2 \beta - a_{25}^p \sin^2 \beta + (a_{24}^p - a_{35}^p) \cos \beta \sin \beta$
a_{45}	$(a_{44}^p - a_{55}^p) \cos \beta \sin \beta$

Table 9.2-2: The reduced mass terms in terms of their counterparts in the cross-sectional principal axes

b_i	Their expression
$b_1(z)$	b_1^p
$b_4(z)$	$b_4^p \cos^2 \beta + b_5^p \sin^2 \beta + 2b_6^p \sin \beta \cos \beta$
$b_5(z)$	$b_5^p \cos^2 \beta + b_4^p \sin^2 \beta - 2b_6^p \sin \beta \cos \beta$
$b_6(z)$	$(b_5^p - b_4^p) \cos \beta \sin \beta + b_6^p (\cos^2 \beta - \sin^2 \beta)$
$b_{13}(z)$	$(b_{14}^p - b_{15}^p) \cos \beta \sin \beta$
$b_{14}(z)$	$b_{14}^p \cos^2 \beta + b_{15}^p \sin^2 \beta - 2b_{13}^p \sin \beta \cos \beta$
$b_{15}(z)$	$b_{15}^p \cos^2 \beta + b_{14}^p \sin^2 \beta + 2b_{13}^p \sin \beta \cos \beta$

9.2-4 Several Comments on the Governing System

In connection with the governing equations of pretwisted rotating beams several comments are in order.

First, in the context of the implemented ply-angle configuration, the governing system, Eqs. (9.2-8) through (9.2-10), involve the flapping-lagging-transverse shear elastic couplings.

However, in addition to this, the pretwist yields a supplementary coupling between the flapping-lagging-transverse shear motions. In this sense, from Tables 9.2-1 and 9.2-2 it is readily seen that for the nonpretwisted beams, the stiffness quantities $a_{23}(z)$, $a_{24}(z)$, $a_{35}(z)$ and $a_{45}(z)$, and the mass terms $b_6(z)$ and $b_{13}(z)$ are zero and, as a result, a number of couplings, within the same categories of motion, become immaterial.

Moreover, for $\beta \to 0$, the stiffness and mass terms are independent of the spanwise coordinate. It is also evident that for doubly-symmetric beams of a rectangular cross-section, when also the rotatory inertia effect is discarded, the stiffnesses $a_{23}(z)$, $a_{24}(z)$, $a_{35}(z)$ and $a_{45}(z)$ as well as the mass terms $b_6(z)$ and $b_{13}(z)$ are zero for any value of the pretwist angle.

With a few exceptions, similar comments are applicable to the classical beam counterpart, as well. In this sense, it is clearly seen that the coupling stiffness quantities induced by pretwist, $a_{23}(z)(= a_{32}(z))$ and mass terms $b_6(z)$ and $b_{13}(z)$ still subsist. However, for doubly symmetric beam cross-sections, implying $a_{22}^p = a_{33}^p$, $b_4^p = b_5^p$ and $b_{14}^p = b_{15}^p$, and in the absence of rotatory inertia that yields $b_6^p = 0$, the two classical bending equations and the associated boundary conditions decouple, for any presetting and pretwist, into two independent groups governing the out of plane and the in-plane bendings.

There is a difference between the boundary conditions at the beam tip, valid for the shearable and unshearable beam. In the latter case, the rotatory inertias and the centrifugal inertia terms appear in the boundary conditions at $z = L$; in the former case they do not.

In connection with the order of the governing system: for both the shearable and unshearable pretwisted rotating beams, the governing equation system is of the eighth order.

The final comment is related to the question of validity of Eqs. (9.1-1), obtained as per the helical multifilament concept of pretwisted rotating blades featuring the bending-bending-transverse shear elastic coupling, concept attributed to Wagner. The answer to this question can be reached by comparing the expression of strains obtained via both Wagner and Washizu approaches. For the latter one we express the 1-D strain measures in the curvilinear coordinate system (η, ζ, z), followed by their transformation to the local orthogonal system (s, n, z).

The related developments, Oh (2004) and Librescu et al. (2005), are presented next. Within the curvilinear coordinates (η, ζ, z), the position vector of a point of the undeformed beam can be defined as

$$\mathbf{r} = x\mathbf{i} + y\mathbf{j} + z\mathbf{k}. \tag{C.9-1}$$

Using (9.2-1a,b) in (C.9-1) one can determine the covariant base vectors

$$\mathbf{g}_1 = \partial\mathbf{r}/\partial\eta = \cos\beta(z)\mathbf{i} + \sin\beta(z)\mathbf{j},$$

$$\mathbf{g}_2 = \partial\mathbf{r}/\partial\zeta = -\sin\beta(z)\mathbf{i} + \cos\beta(z)\mathbf{j}, \tag{C.9-2}$$

$$\mathbf{g}_3 = \partial\mathbf{r}/\partial z = -\beta'(z)(\eta\sin\beta + \zeta\cos\beta)\mathbf{i} + \beta'(z)(\eta\cos\beta - \zeta\sin\beta)\mathbf{j} + \mathbf{k},$$

where, for the sake of convenience, $x^P \Rightarrow \eta$, $y^P \Rightarrow \zeta$, $z^P \Rightarrow z$.

As a result, the components of the covariant symmetric metric tensor $g_{ij}(= \mathbf{g}_i \cdot \mathbf{g}_j)$ of the underformed beam are

$$g_{11} = g_{22} = 1; \; g_{12} = 0; \; g_{13} = -\beta'\zeta, \; g_{23} = \beta'\eta; \; g_{33} = 1 + (\beta')^2(\eta^2 + \zeta^2). \tag{C.9-3}$$

The components of the contravariant metric tensor g^{ij} that are obtained from $g^{ij} g_{im} = \delta_m^j$, where δ_m^j is the Kronecker symbol, are given by

$$g^{11} = 1 + (\beta')^2 \zeta^2; \quad g^{22} = 1 + (\beta')^2 \eta^2;$$
$$g^{12} = -(\beta')^2 \eta \zeta; \quad g^{13} = \beta' \zeta; \quad g^{23} = -\beta' \eta, \quad g^{33} = 1. \tag{C.9-4}$$

Using Eqs. (9.1-3) through (9.1-5) it can readily be seen that the position vector of a point of the deformed beam is

$$\mathbf{R} = [u_0 + \eta \cos \beta - \zeta \sin \beta]\mathbf{i}$$
$$+ [v_0 + \eta \sin \beta + \zeta \cos \beta]\mathbf{j} + \left[z + R_0 + \theta_x \left(y - n\frac{dx}{ds}\right) + \theta_y \left(x + n\frac{dy}{ds}\right)\right]\mathbf{k}. \tag{C.9-5}$$

The covariant base vectors of the deformed beam are

$$G_1 = \partial \mathbf{R}/\partial \eta = \mathbf{i} \cos \beta + \mathbf{j} \sin \beta + [\theta_x \sin \beta + \theta_y \cos \beta]\mathbf{k},$$
$$G_2 = \partial \mathbf{R}/\partial \zeta = -\mathbf{i} \sin \beta + \mathbf{j} \cos \beta + [\theta_x \cos \beta - \theta_y \sin \beta]\mathbf{k}, \tag{C.9-6}$$
$$G_3 = \partial \mathbf{R}/\partial z = (u_0' - \beta'[\eta \sin \beta + \zeta \cos \beta])\mathbf{i} + (v_0' + \beta'[\eta \cos \beta - \zeta \sin \beta])\mathbf{j}$$
$$+ \left[1 + \theta_x' \left(y - n\frac{dx}{ds}\right) + \theta_y' \left(x + n\frac{dy}{ds}\right) + \beta'\theta_x \left(x + n\frac{dy}{ds}\right) - \beta'\theta_y \left(y - n\frac{dx}{ds}\right)\right]\mathbf{k}.$$

Based on Eqs. (C.9-6) one obtain the covariant components of the metric tensor of the deformed space, $G_{ij} = \mathbf{G}_i \cdot \mathbf{G}_j$. Discarding the higher order terms, these are:

$$G_{11} = G_{22} = 1; \quad G_{12} = 0; \quad G_{13} = -\beta'\zeta + u_0' \cos \beta + v_0' \sin \beta + \theta_x \sin \beta + \theta_y \cos \beta;$$
$$G_{23} = \beta'\eta - u_0' \sin \beta + v_0' \cos \beta + \theta_x \cos \beta - \theta_y \sin \beta; \quad G_{33} = 1 + (\beta')^2(\eta^2 + \zeta^2)$$
$$- 2\beta'[yu_0' - xv_0'] + 2\left[\theta_x' \left(y - n\frac{dx}{ds}\right) + \theta_y' \left(x + n\frac{dy}{ds}\right) + \beta'\theta_x \left(x + n\frac{dy}{ds}\right)\right.$$
$$\left. - \beta'\theta_y \left(y - n\frac{dx}{ds}\right)\right] + (u_0')^2 + (v_0')^2. \tag{C.9-7}$$

Based on (C.9-3) and (C.9-7) one can determine the strains $f_{ij} = (G_{ij} - g_{ij})/2$ in the curvilinear coordinates $(\alpha^1, \alpha^2, \alpha^3) \equiv (\eta, \zeta, z)$

$$f_{11} = f_{12} = f_{22} = 0; \quad 2f_{13} = u_0' \cos \beta + v_0' \sin \beta + (\theta_x \sin \beta + \theta_y \cos \beta),$$
$$2f_{23} = -u_0' \sin \beta + v_0' \cos \beta + \theta_x \cos \beta - \theta_y \sin \beta,$$
$$f_{33} = -\beta'[yu_0' - xv_0'] + \frac{1}{2}\left((u_0')^2 + (u_0')^2\right) + \theta_x' \left(y - n\frac{dx}{ds}\right) \tag{C.9-8}$$
$$+ \theta_y' \left(x + n\frac{dy}{ds}\right) + \beta'\theta_x \left(x + n\frac{dy}{ds}\right) - \beta'\theta_y \left(y - n\frac{dx}{ds}\right).$$

In order to express the strain tensor f_{ij} in terms of e_{ij} defined with respect to the local Cartesian coordinate system (s, n, z), one should present a few preliminary results. Defining the position vector $\mathbf{r}_1$ of a point on the underformed mid-contour as (see Fig. C.9-1)

$$\mathbf{r}_1 = \eta(s)\mathbf{g}_1 + \zeta(s)\mathbf{g}_2, \tag{C.9-9}$$

the unit vectors in the direction of coordinates $(y^1, y^2, y^3) \equiv (s, n, z)$, see Fig. C.9-1, are as follows:

$$\mathbf{j}_1 = \frac{d\mathbf{r}_1}{ds} = \frac{d\eta}{ds}\mathbf{g}_1 + \frac{d\zeta}{ds}\mathbf{g}_2, \; \mathbf{j}_2 = \mathbf{j}_1 \times \mathbf{k} = \frac{d\zeta}{ds}\mathbf{g}_1 - \frac{d\eta}{ds}\mathbf{g}_2, \; \text{and} \; \mathbf{j}_3 = \mathbf{k}. \qquad \text{(C.9-10)}$$

Applying the transformation law between e_{ij} and f_{kl}

$$e_{ij} = \frac{\partial \alpha^k}{\partial y^i} \frac{\partial \alpha^l}{\partial y^j} f_{kl}, \qquad \text{(C.9-11)}$$

where

$$\frac{\partial \alpha^k}{\partial y^i} = g^{kl} (\mathbf{j}_i \cdot \mathbf{g}_l). \qquad \text{(C.9-12)}$$

In expanded form the various components of (C.9-12) are expressible as

$$\begin{aligned}
\frac{\partial \alpha^1}{\partial y^1} &= \frac{d\eta}{ds}; & \frac{\partial \alpha^1}{\partial y^2} &= \frac{d\zeta}{ds}; & \frac{\partial \alpha^1}{\partial y^3} &= \beta' \zeta, \\[2mm]
\frac{\partial \alpha^2}{\partial y^1} &= \frac{d\zeta}{ds}; & \frac{\partial \alpha^2}{\partial y^2} &= -\frac{d\eta}{ds}; & \frac{\partial \alpha^2}{\partial y^3} &= -\beta' \eta, \\[2mm]
\frac{\partial \alpha^3}{\partial y^1} &= 0; & \frac{\partial \alpha^3}{\partial y^2} &= 0; & \frac{\partial \alpha^3}{\partial y^3} &= 1,
\end{aligned} \qquad \text{(C.9-13)}$$

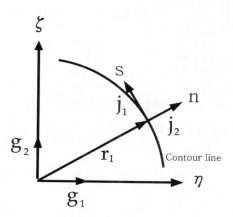

Figure C.9-1: Beam contour-line indicating the coordinates (η, ζ) and (s, n)

and in conjunction with (C.9-8), (C.9-11) and (C.9-13) one can obtain the strain components e_{ij}. These are:

$$\begin{aligned}
e_{33}^{(0)} &\equiv \varepsilon_{zz}^{(0)} = \frac{1}{2}((u'_0)^2 + (v'_0)^2) + y\theta'_x + x\theta'_y, \\[2mm]
e_{33}^{(1)} &\equiv \varepsilon_{zz}^{(1)} = \theta'_y \frac{dy}{ds} - \theta'_x \frac{dx}{ds} + \beta' \theta_x \frac{dy}{ds} - \beta' \theta_y \frac{dx}{ds}, \\[2mm]
e_{13} &\equiv \varepsilon_{sz} = (\theta_y + u'_0)\frac{dx}{ds} + (\theta_x + v'_0)\frac{dy}{ds}, \\[2mm]
e_{23} &\equiv \varepsilon_{nz} = (\theta_y + u'_0)\frac{dy}{ds} - (\theta_x + v'_0)\frac{dx}{ds}.
\end{aligned} \qquad \text{(C.9-14)}$$

Equations (C.9-14) coincide with Eqs. (9.1-1) when the latter ones are specialized for the coupling involving bending-bending-transverse shear. This substantiates the adoption of the helical multifilament approach and the fact that for the present case, the strains defined with respect to the local Cartesian coordinates (s, n, z) are not affected by the pretwist.

9.2-5 Numerical Results and Discussion

Before presenting results revealing the effects of anisotropy coupled with that of the pretwist on dynamics of thin-walled rotating blades, a number of numerical predictions intended to be compared with the ones obtained within the solid beam model of rectangular cross-sections are displayed.

Tables 9.2-3 (see Oh et al., 2003) present the effects of pretwist on the first five natural frequencies of a nonrotating unshearable solid cross-section beam. The beam (see Rosen et al., 1987) is characterized by

$$a_{22} = 2,869.7N.\text{m}^2; a_{33} = 57,393N.\text{m}^2; b_1 = 34.47\text{Kg/m}, \text{ and } L = 3.048\text{m}.$$
(9.2-15)

Table 9.2-3: Effects of the pretwist angle on natural frequencies ω_i (rad/s) of a nonrotating unshearable blade ($R_0 = 0$)

| ω_i | Untwisted blade ($\beta_0 = 0$) | | |
| | Rosen et al. (1987) | | |
	(Exact)	Rosen et al. (1987)	Oh et al. (2003)
$i = 1$	3.45305	3.44898	3.45317
$i = 2$	15.4425	15.4184	15.4429
$i = 3$	21.6399	21.5542	21.6404
$i = 4$	60.5924	60.984	60.6120
$i = 5$	96.7764	96.3913	96.77911
ω_i	Pretwisted blade ($\beta_0 = 40$ deg.)		
	Rosen et al. (1987)	Banerjee (2001)	Oh et al. (2003)
$i = 1$	3.47257	3.47173	3.47182
$i = 2$	13.2740	13.3465	13.3413
$i = 3$	25.2700	25.1707	25.1677
$i = 4$	56.3009	56.3716	56.3485
$i = 5$	103.200	103.263	103.4122

The results obtained within the present model and solution methodology (based on Extended Galerkin Method – see e.g. Librescu et al., 1997) displayed in Tables 9.2-3, reveal excellent agreement with the available exact and approximate predictions by Rosen et al. (1987) and Banerjee (2001).

Moreover, these reveal that the even mode eigenfrequencies decay with the increase of the pretwist, whereas the odd mode eigenfrequencies exhibit an opposite trend. This feature was reported earlier by Rao (1971).

Table 9.2-4 compares the frequency predictions of a pretwisted and unrotating solid beam with the experimental and the theoretical ones. The characteristics of the beam as considered by Subrahmanyam et al. (1981), are

$a_{22}^p = 487.9\text{N.m}^2$, $a_{33}^p = 2.26$ N.m^2, $a_{44}^p = a_{55}^p = 3.076 \times 10^6\text{N.m}^2$, $b_1^p = 0.3447\text{kg/m}$, $b_4^p = 8.57 \times 10^{-8}\text{kg.m}$, $b_5^p = 0.19 \times 10^{-4}\text{kg.m}$, $b_{14}^p = b_{15}^p = 0$, $L = 15.24 \times 10^{-2}\text{m}$. The present displayed predictions are in good agreement with the theoretical and experimental findings.

Table 9.2-4: Comparison of coupled flapping-lagging frequencies ω_i (Hz) of a pretwisted beam ($\beta_0 = 45$ deg., $\Omega = 0$, $R_0 = 0$)

	Subrahmanyam et al. (1981)		Dawson (1968)	Rao (1977)	Carnegie (1959)	Oh et al. (2003)
	Reissner method	Potential Energy process	Rayleigh-Ritz	Galerkin method	Experiments	Extended Galerkin
ω_i						
$i = 1$	61.9	62.0	62.0	61.9	59.0	62.0
$i = 2$	304.7	305.1	301.0	305.0	290.0	305.1
$i = 3$	937.0	955.1	953.0	949.0	920.0	949.0
$i = 4$	1205.1	1214.7	1230.0	1220.0	1110.0	1206.1

Although the equations are valid for a beam of arbitrary closed-cross section, for the sake of illustration a rotating beam modelled as a composite box-beam (see Fig. 9.2-1), characterized by a cross-section ratio $\mathcal{R}(\equiv b/c) = 1.5$ was considered. Unless otherwise specified, the dimensions of the beam and its material properties are listed in Tables 6.4-1 and 6.4-2, respectively.

Figure 9.2-2 depicts the first three coupled eigenfrequencies of the non-rotating beam as a function of the pretwist angle at the beam tip. The trend of variation of ω_i vs. $\beta(L)(\equiv \beta_0)$ as illustrated in this figure, is similar to that featured by a solid beam of similar $\mathcal{R}$, as reported by Subrahmanyam et al. (1981), Slyper (1962) and Carnegie and Thomas (1972). Moreover, as the results reveal, within the first coupled frequency, the increase of the ply-angle yields, with the increase of Ω, increased natural frequencies.

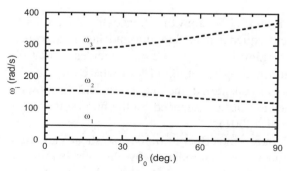

Figure 9.2-2: Variation of first three coupled flapping-lagging natural frequencies vs. pretwist angle, $\theta = 0$ and $\Omega = 0$

However, this trend is reversed for higher frequencies, in the sense that the increase of θ yields larger frequencies at smaller rotational speeds, while with the increase of Ω, the variation of ω_i due to the increase of θ becomes marginal.

Figures 9.2-3a through 9.2-3c display, for selected values of the blade rotational speed, the variations of first three eigenfrequencies as a function of the pretwist angle, for ply-angles $\theta = 0$ and 45 deg. The trend emerging from these plots coincides with that reported by Subrahmanyam and Kaza (1986), for a rotating blade modeled as a solid isotropic beam.

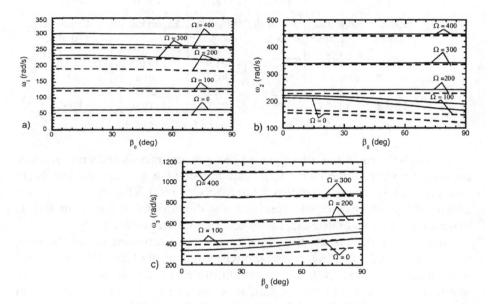

Figure 9.2-3: Variation of first three coupled flapping-lagging natural frequencies vs. pretwist angle for selected rotational speeds Ω (rad/s), and for $\theta = 0$ (- - - - -) and 45 deg.(———), a) First natural frequency; b) Second natural frequency; c) Third natural frequency

Figures 9.2-4a and 9.2-4b show the effect of the hub radius R_0 on the first three decoupled eigenfrequencies in flapping and lagging, respectively. Whereas the trend as emerging from Fig. 9.2-4a coincides to that obtained by Boyce (1956) and confirmed by Lo et al. (1960) where only the flapping motion of rotating blades was considered, the one in Fig. 9.2-4b showing that the lagging eigenfrequencies are little influenced by the hub radius was put into evidence first by Song et al. (2001).

Figure 9.2-5 shows the effect of transverse shear yielding a decay of higher mode eigenfrequencies as predicted by the unshearable nonrotating beam model. The same trend related with the influence of transverse shear was reported for rotating solid beams by Carnegie and Thomas (1972).

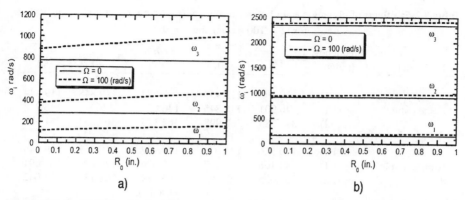

a)   b)

Figure 9.2-4: a) Effect of hub radius on the first three decoupled flapping natural frequencies for both rotating and nonrotating blades, $\theta = 0$. b) Counterpart of Fig. 9.2-4a for the case of lagging natural frequencies

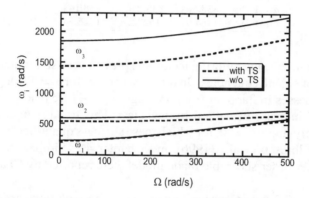

Ω (rad/s)

Figure 9.2-5: Effect of transverse shear on the first three coupled natural frequencies; $\theta = 90$ deg. and $\beta_0 = 90$ deg.

Table 9.2-5 highlights the effects of the chord-to-thickness ratio c/h on the first three natural frequencies ω_i (rad/s), within the shearable and unshearable box-beam models, and for the ply-angle configurations, $\theta = 0$ and 90 deg. (see Song et al., 2001, 2002). Here ω_1 (F1) and ω_3 (F2) are the first consecutive frequencies in flapping, whereas ω_2 (L1) is the first frequency in lagging.

The results show that the decrease of the chord-to-thickness ratio c/h from 50 to 10 (implying, when c is hold fixed, a five times increase of the wall thickness), plays a very modest role in the increase of eigenfrequencies. This trend is due to the inherent large bending stiffnesses featured by the box-beam. As a result, the increase of the wall thickness yields only a slight increase of the global bending stiffness quantities.

Table 9.2-5: Effects of ply-angle and of chord-to-thickness ratio on the natural frequencies ω_i for $\Omega = 100$ (rad/s) and $\beta = 0$

Model	Frequency (rad/s)	10	20	30	40	50
				$\theta = 0$		
	ω_1 (F1)	120.393	119.957	119.874	119.848	119.844
Shearable	ω_2 (L1)	146.901	146.845	146.834	146.83	146.828
	ω_3 (F2)	367.583	362.998	362.125	361.834	361.683
	ω_1 (F1)	120.516	120.074	119.992	119.963	119.949
Nonshearable	ω_2 (L1)	148.166	148.107	148.096	148.093	148.091
	ω_3 (F2)	369.025	364.36	363.489	363.183	363.042
				$\theta = 90$ deg.		
	ω_1 (F1)	265.037	259.471	258.423	258.055	257.855
Shearable	ω_2 (L1)	667.59	667.396	667.359	667.347	667.341
	ω_3 (F2)	1312.12	1287.78	1283.13	1281.5	1280.74
	ω_1 (F1)	272.406	266.265	265.112	264.708	264.52
Nonshearable	ω_2 (L1)	838.604	838.216	838.144	838.119	838.107
	ω_3 (F2)	1564.91	1523.2	1515.34	1512.58	1511.3

In contrast to this, the tailoring technique plays a very significant role toward the increase of eigenfrequencies. In the present case, for $\theta = 90$ deg., maximum flexural stiffnesses in flapping (a_{33}) and lagging (a_{22}) are reached this yielding a remarkable increase of natural frequencies. This trend is in contrast to that occurring at ply-angle $\theta = 0$, for which these stiffness quantities have minimum values (see Librescu et al. (1996) and Chapter 4). Similar trends obtained within a higher-order shearable blade model were reported by Chandiramani et al. (2002).

Finally, Fig. 9.2-6 displays the convergence of the solution carried out via the Extended Galerkin Method, as a function of the considered number of terms. Here, the variation of the coupled fundamental eigenfrequency vs. the rotational speed was considered.

The results of this graph show, on the one hand, the extremely fast convergence of the numerical methodology and, on the other hand, the fact that five terms, yield results in excellent agreement with those obtained by considering more terms.

It should be added that the higher-order shear deformable model presented in Chapters 2 and 3, was applied to the study of free vibration and dynamic response to time-dependent pressure pulses of rotating blades by Chandiramani et al. (2002) and Chandiramani and Librescu (2002).

As a mere remark, due to its evident importance, the dynamic response of blades of helicopter gunships to blast loads was considered a mandatory topic

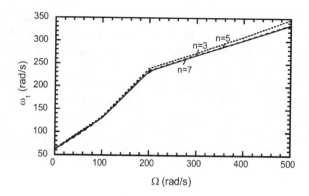

Figure 9.2-6: Convergence of the solution carried out as per the Extended Galerkin Method for the coupled flap-lag fundamental eigenfrequency as a function of the rotational speed ($\theta = 45$ deg.)

of study from the early phases of their design (Coyle and Harris, 1960). There is no doubt that this study continues to be an important topic of research.

9.2-6 Adaptive Rotating Thin-Walled Beams

9.2-6a Basic Equations

In Section 6.6, we presented the concept of adaptive/smart thin-walled beams and described its implementation toward their free vibration feedback control.

In this chapter we will apply this concept to rotating thin-walled beams, and show its efficiency in the context of the free vibration control. A comprehensive review of the state-of-the-art in the area of smart structures technology applied to rotorcraft systems is due to Chopra (2000). In the same context, the works by Büter and Breitbach (2000) and Cesnik and Shin (2001 a,b) deserve to be mentioned.

Here, Case (b) considered in Chapter 6.6-2 related to piezoactuators distribution will be adopted; the piezoactuators are spread over the entire beam span and are bonded/embedded into the host structure as indicated in Fig. 9.2-1.

Based on the converse piezoelectric effect and of the out-of-phase activation, boundary control moments $\tilde{M}_x$ and $\tilde{M}_y$ are piezoelectrically induced at the beam tip. The host structure is considered in the form of a cantilevered single-cell closed cross-section beam featuring the Circumferential Uniform Stiffness Configuration (CUS). As a result, with the exception of the boundary conditions at $z = L$, Eqs. (9.2-10c,d) are modified as:

$$\delta\theta_y : a_{22}(z)\theta'_y + a_{24}(z)(u'_0 + \theta_y) + a_{25}(z)(v'_0 + \theta_x) + a_{23}(z)\theta'_x = \tilde{M}_y,$$

$$\delta\theta_x : a_{33}(z)\theta'_x + a_{34}(z)(u'_0 + \theta_y)a_{35}(z)(v'_0 + \theta_x) + a_{32}(z)\theta'_y = \tilde{M}_x,$$

$$(9.2\text{-}16a,b)$$

while Eqs. (9.2-8), (9.2-9) and (9.2-10) remain unaffected. Note that the stiffness quantities a_{ij} and inertia terms b_i have to be determined in their entirety, in the sense of including in their expression the contributions of both the host structure and the piezoactuators.

In Eqs. (9.2-16), $\tilde{M}_x$ and $\tilde{M}_y$ are the piezoelectrically induced-bending moments expressed by Eqs. (6.6-1). As in Chapter 6.6-4, two feedback control laws are implemented: the proportional and the velocity feedback ones expressed by Eqs. (6.6-19) and (6.6-20), respectively, and the same discretization procedure of the closed-loop eigenvalue problem as in Chapter 6.6-5 is carried out.

9.2-6b Numerical Simulations

Although the equations are valid for a beam of arbitrary closed-cross section, for the sake of illustration we consider a rotating beam modelled as a composite box-beam (see Fig. 9.2-1), characterized by a cross-sectional ratio $\mathcal{R}(\equiv b/c) = 5$. In addition, unless otherwise specified, the dimensions considered are $L = 80$ in. (2.023m); $c = 10$in. (0.254 m), $b = 2$ in. (50.8×10^{-3} m); $h = 0.4$in. (10.16 $\times 10^{-3}$ m), and $h_p = 0.015$ in. (0.381×10^{-3} m).

Depending on whether the flapping, lagging or flapping-lagging coupled motions are intended to be controlled, the piezoactuator layers (selected to be of PZT-4) should be located on the top and bottom flanges, on the opposite lateral webs of the master structure, or on both the flanges and webs, respectively.

The geometric characteristics of the structure are shown in Fig. 9.2-1; the properties of the PZT-4 piezoceramic have been provided in Table 6.6-1, while the mechanical characteristics in the on-axis configuration of the host structure corresponding to a graphite/epoxy material, are supplied in Table 6.4-2.

Note that, in spite of the extensive work and results devoted to eigenvibration problem of pretwisted rotating and nonrotating blades modeled as solid isotropic beams and incorporating bending-bending cross coupling, (see e.g. Slyper, 1962; Carnegie and Thomas, 1972; Sabuncu, 1985; Lin, 1997; and Balhaddad and Onipede, 1998), to the best of the authors' knowledge, there are no available results associated with both adaptive and non-adaptive anisotropic pretwisted rotating *T WBs*.

Figures 9.2-7 and 9.2-8 highlight the effects of rotational velocity Ω considered in conjunction with the implementation of the velocity feedback control strategy on the decoupled fundamental natural frequencies in flapping and lagging that occur at ply-angle $\theta = 0$ of the host structure. The results reveal that, at moderate rotational speeds, the first flapwise mode exhibits the lowest frequency. However, due to the centrifugal stiffening effect (which is much more significant in the flapping modes), beyond a certain angular speed the lagging frequency becomes the lower of the two. Figure 9.2-7 shows that, while for the unactivated beam a frequency crossing of lagging and flapping modes occurs in the vicinity of $\Omega \cong 150$ rad/s, for the activated beam the crossing is shifted toward larger rotational speeds.

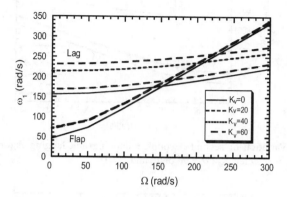

Figure 9.2-7: Variation of first decoupled flapping and lagging natural frequencies vs. Ω for selected K_v; $\theta = 0$ and $\beta = 0$

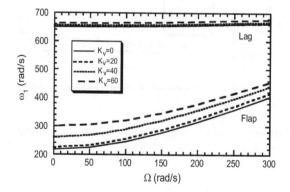

Figure 9.2-8: Variation of the first decoupled flapping and lagging natural frequencies vs. rotational speed for selected K_v; $\theta = 90$ deg. and $\beta = 0$

In Figs. 9.2-7 through 9.2-16 the following nondimensional control gains are used; $K_p \equiv k_p L/(\hat{a}_{33}h_p)$, and $K_v \equiv k_v L/(\hat{a}_{33}h_p)$, where $\hat{a}_{33} \equiv a_{33}|_{\theta=0}$.

Figure 9.2-9 obtained for $\theta = 90$ deg. shows, for the second decoupled flapping and lagging natural frequencies, that the frequency crossings occur at larger values of Ω as compared to their counterparts for $\theta = 0$. This trend is valid for both unactivated and activated rotor blades. For higher modes, see Figs. 9.2-9 and 9.2-10, the frequencies remain separated in the usual range of rotational speeds, in the sense that no frequency crossing occur in that range. Recall that, via velocity feedback control, piezoelectrically induced damping is generated.

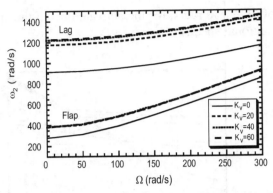

Figure 9.2-9: Variation of second decoupled flapping and lagging natural frequencies vs. rotational speed for selected K_v; $\theta = 90$ deg., and $\beta = 0$

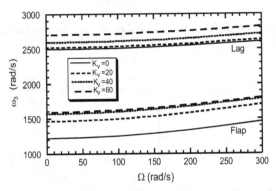

Figure 9.2-10: Variation of third decoupled flapping and lagging natural frequencies vs. rotational speed for selected K_v; $\theta = 90$ deg. and $\beta = 0$

Figures 9.2-11 and 9.2-12 display the induced damping factors associated with the decoupled motions in flapping and lagging, respectively. It is seen that the increase of ζ_r with K_v takes place until a certain values of the feedback gain, specific to each mode number. Beyond that specific values of K_v, referred to as the optimum feedback gain, a sharp drop of ζ_r is experienced.

This trend is consistent with that reported in different contexts by Tzou (1993), Song and Librescu (1999) and Librescu and Na (2001). The comparison of results supplied in Figs. 9.2-11 and 9.2-12 reveals that, in order to get almost the same damping factor, larger feedback gains in lagging than in flapping are needed.

As concerns the proportional feedback control, the obtained results reveal that this control method does not provide any substantial advantage compared to that supplied by velocity feedback control.

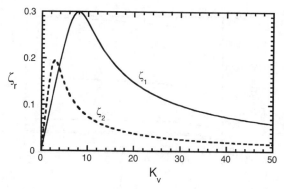

Figure 9.2-11: Piezoelectrically induced damping factors in flapping vs. K_v; $\theta = 0$, $\beta = 0$ and $\Omega = 0$

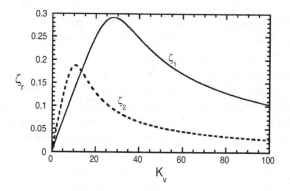

Figure 9.2-12: Counterpart of Fig. 9.2-11 for the damping factor in lagging

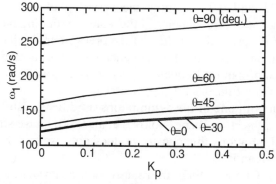

Figure 9.2-13: Variation of the first coupled flapping-lagging natural frequency vs. K_p for selected ply-angles, θ (deg.): $\beta_0 = 90$ deg. and $\Omega = 100$ rad/s

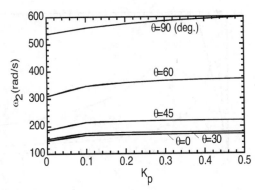

Figure 9.2-14: Variation of the second coupled flapping-lagging natural frequency vs. K_p for selected ply-angles, θ (deg.): $\beta_0 = 90$ deg. and $\Omega = 100$ rad/s

Figures 9.2-13 and 9.2-14 show the effects of the feedback gain K_p and of the ply-angle on the first two coupled frequencies of a pretwisted rotating blade ($\beta_0 = 90$ deg. and $\Omega = 100$ rad/s). The results reveal the relatively modest effect played by the piezoelectric actuation.

Further results on bending vibration feedback control of rotating tapered beams incorporating adaptive capabilities have been presented by Na and Librescu (2000) and more fully in Na et al. (2003). In that context, the concept of the piezoelectrically induced boundary moment control coupled with the implementation of a combined velocity/acceleration feedback control law was used.

The results obtained by Chandiramani et al. (2002, 2003b, 2004) on optimal vibration and dynamic response control of rotating beams modeled as per a higher-order shearable theory should also be indicated. Within these studies both the classical LQR and the instantaneous IOC controls have been implemented.

Figures 9.2-15 and 9.2-16, show the variation of the first two coupled flapping-lagging natural frequencies of the adaptive rotating blade as a function of the ply-angle for prescribed values of the pretwist angle. The significant effect of ply-angle on increasing the frequencies is worth noting. Moreover, the variation with β_0 of natural frequencies is consistent, as it should be, with the trend emerging from Fig. 9.2-2.

In connection with one of the assumptions used here postulating the invariability of the cross-section shape, Volovoi et al. (2001) have shown that for the CUS beam configurations this constraint does not result in overpredictions of related stiffness quantities.

In Chandra and Chopra (1992), the rotating box-beam was modeled to include the bending-torsion elastic coupling via the implementation of the *circumferentially asymmetric stiffness (CAS) configuration*.

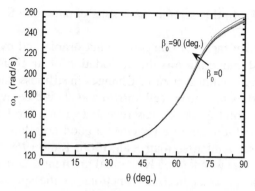

Figure 9.2-15: Variation of the first coupled flapping-lagging natural frequency vs. ply-angle for selected values of the beam tip pretwist angle; $K_p = 0.1$ and $\Omega = 100$ rad/s

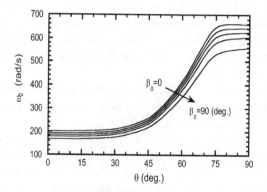

Figure 9.2-16: Counterpart of Fig. 9.2-15 for the second coupled natural frequency

9.3 PRETWISTED ROTATING THIN-WALLED BEAMS OPERATING IN A TEMPERATURE ENVIRONMENT

9.3-1 Preliminaries

For reasons of efficiency involving gas dynamics and weight, the blades of rotating turbomachinery must be thin, yet they must operate in severe thermal environments and at high rotational speeds.

In spite of the evident practical importance of this problem, there are almost no studies devoted to the vibrational analysis of rotating thin-walled beams featuring anisotropy and pretwist and subjected to a temperature field. The extensive survey-papers reviewing in depth the literature devoted to pretwisted (Rosen, 1991), and helicopter (Kunz, 1994) blades, clearly reveal the extreme scarcity of the research related to this topic. The few available results on these issues have been reported by Tomar and Jain (1985), Song et al. (2001b) and Na et al. (2003, 2004). In this section the results obtained by Song et al. (2001b) will be reported.

It is a well known fact that one of the undesirable and by far most serious effects of elevated temperature is the degradation of mechanical properties of constituent materials of the structure. Changes in stiffness due to such changes in material properties are fairly well known and should be evaluated.

As the experimental evidence has revealed, a linear relationship between Young's moduli and temperature provides a good correlation for most engineering materials (see Ambartsumian et al., 1966; Tang, 1969; Noor and Burton, 1992). In such a case, when a steady thermal gradient is considered, the elastic coefficients of the material become functions of the space variables. Consequently, although the considered structure is of uniform thickness, in such circumstances the structure will feature a material-induced variable stiffness (see e.g. Tang, 1969; Tomar and Jain, 1985; Fauconneau and Marangoni, 1970). The goal of this section is to investigate the effect of the *non-homogeneity* of the structure induced by a steady thermal gradient and the accompanied degradation of material properties, on natural frequencies and mode shapes of a pretwisted rotating beam.

The analysis is carried out in the framework of a thin-walled beam structural model of an arbitrary shape cross-section that incorporates non-classical effects similar to those considered in Section 9.2.

In addition, for the already mentioned reasons, the lay-up associated with the circumferentially uniform stiffness configuration (CUS) is implemented. It is also supposed that the rotating beam is exposed to a steady temperature field T featuring a spanwise distribution, thus implying that $T = T(z)$.

9.3-2 Basic Assumptions

The structural model considered here and the adopted systems of coordinates are similar to those involved in Section 9.2. Moreover, the assumptions formulated in Section 9.2-2 will be adopted. Consequently, the case of a straight pretwisted flexible beam of length L rotating with the constant angular velocity Ω normal to the plane of rotation is considered (see Fig. 9.2-1).

We will also assume that the beam is subjected to a steady temperature that is uniform in the s-direction, but featuring a linear distribution in the spanwise z-direction. Corresponding to such a distribution, two scenarios A) and B) (see Fig. 9.3-1) of the temperature variation are considered:

$$\text{Profile A}) :\ T(z) = T_0 z/L, \text{ and Profile B}) :\ T(z) = T_0(1 - z/L).$$
$$(9.3\text{-}1\text{a,b})$$

In Eqs. (9.3-1), $T(z)$ denotes the temperature above a stress-free reference temperature at any point of the beam in its spanwise direction, whereas T_0

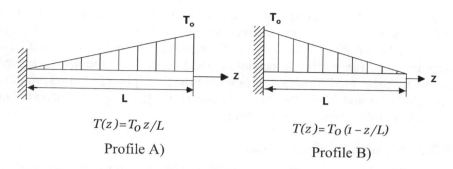

$$T(z)=T_0\, z/L$$

Profile A)

$$T(z)=T_0\,(1-z/L)$$

Profile B)

Figure 9.3-1: Two scenarios of the temperature variation

denotes the temperature excess above the reference temperature at the beam cross sections $z = L$ or $z = 0$, respectively. The temperature dependence of Young's moduli E_i will be considered as (see Ambartsumian et al., 1966; Noor and Burton, 1992),

$$E_i(z) = E_i^0\big(1 - \gamma T(z)\big), \tag{9.3-2}$$

where E_i^0 denote the Youngs' moduli at the reference temperature, while γ is the slope of the variation of E_i with T. It is also assumed (see Ambartsumian et al., 1966; Tang, 1969), that Poisson's ratios are temperature independent. By virtue of Eqs. (9.3-2), (9.2-1), considered in conjunction with those in Table 9.2-1, the expressions of stiffness quantities in terms of their cross-section principal axes (x^p, y^p) evaluated at $T = 0$, and consistent with the CUS ply-angle configuration are supplied in Table 9.3-1.

Table 9.3-1: Expression of stiffness quantities $(a_{ij}(z) = a_{ji}(z))$, as influenced by the temperature field and pretwist angle in terms of their cross-sectional principal axes counterparts $a_{ij}^p(= a_{ji}^p)$, corresponding to $T = 0$

$a_{ij}(z)$	Their expression
a_{22}	$(1 - \gamma T(z))[a_{22}^p \cos^2 \beta + a_{33}^p \sin^2 \beta - 2a_{23}^p \sin \beta \cos \beta]$
a_{33}	$(1 - \gamma T(z))[a_{33}^p \cos^2 \beta + a_{22}^p \sin^2 \beta + 2a_{23}^p \sin \beta \cos \beta]$
a_{44}	$(1 - \gamma T(z))[a_{44}^p \cos^2 \beta + a_{55}^p \sin^2 \beta - 2a_{45}^p \sin \beta \cos \beta]$
a_{55}	$(1 - \gamma T(z))[a_{55}^p \cos^2 \beta + a_{44}^p \sin^2 \beta + 2a_{54}^p \sin \beta \cos \beta]$
a_{23}	$(1 - \gamma T(z))[(a_{22}^p - a_{33}^p) \sin \beta \cos \beta]$
$a_{24} = -a_{35}$	$(1 - \gamma T(z))[-(a_{25}^p + a_{34}^p) \sin \beta \cos \beta]$
a_{25}	$(1 - \gamma T(z))[a_{25}^p \cos^2 \beta - a_{34}^p \sin^2 \beta + (a_{24}^p - a_{35}^p) \cos \beta \sin \beta]$
a_{34}	$(1 - \gamma T(z))[a_{34}^p \cos^2 \beta - a_{25}^p \sin^2 \beta + (a_{24}^p - a_{35}^p) \cos \beta \sin \beta]$
a_{45}	$(1 - \gamma T(z))[(a_{44}^p - a_{55}^p) \cos \beta \sin \beta]$

As is clearly seen, the stiffness quantities feature a spanwise variation induced by both the pretwist and the thermal degradation of elastic properties of the material of the structure.

As concerns the reduced mass coefficients $b_i(z)$, their spanwise variation is induced by only the pretwist effect, and as such, their expressions remain similar to those defined in Table 9.2-2. Excepting for the expression of stiffness quantities a_{ij}, the governing equations and boundary conditions as derived for purely pretwisted rotating beams within the CUS ply-angle configuration, Sections 9.2-3 and 9.2-4, are similar to those arising in this new context. For this reason, these are not repeated here.

9.3-3 Numerical Simulations

In order to put into evidence the effects played by the temperature degradation considered in conjunction with other parameters, such as the pretwist, ply-angle, etc., on free vibration characteristics of rotating blades, we consider the case of a box-beam featuring the geometrical characteristics used in Section 9.2-6. In addition, the graphite/epoxy material, will be considered as the constituent material of the beam. We use the Extended Galerkin Method, so that the study of the eigenvalue problem follows that already developed in previous sections. Figures 9.3-2 through 9.3-4, highlight for the thermal profile A), the effects of the thermal degradation expressed in terms of the parameter $\alpha \equiv \gamma T_0$, where $0 \leq \alpha \leq 1$, and of the rotational speed Ω, on the first three decoupled natural frequencies in flapping and lagging for a blade characterized by zero-pretwist and ply-angle $\theta = 0$.

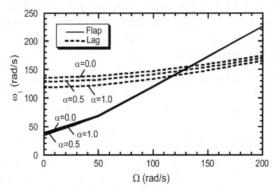

Figure 9.3-2: Variation of first decoupled natural frequencies vs. Ω for selected values of α; $\theta = 0$, $\beta = 0$, temperature profile A)

These plots show that with the increase of Ω, the effects of thermal degradation become less pronounced for the frequencies in flapping than in lagging. In other words, as is readily seen from Fig. 9.3-2, whereas the increase of α yields a decrease of the lagging frequency, the flapping one is only slightly affected by the thermal degradation. In addition, this graph shows that with the increase of the thermal degradation parameter α, the flapping-lagging frequency crossings occur at lower rotational speeds Ω.

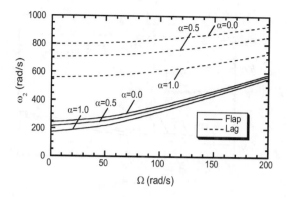

Figure 9.3-3: Counterpart of Fig. 9.3-2 for the second decoupled natural frequencies

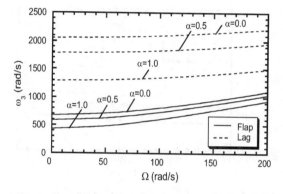

Figure 9.3-4: Counterpart of Fig. 9.3-2 for the third decoupled natural frequencies

For the second and third natural frequencies, the results in Figs. 9.3-3 and 9.3-4, respectively, reveal an accentuation of the trend outlined in Fig. 9.3-2, in the sense that the lagging frequencies become more and more sensitive to the thermal degradation, whereas the flapping ones reveal a much lower sensitivity to this effect.

Figures 9.3-5 through 9.3-7 display the counterparts of Figs. 9.3-2 through 9.3-4, generated for the case of the ply-angle $\theta = 90$ deg. As is clearly seen, due to the larger bending stiffnesses a_{33} and a_{22} experienced at this ply-angle, the frequencies are higher than their counterparts obtained for $\theta = 0$.

In addition, due to this increase of bending stiffnesses, the centrifugal stiffening effect becomes weaker as compared to the case of the ply-angle $\theta = 0$. This is reflected by the very small increase of flapping and lagging frequencies associated with the increase of Ω. These plots, show the deleterious effect of the thermal degradation of material properties on natural frequencies.

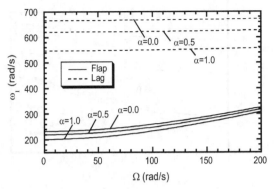

Figure 9.3-5: Variation of first decoupled natural frequencies vs. blade rotational speed Ω for selected values of α; $\theta = 90$ deg., $\beta = 0$, temperature profile A)

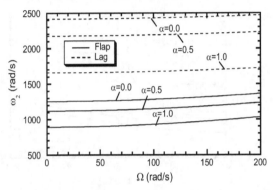

Figure 9.3-6: Counterpart of Fig. 9.3-5 for the second decoupled natural frequencies

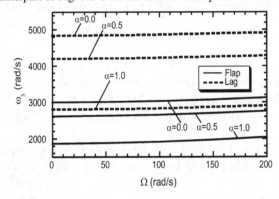

Figure 9.3-7: Counterpart of Fig. 9.3-5 for the third decoupled natural frequencies

Figures 9.3-8 through 9.3-10 depict the influence of the thermal degradation and that of the ply-angle on the first three natural coupled frequencies of the non-rotating blade. These graphs, display the implications of the two temperature profile variations A) and B).

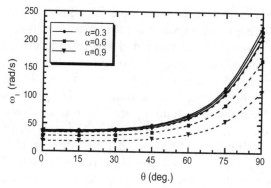

Figure 9.3-8: First coupled flap-lag natural frequency vs. θ for selected values of α and for both temperature scenarios ——— temperature profile A), - - - - - temperature profile B)

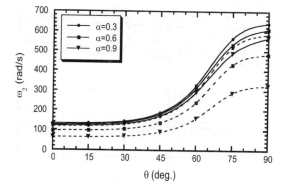

Figure 9.3-9: Counterpart of Fig. 9.3-8 for the second coupled flap-lag natural frequency

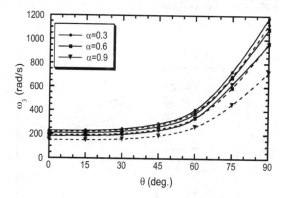

Figure 9.3-10: Counterpart of Fig. 9.3-8 for the third coupled flap-lag natural frequency

The results reveal that the temperature profile B) has more severe implications upon the decay of natural frequencies than the temperature profile A). In addition, the results show that the increase of the ply-angle, yields an increase of natural frequencies, but at the same time the effects of thermal degradation become more prominent. Moreover, this trend becomes more pronounced in the case of the thermal profile B) as compared to its counterpart A). By comparing the variation of coupled frequencies as emerging from Figs. 9.3-8 through 9.3-10 with their decoupled counterparts obtained for $\theta = 0$ (Figs. 9.3-2 through 9.3-4), and $\theta = 90$ deg., (Figs. 9.3-5 through 9.3-7), one can conclude that the coupled natural frequencies ω_1 and ω_3 are flapping dominated, whereas ω_2 is lagging dominated.

Figures 9.3-11 through 9.3-13 display the first three mode eigenfrequencies for the non-rotating blade as a function of the pretwist angle β_0 at the beam tip, with consideration of the thermoelastic degradation factor α as a parameter.

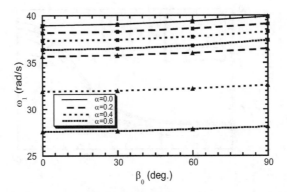

Figure 9.3-11: Variation of the first coupled natural frequency vs. pretwist angle, for selected α, $(\Omega = 0, \theta = 0)$, ──■── temperature profile A); ──▲── temperature profile B)

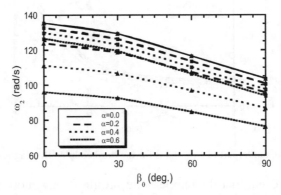

Figure 9.3-12: Variation of the second coupled natural frequency vs. pretwist angle, for selected α, $(\Omega = 0, \theta = 0)$, ──■── temperature profile A); ──▲── temperature profile B)

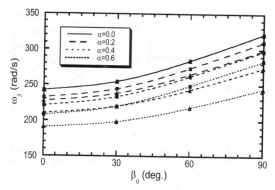

Figure 9.3-13: Variation of the third coupled natural frequency vs. pretwist angle, for selected α, ($\Omega = 0$, $\theta = 0$), —■— temperature profile A); —▲— temperature profile B)

The results involving both temperature profiles A) and B) reveal that: (a) for the non-degraded structure, i.e. for $\alpha = 0$, the trend of variation is similar to that reported by Slyper (1962), Subrahmanyam et al. (1981) and Dokumaci et al. (1967), (b) the thermal degradation of elastic properties has a strong effect on eigenfrequencies, in the sense that with the increase of α, a decrease of eigenfrequencies is experienced, and (c) the strongest decrease in the natural frequencies is featured in the case of the temperature profile B).

Figures 9.3-14 through 9.3-16 depict the variation of the three mode eigenfrequencies vs. the rotating speed, for the temperature profile B) and a prescribed pretwist $\beta_0 = 45$ deg., for two values of the thermal degradation factor α, and for selected values of the ply- angle.

These plots reveal the following features: (a) with the increase of the rotational speed, due to the centrifugal stiffening effect, the eigenfrequencies increase; (b) due to the pretwist, even for ply-angles $\theta = 0$; 90 deg., the flapping and lagging modes are coupled.

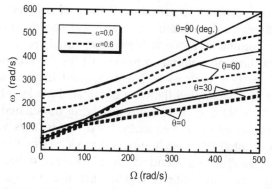

Figure 9.3-14: Variation of the first coupled natural frequency vs. rotational speed for the temperature profile B), $\beta_0 = 45$ deg.

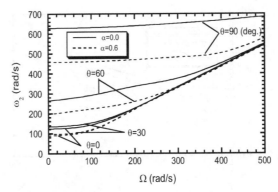

Figure 9.3-15: Variation of the second coupled natural frequency vs. rotational speed for the temperature profile B), $\beta_0 = 45$ deg.

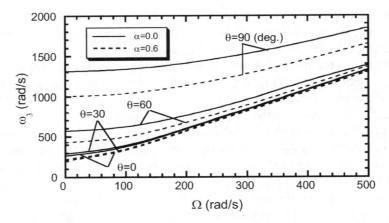

Figure 9.3-16: Counterpart of Fig. 9.3-15 for the third coupled natural frequency

For these ply-angles, as the rotational speed is increased, changes in their nature occur in the sense of: (i) for $\theta = 0$ the first coupled frequence mode evolves from a flapwise dominated mode at lower speeds to one which is basically a lagging dominated mode at higher rotational speeds, and for $\theta = 90$ deg., it remains a flapping dominated mode; (ii) for the second coupled mode frequency, for $\theta = 0$ it starts as a first decoupled lagging dominated mode and becomes the first decoupled flapwise dominated mode at higher rotational speeds, and for $\theta = 90$ deg. it remains a lagging dominated mode, whereas (iii) for the third coupled mode frequency, for $\theta = 0$ it starts and is maintained as the second decoupled flapwise dominated mode, a trend that remains valid also for $\theta = 90$ deg. A similar trend restricted to metallic beams was reported by Rosen (1991). Last, but not least, (iv) the directional property of composite material structures can play an important role in enhancing the free vibration response of rotating blades and alleviate the thermal degradation effect.

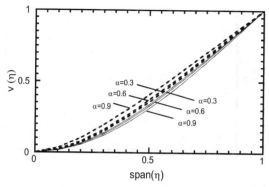

Figure 9.3-17: Variation of the first decoupled mode shape in flapping for selected values of α; $\theta = 0$, $\beta = 0$, $\Omega = 0$. ——— temperature profile A), - - - - - temperature profile B)

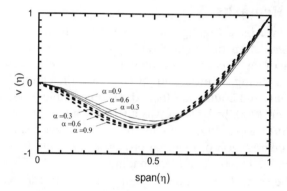

Figure 9.3-18: Counterpart of Fig. 9.3-17 for the second decoupled mode shape in flapping

Figures 9.3-17 through 9.3-19 display the variation of the decoupled first three mode shapes in flapping as influenced by α, for a non-rotating beam, with zero pretwist and for both thermal gradient scenarios. The modes are normalized to unity at $z = L$.

These plots reveal that the temperature profile B) exerts a stronger influence on the mode shapes as compared to that of the temperature profile A). For the same conditions, the lagging mode shapes show a similar behavior to their flapping mode counterparts. Consideration in this context of $\Omega \neq 0$, results in similar trends as those corresponding to $\Omega = 0$. For this reason, no additional plots associated with this case are displayed here.

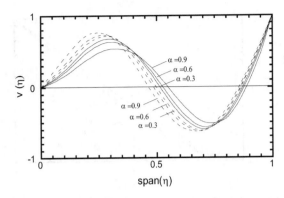

Figure 9.3-19: Counterpart of Fig. 9.3-17 for the third decoupled mode shape in flapping

9.3-4 Final Remarks

In this section, we have highlighted the strong influence played by thermo-mechanical degradation on eigenvibration response. It was also shown that implementation of the directional property of composite material structures can contribute to enhancing, without weight penalties, the eigenvibration response of rotating blades, and counteract their thermomechanical degradation. In the paper by Na et al. (2003b, 2004) it was also shown that via implementation of a feedback control capability, one can alleviate and even suppress the effects of the thermal degradation of material properties on both the eigenfrequencies and on the dynamic response to blast loading.

In spite of the special case of a box-beam and of a linear spanwise variation of the temperature as considered in the numerical simulations, the approach is quite general, in the sense of the possibility of considering a more complex cross-sectional shape, as well as an arbitrary spanwise temperature variation, such as

$$\text{Profile A)} \Rightarrow T(z) = T_0 z^n / L^n,$$
$$\text{Profile B)} \Rightarrow T(z) = T_0 (1 - z^n / L^n), \qquad (9.3\text{-}3\text{a,b})$$

where n is a parameter to be accommodated with the character of the problem at hand.

9.4 EFFECTS OF PRESETTING ON COUPLED VIBRATIONS OF ROTATING BLADES

9.4-1 Preliminaries

In spite of the extensive work devoted to the dynamic analysis of rotating blades, there is no structural model encompassing the basic features of advanced fil-amentary composite structural systems, such as directionality and transverse shear, as well as the presetting and the pretwist effects. In this chapter we investigate the coupled flapping-lagging-transverse shear vibration of a pretwisted rotating composite thin-walled beam mounted on a rigid hub of radius R_0 at a setting angle γ, and featuring the previously mentioned effects.

To this end, we consider a straight pretwisted flexible beam of length L mounted on a rigid hub of radius R_0, rotating at the constant angular velocity Ω as shown in Fig. 9.4-1. The beam is allowed to vibrate flexurally in a plane making an angle γ, referred to as setting angle with the plane of rotation. The origin of the rotating system of coordinates (x, y, z) is located at the blade root, at an offset R_0 from the rotation axis. Besides the rotating coordinates (x, y, z), we also define the local coordinates (x^P, y^P, z^P), where x^P and y^P are the *principal axes* of an arbitrary beam cross-section that rotate around the longitudinal beam axis with the pretwisted cross-sections.

The two coordinate systems are related by the following transformation formulae:

$$
\begin{aligned}
x &= x^P \cos(\gamma + \beta(z)) - y^P \sin(\gamma + \beta(z)), \\
y &= x^P \sin(\gamma + \beta(z)) + y^P \cos(\gamma + \beta(z)), \\
z &= z^P,
\end{aligned}
\tag{9.4-1a-c}
$$

where, as in Eqs. (9.2-1), $\beta(z)(\equiv \beta_0 z/L)$ denotes the pretwist angle of a current beam cross-section where β_0 is the pretwist at the beam tip.

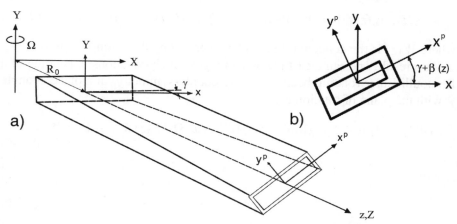

Figure 9.4-1: a) Geometry of the beam; b) Cross-section of the beam with pretwist and presetting angles

All the notations and developments that were displayed in Sections 9.1 and 9.2 remain valid. For the present case, and in the context of the *Circumferentially Uniform Stiffness Configuration*, the governing equations involving the flapping-lagging coupling, are similar to those derived in Section 9.2, where only the effect of the pretwist was included. The single difference involves the expressions for the stiffness and mass terms, $a_{ij}(z)$ and $b_i(z)$ as displayed in Tables 9.2-1 and 9.2-2, respectively. For the present case that involves both the presetting and pretwist, the replacement

$$\beta(z) \rightarrow \beta(z) + \gamma, \tag{9.4-2}$$

should be carried out in the expressions of $a_{ij}(z)$ and $b_i(z)$ contained in Tables 9.2-1 and 9.2-2.

9.4-2 A Special Case

We consider the special case of unshearable beam model featuring doubly symmetry of its cross-sections.

For this case, implying $a_{22}^p = a_{33}^p$; $b_4^p = b_5^p$ and $b_{14}^p = b_{15}^p$, and in the absence of rotatory inertia (yielding $b_6^p = 0$), the two classical bending equations and the associated boundary conditions exactly decouple into two independent groups, each of these governing, for any pretwist/presetting angle, the bending in flapping and lagging. Consequently, in this special case

$$a_{22} = a_{33} \equiv S; \quad b_4 = b_5 \text{ and } b_{15} = b_{14}. \tag{9.4-3a-c}$$

From the system of governing equations (9.2-12) through (9.2-14), specialized to the case of zero pretwist, one can express u_0 and v_0 as

$$[u_0(z,t); \ v_0(z,t)] = W(z)[\sin\gamma, \cos\gamma]\exp(i\omega t), \qquad [i = \sqrt{-1}] \tag{9.4-4}$$

where $W(z)$ is the deflection in the plane defined by the setting angle γ.

As a result, one can recast the dynamic governing equations in bending of a Bernoulli-Euler rotating beam featuring doubly-symmetry, in a plane at angle γ with the plane of rotation as:

$$[SW'']'' + \{[b_5 + \delta_n b_{15}](\omega^2 + \Omega^2)W'\}' - b_1\Omega^2[R(z)W']' - b_1\omega_1^2 W = 0, \tag{9.4-5}$$

where

$$\omega_1^2 = \omega^2 + \Omega^2 \sin^2\gamma, \tag{9.4-6}$$

and the boundary conditions:

$$W = W' = 0, \qquad \text{at } z = 0 \tag{9.4-7a,b}$$

and

$$[SW'']' + [b_5 + \delta_n b_{15}](\omega^2 + \Omega^2)W' = 0, \qquad SW'' = 0. \qquad \text{at } z = L \tag{9.4-8a,b}$$

These equations specialized for the case of the absence of rotatory inertia, reduce to those supplied by Hodges (1981).

It should be noticed that for $\gamma = \pi/2$ and $\gamma = 0$, the previous equations reduce to the rotating beam experiencing the motion in lagging and flapping, respectively.

Equation (9.4-6) shows that the flapping frequency is $\omega_1^2 \Longrightarrow \tilde{\omega}^2$, while the vibration frequencies at any γ expressed in terms of that at $\gamma = 0$ is

$$\omega^2 = \tilde{\omega}^2 - \Omega^2 \sin^2 \gamma. \qquad (9.4\text{-}9)$$

This expression reveals that the frequencies at any setting angle are lower than those corresponding to the purely flapping motion counterpart.

Applying the Rayleigh-quotient procedure to the governing Eq. (9.4-5) considered in conjunction with the boundary conditions and Eq. (9.2-11), yields the natural frequency at any setting angle where $W = W(z)$ should fulfil the boundary conditions, Eqs. (9.4-7,8).

$$\omega^2 = \frac{\int_0^L \left\{ S(W'')^2 + \Omega^2 \left[b_1 R(z)(W')^2 - \underbrace{(b_5 + \delta_n b_{15})(W')^2}_{} - b_1 W^2 \sin^2 \gamma \right] \right\} dz}{\int_0^L \left\{ \underbrace{(b_5 + \delta_n b_{15})\,(W')^2}_{\cdots} + b_1 W^2 \right\} dz},$$

$$(9.4\text{-}10)$$

From this equation one can easily conclude that rotatory inertia yields a diminution of fundamental frequency, whereas the increase of the rotor hub radius results in frequency increase. As concerns the setting angle, its increase yields a decrease of the fundamental frequency. These conclusions are in excellent agreement with those outlined by Lo and Renbarger (1952). The results also reveal that for the non-rotating blade, the effects of the hub radius and presetting angle on eigenfrequency are immaterial.

9.4-3 Comparisons with Available Numerical Predictions

As pointed out in the previous sections, the equations derived for rotating *thin-walled beams* are similar to the ones corresponding to a *solid beam* model. The difference occurs only in the *proper* expression of cross-sectional stiffness quantities and mass terms. Based on this feature, in order to validate both the solution method and the structural model presented here, comparisons with a number of results available in the literature are supplied.

Table 9.4-1: Effect of setting angle on natural frequency parameter $\bar{\omega}_i$ of rotating cantilever shearable beams ($\bar{\Omega} = 10$ and $\bar{R}_0 = 3$)

$\bar{\omega}_i$		$\eta = 0.1,$ $\mu = 0.030588,$			$\eta = 0.025$ $\mu = 0.0076f47$		
		$\gamma = 0$	$\gamma = 45$ deg.	$\gamma = 90$ deg.	$\gamma = 0$	$\gamma = 45$ deg.	$\gamma = 90$ deg.
$i = 1$	Oh et al.	23.0362	21.973	20.853	23.510	22.431	21.298
	W.	23.050	21.987	20.867	23.524	22.446	21.313
	Y.	23.037	21.974	20.850	23.514	22.436	21.302
	L.& L.	22.938	21.873	20.753	23.491	22.411	21.277
$i = 2$	Oh et al.	45.429	45.195	44.957	56.062	55.653	55.240
	W.	45.598	45.359	45.115	56.105	55.696	55.284
	Y.	45.428	45.194	44.955	56.072	55.662	55.250
	L. & L.	44.781	44.550	44.315	55.984	55.575	55.162
$i = 3$	Oh et al.	66.866	66.775	66.681	97.002	96.782	96.562
	W.	67.716	67.619	67.520	97.188	96.968	96.747
	Y.	66.854	66.793	66.668	97.011	96.792	96.570
	L. & L.	66.287	66.200	66.109	96.913	96.693	96.473
$i = 4$	Oh et al.	72.315	72.148	71.985	144.194	144.052	143.911
	W.	73.076	72.914	72.756	144.490	144.349	144.108
	L.	72.313	72.146	71.982	143.815	143.673	143.531
	L. & L.	71.967	71.792	71.620	143.71	143.57	143.43

Herein W. $\rightarrow$ Wang et al. (1976); Y. $\rightarrow$ Yokoyama (1988); L. & L. $\rightarrow$ Lee and Lin (1994); Oh et al. $\rightarrow$, Oh, Song and Librescu (2003a)

Tables 9.4-1 through 9.4-3, display within the assumptions of doubly-symmetric cross-section, comparisons with a number of previously obtained results. In this sense, in Table 9.4-1, the effects of the setting angle and of selected rotational speeds on the first four natural frequencies of shearable beams, are compared to those obtained in the papers by Wang et al. (1976), Yokoyama (1988), and Lee and Lin (1994), where various solution methodologies have been used.

In these numerical simulations the following dimensionless parameters of the rotational speed, natural frequency, and hub radius have been used, and are listed in the same sequence:

$$(\bar{\Omega}^2, \bar{\omega}^2) = (\Omega^2, \omega^2)\,(b_1 L^4/a_{33}); \quad \bar{R}_0 = R_0/L. \tag{9.4-11}$$

The results reveal the excellent agreement, especially with the predictions by Yokoyama (1988) obtained via FEM.

The trend of variation of natural bending frequencies with the setting angle is consistent with that emerging from Eq. (9.4-10), in the sense that the increase of γ yields a decrease of bending natural frequencies, this being valid for any mode number.

Table 9.4-2: Natural frequency parameter $\bar{\omega}_1$ of rotating unshearable cantilever beams ($\gamma = 90$ deg., $\beta_0 = 0$)

			$\bar{\omega}_1$			
$\bar{\Omega}$	$\bar{R}_0 = 0$				$\bar{R}_0 = 1$	
	Yokoyama (1988)	Putter and Manor (1978)	Oh et al. (2003a)	Yokoyama (1988)	Putter and Manor (1978)	Oh et al. (2003a)
0	3.516	——	3.51602	3.516	——	3.51602
2	3.622	3.6118	3.62179	4.401	4.4005	4.40054
5	4.074	4.0739	4.07390	7.412	7.4115	7.41130
10	5.050	5.0490	5.04889	13.261	13.2580	13.2576
20	6.794	6.7757	6.77927	25.318	25.2881	25.2992
50	10.899	10.4806	10.7630	61.881	61.6408	61.8446

Within the assumptions of doubly-symmetrical cross-section and of the unshearable beam model, and for zero pretwisted rotating beams, Tables 9.4-2 and 9.4-3 display comparisons of lagging ($\gamma = 90$ deg.) frequencies with a number of available results from the literature. The results are compared with those by Yokoyama (1988) and Putter and Manor (1978), and an excellent agreement is remarked. In addition, in this table $\eta \equiv (b_4 + \delta_n b_{14})/(b_1 L^2)$, $\mu \equiv a_{33}/(L^2 a_{55})$.

In Tables 9.4-4 comparisons of frequency predictions obtained for a rotating unshearable blade are supplied; the effect played by the presetting angle clearly emerges. The beam characteristics are those listed in Eq. (9.2-15). Moreover in contrast to the case considered in Table 9.4-1, characterized by $a_{22} = a_{33}$, $b_4 = b_5$, the findings in Table 9.4-4 reveal that for blades featuring $a_{22} \neq a_{33}$ and $b_4 \neq b_5$, the increase of the setting angle yields, depending on the odd or even mode number, either a decrease or increase of natural frequencies, respectively. At the same time, from Table 9.2-3, in the same conditions characterizing the blade as in Table 9.4-4, the increase of the pretwist angle exerts an opposite effect to that generated by the increase of the setting angle.

Table 9.4-3: Natural frequency parameter $\bar{\omega}_2$ of rotating unshearable cantilever beams ($\gamma = 90$ deg., $\beta_0 = 0$)

			$\bar{\omega}_2$			
$\bar{\Omega}$	$\bar{R}_0 = 0$				$\bar{R}_0 = 1$	
	Yokoyama (1988)	Putter and Manor (1978)	Oh et al. (2003a)	Yokoyama (1988)	Putter and Manor (1978)	Oh et al. (2003a)
0	22.036	——	22.0344	22.036	——	22.0344
2	22.528	22.5263	22.5263	23.282	23.2803	23.2802
5	24.952	24.9500	24.9500	28.926	28.9238	28.9240
10	32.123	32.1197	32.1198	43.237	43.2267	43.2282
20	51.372	51.3531	51.3573	76.659	76.5942	76.6353
50	116.417	116.1996	116.359	182.386	181.9361	182.316

Table 9.4-4: Influence of presetting angle on natural frequencies ω_i (rad/s) of a rotating unshearable blade with zero pretwist ($\beta_0 = 0$, $\Omega = 300$ rpm)

ω_i	$\gamma = 0$		$\gamma = 22.5$ deg.	
	Rosen et al. (1987)	Oh et al. (2003a)	Rosen et al. (1987)	Oh et al. (2003a)
$i = 1$	19.711	19.7193	18.536	18.0769
$i = 2$	32.488	32.4997	33.168	33.4381
$i = 3$	81.585	81.7388	80.620	80.7609
$i = 4$	121.33	121.531	121.97	122.183
$i = 5$	141.62	142.132	141.11	141.618

In this sense, as it clearly appears, depending on the mode number, i.e. odd or even, the increase of the pretwist angle is accompanied by an increase or decrease of the natural frequency, respectively. In addition to these conclusions, the very good agreement with the results obtained by Rosen et al. (1987) should be remarked.

9.4-4 Results and Discussion

For the sake of illustration, we consider a rotating beam modelled as a composite box-beam (see Fig. 9.4-1) featuring the same geometrical and material characteristics as in Section 9.2.

The results reveal a continuous increase of natural frequencies that accompanies the increase of the ply-angle. Moreover, consistent with the results displayed in Table 9.4-4, depending on the odd or even mode number, the increase of the presetting angle yields either a decrease or increase of natural frequencies, respectively.

Figures 9.4-2 and 9.4-3 display the variation of the first two natural frequencies of the rotating beam ($\Omega = 100$ rad/s), as a function of the ply-angle, for selected values of the setting angle, and for the pretwist, $\beta_0 = 90$ deg.

Related to these results, one should remark from Figs. 9.4-4 and 9.4-5 that the increase of the rotational speed yields an increase of natural frequencies as compared to those previously displayed, where a lower angular speed was considered. Moreover, as appears from the latter two figures, at larger angular speeds, the effect of the ply-angle is more individualized for each of the considered setting angles than in the case of lower angular speeds, Figs. 9.4-2 and 9.4-3.

Figures 9.4-6 through 9.4-8, display in succession the variations of the first three coupled natural frequencies, as a function of the presetting angle, for selected values of the rotational speed. Again, the effect of setting angle on natural frequencies of nonrotating beam is immaterial.

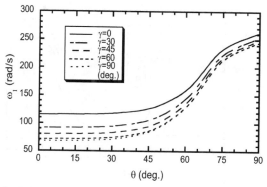

Figure 9.4-2: Variation of the first coupled flapping-lagging natural frequency vs. ply angle for different setting angles ($\beta_0 = 90$ deg., $\Omega = 100$ rad/s, $\overline{R}_0 = 0.1$)

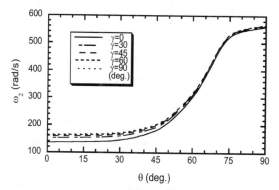

Figure 9.4-3: Variation of the second coupled flapping-lagging natural frequency vs. ply angle for different setting angles ($\beta_0 = 90$ deg., $\Omega = 100$ rad/s, $\overline{R}_0 = 0.1$)

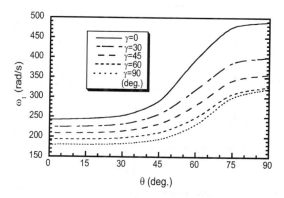

Figure 9.4-4: The counterpart of Fig. 9.4-2 for $\Omega = 400$ rad/s

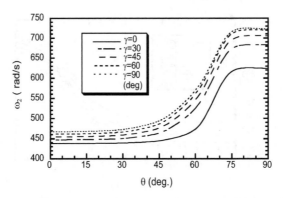

Figure 9.4-5: The counterpart of Fig. 9.4-3 for $\Omega = 400$ rad/s

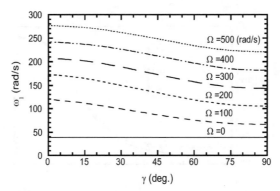

Figure 9.4-6: First coupled flapping-lagging natural frequency vs. presetting angle for selected rotational speeds ($\theta = 0$, $\beta_0 = 30$ deg., $\overline{R}_0 = 0.1$)

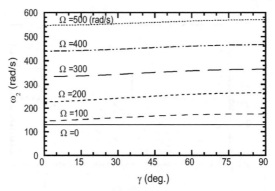

Figure 9.4-7: Second coupled flapping-lagging natural frequency vs. γ for selected rotational speeds ($\theta = 0$, $\beta_0 = 30$ deg., $\overline{R}_0 = 0.1$)

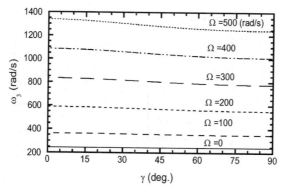

Figure 9.4-8: Third coupled flapping-lagging natural frequency vs. γ for selected rotational speeds ($\theta = 0$, $\beta_0 = 30$ deg., $\overline{R}_0 = 0.1$)

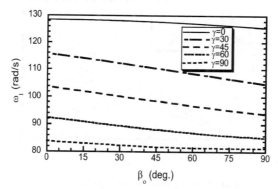

Figure 9.4-9: Variation of the first coupled flapping-lagging natural frequency vs. pretwist angle for selected γ (deg.) of the rotating beam ($\theta = 45$ deg., $\Omega = 100$ rad/s, $\overline{R}_0 = 0.1$)

On the other hand, the trend of variation of various mode frequencies as a function of the presetting angle and angular velocity remains similar to that previously emphasized. In these numerical simulations, the considered pretwist angle was $\beta_0 = 30$ deg.

In Figs. 9.4-9 and 9.4-10, for selected values of the setting angle there are displayed in succession the variations of the first two natural frequencies, as a function of the pretwist angle. For zero presetting angle, the trend of variation of natural frequencies with that of the pretwist angle, coincides with that presented in Section 9.2-5.

Figures 9.4-11 and 9.4-12 show the implications of the hub radius, coupled with that of the rotating speed on the fundamental coupled flapping-lagging natural frequency, for the beam featuring the setting angles $\gamma = 0$ and $\gamma = 90$ deg., and for the ply-angles $\theta = 0$ and $\theta = 45$ deg., respectively. Due to the fact that in these considered cases, as the pretwist is $\beta_0 = 90$ deg., the flapping-lagging coupling is present also for $\gamma = 0$ and $\theta = 0$.

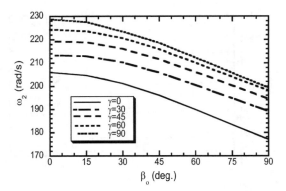

Figure 9.4-10: The counterpart of Fig. 9.4-9 for the second coupled natural frequency

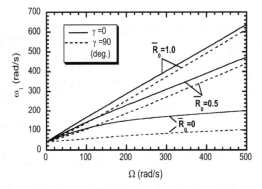

Figure 9.4-11: Variation of the first coupled flapping-lagging natural frequency vs. the rotational speed for different hub radii and two setting angles ($\beta_0 = 90$ deg., $\theta = 0$)

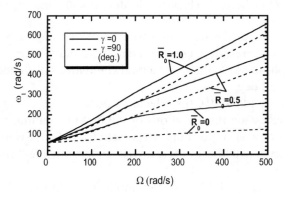

Figure 9.4-12: The counterpart of Fig. 9.4-11 for the ply-angle $\theta = 45$ deg.

The results reveal that the differences between the frequencies corresponding to $\gamma = 0$ and $\gamma = 90$ deg. tend to decay with the increase of the hub radius. This trend is consistent with that reported in the papers by Kumar (1974). At the same time, the results show that for the same $\bar{R}_0$, the increase of the ply-angle tends to accentuate the difference between the frequencies corresponding to $\gamma = 0$ and $\gamma = 90$ deg.

Figures 9.4-13 through 9.4-15 further underline the considerable role that the tailoring technique can play in the increase, without weight penalties, of the coupled flapping-lagging natural frequencies of rotating beams. This renders the composite material systems overwhelmingly superior to the metallic structures deserves to be highlighted again.

In addition, these plots reveal a number of trends related to the implications of the setting angle on each mode frequency, trends that have already been outlined in the previously displayed numerical simulations.

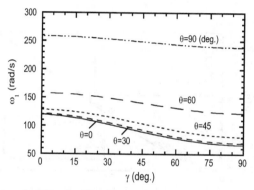

Figure 9.4-13: First coupled flapping-lagging natural frequency vs. γ for selected ply-angles $(\beta = 0, \Omega = 100 \text{ rad/s}, \bar{R}_0 = 0.1)$

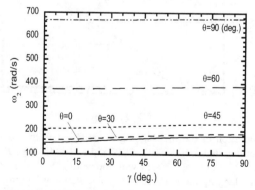

Figure 9.4-14: The counterpart of Fig. 9.4-13 for the second coupled flapping-lagging natural frequency

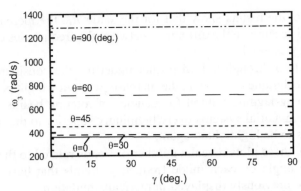

Figure 9.4-15: The counterpart of Fig. 9.4-13 for the third coupled flapping-lagging natural frequency

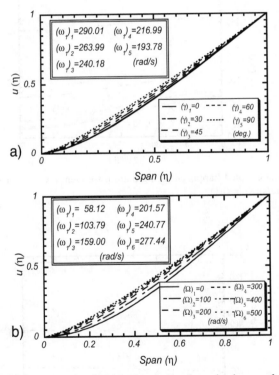

Figure 9.4-16: a) Variation of the first lagging mode shape in the coupled flapping-lagging motion for various setting angles ($\beta = 0$, $\Omega = 400$ rad/s, $\theta = 45$ deg., $\overline{R}_0 = 0.1$). The coupled natural frequencies $(\omega_1)_i$ (rad/s) associated with the setting angles are indicated in the inset. b) Variation of the first lagging mode shape in coupled flapping-lagging for various rotational speeds $(\Omega)_i$ (rad/s), ($\beta = 0$, $\gamma = 45$ deg., $\theta = 45$ deg., $\overline{R}_0 = 0.1$). The natural frequencies $(\omega_1)_i$ (rad/s) associated with each of the rotational speeds $(\Omega)_i$ (rad/s) are indicated in the inset

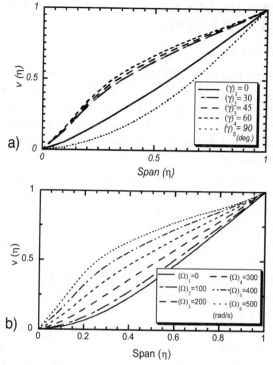

Figure 9.4-17: a) Counterpart of Fig. 9.4-16a for the first flapping mode shape. b) Counterpart of Fig. 9.4-16b for the first flapping mode shape

Finally, Figs. 9.4-16 and 9.4-17, provide the first lagging and first flapping normalized mode shapes, respectively, for selected values of the setting angle or of the rotating speed. These plots show that the flapping mode is more strongly affected by the setting angle/rotating speed than the lagging mode.

BIBLIOGRAPHICAL COMMENTS

In the paper by Na et al. (2004), where a similar structural model to that in Section 9.4 was used, further results related with the influence of the non-uniformity of the beam cross-section and of feedback control on their free/forced vibration have been reported.

For rotating isotropic solid beams featuring the taper in both the chord and height directions, issues related with the influence of shear deformation, hub radius and setting angle on bending free vibration characteristics have been analyzed via the use of the FEM by Bazoune et al. (1999) and Khulief and Bazoune (1992).

9.5 ROTATING BLADES MADE-UP OF FUNCTIONALLY GRADED MATERIALS

9.5-1 Preliminaries

Functionally graded materials (FGMs) for high-temperature structural applications are a new generation of microscopically inhomogeneous composites, whose thermomechanical properties vary smoothly and continously in predetermined directions throughout the body of the structure. This feature is achieved by gradually varying the volume fraction of constituent materials, usually ceramics and metals.

The ceramic in a FGM offers thermal barrier effects and protects the metal from corrosion and oxidation, while the FGM is toughened and strengthened by the metallic composition. As a result, these materials are able to withstand high temperature gradients without structural failures.

In contrast to standard laminated composite structures whose material properties are piecewise constant through their thickness, and, as a result, are exposed to adverse interface effects that can yield the failure of the structure, the FGMs, due to the continuous variation of their material properties, are not exposed to such damaging effects.

In addition to the research work devoted to modeling of 3-D FGM media (see e.g. Aboudi et al., 1996, 1999; Pindera and Aboudi, 2000; Berezovski et al., 2003), the studies involving thin-walled structures made of FGMs have been mainly devoted to beams, plates and shells (see in this sense Tanigawa, 1992; Noda and Jin, 1993; Birman, 1995; Reddy and Chin, 1998; Loy et al., 1999; Praveen and Reddy, 1998; Gong et al., 1999; Reddy, 2000; Pradhan et al., 2000; Cheng and Batra, 2000; Shen, 2002; Sankar, 2001; Sankar and Tzeng, 2002; Vel and Batra, 2002; Hilton, 2005). In spite of its evident practical importance, with the exception of the results by Oh et al. (2003b-e) that will also be supplied here, there is no research work related to the modeling and behavior of rotating blades operating in a high temperature environment and made of functionally graded materials. This chapter is devoted to this topic.

We consider a straight, pretwisted tapered thin-walled beam rotating with a constant angular velocity and exposed to a steady temperature field experiencing a gradient through the blade thickness. It is assumed that the blade is made of functionally graded materials whose properties vary continuously across the blade thickness and that the material properties are temperature-dependent.

9.5-2 Structural Model. Basic Assumptions

We consider a straight, tapered and pretwisted thin-walled rotating beams mounted on a rigid hub of radius R_0 at a setting angle γ. The origin of the rotating axis system (x, y, z) is located at the blade root at an offset R_0 from the rotation axis (see Fig. 9.4-1). As in Sections 9.3 and 9.4, also in this context, we use the same coordinate systems.

It is assumed that the beam is tapered in both the chordwise and hight directions. As a result, the distributions along the blade span of the chord $c(z)$ and height $b(z)$ of the mid-line cross-section profiles are expressed as

$$c(z) = \left[1 + \frac{z}{L}(\alpha_c - 1)\right] c_R, \quad b(z) = \left[1 + \frac{z}{L}(\alpha_h - 1)\right] b_R. \qquad (9.5\text{-}1a,b)$$

Here $\alpha_c = c_T/c_R$ and $\alpha_h = b_T/b_R$ define the taper ratios of the chord and height, respectively, where T and R identify the blade characteristics at the tip and root cross-sections. For further use, we define the parameter

$$S(z) \equiv b(z)/c(z). \qquad (9.5\text{-}2)$$

when $\alpha_h = \alpha_c$, $S(z) \Rightarrow S = b_R/c_R$.

Equation (9.5-1) shows that for $(\alpha_c, \alpha_h) < 1$ there is a decay of the chord and height toward the blade tip, whereas when $(\alpha_c, \alpha_h) > 1$, the opposite trend occurs, i.e. the decay is toward the blade root.

9.5-3 3-D Constitutive Equations

The design of advanced turbine blades must meet high efficiency and high reliability standards, regardless of the severe temperature gradients under which these should operate. To this end, functionally graded ceramic-metal based materials are best suited for this role.

Since these materials are isotropic, the corresponding thermoelastic constitutive law adapted to the case of thin-walled structures is expressed as

$$\begin{Bmatrix} \sigma_{ss} \\ \sigma_{zz} \\ \sigma_{zn} \\ \sigma_{ns} \\ \sigma_{sz} \end{Bmatrix} = \begin{bmatrix} Q_{11} & Q_{12} & 0 & 0 & 0 \\ Q_{12} & Q_{11} & 0 & 0 & 0 \\ 0 & 0 & Q_{44} & 0 & 0 \\ 0 & 0 & 0 & Q_{55} & 0 \\ 0 & 0 & 0 & 0 & Q_{66} \end{bmatrix} \begin{Bmatrix} \varepsilon_{ss} \\ \varepsilon_{zz} \\ \varepsilon_{zn} \\ \varepsilon_{ns} \\ \varepsilon_{sz} \end{Bmatrix} - \begin{Bmatrix} \hat{\alpha}\Delta T \\ \hat{\alpha}\Delta T \\ 0 \\ 0 \\ 0 \end{Bmatrix}.$$

$$(9.5\text{-}3)$$

Here, the reduced thermoelastic coefficients are defined as:

$$Q_{11} = \frac{E}{1 - v^2}, \quad Q_{12} = \frac{Ev}{1 - v^2}, \quad Q_{66} = \frac{E}{2(1 + v)} \ (\equiv G),$$

$$Q_{44} = Q_{55} = k^2 G, \quad \hat{\alpha} = \frac{1}{1 - v}\alpha, \qquad (9.5\text{-}4a\text{-}e)$$

where E and v are the Young's modulus and Poisson's ratio, respectively, k^2 is the transverse shear correction factor, $\Delta T(s, z, n)$ is the steady-state temperature rise from that of the stress free state, while α is the thermal expansion coefficient.

For a ceramic/metal FGM, the material properties vary continuously across the blade thickness according to the law (see Reddy and Chin, 1998; Reddy, 2000)

$$E(n) = (E_c - E_m)\left(\frac{2n+h}{2h}\right)^K + E_m; \quad v(n) = (v_c - v_m)\left(\frac{2n+h}{2h}\right)^K + v_m;$$

$$\rho(n) = (\rho_c - \rho_m)\left(\frac{2n+h}{2h}\right)^K + \rho_m; \quad \alpha(n) = (\alpha_c - \alpha_m)\left(\frac{2n+h}{2h}\right)^K + \alpha_m,$$

$$(9.5\text{-}5a\text{-}d)$$

where subscripts m and c identify the quantities affiliated to metal and ceramic, respectively, while K $(0 \leq K \leq \infty)$ plays the role of a volume fraction parameter.

Equations (9.5-5), show that for $n = h/2$, $E \rightarrow E_c$, $v \rightarrow v_c$ and $\rho \rightarrow \rho_c$, while for $n = -h/2$, $E \rightarrow E_m$, $v \rightarrow v_m$ and $\rho \rightarrow \rho_m$.

This means that the material properties vary continously from the top surface of the blade where the material is entirely ceramic, to fully metal at the bottom surface. Equations (9.5-5) show that, starting with $K = 0$, for which the blade is entirely from ceramic, the increase of the volume fraction parameter K yields a continuous increase of the metal content to the detriment of the ceramic.

As a result, the compositional material profile of the FGM blade is governed by the specific value of the volume fraction parameter K. As an example, consider the case when $K = 1$. In such a case, $P(n) = (P_c + P_m)/2 + n(P_c - P_m)/h$ where P is one of the generic properties of the material. Again, for $n = h/2$ and $n = -h/2$, the blade properties belong to ceramic and metal, respectively. It is also evident, that on the blade middle- surface, i.e. when $n = 0$ whereas for $n > 0$ and $n < 0$, the blade material is dominated by the ceramic and metal phases, respectively. It is assumed that the blade is subjected to a steady-state 1-D temperature distribution through its thickness. As an alternative procedure, Mori-Tanaka mean field scheme can be applied to evaluate the locally effective material properties (see Cheng and Batra, 2000; Vel and Batra, 2002).

$$P(0) = (P_c + P_m)/2, \tag{9.5-6}$$

The steady-state equation of the 1-D heat transfer is expressed by

$$\frac{d}{dn}\left[\kappa(n)\frac{dT}{dn}\right] = 0, \tag{9.5-7}$$

where the boundary conditions are $T(h/2) = T_t$ and $T(-h/2) = T_b$; $\kappa(n)$ is the thermal conductivity in the thickness direction, while T_t and T_b denote the temperatures at the top and bottom surfaces of the FGM blade, respectively.

The solution of (9.5-7) in conjunction with the conditions on the bounding surfaces yields the steady-state temperature distribution $T(n)$

$$T(n) = T_b\left[1 + \frac{\lambda}{\int_{-h/2}^{h/2} \frac{1}{\kappa(n)}dn} \int_{-h/2}^{n} \frac{1}{\kappa(n)}dn\right], \qquad (9.5\text{-}8)$$

where

$$\lambda = \frac{T_t - T_b}{T_b}, \qquad (9.5\text{-}9)$$

represents a measure of the normalized temperature gradient across the blade wall thickness.

Throughout the numerical simulations, we take $T_b = 300°$ K. The thermal conductivity $\kappa(n)$ of the FGM is assumed to vary according to the law

$$\kappa(n) = (\kappa_c - \kappa_m)\left(\frac{2n + h}{2h}\right)^K + \kappa_m, \qquad (9.5\text{-}10)$$

where the thermal conductivities κ_c and κ_m are considered to be temperature-independent.

However, the remaining properties of the FGM are assumed to be temperature- dependent and to vary according to the law obtained from experiments (see Touloukian, 1967). These are expressed in generic form as

$$P(n) = P_0(P_{-1}/T + 1 + P_1 T + P_2 T^2 + P_3 T^3), \qquad (9.5\text{-}11)$$

where P_0, P_{-1}, P_1, P_2 and P_3 are constants in the cubic fit of the specific temperature-dependent material property, and T (in Kelvin) is the environmental temperature.

9.5-4 Governing System

As in the previous sections, Hamilton's principle is used to derive the equations of motion and boundary conditions of rotating blades. Due to the involvement of the temperature field, extra terms associated with it will appear in the governing equation system. For this reason, the equations featuring the flapping-lagging-transverse shear coupling are supplied here.

Since the structure composed of FGMs is isotropic, a number of stiffness coefficients such as a_{42}, a_{52} and a_{53} disappear. As a result, Eqs. (9.2-8) become

Governing System:

$$\delta u_0 : \left[a_{44}(z)(u_0' + \theta_y) + a_{45}(z)(v_0' + \theta_x)\right]' - b_1\ddot{u}_0 + \underline{b_1 u_0 \Omega^2}$$
$$+ \underline{\Omega^2[R(z)u_0']'} + p_x = 0,$$

$$\delta v_0 : \left[a_{55}(z)(v_0' + \theta_x) + a_{54}(z)(u_0' + \theta_y)\right]' - b_1\ddot{v}_0 + \underline{\Omega^2[R(z)v_0']'} + p_y = 0,$$

$$\delta\theta_y : \left[a_{22}(z)\theta_y' + a_{23}(z)\theta_x'\right]' - a_{44}(z)(u_0' + \theta_y) - a_{45}(z)(v_0' + \theta_x)$$
$$- \left(b_5(z) + \delta_n b_{15}(z)\right)(\underset{\cdots}{\ddot{\theta}_y} \underline{- \Omega^2\theta_y})$$
$$- \left(b_6(z) - \delta_n b_{13}(z)\right)(\underset{\cdots}{\ddot{\theta}_x} \underline{- \Omega^2\theta_x}) + m_y = (M_y^T)',$$

$$\delta\theta_x : \left[a_{33}(z)\theta_x' + a_{32}(z)\theta_y'\right]' - a_{55}(z)(v_0' + \theta_x) - a_{54}(z)(u_0' + \theta_y)$$
$$- \left(b_4(z) + \delta_n b_{14}(z)\right)(\underset{\cdots}{\ddot{\theta}_x} \underline{- \Omega^2\theta_x})$$
$$- \left(b_6(z) - \delta_n b_{13}(z)\right)(\underset{\cdots}{\ddot{\theta}_y} \underline{- \Omega^2\theta_y})$$
$$+ m_x = (M_x^T)'. \tag{9.5-12a-d}$$

Here, p_x, p_y, m_x and m_y are the external loads and moments that are assumed to be functions of both the spanwise and time coordinates. If the blade is clamped at $z = 0$ and free at $z = L$, the boundary conditions are

at $z = 0$:

$$u_0 = v_0 = \theta_y = \theta_x = 0, \tag{9-5-13a-d}$$

and at $z = L$:

$$\delta u_0 : a_{44}(L)(u_0' + \theta_y) + a_{45}(L)(v_0' + \theta_x) = 0,$$
$$\delta v_0 : a_{55}(L)(v_0' + \theta_x) + a_{54}(L)(u_0' + \theta_y) = 0,$$
$$\delta\theta_y : a_{22}(L)\theta_y' + a_{23}(L)\theta_x' = M_y^T(L), \tag{9.5-14a-d}$$
$$\delta\theta_x : a_{33}(L)\theta_x' + a_{32}(L)\theta_y' = M_x^T(L).$$

These equations can address the static and dynamic response of rotating blades exposed to external thermomechanical loads.

In these equations,

$$R(z) = -\int_L^z b_1(z)(R_0 + z)dz. \tag{9.5-15}$$

For a uniform cross-section blade Eq. (9.5-15) reduces to Eq. (9.2-11).

In addition, the 1-D thermal loading terms that play the role of bending moments about the axes x and y, are given by

$$M_x^T(z) = \oint \left[y\hat{N}_{zz}^T - \frac{dx}{ds}\hat{L}_{zz}^T \right] ds, \text{ and } M_y^T(z) = \oint \left[x\hat{N}_{zz}^T + \frac{dy}{ds}\hat{L}_{zz}^T \right] ds,$$

(9.5-16a,b)

respectively, where

$$\hat{N}_{zz}^T(s, z) = \left(1 - \frac{A_{12}}{A_{11}} \right) N_{zz}^T; \quad \hat{L}_{zz}^T = L_{zz}^T - \frac{B_{12}}{A_{11}} N_{zz}^T,$$

(9.5-17a,b)

while

$$(N_{zz}^T, \ L_{zz}^T) = \int_{-h/2}^{h/2} \Delta T (Q_{11} + Q_{12})\hat{\alpha}(1, n)dn.$$

(9.5-17c)

The stiffness quantities $a_{ij}(z)$ intervening in the governing system, are obtainable from those supplied in Table 9.2-1, where the change $\beta \rightarrow \beta(z) + \gamma$ should be carried out.

For the present case the stiffness quantities a_{ij}^p that are non-zero are provided next. In addition, in the definition of mass terms, consistent with (9.5-5c), we have

$$(m_0; m_2) = \int_{-h/2}^{h/2} [(\rho_c - \rho_m)\left(\frac{2n + h}{2h} \right)^K (1; n^2) + \rho_m(1; n^2)]dn. \quad (9.5\text{-}18)$$

Table 9.5-1: Expressions of a_{ij}^p

a_{ij}^p	Their expression
a_{22}^p	$\oint \left[K_{11}x^2 + K_{44}\left(\dfrac{dy}{ds} \right)^2 \right]ds$
a_{33}^p	$\oint \left[K_{11}y^2 + K_{44}\left(\dfrac{dx}{ds} \right)^2 \right]ds$
a_{44}^p	$\oint \left[K_{22}\left(\dfrac{dx}{ds} \right)^2 + A_{44}\left(\dfrac{dy}{ds} \right)^2 \right]ds$
a_{55}^p	$\oint \left[K_{22}\left(\dfrac{dy}{ds} \right)^2 + A_{44}\left(\dfrac{dx}{ds} \right)^2 \right]ds$

9.5-5 Validation of the Model and of the Solution Methodology

The equations derived for rotating tapered pretwisted thin-walled beams are similar to those corresponding to a solid beam. The difference occurs only in the proper expression of cross sectional stiffness and mass terms. In order to validate the present model, comparisons with a number of results available in the literature that have been carried out by using a solid beam model will be presented.

Two groups of comparisons are supplied. The first group involves the free vibration problem of a chordwise tapered rotating blade. The blade material is assumed to be isotropic and the temperature effects are neglected. In Table 9.5-2 the following dimensionless parameters have been considered:

$$\bar{\omega}_i^2 = \frac{b_1[0]L^4}{a_{33}[0]}\omega^2, \quad \bar{\Omega}^2 = \frac{b_1[0]L^4}{a_{33}[0]}\Omega^2, \quad \frac{a_{55}[\eta]}{a_{55}[0]} = \frac{b_1[\eta]}{b_1[0]} = 1 - \alpha_c\eta,$$

(9.5-19a-d)

$$\frac{b_4[\eta] + b_{14}[\eta]}{b_4[0] + b_{14}[0]} = \frac{a_{33}[\eta]}{a_{33}[0]} = (1 - \alpha_c\eta)^3.$$

The comparisons concern both the shearable and the classical model counterpart, and in both cases, very good agreements were reached. Note that the predictions by Hodges and Rutkowski (1981) have been obtained via FEM.

Table 9.5-2: Comparisons of dimensionless flapping natural frequencies for a tapered untwisted rotating beam ($\alpha_c = 0.5$, $\alpha_h = 1$, $R_0 = 0$, $\lambda = 0$, $\beta_0 = 0$)

		Timoshenko		Bernoulli-Euler			
		$\dfrac{b_4[0] + \delta_n b_{14}[0]}{b_1[0]L^2} = 0.0064$		$\dfrac{b_4[0] + \delta_n b_{14}[0]}{b_1[0]L^2} = 0$			
$\bar{\Omega}$	$\bar{\omega}_i$	$\dfrac{a_{33}[0]}{a_{55}[0]L^2} = 0.01958$		$\dfrac{a_{33}[0]}{a_{55}[0]L^2} = 0$			
		L. & L.	Present	L. & L.	L. & K.	H. & R.	Present
	$\bar{\omega}_1$	3.6500	3.6499	3.8238	3.8238	3.8242	3.8238
	$\bar{\omega}_2$	15.022	15.022	18.317	18.317	18.320	18.317
0	$\bar{\omega}_3$	32.785	32.784	47.265	47.264	47.271	47.274
	$\bar{\omega}_4$	53.341	53.379	90.450	–	–	91.865
	$\bar{\omega}_1$	4.8866	4.8865	5.0927	5.0927	5.0927	5.0927
	$\bar{\omega}_2$	16.460	16.460	19.684	19.684	19.684	19.684
3	$\bar{\omega}_3$	34.458	34.457	48.619	48.619	48.620	48.629
	$\bar{\omega}_4$	55.358	55.395	91.822	–	–	93.201
	$\bar{\omega}_1$	6.4712	6.4711	6.7434	6.7345	6.7432	6.7434
	$\bar{\omega}_2$	18.744	18.744	21.905	21.905	21.905	21.906
5	$\bar{\omega}_3$	37.224	37.223	50.934	50.934	49.646	50.945
	$\bar{\omega}_4$	58.730	58.744	94.206	–	–	95.529
	$\bar{\omega}_1$	10.991	10.991	11.502	11.502	11.501	11.502
	$\bar{\omega}_2$	26.928	26.928	30.183	30.183	30.177	30.183
10	$\bar{\omega}_3$	47.883	47.884	60.564	60.564	60.555	60.582
	$\bar{\omega}_4$	71.989	72.020	104.61	–	–	105.72

L. & L. → Lee and Lin (1994); L. & K. → Lee and Kuo (1992); H. & R. → Hodges and Rutkowski (1981)

The second group of comparisons involves the case of a metallic pre-twisted rotating beam. The comparisons are summarized in Table 9.5-3 for selected values of β_0 and of the dimensionless rotational speed $\bar{\Omega}$. The comparisons show an excellent agreement with the results obtained elsewhere, and with the ones based on the present model and solution methodology.

Table 9.5-3: Effects of the pretwist angle on natural frequencies of a rotating blade $(\bar{R}_0 = 0, \ \gamma = 0)$

$\bar{\Omega}$	$\bar{\omega}_i$	$\beta_0 = 0$		$\beta_0 = 30$ deg.		$\beta_0 = 60$ deg.		$\beta_0 = 90$ deg.	
		S.& K.	Present	S.& K.	Present	S.& K.	Present	S.& K.	Present
0	$\bar{\omega}_1$	3.5160	3.5160	3.5245	3.5245	3.5496	3.5496	3.5900	3.5900
	$\bar{\omega}_2$	7.0320	7.0320	6.9585	6.9586	6.7595	6.7597	6.4847	6.4850
	$\bar{\omega}_3$	22.034	22.034	22.339	22.339	23.210	23.209	24.531	24.531
	$\bar{\omega}_4$	44.068	44.069	42.896	42.898	40.306	40.310	37.457	37.465
	$\bar{\omega}_5$	61.695	61.715	63.423	63.437	67.627	67.666	72.973	73.062
0.5	$\bar{\omega}_1$	4.0049	4.0049	4.0082	4.0082	4.0183	4.0183	4.0354	4.0354
	$\bar{\omega}_2$	7.0742	7.0742	7.0064	7.0065	6.8227	6.8229	6.5692	6.5695
	$\bar{\omega}_3$	22.484	22.484	22.780	22.780	23.627	23.627	24.917	24.916
	$\bar{\omega}_4$	44.260	44.261	43.108	43.110	40.551	40.555	37.730	37.738
	$\bar{\omega}_5$	62.141	62.160	63.847	63.830	68.016	68.054	73.333	73.420
1.0	$\bar{\omega}_1$	5.1916	5.1916	5.1824	5.1824	5.1552	5.1552	5.1120	5.1121
	$\bar{\omega}_2$	7.1919	7.1919	7.1461	7.1462	7.0096	7.0097	6.8250	6.8253
	$\bar{\omega}_3$	23.783	23.783	24.055	24.055	24.839	24.838	26.041	26.040
	$\bar{\omega}_4$	44.830	44.831	43.735	43.737	41.273	41.276	38.533	38.540
	$\bar{\omega}_5$	63.457	63.475	65.103	65.076	69.171	69.206	74.400	74.483
3.0	$\bar{\omega}_1$	8.2496	8.2496	8.2156	8.2155	8.1195	8.1194	7.9774	7.9776
	$\bar{\omega}_2$	11.740	11.740	11.749	11.748	11.772	11.772	11.804	11.803
	$\bar{\omega}_3$	34.699	34.700	34.834	34.834	35.233	35.232	35.872	35.871
	$\bar{\omega}_4$	50.514	50.514	49.804	49.806	48.078	48.081	46.044	46.050
	$\bar{\omega}_5$	75.952	75.960	77.191	77.175	80.470	80.492	84.936	84.989
5.0	$\bar{\omega}_1$	9.6262	9.6260	9.6137	9.6135	9.5774	9.5773	9.5203	9.5208
	$\bar{\omega}_2$	18.707	18.706	18.709	18.710	18.714	18.715	18.723	18.724
	$\bar{\omega}_3$	49.635	49.637	49.637	49.639	49.658	49.659	49.742	49.742
	$\bar{\omega}_4$	60.266	60.267	59.896	59.900	58.962	58.966	57.827	57.838
	$\bar{\omega}_5$	95.733	95.738	96.623	96.679	99.085	99.114	102.60	102.65

(header spanning: $b/c = 0.5$)

In Table 9.5-3 the parameters are defined as: $\bar{\Omega} = \dfrac{\Omega}{3.51602\lambda_1}, \ \bar{\omega}_i = \dfrac{\omega_i}{\lambda_1}, \lambda_1 = \sqrt{a_{22}/b_1 L^4}$

S. & K. $\Rightarrow$ Subrahmanyam and Kaza (1986)

9.5-6 Numerical Results and Discussion

The turbine blade is modeled as a tapered thin-walled beam of rectangular cross-section with geometric characteristics

$$R_0 = 1.3\text{m}, \ L = 1.52\text{m}, \ c_R = 0.257\text{m}, \ b_R = 0.0827\text{m}, \ h = 0.01654\text{m}$$

For box-beam blades that are non-uniform in the height and width directions, the inertia terms b_i^p are

$$[b_1^p(z); b_4^p(z); b_5^p(z); b_6^p] = m_0[2c(z) + 2c(z)S(z); c^3(z)S^2(z)(3 + S(z))/6;$$
$$c^3(z)(1 + 3S(z))/6; 0], \qquad (9.5\text{-}20a,b)$$
$$[b_{13}^p(z); b_{14}^p(z); b_{15}^p(z)] = m_2[0; 2c(z); 2c(z)S(z)],$$

The blade is made up of FGMs with two distinct material phases, that is ceramic, [Silicon Nitride (SN)], and metal, [Stainless Steel (SS)], featuring temperature-dependent properties. The material properties denoted generically as P, can be expressed as a function of temperature in the form of Eq. (9.5-11). For the two basic constituents, the coefficients P_i are provided in Table 9.5-4.

Table 9.5-4: Material properties of FGM constituents*

Material Properties	Material	P_0	P_{-1}	P_1	P_2	P_3
$E(N/m^2)$	SN	348.43×10^9	0	-3.070×10^{-4}	2.160×10^{-7}	-8.946×10^{-11}
	SS	201.04×10^9	0	3.079×10^{-4}	-6.534×10^{-7}	0
ν	SN	0.2400	0	0	0	0
	SS	0.3262	0	-2.002×10^{-4}	3.797×10^{-7}	0
$\rho(kg/m^3)$	SN	2370	0	0	0	0
	SS	8166	0	0	0	0

*The properties are evaluated at $T = 300° K$

This trend is stronger for larger temperature gradients than in that of the uniform through-the-blade thickness temperature distribution. Therefore, the proper selection of K can lead to the increase of the bending stiffness.

Figure 9.5-1 displays the variation of bending stiffness quantities a_{22} and $0.1a_{33}$, as a function of the volume fraction parameter, for two values of the temperature gradient; the bending stiffnesses a_{22} and a_{33} decrease with the increase of K from $K = 0$ (full ceramic) towards $K = 10$ (full metal).

Figures 9.5-2 and 9.5-3 show for non-pretwisted and non-rotating blades the variations of the first and second decoupled natural frequencies in flapping and lagging, with that of the temperature gradient, for selected values of the volume fraction parameter K.

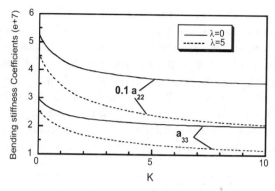

Figure 9.5-1: Variation of bending stiffness quantities with that of the volume fraction parameter K, for two selected values of λ ($\alpha_c = \alpha_h = 1$)

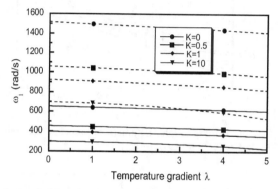

Figure 9.5-2: First decoupled flapping (——— ω_{F1}) and lagging (- - - - - ω_{L1}) natural frequencies with temperature gradient for selected values of K ($\beta_0 = \Omega = \gamma = 0$, $\alpha_c = \alpha_h = 1.2$)

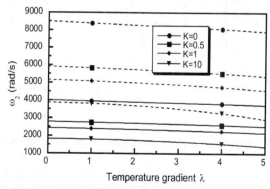

Figure 9.5-3: Counterpart of Fig. 9.5-2 for the second decoupled flapping (——— ω_{F2}) and lagging (- - - - - ω_{L2}) natural frequencies

As a general trend, the increase of the thermal gradient λ results in a decay of the eigenfrequencies. This trend is stronger for the lagging frequencies than for the flapping ones. A similar trend was found in Section 9.3. However, the strongest variation is due to the variation of the volume fraction K. In this sense, the eigenfrequencies are the largest ones in ceramic blades and decay with the increase of K, towards the full metallic blades. Moreover, the decay is steeper for the lagging frequencies, than for the flapping ones.

The counterparts of Figs. 9.5-2 and 9.5-3 for the coupled bending-bending natural frequencies, occurring for pretwisted blades are displayed in Figs. 9.5-4 and 9.5-5. The results reveal a similar trend of variation of natural frequencies with respect to the variation of K and λ, as for the decoupled natural frequencies.

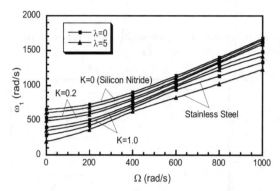

Figure 9.5-4: First flapping-lagging coupled natural frequencies vs. rotational speed for selected values of K, ($\beta_0 = 58$ deg., $\gamma = 0$, $\alpha_c = \alpha_h = 1.2$)

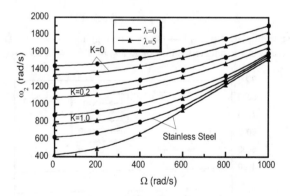

Figure 9.5-5: Counterpart of Fig. 9.5-4 for the second coupled natural frequency

However, the latter results reveal that the strongest decay of natural frequencies with the increase of λ occurs for increased values of K approaching the metallic blade.

Figures 9.5-6 and 9.5-7 display, for untwisted tapered blades, the variations of lagging and flapping natural frequencies with the rotating speed. The variation of the lagging and flapping frequencies with the rotational speed reveal a trend that was already reported. In addition, the results show that for the first natural frequencies, for tapers $\alpha_c < 1$, there is an increase of eigenfrequencies in lagging and flapping as compared to the blades characterized by $\alpha_c \geq 1$.

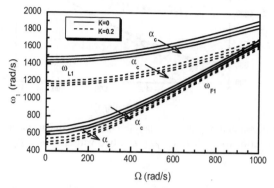

Figure 9.5-6: First decoupled flapping and lagging natural frequencies vs. rotational speed for selected chord taper ratios ($\alpha_c = 0.7$; 1; 1.3) and two values of K, ($\beta_0 = 0$, $\gamma = 0$, $\alpha_h = 1$, $\lambda = 5$)

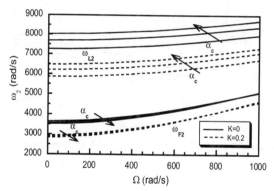

Figure 9.5-7: Counterpart of Fig. 9.5-6 for the second decoupled flapping and lagging natural frequencies

However, for the flapping frequencies, this trend tends to be attenuated with the increase of Ω. For the second frequencies in flapping and lagging, there is a more complex variation with the chord taper ratio.

In this sense, for the uniform tapered blades, corresponding to $\alpha_c \geq 1$, the lagging frequencies are larger than those corresponding to $\alpha_c < 1$, a trend that occurs irrespective of the rotational speed. However, at relatively lower rotational speeds, the flapping frequencies for tapers $\alpha_c \leq 1$ are slightly larger than those corresponding to $\alpha_c > 1$, a trend that is reversed once Ω increases.

Figures 9.5-8 and 9.5-9 show the effect of the height-taper ratio α_h on flapping and lagging frequencies. In this case, as is readily seen, the trend of variation with α_h is more complex than that with α_c.

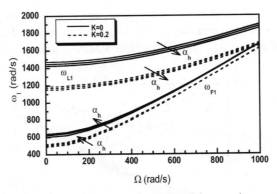

Figure 9.5-8: First decoupled flapping and lagging natural frequencies vs. rotational speed for various height taper ratios ($\alpha_h = 0.7; 1; 1.3$), and two values of K, ($\beta_0 = 0$, $\gamma = 0$, $\alpha_c = 1$, $\lambda = 5$)

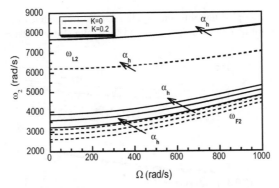

Figure 9.5-9: Counterpart of Fig. 9.5-8 for the second decoupled flapping and lagging natural frequencies

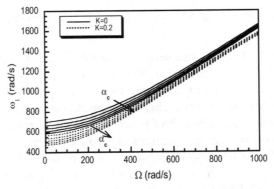

Figure 9.5-10: First natural frequency vs. rotational speed for various chord taper ratios $\alpha_c = 0.6; 0.8; 1; 1.3; 1.4$ and two values of K, ($\beta_0 = 0$, $\gamma = 0$, $\alpha_h = 1$, $\lambda = 5$)

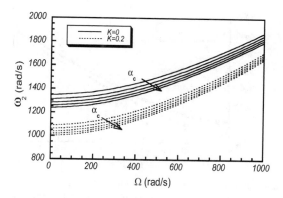

Figure 9.5-11: The counterpart of Fig. 9.5-10 for the second natural frequency

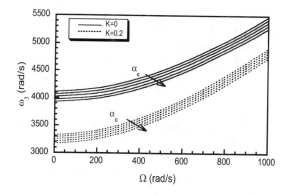

Figure 9.5-12: The counterpart of Fig. 9.5-10 for the third natural frequency.

Note that in all cases (see Figs. 9.5-10 through 9.5-12), the chord taper $\alpha_c \leq 1$ plays a more beneficial influence on coupled frequencies than for $\alpha_c > 1$, a trend that remains valid at any Ω.

An opposite effect (see Figs. 9.5-13 through 9.5-15) is played by the height taper, in the sense that the tapered blades ($\alpha_h \geq 1$), experience larger natural frequencies than those corresponding to $\alpha_h < 1$.

The pretwist plays an alternating effect on the natural frequencies: for the odd mode frequencies the increase of the pretwist is beneficial, in the sense of their increase, while for the even mode frequencies it is not.

The effect of pretwist coupled with that of taper on the FGM nonrotating blades is supplied in Figs. 9.5-16 through 9.5-18.

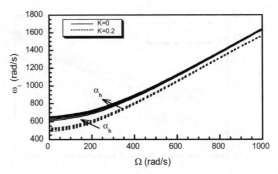

Figure 9.5-13: First natural frequency vs. rotational speed for selected height taper ratios (α_h = 0.6; 0.8; 1; 1.2; 1.4) and two values of K, (β = 90 deg., γ = 0, α_c = 1, λ = 5)

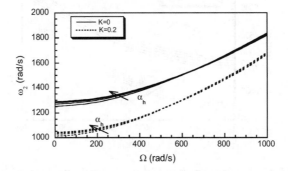

Figure 9.5-14: Counterpart of Fig. 9.5-13 for the second natural frequency

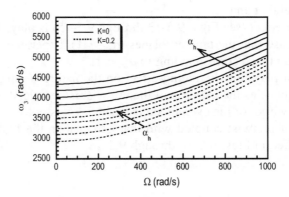

Figure 9.5-15: Counterpart of Fig. 9.5-13 for the third natural frequency

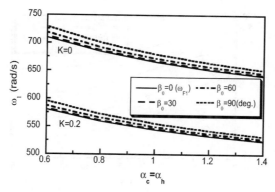

Figure 9.5-16: First natural frequency vs. taper ratio for selected pretwist angles, (for $\beta_0 = 0$, $\omega_1 \rightarrow \omega_{F1}$) and for $K = 0$ and $K = 0.2$ ($\Omega = 0$, $\lambda = 0$)

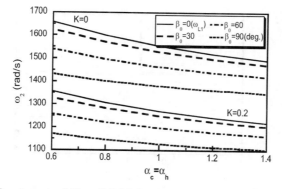

Figure 9.5-17: Counterpart of Fig. 9.5-16 for the second natural frequency (for $\beta_0 = 0$, $\omega_2 = \omega_{L_1}$)

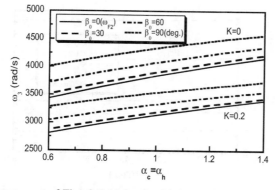

Figure 9.5-18: Counterpart of Fig. 9.5-16 for the third natural frequency (for $\beta_0 = 0$, $\omega_3 = \omega_{F2}$)

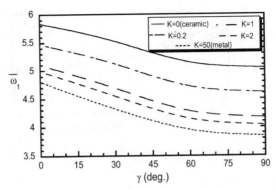

Figure 9.5-19: Variation of the first natural frequency vs. presetting angle, for selected values of the parameter K ($\bar{\Omega} = 3$, $\lambda = 0$, $\beta_0 = 45$ deg.)

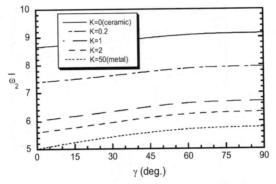

Figure 9.5-20: Counterpart of Fig. 9.5-19 for second natural frequency vs. presetting angle, for selected values of K

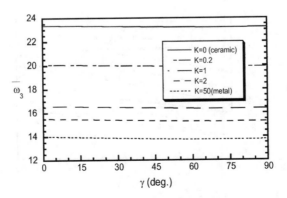

Figure 9.5-21: Counterpart of Fig. 9.5-19 for the third natural frequency vs. presetting angle, for selected values of K

On the other hand, with the simultaneous increase of the taper in chord and height, the first two coupled natural frequencies decrease for $\alpha_c(=\alpha_h) > 1$ and increase for $\alpha_c(=\alpha_h) \leq 1$, while the third one increases.

Until this point, in the numerical simulations, the effect of the presetting angle was neglected. The next figures highlight its influence. Figures 9.5-19 through 9.5-21 show in succession the variations of the first three dimensionless natural frequencies $\overline{\omega}_i$, as a function of the setting angle, for a prescribed value of the rotational speed, $\overline{\Omega}$, pretwist angle, and for selected values of K. In the numerical simulations the following dimensionless expressions of the rotational speed, and natural frequencies have been used: $(\overline{\Omega}^2, \overline{\omega}_i^2) = (\Omega^2, \omega_i^2)b_1 L^4/a_{33}$.

The results reveal that, depending on the odd or even mode number, the increase of the presetting angle yields either a decrease or increase of natural frequencies. These plots show the strong effect played by the volume fraction parameter K: the eigenfrequencies are the largest ones for ceramic blades ($K = 0$), and decay with the increase of K, toward the full metallic blades. Due to the fact that in the case considered here, both the pretwist and presetting are different of zero, the frequencies are coupled, in the sense that these represent the flapping-lagging coupled natural frequencies. Figures 9.5-22 through 9.5-24, depict the variation of the first three coupled natural frequencies vs. the dimensionless rotating speed, for two values of the setting angle: $\gamma = 0$ (corresponding to the motion in flapping) and $\gamma = 90$ deg. (corresponding to the motion in lagging), and for selected values of the parameter K. These plots, reveal the beneficial effect consisting of the increase of natural frequencies with the increase of $\overline{\Omega}$.

The effect of the presetting angle, is similar to that shown in the previous three figures: for odd mode numbers the frequencies decay with the increase of γ, while for even mode numbers, the opposite effect takes place.

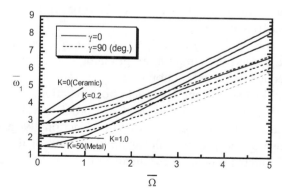

Figure 9.5-22: Variation of the first natural frequency of the uniform blade vs. the rotational speed for selected values of K ($\beta_0 = 45$ deg., $\lambda = 0$)

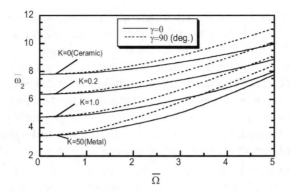

Figure 9.5-23: Counterpart of Fig. 9.5-22 for the second natural frequency

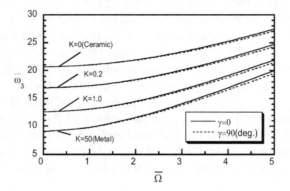

Figure 9.5-24: Counterpart of Fig. 9.5-22 for the third natural frequency

These figures show that while the decrease of the first natural frequency with the increase of the volume fraction parameter K tends to be attenuated once the rotational speed increases, this trend tends to become less accentuated for larger mode eigenfrequencies.

Finally, Figs. 9.5-25 through 9.5-27 show in succession for selected values of the setting angle and the temperature gradient λ, the variations of the first three natural frequencies as a function of K.

Note that within the present approach, for the effective material properties the simple rule of mixtures was used. In order to validate its accuracy, the first three natural frequencies of a rotating functionally graded beam, obtained via the use of the present rule of mixtures and of the more elaborate one by the Mori-Tanaka scheme (see Cheng and Batra, 2000), are compared in Table 9.5-5.

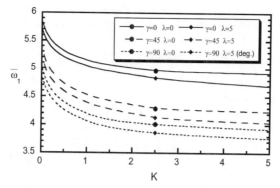

Figure 9.5-25: Variation of the first dimensionless frequency of uniform blades as a function of K for selected setting angles ($\beta_0 = 45$ deg., $\bar{\Omega} = 3$)

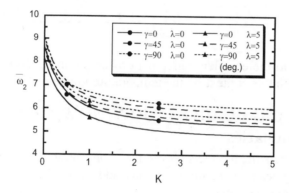

Figure 9.5-26: Counterpart of Fig. 9.5-25 for the second natural frequency

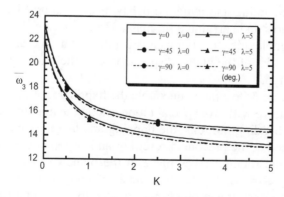

Figure 9.5-27: Counterpart of Fig. 9.5-25 for the third natural frequency

Table 9.5-5: Comparison of dimensionless natural frequencies determined via the simple rule of mixtures and the Mori-Tanaka (M-T) scheme for selected values of the temperature gradient and the parameter K, $(\overline{\Omega} = 3,\ \beta_0 = 0,\ \gamma = 90$ deg, $\overline{R}_0 = 1.3)$

Volume fraction parameter	Temperature gradient	$\lambda = 0$			$\lambda = 2$		
	Dimensionless frequencies	$\overline{\omega}_1$	$\overline{\omega}_2$	$\overline{\omega}_3$	$\overline{\omega}_1$	$\overline{\omega}_2$	$\overline{\omega}_3$
Fully Ceramic	Present	5.020	9.335	22.571	4.955	9.148	22.101
	M-T	5.020	9.335	22.571	4.955	9.148	22.101
$K = 0.5$	Present	4.334	7.313	17.420	4.293	7.188	17.099
	M-T	4.324	7.282	17.340	4.282	7.154	17.013
$K = 1.0$	Present	4.157	6.774	16.038	4.122	6.667	15.761
	M-T	4.148	6.746	15.964	4.111	6.636	15.678
Fully Metal	Present	3.839	5.811	13.543	3.811	5.732	13.327
	M-T	3.839	5.811	13.543	3.811	5.732	13.327

In addition to the excellent agreement of predictions, the results show the effects played in this context by the temperature gradient, λ.

9.6 ROTATING BLADES CARRYING A TIP MASS

9.6-1 Preliminaries

In spite of the beneficial effects played by a concentrated mass mounted at the tip of a rotating beam, there are few studies addressing the vibrational behavior of such structural systems.

Among the studies related to this topic, the pioneering one due to Boyce and Handelman (1961) should be mentioned. We now present several basic issues related to the modeling and dynamic behavior of rotating blades modelled as thin-walled beams and carrying a tip mass. However, in order to eliminate the unavoidable complexities involving consideration of a general anisotropy of the constituent material of the blade, a special one of the transverse-isotropy type will be considered.

This will enable us to exactly decouple the flapping motion from the lagging one. As a result, we will give special emphasis on the dynamic behavior of the two decoupled bending motions, the flapping and lagging ones.

It is assumed that the precone, presetting and sweep angles of the blade are zero, and that the rotation takes place in the plane (X, Z) with the constant angular velocity $\mathbf{\Omega}(\equiv \Omega \mathbf{J} = \Omega \mathbf{j})$, the spin axis being along the Y-axis (see Fig. 9.6-1). In this context, this problem was treated by Song and Librescu (1999).

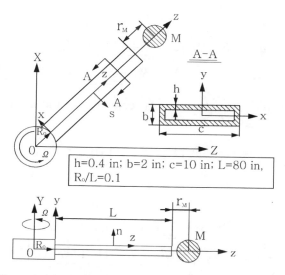

Figure 9.6-1: Geometry of the rotating beam with a tip mass

9.6-2 Kinematics, Equations of Motion and Boundary Conditions

Hamilton's principle will be used to generate the equations of motion and the associated boundary conditions. First, in order to evaluate the quantities entering the energy functional, the position vectors, $\mathbf{R}\ (\equiv \mathbf{R}(x, y, z; t))$ and $\mathbf{R}_M$ of a deformed point of the beam and of the tip mass centroid are defined, respectively,

$$\mathbf{R} = \mathbf{R}_o + \mathbf{r} + \Delta; \qquad \mathbf{R}_M = (R_o + L + r_M)\,\mathbf{k} + \Delta_o. \qquad (9.6\text{-}1a,b)$$

In Eq. (9.6-1), $\mathbf{r}\ (\equiv x\mathbf{i} + y\mathbf{j} + z\mathbf{k})$ denotes the position vector measured in the rotating axis system; $\mathbf{R}_o = R_o\mathbf{K}$ while $\Delta\ (\equiv u\mathbf{i} + v\mathbf{j} + w\mathbf{k})$ and $\Delta_o \equiv (\hat{u}_o\mathbf{i} + \hat{v}_o\mathbf{k} + \hat{w}_o\mathbf{k})$ denote the displacement vectors of the points of the blade and of the tip mass centroid, respectively, where, by virtue of the assumption that the displacement components of the tip mass centroid coincide with those of the blade tip, we have, $(\hat{u}_o, \hat{v}_o, \hat{w}_o) \equiv (u_o, v_o, w_o)\,|_{z=L}$; r_M denotes the offset between the beam tip and centroid of the tip mass, while L denotes the beam span. We will also make use of the equations expressing the time derivatives of unit vectors, Eq. (9.1-6), of expressions of components R_i of the position vector of a deformed point of the blade and tip mass, and, as a result, of expressions of the velocity V_i and acceleration a_i components. The latter are

$$V_1 = \dot{u} + (R_o + z + w)\,\Omega; \qquad V_2 = \dot{v}; \qquad V_3 = \dot{w} - (x + u)\,\Omega, \qquad (9.6\text{-}2a\text{-}c)$$

and

$$a_1 = \ddot{u} + \underline{2\dot{w}\Omega} - \underline{(x+u)\,\Omega^2}; \quad a_2 = \ddot{v},$$

$$a_3 = \ddot{w} - \underline{2\dot{u}\Omega} - \underline{(R_o + z + w)\,\Omega^2}, \qquad (9.6\text{-}3a\text{-}c)$$

respectively, whereas the velocity and acceleration components of the tip mass centroid are expressed as

$$(V_1)_M = \left(R_o + L + r_M + \hat{w}_o\right)\Omega + \dot{\hat{u}}_o; \quad (V_2)_M = \dot{\hat{v}}_o; \quad (V_3)_M = \dot{\hat{w}}_o - \hat{u}_o\Omega \qquad (9.6\text{-}4a\text{-}c)$$

and

$$(a_1)_M = \left(\ddot{\hat{u}}_o + \underline{2\dot{\hat{w}}_o\Omega} - \underline{\underline{\Omega^2\hat{u}_o}}\right); \quad (a_2)_M = \ddot{\hat{v}}_0,$$

$$(a_3)_M = \left[\ddot{\hat{w}}_o - \underline{2\dot{\hat{u}}_o\Omega} - \underline{\underline{\left(R_o + L + r_M + \hat{w}_o\right)\Omega^2}}\right], \qquad (9.6\text{-}5a\text{-}c)$$

respectively. Single and double superposed lines identify the Coriolis and centrifugal acceleration terms, respectively.

In addition, in Eqs. (9.6-2) and (9.6-3), u, v and w denote the components of the 3-D displacement vector defined by Eqs. (2.1-6) and (2.1-43).

Following the procedure described in Section 9.1, in conjunction with the previously displayed equations, from Hamilton's principle we obtain the equations of motion and the associated boundary conditions of rotating thin-walled beams carrying a tip mass. These are

Equations of motion:

$$\delta u_o: \ Q_x' + \left(T_z u_o'\right)' - b_1\left(\ddot{u}_o + \underline{2\dot{w}_o\Omega} - \underline{u_o\Omega^2}\right) + p_x = 0,$$

$$\delta v_o: \ Q_y' + \left(T_z v_o'\right)' - b_1\ddot{v}_o + p_y = 0,$$

$$\delta w_o: \ T_z' - b_1\left[(\ddot{w}_o - \underline{2\dot{u}_o\Omega}) - \underline{(R_o + z + w_o)\,\Omega^2}\right] + p_z = 0,$$

$$\delta\theta_y: \ M_y' - Q_x - (b_5 + \delta_n b_{15})\,(\ddot{\theta}_y - \underline{\Omega^2\theta_y}) + m_y = 0, \qquad (9.6\text{-}6a\text{-}f)$$

$$\delta\theta_x: \ M_x' - Q_y - (b_4 + \delta_n b_{14})\,(\ddot{\theta}_x - \underline{\Omega^2\theta_x}) - 2b_4\Omega\dot{\phi} + m_x = 0,$$

$$\delta\phi: \ \underline{B_\omega''} + M_z' + \left(T_r\phi'\right)' - (b_4 + b_5)\ddot{\phi} + 2b_4\Omega\dot{\theta}_x + \underline{(b_4 - b_5)\,\Omega^2\phi}$$

$$+ (b_{10} + \delta_n b_{18})\left(\underline{\ddot{\phi}''} - \underline{\underline{\Omega^2\phi''}}\right) + m_z + \underline{m_\omega'} = 0,$$

The static boundary conditions:

$$\delta u_o: \quad Q_x + T_z u_o' + M\left(\ddot{u}_o + \underline{2\Omega\dot{w}_o} - \underline{\Omega^2 u_o} - \underline{r_M\ddot{\theta}_y}\right) = \underset{\sim}{Q}_x,$$

$$\delta v_o: \quad Q_y + T_z v_o' + M\left(\ddot{v}_o - \underline{r_M\ddot{\theta}_x}\right) = \underset{\sim}{Q}_y,$$

$$\delta w_o: \quad T_z + M\left[\ddot{w}_o - \underline{2\Omega\dot{u}_o} - \underline{(R_o + L + r_M + w_o)\Omega^2}\right] = \underset{\sim}{T}_z,$$

$$\delta\theta_y: \quad M_y + \underline{I_{M_y}\ddot{\theta}_y} - \underline{Mr_M\ddot{u}_o} = \underset{\sim}{M}_y,$$

$$\delta\theta_x: \quad M_x + \underline{I_{M_x}\ddot{\theta}_x} - \underline{Mr_M\ddot{v}_o} = \underset{\sim}{M}_x, \qquad\qquad (9.6\text{-}7\text{a-g})$$

$$\delta\phi: \quad \underline{B_\omega'} + M_z + T_r\phi' + (b_{10} + \delta_n b_{18})\left(\underline{\ddot{\phi}'} - \underline{\underline{\Omega^2\phi'}}\right) + \underline{I_{M_z}\ddot{\phi}} = \underset{\sim}{M}_z,$$

$$\delta\phi': \quad \underline{B_\omega} = \underline{\underset{\sim}{B}_\omega}.$$

The geometric boundary conditions:

$$u_o = \underset{\sim}{u}_o; \quad v_o = \underset{\sim}{v}_o; \quad w_o = \underset{\sim}{w}_o; \quad \theta_y = \underset{\sim}{\theta}_y; \quad \theta_x = \underset{\sim}{\theta}_x; \quad \phi = \underset{\sim}{\phi}; \quad \underline{\phi'} = \underline{\underset{\sim}{\phi'}}$$

$$(9.6\text{-}8\text{a-g})$$

The quantities related to the mass moments of inertia of the tip mass are defined in Table 9.6-1. These quantities, as well as the tip mass appear only in the boundary conditions at the beam tip, and are identified by the dotted line •••• . For non-rotating beams, a special case of previously obtained equations have been derived by Song et al. (1994).

Table 9.6-1: Definitions of the mass moments of inertia of the tip mass

Quantity	Definition
I_{M_x}	$M\left(\bar{k}_x^2 + \bar{r}_M^2\right)L^2$
I_{M_y}	$M\left(\bar{k}_y^2 + \bar{r}_M^2\right)L^2$
I_{M_z}	$M\bar{k}_z^2 L^2$

Here, $\left(\bar{k}_x, \bar{k}_y, \bar{k}_z, \bar{r}_M\right) \equiv \left(k_x, k_y, k_z, r_M\right)/L$, where k_x, k_y and k_z denote the radii of gyration about the *x, y,* and *z*-axes, respectively. As in Section 9.1, one obtains

$$T_z(z) = b_1 \Omega^2 [R_o (L - z) + \frac{1}{2} (L^2 - z^2)]$$

$$+ \underline{M \Omega^2 (R_o + L + r_M)} \quad , \tag{9.6-9}$$

and

$$T_r = T_z \hat{I}_p. \tag{9.6-10}$$

Note that the equations of motion, Eqs. (9.6-6), and boundary conditions, Eqs. (9.6-7) and (9.6-8), are independent of the constitutive equations, and are therefore valid for any type of anisotropic material.

9.6-3 The Governing System

9.6-3a Transverse Shear Effect Included

The governing system of rotating blades can be represented in terms of displacement quantities. This can be done by replacing in the equations of motion the stress-resultants and stress-couples in terms of displacements.

Note that even if Coriolis' effect is neglected, the governing equations for a general type of anisotropy of the layer materials, are fully coupled.

However, for the type of anisotropy considered here (i.e. transverse-isotropy, the surface of isotropy being parallel to the beam mid-surface), the governing equations obtained from Eqs. (9.6-6a,d), (9.6-6b,e), (9.6-6c) and (9.6-6f) are uncoupled. This means that the governing system splits exactly into four uncoupled subsystems, governing the coupled shear-chordwise bending, the shear-flapwise bending, as well as the extension and the twist motions, respectively. These sub-systems of equations are labeled as (A), (B), (C) and (D), respectively.

Since the analysis is confined to the lagging and flapping vibrations only, the pertinent governing equations are given explicitly as *the equations involving the chordwise bending-shear coupling (i.e. system (A))* (see Song and Librescu, 1999):

The governing equations

$$\delta u_o : \quad a_{44} \left(u_o'' + \theta_y' \right) + \Omega^2 \left\{ \left[\!\left[b_1 \left[R_o(L - z) + \frac{1}{2} (L^2 - z^2) \right] \right.\right.$$

$$+ \underline{M (R_o + L + r_M)} \left.\left.\right]\!\right] u_o' \right\}'$$

$$- 2b_1 \Omega \dot{w}_o + b_1 \Omega^2 u_o - b_1 \ddot{u}_o + p_x = 0,$$

$$\delta \theta_y : \quad a_{22} \theta_y'' - a_{44} \left(u_o' + \theta_y \right)$$

$$- (b_5 + \delta_n b_{15})(\ddot{\theta}_y - \Omega^2 \theta_y) + m_y = 0. \qquad (9.6\text{-}11\text{a,b})$$

and the *boundary conditions*

At $z = 0$:

$$u_o = \theta_y = 0, \qquad\qquad (9.6\text{-}12\text{a,b})$$

and at $z = L$:

$$\delta u_o : \quad a_{44} \left(u_o' + \theta_y \right) + M\Omega^2 \left(R_o + L + r_M \right) u_o'$$

$$+ M \left(\ddot{u}_o + 2\Omega \dot{w}_o - \Omega^2 u_o - r_M \ddot{\theta}_y \right) = 0,$$

$$\delta \theta_y : \quad a_{22} \theta_y' + I_{M_y} \ddot{\theta}_y - M r_M \ddot{u}_o = 0. \qquad (9.6\text{-}13\text{a,b})$$

The equations involving the flapwise bending-shear coupling (i.e. system (B)) are

$$\delta v_o : \quad a_{55} \left(v_o'' + \theta_x' \right) + \Omega^2 \left\{ \left[\! \left[b_1 \left[R_o(L - z) + \frac{1}{2}(L^2 - z^2) \right] \right. \right. \right.$$

$$\left. \left. \left. + M\left(R_o + L + r_M \right) \right] \! \right] v_o' \right\}' - b_1 \ddot{v}_o + p_y = 0, \qquad (9.6\text{-}14\text{a,b})$$

$$\delta \theta_x : \quad a_{33} \theta_x'' - a_{55} \left(v_o' + \theta_x \right) - (b_4 + \delta_n b_{14}) \left(\ddot{\theta}_x - \Omega^2 \theta_x \right)$$

$$- 2b_4 \Omega \dot{\phi} + m_x = 0,$$

and the boundary conditions:

At $z = 0$:

$$v_o = \theta_x = 0, \qquad\qquad (9.6\text{-}15\text{a,b})$$

and at $z = L$:

$$\delta v_o : \quad a_{55}(v_o' + \theta_x) + \underline{M\Omega^2(R_o + L + r_M)v_o'}$$

$$+ M(\ddot{v}_o - r_M\ddot{\theta}_x \quad) = 0,$$

$$\delta\theta_x : \quad a_{33}\theta_x' + I_{M_x}\ddot{\theta}_x \quad - \quad Mr_M\ddot{v}_o = 0. \qquad (9.6\text{-}16\text{a,b})$$

9.6-3b Transverse Shear Effect Neglected

Elimination of $a_{44}(u_o' + \theta_y)$ and $a_{55}(v_o' + \theta_x)$ in the group of equations (A) and (B), respectively, followed by consideration of $\theta_x \Rightarrow -v_o'$ and $\theta_y \Rightarrow -u_o'$, results in their non-shearable counterpart, expressed as

Classical counterpart of the equation group (A):

$$\delta u_o : \quad a_{22}u_o'''' - \Omega^2\left\{\left[\!\!\left[b_1\left[R_o(L - z) + \frac{1}{2}(L^2 - z^2)\right]\right.\right.\right.$$

$$+ \; M(R_o + L + r_M) \left.\!\!\right]u_o'\right\}' + 2b_1\Omega\dot{w}_o - b_1\Omega^2 u_o \qquad (9.6\text{-}17)$$

$$+ \; b_1\ddot{u}_o - (b_5 + \delta_n b_{15})(\ddot{u}_o'' - \Omega^2 u_o'') - p_x - m_y' = 0,$$

and the boundary conditions:

At $z = 0$:

$$u_o = u_o' = 0, \qquad\qquad (9.6\text{-}18\text{a,b})$$

and at $z = L$:

$$\delta u_o : \quad a_{22}u_o''' - M\Omega^2(R_o + L + r_M)u_o' - M(\ddot{u}_o + 2\Omega\dot{w}_o - \Omega^2 u_o$$

$$+ \; r_M\ddot{u}_o') - (b_5 + \delta_n b_{15})(\ddot{u}_o' - \Omega^2 u_o') - m_y = 0,$$

$$\qquad\qquad (9.6\text{-}19\text{a,b})$$

$$\delta u_o' : \quad a_{22}u_o'' + I_{M_y}\ddot{u}_o' + Mr_M\ddot{u}_o = 0.$$

Classical counterpart of the equation group (B):

$$\delta v_o : \quad a_{33} v_o'''' - \Omega^2 \left\{ \left[b_1 \left[R_o (L - z) + \frac{1}{2}(L^2 - z^2) \right] \right. \right.$$

$$\left. \left. + \; M(R_o + L + r_M) \right] v_o' \right\}'$$

$$- (b_4 + \delta_n b_{14})(\; \ddot{v}_o'' - \Omega^2 v_o'') + b_1 \ddot{v}_o - p_y - m_x' = 0, \qquad (9.6\text{-}20)$$

and the boundary conditions:

At $z = 0$:

$$v_o = v_o' = 0, \qquad\qquad (9.6\text{-}21\text{a,b})$$

and at $z = L$:

$$\delta v_o : \quad a_{33} v_o''' - \; M\Omega^2 (R_o + L + r_M \;) v_o' - \; M(\ddot{v}_0 + \; r_M \ddot{v}_o' \;)$$

$$- (b_4 + \delta_n b_{14}) (\; \ddot{v}_o' - \Omega^2 v_o') + 2\, b_4 \Omega \dot{\phi} - m_x = 0, \qquad (9.6\text{-}22\text{a,b})$$

$$\delta v_o' : \quad a_{33} v_o'' + \; I_{M_x} \ddot{v}_o' \; + \; Mr_M \ddot{v}_o = 0.$$

Note that the lagging-extension and the flapping-twist couplings induced by the Coriolis effect, occur within the *classical* beam model as well.

Consistent with the number of four boundary conditions, the governing equations associated with lagging and flapping are each of fourth order, a feature that holds valid for both shear-deformable and non-shearable rotating beams.

9.6-4 Numerical Simulations

To validate both the accuracy of the solution method and the structural model, comparisons with a number of results scattered in the specialized literature will be presented. Two ways of solving the eigenvalue problems namely, an exact one based on the Laplace transform technique, and the Extended Galerkin Method show an excellent agreement. Tables 9.6-2a,b compare the results associated with the fundamental and second flapping frequencies obtained for the non-shearable beam model for selected values of $\overline{\Omega}(\equiv \Omega(b_1 L^4/a_{33}))$ and $m_t(\equiv M/b_1 L)$. We set $r_o(\equiv R_o/L) = 0$; $r(\equiv (b_4 + \delta_n b_{14})/(b_1 L^2)) = 0$ and $\tilde{I}_{M_x}(\equiv I_{M_x}/b_1 L^3) = 0$.

Table 9.6-2a: Dimensionless fundamental flapping frequency, $\overline{\omega}_1$, for selected values of the angular speed $\overline{\Omega}$ and tip mass m_t

m_t	$\overline{\Omega}$	H.	Y. & M.	Y.	P.	H. & R.	W.	Present
0	0	3.51602	3.5160		3.5160	3.5162	3.5160	3.51602
1			1.5573	1.5573			1.5573	1.5573
0	3	4.79644	4.7973		4.7964	4.79728	4.7973	4.7973
1			3.5823	3.5823			3.5823	3.5823
0	6	7.35614	7.3604		7.3561	7.36037	7.3604	7.3604
1			6.5092	6.5091			6.5090	6.5098
0	9	10.2192	10.2258			10.2257	10.2257	10.2257
1			9.4904	9.4903			9.4899	9.4907
0	12	12.1634	12.1706		12.1634	12.1702	12.1702	12.1702
1			12.4827				12.4814	12.4848

H. → Hodges (1981); Y & M. → Yokayama and Markiewicz (1993); P. → Peters (1973); H. & R. → Hodges and Rutkowski (1981); W. → Wright et al. (1982); Y. → Yoo et al. (2001).

The dimensionless natural frequencies $\overline{\omega}_i^2 (\equiv \omega_i^2 (b_1 L^4/a_{33}))$ were obtained by Peters (1973) and Hodges (1981) via the asymptotic methods; by Yokoyama and Markiewicz (1993) and Hodges and Rutkowski (1981), via the FEM; by Wang et al. (1976), Du et al. (1994) and Wright et al. (1982), via the Legendre polynomials, power series and the Frobenius technique, respectively; Yoo et al. (2001) via a numerical method.

Figure 9.6-2 shows the dependence of the fundamental eigenfrequency in flapping and lagging on the beam rotation speed ratio for selected values of the tip mass. As in Du et al. (1994), $r = r_0 = \tilde{I}_{M_x} = \tilde{I}_{M_y} = 0$. The results show that for the non-rotating beam, the higher the tip mass the lower the frequencies. Also for the flapping motion, with the increase of the spin rate, the curves corresponding to different tip mass ratios asymptotically approach the line $\overline{\omega}_1 = \overline{\Omega}$. By contrast, with the increase of the rotational speed, the first lagging frequency exhibits a very reduced increase, and in addition, with the increase of the tip mass and $\overline{\Omega}$, the associated curves remain parallel to each other. This trend has been reported previously by Boyce and Handelman (1961), Kumar (1974), Hoa (1979) and Bhat (1986).

Table 9.6-2b: Dimensionless flapping frequency, $\bar{\omega}_2$, for selected values of the angular speed $\bar{\Omega}$ and tip mass m_t

m_t	$\bar{\Omega}$	Y.& M.	P.	Y.	W.	H.& R.	Present
0	0	22.0352	22.0345		22.0345	22.0345	22.0344
1		16.2504		16.2527	16.2500		16.2501
0	3	23.3210	23.3188		23.3203	23.3203	23.3203
1		20.3507	20.3524	20.3524			20.3504
0	6	26.8098	26.7933		26.8091	26.8091	26.8091
1		29.2926		29.2933	29.2917		29.2914
0	9	31.7716	29.2933		31.7705	31.7705	31.7705
1		39.8082		39.8079			39.8051
0	12	37.6050	37.6031		37.6031	37.6033	
1		50.8684			50.8594		50.8613

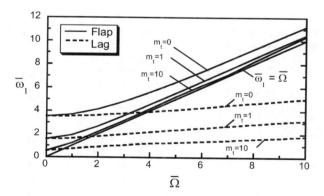

Figure 9.6-2: Fundamental flapping and lagging frequencies vs. rotating speed for mass ratios $(m_t = 0, 1, 10)$ (Euler-Bernoulli model), $(r = 0,\ r_o = 0,\ \tilde{I}_{M_y}(\equiv I_{M_y}/(b_1 L^3)) = \tilde{I}_{M_x} = 0)$

For a rotating non-shearable beam and for selected tip mass ratios, Table 9.6-3 compares the first and second flapping natural frequency predictions obtained by Lee (1993) via a numerical procedure, with those derived by Song and Librescu (1999). As before, $r_o = 0, r = 0$ and $\tilde{I}_{M_x} = 0$.

Table 9.6-3: The first two natural flapping frequencies $\overline{\omega}_i$ for two values of $\overline{\Omega}$ and selected mass ratios m_t

m_t	$\overline{\Omega} = 1$				$\overline{\Omega} = 100$			
	$\overline{\omega}_1$		$\overline{\omega}_2$		$\overline{\omega}_1$		$\overline{\omega}_2$	
	Lee (1993)	Present	Lee (1993)	Present	Lee (1993)	Present	Lee (1993)	Present
0	3.6816	3.6816	22.181	22.1810	101.31	101.490	248.67	248.994
0.02	3.5548	3.5548	21.405	21.4048	101.27	101.449	249.24	249.558
0.04	3.4414	3.4414	20.789	20.7889	101.23	101.388	250.83	251.184
0.05	2.2930	2.2930	17.227	17.2266	100.90	100.993	322.58	323.326
1	1.9017	1.9017	16.757	16.557	100.80	100.608	392.32	393.483
2	1.5902	1.5902	16.725	16.7248	100.72	100.922	504.66	506.086
5	1.3222	1.3223	17.490	17.4896	100.67	99.7214	746.95	749.473
10	1.2052	1.2052	18.972	18.9724	100.65	101.325	1031.4	1034.54

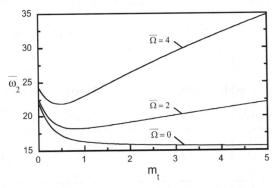

Figure 9.6-3: Variation of second flapping frequencies with mass ratio m_t for different rotating speeds (Euler-Bernoulli model) $(r = 0, r_o = 0, \tilde{I}_{M_x} = 0)$

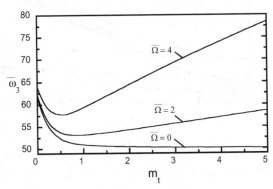

Figure 9.6-4: Counterpart of Fig. 9.6-3 for the third flapping frequencies vs. mass ratio m_t for different rotating speeds (Euler-Bernoulli model)

Figures 9.6-3 and 9.6-4 depict the variation of $\overline{\omega}_n$ $(n = 2, 3)$ vs. the tip mass ratio for selected values of the rotational speed. The displayed results reveal: (a) with the increase of the tip mass for zero rotational speed, the natural frequencies decrease, and (b) for higher mode frequencies, and in contrast to the fundamental frequency, there is a faster increase with the increase of rotational speeds and of the tip mass.

This trend is consistent with that reported by Kumar (1974). A combined plot depicting $\overline{\omega}_n (n = 2, 3)$ as a function of $\overline{\Omega}$, for selected values of the tip mass ratio is presented in Fig. 9.6-5. The displayed results reveal that for higher mode flapping frequencies there is a cross-over between the frequencies of beams carrying different tip masses; it takes place first for the larger values of m_t. A similar trend was reported previously by Boyce and Handelman (1961), Hoa (1979) and Bhat (1986).

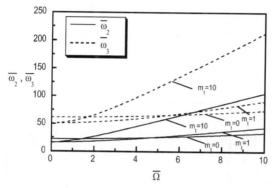

Figure 9.6-5: Second and third flapping frequencies vs. rotating speed for mass ratio $m_t = 0, 1; 10$, (Euler- Bernoulli model), $(r = 0, r_o = 0, \tilde{I}_{M_x} = 0)$

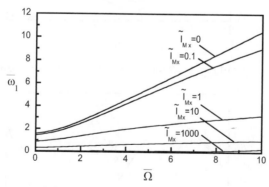

Figure 9.6-6: Flapping fundamental frequency vs. rotating speed for selected values of the moment of inertia of the tip mass (Euler-Bernoulli model) $(m_t = 1, r = 0, r_o = 0)$

Figure 9.6-6 displays the dependence of the fundamental flapping frequency vs. the rotational beam speed $\overline{\Omega}$ for selected values of the moment of inertia of the tip mass $\tilde{I}_{M_x}$. For smaller values of moment of inertia, $\overline{\omega}_1$ increases with $\overline{\Omega}$.

However for large moments of inertia $\tilde{I}_{M_x}$, $\overline{\omega}_1$ tends to zero for any rotational speed. This trend coincides with that reported by Bhat (1986). Figures 9.6-7 through 9.6-9 depict, for different rotational speeds, the first three normalized flapping mode shapes.

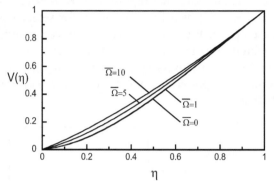

Figure 9.6-7: Influence of rotating speed on the first flapping mode shape ($m_t = 0$, $r_o = 0.1$)

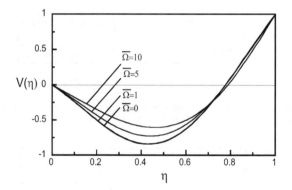

Figure 9.6-8: Counterpart of Fig. 9.6-7 for the second flapping mode shape of rotating speed

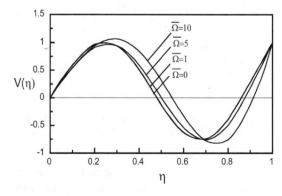

Figure 9.6-9: Counterpart of Fig. 9.6-7 for the third flapping mode shape

The results show that with the increase of $\overline{\Omega}$ the nodal points are shifted towards the tip of the blade; this trend becomes stronger as the mode number increases.

The influence of the tip mass on the first three normalized flapping mode shapes of a non-rotating beam appears clearly from Figs. 9.6-10 through 9.6-12. These graphs, show that the tip mass exerts a great influence on the mode shapes. Among others, for any $m_t \neq 0$, the number of nodal points is reduced by one when comparing with the counterpart case corresponding to the absence of the tip mass.

For a beam equipped with a tip mass ($m_t \neq 0$), the influence of $\tilde{I}_{M_x}$ on the first three mode shapes emerges from Figs. 9.6-13 through 9.6-15; $\tilde{I}_{M_x}$ exerts the strongest influence on the second mode shape.

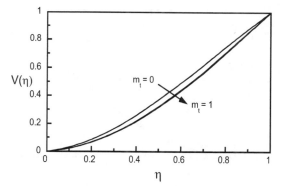

Figure 9.6-10: Influence of the tip mass ratio on the first flapping mode shape of a non-rotating blade ($\overline{\Omega} = 0$, $\tilde{I}_{M_x} = 0.1$ for $m_t \neq 0$, $\tilde{I}_{M_x} = 0$ for $m_t = 0$)

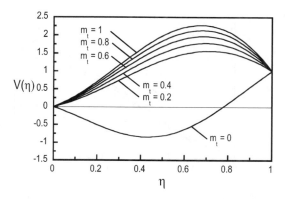

Figure 9.6-11: Counterpart of Fig. 9.6-10 for the second flapping mode shape

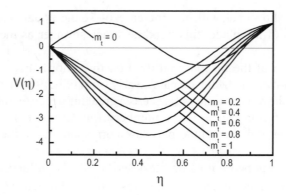

Figure 9.6-12: Counterpart of Fig. 9.6-10 for the third flapping mode shape

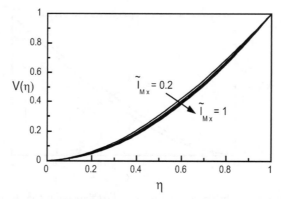

Figure 9.6-13: Influence of the moment of inertia of the tip mass, on the first flapping mode shape of a non-rotating beam ($\overline{\Omega} = 0$, $m_t = 0.5$)

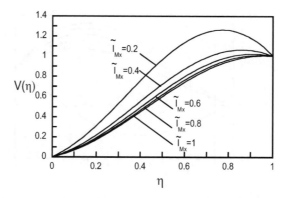

Figure 9.6-14: The counterpart of Fig. 9.6-13 for the second flapping mode shape

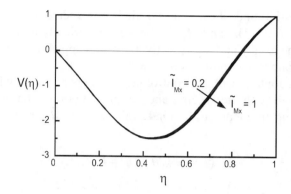

Figure 9.6-15: The counterpart of Fig. 9.6-13 for the third flapping mode shape

9.6-5 Influence of the Induced Strain Actuation

9.6-5a General Considerations

Again consider that the piezoactuators are spread over the entire beam span and the actuation is out-of-phase; bending moments $\tilde{M}_x$ and $\tilde{M}_y$ are piezoelectrically induced at the beam tip, their expression being given by Eqs. (6.6-1). As a result, only the equations at the beam tip are affected by the presence of the induced bending moments; for the shearable beams, from the governing system associated with the lagging motion, Eq. (9.6-11) through (9.6-13), only the boundary condition (9.6-13b) at $z = L$ is changed:

$$\delta\theta_y : a_{22}\theta_y' + \underline{I_{My}\ddot{\theta}_y} - Mr_M\ddot{u}_0 = \tilde{M}_y, \tag{9.6-23}$$

while for the unshearable beam only Eq. (9.6-19b) at $z = L$ is modified as

$$\delta u_0' : a_{22}u_0'' + \underline{I_{My}\ddot{u}_o'} + Mr_m\ddot{u}_0 = -\tilde{M}_y. \tag{9.6-24}$$

Similarly, from the equations in flapping, Eqs. (9.6-14) through (9.6-16), the only boundary condition at $z = L$, (9.6-16b) is changed:

$$\delta\theta_x : a_{33}\theta_x' + \underline{I_{Mx}\ddot{\theta}_x} - Mr_M\ddot{v}_0 = \tilde{M}_x, \tag{9.6-25}$$

while for the unshearable beam counterpart, from the Eqs. (9.6-20) through (9.6-22), Eq. (9.6-22b) becomes:

$$\delta v_0' : a_{33}v_0'' + \underline{I_{Mx}\ddot{v}_0'} + Mr_M\ddot{v}_0 = -\tilde{M}_x. \tag{9.6-26}$$

In Eqs. (9.6-23) and (9.6-25) and in their non-shearable counterparts, Eqs. (9.6-24) and (9.6-26), $\tilde{M}_y$ and $\tilde{M}_x$ are the piezoelectrically induced bending moments about the axes y and x, respectively, which appear as non-homogeneous terms in the boundary conditions at $z = L$.

Within the adopted feedback control law and depending on whether the control is associated with the lagging and flapping motions, the piezoelectrically induced bending moments at the blade tip, $\tilde{M}_y$ and $\tilde{M}_x$, respectively, are expressed as

$$\tilde{M}_y(L) = k_v \dot{\theta}_y(L) + k_a \ddot{\theta}_y(L),$$
$$\tilde{M}_x(L) = k_v \dot{\theta}_x(L) + k_a \ddot{\theta}_x(L). \qquad (9.6\text{-}27\text{a,b})$$

Here, k_v and k_a denote the velocity and acceleration feedback gains, respectively. In this case, the eigenvalue problems involving the equations (9.6-11), (9.6-12), (9.6-13a) and (9.6-23); the Eqs. (9.6-14), (9.6-15), (9.6-16a) and (9.6-25), and (9.6-27) lead to the associated closed-loop eigenvalue problems in lagging and flapping, respectively.

9.6-5b Results and Discussion

For the free vibration problem, it is necessary to solve the closed-loop eigenvalue problem; the unknown displacement variables, u_0, v_0, θ_x and θ_y are represented in a generic form as

$$\mathcal{F}(z, t) = \bar{\mathcal{F}}(z) \exp(\lambda t). \qquad (9.6\text{-}28)$$

Use of representation (9.6-28) in equations associated with flapping and the lagging motions results in two different eigenvalue problems in terms of $\bar{v}_0(z)$ and $\bar{\theta}_x(z)$ on one hand, and in terms of $\bar{u}_0(z)$ and $\bar{\theta}_y(z)$, on the other hand. The discretization of the differential eigenvalue problems in the spatial domain was done via the Extended Galerkin Method which is carried out directly in Hamilton's principle; $\bar{u}_0(z)$, $\bar{v}_0(z)$, $\bar{\theta}_x(z)$ and $\bar{\theta}_y(z)$ are expanded in series of trial functions satisfying the essential boundary conditions. For $k_v \neq 0$, the solution of the algebraic eigenvalue problem yields the closed-loop eigenvalues

$$(\lambda_r, \bar{\lambda}_r) = \sigma_r \pm i\omega_{dr} \qquad r = 1, 2, \dots n, \qquad (9.6\text{-}29)$$

which depend on the feedback control gain k_v, where σ_r is a measure of the damping in the rth mode, while ω_{dr} is the rth frequency of damped oscillations.

Before presenting the numerical results, one should mention that when the acceleration feedback control law is implemented, for *unshearable beams* one can obtain an explicit form of closed-loop natural frequencies.

Applying Rayleigh quotient procedure to the governing equations in conjunction with the associated boundary conditions and Eqs. (9.6-27) specialized by considering $k_v = 0$, one obtain the closed-loop natural frequencies in flapping and lagging:

$$\omega_{flap}^2 = \frac{\int_0^L [a_{33}(\bar{v}_0'')^2 + \Omega^2 P(z)(\bar{v}_0')^2 - I_{xx}\Omega^2(\bar{v}_0')^2]dz}{\int_0^L [b_1\bar{v}_0^2 + I_{xx}(\bar{v}_0')^2]dz + (I_{Mx} - k_a)(\bar{v}_0'(L))^2 + M\bar{v}_0^2(L)},$$

(9.6-30a,b)

and

$$\omega_{lag}^2 = \frac{Q}{S},$$

where

$$Q = \int_0^L [a_{22}(\bar{u}_0'')^2 + \Omega^2 P(z)(\bar{u}_0')^2 - b_1\Omega^2(\bar{u}_0)^2 - I_{yy}\Omega^2(\bar{u}_0')^2]dz$$
$$- M\Omega^2\bar{u}_0(L),$$

(9.6-31a,b)

$$S = \int_0^L [b_1\bar{u}_0^2 + I_{yy}(\bar{u}_0')]dz + (I_{My} - k_a)(\bar{u}_0'(L))^2 + M\bar{u}_0^2(L).$$

In the previous equations

$$P(z) = b_1 R(z) + M R_t,$$

while

(9.6-31c,d)

$$R(z) = R_0(L - z) + \frac{1}{2}(L^2 - z^2) \text{ and } R_t = R_0 + L + r_m.$$

Equations (9.6-30) show that the increase of acceleration feedback gain k_a results in an increase of flapping and lagging natural frequencies. The influence of the other parameters on the natural frequencies appears clearly from the same expressions. Special cases of Eqs. (9.6-30a,b) derived for the case of the non-activated isotropic blades without tip mass can be found in the works by Isakson and Eisley (1960), Johnson (1980) and Peters (1995).

Figures 9.6-16 through 9.6-18 supply results on the variation of dimensionless natural damped frequencies $\bar{\omega}_r$ ($r = 1, 2, 3$), of the lagging motion versus the dimensionless feedback gain $K_v (\equiv k_v/(b_1 a_{22})^{1/2})$, for the first three modes of the non-rotating beam and the absence of the tip mass, for both shearable and non-shearable beams. The results reveal that the increase of the velocity feedback gain results in an increase of the natural frequencies until a fixed value of K_v which depends on the mode number. Beyond that value of K_v, with further increase of the feedback gain, the natural frequencies exhibit a very slight variation. These plots also show that the classical Euler-Bernoulli beam model overestimates the adaptive capability as compared to that of shearable beams. Note that the maximum values of K_v until which the frequencies feature an increase, diminish with the mode number.

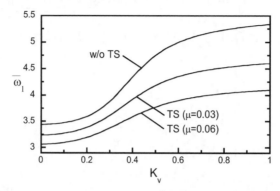

Figure 9.6-16: Influence of the velocity feedback gain and of transverse shear on the fundamental lagging natural frequency ($\mu = a_{22}/a_{44}L^2$, $\bar{\Omega} = 0$, $m_t = 0$)

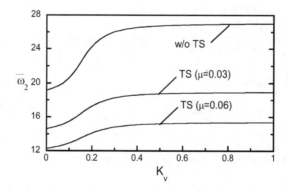

Figure 9.6-17: The counterpart of Fig. 9.6-16 for the second lagging natural frequency

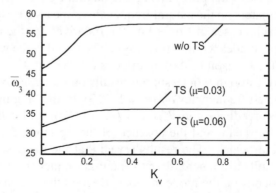

Figure 9.6-18: The counterpart of Fig. 9.6-16 for the third lagging natural frequency

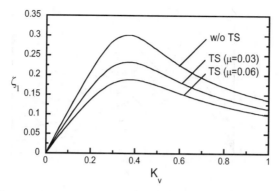

Figure 9.6-19: Influence of the velocity feedback gain and of transverse shear on the piezoelectrically induced first mode damping ratio in flapping, $(\bar{\Omega} = m_t = 0)$

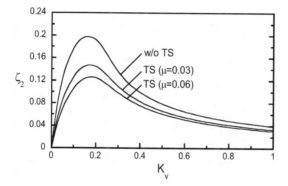

Figure 9.6-20: The counterpart of Fig. 9.6-19 for the induced second mode damping ratio

Figures 9.6-19 through 9.6-21 depict the induced damping factor ζ_r versus the feedback gain K_v, for both the classical and shearable beam models; the induced damping increases with the increase of K_v until a certain value, beyond which a sharp drop in ζ_r is experienced. This trend is similar with that reported by Tzou and Zhong (1991) and Tzou (1993). The limiting values of K_v at which the maxima for ζ_r occur are consistent with those at which $\bar{\omega}_r$ do not exhibit any further increase with the increase of K_v. These graphs, as well as the previous ones, reveal that the classical beam model overpredicts inadvertently both the eigenfrequencies and the induced damping.

Figures 9.6-22 through 9.6-24 depict the variation of closed-loop flapping frequencies $\bar{\omega}_i$ versus the dimensionless feedback gain $K_v (\equiv k_v / (b_1 a_{33})^{1/2})$, for the rotating blade, $(\bar{\Omega} = 1)$, and selected values of the mass ratio m_t; the increase of the m_t results in a decrease of damped frequencies.

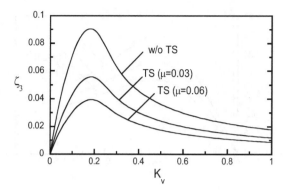

Figure 9.6-21: The counterpart of Fig. 9.6-19 for the induced third mode damping ratio

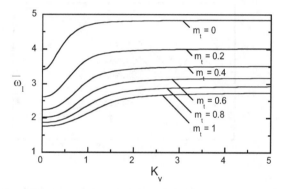

Figure 9.6-22: Flapping fundamental damped eigenfrequency vs. velocity feedback control gain K_v for selected mass ratios m_t, $(\bar{\Omega} = 1)$

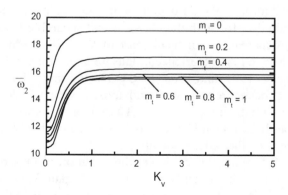

Figure 9.6-23: The counterpart of Fig. 9.6-22 for the second damped flapping eigenfrequency

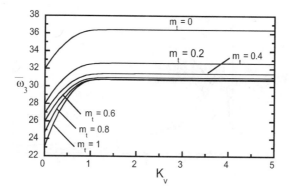

Figure 9.6-24: The counterpart of Fig. 9.6-22 for the third damped flapping eigenfrequency

Moreover, the limiting values of K_v beyond which the frequencies do not exhibit any further rise, increase with the increase of $m_t (\equiv M/m_b)$. Also, with the increase of the mode number, beyond a certain value of m_t, the natural frequencies are less sensitive to the increase of m_t.

Figures 9.6-25 through 9.6-27 depicting the variation of the induced damping ratio in flapping versus the feedback gain K_v, for $\bar{\Omega} = 1$ and selected values of m_t, reveal a different trend, in the sense that the increase of m_t results in an increase of ζ_r. However, the plots reveal that this increase occurs until a limiting value of K_v, beyond which a sharp drop of ζ_r is experienced.

Finally, Figs. 9.6-28 through 9.6-30 depict the dependence of $\bar{\omega}_n (n = \overline{1,3})$ vs. $\bar{\Omega}$ for selected values of the dimensionless accelerated feedback gain $K_a (\equiv k_a/b_1 L^2)$; the closed-loop natural frequencies increase with the acceleration feedback gain, a trend that emerges also from Eqs. (9.6-30). For higher mode frequencies, for $K_a = 0.05$ a substantial increase of natural frequencies is obtained. However, with the increase of K_a beyond that value, there is no further increase of the natural frequency.

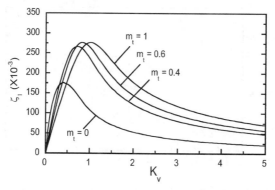

Figure 9.6-25: Piezoelectrically induced first mode damping ratio in flapping vs. velocity feedback gain K_v for selected values of m_t, $(\bar{\Omega} = 1)$

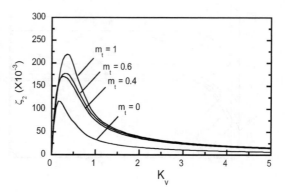

Figure 9.6-26: The counterpart of Fig. 9.6-25 for the second mode

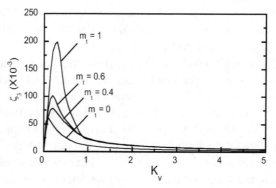

Figure 9.6-27: The counterpart of Fig. 9.6-25 for the third mode piezoelectrically induced damping ratio

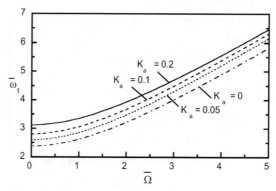

Figure 9.6-28: Dimensionless closed-loop flapping natural frequency $\bar{\omega}_1$ vs. dimensionless rotational speed for selected values of K_a ($m_t = 0.2$)

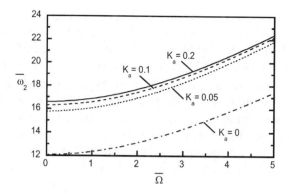

Figure 9.6-29: The counterpart of Fig. 9.6-28 for the closed-loop natural frequency $\bar{\omega}_2$

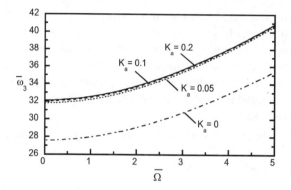

Figure 9.6-30: The counterpart of Fig. 9.6-28 for the closed-loop natural frequency $\bar{\omega}_3$

9.7 PRETWISTED ROTATING BLADES FEATURING EXTENSION-TWIST ELASTIC COUPLING

9.7-1 Preliminaries

We have not yet addressed an important issue, namely that of the modeling and vibration of pretwisted rotating blades built-up as composite thin-walled beams and featuring the twist-extension elastic coupling. In spite of its importance in the design of helicopter and turbomachinery blades, the large majority of studies devoted to this topic have considered standard isotropic structural models.

Within the context of an Bernoulli-Euler isotropic solid beam, a pioneering work was accomplished by Houbolt and Brooks (1958). Their search was followed by those of Rosen and Friedmann (1978), Rosen (1978), Washizu (1964), Tsuiji (1976), Hodges (1980), Krenk (1984), Kosmatka (1992). Studies emphasizing the implications of a number of important effects, such as warping restraint, tennis-racket, and tension-twist coupling on vibration of rotating metallic solid beams have been carried-out by Kaza and Kielb (1984) and Subrahmanyam and Kaza (1985), while for thin-walled metallic beams of open cross-section by McGee (1992). See also the survey-work by Rosen (1991).

In addition to the overall great interest in the study of vibration of pretwisted rotating blades modeled as composite thin-walled beams featuring twist-extension elastic coupling, the works done by Bauchau et al. (1986), Nixon (1987, 1989, 1992), and Lake et al. (1992) show that one can use the tailoring technique to improve tilt-rotor performance by determining the optimum blade twist in both the helicopter and airplane flight modes. As per an advanced composite structural model, this problem was studied by Oh et al. (2004, 2005), and here the basic results will be supplied.

However, before plunging into the development of this section, one should recall the already indicated fact (see the comment at the end of Section 9.2-2), namely that two approaches of the problem are in use. One of these attributed to Wagner, is based on the widely accepted helical multifilament concept applied among others by Houbolt and Brooks (1958). Within the other one related with the name of Washizu (1964), the tensorial approach enabling one of converting the quantities evaluated in the curvilinear (η, ζ, z) system to the local Cartesian one (s, n, z) was used.

The developments associated with the former approach are presented next.

9.7-2 Kinematics

Consider an initially twisted thin-walled beam of length L mounted on a rigid hub of radius R_0 rotating at the constant angular velocity Ω. Three coordinate systems are defined, the inertial one (X, Y, Z), attached to the center of the hub O (see Fig. 9.4-1a), the curvilinear coordinate system $(x^p (\equiv \eta), y^p (\equiv \zeta), z)$, and the (x, y, z) system. The axes η and ζ are the principal axes of the beam cross-section that lie in the cross-sectional planes of the beam, with the origin at the axis Z. These are orthogonal and rotate about the longitudinal axis with the pretwisted cross-section. The third system (x, y, z), rotates with the blade, its origin being located at the blade root, at an offset R_0 from the rotation axis. The z-axis is coincident with the inertial Z-axis.

Finally, we define a local Cartesian system coordinate (s, n, z), where s is the circumferential coordinate, while n the thickness one, where $|n| \leq h/2$.

Note that in the kinematical equations, only the quantities related to twist and extension will be retained. Due to the assumed non-uniformity of the beam in the spanwise direction, x and y become functions of both the circumferential and spanwise coordinates, s and z, respectively, that is $x \equiv x(s, z)$ and $y \equiv y(s, z)$; the primary and secondary warping functions will exhibit a similar dual dependence.

In order to get pertinent kinematic results to be used later, we will follow the approach by Houbolt and Brooks (1958), and we will consider the configuration of the pretwisted beam before and after deformation (see Figs. 9.7-1a,b, respectively). The undeformed position of a fiber identified by point P of the beam cross-section in terms of the distances η and ζ along the principal axes, and its rate of change with respect to z are:

$$x(s, z) = \eta(s) \cos \beta(z) - \zeta(s) \sin \beta(z),$$
$$y(s, z) = \eta(s) \sin \beta(z) + \zeta(s) \cos \beta(z), \qquad (9.7\text{-}1a,b)$$

and

$$x' = -\beta' y; \quad y' = \beta' x, \qquad (9.7\text{-}1c,d)$$

respectively.

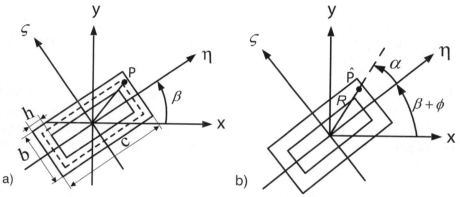

Figure 9.7-1: Box-beam cross-section geometry. (a) Box-beam cross-section before deformation. (b) Box beam cross-section after deformation

Now, the position of the fiber after deformation, involving only the elastic twist and extension, identified by point $\hat{P}$, see Fig. 9.7-1b, is

$$
\begin{aligned}
x_1 &= R \cos(\beta + \phi + \alpha) = R[\cos(\beta + \phi) \cos \alpha - \sin(\beta + \phi) \sin \alpha] \\
&= \eta \cos(\beta + \phi) - \zeta \sin(\beta + \phi), \\
y_1 &= R \sin(\alpha + \beta + \phi) = \eta \sin(\beta + \phi) + \zeta \cos(\beta + \phi), \qquad (9.7\text{-}2a\text{-}c) \\
z_1 &= z + R_0 + w = z + R_0 + w_0(z, t) - [\overline{F} - nr_t]\phi'.
\end{aligned}
$$

From (9.7-2) one can obtain, in conjunction with (9.7-1), the rate of change of x_1, y_1 and z_1 with respect to z

$$
\begin{aligned}
x_1' &= -(\beta' + \phi')[\eta \sin(\beta + \phi) + \zeta \cos(\beta + \phi)], \\
y_1' &= (\beta' + \phi')[\eta \cos(\beta + \phi) - \zeta \sin(\beta + \phi)], \qquad (9.7\text{-}3a\text{-}c) \\
z_1' &= 1 + w_0' - [\overline{F} - nr_t]\phi''.
\end{aligned}
$$

From these equations, one can determine the longitudinal strain. To this end, the expressions of dS/dz and ds/dz are needed.

From Eqs. (9.7-3), we have

$$\left(\frac{dS}{dz}\right)^2 = (x_1')^2 + (y_1')^2 + (z_1')^2$$

$$= 1 + 2\left\{w_0' - (\bar{F} - nr_t)\phi''\right\} + (\beta' + \phi')^2(\eta^2 + \zeta^2). \qquad (9.7\text{-}4)$$

As concerns the original length (ds/dz), this can be obtained directly from (9.7-4) by letting $\phi' = 0$, $\phi'' = 0$ and $w_0' = 0$.

Thus,

$$\left(\frac{ds}{dz}\right)^2 = 1 + (\eta^2 + \zeta^2)(\beta')^2. \qquad (9.7\text{-}5)$$

As a result, with $(\eta^2 + \zeta^2)(\beta')^2 \ll 1$ that according to Houbolt and Brooks (1958) is less than 0.03 to 0.04, for the longitudinal strain $\epsilon_{zz} = (dS - ds)/ds = dS/ds - 1$, one obtains its expression as

$$\epsilon_{zz} = w_0' - (\bar{F} - nr_t)\phi'' + \frac{1}{2}(\eta^2 + \zeta^2)[(\phi')^2 + 2\beta'\phi']. \qquad (9.7\text{-}6)$$

The last term in the brackets represens the contribution of the pretwist to the axial strain, and is essential in the analysis of pretwisted beams. The expression of ε_{zz} can be cast as

$$\epsilon_{zz} = \epsilon_{zz}^{(0)} + n\epsilon_{zz}^{(1)}, \qquad (9.7\text{-}7a)$$

where

$$\epsilon_{zz}^{(0)} = \frac{1}{2}(\eta^2 + \zeta^2)[(\phi')^2 + 2\beta'\phi'] + w_0' - \bar{F}\phi'', \qquad (9.7\text{-}7b,c)$$

and

$$\epsilon_{zz}^{(1)} = r_t\phi''.$$

In the next developments a more elaborate expression of the components of the acceleration of twisted rotating blade is needed. To this end, one should use Eqs. (9.7-2a-c) and the expressions of displacements resulting from Eqs. (9.7-1) through (9.7-2), provided by

$$u(x, y, z; t) = x_1 - x \equiv x(\cos\phi - 1) - y\sin\phi,$$
$$v(x, y, z; t) = y_1 - y \equiv x\sin\phi + y(\cos\phi - 1),$$

and

$$w(x, y, z; t) = z_1 - (z + R_0) \equiv w_0 - (\bar{F} - nr_t)\phi'. \qquad (9.7\text{-}8a\text{-}c)$$

Using in Eq. (9.1-7b) the representations of displacements u, v and w supplied by Eqs. (9.7-8), one gets the components of the acceleration vector

$$
\begin{aligned}
a_x &= -\ddot{\phi}(\eta N + \zeta M) - \dot{\phi}^2(\eta M - \zeta N) \\
&\quad + \underline{2\Omega[\dot{w}_0 - \dot{\phi}'(\bar{F} - nr_t)]} - \underline{\underline{\Omega^2(\eta M - \zeta N)}}, \\
a_y &= -\ddot{\phi}(\zeta N - \eta M) - \dot{\phi}^2(\eta N + \zeta M), \\
a_z &= \ddot{w}_0 - \ddot{\phi}'(\bar{F} - nr_t) + \underline{2\Omega\dot{\phi}(\eta N + \zeta M)} \\
&\quad - \underline{\underline{\Omega^2[R_0 + z + w_0 - \phi'(\bar{F} - nr_t)]}},
\end{aligned}
\qquad (9.7\text{-}9a\text{-}c)
$$

where, for brevity

$$
M \equiv \cos(\beta + \phi) \text{ and } N \equiv \sin(\beta + \phi), \qquad (9.7\text{-}9d,e)
$$

and where the terms underscored by one and two supperposed lines are associated with the Coriolis and centrifugal accelerations, respectively.

9.7-3 The Equations of Motion and Boundary Conditions

In order to derive the equations of motion and the associated boundary conditions, the extended Hamilton's principle expressed by Eq. (9.1-41) is used. First we obtain the various energy quantities.

For this case, we have

$$
\int_{t_0}^{t_1} \delta \mathcal{K} dt = \int_{t_0}^{t_1} \int_{\tau} (\dot{\mathbf{R}} \cdot \delta \dot{\mathbf{R}}) d\tau dt = -\int_{t_0}^{t_1} dt \int_{\tau} \rho \ddot{\mathbf{R}} \cdot \delta \mathbf{R} d\tau. \qquad (9.7\text{-}10)
$$

Having in view Eqs. (9.7-8) and (9.7-9) in (9.7-10), and carrying out the integration across the wall thickness, we find the three terms in the integrand of Eq. (9.7-10) as

$$
\int_{t_0}^{t_1} dt \int_{\tau} \rho a_x \delta u d\tau
$$

$$
= -\int_{t_0}^{t_1} dt \int_{\tau} \rho \{[-\ddot{\phi}(\eta N + \zeta M) - \dot{\phi}^2(\eta M - \zeta N)
$$

$$
+ 2\Omega[\dot{w}_0 - \dot{\phi}'(\bar{F} - nr_t)] - \Omega^2(\eta M - \zeta N)](\eta N + \zeta M)\delta\phi\} d\sigma dz
$$

$$
= -\int_{t_0}^{t_1} dt \int_0^L \oint \{-m_0\ddot{\phi}(\eta N + \zeta M)^2
$$

$$- m_0 \dot{\phi}^2 (\eta M - \zeta N)(\eta N + \zeta M) + 2\Omega[\dot{w}_0(\eta N + \zeta M)$$
$$- \dot{\phi}'(m_0 \bar{F} - m_2 r_t)(\eta N + \zeta M)] - m_0 \Omega^2[(\eta^2 - \zeta^2)MN$$
$$+ \eta\zeta(M^2 - N^2)]\}\delta\phi\, ds\, dz,$$

$$\int_{t_0}^{t_1} dt \int_{\tau} \rho a_y \delta v\, d\tau$$
$$= \int_{t_0}^{t_1} dt \int_{\sigma} \int_0^L \rho\{[-\ddot{\phi}(\zeta N - \eta M) - \dot{\phi}^2\,(\eta N + \zeta M)](\eta M$$
$$- \zeta N)\}\delta\phi\, d\sigma\, dz = \int_{t_0}^{t_1} dt \int_0^L \oint [m_0\ddot{\phi}(\eta M - \zeta N)^2$$
$$- m_0\dot{\phi}^2\big((\eta^2 - \zeta^2)MN + \eta\zeta(M^2 - N^2)\big)]\delta\phi\, ds\, dz,$$

and   (9.7-11a-c)

$$\int_{t_0}^{t_1} dt \int_{\tau} \rho a_z \delta w\, d\tau$$
$$= \int_{t_0}^{t_1} dt \int_0^L \oint \{[m_0(\ddot{w}_0 - \Omega^2(R_0 + z + w_0))]\delta w_0$$
$$+ [(m_0\bar{F}^2 + m_2 r_t^2)\ddot{\phi}' - \phi'(m_0\bar{F}^2 + m_2 r_t^2)\Omega^2]\delta\phi'\}ds\, dz.$$

In these equations, $d\tau = d\sigma\, dz (= ds\, dn\, dz)$. Adding these three equations, collecting the terms associated with the same variations $\delta\phi$ and δw_0, performing the integration along the entire cross-section beam circumference, integrating by parts as to relieve $\delta\phi$ of any differentiation with respect to the z-variable, and recalling that the product of inertia with respect to the principal axes is zero,

i.e. $\oint \zeta\eta\, ds = 0$, we write Eq. (9.7-10) as

$$\int_{t_0}^{t_1} \delta \mathcal{K}\, dt = -\int_{t_0}^{t_1} \Big\{ \int_0^L \big[b_1[\ddot{w}_0 - (R_0 + z + w_0)\Omega^2]\delta w_0$$
$$+ \{I_p\ddot{\phi} + (I_{\zeta\zeta}^0 - I_{\eta\eta}^0)\Omega^2 MN - [I_{ww}(\ddot{\phi}' - \Omega^2\phi')]'\}\delta\phi\big]dz$$
$$+ \big[I_{ww}(\ddot{\phi}' - \Omega^2\phi')\delta\phi\big]_0^L \Big\}dt.$$ (9.7-12)

The cross-sectional quantities are supplied in Table 9.7-1; (m_0, m_2) defined by Eq. (4.1-5) denote the reduced mass terms per unit beam length. For beams with zero pretwist $I_p \to b_4 + b_5$; $I_{\zeta\zeta}^0 \to b_4$, $I_{\eta\eta}^0 \to b_5$; I_{ww} stands for $b_{10} + \delta_n b_{18}$.

Table 9.7-1: Cross-section quantities

Quantity	$I_p(\equiv I_{\zeta\zeta}^0 + I_{\eta\eta}^0)$	$\bar{I}_p$	I_{ps}	$I_{\eta\eta}^0$
Definition	$m_0 \oint (\eta^2 + \zeta^2)ds$	I_p/m_0	$\oint (\eta^2 + \zeta^2)^2 ds$	$m_0 \oint \zeta^2 ds$
Quantity	$I_{\zeta\zeta}^0$	I_{ww}	$\mathcal{B}$	$S(z)$
Definition	$m_0 \oint \eta^2 ds$	$m_0 \oint \bar{F}^2 ds + m_2 \oint r_t^2 ds$	$K_{11}[I_{ps} - \bar{I}_p^2/S]$	$\oint ds$

On the other hand, for the coupled twist-extension motion, $\delta\mathcal{W}$ is expressed as follows:

$$\delta\mathcal{W} = \frac{1}{2} \oint \int_0^L \left(N_{zz}\delta\varepsilon_{zz}^{(0)} + L_{zz}\delta\varepsilon_{zz}^{(1)} + N_{sz}\delta\gamma_{sz}\right)ds\,dz, \qquad (9.7\text{-}13)$$

where $\varepsilon_{zz}^{(0)}$ and $\varepsilon_{zz}^{(1)}$ are provided by Eqs. (9.7-7b,c), while

$$\gamma_{sz} = \psi(s, z)\phi'(z). \qquad (9.7\text{-}14)$$

In light of these latter equations and of the definitions of 1-D stress measures, Eqs. (3.1-17), one can reduce Eq. (9.7-13) to

$$\delta\mathcal{W} = \int_0^L \left[T_z\delta w_0' - B_w\delta\phi'' + T_r\phi'\delta\phi' + M_z\delta\phi'\right]dz, \qquad (9.7\text{-}15)$$

where

$$T_z(z, t) = \oint N_{zz}ds,$$

$$T_r(z, t) = \oint N_{zz}(\eta^2 + \zeta^2)ds,$$

$$B_w(z, t) = \oint [\bar{F}(s, z)N_{zz} - r_t(s, z)L_{zz}]ds, \qquad (9.7\text{-}16\text{a-d})$$

$$M_p(z, t) = \oint [N_{zz}(\eta^2 + \zeta^2)\beta' + \psi(s, z)N_{sz}]ds.$$

T_z and T_r stand for 1-D extensional stress resultant measures, while B_w and M_p denote the warping moment and the twist moment, respectively.

In these equations, the shell stress resultants N_{zz} and N_{sz} whose expressions have been obtained in a general context, Eqs. (A.102), become for the present case:

$$\begin{aligned} N_{zz} &= K_{11}\varepsilon_{zz}^{(0)} + K_{13}\phi' + K_{14}\varepsilon_{zz}^{(1)}, \\ N_{sz} &= K_{21}\varepsilon_{zz}^{(0)} + K_{23}\phi' + K_{24}\varepsilon_{zz}^{(1)}, \end{aligned} \qquad (9.7\text{-}17\text{a,b})$$

where K_{ij} are 2-D stiffness quantities given by Eqs. (A.92). In Eq. (9.7-16d), $\psi(s, z)$ is the torsional function defined by Eq. (2.1-44c), while $r_t(s, z)$ is defined by Eq. (2.1-11b). Due to the double dependence, on s and z of the primary and secondary warping functions, the use in this context of the free warping twist model is precluded.

Using in Eq. (9.7-16a) the representation of N_{zz}, Eq. (9.7-17a), considered in conjunction with (9.7-7), yields

$$
\begin{aligned}
T_z &= \oint N_{zz}ds \\
&= \oint \left\{ K_{11}\left[w_0' - \bar{F}\phi'' + (\eta^2 + \varsigma^2)\left(\beta'\phi' + \frac{1}{2}(\phi')^2 \right) \right] + K_{13}\phi' \right\}ds,
\end{aligned}
$$
(9.7-18a)

or, in light of the CUS configuration,

$$
T_z = K_{11}\left[Sw_0' + \bar{I}_p\left(\beta'\phi' + \frac{1}{2}(\phi')^2 \right) \right] + K_{13}S\phi'. \tag{9.7-18b}
$$

Similarly, Eq. (9.7-16b) in conjunction with (9.7-18b) reduces to

$$
T_r = \frac{T_z \bar{I}_p}{S} + B\left[\beta'\phi' + \frac{1}{2}(\phi')^2 \right]. \tag{9.7-19}
$$

Similar steps applied to Eq. (9.7-16d), considered in conjunction with Eqs. (9.7-17) yields:

$$
\begin{aligned}
M_p &= \oint \left\{ (\eta^2 + \varsigma^2)\beta'\left[K_{11}\epsilon_{zz}^{(0)} + K_{13}\phi' + K_{14}\epsilon_{zz}^{(1)} \right] \right. \\
&\quad \left. + \psi(s, z)\left[K_{21}\varepsilon_{zz}^{(0)} + K_{23}\phi' + K_{24}\varepsilon_{zz}^{(1)} \right] \right\}ds.
\end{aligned}
$$
(9.7-20a)

In view of Eqs. (9.7-7b,c), and consistent with the CUS configuration, its final form becomes

$$
M_p = a_{71}w_0' + a_{77}\phi'. \tag{9.7-20b}
$$

In a similar way, for the present case Eq. (A102b) yields

$$
L_{zz} = K_{41}\varepsilon_{zz}^{(0)} + K_{43}\phi' + K_{44}\varepsilon_{zz}^{(1)}, \tag{9.7-21}
$$

and the expression of B_w given by (9.7-16c) becomes

$$
B_w = -a_{66}\phi'', \tag{9.7-22}
$$

where a_{66} is the warping stiffness whose expression is recorded in Table 9.7-2.

Table 9.7-2: Expression of stiffness quantities

$a_{ij} = a_{ji}$	Definition
a_{11}	$\oint K_{11} ds$
$a_{17} \ (\equiv a_{71}^\beta + a_{71}^0)$	$\underbrace{K_{11}\beta' \bar{I}_p}_{a_{71}^\beta} + \underbrace{K_{13}S}_{a_{71}^0}$
a_{66}	$\oint [K_{11}\bar{F}^2 - 2K_{14}\bar{F}r_t + K_{44}r_t^2]ds$
a_{67}	$\oint [K_{13}\bar{F} - K_{43}r_t]ds$
$a_{77} \ (\equiv a_{77}^0 + (a_{77}^\beta)_1 + (a_{77}^\beta)_2)$	$\oint \underbrace{\psi(s,z)K_{23}ds}_{a_{77}^0} + \underbrace{(K_{13} + \psi(s,z)K_{21})\bar{I}_p\beta'}_{(a_{77}^\beta)_1} + \underbrace{K_{11}I_{ps}(\beta')^2}_{(a_{77}^\beta)_2}$

Before using Hamilton's principle, one should carry out integration by parts in Eq. (9.7-15) as to relieve the virtual displacements of any differentiation. This yields:

$$\int_{t_0}^{t_1} \int_0^L \delta \mathcal{W} dz dt = \int_{t_0}^{t_1} \left\{ \int_0^L -[T'_z \delta w_0 + (B''_\omega + (T_r \phi')' + M'_z)\delta\phi]dz \right.$$
$$\left. + [T_z \delta w_0 - B_\omega \delta\phi' + (B'_\omega + T_r\phi' + M_p)\delta\phi]\Big|_0^L \right\} dt. \tag{9.7-23}$$

Replacing in Hamilton's principle, Eq. (9.1-41), Eqs. (9.7-12) and (9.7-23), terms associated with the virtual work of external loads (specialized for zero edge-loads), collecting the terms associated with the same virtual displacements (considered to be independent and arbitrary within $[0, L]$ and $[t_0, t_1]$), by setting their coefficients to zero, we find the equations of motion.

• *Equations of motion:*

$$\delta w_0 : \ T'_z - b_1 \ddot{w}_0 + \underline{b_1 \Omega^2 (R_0 + z + w_0)} + p_z = 0,$$

$$\delta\phi : \ B''_w + [T_r \phi']' + M'_z - I_p \ddot{\phi} + \underset{\sim\sim\sim\sim\sim\sim\sim\sim\sim}{(I_{\eta\eta}^0 - I_{\zeta\zeta}^0)\Omega^2 MN}$$

$$+ \ [\underline{I_{ww}}(\ddot{\phi}' - \Omega^2\phi')]' + m_z + \ \underline{m'_w} = 0, \tag{9.7-24a,b}$$

and *the boundary conditions for cantilevered blades*:

at $z = 0$:

$$\phi = \underline{\phi'} = w_0 = 0, \qquad (9.7\text{-}25\text{a-c})$$

and at $z = L$:

$$T_z = 0; \quad B_w = 0; \quad B'_w + M_z + T_r\phi' + \underline{\underline{I_{ww}(\ddot{\phi}' - \Omega^2\phi')}} = 0.$$

$$(9.7\text{-}26\text{a-c})$$

Both equations of motion and boundary conditions, Eqs. (9.7-24) through (9.7-26), coincide with those extracted, for the present case, from Eqs. (9.1-14), (9.1-15) and (9.1-16). In these equations the various effects are identified as indicated in Sect. 9.1-4. The three boundary conditions at each edge requires that the governing system should be of the sixth order.

For small twist angles, $\sin\phi \Rightarrow \phi$ and $\cos\phi \rightarrow 1$, so that tennis-racket term is

$$(I^0_{\eta\eta} - I^0_{\zeta\zeta})\Omega^2 MN \rightarrow (I^0_{\eta\eta} - I^0_{\zeta\zeta})\Omega^2[\sin\beta\cos\beta + \phi\cos 2\beta]. \qquad (9.7\text{-}27)$$

In addition, by virtue of Eq. (9.7-19), the second term in Eq. (9.7-24b) can be expressed in terms of T_z as:

$$[T_r\phi']' = T'_r\phi' + T_r\phi'' = \frac{T'_z\bar{I}_p}{S}\phi' + \frac{T_z\bar{I}_p}{S}\phi''. \qquad (9.7\text{-}28)$$

Now, in conjunction with these last equations, Eq. (9.7-24b) becomes

$$\delta\phi: \quad - \underline{[a_{66}\phi'']''} + \left(\frac{T_z\bar{I}_p}{S}\phi'\right)' + [a_{71}w'_0 + a_{77}\phi']' - I_p\ddot{\phi}$$

$$+ \underwave{(I^0_{\eta\eta} - I^0_{\zeta\zeta})\Omega^2[\sin\beta\cos\beta + \phi\cos 2\beta]} + [I_{ww}(\underline{\ddot{\phi}'} - \underline{\underline{\Omega^2\phi'}})]'$$

$$+ m_z + m'_w = 0, \qquad (9.7\text{-}29)$$

whereas Eq. (9.7-24a) remains unchanged.

In order to reduce the governing equations to a single equation, w_0 should be exactly eliminated in (9.7-24a,b). To this end, w'_0 is extracted from Eq. (9.7-18b), thus yielding

$$w'_0 = \frac{T_z}{K_{11}S} - \frac{\bar{I}_p}{S}\left(\beta'\phi' + \frac{1}{2}\phi'^2\right) - \frac{K_{13}}{K_{11}}\phi'. \qquad (9.7\text{-}30)$$

Replacing (9.7-20b), (9.7-22) and (9.7-28) considered in conjunction with (9.7-30) into the equation of motion (9.7-24b) and boundary condition (9.7-26b,c), one obtains *the governing equation*

$$
-\ [a_{66}\phi'']'' + \left(\frac{T_z I_p}{S}\phi'\right)' + (B\beta'^2\phi')' + \left[\left(a_{77}^0 - \frac{K_{13}a_{71}^0}{K_{11}}\right)\phi'\right]' - I_p\ddot{\phi}
$$

$$
+\ (I_{\eta\eta}^0 - I_{\zeta\zeta}^0)\Omega^2\phi\cos 2\beta\ + [I_{ww}(\ddot{\phi}' - \Omega^2\phi')]' + m_z + m_w'
$$

$$
= -\left(\frac{a_{71}T_z}{K_{11}S}\right)', \tag{9.7-31}
$$

and the boundary conditions:

At $z = 0$:

$$
\phi = \phi' = 0, \tag{9.7-32a,b}
$$

and at $z = L$:

$$
a_{66}\phi'' = 0, \tag{9.7-32c,d}
$$

$$
-\ (a_{66}\phi'')' + B\beta'^2\phi' + \left(a_{77}^0 - \frac{K_{13}a_{71}^0}{K_{11}}\right)\phi' + I_{ww}(\ddot{\phi}' - \Omega^2\phi')
$$

$$
+\ \frac{a_{71}}{K_{11}S}T_z = 0.
$$

As concerns T_z, the neglect of the axial inertia term in Eq. (9.7-24a), followed by its integration with respect to z, in conjunction with the homogeneous BCs at $z = L$, yields

$$
T_z(z) = \int_z^L b_1\Omega^2(R_0 + z)dz. \tag{9.7-33}
$$

In view of (9.7-33), the governing equation (9.7-31) and the boundary conditions are expressed in term of the unknown function $\phi(z, t)$. For the ply-angles $\theta = 0, 90$ deg., the terms underscored by a solid line become zero.

One should remark the full agreement of the isotropic counterpart of these equations, with those obtained by Houbolt and Brooks (1958). The same is valid in connections with the equations derived by Kaza and Kielb (1984) and Subrahmanyam and Kaza (1985) for a solid beam counterpart, and with those contained in McGee (1992).

Equations (9.7-31) and (9.7-32) in conjunction with Eq. (9.7-33) govern the motion of pretwisted rotating thin-walled beams featuring the twist-extension elastic coupling. For uniform cross-section beams, the primary and secondary warping functions depend on the circumferential coordinate only; for the free warping case, the governing equation becomes of the second order, and as a result, a single boundary condition at each edge is required.

Notice that no specific variation of the pretwist angle β along the beam span has been prescribed yet. Usually, a linear variation is considered, such as $\beta(z) = \beta_0 z/L$, where β_0 is the pretwist at the beam free end.

To illustrate the implications of each term in Eq. (9.7-31), it is useful to define their significance. In the order of their appearance, these are related to the warping restraint (WR); tension-torsion coupling (TT); pretwist (PT); Saint-Venant torsion (FW); appearing in the same bracket with a term induced by the anisotropy of the material; inertia in twist; tennis-racket (TR), and the inertial warping appearing in the same bracket with the centrifugal warping term, and the distributed twist moment loadings. The term on the right-hand side of Eq. (9.7-31) is an external twist moment due to the centrifugal torsion and pretwist.

9.7-4 Validation of the Structural Model and Solution Methodology

We can validate the present solution methodology based on EGM by comparing our predictions with those generated for a solid beam.

Tables 9.7-3 and 9.7-4 supply comparisons of the non-rotating eigenfrequencies obtained within the present approach with those obtained for isotropic pretwisted beams by Kaza and Kielb (1984). The relationship between the parameters by Kaza and Kielb (1984) and the present ones are as follows:

$$\frac{a_{77}^0}{a_{66}}L^2 = \frac{24}{1+\nu}(L/c)^2; \quad \frac{\mathcal{B}}{a_{66}}L^2 = \frac{4}{5}(c/b)^2; \quad \frac{I_{ww}}{a_{66}}L^2 = \frac{\rho}{E}L^2; \quad \frac{I_p L^4}{a_{66}}$$

$$= \frac{12\rho L^4}{Eb^2} \text{ and } \bar{\omega}_i = \omega_i/\omega_{on} \text{ where } \omega_{on} = \frac{n\pi}{L}\frac{b}{c}\sqrt{G/\rho}$$

The results presented in these tables reveal an excellent agreement.

9.7-5 Results and Discussion

Although the equations are valid for a beam of arbitrary closed-cross section, for illustration purposes we consider the case of a rotating beam modelled as a composite box-beam (see Fig. 9.7-1), characterized by a cross-section ratio $\mathcal{R}(\equiv c/b) = 5$. We take $\bar{R} = R_0/L = 0.1$; $L/c = 8$; $c = 10$ in. (0.254m);

$b = 2$ in. (50.8×10^{-3}m); $h = 0.4$in. (10.16×10^{-3}m). The mechanical characteristics of the beam as considered in the numerical simulations correspond to the graphite/epoxy material.

Table 9.7-3: Comparison of dimensionless torsional frequencies for nonrotating pretwisted beam ($b/c = 0.05$)

L/c	Mode	$\beta_0 = 0$		15 deg.		30 deg.	
		K.& K.	Oh et al.	K.& K.	Oh et al.	K. & K.	Oh et al.
2	$\bar{\omega}_1$	1.1443	1.1443	1.2785	1.2803	1.6099	1.6154
	$\bar{\omega}_2$	1.2436	1.2442	1.3704	1.3728	1.6869	1.6931
	$\bar{\omega}_3$	1.4281	1.4303	1.540	1.5447	1.8309	1.8387
4	$\bar{\omega}_1$	1.0656	1.0659	1.1014	1.1020	1.2018	1.2040
	$\bar{\omega}_2$	1.0959	1.0963	1.1307	1.1317	1.2289	1.2313
	$\bar{\omega}_3$	1.1536	1.1544	1.1868	1.1882	1.2811	1.2838
6	$\bar{\omega}_1$	1.0421	1.0428	1.0582	1.0586	1.1050	1.1069
	$\bar{\omega}_2$	1.0562	1.0573	1.0721	1.0736	1.1184	1.1205
	$\bar{\omega}_3$	1.0838	1.0849	1.0993	1.1006	1.1444	1.1467
8	$\bar{\omega}_1$	1.0310	1.0322	1.0401	1.0421	1.0669	1.0688
	$\bar{\omega}_2$	1.0391	1.0412	1.0481	1.0503	1.0747	1.0776
	$\bar{\omega}_3$	1.0550	1.0569	1.0639	1.0660	1.0901	1.0928
10	$\bar{\omega}_1$	1.0245	1.0276	1.0303	1.0335	1.0476	1.0512
	$\bar{\omega}_2$	1.0297	1.0327	1.0355	1.0387	1.0527	1.0563
	$\bar{\omega}_3$	1.0400	1.0430	1.0458	1.0489	1.0628	1.0663

K. & K. → Kaza and Kielb (1984); Oh et al. (2005)

Table 9.7-4: Comparison of dimensionless torsional frequencies for nonrotating pretwisted beam ($b/c = 0.2$)

L/c	Mode	$\beta_0 = 0$ deg.		15 deg.		30 deg.	
		K.& K.	Oh et al.	K.& K.	Oh et al.	K. & K.	Oh et al.
2	$\bar{\omega}_1$	1.1443	1.1452	1.1532	1.1542	1.1795	1.1807
	$\bar{\omega}_2$	1.2436	1.2527	1.2520	1.2613	1.2767	1.2807
	$\bar{\omega}_3$	1.4281	1.4589	1.4354	1.4666	1.4573	1.4893
4	$\bar{\omega}_1$	1.0656	1.0657	1.0679	1.0688	1.0747	1.0753
	$\bar{\omega}_2$	1.0959	1.0985	1.0981	1.1006	1.1047	1.1074
	$\bar{\omega}_3$	1.1536	1.1607	1.1557	1.1629	1.1620	1.1693
6	$\bar{\omega}_1$	1.0421	1.0435	1.0431	1.0439	1.0462	1.0474
	$\bar{\omega}_2$	1.0562	1.0582	1.0572	1.0593	1.0602	1.0623
	$\bar{\omega}_3$	1.0838	1.0877	1.0847	1.0887	1.0876	1.0916
8	$\bar{\omega}_1$	1.0310	1.0325	1.0315	1.0336	1.0333	1.0357
	$\bar{\omega}_2$	1.0391	1.0417	1.0396	1.0422	1.0413	1.0438
	$\bar{\omega}_3$	1.0550	1.0585	1.0555	1.0591	1.0572	1.0608
10	$\bar{\omega}_1$	1.0245	1.0272	1.0248	1.0284	1.0259	1.0287
	$\bar{\omega}_2$	1.0297	1.0332	1.0301	1.0334	1.0312	1.0364
	$\bar{\omega}_3$	1.0400	1.0440	1.0404	1.0444	1.0415	1.0473

K. & K. → Kaza and Kielb (1984); Oh et al. (2005)

The first three figures, Figs. 9.7-2 through 9.7-4, display the first three eigenfrequencies as a function of the variation of pretwist angle at the blade tip, and for selected values of the rotational speed; the ply-angle of the constituent materials is $\theta = 0$. These plots show the strong centrifugal stiffening effect played by the tension-torsion coupling term in all three eigenfrequencies.

As concerns the pretwist, its effect on the fundamental mode frequency is quite different from the other ones, in the sense that with the increase of β_0 and Ω, that yields an increase of the softening tennis-racket term, the fundamental frequency exhibits a decrease. In contrast to this trend, for the remaining frequencies, and for any Ω, with the increase of the pretwist, this destiffening effect almost disappears. This trend was reported by Subrahmanyam and Kaza (1985), McGee (1992) and Rosen (1991), for isotropic blades. This implies that for higher modes, the natural frequencies are more strongly influenced by the tension-torsion coupling term than by the tennis-racket effect.

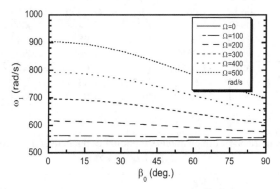

Figure 9.7-2: Variation of the first torsional natural frequency with pretwist angle for selected rotational speeds, $(\theta = 0)$

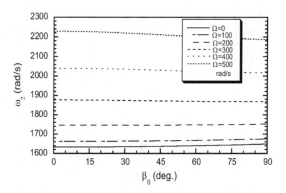

Figure 9.7-3: Counterpart of Fig. 9.7-2 for the second torsional natural frequency

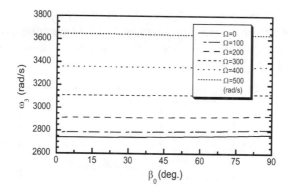

Figure 9.7-4: Counterpart of Fig. 9.7-2 for the third torsional natural frequency

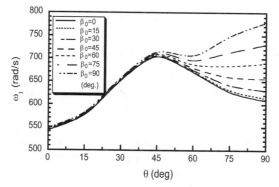

Figure 9.7-5: First natural frequency vs. ply-angle for selected pretwist angles ($\Omega = 0$)

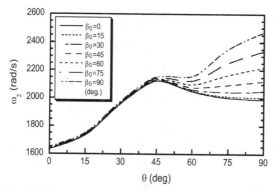

Figure 9.7-6: The counterpart of Fig. 9.7-5 for the second natural frequency

Figures 9.7-5 through 9.7-7 depict the variation of the first three mode eigenfrequencies with ply-angle for selected values of the pretwist angle. These plots clearly show that until $\theta = 45$ deg., there is a continuous increase of non-rotating eigenfrequencies, and that the pretwist plays a marginal role in that increase. However, beyond $\theta = 45$ deg., the pretwist plays a more nuanced role, in the sense that until $\beta_0 \approx 45$ deg. there is a drop of eigenfrequencies with the increase of θ, while for $\beta_0 > 45$ deg., after a slight decrease, an increase with the increase of θ is experienced.

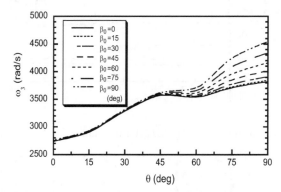

Figure 9.7-7: The counterpart of Fig. 9.7-5 for the third natural frequency

This trend is explainable by examining Tables 9.7-5 and 9.7-6 which display the variations of the warping stiffness a_{66}, and of the stiffness $(a_{77}^0 - a_{71}^0 K_{13}/K_{11})$ with θ, as well as that of the term $\mathcal{B}(\beta')^2$ as a function of θ and β_0. These tables show that the fourth and third term in Eqs. (9.7-31) and (9.7-32d) are responsible for that drop in eigenfrequencies, whereas the lower values of the third term in Eq. (9.7-31), as compared to those generated by the fourth term, are responsible for the marginal contribution of β_0 until $\theta = 45$ deg.

Table 9.7-5: Variation of some selected stiffness quantities with that of the ply-angle

θ (deg.)	Stiffness	a_{66}	a_{77}^0	$a_{71}^0(K_{13}/K_{11})$	$a_{77}^0 - a_{71}^0 K_{13}/K_{11}$
0		54.0	34439	0	34439
15		55.6	38581	175.5	38406
30		67.4	53244	3343.5	49902
60		424.8	152936	103035	49901
75		1499.5	146218	107812	38406
90		2028.1	34439	0	34439

Table 9.7-6: Variation of the pretwist term with the pretwist and the ply-angle

θ (deg.)	β_0 (deg.)	$\mathcal{B}(\beta_0/L)^2$						
		0	15	30	45	60	75	90
0	0		19.32	77.30	173.9	309.2	483.1	695.7
15	0		19.80	79.18	178.2	316.71	494.86	712.59
30	0		23.05	92.21	207.48	368.85	576.33	829.92
45	0		41.12	164.49	370.11	657.98	1028.1	1480.5
60	0		145.27	581.07	1307.4	2324.4	3631.7	5229.6
75	0		534.14	2136.6	4807.2	8546.3	13354	19229
90	0		72578	2903.1	6532	11613	18144	26128

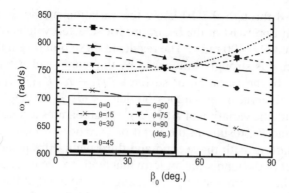

Figure 9.7-8: First natural frequency vs. pretwist for selected ply-angles (Ω = 300 rad/s)

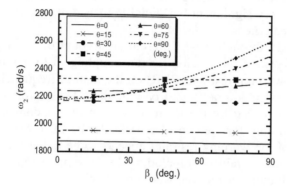

Figure 9.7-9: The counterpart of Fig. 9.7-8 for the second natural frequency

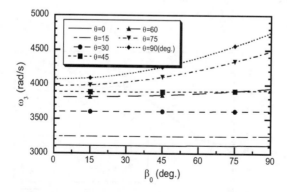

Figure 9.7-10: The counterpart of Fig. 9.7-8 for the third natural frequency

Figures 9.7-8 through 9.7-10 show the variation of the first three natural frequencies with pretwist at the beam tip, for selected values of the ply-angle θ, and for a fixed $\Omega = 300$ rad/s. The results in these plot reveal that: a) the first natural frequency is strongly influenced by the tennis-racket term that plays a softening role, b) the variation of stiffness $(a_{77}^0 - a_{71}^0 K_{13}/K_{11})$ (that exhibits a decay in the interval of ply-angles 60 deg. $< \theta < 90$ deg.), considered in conjunction with the variation of $\mathcal{B}(\beta_0/L)^2$, (that features a continuous increase with both β_0 and θ), results in the trend of variation of ω_1 as emerging from Fig. 9.7-8. Starting with the second and third natural frequencies, due to the softening effect of the tennis-racket term that tends to decay, and of the tension-torsion effect that becomes stronger, a change of the trend of the variation of ω_2 and ω_3 vs. β_0 is experienced, see Figs. 9.7-9 and 9.7-10. In this sense, there is no longer a decrease of the respective natural frequencies with the variation of β_0, but, starting with $\theta = 60$ deg. there is instead, an increase of these. These plots show the very significant beneficial influence of the directional property of composite materials.

Figures 9.7-11 through 9.7-13 present an account of the implications of a number of effects, of the tension-torsion coupling (TT); tennis-racket (TR), pretwist (PT), warping restraint (WR) on the natural frequencies vs. the ply-angle. We take $\Omega = 300$ rad/s, and $\beta_0 = 90$ deg. These plots show the severe softening effect played by the tennis-racket on the fundamental natural frequency. In addition, for the considered slender beam, for $0 < \theta < 45$ deg., the influence of the warping restraint appears to be rather weak, in the sense that the predictions obtained by including the warping restraint are in excellent agreement with those of the free warping model. However, an increase of its influence is manifested for higher mode frequencies and beyond $\theta = 45$ deg. The hardening effect of the tension-torsion effect at any θ and for any mode frequency appears clearly in these graphs.

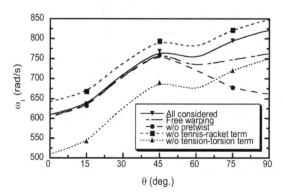

Figure 9.7-11: Variation of first natural frequency with ply-angle ($\Omega = 300$ rad/s; $\beta_0 = 90$ deg)

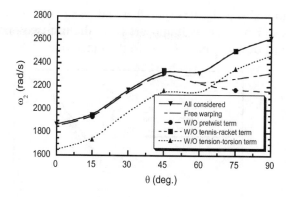

Figure 9.7-12: Counterpart of Fig. 9.7-11 for the second natural frequency

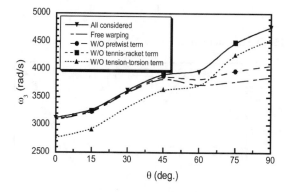

Figure 9.7-13: Counterpart of Fig. 9.7-11 for the third natural frequency

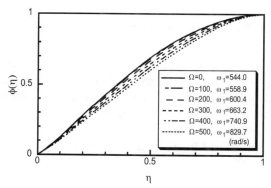

Figure 9.7-14: Variation of first torsional mode shape for different rotational speeds ($\theta = 0$; $\beta_0 = 45$ deg.)

Figures 9.7-14 through 9.7-16, provide in succession the three normalized mode shapes of a beam characterized by $\theta = 0$ and $\beta_0 = 45$ deg., for a number of selected rotational speeds. It is seen that Ω does not change too significantly the mode shapes. However, as it appears from the last two ones, the increase of Ω yields a shift of nodal points toward the blade tip.

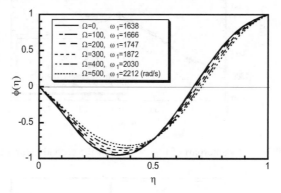

Figure 9.7-15: The counterpart of Fig. 9.7-14 for the second torsional mode shape

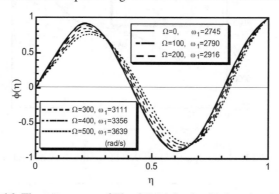

Figure 9.7-16: The counterpart of Fig. 9.7-14 for the third torsional mode shape

Note that in the numerical simulations the membrane shear stiffness G_{sz} is uniform along the beam cross-section contour. When, due to the specific type of construction (e.g. multilayered beams), this condition is not fulfilled, the torsional function related to the primary warping function has to involve the membrane shear non-uniformity. The implications of the violation of this requirement was discussed by Jung et al. (1999) and more fully by Qin and Librescu (2001, 2002).

9.7-6 A Few Comments about Washizu's Approach

The developments of this section are based on Wagner's approach. It is however worthy to infer if Wagner approach of pretwisted rotating blades featuring extension-twist coupling would provide similar or different results from those obtainable via the use of Washizu's approach.

In Section 9.2-4 it was already shown that for pretwisted rotating blades featuring bending-bending-transverse shear coupling, the basic equations obtained through both approaches are identical.

Since the procedure of proving the similarity of results for this case is close to that developed in Section 9.2-4, here we limit ourselves to provide only the basic results. For more details, see Oh et al. (2005).

As concerns the basic vector, and the components of covariant and contravariant metic tensors, their expressions are similar to those provided by Eqs. (C.9-2) through (C.9-4). On the other hand, by using Eqs. (9.7-2), the position vector of a point of the beam featuring elastic extension and twist is

$$
\begin{aligned}
\mathbf{R} =&[\eta \cos(\beta + \phi) - \zeta \sin(\beta + \phi)]\mathbf{i} \\
&+ [\eta \sin(\beta + \phi) + \zeta \cos(\beta + \phi)]\mathbf{j} \\
&+ [z + w_0 - (\overline{F} - nr_t)\phi']\mathbf{k}.
\end{aligned} \tag{9.7-34}
$$

Then, the covariant base vectors of the deformed space are

$$
\mathbf{G}_1 = \partial\mathbf{R}/\partial\eta = \cos(\beta + \phi)\mathbf{i} + \sin(\beta + \phi)\mathbf{j} - \frac{\partial}{\partial\eta}[\overline{F} - nr_t]\phi'\mathbf{k},
$$

$$
\mathbf{G}_2 = \partial\mathbf{R}/\partial\zeta = -\sin(\beta + \phi)\mathbf{i} + \cos(\beta + \phi)\mathbf{j} - \frac{\partial}{\partial\zeta}[\overline{F} - nr_t]\phi'\mathbf{k},
$$

$$
\tag{9.7-35a-c}
$$

$$
\begin{aligned}
\mathbf{G}_3 = \partial\mathbf{R}/\partial z =& -(\beta + \phi)'[\eta \sin(\beta + \phi) + \zeta \cos(\beta + \phi)]\mathbf{i} \\
&+ (\beta + \phi)'[\eta \cos(\beta + \phi) - \zeta \sin(\beta + \phi)]\mathbf{j} \\
&+ [1 + w_0' - (\overline{F} - nr_t)\phi'' - (\overline{F} - nr_t)'\phi']\mathbf{k}.
\end{aligned}
$$

For uniform cross-section beams, the last term in Eq. (9.7-35c) and in the reminder of this section is taken to be zero. Proceeding as in Section 9.2-4, we can determine the strain components in the curvilinear coordinate system $(\eta, \zeta, z) (\equiv (\alpha^1, \alpha^2, \alpha^3))$ as:

$$
f_{11} = 0; \quad f_{12} = 0; \quad 2f_{13} = -\phi'\zeta - \left(\frac{\partial\overline{F}}{\partial\eta} - n\frac{\partial r_t}{\partial\eta}\right)\phi';
$$

$$
2f_{23} = \phi'\eta - \left(\frac{\partial\overline{F}}{\partial\zeta} - n\frac{\partial r_t}{\partial\zeta}\right)\phi'; \quad f_{22} = 0, \tag{9.7-36}
$$

$$
f_{33} = w_0' - (\overline{F} - nr_t)\phi'' + \left(\beta'\phi' + \frac{1}{2}(\phi')^2\right)(\eta^2 + \zeta^2).
$$

In order to determine the strain components e_{ij} in the orthogonal Cartesian system $(s, n, z)(\equiv y^1, y^2, y^3)$, the procedure developed in Section 9.2-4 is used. As a matter of fact, Eqs. (C.9-12) through (C.9-13) remain valid also for this case.

Finally, we get the following expression of ϵ_{33} in the (s, n, z) coordinates:

$$e_{33} \equiv \varepsilon_{zz} = w_0' - \overline{F}\phi'' + \frac{1}{2}(\eta^2 + \zeta^2)(\phi')^2$$

$$- \beta'\phi'\left(\eta\frac{\partial\overline{F}}{\partial\zeta} - \zeta\frac{\partial\overline{F}}{\partial\eta}\right) + nr_t\phi''. \qquad (9.7\text{-}37)$$

To render explicitly the derivatives of $\overline{F}$ in (9.7-37), we take as in the case of a solid beam, $\overline{F} = -\eta\zeta$.

In this case, one obtains for ε_{zz}

$$\varepsilon_{zz} = w_0' - \overline{F}\phi'' + \frac{1}{2}\left[(\eta^2 + \zeta^2)(\phi')^2 + 2\beta'\phi'(\eta^2 - \zeta^2)\right] + nr_t\phi''. \quad (9.7\text{-}38)$$

For $\eta^2 \gg \zeta^2$ one can see that Eqs. (9.7-7) obtained via Wagner's approach coincide with the present ones. As concerns γ_{sz} and γ_{nz}, these are obtained as

$$e_{13}(\equiv \gamma_{sz}) = 2\frac{A_c}{S}\phi', \quad e_{23}(\equiv \gamma_{nz}) = 0. \qquad (9.7\text{-}39)$$

In order to get a sense of the amount of difference incurred when Washizu's and Wagner's approaches are used, in Table 9.7-3 one compares the first natural frequency obtained within the two approaches. The blade is considered to be in the form of a box-beam characterized by $c = 10$in., $h = 0.2$in., $\Omega = 300$ rad/s.

The results obtained for various β_0, ply-angles θ, and cross-section aspect-ratios $\mathcal{R}(\equiv c/b)$ are displayed in Table 9.7-7.

Table 9.7-7: Comparison of predictions obtained via Wagner's and Washizu's approaches

β_0 (deg.)	θ (deg.)	$\mathcal{R} = 0.5$ Wagner	$\mathcal{R} = 0.5$ Washizu	$\mathcal{R} = 0.1$ Wagner	$\mathcal{R} = 0.1$ Washizu	$\mathcal{R} = 0.2$ Wagner	$\mathcal{R} = 0.2$ Washizu	$\mathcal{R} = 0.5$ Wagner	$\mathcal{R} = 0.5$ Washizu
	0	478.95	478.95	549.28	549.28	695.63	695.63	950.74	950.74
0	45	503.14	503.14	614.21	614.21	832.86	832.86	1196.7	1196.7
	90	490.99	490.99	577.76	577.76	747.69	747.69	1002.1	1002.1
	0	450.90	450.90	526.60	526.61	680.17	680.20	944.49	944.70
30	45	477.70	477.70	594.91	594.92	820.52	820.58	1191.9	1192.2
	90	492.48	492.53	580.09	580.40	750.10	751.45	1003.0	1009.1
	0	378.78	378.79	470.39	470.43	643.26	643.42	929.98	930.65
60	45	414.20	414.21	548.54	548.61	791.80	792.07	1881.0	1182.1
	90	513.75	513.96	601.63	602.80	767.70	772.85	1010.2	1034.1

The agreement between the predictions obtained via the two approaches appears to be exceptional for both thin ($\mathcal{R} = 0.05$) and rather thick ($\mathcal{R} = 0.5$) cross-section blades. This is the explanation why in Section 9.7 Wagner's approach was applied only.

These results modify the conclusions conjectured in Kaza and Kielb (1984) according to which Wagner's approach is applicable to *thin* cross-section beams only.

It should be indicated here that within the geometrically nonlinear formulation, a theory of naturally curved and twisted composite beams developed along Washizu's concept is due to Bauchau and Hong (1987).

9.8 A FEW RESULTS ON GEOMETRICALLY NONLINEAR ROTATING BEAMS

9.8-1 Preliminaries

The linear model of rotating beams is of invaluable importance towards determining, among others, of free vibration and dynamic response to time-dependent external excitation. However, in many important problems, a nonlinear model is needed.

Such a model can be instrumental in related stability and aeroelastic analyses (see e.g. Friedmann, 1977, 1983; Hong and Chopra, 1985), in a more accurate evaluation of forced response, as well as in the possibility to capture the effect of steady/unsteady external loads in the evaluation of natural frequencies and of various instabilities. In the next developments, a few results reported by Librescu et al. (2005) on the modeling and behavior of composite rotating blades, in general, and of those made up of FGMs, in particular, are presented.

9.8-2 Kinematics

Toward the modeling of geometrically nonlinear theory of rotating beams a number of statements are postulated:

(i) The flexural displacements u and v in the x and y directions are small but finite, while the twist ϕ of the cross-section can be arbitrarily large.

(ii) The axial displacement w is much smaller than u or v so that products of the derivatives of w can be neglected in the strain-displacement relations.

(iii) Strains are small so that a linear constitutive law can be used to relate the second Piola-Kirchhoff stress tensor to Green-Lagrange's strain tensor.

(iv) The effects of pretwist, presetting and of initial geometric imperfections are not included.

In this case, representation of displacements u and v is that provided by Eqs. (2.2-1) in which one considers that $x_P = y_P = 0$. For the sake of completeness these are recorded next

$$u(x, y, z, t) = u_0(z, t) - x(1 - \cos \phi(z, t)) - y \sin \phi(z, t), \quad \text{(9.8-1a)}$$

$$v(x, y, z, t) = v_0(z, t) + x \sin \phi(z, t) - y(1 - \cos \phi(z, t)), \quad \text{(9.8-1b)}$$

where ϕ is the elastic twist, positive counterclockwise, while

$$w(x, y, z, t) = w_0(z, t) + y\theta_x(z, t) + x\theta_y(z, t) - \phi'(z, t)[\overline{F}(s) - nr_t(s)].$$
$$\text{(9.8-1c)}$$

As a result, the 3-D strains expressed in terms of 1-D strain measures assume the representation

$$\epsilon_{zz}(s, z, n, t) = \epsilon_{zz}^{(0)} + n\epsilon_{zz}^{(1)} + n^2\epsilon_{zz}^{(2)}, \quad \text{(9.8-2a)}$$

$$\gamma_{sz}(s, z, n, t) = \gamma_{sz}^{(0)} + n\gamma_{sz}^{(1)}; \quad \gamma_{nz}(s, z, n, t) = \gamma_{nz}^{(0)}, \quad \text{(9.8-2b,c)}$$

where the detailed expressions of component terms are given in Eqs. (2.2-9), (2.2-12) and (2.2-13).

The previously displayed equations include the effect of transverse shear deformations. When this effect is neglected, implying adoption of Bernoulli-Euler beam model,

$$\theta_x(z, t) = u_0' \sin \phi - v_0' \cos \phi; \quad \theta_y(z, t) = -v_0' \sin \phi - u_0' \cos \phi. \quad \text{(9.8-3a,b)}$$

9.8-3 Equations of Motion and Boundary Conditions

The extended Hamilton's principle will be used to derive the equations of motion and the associated boundary conditions. To this end, the expressions of the strain energy, Eq. (3.1-16), of kinetic energy obtained by using Eqs. (9.8-1), and (9.1-7) in (9.1-8), and of expression of work performed by the external loading, Eq. (3.1-29), are replaced in Hamilton's functional.

Application of the standard procedure yields the equations of motion and boundary conditions of rotating beams exposed to a temperature field $T(n)$ as provided next.

The equations of motion:

$$\delta u_0 : (T_z u_0')' + \left\{ \left[Q_x - (M_x - M_x^T)\phi' \right] \cos \phi \right.$$

$$\left. - \left[Q_y + (M_y - M_y^T)\phi' \right] \sin \phi \right\}' + p_x(z, t) - m_1 = 0,$$

$$\delta v_0 : \; (T_z v_0')' + \Bigg\{ \Big[Q_x - (M_x - M_x^T)\phi' \Big] \sin \phi$$

$$+ \Big[Q_y + (M_y - M_y^T)\phi' \Big] \cos \phi \Bigg\}' + p_y(z, t) - m_2 = 0,$$

$$\delta w_0 : \; T_z' - m_3 = 0,$$

$$\delta \theta_x : \; M_x' - Q_y + m_x - m_4 = (M_x^T)',$$

$$\delta \theta_y : \; M_y' - Q_x + m_y - m_5 = (M_y^T)', \qquad\qquad (9.8\text{-}4a\text{-}f)$$

$$\delta \phi : \; B_w'' + M_z' + (T_r \phi')'$$

$$- \Bigg\{ \Big[M_x - M_x^T)u_0' - (M_y - M_y^T)v_0' \Big] \cos \phi + \Big[(M_x - M_x^T)v_0'$$

$$+ (M_y - M_y^T)u_0' \Big] \sin \phi \Bigg\}' + \Bigg\{ \Big[(M_x - M_x^T)v_0'\phi'$$

$$+ (M_y - M_y^T)u_0'\phi' + Q_y u_0' - Q_x v_0' \Big] \cos \phi$$

$$+ \Big[- (M_x - M_x^T)u_0'\phi' + (M_y - M_y^T)v_0'\phi'$$

$$+ Q_y v_0' + Q_x u_0' \Big] \sin \phi \Bigg\} - m_6 = 0.$$

The boundary conditions at $z = 0$:

$$u_0 = v_0 = w_0 = \theta_x = \theta_y = \phi = \phi' = 0, \qquad\qquad (9.8\text{-}5a\text{-}g)$$

and at $z = L$:

$$\delta u_0 : \; (T_z u_0') + \Bigg\{ \Big[Q_x - (M_x - M_x^T)\phi' \Big] \cos \phi$$

$$- \Big[Q_y + (M_y - M_y^T)\phi' \Big] \sin \phi \Bigg\} = 0,$$

$$\delta v_0 : \; (T_z v_0') + \Bigg\{ \Big[Q_x - (M_x - M_x^T)\phi' \Big] \sin \phi$$

$$+ \Big[Q_y + (M_y - M_y^T)\phi' \Big] \cos \phi \Bigg\} = 0,$$

$$\delta w_0 : \; T_z = 0, \qquad\qquad (9.8\text{-}6a\text{-}g)$$

$$\delta \theta_x : \; M_x = M_x^T,$$

$$\delta \theta_y : \; M_y = M_y^T,$$

$\delta\phi:$ $\underline{B'_w} + M_z + (T_r\phi')$

$$-\left\{\left[(M_x - M_x^T)u'_0 - (M_y - M'_y)v'_0\right]\cos\phi + \left[(M_x - M_x^T)v'_0\right.\right.$$

$$\left.\left. + (M_y - M_y^T)u'_0\right]\sin\phi\right\} - m_7 = 0,$$

$\delta\phi':$ $\underline{B_w} = 0.$

In these equations, the 1-D stress resultant/couple measures are expressed by Eqs. (3.1-17a-h). Herein, however, the notation T_r that is standard in the helicopter literature replaces Λ_z appearing in Eq. (3.1-17h). In addition, the 1-D thermal loading terms are provided by Eqs. (9.5-16).

The inertia and gyroscopic terms appearing in the equations of motion and boundary conditions are expressed as

$$m_1 = b_1\left[\ddot{u}_0 + \underline{2\Omega\dot{w}_0} - \Omega^2 u_0\right],$$
$$m_2 = b_1\ddot{v}_0,$$
$$m_3 = b_1\left\{\ddot{w}_0 - \left[\underline{2\Omega\dot{u}_0} + \Omega^2(R_0 + z + w_0)\right]\right\},$$
$$m_4 = (b_4 + b_{14})\left\{\ddot{\theta}_x + \left[-\Omega^2\theta_x + \underline{2\Omega(\cos\phi)\dot{\phi}}\right]\right\}, \qquad (9.8\text{-}7\text{a-g})$$
$$m_5 = (b_5 + b_{15})\left\{\ddot{\theta}_y + \left[-\Omega^2\theta_y + \underline{2\Omega(\sin\phi)\dot{\phi}}\right]\right\},$$
$$m_6 = (b_4 + b_5 + b_{14} + b_{15})\ddot{\phi} - (b_{10} + b_{18})(\ddot{\phi}'' - \Omega^2\phi'')$$

$$- \left\{\underline{2\Omega\left[(b_4 + b_{14})\,\dot{\theta}_x\cos\phi + (b_5 + b_{15})\dot{\theta}_y\sin\phi\right]} - \Omega^2\underwave{\left[(b_4 + b_{14})\right.}\right.$$

$$\left.\left. - \underwave{(b_5 + b_{15})\right]\cos\phi\,\sin\phi}\right\},$$
$$m_7 = -(b_{10} + b_{18})\left[\ddot{\phi}' - \Omega^2\phi'\right].$$

At this point, using expressions (3.1-17) of 1-D stress-resultant/couples in conjunction with Eqs. (3.1-15), (9.8-2) and (A.89), one obtain the of 1-D stress resultant/couple in terms of 1-D displacement quantities whose expressions are supplied in Table 9.8-1.

The stiffness quantities a_{ij} are provided in Table 4.3-1, while the few ones not included there are provided next:

$$a_{19} = \oint(K_{11}(x^2 + y^2)/2 + K_{14}r_n(s) + K_{16}/2)ds,$$

Table 9.8-1: Expression of 1-D stress resultant/couple

Quantity	Its expression
T_z	$a_{11}\left[w_0' + \dfrac{1}{2}\left((u_0')^2 + (v_0')^2\right)\right] + a_{19}(\phi')^2,$
T_r	$a_{81}\left[w_0' + \dfrac{1}{2}((u_0')^2 + (v_0')^2)\right] + a_{89}(\phi')^2,$
M_x	$a_{33}[\theta_x' - (u_0'\cos\phi + v_0'\sin\phi)\phi'],$
M_y	$a_{22}[\theta_y' - (u_0'\sin\phi - v_0'\cos\phi)\phi'],$
Q_x	$a_{44}(\theta_y + u_0'\cos\phi + v_0'\sin\phi),$
Q_y	$a_{55}[\theta_x - u_0'\sin\phi + v_0'\cos\phi],$
B_ω	$-a_{66}\phi'',$
M_z	$a_{77}\phi',$

$$a_{81} = \oint (K_{11}(x^2 + y^2) + K_{61} + 2K_{41}r_n(s))ds, \qquad (9.8\text{-}8a\text{-}c)$$

$$a_{89} = \oint \left\{[K_{11}(x^2 + y^2)/2 + K_{61} + 2K_{14}r_n](x^2 + y^2)\right.$$
$$\left. + K_{66}/2 + 2K_{44}r_n^2\right\}ds.$$

Based on Eqs. (9.8-7) and of the expressions in Table 9.8-1 one can represent the equations of motion and boundary conditions, Eqs. (9.8-4) and (9.8-6), respectively, in terms of 1-D displacement measures. Apart from the fact that the system of governing equations and boundary conditions obtainable in this way include higher order nonlinearities, it results to be fully coupled.

In this form, the structural model accounts for a number of nonclassical effects, such as the anisotropy of the constituent materials, transverse shear and warping inhibition. Due to its high intricacy, a simplified formulation of the problem is presented next.

- *Simplifications of the governing equation system.*

  Equation (9.8-4c) considered in conjunction with (9.8-7c) where one stipulates that the longitudinal acceleration $\ddot{w}_0$ is negligibly small, yields

$$T_z' = -b_1[2\Omega\dot{u}_0 + \Omega^2(R + z)]. \qquad (9.8\text{-}9)$$

Its integration in conjunction with the boundary condition (9.8-6c) yields

$$T_z = b_1\Omega^2 R(z) + \underline{2b_1\Omega \int_z^L \dot{u}_0 dz}, \qquad (9.8\text{-}10a)$$

where

$$R(z) = R_0(L - z) + 0.5\,(L^2 - z^2), \qquad (9.8\text{-}10\text{b})$$

R_0 denoting the hub radius.

T_z extracted from Table 9.8-1 in the form

$$T_z/a_{11} = w_0' + \frac{1}{2}\big((u_0')^2 + (v_0')^2\big) + (a_{19}/a_{11})(\phi')^2, \qquad (9.8\text{-}11)$$

in conjunction with (9.8-10a) yields

$$w_0' = \frac{b_1\Omega^2 R(z)}{a_{11}} + \frac{2b_1\Omega}{a_{11}} \int_z^L \dot{u}_0 dz$$
$$- \frac{1}{2}\big((u_0')^2 + (v_0')^2\big) - (a_{19}/a_{11})(\phi')^2, \qquad (9.8\text{-}12)$$

wherefrom one gets

$$\dot{w}_0 = \frac{2b_1\Omega}{a_{11}} \int_z^L \int_0^z \ddot{u}_0 d\zeta\, dz$$
$$- \int_0^z (u_0'\dot{u}_0' + v_0'\dot{v}_0')dz - 2(a_{19}/a_{11}) \int_0^z \phi'\dot{\phi}' dz, \qquad (9.8\text{-}13)$$

where ζ is a dummy variable associated with the coordinate z.

By virtue of (9.8-12), T_r results as

$$T_r = \frac{a_{81}}{a_{11}} \left[b_1\Omega^2 R(z) + 2b_1\Omega \int_z^L \dot{u}_0 dz \right]$$
$$+ (a_{89} - a_{81}a_{19}/a_{11})(\phi')^2. \qquad (9.8\text{-}14)$$

This expression of $\dot{w}_0$ will be used in the inertia term m_1, while T_z and T_r as provided by Eq. (9.8-10) and (9.8-14), in Eqs. (9.8-4) and (9.8-5).

A simplification of the governing equations can be achieved by considering in Eqs. (9.8-4) and (9.8-5) the twist small but finite, implying $\sin\phi \approx \phi$ and $\cos\phi \approx 1$. In such a context, the explicit form of the obtained governing equations was supplied in Librescu et al. (2005).

Since the equations should be considered in conjunction with (9.8-10) and (9.8-14), the governing system results in an integro-partial differential nonlinear system of fully coupled equations. One possible approach for simplifying the formulation is to split the coupled equations into uncoupled groups. Some results along this way will be presented next.

9.8-4 Special Cases

• *The case of the high stiffness in twist*
In this case one can consider $\phi \to 0$. As a result, the equations governing the flapping-lagging-transverse shear coupled motion are obtained as:

$$\delta u_0 : b_1\Omega^2[R(z)u_0']' + 2b_1\Omega\underbrace{\left[u_0'\int_z^L \ddot{u}_0 dz\right]'}_{A\quad A\quad A} + a_{44}[\theta_y + u_0']' - b_1\ddot{u}_0$$

$$- 2b_1\Omega\left\{\frac{2b_1\Omega}{a_{11}}\int_z^L\int_0^z \ddot{u}_0 d\zeta dz - \int_0^z (u_0'\ddot{u}_0 + v_0'\ddot{v}_0)dz\right\}$$

$$+ b_1\Omega^2 u_0 + p_x(z, t) = 0,$$

$$\delta v_0 : b_1\Omega^2[R(z)v_0']' + 2b_1\Omega\underbrace{\left[v_0'\int_z^L \ddot{u}_0 dz\right]'}_{A\quad A\quad A} + a_{55}(\theta_x + v_0')'$$

$$- b_1\ddot{v}_0 + p_y(z, t) = 0, \qquad\qquad (9.8\text{-}15\text{a-d})$$

$$\delta\theta_x : a_{33}\theta_x'' - a_{55}(\theta_x + v_0') - (b_4 + b_{14})[\ddot{\theta}_x - \Omega^2\theta_x] + m_x(z, t) = (M_x^T)',$$

$$\delta\theta_y : a_{22}\theta_y'' - a_{44}(\theta_y + u_0') - (b_5 + b_{15})[\ddot{\theta}_y - \Omega^2\theta_y] + m_y(z, t) = (M_y^T)'.$$

and the boundary conditions at $z = 0, L$:

$$\delta u_0 \; [a_{44}[\theta_y + u_0']] = 0,$$
$$\delta v_0 \; [a_{55}[\theta_x + v_0']] = 0, \qquad\qquad (9.8\text{-}16\text{a-d})$$
$$\delta\theta_x \; [a_{33}\theta_x' - M_x^T] = 0,$$
$$\delta\theta_y \; [a_{22}\theta_y' - M_y^T] = 0,$$

where at $z = 0$ the boundary conditions are entirely kinematic, while at $z = L$, entirely static.

The above governing system corresponds to the shearable structural model. As a sub-case of it, the *Bernoulli-Euler* counterpart of Eqs. (9.8-15) and (9.8-16) becomes

$$\delta u_0 : a_{22}u_0'''' - b_1\Omega^2[R(z)u_0']' - 2b_1\Omega\left[u_0'\int_z^L \dot{u}_0 dz\right]'$$

$$- (b_5 + b_{15})(\ddot{u}_0 - \Omega^2 u_0)'' + b_1(\ddot{u}_0 - \Omega^2 u_0)$$

$$+ [(2b_1\Omega)^2/a_{11}]\int_z^L \int_0^z \ddot{u}_0 d\zeta dz$$

$$- 2b_1\Omega\int_0^z (u_0'\dot{u}_0' + v_0'\dot{v}_0')dz - p_x - m_y' + (M_y^T)'' = 0, \quad (9.8\text{-}17a,b)$$

$$\delta v_0 : a_{33}v_0'''' - b_1\Omega^2[R(z)v_0']' - 2b_1\Omega\left[v_0'\int_z^L \dot{u}_0 dz\right]'$$

$$- (b_4 + b_{14})(\ddot{v}_0'' - \Omega^2 v_0'') + b_1\ddot{v}_0 - p_y - m_x' + (M_x^T)'' = 0.$$

The associated boundary conditions at $z = 0, L$ are

$$\delta u_0\left[-(b_5 + b_{15})(\ddot{u}_0 - \Omega^2 u_0)' - m_y + (M_y^T)'\right] = 0,$$

$$\delta v_0\left[-(b_4 + b_{14})(\ddot{v}_0 - \Omega^2 v_0)' - m_x + (M_x^T)'\right] = 0, \quad (9.8\text{-}18a\text{-}d)$$

$$\delta v_0'\left[a_{33}v_0'' + M_x^T\right] = 0,$$

$$\delta u_0'\left[a_{22}u_0'' + M_y^T\right] = 0.$$

In the previously displayed equations the terms underscored by —△—△—△— are associated with the nonlinear Coriolis effect.

The system of equations from Hodges and Ormiston (1976) obtained in the case of Euler-Bernoulli solid beam model, specialized for zero precone, is similar to those displayed previously (Eqs. 9.8-17), wherein the thermal terms and rotatory inertias are removed.

The case corresponding to inextensible beams, implying $a_{11} \rightarrow \infty$, can readily be obtained.

● *The case of twist motion*
Replacing (9.8-14) in the equations of twist and discarding higher-order nonlinear terms results in the governing equation:

$$\delta\phi : - \underline{a_{66}\phi''''} + a_{77}\phi'' + \frac{a_{81}}{a_{11}}b_1\Omega^2\big[R(z)\phi'\big]'$$

$$+ 2b_1\Omega\left(\frac{a_{81}}{a_{11}}\right)\left[\phi'\int_z^L \dot{u}_0 dz\right]' \quad - \big[a_{33}\theta_x'u_0' - a_{22}\theta_y'v_0'\big]'$$

$$+ a_{55}(\theta_x + v_0')u_0' - a_{44}(\theta_y + u_0')v_0'$$

$$+ (M_x^T u_0' - M_y^T v_0')' + (M_x^T v_0' + M_y^T u_0')'\phi \qquad (9.8\text{-}19)$$

$$- (b_4 + b_5 + b_{14} + b_{15})\ddot{\phi} + (b_{10} + b_{18})(\ddot{\phi}'' - \Omega^2\phi'')$$

$$+ 2\Omega[(b_4 + b_{14})\dot{\theta}_x \cos\phi + (b_5 + b_{15})\dot{\theta}_y \sin\phi)]$$

$$+ \Omega^2[\ (b_4 + b_{14}) - (b_5 + b_{15})]\cos\phi\sin\phi = 0.$$

The boundary conditions at $z = 0, L$ are:

$$\delta\phi\left[- \underline{a_{66}\phi'''} + a_{77}\phi' - \big[a_{33}\theta_x'u_0' - a_{22}\theta_y'v_0'\big] + (M_x^T v_0' + M_y^T u_0')\phi\right.$$

$$\left. + (b_{10} + b_{18})[\ddot{\phi}' - \Omega^2\phi']\right] = 0,$$

$$\delta\phi'\left[\ \underline{a_{66}\phi''}\ \right] = 0. \qquad (9.8\text{-}20a,b)$$

One should remark that in this form, the twist equations are coupled with terms belonging to flapping, lagging and transverse shear. Discarding nonlinear Coriolis effect, and assuming large bending stiffnesses resulting in $v_0 \to 0$ and $u_0 \to 0$, one obtains the governing equation

$$\delta\phi : - \underline{a_{66}\phi''''} + a_{77}\phi'' + \frac{a_{81}}{a_{11}}b_1\Omega^2[R(z)\phi']' - (b_4 + b_5 + b_{14} + b_{15})\ddot{\phi}$$

$$+ (b_{10} + b_{18})[\ddot{\phi} - \Omega^2\phi]'' + \Omega^2[\ (b_4 + b_{14}) - (b_5 + b_{15})]\ \phi = 0,$$

$$(9.8\text{-}21)$$

and the boundary conditions at $z = 0, L$:

$$\delta\phi\left[- \underline{a_{66}\phi'''} + a_{77}\phi' + (b_{10} + b_{18})[\ddot{\phi}' - \Omega^2\phi']\right] = 0,$$

$$\delta\phi'\left[\ \underline{a_{66}\phi''}\ \right] = 0. \qquad (9.8\text{-}22a,b)$$

In spite of the absence of thermal terms in Eqs. (9.8-21) and (9.8-22), its effect is however present in the stiffness quantities that are affected by the thermal degradation.

In the case of the discard of the bending-twist coupling, for metallic blades, the free warping version of the present governing equation coincides with its counterpart from Hodges and Ormiston (1976) and Subrahmanyan et al. (1987). Notice that in these indicated works the boundary conditions have not been supplied.

9.8-5 A Few Results

Based on previously supplied equations, one can determine the non-linear response to external and thermal loads that can be constant or time-dependent.

For aeroelastic studies, the unsteady aerodynamic loads have to be included. In these cases, one can arrive to an eigenvalue problem and the nature of eigenvalues, real or complex-conjugate, can provide the proper conditons of the occurrence of the instability by flutter, and the influence in this respect of thermomechanical loads.

In order to get the eigenvalue problem, the idea is to linearize the resulting nonlinear equations about an equilibrium state by the usual practice that consists of assuming that the dependent variables are the sum of a steady-state part and a (small) perturbation part, whose squares and higher powers are discarded after the substitution of their representations in the governing equations.

This process leads to two sets of differential equations, one set for the steady-state quantities, and another set for the disturbance quantities. However, the latter set of equations contains information reached from the solution of the former set. In the context of plates/shells the approach was developed by Librescu et al. (1996 c,d). Assuming that in this set of equations expressed in terms of disturbance quantities, the considered loads are time-dependent, a Fourier transform can convert this system of equations from the time-domain to the frequency-domain, and, in this way, is possible to study the system from the point of view of its stability. This can be accomplished by using the analytical criterion described in Librescu (1975).

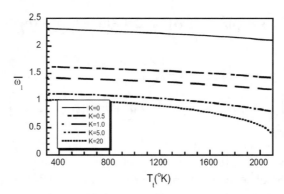

Figure 9.8-1: The effect of the temperature gradient of the rotating blade on the first dimensionless natural frequency for selected values of K ($\overline{\Omega} = 0.1$; $p_m = 1$)

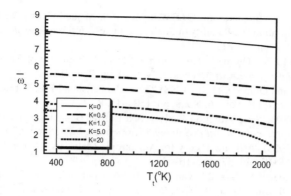

Figure 9.8-2: Counterpart of Fig. 9.8-1 for the second dimensionless natural frequency

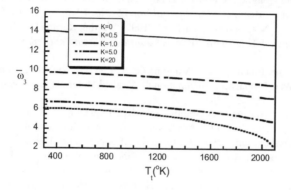

Figure 9.8-3: Counterpart of Fig. 9.8-1 for the third dimensionless natural frequency

Using this methodology, in Figs. 9.8-1 through 9.8-3 there were obtained the first three dimensionless eigenfrequencies as influenced by the static variation of the temperature on the top-face of the rotating blade whose cross-section is in the form of a box-beam. In these plots $\bar{\omega}_i = \omega_i/\omega_1$ and $\bar{\Omega} = \Omega/\omega_1$, where $\omega_1 = 496.73$ rad/s is the first natural frequency of the blade from full metal determined for conditions $\lambda = 0$ and $\mathcal{R} = 1$, where $p_m = p_y L^3/a_{33}^m$. The results show that the blade from full ceramic experiences the highest natural frequency and is not so susceptible to thermal degradation as the metallic blades that feature, in addition, the lowest natural frequency.

REFERENCES

Aboudi, J., Pindera, M. and Arnold, S. M. (1996) "Thermoelastic Theory for the Response of Materials Functionally Graded in Two Directions," *International Journal of Solids and Structures*, Vol. 33, No. 7, pp. 931–966.

Aboudi, J., Pindera, M. J. and Arnold, S. M. (1999) "Higher-Order Theory for Functionally Graded Materials," *Composites: Part B*, 30, pp. 777–832.

Ambartsumian, S. A., Bagdasarian, G. E. Durgarian, S. M. and Gnuny, V. T. (1966) "Some Problems of Vibration and Stability of Shells and Plates," *International Journal of Solids and Structures*, Vol. 2, pp. 59–81.

Anderson, G. L. (1975) "On the Extensional and Flexural Vibrations of Rotating Bars," *International Journal of Nonlinear Mechanics*, Vol. 10, pp. 223–236.

Balhaddad, A. S. and Onipede, D. Jr. (1998) "Three-Dimensional Free Vibration of Pretwisted Beams," *AIAA Journal*, Vol. 36, No. 8, pp. 1524–1528.

Banerjee, J. R. (2001), "Free Vibration Analysis of a Twisted Beam Using the Dynamic Stiffness Method," *International Journal of Solids and Structures*, Vol. 38, pp. 6703–6722.

Bauchau, O. A. (1985) "A Beam Theory for Anisotropic Materials," *Journal of Applied Mechanics, Trans. ASME*, Vol. 52, pp. 416–422.

Bauchau, O. A., Loewy, R. G. and Bryan, P. S. (1986) "Approach to Ideal Twist Distribution in Tilt Rotor VTOL Blade Designs," *RTC Report No. D-86-2*, Rensselaer Polytechnic Institute, Troy, NY July.

Bauchau, O. A. and Hong, C. H. (1987) "Large Displacement Analysis of Naturally Curved and Twisted Composite Beams," *AIAA Journal* Vol. 25, No. 10, pp. 1469–1475.

Bazoune, A., Khulief, Y. A. and Stephen, N. G. (1999) "Further Results for Modal Characteristics of Rotating Tapered Timoshenko Beams," *Journal of Sound and Vibration*, Vol. 219, 1, pp. 157–174.

Berezovski, A., Engelbrecht, J. and Maugin, G. A. (2003) "Numerical Simulation of Two-Dimensional Ware Propagation in Functionally Graded Materials, *European Journal of Mechanics A/Solids*, 22, pp. 257–265.

Bhat, R. B. (1986) "Transverse Vibrations of a Rotating Uniform Cantilever with Tip Mass as Predicted by Using Beam Characteristics Orthogonal Polynomials in the Rayleigh-Ritz Method," *Journal of Sound and Vibration*, Vol. 105, No. 2, pp. 199–210.

Bhuta, P. G. and Jones, J. P. (1965) "On Axial Vibrations of a Whirling Bar," *Journal of the Acoustical Society of America*, Vol. 35, No. 2, February, pp. 217–221.

Bielawa, R. L. (1992) *Rotary Wing Structural Dynamics and Aeroelasticity*, AIAA Education Series, AIAA, Inc., Washington, D.C.

Birman, V. (1995) "Buckling of Functionally Graded Hybrid Composite Plates," *Proceedings of the 10th Conference on Engineering Mechanics*, Vol. 2, Boulder, CO, pp. 1199–1202.

Boyce, W. E. (1956) "Effect of Hub Radius on the Vibration of a Uniform Bar," *Journal of Applied Mechanics, Trans. ASME*, Vol. 23, pp. 287–290.

Boyce, W. E. and Handelman, G. H. (1961) "Vibration of Rotating Beams with Tip Mass," *Zeitschrift für Agew. Math. and Physik*, XII, 5, 369–392.

Büter, A. and Breitbach E. (2000) "Adaptive Blade Twist-Calculations and Numerical Results," *Aerospace Science and Technology*, 4, pp. 309–319.

Carnegie, W. (1959) "Vibrations of Pre-twisted Cantilever Blading," *Proc. Inst. Mech. Engrs. London*, Vol. 173, pp. 343–362.

Carnegie, W. and Thomas, J. (1972) "The Effects of Shear Deformation and Rotary Inertia on the Lateral Frequencies of Cantilever Beams in Bending," *Journal of Engineering for Industry, Trans. ASME*, Vol. 94, pp. 267–278.

Cesnik, C. E. S. and Shin, S. J. (2001a) "On the Twist Performance of a Multiple-Cell Active Helicopter Blade," *Smart Materials and Structures*, Vol. 10, pp. 53–61.

Cesnik, C. E. S. and Shin, S. J. (2001b) "On The Modeling of Integrally Activated Helicopter Blades," *International Journal of Solids and Structures*, Vol. 38, pp. 1765–1789.

Chandra, R. and Chopra, I. (1992) "Experimental-Theoretical Investigation of the Vibration Characteristics of Rotating Composite Box Beams," *Journal of Aircraft*, Vol. 29, No. 4, pp. 657–664.

Chandiramani, N. K., Librescu, L. and Shete, C. D. (2002) "On The Free-Vibration of Rotating Composite Beams Using a Higher-Order Transverse Shear Formulation," *Aerospace Science and Technology*, Vol. 6, pp. 545–561.

Chandiramani, N. K. and Librescu, L. (2002) "Optimal Vibration Control of a Rotating Shearable Blade Using Distributed Piezoelectric Sensing and Actuation," *Smart Structures and Materials; Modeling Signal Processing and Control*, (V. S. Rao Ed.) *Proceedings of SPIE*, Vol. 4693, SPIE, pp. 451–460.

Chandiramani, N. K., Shete, C. D. and Librescu, L. (2003a) "Vibration of Higher-Order Shearable Pretwisted Rotating Composite Blades," *International Journal of Mechanical Sciences*, Vol. 45, pp. 2017–2041.

Chandiramani, N. K., Shete, C. D. and Librescu, L. (2003b) "Optimal Control of Pretwisted Shearable Smart Composite Rotor Blades," *AIAA-2003-1540, 44th AIAA/ASME/ASCE/AHS Structures, Structural Dynamics and Materials Conference*, Norfolk, VA, April 7–10.

Chandiramani, N. K., Librescu, L., Saxena, V. and Kumar, A. (2004) "Optimal Vibration Control of a Rotating Composite Beam with Distributed Piezoelectric Sensing and Actuations," *Smart Materials and Structures*, Vol. 13, pp. 433–442.

Chelu, P. and Librescu, L. (2005) "Dynamic Response of Spinning Thin-Walled Composite Booms Exposed to Solar Radiation Using Wavelet-Galerkin Method," *Proceedings of the Sixth International Congress on Thermal Stresses, TS 2005*, Vienna University of Technology, F. Ziegler, R. Heuer and C. Adam (Eds), Vol. 2, pp. 459–462, TU Wien, May 2005.

Cheng, Z.-W. and Batra, R. C. (2000) "Three-Dimensional Thermoelastic Deformation of a Functionally Graded Elliptic Plate," *Composites, Part B*, Vol. 31, pp. 97–100.

Chopra, I. (2000) "Status of Application of Smart Structures Technology to Rotorcraft Systems," *Journal of the American Helicopter Society*, Vol. 45, No. 4, pp. 228–252.

Coyle, T. and Harris, W. R. (1960) "The Effects of Blast Against Rotor Blades," (V) *BRL, Technical Note No. 1342*, Sept. 1960, AD 370885.

Dawson, B. (1968), "Coupled Bending-Bending Vibrations of Pretwisted Cantilever Blading Treated by Rayleigh-Ritz Energy Method," *Journal of Mechanical Engineering Science*, Vol. 10, No. 5, pp. 381–388.

Dokumaci, E., Thomas, J. and Carnegie, W. (1967) "Matrix Displacement Analysis of Coupled Bending-Bending Vibration of Pretwisted Blading," *Journal of Mechanical Engineering Science*, Vol. 9, No. 4, pp. 247–254.

Du, H., Lim, M. K. and Liew, K. M. (1994) "A Power Series Solution for Vibration of a Rotating Timoshenko Beam," *Journal of Sound and Vibration*, Vol. 175, No. 4, pp. 505–523.

Eick, C. D. and Mignolet, M. P. (1995, 1996) "Vibration and Buckling of Flexible Rotating Beams," *AIAA Journal*, Vol. 33, No. 3, March. pp. 528–538, and *AIAA Journal*, Vol. 34, No. 3, pp. 641–643.

Ewins, D. J. and Henry, R. (1992) "Structural Dynamic Characteristics of Individual Blades," *Vibration and Rotor Dynamics*, von Kármán Institute for Fluid Dynamics, Lecture Series 1992-06, September, pp. 14.1 – 14.27.

Fauconneau, G. and Marangoni, R. D. (1970) "Effect of a Thermal Gradient on the Natural Frequencies of a Rectangular Plate," *International Journal Mechanical Sciences*, Vol. 12, pp. 113–122.

Flax, A. H. (1996) "Comments on Vibration and Buckling of Flexible Rotating Beams," *AIAA Journal*, Vol. 34, No. 3, March. pp. 640–641.

Friedmann, P. (1977) "Recent Developments in Rotaway-Wing Aeroelasticity," *Journal of Aircraft*, Vol. 14, No. 11, pp. 1027–1041.

Friedmann, P. (1983) "Formulation and Solution of Rotary-Wing Aeroelastic Stability and Response Problems," *Vertica*, Vol. 7, No. 2, pp. 101–141.

Gern, F. H. and Librescu, L. (1998) "Effects of Externally Mounted Stores on Flutter Character-istics of Advanced Swept Cantilevered Aircraft Wings," *Aerospace Science and Technology*, Vol. 2, No. 5, pp. 321–333.

Gern, F. H. and Librescu, L. (1999) "Aeroelatic Tailoring of Advanced Aircraft Wings Carrying External Stores," *Atti della Accademia delle Scienze di Torino, Classe di Scienze Fisiche, Mathematiche's Naturali, Quaderni*, 1, pp. 201–219, (Issue devoted to Placido Cicala).

Gern, F. H. and Librescu, L. (2000) "Aeroelastic Tailoring of Composite Wings Exhibiting Nonclassical Effects and Carrying External Stores," *Journal of Aircraft*, Vol. 37, No. 6, pp. 1097–1004.

Gern, F. H. and Librescu, L. (2001) "Static and Dynamic Aeroelasticity of Advanced Aircraft Wings Carrying Exernal Stores," *AIAA Journal*, Vol. 36, No. 7, pp. 1121–1129.

Giurgiutiu, V. and Stafford, R. O. (1977) "Semi-Analytic Methods for Frequencies and Mode Shapes of Rotor Blades," *Vertica*, Vol. 1, pp. 291–306.

Gong, S. W., Lam, K. Y. and Reddy, J. N. (1999) "The Elastic Response of Functionally Graded Cylindrical Shells to Low-Velocity Impact," *International Journal of Impact Engineering*, Vol. 22, No. 4, pp. 397–417.

Hilton, H. H. (2005) "Optimum Linear and Nonlinear Viscoelastic Functionally Graded Materials-Characterizations and Analysis," *Composites Part A: Manufacturing and Applied Sciences*, (in press).

Hoa, S. V. (1979) "Vibration of a Rotating Beam with Tip Mass," *Journal of Sound and Vibration*, Vol. 67, No. 2, pp. 369–381.

Hodges, D. H. (1977) "On the Extensional Vibrations of Rotating Bars," *International Journal of Nonlinear Mechanics*, Vol. 12, pp. 293–296.

Hodges, D. H. (1980), "Torsion of Pretwisted Beams Due to Axial Loading," *ASME Journal of Applied Mechanics*, Vol. 47, pp. 393–397.

Hodges, D. H. (1981) "An Approximate Formula for the Fundamental Frequency of a Uniform Rotating Beam Clamped off the Axis of Rotation," *Journal of Sound and Vibration*, Vol. 77, No. 1, pp. 11–18.

Hodges, D. H. and Ormiston, R. A. (1976) "Stability of Elastic Bending and Torsion of Uniform Cantilever Rotor Blades in Hover with Variable Structural Coupling", NASA TND-8192, April.

Hodges, D. H., Rutkowski, M. J. (1981), "Free-Vibration Analysis of Rotating Beams by a Variable-Order Finite Element Method," *AIAA Journal*, Vol. 19, No. 11, pp. 1459–1466.

Hodges, D. H. (1990) "Review of Composite Rotor Blade Modeling," *AIAA Journal*, Vol. 28, No. 3, pp. 561–565.

Hong, C. H. and Chopra, I. (1985) "Aeroelastic Stability of a Composite Rotor Blade," *Journal of American Helicopter Society*, Vol. 30, No. 2, pp. 57–67.

Houbolt, J.C. and Brooks, G. W. (1958) "Differential Equations of Motion for Combined Flap-wise Bending, Chordwise Bending, and Torsion of Twisted Nonuniform Rotor Blades," *NACA TR 1346*.

Isakson, G. and Eisley, J. G. (1960) *"Natural Frequencies in Bending of Twisted Rotating Blades,"* NASA TN D-371.

Johnson, W. (1980) *Helicopter Theory*, Princeton University Press.

Jung, S. N., Nagaraj, V. T. and Chopra, I. (1999) "Assessment of Composite Rotor Blade: Mod-eling Techniques," *Journal of the American Helicopter Society*, Vol. 44, No. 3, pp. 188–205.

Jung, S. N., Nagaraj, V. T. and Chopra, I. (2001) "Refined Structural Dynamics Model for Composite Rotor Blades," *AIAA Journal*, Vol. 39, No. 2, pp. 339–348.

Karpouzian G. and Librescu, L. (1994) "A Comprehensive Model of Anisotropic Composite Aircraft Wings and Its Use in Aeroelastic Analyses," *Journal of Aircraft*, May–June, Vol. 31, No. 3, pp. 702–712.

Karpouzian G. and Librescu, L. (1995) "Exact Flutter Solution of Advanced Composite Swept Wings in Various Flight Speed Regimes," *AIAA Paper 95-1382, Proceedings of the 36th AIAA/ASME/ASCE/AHS/ASC Structures, Structural Dynamics and Materials Conference*, New Orleans, LA, April 10–12.

Karpouzian, G. and Librescu, L. (1996) "Non-Classical Effects on Divergence and Flutter of Anisotropic Swept Aircraft Wings," *AIAA Journal*, Vol. 34, No. 4, April, pp. 786–794.

Kaza, K. B. and Kielb, R. E. (1984) "Effects of Warping and Pretwist on Torsional Vibration of Rotating Beams," *Journal of Applied Mechanics, Trans. ASME*, Vol. 51, Dec., pp. 913–920.

Khulief, Y. A. and Bazoune, A. (1992) "Frequencies of Rotating Tapered Timoshenko Beams with Different Boundary Conditions," *Computers & Structures*, Vol. 42, No. 5, pp. 781–795.

Kosmatka, J. B. (1992) "Extension-Bend-Twist Coupling Behavior of Nonhomogeneous Anisotropic Beams with Initial Twist," *AIAA Journal*, Vol. 30, No. 2, pp. 519–527.

Krenk, S. (1984) "A Linear Theory for Pretwisted Elastic Beams," *ASME Trans., Journal of Applied Mechanics*, Vol. 50, pp. 137–142.

Kumar R. (1974) "Vibration of Space Booms Under Centrifugal Force Field," *Canadian Aeronautics and Space Institute (CASI) Trans.* Vol. 7, pp. 1–5.

Kunz, D. L. (1994), "Survey and Comparison of Engineering Beam Theories for Helicopter Rotor Blades," *Journal of Aircraft*, Vol. 31, No. 3, pp. 473–479.

Kvaternik, R. G., White, W. F. Jr. and Kaza, K. R. V. (1978) "Nonlinear Flap-Lag-Axial Equations of a Rotating Beam with Arbitrary Precone Angle," *AIAA Paper No. 78-491*.

Lake, R. C., Nixon, M. W., Wilbur, M. L. Singleton, J. D. and Mirick, R. H. (1992) "A Demonstration of Passive Blade Twist Control Using Extension-Twist Coupling," Paper AIAA-92-2468-CR, *SDM Conference*, Dallas, Texas.

Lee, H. P. (1993) "Vibration of an Inclined Rotating Cantilever Beam With Tip Mass," *Journal of Vibration and Acoustics, Trans. ASME*, 115, July, pp. 241–245.

Lee, S. Y. and Kuo, Y. H., (1992) "Bending Vibrations of a Rotating Non-Uniform Beam with an Elastically Restrained Root," *Journal of Sound and Vibration*, Vol. 154, No. 3, pp.441–451.

Lee, S. Y. and Lin, S. M. (1994) "Bending Vibration of Rotating Nonuniform Timoshenko Beams with an Elastically Restrained Root," *Journal of Applied Mechanics*, Trans. ASME, Paper No. 94-WA/APM-8.

Leissa, A. and Co, C. M. (1984) "Coriolis Effects on the Vibration of Rotating Beams and Plates," *Proceedings of the 12th Southeastern Conference on Theoretical and Applied Mechanics*, Callaway Gardens, Pine Mountain, GA, May 10–11, pp. 508–513.

Librescu, L. (1975) *Elastostatics and Kinetics of Anisotropic and Heterogeneous Shell-Type Structures*, Noordhoff International Publishing, Leyden, Netherlands, pp. 560–598.

Librescu, L. and Song, O. (1991) "Behavior of Thin-Walled Beams Made of Advanced Composite Materials and Incorporating Non-Classical Effects," *Applied Mechanics Reviews*, Vol. 44, No. 11, Part 2, November, pp. 174–180.

Librescu, L. and Thangjiham, S. (1991) "Analytical Studies on Static Aeroelastic Characteristics for Composite Forward-Swept Wing Aircraft," *Journal of Aircraft*, Vol. 28, No. 2, pp. 151–157.

Librescu, L. and Song, O. (1992) "On the Static Aeroelastic Tailoring of Composite Aircraft Swept Wings Modelled as Thin-Walled Beam Structures," *Composites Engineering*, Vol. 2, No. 5–7, pp. 497–512.

Librescu, L., Song, O. and Rogers, C. A. (1993) "On Adaptive Vibration Behavior of Cantilevered Structures Modelled as Composite Thin-Walled Beams," *International Journal of Engineering Science*, Vol. 31, No. 5, pp. 775–792.

Librescu, L., Meirovitch, L. and Song, O. (1996a) "Refined Structural Modeling for Enhancing Vibrational and Aeroelastic Characteristics of Composite Aircraft Wings," *La Recherche Aerospatiale*, 1, pp. 23–35.

Librescu, L., Meirovitch, L. and Song, O. (1996b) "Integrated Structural Tailoring and Control Using Adaptive Materials for Advanced Aircraft Wings," *Journal of Aircraft*, Vol. 22, No. 1, Jan.– Feb., pp. 203–213.

Librescu, L., Lin, W., Nemeth, M. P. and Starnes, Jr., J. H., (1996c) "Vibration of Geometrically Imperfect Panels Subjected to Thermal and Mechanical Loads," *Journal of Spacecraft and Rockets*, Vol. 33, No. 2, March–April, pp. 285–291.

Librescu, L., Lin, W., Nemeth, M. P. and Starnes, Jr., J. H., (1996d) "Frequency-Load Interaction of Geometrically Imperfect Curved Panels Subjected to Heating," *AIAA Journal*, Vol. 34, No. 1, pp. 166–177.

Librescu, L., Meirovitch, L. and Na, S. S. (1997) "Control of Cantilevers Vibration via Structural Tailoring and Adaptive Materials," *AIAA Journal*, Vol. 35, No. 8, August pp. 1309–1315.

Librescu, L., Song, O. and Kwon, H. D. (1999) "Vibration and Stability Control of Gyroelastic Thin-Walled Beams via Smart Materials Technology," in *Smart Structures*, J. Holnicki-Szulk and J. Rodellar, (Eds.), Kluwer Academic Publication, pp. 163–172.

Librescu, L. and Na, S. S. (2001) "Active Vibration of Thin-Walled Tapered Beams Using Piezoelectric Strain Actuation," *Journal of Thin-Walled Structures*, Vol. 39, No. 1, pp. 65–68.

Librescu, L., Oh, S-Y. and Song, O. (2003) "Spinning Thin-Walled Beams Made of Functionally Graded Materials: Modeling, Vibration and Instability," *European Journal of Mechanics A/Solids*, Vol. 23, No. 3, pp. 499–515.

Librescu, L., Oh, S-Y. and Song, O. (2005) "Thin-Walled Beams Made of Functionally Graded Materials and Operating in a High Temperature Environment: Vibration and Stability," *Journal of Thermal Stresses*, Vol. 28, Nos. 6–7, pp. 649–712.

Lin, S. M. (1997) "Vibration of Elastically Restrained Nonuniform Beams with Arbitrary Pretwist," *AIAA Journal*, Vol. 35, No. 11, pp. 1681–1687.

Lo, H. and Renbarger, J. L. (1952) "Bending Vibrations of a Rotating Beam," *Proceedings of the First U. S. National Congress of Applied Mechanics*, New York, N.Y., pp. 75–79.

Lo, H., Goldberg, J. E. and Bogdanoff, J. L. (1960) "Effect of Small Hub-Radius Change on Bending Frequencies of a Rotating Beam," *Journal of Applied Mechanics*, Trans. ASME, Vol. 27, September, pp. 548–550.

Loy, C. T., Lam, K. Y. and Reddy, J. N. (1999) "Vibration of Functionally Graded Cylindrical Shells," *International Journal of Mechanical Sciences*, Vol. 41, No. 3, pp. 309–324.

Mansfield, E. H. and Sobey, A. J. (1979) "The Fibre Composite Helicopter Blade — Part I: Stiffness Properties: Part II: Prospects for Aeroelastic Tailoring," *Aeronautical Quarterly*, Vol. 30, No. 2, pp. 413–449.

McGee, O. G. (1992) "Influence of Warping-Pretwist Coupling on the Torsional Vibration of Centrifugally-Stressed Cantilevers with Thin-Walled Open Profiles," *Computers & Structures*, Vol, 42, No. 2, pp. 175–195.

Na, S. S. and Librescu (2000) "Modeling and Vibration Feedback Control of Rotating Tapered Beams Incorporating Adaptive Capabilities," Recent Advanced in Solids and Structures-2000, PVP-Vol. 415, H. H. Chung and Y. W. Kwon (Eds.), ASME, New York, pp. 35–43.

Na, S. S., Librescu, L. and Shim J-K. (2003a) "Modeling and Bending Vibration Control of Nonuniform Thin-Walled Rotating Beams Incorporating Adaptive Capabilities," *International Journal of Mechanical Sciences*, Vol. 45, No. 8, pp. 1347–1367.

Na, S. S., Librescu, L. and Jung, H. (2003b) "Vibration Control of Rotating Composite Thin-Walled Beams in a Temperature Environment," in *Proceedings of the 5th International Congress on Thermal Stresses and Related Topics*, Blacksburg, VA, June 8–11, L. Librescu and P. Marzocca (Eds.), Vol. 2, WA-6-4-(1-4).

Na, S. S. and Librescu, L. and Jung, H. (2004) "Dynamics and Active Bending Vibration Control of Turbomachinery Rotating Blades Featuring Temperature-Dependent Material Properties," *Journal of Thermal Stresses*, Vol. 24, pp. 625–644.

Na, S. S., Librescu, L., Rim, S. and Jeong, I-J.,(2004) "Free Vibration and Control of Composite Non-Uniform Thin-Walled Beams Featuring Bending-Bending Elastic Coupling," *Proceeding of the IMECE 2004, ASME International Mechanical Engineering Congress*, November 13–19, 2004, Anaheim, Califormia.

Nagaraj, V. T. and Sahu, N. (1982) "Torsional Vibration of Non-Uniform Rotating Blades with Attachment Flexibility," *Journal of Sound and Vibration*, Vol. 80, No. 3, pp. 401–411.

Nixon, M. W. (1987) "Extension-Twist Coupling of Composite Circular Tubes with Application to Tilt Rotor Blade Design," *28th Structure, Structural Dynamics and Materials Conference*, April 6–8, Monterey, CA, AIAA Paper No. 87-0772, pp. 295–303.

Nixon, M. W. (1989) "Analytical and Experimental Investigations of Extension-Twist-Coupled Structures," George Washington University Masters Thesis, Hampton, VA.

Nixon, M. W. (1992) "Parameter Studies for Tiltrotor Aeroelastic Stability in High-Speed Flight," *AIAA-92-5568-CR*.

Noda, N. and Jin, Z. H. (1993) "Thermal Stress Intensity Factors for a Crack in a Strip of a Functionally Gradient Material," *International Journal of Solids and Structures*, Vol. 30, pp. 1039–1056.

Noor, A. K. and Burton, W. S. (1992) "Computational Models for High-Temperature Multilayered Composite Plates and Shells," *Applied Mechanics Reviews*, Vol. 45, No. 10, pp. 414–446.

Oh, S-Y., Song, O. and Librescu, L. (2003a) "Effects of Pretwist and Presetting on Coupled Bending Vibrations of Rotating Composite Beams," *International Journal of Solids and Structures*, Vol. 40, pp. 1203 – 1224.

Oh, S-Y., Librescu, L. and Song, O. (2003b) "Thin-Walled Rotating Blades Made of Functionally Graded Materials: Modeling and Vibration Analysis," *AIAA 2003-1541, 44th AIAA/ASME/ASCE/AHS Structures, Structural Dynamics, and Materials Conference*, Norfolk, VA April 7–10.

Oh, S-Y., Librescu, L. and Song, O. (2003c) "Thin-Walled Rotating Blades Made of Functionally Graded Materials: Thermoelastic Modeling and Vibration Analysis," in *Thermal Stresses 03*, Vol. 1, MA-2-5-1-MA-2-5-4, L. Librescu and P. Marzocca (Eds.), Virginia Tech, Blacksburg, VA, USA.

Oh, S-Y., Librescu, L. and Song, O. (2003d) "Thermoelastic Modeling and Vibration of Functionally Graded Thin-Walled Rotating Blades,"*AIAA Journal*, Vol. 41, No. 10, pp. 2051–2060.

Oh, S-Y., Librescu, L. and Song, O. (2003e) "Vibration of Turbomachinery Rotating Blades Made-Up of Functionally Graded Materials and Operating in a High Temperature Field," *Acta Mechanica*, Vol. 166, pp. 69–87.

Oh, S-Y., Librescu, L. and Song, O. (2004) "Thin-Walled Rotating Composite Blades Featuring Extension-Twist Elastic Coupling," *AIAA-2004-2049*, 45th AIAA/ASME/ASCE/AHS/ASC Structures, Structural Dynamics & Materials Conference, Palm Springs, CA April 19–22.

Oh, S-Y., Librescu, L. and Song, O. (2005) "Modeling and Vibration of Thin-Walled Rotating Composite Blades Featuring Extension-Twist Elastic Coupling," *The Aeronautical Journal*, Vol. 109, May, No. 1095, pp. 233–246.

Oh, S-Y. (2004) Personal Communication.

Palazotto, A. N. and Linnemann, P.E. (1991) "Vibration and Buckling Characteristics of Composite Cylindrical Panels Incorporating the Effects of a Higher Order Shear Theory," *International Journal of Solids and Structures*, Vol. 28, No. 3, pp. 341–361.

Peters, D. A. (1973) "An Approximate Solution for the Free Vibrations of Rotating Uniform Cantilever Beams," NASA TM X-62, 299, September.

Peters, D. A. (1995) "Aeroelastic Response of Rotorcraft," in a *Modern Course in Aeroelasticity*, (Third Edition), Edited by E. H. Dowell, Kluwer Academic Publishers, Dordrecht, The Netherlands.

Pindera, M-J. and Aboudi, J. (2000) "A Coupled Higher-Order Theory for Cylindrical Structural Components with Bi-Directionally Graded Microstructures," *NASA CR 210350*, NASA-Glenn Research Center, Cleveland, OH.

Pradhan, S. C., Loy, C. T., Lam, K. Y. and Reddy, J. N. (2000) "Vibration Characteristics of Functionally Graded Cylindrical Shells Under Various Boundary Conditions," *Journal of Applied Acoustics*, Vol. 61, No. 1, pp. 119–129.

Praveen, G. N. and Reddy, J. N. (1998) "Nonlinear Transient Thermoelastic Analysis of Functionally Graded Ceramic-Metal Plates," *International Journal of Solids Structures*, Vol. 35, No. 33, pp. 4457–4476.

Putter, S. and Manor, H. (1978), "Natural Frequencies of Radial Rotating Beams," *Journal of Sound and Vibrations*, Vol. 56, No. 2, pp. 175–185.

Qin, Z. and Librescu, L. (2001) "Static and Dynamic Validations of a Refined Thin-Walled Composite Beam Model," *AIAA Journal*, Vol. 39, No. 12, pp. 2422–2424.

Qin, Z. and Librescu, L. (2002) "On a Shear-Deformable Theory of Anisotropic Thin-Walled Beams: Further Contribution and Validations," *Composite Structures*, Vol. 56, No. 4, pp. 345–358.

Rand, O. (1991) "Periodic Response of Thin-Walled Composite Helicopter Rotor Blades," *Journal of the American Helicopter Society*, Vol. 36, No. 4, pp. 3–11.

Rand, O. (1996) "Analysis of Composite Rotor Blades," in *Numerical Analysis and Modelling of Composite Materials*, J. W. Bull, Editor, Blackie Academic and Professional, Chapman and Hall, pp. 1–26.

Rao, J. S. (1971) "Flexural Vibration of Pre-Twisted Beams of Rectangular Cross-Section," *Journal of the Aeronautical Society of India*, Vol. 23, No. 1, pp. 62–64.

Rao, J. S. (1977), "Coupled Vibrations of Turbomachine Blades," *The Shock Vibration Bulletin*, Vol. 47, pp. 107–125.

Rao, J. S. (1991) *Turbomachine Blade Vibration*, John Wiley & Sons, New York, Chickester, Brisbane, Toronto, Singapore.

Reddy, J. N. and Chin, C. D. (1998) "Thermomechanical Analysis of Functionally Graded Cylinders and Plates," *Journal of Thermal Stresses*, Vol. 21, pp. 593–626.

Reddy, J. N. (2000) "Analysis of Functionally Graded Plates," *International Journal of Numerical Meth. Engr.*, No. 47, pp. 663–684.

Rehfield, L. W. (1985) "Design Analysis Methodology for Composite Rotor Blades," *AFWAL-TR-85-3094*, June, pp. (V(a)-1) – (V(a)-15).

Rehfield, L. W. and Atilgan, A. R. (1989) "Toward Understanding the Tailoring Mechanisms for Thin-Walled Composite Tubular Beams," in *Proceeding of the First USSR-USA Symposium on Mechanics of Composite Materials*, Riga, Latvia, USSR, May 23–26, 1989, S. W. Tsai, J. M. Whitney, T. W. Choi and R. M. Jones (Eds.), ASME, pp. 187–196.

Rosen, A. and Friedmann, P. (1978) "Nonlinear Equations of Equilibrium for Elastic Helicopter or Wind Turbine Blades Undergoing Moderate Deformation," University of California at Los Angeles, School of Engineering and Applied Science Report, *UCLA-ENG-7718, DOE/NASA/3082-78/1, NASA CR-159478*.

Rosen, A. (1978, 1980) "The Effect of Initial Twist on the Torsional Rigidity of Beams-Another Point of View," *Technion, Department of Aeronautical Engineering, TAE Report No. 360*, published also in 1980, *Journal of Applied Mechanics, ASME* Vol. 47, pp. 389–393.

Rosen, A., Loewy, R. G., Mathew, M. B. (1987), "Use of Twisted Principal Coordinates and Non-Physical Coordinates in Blade Analysis, *Vertica*, Vol. 11, 541–572.

Rosen, A. (1991) "Structural and Dynamic Behavior of Pretwisted Rods and Beams," *Applied Mechanics Reviews*, Vol. 44, No. 12, art 1, pp. 483–515.

Sabuncu, M. (1985) "Coupled Vibration Analysis of Blades with Angular Pretwist of Cubic Distribution," *AIAA Journal*, Vol. 23, No. 9, pp. 1424–1430.

Sankar, B. V. (2001) "An Elasticity Solution for Functionally Graded Beams," *Composites Science and Technology*, Vol. 61, pp. 689–696.

Sankar, B. V. and Tzeng, T. J. (2002) "Thermal Stresses in Functionally Graded Beams," *AIAA Journal*, Vol. 40, No. 6, pp. 1228–1232.

Shen, H. -S. (2002) "Postbuckling Analysis of Axially-Loaded Functionally Graded Cylindrical Shells in Thermal Environments," *Composites Science and Technology*, Vol. 62, pp. 977–987.

Slyper, H. A. (1962) "Coupled Bending Vibration of Pretwisted Cantilever Beams," *Journal of Mechanical Engineering Sciences*, Vol. 4, No. 4, pp. 365–379.

Song, O. and Librescu, L. (1993) "Free Vibration of Anisotropic Composite Thin-Walled Beams of Closed Cross-Section Contour," *Journal of Sound and Vibration*, Vol. 167, No. 1, pp. 129–147.

Song, O., Librescu, L. and Rogers, C. A. (1994) "Adaptive Response Control of Cantilevered Thin-Walled Beams Carrying Heavy Concentrated Masses," *Journal of Intelligent Materials Systems and Structures*, Vol. 5, No. 1, January, pp. 42–48.

Song, O. and Librescu, L. (1995) "Bending Vibration of Cantilevered Thin-Walled Beams Subjected to Time-Dependent External Excitations," *Journal of the Acoustical Society of America*, Vol. 98, No. 1, pp. 313–319.

Song, O. and Librescu, L. (1996) "Bending Vibrations of Adaptive Cantilevers with External Stores," *International Journal of Mechanical Sciences*, Vol. 28, No. 5, pp. 483–498.

Song, O. and Librescu, L. (1997) "Structural Modeling and Free Vibration Analysis of Rotating Composite Thin-Walled Beams," *Journal of the American Helicopter Society*, Vol. 42, No. 4, pp. 358–369.

Song, O. and Librescu, L. (1999) "Modeling and Dynamic Behavior of Rotating Blades Carrying a Tip Mass and Incorporating Adaptive Capabilities," *Acta Mechanica*, Vol. 134, pp. 169–197.

Song, O., Librescu, L. and Oh, S-Y. (2001a) "Vibration of Pretwisted Adaptive Rotating Blades Modeled as Anisotropic Thin-Walled Beams," *AIAA Journal*, Vol. 39, No. 2, February, pp. 285–295.

Song, O., Librescu, L. and Oh, S.-Y (2001b) "Dynamic of Pretwisted Rotating Thin-Walled Beams Operating in a Temperature Environment," *Journal of Thermal Stresses*, Vol. 24, No. 3, pp. 255–279.

Song, O, Oh, S-Y. and Librescu, L. (2002) "Dynamic Behavior of Elastically Tailored Rotating Blades Modeled as Pretwist Thin-Walled Beams and Incorporating Adaptive Capabilities," *International Journal of Rotating Machinery*, Vol. 8, No. 1.

Stafford, R. O. and Giurgiutiu, V. (1975) "Semi-Analytic Methods for Rotating Timoshenko Beams," *International Journal of Mechanical Sciences*, Vol. 17, pp. 719–727.

Stemple, A. D. and Lee, S. W. (1989) "Finite Element Modeling for Composite Beams Undergoing Large Deflections with Arbitrary Cross Sectional Warping," *Int. of J. Numerical Methods in Engineering*, Vol. 28, No. 9, pp. 2143–2160.

Subrahmanyam, K. B., Kulkarni, S. V. and Rao, J. S. (1981) "Coupled Bending-Bending Vibrations of Pretwisted Cantilever Blading Allowing for Shear Deformation and Rotary Inertia by Reissner Method," *International Journal of Mechanical Sciences*, Vol. 23, pp. 517–530.

Subrahmanyam, K. B. and Kaza, K. R. V. (1985) "Finite Difference Analysis of Torsional Vibrations of Pretwisted, Rotating, Cantilever Beams with Effects of Warping," *Journal of Sound and Vibration*, Vol. 99, No. 2, pp. 213–224.

Subrahmanyam, K. B. and Kaza, K. R. V. (1986) "Vibration and Buckling of Rotating Pretwisted, Preconed Beams Including Coriolis Effects," *Journal of Vibration, Acoustics, Stress and Reliability in Design, Trans. ASME*, Vol. 108, April, pp. 140–149.

Subrahmanyam, K. B., Kaza, K. R. V., Brown, G. V. and Lawrence, C. (1987) "Nonlinear Vibration and Stability of Rotating Pretwisted, Preconed Blades Including Coriolis Effects," *Journal of Aircraft*, Vol. 24, No. 5, pp. 342–352.

Tang, S. (1969) "Natural Vibration of Isotropic Plates with Temperature-Dependent Properties," *AIAA Journal*, Vol. 7, No. 4, pp. 725–727.

Tanigawa, Y. (1992) "Theoretical Approach of Optimum Design for a Plate of Functionally Gradient Materials Under Thermal Loading," *Thermal Shock and Thermal Fatigue Behavior of Advanced Ceramics, NATO ASI Series E*, Vol. 241, pp. 171–180.

Tomar, J. A. and Jain, R. (1985) "Thermal Effect on Frequencies of Coupled Vibrations of Pretwisted Rotating Beams," *AIAA Journal*, Vol. 23, No. 8, pp. 1293–1296.

Touloukian, Y. S. (1967) *Thermophysical Properties of High Temperature Solid Materials*, Macmillan, New York.

Tsuiji, T. (1976) "Torsion of Pretwisted Thin-Walled Beams," *Theoretical and Applied Mechanics*, Vol. 26, University of Tokyo Press, pp. 75–80.

Tzou, H. S. and Zhong, J. R. (1991) "Adaptive Piezoelectric Shell Structures: Theory and Experiments," *32nd AIAA SDM Conference*, Baltimore, Maryland, April 8–12, Paper No. AIAA-91-1238.

Tzou, H. S. (1993) *Piezoelectric Shells, Distributed Sensing and Control of Continua*, Kluwer Academic Publ., Dordrecht/Boston/London.

Vel, S. S. and Batra, R. C., (2002) "Exact Solution for Thermoelastic Deformations of Functionally Graded Thick Rectangular Plates," *AIAA Journal*, Vol. 40, No. 7, pp. 1421–1433.

Venkatesan, C. and Nagaraj, C. T.,(1981) "On the Axial Vibrations of Rotating Bars," *Journal of Sound and Vibration*, Vol. 74, No. 1, pp. 143–147.

Volovoi, V. V., Hodges, D. H., Cesnik, C. E. S. and Popescu, B., (2001) "Assessment of Beam Modeling Methods for Rotor Blades Applications," *Mathematical and Computer Modelling*, Vol. 33, Nos. 10–11, pp. 1099–1112.

Vorob'ev, I. S. (1988) *Vibrations of Turbomachinery Blades* (In Russian)," Kiev, Naukova, Dumka.

Wang, J. T. S., Mahrenholtz, O., Böhm, J. (1976), "Extended Galerkin's Method for Rotating Beam Vibrations Using Legendre Polynomials, *Solid Mechanics Archives*, Vol. 1, pp. 341–356.

Washizu, K. (1964) "Some Considerations on a Naturally Curved and Twisted Slender Beams," *Journal of Mathematics and Physics*, Vol. 43, No. 2, pp. 111–116.

Wright, A. D., Smith, C. E., Thresher, R. W. and Wang, J. L. C., (1982) "Vibration Modes of Centrifugally Stiffened Beams," *Journal of Applied Mechanics, Trans. ASME*, Vol. 49, March, pp. 197–202.

Yokoyama, T. (1988), "Free Vibration Characteristics of Rotating Timoshenko Beams," *International Journal of Mechanical Sciences*, Vol. 30, No. 10, pp.743–755.

Yokoyama, T. and Markiewicz, M. (1993) "Flexural Vibrations of Rotating Timoshenko Beam with Tip Mass," Asia-Pacific Vibration Conference, Kitakyushu, Japan, Nov. pp. 382–387.

Yoo, H. H., Kwak, J. Y. and Chung, J. (2001) "Vibration Analysis of Rotating Pre-Twisted Blades with a Concentrated Mass," *Journal of Sound and Vibration*, Vol. 240, No. 5, pp. 891–908.

Young, M. I. (1973) "The Influence of Pitch and Twist on Blade Vibration," *Journal of Aircraft*, Vol. 10, No. 6, pp. 383–384.

Chapter 10

SPINNING THIN-WALLED ANISOTROPIC BEAMS

10.1 UNTWISTED BEAMS
10.1-1 Introduction

Rotating structures with angular velocity vector (referred to as the spin rate vector), parallel to their longitudinal axis plays an important role in modern technology.

They are found in gas turbines for higher-power aircraft engines, in helicopter drive applications, in space structures such as satellite booms, as well as in the modern machinery, such as in the cutting tools used in boring and milling operations.

The analysis of spinning structures differs from that of their non-rotating counterpart; Coriolis and centripetal effects have to be included.

The incorporation of the new composite materials for future high-performance spinning structural systems constitutes a natural trend which is likely to grow in the years ahead. However, in order to exploit their capabilities, a better understanding of the implications of elastic couplings generated by their anisotropy, is needed. This chapter provides a comprehensive derivation and solution of the dynamics of spinning structures modeled as thin-walled beams.

Although a conservatively loaded structure subject to a constant axial angular rotation falls into the category of gyroscopic conservative systems, it can feature, in some conditions, the instabilities by divergence and flutter, the latter being peculiar to non-conservative systems. We assume that the structure, modelled as a thin-walled beam, encompasses non-classical effects such as transverse shear, warping inhibition, rotatory inertia, Coriolis acceleration and anisotropy of its constituent materials.

To achieve an increase of the spin rate without the occurrence of any instability, a structural tailoring analysis is pursued. However, to avoid the inherent complexities related to the use of the fully coupled elastic system, a *circumferentially uniform stiffness configuration* is used so that the governing equations split exactly into two independent subsystems, one involving the transverse bending-lateral bending-transverse shear coupling, and the other the twist-extension coupling; the pertinent equations are supplied in Chapter 4. Herein we consider the former type of structural coupling.

This problem has been approached in the framework of the solid beam theory, i.e., for Euler-Bernoulli beams by Chen and Liao (1991), Liao and Dang (1992), Gulick (1994), Filipich et al. (1987, 1989), Johnston and Thornton (1996), and for Timoshenko beams by Curti et al. (1992), Sturia and Argento (1996), Argento (1995), Argento and Scott (1992, 1995), Kim and Bert (1993), Bert and Kim (1995a,b), Kim et al. (1999, 2001), Zu and Han (1992), Zu and Melianson (1998), Chang et al. (2004); for circular cylindrical shells (see e.g. Zohar and Aboudi, 1973; Saito and Endo, 1986; Huang and Soedel, 1988; Rand and Stavsky, 1991; Chen et al., 1993; Tylikowski, 1996; Hua and Lam, 1998) and for a circular conical shell by Lam and Hua (1997) and Sivadas (1995). With few exceptions, the analysis was carried for metallic structures only. The pioneering work by Bauchau (1981) in the modeling of the advanced composite shaft should be mentioned. Studies addressing various issues of spinning thin-walled beams have been carried out by Song and Librescu (1997, 1998), Song et al. (2000, 2001, 2002), Librescu et al. (1999, 2003, 2005), Chen and Peng (1998), Chang et al. (2004), Oh et al. (2005), and within the structural model developed by Song and Librescu (1993), by Ko and Kim (2003).

In addition to tailoring, active feedback control based on piezoelectric induced actuation will be considered.

10.1-2 Coordinate Systems and Basic Assumptions

Consider a straight untwisted flexible beam of length L, rotating about its longitudinal z-axis at a constant rate Ω (see Fig. 10.1-1).

Consider two sets of coordinates, an inertial one $OXYZ$, and a body attached rotating frame of reference $Oxyz$ with the common origin O, located in the geometric center (coinciding with the elastic center of the beam). At $t = 0$, the axes of the two systems coincide, while in the undeformed configuration, the body-fixed and inertial coordinates Oz and OZ coincide at any time t. Associated with the coordinate systems (x, y, z) and (X, Y, Z), one define the unit vectors $(\mathbf{i}, \mathbf{j}, \mathbf{k})$ and $(\mathbf{I}, \mathbf{J}, \mathbf{K})$, respectively. In addition, we define a local coordinate system, (n, s, z); the spin rate vector is $\boldsymbol{\Omega} = \Omega\mathbf{k}(\equiv \Omega\mathbf{K})$ with $\dot{\boldsymbol{\Omega}} = 0$.

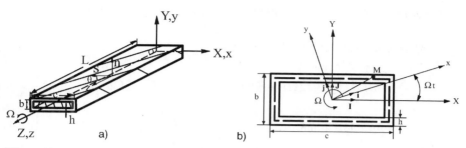

Figure 10.1-1: (a) Thin-walled beam featuring CUS configuration, (b) cross-section of the beam

Consider a single-cell thin-walled beam, of arbitrary cross-sectional shape. The following assumptions are adopted: (i) the original cross-section of the beam is preserved, (ii) transverse shear effects are incorporated, (iii) the constituent material of the structure features anisotropic properties, with a special lay-up inducing flapping-lagging coupling and (iv) the centroidal axis coincides with the spin axis of the beam.

10.1-3 Kinematics

The position vector of a generic point $M(x, y, z)$ belonging to the deformed structure is:

$$\mathbf{R}(x, y, z; t) = (x + u)\mathbf{i} + (y + v)\mathbf{j} + (z + w)\mathbf{k}, \tag{10.1-1}$$

where x, y and z are the Cartesian coordinates in its undeformed state. Since the spin rate was assumed to be constant and having in view Eqs. (R.8) from Section 9.1-3, the velocity and acceleration of a generic point of the spinning blade are

$$\dot{\mathbf{R}} = \left[\dot{u} - \Omega(y + v)\right]\mathbf{i} + \left[\dot{v} + \Omega(x + u)\right]\mathbf{j} + \dot{w}\mathbf{k}, \tag{10.1-2}$$

and

$$\ddot{\mathbf{R}} = \left[\ddot{u} - 2\Omega\dot{v} - \underline{(x + u)\Omega^2}\right]\mathbf{i} + \left[\ddot{v} + 2\Omega\dot{u} - \underline{(y + v)\Omega^2}\right]\mathbf{j} + \ddot{w}\mathbf{k}. \tag{10.1-3}$$

The displacements u, v and w are given by Eqs. (2.1-6) and (2.1-28).

10.1-4 Governing Equations

Hamilton's principle (see Eq. (3.1-1)) will be used to derive the equations of motion, and the associated boundary conditions of spinning beams. The only term peculiar to the problem at hand is $\int_{t_0}^{t_1} \delta\mathcal{K}dt$.

By virtue of Eqs. (10.1-1) and (10.1-3), this term reduces to

$$\int_{t_0}^{t} \delta \mathcal{K} dt \left(\equiv - \int_{t_0}^{t_1} dt \int_{\tau} \rho \ddot{\mathbf{R}} \cdot \delta \mathbf{R} d\tau \right)$$

$$= - \int_{t_0}^{t_1} dt \int_{\tau} \{ [\ddot{u} - 2\Omega \dot{v} - \Omega^2 (x + u)] \delta u + [\ddot{v} + 2\Omega \dot{u}$$

$$- (y + v)\Omega^2] \delta v + \ddot{w} \delta w \} \rho d\tau = - \int_{t_0}^{t_1} dt \int_{\tau} \Big\{ [\ddot{u}_0 - \underline{y\ddot{\phi}} \qquad (10.1\text{-}4)$$

$$- 2\Omega(\dot{v}_0 - \underline{x\dot{\phi}}) - \Omega^2(x + u_0 - \underline{y\phi})](\delta u_0 - \underline{y\delta\phi})$$

$$+ [\ddot{v}_0 + \underline{x\ddot{\phi}} + 2\Omega(\dot{u}_0 - \underline{y\dot{\phi}}) - (y + v_0 + \underline{x\phi})\Omega^2](\delta v_0 + \underline{x\delta\phi})$$

$$+ \left[\underline{\ddot{w}_0} + \left(y - n\frac{dx}{ds} \right)\ddot{\theta}_x + \left(x + n\frac{dy}{ds} \right)\ddot{\theta}_y - \underline{\ddot{\phi}'(\overline{F} - nr_t)} \right] \delta \left[\underline{w_0} \right.$$

$$\left. + \theta_x \left(y - n\frac{dx}{ds} \right) + \theta_y \left(x + n\frac{dy}{ds} \right) - \underline{\phi'(\overline{F} - nr_t)} \right] \Big\} \rho d\tau.$$

Since only the bending-bending coupled motion is considered, the underscored terms in Eq. (10.1-4) that belong to the twist and extension will be discarded.

Including the effect of a longitudinal compressive dead force P via the developments by Bhaskar and Librescu (1995), we obtain the following governing equations involving bending-bending-transverse shear coupling achievable via skewing the ply-angle according to the law $\theta(y) = \theta(-y)$ and $\theta(x) = \theta(-x)$:

$$\delta u_0 : \quad a_{43}\theta_x'' + a_{44}(u_0'' + \theta_y') - Pu_0'' = b_1\ddot{u}_0 - 2b_1\Omega\dot{v}_0 - \underline{b_1 u_0 \Omega^2},$$

$$\delta v_0 : \quad a_{52}\theta_y'' + a_{55}(v_0'' + \theta_x') - Pv_0'' = b_1\ddot{v}_0 + 2b_1\Omega\dot{u}_0 - \underline{b_1 v_0 \Omega^2},$$

$$\delta\theta_y : \quad a_{22}\theta_y'' + a_{25}(v_0'' + \theta_x') - a_{44}(u_0' + \theta_y) - a_{43}\theta_x' = (b_5 + \delta_n b_{15})\ddot{\theta}_y \;,$$

$$\delta\theta_x : \quad a_{33}\theta_x'' + a_{34}(u_0'' + \theta_y') - a_{55}(v_0' + \theta_x) - a_{52}\theta_y' = (b_4 + \delta_n b_{14})\ddot{\theta}_x \;,$$

$$(10.1\text{-}5a\text{-}d)$$

and the *boundary conditions* at $z = 0, \; L$:

$$\delta u_0 : \quad Q_x = \underline{Q}_x \text{ or } u_0 = \underline{u}_0,$$

$$\delta v_0 : \quad Q_y = \underline{Q}_y \text{ or } v_0 = \underline{v}_0,$$

$$\delta\theta_y : \quad M_y = \underline{M}_y \text{ or } \theta_y = \underline{\theta}_y, \qquad (10.1\text{-}6a\text{-}d)$$

$$\delta\theta_x : \quad M_x = \underline{M}_x \text{ or } \theta_x = \underline{\theta}_x.$$

The static version of homogeneous boundary conditions reads:

$$\delta u_0 : \quad a_{43}\theta_x' + a_{44}(u_0' + \theta_y) - Pu_0' = 0,$$
$$\delta v_0 : \quad a_{52}\theta_y' + a_{55}(v_0' + \theta_x) - Pv_0' = 0,$$
$$\delta\theta_y : \quad a_{22}\theta_y' + a_{25}(v_0' + \theta_x) = 0, \qquad (10.1\text{-}7\text{a-d})$$
$$\delta\theta_x : \quad a_{33}\theta_x' + a_{34}(u_0' + \theta_y) = 0.$$

The terms in the governing equations underscored by a solid line, by two superposed lines and by a dotted line are associated with the Coriolis acceleration, the centrifugal acceleration and with the rotatory inertia effect, respectively. These terms contribute further to the bending-bending elastic coupling.

10.1-5 Special Cases

Two special cases, are reported here:

(i) an orthotropic axisymmetric beam with the principal axes of orthotropy parallel to the geometrical axes. The coupling stiffnesses $a_{43} = a_{52} = 0$, and the axisymmetry implies

$$a_{22} = a_{33} \equiv A; \ a_{44} = a_{55} \equiv B, \ b_4 + \delta_n b_{14} = b_5 + \delta_n b_{15} \equiv C.$$
$$(10.1\text{-}8\text{a-c})$$

In terms of the complex displacement variables

$$U = u_0 + i v_0; \ \Theta = \theta_y + i\theta_x, \ (i = \sqrt{-1}) \qquad (10.1\text{-}9\text{a,b})$$

the governing equations reduce to

$$B(U'' + \Theta') - PU'' - b_1(\ddot{U} + 2i\Omega\dot{U} - U\Omega^2) = 0,$$
$$A\Theta'' - B(U' + \Theta) - C\ddot{\Theta} = 0, \qquad (10.1\text{-}10\text{a,b})$$

while the homogeneous boundary conditions at $z = 0, L$ become

$$B(U' + \Theta) - PU' = 0 \text{ or } U = 0; \ A\Theta' = 0 \text{ or } U' = 0. \quad (10.1\text{-}11\text{a,b})$$

The stiffness quantity B associated with the transverse shear effect couples the two governing equations. Solutions of the eigenvalue problem based on this complex representation have been supplied e.g. by Han and Zu (1992) and Zu and Han (1992).

(ii) a non-shear deformable beam model. Elimination from Eqs. (10.1-5) and (10.1-7) of the quantities $a_{44}(u_0'' + \theta_y')$ and $a_{55}(v_0'' + \theta_x')$, operation followed by the replacements $\theta_x \rightarrow -v_0'$ and $\theta_y \rightarrow -u_0'$ stating the absence of transverse shear, yields the governing equations:

$$\delta u_0 : \; a_{22}u_0'''' + Pu_0'' - \underdot{(b_5 + \delta_n b_{15})\ddot{u}_0''} + b_1(\ddot{u}_0 - 2\Omega\dot{v}_0 - \underline{\underline{u_0\Omega^2}}) = 0,$$

$$\text{(10.1-12a,b)}$$

$$\delta v_0 : \; a_{33}v_0'''' + Pv_0'' - \underdot{(b_4 + \delta_n b_{14})\ddot{v}_0''} + b_1(\ddot{v}_0 + 2\Omega\dot{u}_0 - \underline{\underline{v_0\Omega^2}}) = 0.$$

and the boundary conditions. Their homogeneous counterpart is:

$$a_{22}u_0''' - \underdot{(b_5 + \delta_n b_{15})\ddot{u}_0'} + Pu_0' = 0 \; \text{ or } \; u_0 = 0,$$

$$a_{33}v_0''' - \underdot{(b_4 + \delta_n b_{14})\ddot{v}_0'} + Pv_0' = 0 \; \text{ or } \; v_0 = 0,$$

$$a_{22}u_0'' = 0 \; \text{ or } \; u_0' = 0, \qquad\qquad \text{(10.1-13a-d)}$$

$$a_{33}v_0'' = 0 \; \text{ or } \; v_0' = 0.$$

Several remarks on the governing equations are in order: the governing equations of shearable thin-walled beams, Eqs. (10.1-5), and their non-shear deformable counterparts, Eqs. (10.1-12), feature the same order (eight), and thus, four boundary conditions have to be prescribed at each end $z = 0, L$ of the beam. These equations are formally similar to those for a solid beam (see e.g. Likins et al., 1973; Filipich et al., 1987). Equations (10.1-12) reveal that in this special case, the coupling arises only via Coriolis acceleration effect.

With the complex displacement variable $U(\equiv u_0 + i v_0)$, the governing system, Eqs. (10.1-12) and (10.1-13), becomes

$$AU'''' + PU'' - \underdot{C\ddot{U}''} + b_1(\ddot{U} + 2i\Omega\dot{U} - U\Omega^2) = 0, \qquad \text{(10.1-14)}$$

whereas the homogeneous boundary conditions reduce to

$$AU''' + PU' - \underdot{C\ddot{U}'} = 0 \text{ or } U = 0,$$

$$AU'' = 0 \text{ or } U' = 0. \qquad\qquad \text{(10.1-15a,b)}$$

To study the associated eigenvalue problem, a modal analysis can be used, see Likins et al. (1973), Filipich et al. (1987, 1989), and Lee et al. (1988). The similarity of equations of type (10.1-12) with those corresponding to pipes conveying fluid was remarked by Païdoussis (1998).

10.1-6 Solution of the Eigenvalue Problem of Gyroscopic Systems

Represent the displacement functions in the form

$$u_0(z,t) = \mathbf{U}^T(z)\,\mathbf{q}_u(t); \quad v_0(z,t) = \mathbf{V}^T(z)\mathbf{q}_v(t),$$
$$\theta_x(z,t) = \mathbf{X}^T(z)\,\mathbf{q}_x(t); \quad \theta_y(z,t) = \mathbf{Y}^T(z)\mathbf{q}_y(t), \qquad (10.1\text{-}16\text{a-d})$$

where

$$\mathbf{U} \equiv \{u_1, u_2 \ldots v_N\}^T; \quad \mathbf{V} \equiv \{v_1, \ v_2, \ldots v_N\}^T,$$
$$\mathbf{X} \equiv \{X_1, X_2, \ldots X_N\}^T; \quad \mathbf{Y} \equiv \{Y_1, \ Y_2, \ldots Y_N\}^T, \qquad (10.1\text{-}17)$$

and

$$\mathbf{q}_u \equiv \{q_1^u, \ q_2^u \ldots q_N^u\}^T; \quad \mathbf{q}_v \equiv \{q_1^v, q_2^v \ldots q_N^v\}^T,$$
$$\mathbf{q}_x \equiv \{q_1^x, q_2^x \ldots q_N^x\}^T; \quad \mathbf{q}_y \equiv \{q_1^y, q_2^y \ldots q_n^y\}^T, \qquad (10.1\text{-}18)$$

denote the vectors of trial functions, and of the generalized coordinates, respectively. Replacing representations (10.1-16) in the variational principle considered in conjunction with Eqs. (10.1-4), and carrying out the indicated variations and the required integration, yield the equation governing the motion of the gyroscopic system.

$$\mathbf{M}\ddot{\mathbf{q}}(t) + \mathbf{G}\dot{\mathbf{q}}(t) + \mathbf{K}\mathbf{q}(t) = 0. \qquad (10.1\text{-}19)$$

Here $\mathbf{M}$ and $\mathbf{K}$ are the symmetric mass and stiffness matrix, respectively, $\mathbf{G}$ is the skew symmetric gyroscopic matrix, while

$$\mathbf{q} = \{\mathbf{q}_u^T, \ \mathbf{q}_v^T, \ \mathbf{q}_x^T, \ \mathbf{q}_y^T\}^T, \qquad (10.1\text{-}20)$$

is the overall $4N \times 1$ vector of generalized coordinates.

Using the method presented in Chapter 6, we express Eq. (10.1-19) in state-space form. Upon defining the $8N \times 1$ state vector $\mathbf{X} = \{\mathbf{q}^T, \ \dot{\mathbf{q}}^T\}^T$ and adjoining the identity $\dot{\mathbf{q}} = \dot{\mathbf{q}}$, Eq. (10.1-19) reduces to

$$\dot{\mathbf{X}}(t) = \mathbf{A}\mathbf{X}(t). \qquad (10.1\text{-}21)$$

In Eq. (10.1-21) the $8N \times 8N$ state matrix $\mathbf{A}$ is given by

$$\mathbf{A} = \begin{bmatrix} \mathbf{0} & | & \mathbf{I} \\ ----- & | & ----- \\ -\mathbf{M}^{-1}\mathbf{K} & | & -\mathbf{M}^{-1}\mathbf{G} \end{bmatrix}, \qquad (10.1\text{-}22)$$

while $\mathbf{I}$ is the identity $4N \times 4N$ matrix. Expressing $\mathbf{X}(t)$ in Eq. (10.1.21) as

$$\mathbf{X}(t) = \mathbf{Z}\exp(\lambda t), \qquad (10.1\text{-}23)$$

where $\mathbf{Z}$ is a constant vector and λ a constant valued quantity, both generally complex, we obtain a standard eigenvalue problem

$$(\mathbf{Z} - \lambda\mathbf{I})\mathbf{X} = \mathbf{0}, \tag{10.1-24}$$

that can be solved for the eigenvalues λ_r and eigenvectors $\mathbf{X}_r$ ($r = 1, 2, \ldots 8N$).

If $\mathbf{K}$ is positive definite then the eigenvalues appear as purely imaginary pairs $\lambda_r = \pm i\omega_r$, where ω_r are the rotating (whirling) frequencies. The motion is purely oscillatory. If $\mathbf{K}$ is positive semi-definite, then at least on eigenvalue will be zero, implying divergence. If $\mathbf{K}$ is negative definite, the eigenvalues will appear in complex conjugate pairs with at least one having a positive real part; the instability is of the flutter type.

10.1-7 Closed-Form Solutions

In some special cases, closed-form solutions can be obtained. Consider a spinning thin-walled beam subjected at its free end to a longitudinal compressive dead force. For simplicity, consider an axisymmetric non-shearable beam clamped at $z = 0$ and free at $z = L$; we use Eqs. (10.1-14) and (10.1-15). Expressing in these equations

$$U(z, t) = \Phi(z)\exp(i\omega t), \tag{10.1-25}$$

where $\Phi(z)$ and ω are complex, one obtains

$$A\Phi'''' + P\Phi'' - b_1(\omega^2 + \underline{\underline{\Omega^2}})\Phi - \underline{2b_1\Omega\omega\Phi} + \underset{\cdots\cdots}{C\omega^2\Phi''} = 0, \tag{10.1-26}$$

and the boundary conditions:

at $z = 0$,

$$\Phi = 0, \quad \Phi' = 0, \tag{10.1-27}$$

and at $z = L$

$$A\Phi''' + P\Phi' + \underset{\cdots\cdots}{C\omega^2\Phi'} = 0; \quad A\Phi'' = 0. \tag{10.1-28}$$

Multiplying Eq. (10.1-26) by $\overline{\Phi}$, the complex conjugate of Φ, integrating the resulting equation over the beam span length, and using the boundary conditions (10.1-27) and (10.1-28), one obtains the eigenvalue problem:

$$(K\omega^2 + G\omega + Q)\Phi = 0. \tag{10.1-29}$$

The coefficients K, G, and Q associated with the kinetic energy, conservative gyroscopic energy and potential energy of the system, respectively, are given in

Table 10.1-1. The condition for neutral flutter instability requires ω to be real which implies the fulfillment of the condition $G^2 - 4KQ = 0$. This condition that represents a statement of coalescence of two eigenfrequencies, provides the compressive load yielding the flutter of the spinning thin-walled beam

$$P_{Flutter} = S_1/S_2, \tag{10.1-30}$$

where S_1 and S_2 are supplied in Table 10.1-2.

Table 10.1-1: Expression of the coefficients appearing in Eq. (10.1-29)

Coefficients	Their expressions
K	$b_1 \int_0^L \Phi \overline{\Phi} dz + C \int_0^L \Phi' \overline{\Phi}' dz \cdots$
G	$2b_1 \Omega \int_0^L \Phi \overline{\Phi} dz,$
Q	$b_1 \Omega^2 \int_0^L \Phi \overline{\Phi} dz + P \int_0^L \Phi' \overline{\Phi}' dz - A \int_0^L \Phi'' \overline{\Phi}'' dz$

Table 10.1-2: Expression of S_i ($i = 1, 2$) appearing in Eq. (10.1-30)

S_i	Their expressions
S_1	$b_1 A \int_0^L \Phi \overline{\Phi} dz \int_0^L \Phi'' \overline{\Phi}'' dz - Cb_1 \Omega^2 \int_0^L \Phi' \overline{\Phi}' dz \int_0^L \Phi \overline{\Phi} dz \cdots$ $+ CA \int_0^L \Phi' \overline{\Phi}' dz \int_0^L \Phi'' \overline{\Phi}'' dz$
S_2	$b_1 \int_0^L \Phi \overline{\Phi} dz \int_0^L \Phi' \overline{\Phi}' dz + C \int_0^L \Phi' \overline{\Phi}' dz \int_0^L \Phi' \overline{\Phi}' dz$

In the absence of rotatory inertia terms, i.e. when $C = 0$, the flutter conditions become

$$P_{flutter} = \frac{A \int_0^L \Phi'' \overline{\Phi}'' dz}{\int_0^L \Phi' \overline{\Phi}' dz},$$

$$\omega_{flutter} = -\frac{G}{2K} \neq 0. \tag{10.1-31}$$

Equation (10.1-30) and Table 10.1-2, show that for non-zero rotatory inertia (i.e. $C \neq 0$), the flutter conditions involving $P_{fluttter}$ and $\omega_{flutter}$ include also Ω; when it is discarded, Eq. (10.1-31) shows that $P_{flutter}$ is independent of Ω.

A similar result for a solid beam was reported by Shieh (1971). These results reveal that due to gyroscopic forces, flutter instability may occur, even though a conservative system can otherwise lose stability by divergence only.

Equation (10.1-29) also gives the condition for the occurrence of the divergence instability:

$$Q = 0. \qquad (10.1\text{-}32)$$

For the divergence instability the eigenvalues and eigenfunctions are real, thus

$$P_{div} = \frac{A \int_0^L (\Phi'')^2 dz - b_1 \Omega^2 \int_0^L \Phi^2 dz}{\int_0^L (\Phi')^2 dz}, \qquad (10.1\text{-}33)$$

or

$$\Omega^2_{div} = \frac{A \int_0^L (\Phi'')^2 dz - P \int_0^L (\Phi')^2 dz}{b_1 \int_0^L \Phi^2 dz}. \qquad (10.1\text{-}34)$$

Equation (10.1-33) reveals that with the increase of Ω^2, P_{div} decreases; see the numerical results in Chen and Liao (1991) for a solid beam. In the absence of the compressive end load, Eq. (10.1-34) gives

$$\Omega^2_{div} = \frac{A \int_0^L (\Phi'')^2 dz}{b_1 \int_0^L \Phi^2 dz}. \qquad (10.1\text{-}35)$$

Note that for an *unloaded* spinning untwisted thin-walled beams, the instability can occur only by divergence. In such a case, a spinning beam with equal flexural stiffnesses is stable for any angular velocities of rotation, except for certain critical values given by (10.1-35). For *loaded* spinning beams, the instability can occur also by flutter; see Huseyin and Plaut (1974, 1975), Huseyin (1978), Shieh (1971, 1982) and Lee and Yun (1996).

10.1-8 Comparisons with Available Predictions

The equations governing the coupled bending vibrations of pretwisted *thin-walled beam* cantilevers are formally similar to those for a *solid beam* (see e.g. Chen and Liao, 1991; Liao and Dang, 1992). The differences lie in the expression of cross-sectional stiffnesses and mass terms. For this reason, use of dimensionless parameters in which these quantities are absorbed, will make it possible to obtain universal results valid for both solid and thin-walled spinning beams. We compare results scattered throughout the literature for solid beams.

One of these results is related to the variation of natural frequencies as function of the spinning speed of an unshearable isotropic beam.

Figure 10.1-2 depicts the dependence of rotating frequency $\overline{\omega}_1 (\equiv \omega_1/\omega_0)$ as a function of the spin rate $\overline{\Omega}(\equiv \Omega/\omega_0)$, for a beam characterized by the ratio of the principal flexural stiffnesses $\mathcal{R}^2 (\equiv a_{33}/a_{22}) = (b/c)^2 = 1$ and 0.25, where $\omega_0^2 (\equiv \sqrt{a_{22}a_{33}}/\rho AL^4)$, A denoting the cross-sectional beam area.

When $\mathcal{R} = 1$, and $\overline{\Omega} = 0$ the flapping and lagging frequencies in each mode coincide, whereas, for $\mathcal{R}^2 = 0.25$, the non-rotating bending frequencies in flapping and lagging do not coincide. For $\mathcal{R} = 1$, with the increase of $\overline{\Omega}$, there is a bifurcation of natural frequencies, resulting in the upper ($\overline{\omega}_{1U}$) and lower frequency branches ($\overline{\omega}_{1L}$) (see Chen and Liao, 1991; Filipich et al., 1987); This reverts to the conclusion that due to the effect of gyroscopic Coriolis forces, two distinct frequency branches of free bending vibration are generated. The spin rate at which the lowest rotating natural frequency branch becomes zero is referred to as the critical spinning speed, and corresponds to the divergence instability.

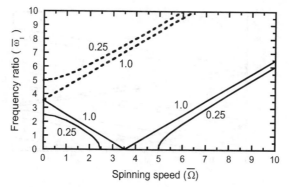

Figure 10.1-2: Variation of the dimensionless fundamental frequency as a function of the dimensionless spin speed for a box-beam of $\mathcal{R}^2 (\equiv a_{33}/a_{22}) = 0.25$ and 1 indicated on the graph: $\overline{\omega}_1 = \omega_1/\omega_0; \overline{\Omega} = \Omega/\omega_0$; ———, $\overline{\omega}_{1L}$; – – – –, $\overline{\omega}_{1U}$

For a beam with square cross-section, there a single critical spinning speed corresponding to each frequency mode number. For a beam with rectangular cross-section (for the present case $\mathcal{R}^2 = 0.25$), there is a whole domain of critical spinning speeds, bordered by $(\overline{\Omega}_{cr})_1$ and $(\overline{\Omega}_{cr})_2$, for which the dynamic system becomes unstable. Herein $(\overline{\Omega}_{cr})_1$ and $(\overline{\Omega}_{cr})_2$, denote the lower and upper bounds of the divergence instability domain (see Fig. 10.1-2). These results coincide with those reported by Filipich et al.(1987) and Chen and Liao (1991) obtained via the use of various numerical procedures.

In Table 10.1-3, comparisons of the theoretical and experimental natural frequencies reported by Slyper (1962) and Sabuncu (1985) for a uniform cantilever, with the present ones, reveal an excellent agreement.

Table 10.1-3: Frequency ratios $\overline{\omega}_i$ for a straight uniform cantilever ($a_{22}/a_{33} = 64$, $L = 35.56$ cm)

Mode No.	Slyper (1962)			Sabuncu (1985)	Present
	Standard	Stodola	Experimental	(FEM)	(EGM)
1	1.00	1.00	1.00	1.00	1.00
2	6.28	6.29	6.09	6.26	6.27
3	17.57	17.74	17.11	17.57	17.55
4	34.38	35.16	33.59	34.60	34.33

10.1-9 Numerical Simulations

Except for a few special cases, we must explore the dependence of natural rotating frequencies upon the spin rate by conducting parametric studies.

All the displayed results are recorded in terms of the spin speed Ω and natural frequencies ω_i normalized by the fundamental natural frequency $\hat{\omega}$ of the non-rotating beam counterpart: $\overline{\Omega} (\equiv \Omega/\hat{\omega})$, and $\overline{\omega}_i (\equiv \omega_i/\hat{\omega})$.

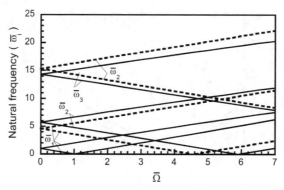

Figure 10.1-3: Rotating frequency-spin rate interaction for a beam of $\mathcal{R} = 1$ and $\theta = 0$ (———) and 90 deg. (.........)

Here $\hat{\omega} = 159.626$ rad/s corresponds to a square cross-section beam with clamped-free boundary conditions and constructed from graphite-epoxy with ply-angle $\theta = 0$. All the numerical illustrations are carried out for rectangular box-beam of fixed dimensions $c = 4$, (10.2cm) $L = 40$in. (1.2m) and $h = 0.4$in. (1.02cm). We consider the unloaded beam; $P = 0$. Figure 10.1-3 shows the dependence $\overline{\omega}_i$ ($i = 1, 2, 3$) vs. $\overline{\Omega}$ for a square cross-section beam $\mathcal{R}(\equiv b/c)$ =1, whose material (graphite-epoxy) involves the ply-angles $\theta = 0$ and 90 deg; the beam with ply-angle $\theta = 90$ deg. has a considerably larger $\overline{\Omega}_{cr}$ than one with $\theta = 0$.

Figure 10.1-4 shows the effects of the ply-angle on the frequency-spin rate interaction of a square cross-section beam; for each ply-angle there is a specific critical spinning speed and the minimum and maximum ones occur for $\theta = 0$ and $\theta = 90$ deg., respectively.

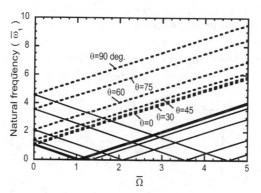

Figure 10.1-4: The influence of the ply-angle on fundamental frequency-spin rate interaction for a beam of $\mathcal{R} = 1$: ——, $\overline{\omega}_L$; ...., $\overline{\omega}_U$

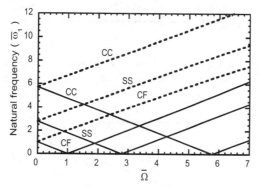

Figure 10.1-5: The influence of boundary conditions on fundamental rotating frequency-spin rate interaction for a beam of $\mathcal{R} = 1$, $\theta = 0$: ——, $\overline{\omega}_L$; ...., $\overline{\omega}_U$

Figure 10.1-5 shows the influence of three different combinations of boundary conditions on the frequency-spin rate interaction; the minimum $\overline{\Omega}_{cr}$ occurs for the clamped-free (CF) beam, the maximum for the clamped-clamped (CC) beam, the intermediate one for the beam simply supported (SS) at both ends.

Figures 10.1-6 through 10.1-8 reveal the effects of non-symmetry of the beam cross-section (measured in terms of the parameter $\mathcal{R}$), on frequency-spin rate interaction; for a beam with $\mathcal{R} \neq 1$, the non-rotating bending frequencies in flapping and lagging do not coincide. In this case, instead of a critical spinning speed, there is a whole domain of critical spinning speeds for which the dynamic system becomes unstable.

Moreover, Fig. 10.1-8 reveals that the domain of instability can be shifted toward larger spin speeds by tailoring the structure; however, this shift of the instability domain towards larger spin rates is paid for by an enlargement of the domain of instability. Figure 10.1-7 shows that the shift of the domain of instability towards larger spin speeds can arise from different boundary conditions. Figure 10.1-6 shows that in contrast to the case $\mathcal{R} < 1$ when the domain of instability is shifted toward lower spin rates as compared to that of $\mathcal{R} = 1$, for the beam cross-sections characterized by $\mathcal{R} > 1$, the domain of instability is shifted toward larger spin rates. However, the *extent* of the *domain* of instability in the latter case is similar to that occurring for beams featuring $\mathcal{R} < 1$.

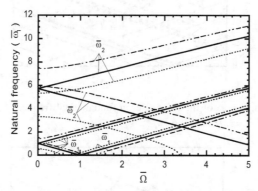

Figure 10.1-6: The influence of the non-symmetry of the beam cross-section measured in terms of $\mathcal{R}$ on rotating frequency-spin rate interaction, $\theta = 0$: —— , $\mathcal{R} = 1$; ...... , $\mathcal{R} = 0 \cdot 5$; —— ··· —— , $\mathcal{R} = 1 \cdot 5$

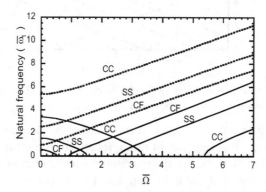

Figure 10.1-7: The influence of boundary conditions on fundamental rotating frequency-spin rate interaction of a beam characterized by $\mathcal{R} = 0 \cdot 5$ and $\theta = 0$: —— , $\overline{\omega}_L$; ......, $\overline{\omega}_U$

10.1-10 Concluding Remarks

Structural tailoring can enhance the behavior of spinning thin-walled beams by increasing the spinning speed and shifting the domain of divergence instability towards larger spin rates. In addition, the various boundary conditions and the cross-section parameter can play an important role on the dynamic behavior of the spinning system in general, and on the instability by divergence, in particular.

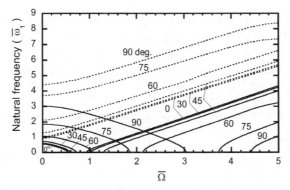

Figure 10.1-8: The influence of ply-angle on fundamental rotating frequency-spin rate interaction for a beam of $\mathcal{R} = 0.5$; ——, $\overline{\omega}_L$; ......, $\overline{\omega}_U$

10.2 VIBRATION AND STABILITY OF SPINNING CIRCULAR CYLINDRICAL SHAFT

10.2-1 Preliminaries

The issues considered in this section can be viewed as a special case of those addressed in Section 10.1. However, the wide applications of the circular cylindrical shaft in the modern technology justifies in full a special study devoted to the vibration and stability of this structural system.

10.2-2 Governing Equations

Consider a straight flexible thin-walled beam of length L and of uniform circular cylindrical cross-sections (see Figs. 10.2-1a,b). Let R be the beam radius of the mid-line cross-section whose points are expressible as

$$x = -R\sin(s/R), \qquad y = R\cos(s/R). \qquad (10.2\text{-}1)$$

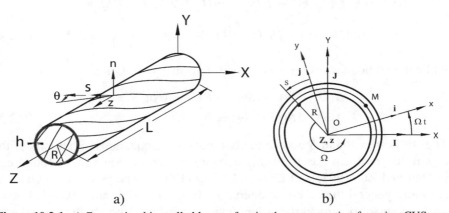

a)   b)

Figure 10.2-1: a) Composite thin-walled beam of a circular cross-section featuring CUS configuration; b) Inertial (X, Y, Z) and body fixed (x, y, z) coordinate systems

This equation shows that for this case the warping quantities, $\overline{F}(s)$ and $\overline{\overline{F}}(n, s)$ in Eqs. (2.1-44) and (2.1-11b) are both zero.

The governing equations and boundary conditions, are (10.1-5) and (10.1-6), respectively, where the stiffnesses $a_{ij}(= a_{ji})$ and reduced mass terms. are provided in Table 10.2-1.

Table 10.2-1: Expression of stiffness quantities and mass terms

$a_{ij}(= a_{ji})$	Their Expressions
$a_{22} = a_{33} \equiv A$	$\pi R[K_{11}R^2 + 2K_{14}R + K_{44}]$
$a_{44} = a_{55} \equiv B$	$\pi R[K_{22} + A_{44}]$
$a_{25} = -a_{34} \equiv S$	$\pi R[RK_{12} + K_{24}]$
$b_4 + \delta_n b_{14} = b_5 + \delta_n b_{15} \equiv C$	$\pi R[m_0 R^2 + \delta_n m_2]$

10.2-3 Two Alternative Representations of Governing Equations

Two complex representations of governing equations are supplied. The former one is associated with the *shearable rotating shaft* with circular cross-section

$$a_{22} = a_{33} \equiv A; \ a_{44} = a_{55} \equiv B,$$
$$a_{25} = -a_{34} \equiv S, \ b_4 + \delta_n b_{14} = b_5 + \delta_n b_{15} \equiv C, \quad (10.2\text{-}2)$$

where A, B, C and S are displayed in Table 10.2-1. Taking

$$U = u_0 + iv_0; \ \Theta = \theta_y + i\theta_x, \quad (i = \sqrt{-1}) \quad (10.2\text{-}3)$$

we can write the governing equations as

$$B(U'' + \Theta') + iS\Theta'' - PU'' - b_1(\ddot{U} + \underline{2i\Omega\dot{U} - U\Omega^2}) = 0,$$
$$A\Theta'' - B(U' + \Theta) - iS(U'' + \Theta') - iS\Theta' - C\underset{\cdots}{\ddot{\Theta}} = 0, \quad (10.2\text{-}4)$$

while the homogeneous boundary conditions at $z = 0, L$ are:

$$B(U' + \Theta) + iS\Theta' - PU' = 0 \text{ or } U = 0,$$
$$A\Theta' - iS(U' + \Theta) = 0 \text{ or } U' = 0. \quad (10.2\text{-}5)$$

The stiffnesses B and S couple the two governing equations. Solutions of the eigenvalue problem based on this complex representation can be found e.g. in Han and Zu (1992) and Liao and Dang (1992). The governing equations are more general than their counterparts, Eqs. (10.1-10) and (10.1-11): they apply to anisotropic materials; the previous ones are valid for the special case of orthotropic materials.

For the latter case, corresponding to the non-shearable shaft, the governing equations are

$$AU'''' + PU'' - C\ddot{U}'' + b_1(\ddot{U} - \underline{\Omega^2 U + 2i\Omega\dot{U}}) = 0, \qquad (10.2\text{-}6)$$

and the homogeneous boundary conditions reduce to

$$AU''' + PU' - C\ddot{U}' = 0, \text{ or } U = 0$$

$$AU'' = 0, \text{ or, } U' = 0. \qquad (10.2\text{-}7)$$

10.2-4 The Eigenvalue and Instability of the Spinning Shaft

As before, we use the Extended Galerkin Method in conjunction with representations of displacement quantities, Eqs. (10.1-16) to obtain a standard eigenvalue problem as given by Eq. (10.1-24). Depending upon the spin speed and compressive load that are involved in the matrix K, and implicitly in the matrix A, as discussed in Section 10.1-6, the eigenvalues can correspond to purely oscillatory motion, and to divergence and flutter instabilities.

In the numerical simulations, the divergence and flutter instabilities will be analyzed for the following three cases of boundary conditions:

a) Clamped at $z = 0$: $\quad u_0 = v_0 = \theta_y = \theta_x = 0,$
   Free at $z = L$: $\quad\quad Q_x = Q_y = M_x = M_y = 0,$

b) Simply-supported at both $z = 0, L$: $\quad u_0 = v_0 = M_x = M_y = 0,$

c) Clamped at both $z = 0, L$: $\quad u_0 = v_0 = \theta_x = \theta_y = 0. \qquad (10.2\text{-}8)$

10.2-5 Results

Figures 10.2-2 through 10.2-4 show the variation of natural frequencies $\overline{\omega}_i (\equiv \omega_i/\hat{\omega})$ versus the spin speed $\overline{\Omega}(\equiv \Omega/\hat{\omega})$, for selected ply-angles and various boundary conditions, for the unloaded beam. The normalizing factor $\hat{\omega} = 138.85$ rad/s is the fundamental frequency of a non-spinning graphite-epoxy beam characterized by $\theta = 0$ and $P = 0$.

The beam geometrical dimensions are: $R = 5$in. (0.127m), $L = 80$in. (2.032m), and $h = 0.4$in. (1.02cm). For $\overline{\Omega} = 0$, that is in the absence of gyroscopic effects, the system is characterized, for each θ, by a single fundamental frequency. With the increase of the ply-angle θ, the non-rotating natural frequencies increase. This trend is attributed to the increase of the bending stiffnesses $a_{22} = a_{33}(\equiv A)$ that is associated with the increase of θ (see Fig. 10.2-2).

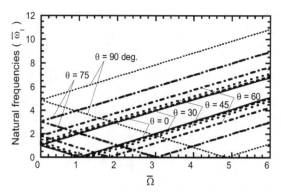

Figure 10.2-2: Variation of upper and lower fundamental frequency branches with spin speed $\overline{\Omega}$ for selected values of the ply-angle [$\overline{P} = 0$, boundary conditions a)]

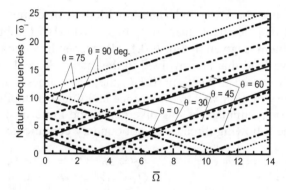

Figure 10.2-3: Counterpart of Fig. 10.2-2 for the boundary conditions b)

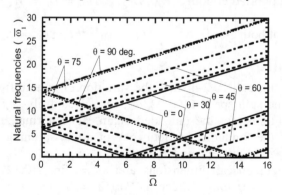

Figure 10.2-4: Counterpart of Fig. 10.2-2 for the boundary conditions c)

As soon as the rotation starts, and the gyroscopic forces are generated, there is a bifurcation of natural frequencies. This reverts to the conclusion that due the effect of the gyroscopic Coriolis force, two distinct frequencies of free bending vibration are produced. The minimum spin rate $\overline{\Omega} = \overline{\Omega}_{cr}$ at which the lowest rotating natural frequency becomes zero-valued, corresponds to the divergence instability. As before, at each ply-angle there is a specific critical spinning speed; the minimum and maximum ones occur for $\theta = 0$ and $\theta = 90$ deg., respectively.

These figures also show that for the clamped-free beam, (Case a), both the critical spinning speed and the eigenfrequencies are, for the same ply-angle, smaller than those corresponding to the boundary conditions b) and c). Figure 10.2-2 reveals that, for the boundary conditions a), the increase of the ply-angle yields a strong and continuous increase of both divergence spin speed $\overline{\Omega}_{cr}$ and of eigenfrequencies $\overline{\omega}_i$. For the boundary conditions b) and c), Figs. 10.2-3 and 10.2-4 reveal that beyond the ply-angle $\theta \approx 75$ deg., there are modest increases of the divergence rotating speed and eigenfrequencies.

Figure 10.2-5 show, for selected values of the spin rate $\overline{\Omega}$, the variation of the upper and lower frequency branches that correspond to the boundary conditions a), versus the dimensionless axial load $\overline{P}(\equiv PL^2/\hat{a}_{33})$. Here $\hat{a}_{33} = 3.3832 \times 10^5 \text{N.m}^2$ is the bending stiffness corresponding to $\theta = 0$.

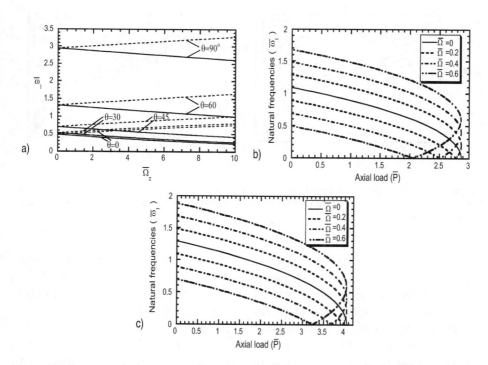

Figure 10.2-5: Variation of upper and lower fundamental frequency branches with increase of axial compressive load, for selected values of the spin speed, boundary conditions a); a) $\theta = 0$, b) $\theta = 30$ deg., c) $\theta = 45$ deg.

These plots show that for each specific value of $\overline{\Omega}$, as the axial compressive load increases, the associated eigenfrequency pairs decrease, and for a specific value of $\overline{P}$, these approach the value of zero. The corresponding value of $\overline{P}$ yielding the frequencies to vanish constitutes the critical axial compressive load (buckling load). This loss of stability is by divergence. These plots show that for one particular ply-angle, the critical axial load, $\overline{P}_{cr}$, corresponding to $\overline{\Omega} = 0$ constitutes an upper-bound, in the sense that for $\overline{\Omega} \neq 0$, the critical compressive loads diminish with the increase of $\overline{\Omega}$.

A closer inspection of the region where the divergence occurs reveals that a slight increase of the compressive load beyond the critical load makes the lower and upper lowest eigenfrequencies become complex conjugate; this slight increase of the compressible load, generates a shift of the instability from divergence to flutter. A similar trend was reported by Ku and Chen (1994), using a finite element method.

The same plots also show that the increase of the ply-angle leads to a significant increase of the critical axial load; also with the increase of the ply-angle, the critical axial compressive load becomes less and less sensitive to the increase of $\overline{\Omega}$.

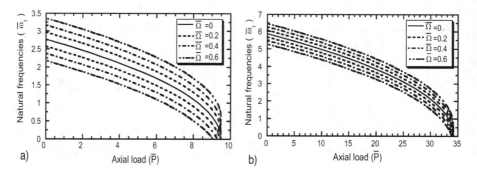

Figure 10.2-6: Variation of upper and lower fundamental frequency with increase of axial compressive load for selected values of the spin speed, a) boundary conditions, b); b) boundary conditions, c)

Figures 10.2-6 a,b are the counterparts of Fig. 10.2-5a, for boundary conditions b) and c), respectively. For these cases, there is a reduction of the sensitivity of the variation of the buckling load with the spin rate. It was also found that the buckling load increases greatly with ply-angle.

Figures 10.2-7 display stability plots of the spinning system in the plane ($\overline{\Omega}$ - $\overline{P}$) for boundary conditions, a) through c), and for selected values of θ. In these plots S denotes the region of stability, D denotes the divergence boundary, while F is that of the flutter instability. For $\overline{\Omega}$ and $\overline{P}$ equal to zero, ω_i^2 are real and positive and the system is stable. With the increase of $\overline{\Omega}$ and/or $\overline{P}$, instabilities by divergence or flutter may occur. Since the system is conservative, initial instability will always be of divergence type, characterized by $\omega_i^2 = 0$, as shown in Figs. 10.2-5 and 10.2-6. The locus of such points in the plane $(\overline{\Omega} - \overline{P})$ defines the divergence instability boundary; in this plane the instability boundary separates two stable regions, or in other words, the divergence occurs, as for box-beams with $\mathcal{R} = 1$, only on this boundary, without the existence of *regions* of divergence instability. In all these plots, the increase of the ply-angle yields a considerable increase of the stability domains. As before, as the boundary conditions change from type a), to types b) and c), there are significant increases in the stability domains. The domain of stability on the right of the divergence instability boundary shows that the gyroscopic effects, increasing with $\overline{\Omega}$, contribute to the increase of the stability domain.

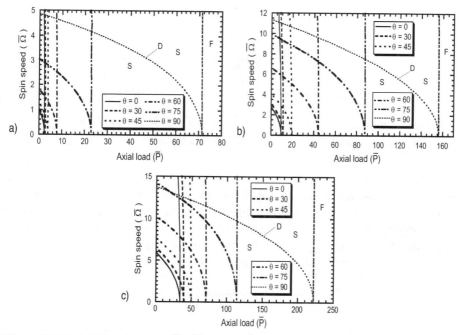

Figure 10.2-7: Stability plot in the $\overline{\Omega}$ - $\overline{P}$ plane displaying the domains of stability, divergence instability boundary, and flutter for selected ply-angles θ (deg). a) Boundary conditions a); b) Boundary conditions b); c) Boundary conditions c)

The same plots there depict the flutter boundary, where two eigenfrequencies ω_j^2 coalesce; for specified ply-angle and boundary conditions, the flutter boundary consists of slightly curved lines emerging at values of $\overline{P}$ slightly lower than $\overline{P}_{\text{div}}$, and at spin rates $\overline{\Omega} > 0$, where $\overline{P}_{\text{div}}$ is the buckling load occuring at $\overline{\Omega} = 0$. In order to get a better understanding of the flutter and divergence instabilities that appear in a very limited region of the variation of $\overline{\Omega}$ and $\overline{P}$,

Table 10.2-2 shows how the instability boundaries change with θ and $\overline{\Omega}$, for boundary conditions a): The flutter instability domain lies to the right of the curved line $\overline{P} = \overline{P}_{\text{flutter}}$ and for any $\overline{\Omega} > 0$. This coincides with the qualitative results obtained by Huseyin and Plaut (1974, 1975), with that reported by Song and Librescu (1998), and is consistent with that emerging from Figs. 10.2-5 through 10.2-6. By contrast, when the rotatory inertia terms are ignored, the flutter boundary does not involve the dependence on $\overline{\Omega}$; in that case the flutter boundary degenerates in a straight line $\overline{P} = \overline{P}_{\text{div}}$, parallel to the $\overline{\Omega}$-axis, see Shieh (1971) and Song and Librescu (1998).

Table 10.2-2: Stability boundaries for selected values of ply-angle.
Boundary conditions a). Rotatory inertias included

$\overline{\Omega}$	$\theta = 0$		$\theta = 45$ deg.		$\theta = 90$ deg.	
	$\overline{P}_{div}$	$\overline{P}_{flutter}$	$\overline{P}_{div}$	$\overline{P}_{flutter}$	$\overline{P}_{div}$	$\overline{P}_{flutter}$
0	2.444	-	4.056	-	71.44	-
1	-	2.421	1.713	4.033	69.19	71.43
2	-	2.353	-	3.965	62.02	71.39
3	-	2.238	-	3.852	48.61	71.33
4	-	2.078	-	3.692	26.90	71.24
5	-	1.872	-	3.488	-	71.13

Figure 10.2-8 highlights the effects of transverse shear on the stability boundaries; in the absence of transverse shear, when the structure is unshearable, a significant increase of the stability domain as compared to the shearable beam case is experienced: the classical (unshearable), structural model inadvertently overestimates the capacity of the rotating shaft to operate without the danger of the occurrence of divergence or flutter instabilities.

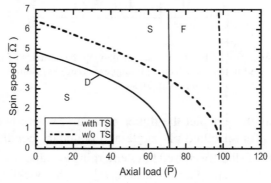

Figure 10.2-8: Effect of transverse shear on the stability boundaries of the rotating shaft [boundary conditions a); $\theta = 90$ deg.]

10.2-6 Concluding Remarks

A rotating shaft can experience instabilities by flutter and divergence. Structural tailoring can enhance their behavior by increasing the spinning speed, and shifting the domains of divergence and flutter instability towards larger spin rates. Neglecting transverse shear effect overestimates the capacity of the system to operate without flutter and divergence instabilities.

10.3 PRETWISTED SPINNING BEAMS

10.3-1 Basic Assumptions

Consider a straight pretwisted flexible beam of length L spinning about their longitudinal z-axis at a constant rate Ω (see Fig. 10.3-1). As in Section 10.1-2, one define two sets of coordinates, an inertial one $OXYZ$, and a rotating frame of reference $Oxyz$ with the common origin O, and consider the spin rate vector $\boldsymbol{\Omega} = \Omega\mathbf{k}(\equiv \Omega\mathbf{K})$ with $\dot{\Omega} = 0$.

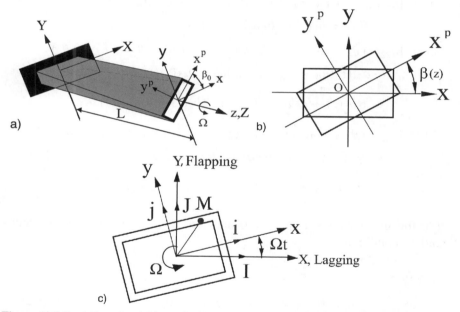

Figure 10.3-1: a) Pretwisted thin-walled beam of closed cross-section. Rotating and inertial coordinate systems. b) Pretwisted beam cross-section; c) Cross-section of the spinning beam

Besides the coordinates (x, y, z), we define the local coordinates (x^p, y^p, z^p), where x^p and y^p denote the principal axes of an arbitrary beam cross-section (see Fig. 10.3-1b).

These two coordinate systems are related by Eqs. (9.2-1), where, as usual, β_o denotes the pretwist at the beam tip. From these equations it becomes apparent that the systems (x^P, y^P) and (x, y) coincide at the beam root ($z = 0$). Define the beam surface Cartesian system (s, z, n), where s and n are the circumferential and thickness coordinates, respectively (see Fig. 10.1-1a). Otherwise, the assumptions coincide with those listed in Section 10.1-2.

10.3-2 Governing System

Having in view that Eqs. (10.1-4) subsist also for the present case, Hamilton's principle, the kinematical and constitutive equations, and the relationships detailed in Section 9.2-3, lead to the following governing equations that include also the effect of a longitudinal compressive dead force P. These equations are:

$$\delta u_0: \ [a_{42}(z)\theta'_y + a_{43}(z)\,\theta'_x + a_{44}(z)\,(u'_0 + \theta_y) + a_{45}(z)\,(v'_0 + \theta_x)]'$$
$$- Pu''_0 = b_1\ddot{u}_0 - \underline{2b_1\Omega\dot{v}_0} - \underline{b_1u_0\Omega^2},$$

$$\delta v_0: \ [a_{52}(z)\,\theta'_y + a_{53}(z)\theta'_x + a_{55}(z)\,(v'_0 + \theta_x) + a_{54}(z)\,(u'_0 + \theta_y)]'$$
$$- Pv''_0 = b_1\ddot{v}_0 + \underline{2b_1\Omega\dot{u}_0} - \underline{b_1v_0\Omega^2}, \qquad (10.3\text{-}1a\text{-}d)$$

$$\delta\theta_y: \ [a_{22}(z)\,\theta'_y + a_{25}(z)\,(v'_0 + \theta_x) + a_{24}(z)(u'_0 + \theta_y) + a_{23}(z)\,\theta'_x]'$$
$$- a_{44}(z)(u'_0 + \theta_y) - a_{43}(z)\theta'_x - a_{45}(z)(v'_0 + \theta_x) - a_{42}(z)\theta'_y$$
$$= \ (b_5(z) + \delta_n b_{15}(z))\,\ddot{\theta}_y \ + \ (b_6(z) - \delta_n b_{13}(z))\,\ddot{\theta}_x,$$

$$\delta\theta_x: \ [a_{33}(z)\,\theta'_x + a_{32}(z)\,\theta'_y + a_{34}(z)\,(u'_0 + \theta_y) + a_{35}(z)(u'_0 + \theta_x)]'$$
$$- a_{55}\,(z)\,(v'_0 + \theta_x) - a_{52}(z)\,\theta'_y - a_{54}\,(z)\,(u'_0 + \theta_y) - a_{53}(z)\theta'_x$$
$$= \ (b_4(z) + \delta_n b_{14}(z))\ddot{\theta}_x \ + \ (b_6(z) - \delta_n b_{13}(z))\ddot{\theta}_y \ .$$

For the blade clamped at $z = 0$ and free at $z = L$, the homogeneous boundary conditions are

At $z = 0$:

$$u_0 = 0, \ v_0 = 0, \ \theta_x = 0, \ \theta_y = 0 \qquad (10.3\text{-}2a\text{-}d)$$

while at $z = L$:

$$\delta u_0: \ a_{42}(L)\,\theta'_y + a_{43}(L)\,\theta'_x + a_{44}(L)\,(u'_0 + \theta_y) + a_{45}(L)\,(v'_0 + \theta_x) - Pu'_0 = 0,$$
$$\delta v_0: \ a_{52}(L)\,\theta'_y + a_{53}(L)\,\theta'_x + a_{55}(L)\,(v'_0 + \theta_x) + a_{54}(L)\,(u'_0 + \theta_y) - Pv'_0 = 0,$$
$$\delta\theta_y: \ a_{22}(L)\theta'_y + a_{24}(L)(u'_0 + \theta_y) + a_{25}(L)\,(v'_0 + \theta_x) + a_{23}(L)\theta'_x = 0,$$
$$\delta\theta_x: \ a_{33}(L)\,\theta'_x + a_{34}(L)(u'_0 + \theta_y) + a_{32}(L)\theta'_y + a_{35}(L)(v'_0 + \theta_x) = 0.$$
$$(10.3\text{-}3a\text{-}d)$$

The coefficients $a_{ij}(z)(=a_{ji}(z))$ and $b_i(z)$ are stiffness and reduced mass terms, respectively; their expressions in terms of their cross-section principal axes (x^p, y^p) are recorded in Tables 9.2-1 and 9.2-2. Equations (10.3-1) and (10.3-3) show that in the context of the ply-angle configuration considered above, the flapwise transverse shear is coupled with the chordwise-bending and the chordwise transverse shear is coupled with the flapwise-bending. In addition to the Coriolis acceleration (whose terms are associated with the gyroscopic effect and are underscored by a solid line), the stiffness quantities $a_{45}(z), a_{23}(z)$ $a_{35}(z)$ and $a_{24}(z)$ and the mass terms $b_6(z)$ and $b_{13}(z)$ which are different from zero only for a pretwisted beam, induce a supplementary coupling between the bending motions in flapping and lagging. However, the expressions for $a_{45}(z), a_{23}(z), a_{24}(z), a_{45}(z)$ as well as $b_6(z)$ and $b_{13}(z)$, for a box-beam of a *square cross-section*, are zero for any value of the pretwist β.

In this case, the flapping-lagging coupling arises independently of the pretwist effect, being a result of the considered ply-angle configuration. Consequently, for a non-spinning pretwisted beam with *square* cross-section, the buckling response should be independent of the pretwist angle β.

10.3-3 Nonshearable Beam Counterpart

The equations for the non-shearable beam counterpart, are obtained by following the method presented for example in Section 9.2-3b. *The associated governing equations* are

$$\delta u_0: \quad [a_{22}(z)u_0'' + a_{23}(z)v_0'']'' + Pu_0'' - [(b_6(z) - \delta_n b_{13}(z))\ddot{v}_0'$$

$$+ (b_5(z) + \delta_n b_{15}(z))\,\ddot{u}_0']' + b_1(z)[\ddot{u}_0 - 2\Omega\dot{v}_0 - \Omega^2 u_0] = 0$$

$$\delta v_0: \quad [a_{33}(z)v_0'' + a_{23}(z)u_0'']'' + Pv_0'' - [(b_4(z) + \delta_n b_{14}(z))\ddot{v}_0' \quad (10.3\text{-}4a,b)$$

$$+ (b_6(z) - \delta_n b_{13}(z))\ddot{u}_0']' + b_1(z)[\ddot{v}_0 + 2\Omega\dot{u}_0 - \Omega^2 v_0] = 0$$

and the *boundary conditions*

at $z = 0$:

$$u_0 = v_0 = u_0' = v_0' = 0 \qquad (10.3\text{-}5a\text{-}d)$$

and at $z = L$:

$$\delta u_0: \quad [a_{22}(L)\,u_0'' + a_{23}(L)v_0'']' + Pu_0' - [\,(b_6(L) - \delta_n b_{13}(L))\ddot{v}_0'$$

$$+ (b_5(L) + \delta_n b_{15}(L))\ddot{u}_0'] = 0,$$

$$\delta v_0 : \quad [a_{33}(L)v_0'' + a_{23}(L)u_0'']' + Pv_0' - [\ (b_6(L) - \delta_n b_{13}(L))\ddot{u}_0'$$

$$+\ (b_4(L) + \delta_n b_{14}(L))\ddot{v}_0'\ ] = 0, \qquad\qquad (10.3\text{-}6a\text{-}d)$$

$$\delta u_0' : \quad a_{22}(L)u_0'' + a_{23}(L)v_0'' = 0,$$

$$\delta v_0' : \quad a_{33}(L)v_0'' + a_{23}(L)u_0'' = 0,$$

Again, flapping-lagging coupling arises from the Coriolis and the pretwist effects. As concerns the solution methodology and determination of instability conditions by divergence and flutter, these issues remain similar to those presented in Section 9.1-7.

10.3-4 Comparisons with Other Results

Consider the free vibration of a pretwisted non-rotating beam of rectangular cross-section, with cross-section ratio $\mathcal{R}^{-1} \equiv c/b = 16$; $b = 0.5$ cm and $L = 100$cm.

Table 10.3-1 compares the normalized frequencies $\overline{\omega}_n (\equiv \omega_n/\omega_0)$ obtained via different numerical procedures, where ω_n are the actual frequencies, and $\omega_0 = 26.083$ rad/s is the lowest frequency of the untwisted beam counterpart.

Table 10.3-1: First three mode eigenfrequencies $\overline{\omega}_n (\equiv \omega_n/\omega_0)$, obtained for selected values of the pretwist angle via various numerical schemes, for selected values of the pretwist angle

Mode	Pretwist angle β_0 (deg.)	Rayleigh-Ritz method (D)	Transform. method (D)	FEM (T&U)	Bernoulli-Euler FEM (L&D)	Timo-shenko FEM (L&D)	Present
	30	1.00	1.00	1.00	1.00	1.00	1.00
1st	60	1.01	1.01	1.01	1.01	1.01	1.01
	90	1.02	1.02	1.02	1.02	1.02	1.02
	30	5.55	5.56	5.56	5.56	5.57	5.56
2nd	60	4.41	4.43	4.42	4.42	4.43	4.42
	90	3.53	3.55	3.54	3.55	3.57	3.54
	30	16.12	16.08	16.06	16.06	16.06	16.19
3rd	60	15.02	14.99	14.98	14.99	15.02	14.98
	90	13.58	13.55	13.53	13.53	13.59	13.50

The table is headed: Frequency Ratio (ω_n/ω_0)

The numerical schemes involve Rayleigh-Ritz and Transformation Methods (Dawson (D), 1968); the Finite Element Method (FEM) for the Euler-Bernoulli and Timoshenko beam models, Tekinalp and Ulsoy (T & U) (1989) and Liao and Dang (L & D) (1992), as well as the present method for shearable beams. The variation of the various mode frequencies with the pretwist angle follows a similar trend to that described by Anliker and Troesch (1963) and Slyper (1962). On the other hand, the results obtained via different numerical procedures agree well.

A final test concerns the buckling of a pretwisted shearable beam with $L = 0.2$m and $c = 2$cm. Table 10.3-2 compares the values of the buckling coefficient $S(\equiv P_{cr}L^2/a^p_{33})$ for selected values of the pretwist angle and cross-section parameter $\mathcal{R}$, predicted via the FEM developed by Liao and Dang (L&D) (1992), with those obtained here; there is an excellent agreement.

Table 10.3-2: Comparison of the buckling coefficients $\hat{P}_{cr}(\equiv P_{cr}L^2/a^p_{33})$ obtained via FEM and the present method, for selected values of the beam tip pretwist angle β_0 and beam cross-section parameter $\mathcal{R}$. Shearable beam model

β_0 (deg.)	$\mathcal{R} = 1.0$		$\mathcal{R} = 0.75$		$\mathcal{R} = 0.5$		$\mathcal{R} = 0.25$	
	(L&D)	Present	(L&D)	Present	(L&D)	Present	(L&D)	Present
0	2.453	2.453	2.453	2.453	2.453	2.453	2.453	2.453
90	2.453	2.453	2.562	2.554	2.624	2.649	2.686	2.669
180	2.453	2.453	2.802	2.794	3.106	3.081	3.304	3.273
270	2.453	2.453	2.963	2.953	3.479	3.439	3.857	3.803
360	2.453	2.453	3.004	2.991	3.596	3.539	4.048	3.973

10.3-5 Numerical Simulation and Discussion

We present a number of cases that highlight the effects of pretwist and compressive load on the vibration and instability of spinning cantilevered thin-walled beams.

The numerical simulations involve a cantilevered graphite-epoxy box-beam of rectangular cross-section.

Figure 10.3-2 presents, for the unloaded beam, the variation of the divergence spin speed as a function of the pretwist angle, for selected values of $\mathcal{R}$. For beams with $\mathcal{R} \neq 1$, the variation of the divergence instability domain has lower and upper bounds, $(\overline{\Omega}_{cr})_1$ and $(\overline{\Omega}_{cr})_2$, respectively, and is represented as a function of the pretwist angle; for $\mathcal{R} = 1$ the spinning divergence speed does not depend on the pretwist angle, and, the critical speed domain reduces to a single critical spin speed; for $\mathcal{R} < 1$ the divergence speed instability domain decreases as the pretwist angle increases. Within the framework of a solid beam model, Liao and Dang (1992) found similar result by using a FEM approach.

Figures 10.3-3a,b show the rotating frequency-spin rate interaction of unloaded box-beams of various cross-section aspect-ratios $\mathcal{R}$. In Fig. 10-3-3b the effect of the pretwist is involved, where the angle of pretwist at the beam tip is $\beta_0 = 45$ deg. It becomes apparent that for the beams whose cross-section is characterized by $\mathcal{R} \neq 1$, with the increase of $\overline{\Omega}$, the lower frequency branch decreases while the upper branch increases. The spin speed at which the lower whirl frequency becomes zero corresponds to the divergence spin speed.

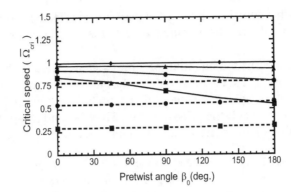

Figure 10.3-2: Variation of the critical spin speed in divergence as a function of the pretwist angle for selected values of $\mathcal{R}$ and $\theta = 0$. The divergence instability domain is extended between $(\Omega_{cr})_1$ and $(\Omega_{cr})_2$: - - -, $(\bar{\Omega}_{cr})_1$; ——, $(\bar{\Omega}_{cr})_2$, ■, $\mathcal{R} = 0.25$; ●.$\mathcal{R} = 0.5$; ▲, $\mathcal{R}$ 0.75; ◆, $\mathcal{R} = 1$

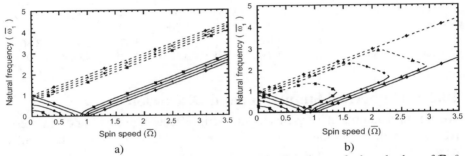

a)   b)

Figure 10.3-3: a) Rotating frequency-spin rate interaction for a beam of selected values of $\mathcal{R}$, $\theta = 0$, and pretwist angle $\beta = 0$: ——, $\bar{\omega}_{1L}$, - - - -, $\bar{\omega}_{1U}$; ■, $\mathcal{R} = 0.25$; ●.$\mathcal{R} = 0.5$; ▲, $\mathcal{R}$ 0.75; ◆, $\mathcal{R} = 1$; b) Rotating frequency-spin rate interaction for a beam of selected values of $\mathcal{R}$, $\theta = 0$, and beam tip pretwist angle $\beta_0 = 45$ deg.: ——, $\bar{\omega}_{1L}$, - - - -, $\bar{\omega}_{1U}$; ■, $\mathcal{R} = 0.25$; ●.$\mathcal{R} = 0.5$; ▲, $\mathcal{R} = 0.75$; ◆, $\mathcal{R} = 1$

Beyond this spin speed the system becomes stable again, and with the further increase of $\bar{\Omega}$, for a certain value of $\bar{\Omega} = \bar{\Omega}_{flutter}$, referred to as the flutter spin speed, two eigenfrequencies coalesce. Beyond $\bar{\Omega}_{flutter}$ the two eigenfrequencies become complex conjugate. Figure 10.3-3a shows that for $\mathcal{R} = 1$, the flutter spin speed is theoretically infinite, while Fig. 10.3-3b shows that it decreases with $\mathcal{R}$. Figures 10.3-3 also show that square cross-section beams, $\mathcal{R} = 1$, for any pretwist angle and spin speed, the occurrence of the flutter instability is precluded.

Figures 10.3-4 through 10.3-8 depict the variation of the *non-rotating* fundamental eigenfrequency $\bar{\omega}_1 (\equiv \omega_1/\hat{\omega}_1)$ vs. the normalized axial dead load, $\bar{P} (\equiv P/\hat{P}_{cr})$, for selected pretwist, ply-angles and cross-section $\mathcal{R}$ parameter, where $\hat{\omega}_1$ and $\hat{P}_{cr} (\equiv P_{cr} L^2/a_{33}^p)$ denote the normalizing fundamental frequency and buckling load, respectively, corresponding to straight ($\beta = 0$), non-spinning ($\Omega = 0$), square cross-section ($\mathcal{R} = 1$) and $\theta = 0$, having the values: $\hat{\omega}_1 = 164.73$ rad/s, $\hat{P}_{cr} = 2.43$, whereas $a_{33}^p = 2.134 \times 10^8$ lb. in.^{2}(6.125×10^5N.m^2).

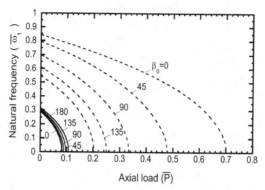

Figure 10.3-4: Dependence of the fundamental natural frequency $\overline{\omega}_1$ in flapping, $\overline{\omega}_F$, and lagging, $\overline{\omega}_L$, on the axial compressive load $\overline{P}$, for selected values of β_0 (deg). ($\mathcal{R} = 0.25$, $\theta = 0$, $\Omega = 0$); ———, $\overline{\omega}_F$, – – –, $\overline{\omega}_L$)

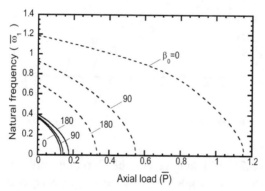

Figure 10.3-5: The counterpart of Figure 10.3-4 for the ply-angle $\theta = 45$ deg, ———, $\bar{\omega}_F$, - - - -, $\bar{\omega}_L$

These plots show that with the increase of the axial compressive load, the fundamental frequencies in flapping and lagging (that for $\mathcal{R} \neq 1$, are different), decrease. The minimum value of the axial (normalized) load for which the frequency vanishes is the critical (buckling) axial compressive load. With the increase of the pretwist angle, for the same compressive load, the flapping mode eigenfrequency increases, while the lagging mode eigenfrequency decreases. These figures also reveal that the cross-section parameter $\mathcal{R}$ and the ply-angle θ can play an important role in enhancing the behavior of the structure by increasing, independently, both the natural frequencies and the buckling load.

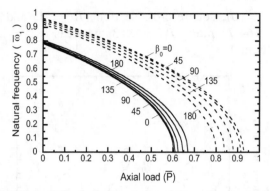

Figure 10.3-6: Dependence of the fundamental natural frequency $\bar{\omega}_1$ on the axial compressive load $\bar{P}$ for selected β_0 (deg.). ($\mathcal{R} = 0.5$, $\theta = 0$; $\Omega = 0$); ——, $\bar{\omega}_F$, - - - -, $\bar{\omega}_L$

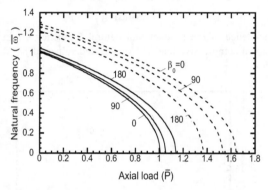

Figure 10.3-7: The counterpart of Fig. 10.3-6 for $\theta = 45$ deg.; —, $\bar{\omega}_F$, - - - -, $\bar{\omega}_L$

By contrast, for $\mathcal{R} = 1$ and $\Omega = 0$, the flapping and lagging fundamental frequencies in each mode coincide; also Fig. 10.3-8 reveals that the pretwist angle does not play any role in frequency-load interaction, but again, the increase of the ply-angle has a beneficial effect on both eigenfrequency and buckling load.

Figure 10.3-9 depicts the lowest normalized buckling load vs. the pretwist angle, for selected values of $\mathcal{R}$. There is an expected result: for $\mathcal{R} = 1$, the buckling load is not affected by the pretwist angle. However, for $\mathcal{R} \neq 1$, the buckling load increases with pretwist, and also with the decrease of $\mathcal{R}$, see Chen and Liao (1991) and Liao and Dang (1992), whose results were obtained for a solid beam via the assumed-mode approximation, and the FEM, respectively.

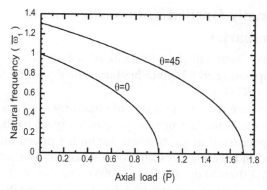

Figure 10.3-8: Dependence of the fundamental frequency $\bar{\omega}_1$ on the compressive axial load for selected values of the pretwist angle, β_0 (deg.) = 0; 45; 90; 135; and 180 ($\mathcal{R} = 1$, θ (deg.) = 0, and 45, $\Omega = 0$)

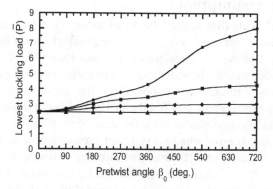

Figure 10.3-9: Variation of the lowest buckling load with the pretwist angle β_0 (deg.) for selected values of the cross-section beam parameter $\mathcal{R}$ ($\theta = 45$ deg., $\Omega = 0$), ——•——, $\mathcal{R} = 0.25$; ——■——, $\mathcal{R} = 0.5$; ——◆——, $\mathcal{R} = 0.75$; ——▲——, $\mathcal{R} = 1$

10.3-6 Concluding Remarks

Pretwisted spinning thin-walled beams can lose their stability by divergence and flutter. In addition to the findings by Shieh (1971, 1982) and Huseyin and Plaut (1974, 1975) on the occurrence of the flutter instability in the case of *loaded* spinning beam, here, the results show that this instability can occur in an unloaded, but pretwisted beam, characterized by a cross-section parameter $\mathcal{R} \neq 1$.

Additional results on spinning thin-walled beams made of functionally graded materials and operating in a temperature field will be provided next.

10.4 Functionally Graded Thin-Walled Beams

10.4-1 Preliminaries

In Section 9.5 we discussed the problems of the modeling and behavior of rotating blades operating in a high temperature field and made of functionally graded materials (FGMs).

Spinning beams are used in a variety of structural systems that operate in a high temperature environment: in aerospace industry, as the booms attached to spin-stabilized spacecraft; in gas turbines for higher power aircraft engines; in cutting tools used in boring and milling operations; in robotic devices. Except for the papers by Librescu et al. (2003, 2005), and Oh et al. (2005), the specialized literature is void of any results related to functionally graded spinning thin-walled beams.

10.4-2 Basic Assumptions

Consider a single-cell thin-walled beam of arbitrary cross-section. The following assumptions are adopted: (i) the original cross-section of the beam is preserved, (ii) transverse shear effects are incorporated, (iii) the concept of functionally graded materials with gradients in metallic-ceramic concentrations across the beam wall thickness is adopted, (iv) the flapwise-chordwise-bending cross coupling is considered, and finally, (v) there is no unbalance, in the sense that the centroidal axis coincides with the axis of rotation.

In the numerical simulations two types of cross-section beams will be considered; (a) rectangular, and (b) uniform circular cylindrical. For brevity, for beams of type (a), the effect of the pretwist will be discarded. For results related to this and other effects on vibration and instability of FGM spinning beams, see Librescu et al. (2003) and Oh et al. (2005).

10.4-3 Governing System

The 1-D constitutive equations are

$$
\left\{ \begin{array}{c} M_y \\ M_x \\ Q_x \\ Q_y \end{array} \right\} = \left[\begin{array}{cccc} a_{22} & 0 & 0 & 0 \\ 0 & a_{33} & 0 & 0 \\ 0 & 0 & a_{44} & 0 \\ 0 & 0 & 0 & a_{55} \end{array} \right] \left\{ \begin{array}{c} \theta_y' \\ \theta_x' \\ u_0' + \theta_y \\ v_0' + \theta_x \end{array} \right\}
$$
$$
+ \left\{ \begin{array}{c} M_y^T \\ M_x^T \\ 0 \\ 0 \end{array} \right\} \tag{10.4-1}
$$

The $a_{ij}(= a_{ji})$ are stiffness coefficients, while M_y^T and M_x^T are 1-D thermal moment terms, given by Eqs. (9.5-16) through (9.5-17). In terms of displacements u_0, v_0, θ_x and θ_y that include also the effect of a longitudinal compressive dead force, the governing equations are:

$$\delta u_0 : a_{44}(u_0' + \theta_y)' - Pu_0'' + p_x(z, t) = b_1\ddot{u}_0 - \underline{2b_1\Omega\dot{v}_0} - b_1 u_0\Omega^2,$$

$$\delta v_0 : a_{55}(v_0' + \theta_x)' - Pv_0'' + p_y(z, t) = b_1\ddot{v}_0 + \underline{2b_1\Omega\dot{u}_0} - b_1 v_0\Omega^2,$$

$$\delta\theta_y : a_{22}\theta_y'' - a_{44}(u_0' + \theta_y) + m_y(z, t) = (b_5 + \delta_n b_{15})\ddot{\theta}_y + (M_y^T)',$$

$$\delta\theta_x : a_{33}\theta_x'' - a_{55}(v_0' + \theta_x) + m_x(z, t) = (b_4 + \delta_n b_{14})\ddot{\theta}_x + (M_x^T)',$$

$$(10.4\text{-}2\text{a-d})$$

and the *boundary conditions* at $z = 0$,

$$u_0 = v_0 = \theta_x = \theta_y = 0, \qquad (10.4\text{-}3\text{a-d})$$

and at $z = L$

$$\delta u_0 : a_{44}(u_0' + \theta_y) - Pu_0' = 0,$$

$$\delta v_0 : a_{55}(v_0' + \theta_x) - Pv_0' = 0,$$

$$\delta\theta_y : a_{22}\theta_y' = M_y^T,$$

$$\delta\theta_x : a_{33}\theta_x' = M_x^T. \qquad (10.4\text{-}4\text{a-d})$$

In Eqs. (10.4-2) the gyroscopic terms generated by Coriolis acceleration are underscored. The 1-D stiffness quantities a_{ij} are provided in Table 9.5-1, whereas for a circular cylindrical beam cross-section as in Table 10.2-1, where for the *FGM* case, K_{12}, K_{14} and K_{24} become zero yielding $a_{25} = a_{34} = 0$. In this case, a_{45} and a_{23}, and the mass terms b_6 and b_{13} become zero-valued quantities as well. These expressions have to be considered in conjunction with Eqs. (9.5-5) and (9.5-6) in which it was assumed that the ceramic/metal FGM composition varies continuously across the beam wall thickness from purely ceramic phase at $n = h/2$ to purely metallic phase at $n = -h/2$.

The temperature $T = T(n)$ is provided by Eq. (9.5-8), while the temperature dependence of mechanical properties of the metallic and ceramic constituents are recorded in Table 9.5-4.

COMMENT

The unshearable version of the governing equations and boundary conditions can be obtained by exactly eliminating the $a_{44}(u_0' + \theta_y)$ and $a_{55}(v_0' + \theta_x)$ terms among the equations (10.4-2) and (10.4-4), and considering afterwards $\theta_y \to -u_0'$ and $\theta_x \to -v_0'$. This way we obtain the following governing equations.

$$\delta u_0 : a_{22}u_0'''' + Pu_0'' + b_1(\ddot{u}_0 - \underline{2\Omega\dot{v}_0} - \Omega^2 u_0) - (b_5 + \delta_n b_{15})\ddot{u}_0''$$
$$+ (M_y^T)'' - p_x - m_y' = 0, \qquad (10.4\text{-}5\text{a,b})$$

$$\delta v_0 : a_{33}v_0'''' + Pv_0'' + b_1(\ddot{v}_0 + \underline{2\Omega\dot{u}_0} - \Omega^2 v_0) - (b_4 + \delta_n b_{14})\ddot{v}_0''$$
$$+ (M_x^T)'' - p_y - m_x' = 0.$$

As concerns the classical *homogeneous* counterpart of the boundary conditions these are:

$$\delta v_0: \quad M_x' = 0; \text{ or } v_0 = 0,$$
$$\delta u_0: \quad M_y' = 0; \text{ or } u_0 = 0,$$
$$\delta u_0': \quad M_y = 0; \text{ or } u_0' = 0,$$
$$\delta v_0': \quad M_x = 0; \text{ or } v_0' = 0.$$

In terms of displacement quantities, the static boundary conditions are:

$$\delta v_0 : a_{33}v_0''' + Pv_0' - (b_4 + \delta_n b_{14})\ddot{v}_0' + (M_x^T)' - m_x = 0,$$
$$\delta u_0 : a_{22}u_0''' + Pu_0' - (b_5 + \delta_n b_{15})\ddot{u}_0' + (M_y^T)' - m_y = 0, \qquad (10.4\text{-}7a\text{-}d)$$
$$\delta u_0' : a_{22}u_0'' + M_y^T = 0,$$
$$\delta v_0' : a_{33}v_0'' + M_x^T = 0.$$

As a mere remark, the governing equations for the shearable and unshearable beam models feature the same order, namely eight, and as a result, in both cases, four boundary conditions should be prescribed at $z = 0, L$.

It should be pointed out that when discarding the rotatory inertia terms these equations coincide with those derived by Rosales and Filipich (1993).

10.4-4 Results

For the rectangular beam cross-section the following geometrical characteristics are used: $L = 80$ in. (2.032 m), $h = 0.005\,L$, $c = 0.05\,L$.

The displayed results are presented in terms of the normalized natural frequencies and spin rate: $\bar{\omega}_i \equiv \omega_i/\hat{\omega}_1$, $\bar{\Omega} \equiv \Omega/\hat{\omega}_1$, where the normalizing fundamental frequency $\hat{\omega}_1$ (= 177.35 rad/s) corresponds to a non-rotating fully-metallic beam of square cross-section, $\mathcal{R} = 1$, in the absence of the thermal gradient, ($\lambda = 0$).

As previously shown, for $\mathcal{R} = 1$, and $\bar{\Omega} = 0$, the flapping and lagging frequencies in each mode coincide. For $\bar{\Omega} \neq 0$, the gyroscopic effect generates a bifurcation of natural frequencies, with upper and lower branches. The spin rate at which the lowest rotating natural frequency becomes zero is referred to as the critical spinning speed, and corresponds to divergence instability. Beyond the critical spinning speed, the frequency corresponding to the lower frequency branch increases with $\bar{\Omega}$.

For a square cross-section beam in the absence of the compressive load, and for selected values of the volume fraction parameter K, Figure 10.4-1 shows the dependence of $\bar{\omega}_1 (\equiv \omega_1/\hat{\omega}_1)$ versus the normalized spin rate $\bar{\Omega}(\equiv \Omega/\hat{\omega}_1)$. The fully metallic beam, ($K = 10$), has the lowest non- rotating natural frequency and the lowest critical divergence spin speed. However, with the decay of K, implying an increase of the ceramic phase in the detriment of the metallic one, a beneficial trend consisting of both the increase of the non-spinning natural frequencies and of the critical spinning speed are resulting. As a limiting case, the fully ceramic beam, ($K = 0$), experiences the highest non-rotating natural frequency and highest critical speed of divergence.

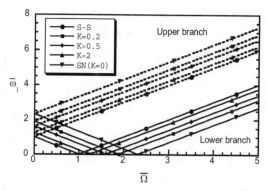

Figure 10.4-1: Variation of the dimensionless natural frequency $\bar{\omega}_1$ vs. dimensionless spinning speed for selected values of the volume fraction parameter K ($\mathcal{R} = 1, \lambda = 0, P = 0$)

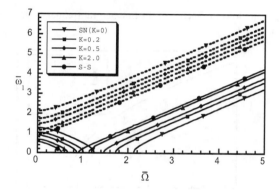

Figure 10.4-2: The counterpart of Fig. 10.4-1 for $\mathcal{R} = 0.5$

For beams free of compressive axial load ($\bar{P} = 0$) and with zero pretwist ($\beta_0 = 0$), the upper and lower frequency branches remain, for any $\bar{\Omega}$, parallel to each other. However, when $\mathcal{R} \neq 1$ (see Fig. 10.4-2) there is a new phenomenon. In contrast to beams of a square cross-section ($\mathcal{R} = 1$), for which there is a single critical spinning speed corresponding to each frequency mode number, for $\mathcal{R} \neq 1$, there is a whole domain of critical spinning speeds for which the systems becomes unstable by divergence. For $\bar{\Omega} = 0$, there are two frequencies, in lagging and flapping, the former being larger than the latter. The influence of K on the frequencies, is the same as for $\mathcal{R} = 1$. However, with the increase of the ceramic phase, the domain of divergence instability is shifted toward larger spin speeds.

Figure 10.4-3 shows the variation of the upper and lower frequency branches, against K, for selected values of the spinning speed. For this case, one assumes that $\mathcal{R} = 1$ and $P = 0$. The results reveal that, as K increases yielding the increase of the metallic phase, a decrease of frequencies corresponding to their upper and lower branches is experienced. For $\bar{\Omega} = 0$, there is a single non-rotating frequency that decreases as K increases.

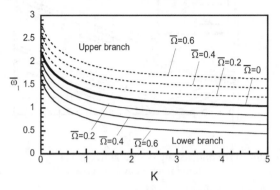

Figure 10.4-3: Upper and lower branches of the first natural frequency vs. K for selected values of Ω ($\mathcal{R} = 1$, $\lambda = 3$, $P = 0$)

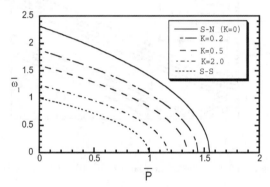

Figure 10.4-4: Variation of the first natural frequency vs. the dimensionless compressive load for selected values of K ($\mathcal{R} = 1$, $\bar{\Omega} = 0$, $\lambda = 0$)

Figure 10.4-4 is associated to a non-spinning beam subjected to a compressive axial load $\bar{P} (\equiv P/P_{cr})$, and characterized by selected values of K. Here, P_{cr} is expressed in terms of the dimensionless buckling coefficient $S (\equiv P_{cr} L^2/(a_{33})_m)$, where $(a_{33})_m = 1.48 \times 10^6 \text{N.m}^2$ denotes the bending stiffness for an untwisted non-rotating fully-metallic beam ($\mathcal{R} = 1$ and $\lambda = 0$). For this case $S = 2.454$. The values of $\bar{P}$ at which the non-spinning frequency becomes zero correspond to the buckling load. Figure 10.4-4 shows that with the decrease of K (i.e. with the increase of the concentration of the ceramic phase in the detriment of the metallic one), the buckling load increases.

Figure 10.4-5 is the counterpart of Fig. 10.4-4 for $\mathcal{R} = 0.5$. For each K there is a domain of the divergence instability bordered by $\bar{P}_{min}$ and $\bar{P}_{max}$. As K decreases, the domain of instability by divergence shifts toward larger values of $\bar{P}$.

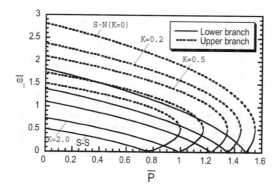

Figure 10.4-5: The counterpart of Fig. 10.4-4 for $\mathcal{R} = 0.5$

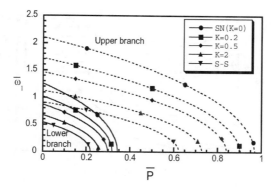

Figure 10.4-6: Variation of the first natural frequency vs. the dimensionless compressive load for selected values of K ($\mathcal{R} = 1$, $\bar{\Omega} = 0.5$, $\lambda = 0$)

A more complex trend occurs for a *spinning* beam that is axially compressed. Figures 10.4-6 through 10.4-8 show that for $\bar{\Omega} \neq 0$ there are two frequency branches. As the axial load $\bar{P}$ increases, two phenomena are experienced: (i) the two frequency branches coalesce this yielding the flutter instability, and (ii) the buckling of the beam occurs for that value of $\bar{P}$ yielding the frequency of the lower branch to become zero.

The plots show that with the increase of the concentration of the ceramic phase, i.e. when $K \to 0$, the loads yielding the buckling and flutter increase.

Figure 10.4-7 reveals that, depending on the relative concentration of the metallic and ceramic phases, various instabilities can be experienced. In this sense, the results of this figure show that the fully ceramic blade experiences, in addition to a domain of static instability, also the flutter instability occuring at that $\bar{P}$ at which the upper and lower frequency branches coalesce. For other $K \neq 0$, only the flutter instability occurs.

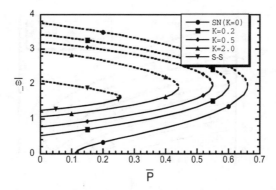

Figure 10.4-7: Variation of the first dimensionless natural frequency vs. the dimensionless compressive load for selected values of K ($\mathcal{R} = 0.5$, $\bar{\Omega} = 2$, $\lambda = 0$)

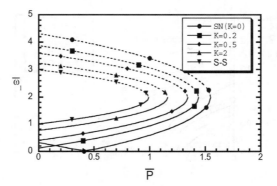

Figure 10.4-8: The counterpart of Fig. 10.4-7 for $\mathcal{R} = 1$

Figure 10.4-8 is the counterpart of Fig. 10.4-7, for $\mathcal{R} = 1$; it shows that: a) for $K = 0$, instead of the *domain* of static instability, there is a single critical value of $\bar{P} = \bar{P}_{div}$, and b) the flutter instabilities for various values of K and $\mathcal{R} = 1$, are larger than those for $\mathcal{R} = 0.5$.

Figures 10.4-9 and 10.4-10, give stability plots of the spinning system in the $\bar{\Omega}$ - $\bar{P}$ plane for selected values of the volume fraction parameter K, and for $\mathcal{R} = 1$ and $\mathcal{R} = 0.5$, respectively. The regions of stability and flutter as well as the boundary/regions of the divergence instability are indicated, Figure 10.4-9, shows that for any $\bar{\Omega}$ and $\bar{P}$ equal to zero, $\bar{\omega}_i^2$ are real and positive and the system is stable. With the increase of $\bar{\Omega}$ and/or $\bar{P}$, divergence or flutter may occur.

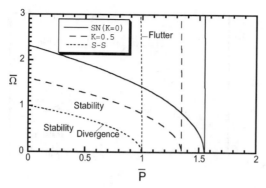

Figure 10.4-9: Stability map in the $\bar{\Omega}$ - $\bar{P}$ plane displaying the domains of stability, divergence and flutter instability, ($\mathcal{R} = 1, \lambda = 0$)

Since the system is conservative, the initial instability will always be of divergence type, characterized by $\bar{\omega}_i^2 = 0$. The locus of such points in the plane $\bar{\Omega}$ - $\bar{P}$ for which $\bar{\omega}_i = 0$ defines the divergence instability boundary; the divergence occurs only on this boundary, without the existence of *regions* of divergence instability. The domain of stability adjacent to the divergence instability boundary and on the right-hand side of it, shows that the gyroscopic effects, increasing with $\bar{\Omega}$, increase the stability domain. The same plots depict the flutter boundary corresponding to $\bar{\Omega}$ and $\bar{P}$ yielding coalescence of the upper and lower frequency pairs; for specified values of K the flutter boundary consists of slightly curved lines emerging at values of $\bar{P}$ slightly lower than $\bar{P}_{div}$, and at spin rates $\bar{\Omega} > 0$, where $\bar{P}_{div}$ is the buckling load obtained at $\bar{\Omega} = 0$. The flutter instability domain lies at the right of the curved line $\bar{P} = \bar{P}_{flutter}$ and for any $\bar{\Omega} > 0$; see also Figs. 10.4-7 and 10.4-8.

Figure 10.4-9 shows that decreasing K, a dramatic increase of the stability domains, and the shift of the divergence and flutter instability boundaries towards larger compressive loads are obtained. Table 10.4-1 summarizes the effects of $\bar{\Omega}$, $\bar{P}$ and K for a cross-section beam characterized by $\mathcal{R} = 1$, on the stability/instability of the beam.

Figure 10.4-10 is the counterpart of Fig. 10.4-9 for $\mathcal{R} = 0.5$ and $K = 0$, i.e. for a fully ceramic blade. In this case, and in contrast to the findings in Fig. 10.4-9, it becomes apparent that increasing $\bar{\Omega}$, for $\bar{P} \neq 0$, one can have either stability, instability by divergence, and again stability. In the latter case, for a rather large value of $\bar{\Omega}$ (that is $\bar{\Omega} > 2$), the increase of $\bar{P}$ can yield after stability, the instability by flutter, while for a lower values of $\bar{\Omega}$, after crossing the domain of divergence instability, with the further increase of $\bar{P}$, the flutter instability can occur.

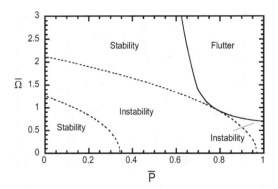

Figure 10.4-10: Stability map in the $\bar{\Omega}$ - $\bar{P}$ plane displaying the domains of stability, divergence and flutter instability ($\mathcal{R} = 0.5$, $K = 0$, $\lambda = 0$)

Table 10.4-1: Stability and instability by divergence and flutter as influenced by the spinning speed, compressive load, and volume fraction parameter ($\mathcal{R} = 1$, $\beta_0 = 0$, $\lambda = 0$)

$\bar{\Omega}$	$K = 0$ (S-N)		$K = 0.2$		$K = 0.5$		$K = 2.0$		$S - S$	
	$\bar{P}_{div}$	$\bar{P}_{fl}$	$\bar{P}_{div}$	$\bar{P}_{fl}$	$\bar{P}_{div}$	$\bar{P}_{fl}$	$\bar{P}_{div}$	$\bar{P}_{fl}$	$\bar{P}_{div}$	$\bar{P}_{fl}$
0	1.5513	–	1.4449	–	1.3482	–	1.1693	–	1.0000	–
0.5	1.4850	1.5512	1.3509	1.4447	1.2266	1.3479	0.9920	1.1689	0.7660	0.9995
1	1.2813	1.5507	1.0616	1.4441	0.8490	1.3471	0.4290	1.1678	–	0.9981
1.5	0.9310	1.5501	0.5560	1.4431	0.1772	1.3457	–	1.1658	–	0.9954
2	0.4171	1.5490	–	1.4416	–	1.3438	–	1.1630	–	0.9918

Note: Indices fl and div denote flutter and divergence, respectively.

To validate the simple rule of mixture used to determine the equivalent thermo-elastic global properties of the FGM, the buckling predictions are compared with those derived via the application of the Mori-Tanaka scheme (see, e.g., Cheng and Batra, 2000). For this purpose a functionally graded beam of a square cross-section featuring selected values of K and various temperature gradients was considered. Table 10.4-2 reveals the excellent agreement of predictions.

The previously reported conclusions on the effects of the volume fraction and of temperature material degradation on vibration and instability of spinning box-beams of cross-section $\mathcal{R} = 1$ remain valid, at least qualitatively, also for the circular cross-section spinning shaft. Figure 10.4-11 shows the effect of the shaft length ratio (= L/R) and of the temperature gradient λ, coupled with that of K on the lower branch of the first natural frequency.

The results of this plot reveal that the increase of the ratio L/R obtained by maintaining L fixed ($L = 80$in), yields a decrease of the lower frequency branch that corresponds to $\bar{\Omega} = \bar{P} = 0.2$, a trend that remains valid for any value of

K. However, for larger L/R, meaning lower values of the radius of the shaft R, with the increase of K there is an attenuation of the frequency decrease. At the same time, the increase of the temperature gradient λ yields a decay of the frequency. As is readily seen from Fig. 10.4-11, with the decrease of R, implying the increase of the parameter L/R, the temperature gradient tends to diminish its influence.

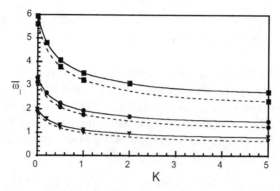

Figure 10.4-11: Variation of the lower branch of the fundamental frequency vs. K for $\bar{\Omega} = \bar{P} = 0.2$, $L = 80$in. (2.032m), for selected values of L/R and two values of the thermal gradient λ ($\lambda = 0$ ——; $\lambda = 4$ - - - -); —▼— $L/R = 16$; —●— $L/R = 10$; —■— $L/R = 5$

Table 10.4-2: Comparisons of the buckling coefficient $S(= P_{cr}L^2/(a_{33})_m)$ based on Mori-Tanaka and the simple rule of mixtures, for selected values of λ and K ($\mathcal{R} = 1$, $\Omega = 0$)

Volume fraction exponent	$\lambda = 0$		$\lambda = 1$		$\lambda = 2$	
	Mixture rule	Mori-Tanaka scheme	Mixture rule	Mori-Tanaka scheme	Mixture rule	Mori-Tanaka scheme
Fully ceramic	3.807	3.807	3.697	3.697	3.605	3.605
$K = 0.2$	3.546	3.583	3.444	3.478	3.349	3.386
$K = 0.5$	3.309	3.357	3.214	3.261	3.115	3.167
$K = 1.0$	3.085	3.133	2.999	3.044	2.895	2.94
Fully metal	2.454	2.454	2.397	2.397	2.275	2.275

Table 10.4-3 displays a comparison of the non-rotating fundamental frequency $\bar{\omega}_1$ ($\equiv \omega_1/\hat{\omega}_1$, where $\hat{\omega}_1 = 375.43$ rad/s), as predicted for a circular cylindrical FGM beam via the simple rule of mixture and the Mori-Tanaka scheme; there is an excellent agreement.

Table 10.4-3: Comparison of dimensionless first natural frequency $\overline{\omega}_1$ determined via the simple rule of mixtures (R-M) and the Mori-Tanaka (M-T) scheme for selected values of λ and K, ($\overline{\Omega} = 0$, $\overline{P} = 0$, $L = 80$in. (2.032m), $h = 0.4$in. (0.01m) $R = 5$in. (0.127m))

	$\lambda = 0$		$\lambda = 3$		$\lambda = 5$	
	R-M	M-T	M-R	M-T	M-R	M-T
$K = 0$ (SN)	2.3145	2.3145	2.2249	2.2249	2.1526	2.1526
$K = 0.2$	1.8920	1.8830	1.8132	1.8010	1.7417	1.7100
$K = 0.5$	1.6127	1.6010	1.5392	1.5233	1.4635	1.4253
$K = 1.0$	1.4075	1.3967	1.3354	1.3206	1.2502	1.2408
SS	1.0000	1.0000	0.8903	0.8903	0.6792	0.6792

Finally, in Table 10.4-4, a comparison of the divergence instability predictions for a circular cylindrical spinning shaft based on the EGM and on the *dynamic stiffness method* used by Curti et al. (1992) is presented. In their work the case of a solid beam was considered. Since the equations for solid and thin-walled beams are similar in form, the following global parameters converted to our notations were considered:

$$a_{44}/a_{22} = a_{55}/a_{33} = 21.82(1/m^2); \quad (b_4 + b_{14})/a_{33} = (b_5 + b_{15})/a_{22},$$
$$= 3.71 \times 10^{-8}(\sec^2); \quad b_1/a_{44} = b_1/a_{55} = 1.09 \times 10^{-7}(\sec^2),$$
$$\text{and } L = 10(m).$$

The results reveal an excellent agreement and that the Euler-Bernoulli model overestimates the critical speed, especially at the high mode numbers.

Table 10.4-4: Comparison of the critical divergence spinning speed based on the Extended Galerkin Method and Dynamic Stiffness Method

	Critical spinning speed Ω (rad/s)			
Mode	Euler-Bernoulli Beam		Timoshenko Beam	
Number	C. et al.	O. et al.	C. et al.	O. et al.
1	64.01	64.01	63.92	63.88
2	256.1	256.1	254.5	254.1
3	1024.2	1024.2	568.4	565.6
4	1600.3	1600.3	999.8	995.9
5	2304.5	2304.5	1540.9	1539.5

C. et al. $\rightarrow$ Curti et al. (1992); O. et al. $\rightarrow$ Oh et al. (2005)

10.5 A BEAM CARRYING A SPINNING TIP ROTOR

10.5-1 Preliminaries

Prompted by the rapid development of the high technology and space exploration, the problems of the modeling, vibration and stability of high performance robotic manipulator systems are likely to constitute topics of an increased attention in the years ahead. In spite of severe environmental conditions under which these devices have to operate, these are required to be light-weight, strong and capable of high precision. Unwanted vibrations and instabilities that can occur, resulting in drastic reductions of the accuracy and precision of their operation and even in their failure, should be addressed in the most appropriate way.

Here a robotic manipulator is modeled as a cantilevered elastic thin-walled beam, carrying at its tip a spinning small rotor. Since it is a gyroscopic system, it can experience in some conditions, the instabilities proper to a non-conservative system, namely the divergence and flutter; we will analyze the conditions generating their occurrence, as well as the factors enabling their postponement.

To this end, the concept of *gyroelastic systems* prompted by D'Eleuterio and Hughes (1987) and Yamanaka et al. (1995, 1996) will be applied.

According to this concept, a gyroelastic system is constituted of a body considered to be continuous in mass, stiffness, and equipped with a large number of small spinning rotors spread over the structure. Moreover, within this concept, also the case of the continuum equipped with discrete spinning rotors can be accommodated. Yamanaka et al. (1996) studied a Bernoulli-Euler solid beam containing a 1-D distribution of gyricity along the beam neutral axis and subjected to a conservative load; within this concept Yamanaka et al. (1995) considered a solid beam carrying a spinning rotor located at the beam tip. Herein, we consider an encompassing structural model equipped with a gyro-device. It consists of an anisotropic thin-walled beam equipped at its free end with a small spinning rotor. This model was developed by Song et al. (2000a,b, 2001c, 2002a) and Librescu et al. (1999). Among the main goals there are those of providing a basis for the modeling, vibrational behavior and stability of the gyroelastic system depicted in Fig. 10.5-1, as a function of the spin rate Ω_z of the rotor, and the conservative compressive force acting along the longitudinal Z-axis. We will investigate the conditions under which, instabilities by divergence (i.e. the static one) and by flutter (i.e. the dynamic one), occur. We will show that the directionality property of the anisotropic materials of the structure can constitute an important tool towards postponement of the occurrence of such instabilities; we consider *circumferentially uniform stiffness configuration* (CUS); as is already known, in this context, the governing equations split exactly into two independent sub-systems, one involving the flapping-lagging-transverse shear coupling, and the other the twist-extension coupling.

10.5-2 Basic Assumptions

Consider a straight flexible thin-walled beam of length L, clamped at $Z = 0$, and free at $Z = L$ where it carries a spinning rigid rotor of mass m (see Fig. 10.5-1). The offset between the beam tip and the centroid of the rotor is denoted as r_m. The rotor is assumed to have an axis of inertial symmetry which is coincident with the longitudinal axis of the beam.

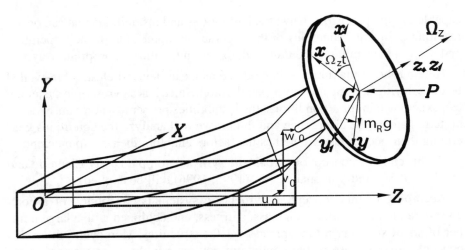

Figure 10.5-1: Thin-walled box-beam with a spinning tip rotor

Two sets of coordinates are considered, an inertial frame $OXYZ$ and a body fixed frame $Gxyz$ attached to the rotor, where G is the rotor mass center; the associated systems of the unit vectors are $\mathbf{I}$, $\mathbf{J}$, $\mathbf{K}$, and $\mathbf{i}$, $\mathbf{j}$, $\mathbf{k}$, respectively. The external force P acting at the beam tip is assumed to remain parallel to the Z-axis of the inertial frame, whereas the spin rate vector of the rotor, $\mathbf{\Omega}_z$ is $\mathbf{\Omega}_z = \Omega_z \mathbf{k}$

The adopted structural model is a thin-walled beam of arbitrary cross-sectional shape; we adopt the assumptions stipulated in Section 10.1-2.

10.5-3 Kinematics

To reduce the 3-D elasticity problem to an equivalent 1-D one, the components of the displacement vector are represented as:

$$u(X, Y, Z; t) = u_0(Z; t) - Y\phi(Z, t); \quad v(X, Y, Z; t) = v_0(Z; t) + X\phi(Z, t),$$

$$w(X, Y, Z; t) = w_0(Z, t) + \theta_X(Z; t)\left[Y(s) - n\frac{dX}{ds}\right] \quad\quad (10.5\text{-}1a\text{-}c)$$

$$+ \theta_Y(Z; t)\left[X(s) + n\frac{dY}{ds}\right] - \phi'(Z; t)[\overline{F}(s) - nr_t(s)].$$

The primes denote differentiation with respect to the longitudinal Z-coordinate. The rotational vector of the beam tip can be expressed as $\Theta = [\theta_X, -\theta_Y, \phi]^T$, while the angular velocity $\mathbf{\Omega}_B$ of the beam cross-section where the rotor is attached, is represented in the absolute system of coordinates through the symbolism of Hughes (1986) as

$$\mathbf{\Omega}_B = \dot{\mathbf{\Theta}} - \frac{1}{2}\tilde{\mathbf{\Theta}}\dot{\mathbf{\Theta}} = \left[\dot{\theta}_X + \frac{1}{2}(\theta_Y\dot{\phi} - \dot{\theta}_Y\phi); -\dot{\theta}_Y \right.$$

$$\left. + \frac{1}{2}(\theta_X\dot{\phi} - \dot{\theta}_X\phi); \dot{\phi} + \frac{1}{2}(\theta_X\dot{\theta}_Y - \dot{\theta}_X\theta_Y) \right]^T. \tag{10.5-2}$$

In Eq. (10.5-2), $\tilde{\mathbf{\Theta}}$ is the skew-symmetric matrix expressed in terms of the elements of the column matrix $\mathbf{\Theta}$, as:

$$\tilde{\mathbf{\Theta}} = \begin{bmatrix} 0 & -\phi & -\theta_Y \\ \phi & 0 & -\theta_X \\ \theta_Y & \theta_X & 0 \end{bmatrix}. \tag{10.5-3}$$

$\mathbf{\Omega}_B$ as expressed by Eq. (10.5-2) has to be evaluated at the beam tip; the terms associated with its effect appear in the boundary conditions at $Z = L$.

10.5-4 Governing Equations

10.5-4a Shearable Beams

For the present structural system Hamilton's principle gives

$$\delta J = \int_{t_0}^{t_1} (\delta \mathcal{W} - \delta \mathcal{K}_B - \delta \mathcal{K}_R - \delta \mathcal{J}_R - \delta \mathcal{J}_A) \, dt = 0, \tag{10.5-4}$$

where $\mathcal{W}$ denotes the strain energy of the elastic beam, $\mathcal{K}_B$ and $\mathcal{K}_R$ denote the kinetic energy of the beam and of the spinning rotor respectively; $\mathcal{J}_R$ and $\mathcal{J}_A$ denote the work done by the spinning rotor and the axial load, respectively, while δ denotes the variation operator.

We need the relationship between the unit vectors $(\mathbf{I}, \mathbf{J}, \mathbf{K})$ and $(\mathbf{i}, \mathbf{j}, \mathbf{k})$:

$$\begin{Bmatrix} \mathbf{I} \\ \mathbf{J} \\ \mathbf{K} \end{Bmatrix} = \mathbf{H} \begin{Bmatrix} \mathbf{i} \\ \mathbf{j} \\ \mathbf{k} \end{Bmatrix}, \tag{10.5-5}$$

where, for small angles of rotation we have

$$\mathbf{H} \equiv h_{ij} = \begin{bmatrix} \cos\Omega_z t - \phi\sin\Omega_z t & -(\sin\Omega_z t + \phi\cos\Omega_z t) & -\theta_Y \\ \sin\Omega_z t + \phi\cos\Omega_z t & \cos\Omega_z t - \phi\sin\Omega_z t & -\theta_X \\ \theta_X\sin\Omega_z t + \theta_Y\cos\Omega_z t & \theta_X\cos\Omega_z t - \theta_Y\sin\Omega_z t & 1 \end{bmatrix}. \tag{10.5-6}$$

The derivation of Eq. 10.5-5 is supplied in Section 10.5-5. By virtue of (10.5-2), the total angular velocity of the rotor can be expressed as

$$\Omega_R = \Omega_B + \Omega_z \mathbf{k}$$

$$= [\dot{\theta}_X + \frac{1}{2}(\theta_Y \dot{\phi} - \dot{\theta}_Y \phi)]\mathbf{I} + [-\dot{\theta}_Y + \frac{1}{2}(\theta_X \dot{\phi} - \dot{\theta}_X \phi)]\mathbf{J}$$

$$+ [\dot{\phi} + \frac{1}{2}(\theta_X \dot{\theta}_Y - \dot{\theta}_X \theta_Y)]\mathbf{K} + \Omega_z \mathbf{k} = \omega_{BX}\mathbf{I} + \omega_{BY}\mathbf{J} + \omega_{BZ}\mathbf{K} + \Omega_z \mathbf{k}$$

$$= (\omega_{BX} h_{11} + \omega_{BY} h_{21} + \omega_{BZ} h_{31})\mathbf{i} + (\omega_{BX} h_{12} + \omega_{BY} h_{22} + \omega_{BZ} h_{32})\mathbf{j}$$

$$+ (\omega_{BX} h_{13} + \omega_{BY} h_{23} + \omega_{BZ} h_{33} + \Omega_z)\mathbf{k} \equiv \omega_{Rx}\mathbf{i} + \omega_{Ry}\mathbf{j} + \omega_{Rz}\mathbf{k}.$$

$$(10.5\text{-}7)$$

Using the expression of the position vector of the rotor

$$\mathbf{R}_R(x, y, z\, t) = u_0(L, t)\mathbf{I} + v_0(L, t)\mathbf{J} + (L + w_0(L, t))\mathbf{K}$$

$$+ x\mathbf{i} + y\mathbf{j} + (r_m + z)\mathbf{k}, \qquad (10.5\text{-}8)$$

one obtains

$$\dot{\mathbf{R}}_R = \dot{u}_0(L, t)\mathbf{I} + \dot{v}_0(L, t)\mathbf{J} + \dot{w}_0(L, t)\mathbf{K} + x\dot{\mathbf{i}} + y\dot{\mathbf{i}} + z\dot{\mathbf{k}}, \qquad (10.5\text{-}9)$$

so that with Eqs. (10.5-5) and (10.5-6), it becomes:

$$\dot{\mathbf{R}}_R = \dot{u}_o \mathbf{I} + \dot{v}_o \mathbf{J} + \dot{w}_o \mathbf{K} + (-y\omega_{Rz} + z\omega_{Ry})\mathbf{i}$$

$$+ (x\omega_{Rz} - z\omega_{Rx})\mathbf{j} + (-x\omega_{Ry} + y\omega_{Rx})\mathbf{k}. \qquad (10.5\text{-}10)$$

Thus the kinetic energy of the rotor is

$$\mathcal{K}_R = \frac{1}{2} \int_{\tau_R} \rho_R (\dot{\mathbf{R}}_R \cdot \dot{\mathbf{R}}_R) \delta_D(Z - L)\, d\tau_R = \frac{1}{2} \{ m_R(\dot{u}_o^2 + \dot{v}_o^2 + \dot{w}_o^2)$$

$$+ J_{xx\dot{y}y}^R (\dot{\theta}_X^2 + \dot{\theta}_Y^2) + J_{zz}^R[(\dot{\phi} + \Omega_z)^2 + 3\Omega_z(\theta_X \dot{\theta}_Y - \dot{\theta}_X \theta_Y)] \}_{Z=L}.$$

$$(10.5\text{-}11)$$

Here, $J_{xx}^R = J_{yy}^R = J_{xxyy}^R = m_R(r_m^2 + \frac{1}{4}k_R^2)$, are the rotor mass moments of inertia, $J_{zz}^R = \frac{1}{2}m_R k_R^2$, is the polar moment of inertia, k_R is the radius of gyration of the rotor, and m_R is the rotor mass, while $\delta_D(\cdot)$ stands for the Dirac's delta function.

In conjunction with Hamilton's condition $\delta v_i = 0$ at t_0, t_1, we have

$$\int_{t_0}^{t_1} \delta \mathcal{K}_R dt \ (\equiv \int_{t_0}^{t} dt \int_{\tau_R} \rho_R \ddot{\mathbf{R}}_R \cdot \delta \mathbf{R}_R d\tau_R)$$

$$= \Big[-m_R \ddot{u}_o \delta u_o - m_R \ddot{v}_o \delta v_o - m_R \ddot{w}_o \delta w_o$$

$$- (J_{xxyy}^R \ddot{\theta}_X - 3\Omega_z J_{zz}^R \dot{\theta}_Y)\delta\theta_X$$

$$- (J_{xxyy}^R \ddot{\theta}_Y + 3\Omega_z J_{zz}^R \dot{\theta}_X)\delta\theta_Y - J_{zz}^R \ddot{\phi}\delta\phi \Big]_{Z=L}. \qquad (10.5\text{-}12)$$

The variation of the work done by the spinning tip rotor is

$$\delta \mathcal{J}_R = [m_R r_m \ddot{\theta}_Y \delta u_o + m_R (r_m \ddot{\theta}_X - g) \delta v_o + m_R r_m (\ddot{v}_o + g) \delta \theta_X$$
$$+ m_R r_m \ddot{u}_o \delta \theta_Y]_{Z=L}, \tag{10.5-13}$$

and that done by the axial dead load is

$$\delta \mathcal{J}_A = -P \int_0^L (u_o'' \delta u_o + v_o'' \delta v_o) \, dZ + P[u_o' \delta u_o + v_o' \delta v_o]_0^L. \tag{10.5-14}$$

The strain energy of the beam $\mathcal{W}_b$ is given by Eq. (3.2-1), and having in view

$$\mathbf{R}_b(X, Y, Z; t) = (X + u) \, \mathbf{I} + (Y + v) \, \mathbf{J} + (Z + w) \mathbf{K},$$
$$\dot{\mathbf{R}}_b(X, Y, Z; t) = \dot{u} \mathbf{I} + \dot{v} \mathbf{J} + \dot{w} \mathbf{K}, \tag{10.5-15a,b}$$

the associated kinetic energy $\mathcal{K}_b$ of the beam in the form of Eq. (3.2-3) is obtained.

The variation of the work done by the gravitational force of the beam is

$$\delta \mathcal{J}_{bg} = -\int_0^L b_1 g \delta v_0 dZ. \tag{10.5-16}$$

As before, to induce the elastic coupling between flapwise-bending and chordwise-bending, we use the ply-angle distribution referred to as *circumferentially uniform stiffness* (CUS) configuration achieved by skewing angle plies with respect to the beam axis according to the law $\theta(Y) = \theta(-Y)$ and $\theta(X) = \theta(-X)$; as usual, θ denotes the dominant ply orientation in the top and bottom, as well as in the lateral walls of the beam, measured from the positive s-axis toward the positive Z-axis.

For this ply-angle configuration, there is an additional system of governing equations and boundary conditions involving the extension-twist coupling; however, in this context these are not used.

In this case, from the variational equation (10.5-4), bearing in mind that δu_0, δv_0, $\delta \theta_X$ and $\delta \theta_Y$ are independent and arbitrary, and setting the coefficients of these variations to zero, yield the equations of motion and the boundary conditions. As concerns the governing equations in terms of displacement quantities about a static equilibrium position, these are similar to those in Eqs. (10.1-5) wherein one should let $\Omega_z = 0$. The boundary conditions at the root cross-section of the beam are identical to Eqs. (10.1-6).

As concerns the boundary conditions at the beam tip, $z = L$, these ones take into account the presence of the spinning tip rotor and are given by

$$\delta u_o : \; a_{43}\theta_X' + a_{44}(u_o' + \theta_Y) - Pu_o' + m_R(\ddot{u}_o - r_m\ddot{\theta}_Y) = 0,$$

$$\delta v_o : \; a_{52}\theta_Y' + a_{55}(v_o' + \theta_X) - Pv_o' + m_R(\ddot{v}_o - r_m\ddot{\theta}_X) = 0,$$

$$\delta\theta_X : \; a_{33}\theta_X' + a_{34}(u_o' + \theta_Y) + J_{xxyy}^R\,\ddot{\theta}_X - 3\Omega_z J_{zz}^R\dot{\theta}_Y - m_R r_m\ddot{v}_o = 0,$$

$$\delta\theta_Y : \; a_{22}\theta_Y' + a_{25}(v_o' + \theta_X) + J_{xxyy}^R\,\ddot{\theta}_Y + 3\Omega_z J_{zz}^R\dot{\theta}_X - m_R r_m\ddot{u}_o = 0.$$

$$(10.5\text{-}17\text{a-d})$$

The terms in Eqs. (10.5-17) underscored by a solid line are associated with the gyroscopic effects; those marked by a dotted line correspond to rotatory inertia terms, and those in the boundary conditions (10.5-17) underscored by —•—•— are associated with the tip rotor. If the spinning rotor were located at a point other then the beam tip, the gyroscopic effects would appear in the governing equations and not in the boundary conditions at $Z = L$.

Note that for $\theta = 0$ and 90 deg., the coupling stiffness quantities a_{43} and a_{52} become zero; in addition, if $\Omega_z = 0$, the governing system and boundary conditions split into two independent sub-systems associated with flapping and lagging motions.

10.5-4b Unshearable Beams

The unshearable version of the governing equations and boundary conditions at the beam root are identical to those in Eqs. (10.1-12) when are specialized for $\Omega_z = 0$, and with (10.1-13), respectively. As concerns the boundary conditions at the beam tip, $Z = L$, these are:

$$\delta u_o : \; a_{22}u_0''' - (b_5 + \delta_n b_{15})\ddot{u}_0' + Pu_0' - m_R(\ddot{u}_0 + r_m\ddot{u}_0') = 0,$$

$$\delta v_0 : \; a_{33}v_0''' - (b_4 + \delta_n b_{14})\ddot{v}_0' + Pv_0' - m_R(\ddot{v}_0 + r_m\ddot{v}_0') = 0,$$

$$\delta u_0' : a_{22}u_0'' + J_{xxyy}^R \ddot{u}_0' + 3\Omega_z J_{zz}^R \dot{v}_0' + m_R r_m \ddot{u}_0 = 0,$$

$$\delta v_0' : a_{33}v_0'' + J_{xxyy}^R \ddot{v}_0' - 3\Omega_z J_{zz}^R \dot{u}_0' + m_R r_m \ddot{v}_0 = 0. \quad (10.5\text{-}18a\text{-}d)$$

In contrast to the case of shearable beams, for the unshearable beam counterpart, the flapping-lagging coupling arises only in the boundary conditions at the beam tip, and is due entirely to the gyroscopic effects.

10.5-5 Explicit Derivation of Eq. (10.5-5)

The matrix **H** given by Eq. (10.5-6) can be obtained by examining the equations describing the coordinate transformations. These include:

• Rotation about the X-axis that involves θ_X

$$\mathbf{T}_{\theta_X} = \begin{bmatrix} 1 & 0 & 0 \\ 0 & \cos\theta_X & \sin\theta_X \\ 0 & -\sin\theta_X & \cos\theta_X \end{bmatrix}.$$

• Rotation about the Y-axis that involves θ_Y

$$\mathbf{T}_{\theta_Y} = \begin{bmatrix} \cos\theta_Y & 0 & \sin\theta_Y \\ 0 & 1 & 0 \\ -\sin\theta_Y & 0 & \cos\theta_Y \end{bmatrix}.$$

• Rotation about Z-axis that involves ϕ

$$\mathbf{T}_\phi = \begin{bmatrix} \cos\phi & \sin\phi & 0 \\ -\sin\phi & \cos\phi & 0 \\ 0 & 0 & 1 \end{bmatrix}.$$

• Rotation about the Z-axis that involves $(\Omega_z t)$

$$\mathbf{T}_{\Omega_z t} = \begin{bmatrix} \cos\Omega_z t & \sin\Omega_z t & 0 \\ -\sin\Omega_z t & \cos\Omega_z t & 0 \\ 0 & 0 & 1 \end{bmatrix}.$$

Based on these partial transformations one can obtain the relationships between the units vectors in the rotating $(x,\ y,\ z)$ and inertial $(X,\ Y,\ Z)$ axes as

$$\begin{Bmatrix} \mathbf{i} \\ \mathbf{j} \\ \mathbf{k} \end{Bmatrix} = \mathbf{F} \begin{Bmatrix} \mathbf{I} \\ \mathbf{J} \\ \mathbf{K} \end{Bmatrix},$$

where

$$\mathbf{F} = \mathbf{T}_{\Omega_z t} \, \mathbf{T}_\phi \, \mathbf{T}_{\theta_Y} \, \mathbf{T}_{\theta_X}. \tag{10.5-19}$$

In an explicit form $\mathbf{F}$ is obtained as:

$$\mathbf{F} = \begin{bmatrix} c\theta_Y c(\phi + \Omega_z t) & \begin{array}{c} c\theta_X s(\phi + \Omega_z t) \\ -s\theta_X s\theta_Y c(\phi + \Omega_z t) \end{array} & \begin{array}{c} s\theta_X s(\phi + \Omega_z t) \\ +c\theta_X s\theta_Y c(\phi + \Omega_z t) \end{array} \\[4mm] -c\theta_Y s(\phi + \Omega_z t) & \begin{array}{c} c\theta_X c(\phi + \Omega_z t) \\ +s\theta_X s\theta_Y s(\phi + \Omega_z t) \end{array} & \begin{array}{c} s\theta_X c(\phi + \Omega_z t) \\ -c\theta_X s\theta_Y s(\phi + \Omega_z t) \end{array} \\[4mm] -s\theta_Y & -c\theta_Y s\theta_X & c\theta_X c\theta_Y \end{bmatrix}, \tag{10.5-20}$$

where the abbreviations $\cos(\cdot) \Longrightarrow c(\cdot)$ and $\sin(\cdot) \Longrightarrow s(\cdot)$ have been used. For small angular displacements,

$$[\sin\theta_X, \, \sin\theta_Y, \, \sin\phi] \Longrightarrow [\theta_X, \, \theta_Y, \, \phi],$$

and

$$[\cos\theta_X, \, \cos\theta_Y, \, \cos\phi] \Longrightarrow [1, 1, 1],$$

$\mathbf{F}$ becomes

$$\mathbf{F} = \begin{bmatrix} c\Omega_z t - \phi s\Omega_z t & s\Omega_z t + \phi c\Omega_z t & \theta_X s\Omega_z t + \theta_Y c\Omega_z t \\ -(s\Omega_z t + \phi c\Omega_z t) & c\Omega_z t - \phi s\Omega_z t & \theta_X c\Omega_z t - \theta_Y s\Omega_z t \\ -\theta_Y & -\theta_X & 1 \end{bmatrix}. \tag{10.5-21}$$

Due to the orthogonality of the transformation matrix, its inverse may be obtained by simply transposing $\mathbf{F}$. In this way one obtains Eq. (10.5-5), where $\mathbf{H} = \mathbf{F}^{-1} \equiv \mathbf{F}^T$ is given by Eq. (10.5-6).

Note that the transformation matrix assumes a sequence of rotations as indicated in Eq. (10.5-19). For another sequence, the tranformation matrix would be different. However, for small rotations, the order of the sequence becomes immaterial.

10.5-6 Solution Methodology

First represent the displacement quantities in the form

$$u_o(Z; t) = \mathbf{U}^T(Z)\mathbf{q}_u(t), \quad v_o(Z; t) = \mathbf{V}^T(Z)\mathbf{q}_v(t),$$
$$\theta_X(Z; t) = \mathbf{X}^T(Z)\mathbf{q}_x(t), \quad \theta_Y(Z; t) = \mathbf{Y}^T(Z)\mathbf{q}_y(t). \tag{10.5-22a-d}$$

Here $\mathbf{U}$, $\mathbf{V}$, $\mathbf{X}$, $\mathbf{Y}$ are the vectors of trial functions that are chosen to satisfy at least the kinematic boundary conditions; $\mathbf{q}_u$, $\mathbf{q}_v$, $\mathbf{q}_x$, $\mathbf{q}_y$ are the vectors of generalized coordinates. Replacing the energy quantities, and the representations (10.5-22) in the variational integral, Eq. (10.5-4), carrying out the indicated variations and the required integrations, one obtains the equation

$$\mathbf{M}\ddot{\mathbf{q}}(t) + \mathbf{G}\dot{\mathbf{q}}(t) + \mathbf{K}\mathbf{q}(t) = \mathbf{0}, \tag{10.5-23}$$

where $\mathbf{M}$ is a real symmetric positive definite mass matrix; $\mathbf{K}$ is the symmetric stiffness matrix, including in its elements the external load P; $\mathbf{G}$ is the real skew-symmetric gyroscopic matrix, while $\mathbf{q} \equiv \{\mathbf{q}_u^T, \mathbf{q}_v^T, \mathbf{q}_x^T, \mathbf{q}_y^T\}^T$, is the overall vector of the generalized coordinates. In the absence of structural damping, and when $\Omega_z = 0$, the matrix $\mathbf{G}$ becomes immaterial.

For synchronous motion,

$$\mathbf{q}(t) = \mathbf{Z}\exp(i\omega t), \tag{10.5-24}$$

where $\mathbf{Z}$ is a constant vector and ω a constant, both generally complex. Following the usual steps, one obtains the eigenvalue problem

$$-\omega^2\mathbf{M}\mathbf{Z} + i\omega\mathbf{G}\mathbf{Z} + \mathbf{K}\mathbf{Z} = \mathbf{0}, \tag{10.5-25}$$

where ω must satisfy the characteristic equation:

$$\Delta(\omega^2, P, \Omega_z) = det| -\omega^2\mathbf{M} + i\omega\mathbf{G} + \mathbf{K}| = 0. \tag{10.5-26}$$

Note that the characteristic polynomial Δ contains even powers of ω only, and, so that the stability will be analyzed in terms of ω^2. Since $\mathbf{K}$ includes the contribution of the external load which can be tensile or compressive, we can have $\mathbf{K} > 0$, $\mathbf{K} = 0$ or $\mathbf{K} < 0$. As a result, one can distinguish three cases:

(1) $\mathbf{K}$ is positive definite. ω_k^2 are real and positive, corresponding to pure oscillatory, or stable motion.

(2) instability by divergence. The condition for its occurrence is $\Delta(\omega^2 = 0, P, \Omega_z) = 0$. The instability by divergence depends on the external load P only, and is the same for both the gyroscopic and non-gyroscopic systems.

(3) $\mathbf{K}$ is negative definite. It is still possible to have stable motion, implying that the gyroscopic effects stabilize an unstable conservative system. However for the same case, the eigenvalues can be complex conjugate with at least one of them having a negative imaginary part, yielding unstable motion of the flutter type. In such a case, as the rotational speed of the rotor Ω_z and the compressive load P increase, the two frequency branches

may coalesce; beyond that point of coalescence, the eigenfrequencies become complex conjugate and correspondingly, bending oscillations with exponentially increasing amplitudes will occur. The values of ω and Ω_z corresponding to the coalescence point are referred to as the flutter frequency ω_{fl}, and the flutter rotational speed $(\Omega_z)_{fl}$, respectively. For each external load P, ω_{fl} and $(\Omega_z)_{fl}$ can be determined.

The entries of the mass $\mathbf{M}(\equiv \mathbf{M}_{4\times4})$, gyroscopic $\mathbf{G}(\equiv \mathbf{G}_{4\times4})$ and stiffness $\mathbf{K}(\equiv \mathbf{K}_{4\times4})$ matrices are provided in Tables 10.5-1 through 10.5-3.

Applying an integration by parts in the expressions of k_{14} and k_{23}, in conjunction with the boundary conditions, one obtains that k_{14} and k_{23} reduce to
$k_{14} = -\int_0^L (a_{44}\mathbf{U}'\mathbf{Y}^T)dZ$ and $k_{23} = -\int_0^L (a_{55}\mathbf{V}'\mathbf{X}^T)dZ$, respectively. These results in conjunction with the ones supplied in Table 10.5-3 yield the conclusion that the stiffness matrix $\mathbf{K}$ is symmetric.

Table 10.5-1: The entries of the mass matrix $\mathbf{M}$

$\mathbf{M} \equiv (m_{ij} = m_{ji})$	Their expression
m_{11}	$-\int_0^L (b_1 \mathbf{U}\mathbf{U}^T)dZ - [m_R\mathbf{U}\mathbf{U}^T]_{Z=L}$
m_{12}	0
m_{13}	0
m_{14}	$[m_R r_m \mathbf{U}\mathbf{Y}^T]_{Z=L}$
m_{22}	$-\int_0^L (b_1 \mathbf{V}\mathbf{V}^T)dZ - [m_R\mathbf{V}\mathbf{V}^T]_{Z=L}$
m_{23}	$[m_R r_m \mathbf{V}\mathbf{X}^T]_{Z=L}$
m_{24}	0
m_{33}	$-\int_0^L (b_4 + \delta_n b_{14})\mathbf{X}\mathbf{X}^T dZ - [J^R_{xxyy}\mathbf{X}\mathbf{X}^T]_{Z=L}$
m_{34}	0
m_{44}	$-\int_0^L (b_5 + \delta_n b_{15})\mathbf{Y}\mathbf{Y}^T dZ - [J^R_{xxyy}\mathbf{Y}\mathbf{Y}^T]_{Z=L}$

Table 10.5-2: The entries different of zero of the gyroscopic matrix $\mathbf{G}$

$\mathbf{G} \equiv (g_{ij})$	Their expression
g_{34}	$[3J^R_{zz}\Omega_z\mathbf{X}\mathbf{Y}^T]_{Z=L}$
g_{43}	$-[3J^R_{zz}\Omega_z\mathbf{Y}\mathbf{X}^T]_{Z=L}$

10.5-7 Results

The numerical illustrations are carried out for a composite box-beam of a rectangular ($b \times c$) cross-section, with $c = 10$in. (0.254m), $h = 0.4$in. (1.02cm), $L = 80$in. (2.032m). The material of the beam is a Graphite/Epoxy composite.

The numerical simulations are carried out in terms of the dimensionless parameters listed in Table 10.5-4.

Table 10.5-3: The entries of the stiffness matrix $\mathbf{K}(\equiv K_{4\times4})$

$\mathbf{K} \equiv (k_{ij})$	Their expression
k_{11}	$\int_0^L (a_{44}\mathbf{U}\mathbf{U}''^T - P\mathbf{U}\mathbf{U}''^T)dZ$ $-[a_{44}\mathbf{U}\mathbf{U}'^T - P\mathbf{U}\mathbf{U}'^T]_{Z=L}$
$k_{12} = k_{21}$	0
k_{13}	$\int_0^L (a_{43}\mathbf{U}\mathbf{X}''^T)dZ - [a_{43}\mathbf{U}\mathbf{X}'^T]_{Z=L}$
k_{14}	$\int_0^L (a_{44}\mathbf{U}\mathbf{Y}'^T)dZ - [a_{44}\mathbf{U}\mathbf{Y}^T]_{Z=L}$
k_{22}	$\int_0^L (a_{55}\mathbf{V}\mathbf{V}''^T - P\mathbf{V}\mathbf{V}''^T)dZ$ $-[a_{55}\mathbf{V}\mathbf{V}'^T - P\mathbf{V}\mathbf{V}'^T]_{Z=L}$
k_{23}	$\int_0^L (a_{55}\mathbf{V}\mathbf{X}'^T)dZ - [a_{55}\mathbf{V}\mathbf{X}^T]_{Z=L}$
k_{24}	$\int_0^L (a_{52}\mathbf{V}\mathbf{Y}''^T)dZ - [a_{52}\mathbf{V}\mathbf{Y}'^T]_{Z=L}$
k_{31}	$\int_0^L (a_{34}\mathbf{X}\mathbf{U}''^T)dZ - [a_{34}\mathbf{X}\mathbf{U}'^T]_{Z=L}$
k_{32}	$-\int_0^L (a_{55}\mathbf{X}\mathbf{V}'^T)dZ$
k_{33}	$\int_0^L (a_{33}\mathbf{X}\mathbf{X}''^T - a_{55}\mathbf{X}\mathbf{X}^T)dZ - [a_{33}\mathbf{X}\mathbf{X}'^T]_{Z=L}$
k_{34}	$\int_0^L (a_{34}\mathbf{X}\mathbf{Y}'^T - a_{52}\mathbf{X}\mathbf{Y}'^T)dZ - [a_{34}\mathbf{X}\mathbf{Y}^T]_{Z=L}$
k_{41}	$-\int_0^L (a_{44}\mathbf{Y}\mathbf{U}'^T)dZ$
k_{42}	$\int_0^L (a_{25}\mathbf{Y}\mathbf{V}''^T)dZ - [a_{25}\mathbf{Y}\mathbf{V}'^T]_{Z=L}$
k_{43}	$\int_0^L (a_{25}\mathbf{Y}\mathbf{X}'^T - a_{43}\mathbf{Y}\mathbf{X}'^T)dZ - [a_{25}\mathbf{Y}\mathbf{X}^T]_{Z=L}$
k_{44}	$\int_0^L (a_{22}\mathbf{Y}\mathbf{Y}''^T - a_{44}\mathbf{Y}\mathbf{Y}^T)dZ - [a_{22}\mathbf{Y}\mathbf{Y}'^T]_{Z=L}$

Table 10.5-4: Dimensionless quantities

Dimensionless quantity	Its expression
$\bar{\omega}_i$	ω_i/ω_0, ($\omega_0 = 164.73$ rad/s)
$\bar{\Omega}_z$	Ω_z/ω_o,
$\bar{P}$	$P/\hat{P}$, ($\hat{P} = 81178.1$ lb.)
$\bar{m}_r$	$m_R/(b_1 L) = 0.5$
$\bar{r}_m$	$r_m/L = 0.2$
$\bar{k}_R$	$k_R/L = 0.2$

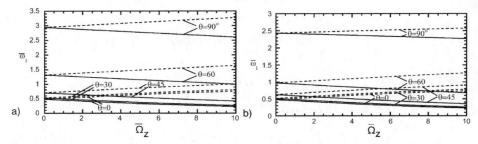

Figure 10.5-2: Influence of the ply-angle of the beam material on the upper and lower branches of the rotating dimensionless fundamental frequency. (a) Unshearable beam model, $\mathcal{R} = 1$, (b) The counterpart of (a) for the shearable beam model

Figures 10.5-2 through 10.5-4 depict, in the absence of the compressive load $\bar{P}$, the dependence of the three normalized eigenfrequencies $\bar{\omega}_i$ ($i = 1, 2, 3$) on the normalized rotor spin speed $\bar{\Omega}_z$. The results are for a square ($\mathcal{R} = 1$) cross-section unshearable beam, and for selected values of the ply-angle θ. For $\bar{\Omega}_z = 0$, i.e. for the non-spinning rotor, because the bending stiffnesses in X and Y directions are equal, the flapping and lagging eigenfrequencies in each mode coincide. As soon as the rotation starts (i.e. for $\bar{\Omega}_z \neq 0$), gyroscopic forces are generated, and there is a bifurcation of rotating eigenfrequencies, with upper and lower frequency branches. The rotor spin rate at which the lowest rotating eigenfrequency becomes zero is the critical spinning speed, $(\bar{\Omega}_z)_{cr}$, and corresponds to the instability by divergence. However, as it will appear later in Fig. 10.5-5, this critical spinning speed is extremely large. These plots show that the tailoring technique can significantly enhance the vibrational behavior of the system; the increase of the ply-angle θ, accompanied by an increase of bending stiffnesses yields the increase of both non-rotating and rotating eigenfrequencies.

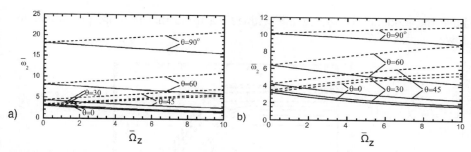

Figure 10.5-3: The counterparts of Figs. 10.5-2a,b, for the second mode dimensionless natural frequency

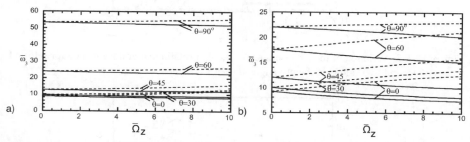

Figure 10.5-4: The counterparts of Figs. 10.5-2a,b for the third mode dimensionless natural frequency

The same figures show that with the increase of the mode number, for fixed spin rates, the distances between the upper and lower frequency branches tend to increase, a trend that is independent of the ply-angle. Figures 10.5-2b through 10.5-4b depict the shearable counterparts of Figs. 10.5-2a through 10.5-4a; classical (unshearable) beam model overestimates both the non-rotating and the rotating frequencies. The beneficial effects induced by the increase of the ply-angle and of $\mathcal{R}$ on both the lower branch of rotating fundamental eigenfrequency and on the rotor spinning divergence speed clearly emerge from this plot.

Figure 10.5-5 shows in logarithmic scale the variation of the lower branch of the first eigenfrequency of the system vs. the rotor spinning rate $\overline{\Omega}_z$, for two values of the ply-angle, and for selected values of the cross-section parameter $\mathcal{R}$.

Figures 10.5-6 through 10.5-8 show the variation of the first three eigenfrequencies vs. $\overline{\Omega}_z$ for selected values of the cross-sectional parameter $\mathcal{R}$, and for a fixed value of the ply-angle $\theta = 45$ deg; for the beams with $\mathcal{R} \neq 1$, the non-rotating frequencies in flapping and lagging are different.

The natural frequencies pairs associated with $\mathcal{R} \neq 1$ are lower than those for $\mathcal{R} = 1$. Moreover, with the increase of the mode number, the eigenfrequencies associated with the lower branch of each frequency pair, instead of decreasing, have the tendency of increasing with $\overline{\Omega}_z$, a trend that is more prominent for lower values of $\mathcal{R}$.

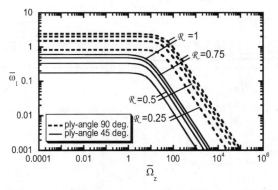

Figure 10.5-5: Variation of the lower branch of the first rotating natural frequency for selected values of $\mathcal{R}$, and two values of the ply-angle, θ (deg.) = 45 and 90

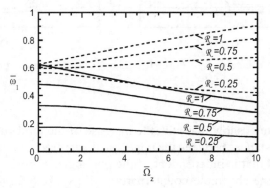

Figure 10.5-6: Influence of $\mathcal{R}$ on the upper and lower branches of the first dimensionless natural frequency, θ = 45 deg.

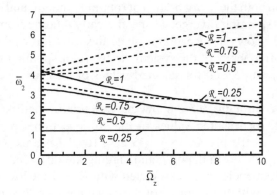

Figure 10.5-7: Counterpart of Fig. 10.5-6 for the second dimensionless natural frequency

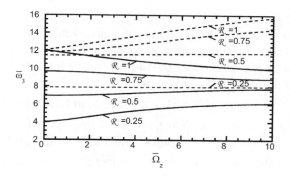

Figure 10.5-8: Counterpart of Fig. 10.5-6 for the third rotating natural frequency

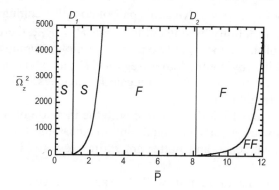

Figure 10.5-9: Stability boundaries for the box-beam with a spinning rotor, ($\mathcal{R} = 1, \theta = 0$)

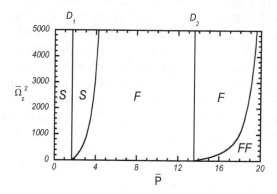

Figure 10.5-10: Counterpart of Fig. 10.5-9 for $\theta = 45$ deg.

Figures 10.5-9 and 10.5-10 supply stability plots in the plane ($\overline{\Omega}_z^2 - \overline{P}$), for $\mathcal{R} = 1$. These were generated by using six trial functions in Eqs. (10.5-22). The domains of stability are marked by S, and those by single and double flutter instability by F and FF, respectively. Since the beam has equal bending stiffnesses in both X and Y directions, the divergence instability under the compressive axial load occurs simultaneously in both X and Y directions; the divergence boundaries marked as D_1 and D_2 can be interpreted as divergence regions of zero thickness. In the load interval, $0 \leq \overline{P} < \overline{P}_{D_1}$, the frequencies ω_k^2 remain positive for any value of $\overline{\Omega}_z^2$: in this load range the system is stable.

For a slight increase of $\overline{P}$ beyond $\overline{P}_{D_1}$ and for low values of $\overline{\Omega}_z^2$, there is a pair of complex conjugate eigenfrequencies, indicating instability by single flutter. As $\overline{\Omega}_z^2$ increases, the gyroscopic effects stabilize the system; all the roots ω_k^2 are positive.

However, at the right of the resulting stability boundary, until $P = \overline{P}_{D_2}$ and for arbitrary $\overline{\Omega}_z^2$, two positive and a pair of complex conjugate frequencies occur, indicating instability by single flutter. For $\overline{P} > \overline{P}_{D_2}$, there are two flutter regions. In one of these, there are two pairs of complex conjugate frequencies, indicating instability by double flutter (region FF); in other, there are two positive values and a pair of complex conjugate frequencies indicating instability by single flutter (region F).

Figures 10.5-9 and 10.5-10 show that with the increase of the ply-angle the stability domains are greatly enlarged, and, at the same time, the instability boundaries by divergence and flutter are shifted towards larger compressive loads $\overline{P}$.

The stability behavior of thin-walled beams of rectangular cross-section ($\mathcal{R} \neq 1$) is much more complex than that for $\mathcal{R} = 1$; in addition to regions of single flutter (F) and divergence (D), there are regions of double (DD), triple (DDD) and quaternary ($DDDD$) divergence and double flutter (FF), as well as instability regions with mixed, single and double divergence and flutter, i.e. DF and DDF, respectively. For a solid isotropic beam with $\mathcal{R} \neq 1$, the complexity of the stability behavior was illustrated by Yamanaka et al. (1996).

Finally, in Fig. 10.5-11, for a similar gyroscopic structural systems considered by Yamamaka et al. (1995), the predictions of stability/instability boundaries obtained by the same authors via Ritz's method, and by the present Extended Galerkin are presented. The predictions reveal an excellent agreement.

10.5-8 Concluding Remarks

The directional properties of composites can be used to improve the stability of a beam carrying a spinning tip rotor. The system can lose its stability by divergence by the separate actions of $\overline{P}$ and $\overline{\Omega}_z$, and by flutter, through the joint action of $\overline{P}$ and $\overline{\Omega}_z$. This implies that in the absence of a compressive load, the system cannot lose its stability by flutter. Gyroscopic effects can stabilize the system.

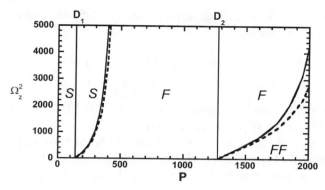

Figure 10.5-11: Validation of the solution methodology based on Yamanaka et al. (1995) gyroscopic model (——) EGM *P* versus (- - - - -) Ritz's method. Stability diagram in the plane $(\Omega_z^2 [(\text{rad/s})^2] - P[N]$

10.6 SMART THIN-WALLED SPINNING BEAMS

10.6-1 Preliminaries

As it was highlighted in the previous sections, the directionality property featured by the advanced composite materials can be used to enhance the behavior of spinning structural systems.

The tailoring technique just discussed is passive: once implemented, the structure is not able to respond as to avoid the occurrence of the resonance, or of any other instability.

As a complementary option, we can apply the feedback control via the adaptive materials technology.

The adaptive capability can be achieved through both the converse and direct piezoelectric effects. In this sense, the capability generated by the former effect, consists of the generation of localized strains in response to an applied voltage, whereas the later obtained via the direct piezoelectric effect, consists of the generation of a charge/voltage as a result of the application of a force.

In this section several results on the enhancement of the vibration and stability of spinning flexible thin-walled beams via implementation of the *adaptive capability* and of the *structural tailoring* will be presented.

We proceed as in Sections 6.6-2 and 6.6-4. The piezoactuators/sensors are spread over the entire beam span and the actuation is out-of-phase; we use the concept of boundary moment control, and the proportional feedback control as described by Eqs. (6.6-19).

Since the piezoelectrically induced bending moments appear in the boundary conditions at the beam tip, we record only the equations that are affected by their presence.

For the *spinning circular shaft*, among Eqs. (10.1-5) and (10.1-7), only Eqs. (10.1-7c, d) are changed to

$$\delta\theta_y : \ a_{22}\theta'_y + a_{25}(v'_0 + \theta_x) - \tilde{M}_y = 0,$$

$$\delta\theta_x : \ a_{33}\theta'_x + a_{34}(u'_0 + \theta_y) - \tilde{M}_x = 0. \qquad (10.6\text{-}1a,b)$$

For the manipulator arm systems, among Eqs. (10.1-5) and (10.5-17), only Eqs. (10.5-17c,d) are affected; the new equations are

Here, $\tilde{M}_x$ and $\tilde{M}_y$ are the piezoelectrically induced boundary bending moments provided by Eqs. (6.6-1), where, for the proportional feedback control, these have to be expressed by Eqs. (6.6-19). In such a context, a closed-loop eigenvalue problem is obtained.

$$\delta\theta_X : \ a_{33}\theta'_X + a_{34}(u'_o + \theta_Y) + \ J^R_{xxyy} \, \ddot{\theta}_X - \frac{3\Omega_z J^R_{zz}\dot{\theta}_Y}{}$$

$$- \ m_R r_m \ddot{v}_o - \tilde{M}_x = 0, \qquad (10.6\text{-}2a,b)$$

$$\delta\theta_Y : \ a_{22}\theta'_Y + a_{25}(v'_o + \theta_X)$$

$$+ \ J^R_{xxyy} \, \ddot{\theta}_Y + \frac{3\Omega_z J^R_{zz}\dot{\theta}_X}{} - \ m_R r_m \ddot{u}_o - \tilde{M}_y = 0.$$

For the two cases under consideration here, $C_{M^a_x}$ and $C_{M^a_y}$ appearing in Eqs. (6.6-19) are expressed as:

$$C_{M^a_x} = e_{31}\left[b\left(1 - \frac{A^*_{11}}{A^*_{12}}\right)(n_+ - n_-) + (n_+^2 - n_-^2) - 2\frac{B^*_{12}}{A^*_{11}}(n_+ - n_-)\right]p_1,$$

$$(10.6\text{-}3a,b)$$

$$C_{M^a_y} = e_{31}\left[c\left(1 - \frac{A^*_{11}}{A^*_{12}}\right)(n_+ - n_-) + (n_+^2 - n_-^2) - 2\frac{B^*_{12}}{A^*_{11}}(n_+ - n_-)\right]p_2,$$

In Eqs. (10.6-3) p_1 and p_2 denote, for the circular cylindrical shaft, the lengths of piezoactuators along the circumferential direction, while for the manipulator arm system, their lengths along the x and y directions, respectively.

Following the procedure described in Section 10.5-6, we obtain an equation similar to Eq. (10.5-25). However, in contrast to that case, the contribution due to the control moment at the beam tip will appear in the matrix **K**.

The conditions of stability/instability discussed in Section 10.5-6, show that the piezoelectric strain actuation can control the free vibration of the system, and prevent loss of stability by divergence and flutter.

10.6-2 Results

10.6-2a Preliminaries

The numerical simulations are carried out for the circular shaft and robotic manipulator described in Sections 10.2-5 and 10.5-7, respectively. The material of the host structure is graphite-epoxy, that of the piezoactuators/sensors is *PZT*-4 piezoceramic.

The distributions of piezoactuators/sensors in both considered cases is supplied in Figs. 10.6-1a and 10.6-1b. For the circular shaft the extension of piezoactuators, p^a, and piezosensors p^s, in the circumference direction are $\pi R/4$ and $\pi R/32$, respectively. In both cases, the dimensionless feedback gain K_p is

$$K_p = \frac{\bar{K}_p L}{\hat{a}_{33} h_p},\tag{10.6-4}$$

where $\hat{a}_{33}$ is the transverse bending stiffness corresponding to the material featuring $\theta = 0$.

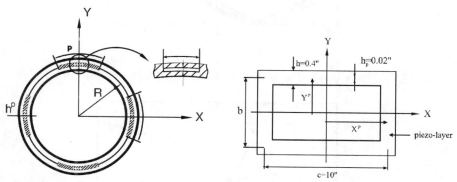

Figure 10.6-1: a) Piezoactuator and piezosensor patch distributions, $(p_x^a = p_y^a = \pi R/4;\ p_x^s = p_y^s = \pi R/32;\ h^p = 0.015in.;\ h^s = 0.005in.)$; b) Cross-section of the adaptive beam

10.6-2b Results for the Circular Shaft

Figures 10.6-2 through 10.6-4 relate the sequence of the ply-angles θ (deg.) = 0, 60, and 90, respectively, and the unloaded beam $(P = 0)$. They show the variation of the dimensionless fundamental natural frequency $\bar{\omega}_1 (\equiv \omega_1/\hat{\omega})$ against the dimensionless spin speed $\bar{\Omega}(\equiv \Omega/\hat{\omega})$, for selected feedback gains. The normalizing factor $\hat{\omega} = 138.85$ rad/s is the fundamental frequency of the non-spinning, unloaded and unactivated beam $(K_p = 0)$, characterized by $\theta = 0$ and $P = 0$.

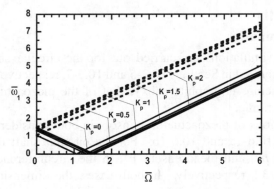

Figure 10.6-2: First natural frequency versus spinning speed for selected values of the proportional feedback gain ($\theta = 0$, $\overline{P} = 0$); ——, lower branch; - - - - -, upper branch

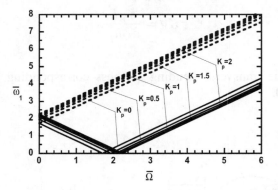

Figure 10.6-3: The counterpart of Fig. 10.6-2 for $\theta = 60$ deg.

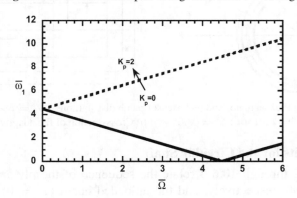

Figure 10.6-4: The counterpart of Fig. 10.6-2 for $\theta = 90$ deg.

For $\overline{\Omega} = 0$, i.e. in the absence of gyroscopic effects, the system is characterized, for each θ, by a single fundamental frequency. With the increase of the ply-angle θ, non-rotating natural frequency increases; the bending stiffnesses $a_{22} = a_{33}(\equiv A)$ increase with the ply-angle θ.

As soon as the rotation starts, gyroscopic forces are generated, and there is a bifurcation of natural frequencies. The effect of the gyroscopic Coriolis force, produces two distinct frequency branches of free bending vibration, the upper and lower frequency branches. The minimum spin rate at which the lowest rotating natural frequency becomes zero corresponds to the divergence instability, $\overline{\Omega}_{cr}$. For each ply-angle there is a specific critical spinning speed and the minimum and maximum ones occur at $\theta = 0$ and $\theta = 90$ deg., respectively.

Figures 10.6-2 through 10.6-4 show that the piezoelectric strain actuation plays an important role in increasing the eigenfrequencies of the non-spinning beam and postponing the divergence whirl instability.

Note that as the ply-angle increases which induces an increase of the bending stiffness, the efficiency of the control via piezoelectric strain actuation decreases.

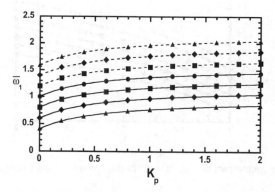

Figure 10.6-5: Upper and lower branches of the first natural frequency versus the proportional feedback gain for selected values of the spinning speed ($\theta = 0$, $\overline{P} = 0$); ———, lower branch; - - - -, upper branch; •, $\overline{\Omega} = 0$; ■, $\overline{\Omega} = 0.2$; ♦, $\overline{\Omega} = 0.4$; ▲, $\overline{\Omega} = 0.6$

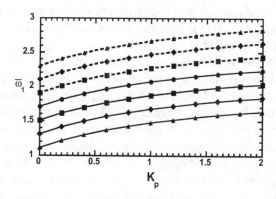

Figure 10.6-6: The counterpart of Fig. 10.6-5 for $\theta = 60$ deg., $\overline{P} = 0$

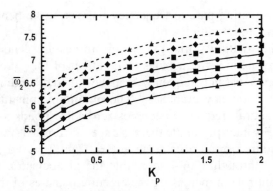

Figure 10.6-7: The counterpart of Fig. 10.6-5 for the second natural frequency: ——, lower branch; - - - -, upper branch; •, $\overline{\Omega} = 0$; ■, $\overline{\Omega} = 0.2$; ♦, $\overline{\Omega} = 0.4$; ▲, $\overline{\Omega} = 0.6$ ($\theta = 0$, $\overline{P} = 0$)

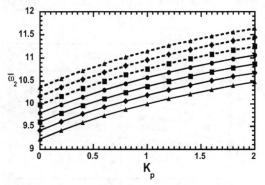

Figure 10.6-8: The counterpart of Fig. 10.6-6 for the second natural frequency ($\theta = 60$ deg., $\overline{P} = 0$)

Figures 10.6-5 and 10.6-6, and 10.6-7 and 10.6-8 show the upper and lower branches of the fundamental and second natural frequencies, respectively, of the unloaded shaft ($P = 0$), as function of the feedback gain, and for selected values of the spinning speed $\overline{\Omega}$.

Figures 10.6-5 and 10.6-6, and their counterparts, Figs. 10.6-7 and 10.6-8, are generated for the ply-angles, in the succession: θ (deg.) $= 0$ and 60. As for a box-beam with $\mathcal{R} = 1$ for $\overline{\Omega} = 0$, there is a single bending frequency, common to flapping and lagging motions, while for $\overline{\Omega} \neq 0$, both the upper and lower frequency branches increase with the increase of the proportional feedback control gain. However, irrespective of the ply-angle, the second mode natural frequency features, for both the upper and lower branches, a higher increase with K_p as compared to the first mode frequency. As concerns the whirl speed, its increase exerts on the upper and lower frequency branches a similar influence to that previously highlighted in Figs. 10.6-2 through 10.6-4.

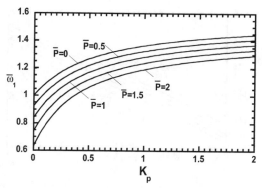

Figure 10.6-9: First non-rotating ($\overline{\Omega} = 0$) natural frequency as a function of the feedback gain, for selected values of the compressive load $\overline{P}$, $\theta = 0$

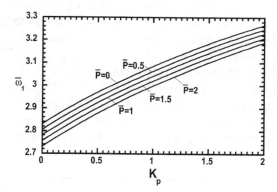

Figure 10.6-10: The counterpart of Fig. 10.6-9 for $\theta = 75$ deg.

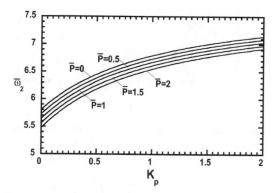

Figure 10.6-11: The counterpart of Fig. 10.6-9 for the second non-rotating natural frequency

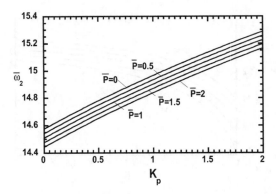

Figure 10.6-12: The counterpart of Fig. 10.6-11 for $\theta = 75$ deg.

Figures 10.6-9 through 10.6-12 show the effect of the compressive axial load $\bar{P}$ and ply-angle on the open/closed loop first and second mode frequencies of the non-rotating shaft. Herein, the dimensionless compressive force is defined as $\bar{P} = PL/\hat{a}_{33}$, where $\hat{a}_{33} = 1.1787 \times 10^8$ lb.in.^{2} (338,323 N.m^2) is the bending normalizing stiffness corresponding to $\theta = 0$.

The results reveal that: (i) the compressive load lowers the eigenfrequencies, (ii) the piezoelectric actuation plays a substantial role in increasing the frequencies and, finally, that (iii) the increase of the ply-angle, is accompanied by both a notable increase of natural frequencies and by a decreasing efficiency of the piezoelectric actuation.

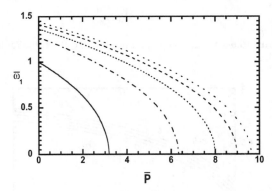

Figure 10.6-13: Variation of the first non-rotating ($\bar{\Omega} = 0$) natural frequency versus the dimensionless compressive load, for selected values of the feedback gain, ($\theta = 0$): ———, $K_p = 0$; — · — · —, $K_p = 0.5$; . . . . , $K_p = 1$; – – – –, $K_p = 1.5$; . . . . . . $\bar{K}_p = 2$

Figures 10.6-13 and 10.6-14 show the variation of the first natural bending frequency of the non-rotating shaft as a function of the compressive axial load, for selected values of the feedback gain; these figures correspond, to the ply-angles $\theta = 0$ and 75 deg., respectively. The value of $\bar{P}$ at which the frequencies become zero, corresponds to the buckling load. This loss of stability is by divergence. These plots show that the feedback actuation plays a significant role in counteracting the detrimental effect of the compressive load by increasing both the eigenfrequencies and the buckling load.

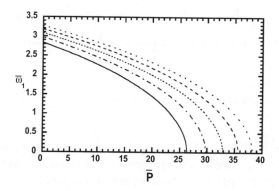

Figure 10.6-14: Counterpart of Fig. 10.6-13 for $\theta = 75$ deg.

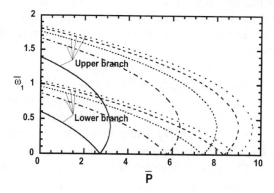

Figure 10.6-15: Variation of the upper and lower frequency branches versus the dimensionless compressive load, for selected values of the feedback gain ($\overline{\Omega} = 0.4$), $\theta = 0$; ———, $K_p = 0$; $-\cdot-\cdot-$, $K_p = 0.5$; $\ldots\ldots$; $K_p = 1$; $----$, $K_p = 1.5$; $\ldots\ldots$ $K_p = 2$

Figures 10.6-15 and 10.6-16 show the variation of the upper and lower frequency branches vs. the axial compressive load, for selected values of the feedback gains and a fixed value of the spinning speed; for fixed values of the ply-angle and feedback gain, the divergence instability occurs at such a buckling load P_{cr} for which the frequency of the lower branch becomes zero.

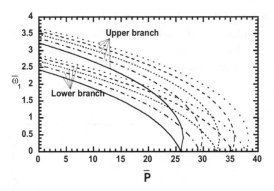

Figure 10.6-16: Counterpart of Fig. 10.6-15 for $\theta = 75$ deg.

As in Section 10.2-5, in the non-adaptive case, near divergence, the increase of the compressive load makes the frequencies of the upper and lower branches to merge, and afterwards to become complex conjugate; this slight increase of the compressive load, generates a shift of the instability from divergence to flutter.

The same plots show that the increase of the ply-angle and of the feedback gain significantly increase the critical axial load. However, with the increase of the ply-angle, that implies an increase of the bending stiffnesses, the buckling load becomes less and less sensitive to the increase of the feedback gain, of $\overline{\Omega}$ and of the ply-angle.

Finally, Figs. 10.6-17 and 10.6-18, display stability plots of the spinning system in the $\overline{\Omega}$ - $\overline{P}$ plane for the ply-angles $\theta = 0$ and $\theta = 45$ deg., respectively In these plots some of the previously reported trends will be highlighted in a more comprehensive context. Herein S and F denote the regions of stability and of flutter instability, respectively, while D denotes the divergence boundary. For $\overline{\Omega}$ and $\overline{P}$ equal to zero, $\overline{\omega}_i^2$ are real and positive and the system is stable. With the increase of $\overline{\Omega}$ and/or $\overline{P}$, instabilities by divergence or flutter may occur. The results supplied in these figures reveal that the increase of ply-angle and of the feedback gain contribute to a dramatic increase of the stability domains, and to the shift of the divergence and flutter instability boundaries towards larger compressive loads. However, as is clearly seen from these plots, increase of the ply- angle yields an unavoidable decay of the efficiency of the piezoelectric feedback actuation.

Figures 10.6-19 and 10.6-20 depict the variation of the normalized open ($K_p = 0$) and closed-loop ($K_p \neq 0$) eigenfrequency ω_1 vs. the normalized rotor spin speed $\overline{\Omega}_z$. The results relate to $P = 0$, $\mathcal{R} = 1$, unshearable beam and selected values of the reduced feedback gain K_p. These figures correspond in sequence to the ply-angles $\theta = 0$ and $\theta = 90$ deg., respectively.

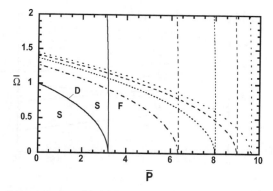

Figure 10.6-17: Stability plot in the $\overline{\Omega}$ - $\overline{P}$ plane displaying the domains of stability, divergence, and flutter instability for $\theta = 0$ and selected values of the feedback gain: ——, $K_p = 0$; —·—·—, $K_p = 0.5$; . . . , $K_p = 1$; — — —, $K_p = 1.5$, - - - -, $K_p = 2$

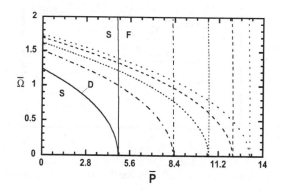

Figure 10.6-18: Counterpart of Fig. 10.6-17 for $\theta = 45$ deg.

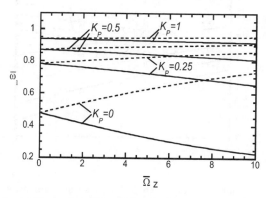

Figure 10.6-19: Upper and lower branches of the rotating open/closed loop dimensionless fundamental frequency. Unshearable beam model, $\mathcal{R} = 1$, $\theta = 0$

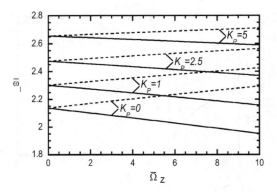

Figure 10.6-20: The counterpart of Fig. 10.6-19 for $\theta = 90$ deg.

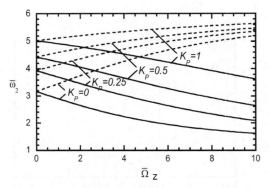

Figure 10.6-21: The counterpart of Fig. 10.6-19 for the second mode open/closed loop dimensionless eigenfrequency

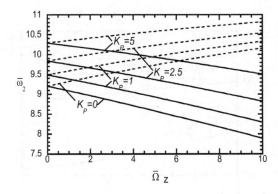

Figure 10.6-22: The counterpart of Fig. 10.6-20 for the second dimensionless eigenfrequency

Table 10.6-1: Stability and instability domains as influenced by the spinning speed, compressive load, ply-angle and piezoelectric induced actuation ($0 \leq \bar{P}_{stable} < \bar{P}_{div}$; $\bar{P}_{div} < \bar{P}_{stable} < \bar{P}_{flutter}$)

$$\theta = 0$$

$\bar{\Omega}$	$K_p = 0$		$K_p = 0.5$		$K_p = 1$		$K_p = 2$	
	$\bar{P}_{div}$	$\bar{P}_{flutter}$	$\bar{P}_{div}$	$\bar{P}_{flutter}$	$\bar{P}_{div}$	$\bar{P}_{flutter}$	$\bar{P}_{div}$	$\bar{P}_{flutter}$
0	3.1969	-	6.3516	-	8.0243	-	9.649	-
0.4	2.726	3.1922	5.78	6.347	7.39	8.0197	8.96	9.6444
0.8	1.23	3.1779	3.98	6.3329	5.43	8.0058	6.84	9.6306
1.2	-	3.154	0.75	6.3095	1.952	7.9826	3.14	9.6078
1.6	-	3.1206	-	6.2767	-	7.9502	-	9.5756
2	-	3.0777	-	6.2345	-	7.9084	-	9.5343

$$\theta = 45 \text{ deg.}$$

$\bar{\Omega}$	$K_p = 0$		$K_p = 0.5$		$K_p = 1$		$K_p = 2$	
	$\bar{P}_{div}$	$\bar{P}_{flutter}$	$\bar{P}_{div}$	$\bar{P}_{flutter}$	$\bar{P}_{div}$	$\bar{P}_{flutter}$	$\bar{P}_{div}$	$\bar{P}_{flutter}$
0	4.8668	-	8.3573	-	10.603	-	13.1833	-
0.4	4.4	4.862	7.816	8.3526	10.011	10.5983	12.533	13.1787
0.8	2.94	4.8478	6.1422	8.3386	8.191	10.5844	10.55	13.165
1.2	0.33	4.824	3.2	8.3151	5.01	10.5612	7.12	13.142
1.6	-	4.7907	-	8.2824	0.29	10.5288	2.08	13.11
2	-	4.7479	-	8.2402	-	10.487	-	13.0687

$$\theta = 90 \text{ deg.}$$

$\bar{\Omega}$	$K_p = 0$		$K_p = 0.5$		$K_p = 1$		$K_p = 2$	
	$\bar{P}_{div}$	$\bar{P}_{flutter}$	$\bar{P}_{div}$	$\bar{P}_{flutter}$	$\bar{P}_{div}$	$\bar{P}_{fl}$	$\bar{P}_{div}$	$\bar{P}_{flutter}$
0	74.2698	-	76.8045	-	79.2299	-	83.7782	-
1	71.33	74.2521	73.832	76.7871	76.23	79.213	80.724	83.7618
2	61.9	74.1989	64.31	76.7351	66.62	79.162	70.95	83.7128
3	44.143	74.1103	46.4	76.6484	48.53	79.077	52.51	83.6311
4	15.21	73.9861	17.2	76.5268	19.1	78.958	22.6	83.5167
5	-	73.8263	-	76.3704	-	78.8047	-	83.3694

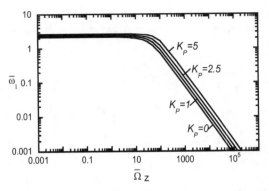

Figure 10.6-23: Variation of the lower branch of the first natural rotating frequency for selected feedback gains, ($\mathcal{R} = 1$). The plot is in logarithmic scale coordinates

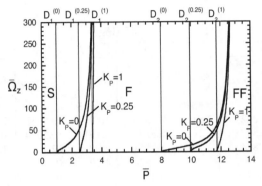

Figure 10.6-24: Influence of piezoelectric strain actuation on stability boundaries ($\mathcal{R} = 1$, $\theta = 0$), $K_p = 0$; 0.25 and 1

For $\bar{\Omega}_z = 0$, having in view that for this case the bending stiffnesses in X and Y directions are equal, the flapping and lagging eigenfrequencies in each mode coincide. As soon as the rotation starts i.e. for $\bar{\Omega}_z \neq 0$, gyroscopic forces are generated, and the rotating eigenfrequencies bifurcate.

In this sense, Figs. 10.6-21 and 10.6-22 that are the counterparts for the second mode natural frequency of the previous two ones, reveal a similar influence played by the ply-angle θ and proportional feedback control. The rotor spin rate at which the lowest rotating eigenfrequency becomes zero-valued corresponds to the divergence instability $(\bar{\Omega}_z)_{cr}$.

In Figure 10.6-23, the variation of the lower branch of the first dimensionless eigenfrequency of the systems vs. the rotor spinning rate $\overline{\Omega}_z$, is depicted in logarithmic scale, for selected values of the feedback gain K_p. The beneficial effects induced by the piezoelectric actuation on both the lower branch of rotating fundamental eigenfrequency and on the rotor spinning divergence speed clearly emerge from this plot.

In Fig. 10.6-24 the stability behavior for the case of a square ($\mathcal{R} = 1$) cross-section beam characterized by the ply-angles θ (deg.) = 0, is supplied. In this figure the domains of stability are marked by S, and those by single and double flutter instability by F and FF, respectively. For this case, the divergence instability under the compressive axial load occurs simultaneously in both X and Y directions. The divergence boundaries marked as D_1 and D_2 are associated with single and double roots, and correspond to the divergence instability regions of zero size. The exponents in brackets accompanying D_1 or D_2 denote the values of K_p used in the control process. In this context $D_1^{(0)}$ and $D_2^{(0)}$ denote divergence boundaries corresponding to the open-loop system.

Similar trends regarding the instability by divergence and flutter to those previously highlighted emerge from this plot.

10.6-3 Concluding Remarks

A few results on the vibrational and stability feedback control of spinning circular shafts and robotic manipulator systems modeled as thin-walled composite beams have been presented. The strong and synergistic effects played by the directional properties of advanced composite materials and of piezoelectric actuation on vibration and stability of considered spinning systems have been highlighted.

Two comments are in order: (a) the directionality property of constituent composite materials can successfully be used to enhance the stability of the system, and (b) the system can lose its stability by divergence by the separate action of $\overline{P}$ and $\overline{\Omega}_z$. However, the lose of stability by flutter is by the action of both $\overline{P}$ and $\overline{\Omega}_z$.

10.7 GEOMETRICALLY NONLINEAR SPINNING BEAMS

10.7-1 Preliminimaries

As in the case of rotating beams, also in the case of the spinning beams, their geometrically nonlinear formulation presents a significant importance.

In the linearized context, in the previous sections of this chapter, it was shown that as the result of the spinning speed, gyroscopic forces are generated.

In some conditions, under the action of these forces, in addition to the instability by divergence, also that by flutter can occur. Although considered to be catastrophic, depending on the nonlinearities that are involved in the structural system, the flutter instability can be also benign (see e.g. Librescu, 1965, 1967, 1975; Librescu et al., 2003, 2004). In the later case, the flutter boundary can be crossed without the catastrophic failure of the structural system, and the failure occurs by fatigue. Consequently, the study of the *character*, benign/catastrophic of the flutter instability boundary is of an enormous importance not only for the aeronautical/aerospace flight vehicles, but also for these structural systems that should operate safely at any Ω, without catastrophic failures.

This is one of the problems that can be addressed within a nonlinear formulation of these structural systems. In the paper by Shaw and Shaw (1989), based on the classical model of a spinning shaft, the issue of simple and double Hopf bifurcation has been investigated.

However, in addition to this problem, other ones that present an equal importance can be addressed within a nonlinear formulation. Several of these issues have been indicated in Section 9.8-1 and remain valid also in this case.

10.7-2 Governing Equations

The nonlinear equations of motion and the boundary conditions presented in Section 9.8-3, remain valid also for the case of spinning beams.

The single difference consists of the dynamic terms m_i that for this case are

$$
\begin{aligned}
m_1 &= b_1[\ddot{u}_0 - 2\Omega\dot{v}_0 - \Omega^2 u_0], \\
m_2 &= b_1[\ddot{v}_0 + 2\Omega\dot{u}_0 - \Omega^2 v_0], \\
m_3 &= b_1\ddot{w}_0, \\
m_4 &= (b_4 + b_{14})\ddot{\theta}_x, \\
m_5 &= (b_5 + b_{15})\ddot{\theta}_y, \\
m_6 &= (b_4 + b_5 + b_{14} + b_{15})\ddot{\phi} - (b_{10} + b_{18})\ddot{\phi}'', \\
m_7 &= -(b_{10} + b_{18})\ddot{\phi}'.
\end{aligned}
\qquad (10.7\text{-}1a\text{-}g)
$$

In such a context, one can see that a full coupling between twist, bending-bending and extension occurs. In order to simplify the problem, the case of thin-walled beams featuring a large twist stiffness, implying that $\phi \to 0$, is considered.

Restricting our study to simply-supported and in-plane immovable boundary conditions, requiring fulfilment of Eqs. (10.2-8b), and of $w_0(0) = w_0(L)$, and assuming that the axial inertia term is negligibly small, i.e. $\ddot{w}_0 = 0$, one obtains from Eqs. (9.8-4c) and the first expression in Table 9.8-1 that

$$
T_z = const. = \frac{a_{11}}{2L}\int_0^L ((u_0')^2 + (v_0')^2)dz. \qquad (10.7\text{-}2)
$$

As a result of (10.7-2), the governing equations become:

$$
\delta u_0 : \frac{a_{11}}{2L}\left[\int_0^L ((u_0')^2 + (v_0')^2)dz\right]u_0'' + a_{44}(\theta_y + u_0')' - b_1(\ddot{u}_0 - 2\Omega\dot{v}_0 \\
- \Omega^2 u_0) + p_x(z, t) = 0,
$$

$$\delta v_0 : \frac{a_{11}}{2L}\left[\int_0^L ((u_0')^2 + (v_0')^2 dz\right]v_0'' + a_{55}(\theta_x + v_0')' - b_1(\ddot{v}_0 + \underline{2\Omega\dot{u}_0}$$

$$- \Omega^2 v_0) + p_y(z, t) = 0, \qquad\qquad (10.7\text{-}3\text{a-d})$$

$$\delta\theta_x : a_{33}\theta_x'' - a_{55}(\theta_x + v_0') - (b_4 + b_{15})\ddot{\theta}_x + m_x(z, t) = (M_x^T)',$$

$$\delta\theta_y : a_{22}\theta_y'' - a_{44}(\theta_y + u_0') - (b_5 + b_{15})\ddot{\theta}_y + m_y(z, t) = (M_y^T)'.$$

and the associated boundary conditions at $z = 0, L$:

$$\delta u_o \left\{\frac{a_{11}}{2L}\left[\int_0^L [(u_0')^2 + (v_0')^2]dz\right]u_0' + a_{44}(\theta_y + u_0')\right\} = 0,$$

$$\delta v_0 \left\{\frac{a_{11}}{2L}\left[\int_0^L [(u_0')^2 + v_0')^2]dz\right]v_0' + a_{55}(\theta_x + v_0')\right\} = 0,$$

$$\delta\theta_x \, [a_{33}\theta_x' - M_x^T] = 0, \qquad\qquad (10.7\text{-}4\text{a-d})$$

$$\delta\theta_y \, [a_{22}\theta_y' - M_y^T] = 0.$$

• *Unshearable structural model*

Eliminating $a_{44}(\theta_y + u_0')$ and $a_{55}(\theta_x + v_0')$ in the previously displayed governing equations and the boundary conditions, one obtain the governing equation

$$\delta v_0 : \quad a_{33}v_0'''' - \frac{a_{11}}{2L}\left[\int_0^L [(u_0')^2 + (v_0')^2]dz\right]v_0''$$

$$- (b_4 + b_{14})\ddot{v}_0'' + b_1\left[\ddot{v}_0 + \underline{2\Omega\dot{u}_0} - \Omega^2 v_0\right]$$

$$- p_y(z, t) - m_x'(z, t) + (M_x^T)'' = 0,$$

$$\delta u_0 : \quad a_{22}u_0'''' - \frac{a_{11}}{2L}\left[\int_0^L [(u_0')^2 + (v_0')^2]dz\right]u_0''$$

$$- (b_5 + b_{15})\ddot{u}_0'' + b_1(\ddot{u}_0 - 2\Omega\dot{v}_0 - \Omega^2 u_0)$$

$$- p_x(z, t) - m_y'(z, t) + (M_y^T)'' = 0, \qquad (10.7\text{-}5\text{a,b})$$

and the associated boundary conditions at $z = 0, L$:

$$\delta u_0 \left\{a_{22}u_0''' - \frac{a_{11}}{2L}\left[\int_0^L [(u_0')^2 + (v_0')^2]dz\right]u_0'\right.$$

$$\left. - (b_5 + b_{15})\ddot{u}_0' - m_y + (M_y^T)'\right\} = 0,$$

$$\delta v_0 \left\{a_{33}v_0''' - \frac{a_{11}}{2L}\left[\int_0^L [(u_0')^2 + (v_0')^2]dz\right]v_0'\right.$$

$$\left. - (b_4 + b_{14})\ddot{v}_0' - m_x + (M_x^T)'\right\} = 0,$$

In the absence of rotatory inertias and of temperature effects the governing equations of the classical beam model coincide with those provided by Shaw and Shaw (1989).

10.7-3 Numerical Results

The previously displayed governing equations reveal their intricacy. While the issue of the charcter of the flutter boundary can be addressed either via determination of the sign of *the first Lyapunov quantity* (see Librescu, 1965, 1967, 1975; Librescu et al., 2003), or by applying center manifold theory (see Shaw and Shaw, 1989), the other issues and their approach have been already indicated in Section 9.8-1.

In Table 10.7-1 there are presented the effects of p_m, K and λ on the first dimensionless eigenfrequency. Needless to say, incorporation of the effect of p_m on the variation of eigenfrequency is not possible without considering the nonlinearities and without approaching the problem in the way indicated in Section 9.8-5.

Table 10.7-1: Effects of nonlinearity on first dimensionless natural frequency of non-spinning beam as a function of K and λ ($\mathcal{R} = 1$, $\overline{\Omega} = 0$)

	$\lambda = 0$			$\lambda = 2$			$\lambda = 4$		
	$p_m = 0$	$p_m = 1$	$p_m = 2$	$p_m = 0$	$p_m = 1$	$p_m = 2$	$p_m = 0$	$p_m = 1$	$p_m = 2$
$K = 0$	1.2639	1.2643	1.2657	1.2264	1.2269	1.2284	1.1190	1.1905	1.1921
$K = 0.5$	0.8883	0.8887	0.8900	0.8598	0.8602	0.8616	0.8240	0.8245	0.8260
$K = 1.0$	0.7777	0.7781	0.7793	0.7511	0.7515	0.7530	0.7120	0.7126	0.7142
$K = 20$	0.5467	0.5472	0.5488	0.5144	0.5150	0.5163	0.4212	0.4223	0.4240

From this table it can be concluded that the pressure has a beneficial effect on the fundamental eigenfrequency. This is due to the fact that it contributes to the increase of geometrical nonlinearities that have a positive influence on structural system response.

REFERENCES

Anliker, M. and Troesch, B. A. (1963) "Coupled Bending Vibration of Pretwist Cantilever Beams," *Zeitschrift für Agewandte Mathematics und Physics (ZAMP)*, Vol. 4, pp. 365–379.

Argento, A. and Scott, R. A. (1992) "Dynamic Response of a Rotating Beam Subjected to an Accelerating Distributed Surface Force," *Journal of Sound and Vibration*, Vol. 157, No. 2, pp. 221–231.

Argento, A. (1995) "A Spinning Beam Subjected to a Moving Deflection Dependent Load, Part I: Response and Resonance," Part II: Parametric Resonance (with Morano, H. L.) *Journal of Sound and Vibration*, Vol. 182, No. 4, pp. 595–615 and pp. 617–622.

Argento, A. and Scott, R. A. (1995) "Elastic Wave Propagating in a Timoshenko Beam Spinning About its Longitudinal Axis," *Wave Motion*, Vol. 21, pp. 67–74.

Bauchau, O. (1981) "Design, Manufacturing and Testing of High Speed Rotating Graphite-Epoxy Shaft," Ph.D. Thesis, Massachusetts Inst. of Technology, Dept. of Aeronautical & Astronautics.

Bert, C. W. and Kim, C. D. (1995a) "Whirling of Composite-Material Drive-shaft Including Bending-Twisting Coupling and Transverse Shear Deformation," *Journal of Vibration and Acoustics*, Trans. ASME, Vol. 117, pp. 17–21.

Bert, C. W. and Kim, C. D. (1995b) "Dynamic Instability of Composite Material Drive Shaft Subjected to Fluctuating Torque and/or Rotational Speed," *Dynamic and Stability of Systems*, Vol. 10, No. 2, pp. 125–147.

Bhaskar, K. and Librescu, L. (1995) "A Geometrically Non-Linear Theory for Laminated Anisotropic Thin-Walled Beams," *International Journal of Engineering Science*, Vol. 33, No. 9, pp. 1331–1344.

Chen, M. L. and Liao, Y. S. (1991) "Vibrations of Pretwisted Spinning Beams under Axial Compressive Loads with Elastic Constraints," *Journal of Sound and Vibration*, Vol. 147, No. 3, pp. 497–513.

Chen, Y., Zhao, H. B., Shen Z. P., Grieger, I. and Kröplin, B. H. (1993) "Vibrations of High Speed Rotating Shells with Calculations for Cylindrical Shells," *Journal of Sound and Vibration*, Vol. 160, pp. 137–160.

Chen, L. W. and Peng, W. K. (1998) "Stability Behavior of Rotating Composite Shafts under Axial Compressive Loads," *Composite Structures*, Vol. 41, 3–4, pp. 253–263.

Chang, C-Y., Chang, M-Y. and Huang, J. C. (2004) "Vibration Analysis of Rotating Composite Shafts Containing Randomly Oriented Reinforcements," *Composite Structures*, Vol. 63, pp. 21–32.

Cheng, Z. W. and Batra, R. C. (2000) "Three-Dimensional Thermoelastic Deformation of a Functionally Graded Elliptic Plate," *Composites, Part B*, Vol. 31, pp. 97–106.

Curti, G., Raffa, F. A. and Vatta, F. (1992) "An Analytical Approach to the Dynamic of Rotating Shafts," *Meccanica*, Vol. 27, pp. 285–292.

Dawson, B. (1968) "Coupled Bending-Bending Vibration of Pretwisted Cantilever Blading Treated by the Rayleigh-Ritz Energy Method," *Journal of Mechanical Engineering Science*, Vol. 10, No. 5.

D'Eleuterio, G. M. and Hughes, P. C. (1987) "Dynamics of Gyroelastic Spacecraft," *Journal Guidance Control and Dynamics*, Vol. 10, No. 4, pp. 401–405.

Filipich, C. P., Maurizi, M. J. and Rosales, M. B. (1987) "Free Vibrations of a Spinning Uniform Beam with Ends Elastically Restrained against Rotation," *Journal of Sound and Vibration*, Vol. 116, No. 3, pp. 475–482.

Filipich, C. P., Maurizi, M. J. and Rosales, M. B. (1989) "A Note on the Free Vibration of a Spinning Beam," *Journal of Sound and Vibration*, Vol. 129, pp. 350–355.

Gulick, D. W. (1994) "Thermally Induced Vibrations of an Axial Boom on a Spin-Stabilized Spacecraft," *AIAA-94-1556 CP*.

Han, R. P. S. and Zu, J. W. Z. (1992) "Modal Analysis of Rotating Shafts. A Body Fixed Axis Formulation Approach," *Journal of Sound and Vibration*, Vol. 156, No. 1, pp. 1–16.

Hua, L. and Lam, K. Y. (1998) "Frequency Characteristics of a Thin Rotating Cylindrical Shell Using the Generalized Differential Quadratic Method," *International Journal of Mechanical Sciences*, Vol. 40, No. 5, pp. 443–459.

Huang, S. C. and Soedel, W. (1988) "On the Forced Vibration of Simply Supported Rotating Cylindrical Shells," *Journal of the Acoustical Society of America*, Vol. 84, No. 1, pp 275–285, July.

Hughes, P. C. (1986) *Spacecraft Attitude Dynamics*, Wiley, New York.

Huseyin, K. (1978) *Vibrational and Stability of Multiple Parameters Systems*, Noordhoof, Alpen Aan Den Rijn, The Netherlands.

Huseyin, K. and Plaut, R. H. (1974–1975) "Transverse Vibrations and Stability of Systems with Gyroscopic Forces," *Journal of Structural Mechanics*, Vol. 3, No. 2, pp. 163–177.

Huseyin, K. and Plaut, R. H. (1975) "Divergence and Flutter Boundaries of Systems under Combined Conservative and Gyroscopic Forces," *Dynamics of Rotors, IUTAM Symposium*, Lyngby, Denmark, August 12–16, 1974, pp. 182–205.

Johnston, J. D. and Thornton, E. A. (1996) "Thermally Induced Response of Radiantly Heated Spinning Spacecraft Booms," *Journal of Thermophysics and Heat Transfer*, Vol. 10, 1, pp. 60–68.

Kim, C. D. and Bert, C. W. (1993) "Critical Speed Analysis of Laminated Composite Hollow Drive Shaft," *Composites Engineering*, Vol. 3, Nos. 7–8, pp. 663–643.

Kim, W., Argento, A. and Scott, R. A. (1999) "Free Vibration of a Rotating Tapered Composite Timoshenko Shaft," *Journal of Sound and Vibration*, Vol. 226, No. 1, pp. 125–147.

Kim, W., Argento, A. and Scott, R. A. (2001a) "Forced-Vibration and Dynamic Stability of a Rotating Tapered Composite Timoshenko Shaft: Bending Motions in End-Milling Operations," *Journal of Sound and Vibration*, Vol. 246, No. 4, pp. 583–600.

Kim, W., Argento, A. and Scott, R. A. (2001b) "Rotating Tapered Composite Shafts: Forced Torsional and Extensional Motions and Static Strength," *Journal of Vibration and Acoustics*, Trans. ASME, Vol. 123, January, pp. 24–29.

Ko, E-E. and Kim, J-H. (2003) "Thermally Induced Vibrations of Spinning Thin-Walled Composite Beam," *AIAA Journal*, Vol. 41, No. 2, pp. 296–303.

Ku, D. M. and Chen, L. W. (1994) "Stability and Whirl Speeds of Rotating Shaft under Axial Loads," *International Journal of Analytical and Experimental Modal Analysis*, Vol. 9, No. 2, pp. 111–123.

Lam, K. Y. and Hua, L. (1997) "Vibration Analysis of a Rotating Circular Conical Shell," *International Journal of Solids and Structures*, Vol. 34, pp. 2183–2197.

Lee, C. W., Katz, R., Ulsoy, A. G. and Scott, R. A. (1988) "Modal Analysis of a Distributed Parameter Rotating Shaft," *Journal of Sound and Vibration*, Vol. 122, pp. 119–130.

Lee, C. W. and Yun, J. S. (1996) "Dynamic Analysis of Flexible Rotors Subjected to Torque and Force," *Journal of Sound and Vibration*, Vol. 192, No. 2, pp. 439–452.

Liao, C. L. and Dang, Y. H. (1992) "Structural Characteristics of Spinning Pretwisted Orthotropic Beams," *Computers & Structures*, Vol. 45, No. 4, pp. 715–731.

Librescu, L., (1965, 1967) "Aeroelastic Stability of Orthotropic Heterogeneous Thin Panels in the Vicinity of the Flutter Critical Boundary," *Journal de Mécanique*, Vol. 4, No. 1, pp. 51–76, (II) *Journal de Mécanique*, Vol. 6, No. 1, pp. 133–152.

Librescu, L. (1975) *Elastostatics and Kinetics of Anisotropic and Heterogeneous Shell Type Structures*, Noordhoff International Publishing, Leyden, Netherlands.

Librescu, L., Song, O. and Kwon, H-D. (1999) "Vibration and Stability Control of Gyroelastic Beams via Smart Materials Technology," *Smart Structures NATO Science Series*, 3. High Technology – Vol. 65, J. Holnicki-Szulc and J. Rodellar (Eds.), Kluwer Academic, Dordrecht, 1999, pp. 163–172.

Librescu, L., Chiocchia, G. and Marzocca, P. (2003a) "Implications of Physical/Aerodynamical Nonlinearities on the Character of Flutter Instability Boundary," *International Journal of Non-linear Mechanics*, Vol. 38, pp. 173–199.

Librescu, L., Oh, S-Y. and Song, O. (2003b) "Spinning Thin-Walled Beams Made of Functionally Graded Materials: Modeling, Vibration and Instability," *European Journal of Mechanics, A/Solids*, Vol. 23, No. 3, pp. 499–515.

Librescu, L., Marzocca, P. and Silva, W. A. (2004) "Linear/Nonlinear Supersonic Panel Flutter in a High Temperature Field," *Journal of Aircraft*, Vol. 41, No. 4, pp. 918–924.

Librescu, L., Oh, S-Y. and Song, O. (2005) "Thin-Walled Beams Made of Functionally Graded Materials and Operating in a High Temperature Environment: Vibration and Stability," *Journal of Thermal Stresses*, Vol. 28, Nos 6–7, pp. 649–712.

Likins, P. W., Barbera, F. J. and Baddeley, V. (1973) "Mathematical Modeling of Spinning Elastic Bodies for Modal Analysis," *AIAA Journal*, Vol. 11, pp. 1251–1258.

Oh, S-Y., Librescu, L. and Song, O. (2005) "Vibration Instability of Functionally Graded Circular Cylindrical Spinning Thin-Walled Beams," *Journal of Sound and Vibration*, Vol. 285, Nos 4–5, pp. 1071–1091.

Païdoussis, M. P. (1998) *Fluid-Structure Interactions. Slender Structures and Axial Flow*, Vol. 1, *Academic Press*, San Diego, London, New York, Boston, Sydney, Tokyo, Toronto.

Rand, O. and Stavsky, Y. (1991) "Free Vibrations of Spinning Composite Cylindrical Shells," *International Journal of Solids and Structures*, Vol. 28, No. 7, pp. 831–843.

Rosales, M. B. and Filipich, C. P. (1993) "Dynamic Stability of a Spinning Beam Carrying an Axial Dead Load," *Journal of Sound and Vibration*, Vol. 163, No. 2, pp. 283–294.

Sabuncu, M. (1985) "Coupled Vibration Analysis of Blades with Angular Pretwist of Cubic Distribution," *AIAA Journal*, Vol. 23, pp. 1424–1430.

Saito, T. and Endo, M. (1986) "Vibration Analysis of Rotating Cylindrical Shells Based on the Timoshenko Beam Theory," *Bulletin of JSME*, Vol. 29, No. 250, pp. 1239–1245, April.

Shaw, J. and Shaw, S. W. (1989) "Instabilities and Bifurcations in a Rotating Shaft," *Journal of Sound and Vibration*, Vol. 132, No. 2, pp. 227–244.

Shieh, R. C. (1971) "Energy and Variational Principles for Generalized (gyroscopic) Conservative Problems," *International Journal of Non-Linear Mechanics*, Vol. 5, pp. 495–509.

Shieh, R. C. (1982) "Some Principles of Elastic Shaft Stability Including Variational Principles," *Journal of Applied Mechanics*, Trans. ASME, Vol. 49, pp. 191–196.

Sivadas, K. R. (1995) "Vibration Analysis of Pre-Stressed Rotating Thick Circular Conical Shell," *Journal of Sound and Vibration*, Vol. 180, pp. 99–109.

Slyper, H. A. (1962) "Coupled Bending Vibrations of Pretwisted Cantilever Beams," *Journal of Mechanical Engineering Science*, Vol. 4, pp 365–379.

Song, O. and Librescu, L. (1993) "Free Vibration of Anisotropic Composite Thin-Walled Beams of Closed Cross-Section Contour," *Journal of Sound and Vibration*, Vol. 167, No. 1, pp. 129–147.

Song, O. and Librescu, L. (1997) "Modelling and Vibration of Pretwisted Spinning Composite Thin-Walled Beams," Paper AIAA-97-1091, part 1, pp. 312-322, Kissimmee, Florida, April 7–10.

Song, O. and Librescu, L. (1998) "Anisotropy and Structural Coupling on Vibration and Instability of Spinning Thin-Walled Beams," *Journal of Sound and Vibration*, Vol. 204, No. 3, pp. 477–494.

Song, O., Jeong, N-H. and Librescu, L. (2000a) "Vibration and Stability of Pretwisted Spinning Thin-Walled Composite Beams Featuring Bending-Bending Electric Coupling," *Journal of Sound and Vibration*, Vol. 237, No. 3, Oct. 26, pp. 513–533.

Song, O., Librescu, L. and Jeong, N. H. (2000b) "Vibration and Stability Control of Spinning Flexible Shaft via Integration of Smart Material Technology," *Adaptive Structures and Material Systems*, ASME 2000 AD-Vol. 60, J. Redmond and J. Main (Eds.), ASME, New York, pp. 443–451.

Song, O., Jeong, N-H. and Librescu, L. (2001a) "Implications of Conservative and Gyroscopic Forces on Vibration and Stability of Elastically Tailored Rotating Shaft Modeled as Composite Thin-Walled Beams," *Journal of the Acoustical Society of America*, Vol. 109, No. 3, pp. 972–981.

Song, O., Kim, J-B. and Librescu, L. (2001b) "Synergistic Implications of Tailoring and Adaptive Materials Technology on Vibration Control of Anisotropic Thin-Walled Beams," *International Journal of Engineering Science*, Vol. 39, No. 1, December, pp. 71–94.

Song, O., Kwon, H-D. and Librescu, L. (2001c) "Modeling, Vibration and Stability of Elastically Tailored Composite Thin-Walled Beams Carrying a Spinning Tip Rotor," *Journal of the Acoustical Society of America*, Vol. 110, Issue 2, August, pp. 877–886.

Song, O., Librescu, L. and Kwon, H-D. (2002a) "Vibration and Stability Control of Robotic Manipulator Systems Consisting of a Thin-Walled Beam and a Spinning Tip Rotor," *Journal of Robotic Systems*, Vol. 19, No. 10, pp. 469–489.

Song, O., Librescu, L. and Jeong, N-H. (2002b) "Vibration and Stability Control of Smart Composite Rotating Shaft Via Structural Tailoring and Piezoelectric Strain Actuators," *Journal of Sound and Vibration*, Vol. 257, No. 3, pp. 503–525.

Sturia, F. A. and Argento, A. (1996) "Free and Forced Vibrations of a Spinning Viscoelastic Beams," *Journal of Vibration and Acoustics*, Trans. ASME, Vol. 118, July, pp. 463–468.

Tekinalp, O. and Ulsoy, A. G. (1989) "Modeling and Finite Element Analysis of Drill Bit Vibration," *Journal of Vibration, Acoustics, Stress and Reliability in Design*, Trans. ASME, Vol. 111, pp. 148–155.

Tylikowski, A. (1996) "Dynamic Stability of Rotating Composite Shafts," *Mechanics Research Communications*, Vol. 23, No. 2, pp. 175–180.

Yamanaka, K., Heppler, G. R. and Huseyin, K. (1995) "The Stability of a Flexible Link with a Tip Rotor and a Compressive Tip Load," *IEEE Trans. Rob. Autom.*, Vol. 11, No. 6, pp. 882–886.

Yamanaka, K., Heppler, G. R. and Huseyin, K. (1996) "Stability of Gyroelastic Beams," *AIAA Journal*, Vol. 34, No. 6, pp. 1270–1278.

Zohar, A. and Aboudi, J. (1973) "The Free Vibrations of a Thin Circular Finite Rotating Cylinder," *International Journal of Mechanical Sciences*, Vol. 15, pp. 269–278.

Zu, J. W. and Han, R. P. S. (1992) "Natural Frequencies and Normal Modes of a Spinning Timoshenko Beam with General Boundary Conditions," *Journal of Applied Mechanics*, Trans. ASME, Vol. 59, June, pp. 197–204.

Zu, J. W. and Melianson, J. (1998) "Natural Frequencies and Normal Modes for Damped Spinning Timoshenko Beam with General Boundary Conditions," *Journal of Applied Mechanics*, Trans. ASME, Vol. 65, September, pp. 770–772.

Chapter 11

THERMALLY INDUCED VIBRATION AND CONTROL OF SPACECRAFT BOOMS

11.1 INTRODUCTION

Space exploration has spurred research on vibration and stability of long spacecraft booms subjected to solar radiation.

Rapid temperature changes at day-night transitions in orbit, generate time dependent bending moments and transverse shear forces that induce oscillations of structural members. These vibrations may jeopardize system operations, and in some conditions, self-excited vibrations of increasing amplitudes, i.e. *thermal flutter*, may lead to the catastrophic failure of the structural system.

In spite of the tremendous research activity devoted to this topic, most studies were accomplished using simplifying assumptions: see Murozono and Sumi (1989, 1991), Sumi et al. (1988, 1990) and Gulick (1994), where the analysis was carried out in the context of a metallic and non-shearable thin-walled structural model with closed cross-sections. The works by Beam (1969), Donohue and Frisch (1969), were accomplished in a standard structural context for open cross-section beams, while in Oguamanam et al. (2004) the model of a cylindrical panel incorporating the geometrical nonlinearities was considered. The survey-paper by Thornton and Foster (1991) and the monograph by Thornton (1996) provide extensive references and discussions on the status of the problem.

Since composite materials are likely to play a great role in the design of space structures, we model the spacecraft boom as a circular thin-walled composite beam of closed cross-section, and include the effects of anisotropy of constituent materials, transverse shear, secondary warping and rotatory inertia. To induce elastic couplings between flapwise-bending and chordwise-bending, we use a special ply-angle distribution achieved via the usual helical winding fiber-reinforced technology.

475

The governing equations and the time-dependent boundary conditions of a boom heated by unidirectional solar radiation at an arbitrary angle of incidence with respect to its undeformed axis, are obtained from generalized Hamilton's principle.

Results display deflection time-history as a function of the material ply-angle, damping factor, and angle of incidence of heat radiation. We discuss thermal flutter instability, and instability feedback control of these structural systems.

With the exception of the work by Song et al. (2001, 2002, 2003a,b), and Chelu and Librescu (2003), there is no other comparable wide-ranging treatment in the specialized literature.

11.2 NON-STATIONARY THERMAL LOADING

A distinguishing feature of thermal radiative transfer is that no medium need be present for radiant interchange to occur; unlike convection and conduction radiant energy passes perfectly through a vacuum.

Consider a thin-walled beam of circular cross-section, radius R, and wall thickness h with a concentrated mass m_{tip} at its free end (see Fig. 11.2-1). The direction of incidence of heat radiation is on the positive y-axis side.

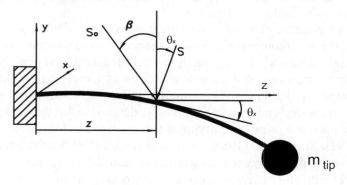

Figure 11.2-1: Boom model and coordinate systems

The boom is subjected to a known incident heat flux S_0 applied at time $t = 0$. The problem is to determine the transient temperature response of the boom. The following assumptions are adopted:

(a) Heat is conducted uniformly only in the circumferential direction, implying that the heat conduction along the beam span is negligible, (b) the thermal conductivity is large, so that the temperature field is assumed to be uniform across the beam thickness, (c) convection heat transfer inside and outside the boom, as well as the radiation emitted and reflected from stars and planets to the object are neglected, (d) the boom radiates as a black body and the heat input at any boom cross section is assumed to be proportional to the cosine of the angle between the direction of heat flux and that defined by the rotation of the normal to the boom mid-line about the x-axis (Lambert's cosine law). By virtue of assumption d), the heat input is

$$S = S_0 \cos(\beta + \theta_x). \qquad (11.2\text{-}1)$$

In Eq. (11.2-1) S_0 denotes the incident heat flux intensity of solar radiation at an angle β with respect to the direction normal to the undeflected boom axis, while S denotes the net radiation heat flux at an arbitrary point of the deflected boom surface along its span.

The dependence of the incident heat flux on $\theta_x(z, t)$, results in the coupling of the thermal and structural problems, and in the dependence of the temperature on the boom spanwise coordinate.

Under these assumptions, the absolute temperature at an arbitrary point of the boom wall features the dependence $T = T(z, \phi, t)$, where z and ϕ are the spanwise and the circumferential coordinates, respectively, see Fig. 11.2-2.

Taking into account both the thermal radiation heating given by Eq. (11.2-1), and of the heat loss by reradiation to space, and using Stefan-Boltzmann law of thermal reradiation, we find the heat balance equation

$$\frac{\partial T}{\partial t} - \frac{k}{\rho c R^2} \frac{\partial^2 T}{\partial \phi^2} + \frac{\sigma \epsilon}{\rho c h} T^4 = \frac{\alpha S_o}{\rho c h} \delta \cos \phi \cos(\beta + \theta_x). \qquad (11.2\text{-}2)$$

Here, k is the circumferential thermal conductivity, ρ and c are the mass density and the specific heat of the material, respectively, t is time, σ is the Stephan-Boltzmann constant, ϵ and α are the surface emissivity and absorptivity, respectively, and $\delta (\equiv \delta(\phi))$ is a tracing function (more exactly a *generalized function*), taking the values 1 or 0 depending on whether the circumferential coordinate ϕ corresponds to the portion of boom surface exposed to radiation, $(-\pi/2 < \phi < \pi/2)$, and to the remaining surface, (i.e. for $\pi/2 < \phi < 3\pi/2$), respectively. Due to the complexity of Eq. (11.2-2), an approximate solution will be obtained.

Since the maximum temperature difference between any two points on the boom surface for any sun orientation is small compared to the steady state mean temperature, the thermodynamic equation of heat-conduction- radiation can be linearized: we write T as

$$T(z, \phi, t) = \bar{T}(z, t) + T_1(z, \phi, t), \qquad (11.2\text{-}3)$$

where $\bar{T}$ is the average absolute temperature assumed to exhibit time dependence in the transient period when it increases from its original value to the steady-state one, while T_1 is the disturbance temperature, where $|T_1| \ll \bar{T}$.

Due to the symmetry of the cross-sectional shape and of the radiational heating profile, it is appropriate to assume that the disturbance temperature around the circumference in a section should be distributed as shown in Fig.

11.2-2, that is temperature rises slightly at the heated surface and drops at the unheated surface.

As a result, the disturbance temperature is expressible as

$$T_1(z, \phi, t) = \hat{T}(z, t) \cos \phi, \tag{11.2-4}$$

where $\hat{T}$ is the maximum disturbance temperature.

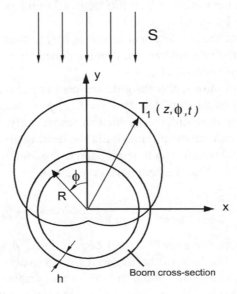

Figure 11.2-2: Distribution of the disturbance temperature in a cross-section of the boom

Expressing $T^4(z, \phi, t)$ as a truncated binomial series expansion about $\bar{T}$ in the form

$$T^4(z, \phi, t) \dot{=} \bar{T}^4(z, t) + 4\bar{T}^3 \hat{T}(z, t) \cos \phi, \tag{11.2-5}$$

and approximating the radiation input function appearing in the right-hand side of Eq. (11.2-2) by using the representation provided by Petrof and Raynor (1968):

$$S_0 \delta \cos \phi \approx S_0 \left(\frac{1}{\pi} + \frac{1}{2} \cos \phi \right), \tag{11.2-6}$$

one obtains from Eq. (11.2-2) two independent equations for the average and disturbance temperatures:

$$\frac{\partial \bar{T}}{\partial t} + \frac{\sigma \epsilon}{\rho c h} \bar{T}^4 = \frac{1}{\pi} \frac{\alpha S_0}{\rho c h} \cos(\beta + \theta_x), \tag{11.2-7}$$

$$\frac{\partial \hat{T}}{\partial t} + \frac{1}{\tau}\hat{T} = \frac{T^*}{\tau}\cos(\beta + \theta_x). \tag{11.2-8}$$

In Eq. (11.2-8), the parameters

$$\tau = \left(\frac{k}{\rho c R^2} + \frac{4\sigma\epsilon\bar{T}^3}{\rho c h}\right)^{-1} = \left[\frac{k}{\rho c R^2} + \frac{4\sigma\epsilon}{\rho c h}\left(\frac{\alpha S_0}{\pi\sigma\varepsilon}\cos\beta\right)^{3/4}\right]^{-1},$$

and

$$T^* = \frac{\alpha S_0}{2\rho c h}\tau, \tag{11.2-9a,b}$$

can be interpreted, respectively, as a measure of the thermal response time of the boom, and the time-independent maximum disturbance temperature when the boom, featuring an orientation perpendicular to the solar flux, is deflected statically. In steady-state conditions, by neglecting the elastic rotation θ_x the *average temperature* determined from (11.2-7) is given by

$$\bar{T} \equiv \bar{T}_{ss} = \left(\frac{1}{\pi}\frac{\alpha S_0 \cos\beta}{\sigma\epsilon}\right)^{1/4}. \tag{11.2-10}$$

For nonzero initial conditions, (11.2-8) gives

$$\hat{T}(z,t) = \hat{T}(z,0)e^{-t/\tau} + \frac{e^{-t/\tau}T^*}{\tau}\int_0^t e^{p/\tau}\cos(\beta + \theta_x)dp, \tag{11.2-11}$$

where p is a dummy time variable. Note that the amplitude of the disturbance temperature as expressed by Eq. (11.2-11) depends nonlinearly on θ_x. Assuming θ_x to be small, one can linearize $\hat{T}(z,t)$, and in conjunction with (11.2-11) one finds that the disturbance temperature depends linearly on the elastic rotation:

$$\hat{T}(z,t) = \hat{T}(z,0)e^{-t/\tau} + \frac{e^{-t/\tau}T^*}{\tau}\int_0^t e^{p/\tau}(\cos\beta - \theta_x\sin\beta)dp. \tag{11.2-12}$$

Equations (11.2-11) and (11.2-12) show that within the coupled thermal-structural analysis we are dealing with a system of equations requiring the simultaneous solution of elastic and thermal equations. Equation (11.2-11) provides the uncoupled counterpart of $\hat{T}$ for zero initial conditions:

$$\hat{T}(t) = T^*(1 - e^{-t/\tau})\cos\beta. \tag{11.2-13}$$

This expression is uniform in the spanwise direction; its steady-state value is $\hat{T} = T^*\cos\beta$. Based on (11.2-11) or (11.2-12) one can perform a coupled thermal-structural analysis, whereas based on (11.2-13) and on its steady-state counterpart, an uncoupled thermal-structural analysis can be carried out.

11.3 THERMALLY INDUCED VIBRATION

We investigate the bending vibration of circular booms under the temperature field expressed by Eq. (11.2-3), considered in conjunction with (11.2-10) and (11.2-11). We know from Chapter 4 that via the implementation of the ply-angle $\theta(y) = \theta(-y)$ and $\theta(x) = \theta(-x)$, i.e. in the context of the Circumferentially Uniform Stiffness Configuration (see Song and Librescu, 1993, and Chapter 4), there is the elastic coupling between the flapwise bending-chordwise bending, and transverse shear.

The terms in the pertinent equations of motion (see Eqs. 3.2-13), that bring a thermal contribution are

$$
\begin{aligned}
\delta u_0 : &\implies Q'_x (\equiv h'_x), \\
\delta v_0 : &\implies Q'_y (\equiv h'_y), \\
\delta \theta_y : &\implies M'_y - Q_x (\equiv m'_y - h_x), \\
\delta \theta_x : &\implies M'_x - Q_y (\equiv m'_x - h_y),
\end{aligned}
\tag{11.3-1a-d}
$$

while in the boundary conditions at $z = L$, (see Eqs. 3.2-14),

$$
\begin{aligned}
\delta u_0 : &\implies Q_x (\equiv h_x), \\
\delta v_0 : &\implies Q_y (\equiv h_y), \\
\delta \theta_y : &\implies M_y (\equiv m_y), \\
\delta \theta_x : &\implies M_x (\equiv m_x).
\end{aligned}
\tag{11.3-2a-d}
$$

In Eqs. (11.3-1) and (11.3-2) the terms in parentheses involve only the temperature, and appear when the 1-D stress-resultants and stress couples indicated outside the parentheses are expressed in terms of the generalized 1-D constitutive equations.

Since the contribution of the temperature to the N_{zn} shear stress resultant is immaterial, Eqs. (3.1-17), (11.3-1), (11.3-2) and (A.61), (A.64) and (A.74) give the time-dependent thermal loading terms

$$
\begin{aligned}
h_x &= -\pi R \mathcal{Q}_3 \hat{T}(z, t), \\
h_y &= 0, \quad m_y = 0, \\
m_x &= \pi R (R \mathcal{Q}_1 + \mathcal{Q}_2) \hat{T}(z, t).
\end{aligned}
\tag{11.3-3a-d}
$$

The coefficients $\mathcal{Q}_i$ are defined as

$$
\mathcal{Q}_1 = \sum_{k=1}^{N} \int_{n_{(k-1)}}^{n_{(k)}} \left(Q_A^{(k)} - \frac{A_{12}}{A_{11}} Q_B^{(k)} \right) dn,
$$

$$\mathcal{Q}_2 = \sum_{k=1}^{N} \int_{n_{(k-1)}}^{n_{(k)}} \left(Q_A^{(k)} n - \frac{B_{12}}{A_{12}} Q_B^{(k)} \right) dn,$$

$$\mathcal{Q}_3 = \sum_{k=1}^{N} \int_{n_{(k-1)}}^{n_{(k)}} \left(Q_C^{(k)} - \frac{A_{16}}{A_{11}} Q_B^{(k)} \right) dn, \qquad (11.3\text{-}4a\text{-}c)$$

where

$$Q_A = \bar{Q}_{12}\alpha_s + \bar{Q}_{22}\alpha_z + Q_{26}\alpha_{sz},$$
$$Q_B = \bar{Q}_{11}\alpha_s + \bar{Q}_{12}\alpha_z + \bar{Q}_{16}\alpha_{sz}, \qquad (11.3\text{-}5a\text{-}c)$$
$$Q_C = \bar{Q}_{16}\alpha_s + \bar{Q}_{26}\alpha_z + \bar{Q}_{66}\alpha_{sz}.$$

In Eqs. (11.3-5)

$$\alpha_s = \alpha_1 \cos^2\theta + \alpha_2 \sin^2\theta,$$
$$\alpha_z = \alpha_2 \cos^2\theta + \alpha_1 \sin^2\theta, \qquad (11.3\text{-}6a\text{-}c)$$
$$\alpha_{sz} = (\alpha_1 - \alpha_2) \sin 2\theta,$$

denote the reduced thermoelastic coefficients in an off-axis system, in terms of those corresponding to the on-axis coordinates, i.e. of the principal axes of orthotropy.

11.4 GOVERNING SYSTEM

11.4-1 Shearable Booms

The governing system of equations and boundary conditions are obtained from Hamilton's principle.

We include the effect of a tip mass simulating a scientific instrument or other pay load that is located at the free end of the boom; the kinetic energy of the tip mass should be added to the kinetic energy of the beam. Its contribution to the Hamilton's energy functional is

$$\int_{t_0}^{t_1} \delta K_{tip} dt = - \int_{t_0}^{t_1} m_{tip} \left[\ddot{u}_0 \delta u_0 + \ddot{v}_0 \delta v_0 \right]_{z=L} dt, \qquad (11.4\text{-}1)$$

where m_{tip} denotes the tip mass.

Following the usual procedure, and using Eqs. (11.3-1) through (11.3-3), we find the governing equations featuring the bending-bending coupling:

$$\delta u_0 : a_{43}\theta_x'' + a_{44}(u_0'' + \theta_y') - h_x' = b_1\ddot{u}_0,$$

$$\delta v_0 : a_{52}\theta_y'' + a_{55}(v_0'' + \theta_x') = b_1\ddot{v}_0,$$

$$\delta\theta_y : a_{22}\theta_y'' + a_{25}(v_0'' + \theta_x') - a_{44}(u_0' + \theta_y)$$

$$- a_{43}\theta_x' + h_x = (b_5 + \delta_n b_{15})\ddot{\theta}_y \quad, \tag{11.4-2a-d}$$

$$\delta\theta_x : a_{33}\theta_x'' + a_{34}(u_0'' + \theta_y') - a_{55}(v_0' + \theta_x)$$

$$- a_{52}\theta_y' - m_x' = (b_4 + \delta_n b_{14})\ddot{\theta}_x \quad.$$

In addition, the boundary conditions are

at $z = 0$:

$$u_0 = v_0 = \theta_y = \theta_x = 0, \tag{11.4-3a-d}$$

and at $z = L$:

$$\delta u_0 : a_{43}\theta_x' + a_{44}(u_0' + \theta_y) + \delta_m\, m_{tip}\,\ddot{u}_0 = h_x,$$

$$\delta v_0 : a_{52}\theta_y' + a_{55}(v_0' + \theta_x) + \delta_m\, m_{tip}\,\ddot{v}_0 = 0,$$

$$\delta\theta_y : a_{22}\theta_y' + a_{25}(v_0' + \theta_x) = 0,$$

$$\delta\theta_x : a_{33}\theta_x' + a_{34}(u_0' + \theta_y) = m_x. \tag{11.4-4a-d}$$

The tracer δ_m takes the values 1 or 0 depending on whether or not the tip mass is considered. Note that at $z = L$, the boundary conditions include nonhomogeneous, time-dependent terms.

Concerning the expressions of stiffness and inertia terms, a_{ij} and b_i, for the circular cross-section beam are given in Table 10.2-1.

11.4-2 Unshearable Booms

Following the usual procedure we find that the unshearable system decouples into two systems: transverse bending and lateral bending.

The former is given by

$$a_{33}v_0'''' + m_x'' - (b_4 + \delta_n b_{14})\ddot{v}_0'' + b_1\ddot{v}_0 = 0, \tag{11.4-5}$$

where the boundary conditions are

at $z = 0$:

$$v_0 = v_0' = 0, \tag{11.4-6a,b}$$

and at $z = L$:

$$a_{33}v_0'' + m_x = 0,$$

$$a_{33}v_0''' + m_x' - (b_4 + \delta_n b_{14})\ddot{v}_0' - \delta_m m_{tip}\ddot{v}_0 = 0. \qquad (11.4\text{-}7a,b)$$

Again the boundary conditions at $z = L$ are time-dependent.

In the absence of rotatory inertia terms, and for an isotropic material boom, Eqs. (11.4-5) through (11.4-7) are similar to those obtained by Murozono and Sumi (1989).

The equations for lateral bending are

$$a_{22}u_0'''' - (b_5 + \delta_n b_{15})\ddot{u}_0'' + b_1\ddot{u}_0 = 0, \qquad (11.4\text{-}8)$$

while the boundary conditions are:

at $z = 0$:

$$u_0 = u_0' = 0, \qquad (11.4\text{-}9a,b)$$

and at $z = L$:

$$u_0'' = 0,$$

$$a_{22}u_0''' - (b_5 + \delta_n b_{15})\ddot{u}_0' - \delta_m m_{tip}\ddot{u}_0 = 0. \qquad (11.4\text{-}10a,b)$$

These equations show that *in the context of the non-shearable beam model,* only the decoupled equation related to transverse bending should be considered in the analysis.

However, as Eqs. (11.4-2)–(11.4-4) reveal, the lateral motion becomes relevant only when it is coupled with the transversal one.

11.4-3 Alternative Form of Equations for Transverse Bending

Equations (11.4-5)–(11.4-7) can be brought to a more explicit form by using the dimensionless parameters

$$\eta \equiv z/L; \quad \bar{t} \equiv t/\tau \text{ (or } p/\tau); \quad V \equiv v_0/L. \qquad (11.4\text{-}11)$$

Considered in conjunction with (11.2-11), one obtains the governing equation

$$\frac{\partial^4 V}{\partial \eta^4} + a_1^2\frac{\partial^2 V}{\partial \bar{t}^2} - a_2^2\frac{\partial^4 V}{\partial \eta^2 \partial \bar{t}^2} + a_3 e^{-\bar{t}}\int_0^{\bar{t}} e^{\bar{t}}\frac{\partial^3 V}{\partial \eta^3}\sin\beta d\bar{t} = 0, \qquad (11.4\text{-}12)$$

and the boundary conditions:

$$V(0, \bar{t}) = 0; \quad \frac{\partial V(0, \bar{t})}{\partial \eta} = 0, \tag{11.4-13a,b}$$

$$\frac{\partial^2 V(1, \bar{t})}{\partial \eta^2} + a_3 e^{\mp} \int_0^{\bar{t}} e^{\bar{t}} \left[\cos \beta + \frac{\partial V(1, \bar{t})}{\partial \eta} \sin \beta \right] d\bar{t} = 0, \tag{11.4-14a,b}$$

$$\frac{\partial^3 V(1, \bar{t})}{\partial \eta^3} - a_2^2 \frac{\partial^3 V(1, \bar{t})}{\partial \eta \partial \bar{t}^2} + a_3 \, e^{-\bar{t}} \int_0^{\bar{t}} e^{\bar{t}} \sin \beta \frac{\partial^2 V(1, \bar{t})}{\partial \eta^2} d\bar{t}$$

$$- \delta_m \mu \, a_1^2 \frac{\partial^2 V(1, \bar{t})}{\partial \bar{t}^2} = 0.$$

Herein,

$$a_1 \equiv \sqrt{\frac{b_1}{a_{33}}} \frac{L^2}{\tau}; \quad a_2 \equiv \sqrt{\frac{b_4 + \delta_n b_{14}}{a_{33}}} \frac{L}{\tau},$$

$$a_3 \equiv \frac{L}{a_{33}} \pi R (R \mathcal{Q}_1 + \mathcal{Q}_2) T^*; \quad \mu \equiv \frac{m_{tip}}{b_1 L}. \tag{11.4-15a-d}$$

The parameter a_2 is related to rotatory inertia, a_3 to the amount of radiation heat input, while $1/a_1$, is the ratio of the characteristic thermal time to the characteristic mechanical time; it is equivalent to the nondimensional parameter B used by Boley (1956), and Boley and Barber (1957) to study the effects of inertia in beams and plates subjected to sudden heat inputs.

Equations (11.4-12) through (11.4-14) contain as a special case those obtained by Murozono and Sumi (1989).

11.5 SOLUTIONS FOR THERMALLY INDUCED VIBRATIONS

11.5-1 Preliminaries

Equations (11.4-2) through (11.4-4) present unavoidable difficulties: the boundary conditions at $z = L$ are non-homogeneous and time-dependent; the method developed by Mindlin and Goodman (1950) is suitable.

Murozono and Sumi (1989) applied Laplace Transform Technique to equations (11.4-12) through (11.4-14). We will apply an approximate method prompted by Boley (1956) and Boley and Barber (1957), and used in the present context by Thornton (1996); we call it Boley-Barber-Thornton approach (BBT).

Later, we present a more general solution methodology based on EGM to vibration and instability feedback control of spacecraft booms.

Suppose the boom is suddenly heated by radiation at the initial time. One can express the thermally induced dynamic response as a superposition of the quasi-static non-oscillatory response and a disturbance:

$$
\begin{aligned}
u_0(z, t) &= \hat{u}_0(z, t) + \bar{u}_0(z, t), \\
v_0(z, t) &= \hat{v}_0(z, t) + \bar{v}_0(z, t), \\
\theta_y(z, t) &= \hat{\theta}_y(z, t) + \bar{\theta}_y(z, t), \\
\theta_x(z, t) &= \hat{\theta}_x(z, t) + \bar{\theta}_x(z, t),
\end{aligned}
\tag{11.5-1a-d}
$$

where the overcarets and overbars identify the quasi-static and disturbance terms respectively; we examine each part separately.

11.5-2 Quasi-Static Solution

We assume very slow motion of the boom. Consistent with it, the inertia terms are neglected; the thermal loading terms expressed by Eqs. (11.3-3), are determined in conjunction with Eq. (11.2-13).

The quasi-static equilibrium shape of the boom is determined by

$$
\begin{aligned}
\delta \hat{u}_0 &: a_{43}\hat{\theta}_x'' + a_{44}(\hat{u}_0'' + \hat{\theta}_y') = 0, \\
\delta \hat{v}_0 &: a_{52}\hat{\theta}_y'' + a_{55}(\hat{v}_0'' + \hat{\theta}_x') = 0, \\
\delta \hat{\theta}_y &: a_{22}\hat{\theta}_y'' + a_{25}(\hat{v}_0'' + \hat{\theta}_x') - a_{44}(\hat{u}_0' + \hat{\theta}_y) - a_{43}\hat{\theta}_x' + h_x = 0, \\
\delta \hat{\theta}_x &: a_{33}\hat{\theta}_x'' + a_{34}(\hat{u}_0'' + \theta_y') - a_{55}(\hat{v}_0' + \hat{\theta}_x) - a_{52}\hat{\theta}_y' = 0,
\end{aligned}
\tag{11.5-2a-d}
$$

and of the boundary conditions

at $z = 0$:

$$
\hat{u}_0 = \hat{v}_0 = \hat{\theta}_y = \hat{\theta}_x = 0,
\tag{11.5-3a-d}
$$

and at $z = L$:

$$
\begin{aligned}
\delta \hat{u}_0 &: a_{43}\hat{\theta}_x' + a_{44}(\hat{u}_0' + \hat{\theta}_y) - h_x = 0, \\
\delta \hat{v}_0 &: a_{52}\hat{\theta}_y' + a_{55}(\hat{v}_0' + \hat{\theta}_x) = 0, \\
\delta \hat{\theta}_y &: a_{22}\hat{\theta}_y' + a_{25}(\hat{v}_0' + \hat{\theta}_x) = 0, \\
\delta \hat{\theta}_x &: a_{33}\hat{\theta}_x' + a_{34}(\hat{u}_0' + \hat{\theta}_y) - m_x = 0.
\end{aligned}
\tag{11.5-4a-d}
$$

Based on these equations, one obtains the quasi-static solution as

$$
\begin{aligned}
\hat{\theta}_x(z, t) &= \frac{1}{A}(a_{44}m_x - a_{34}h_x)z; \quad \hat{\theta}_y = 0, \\
\hat{u}_0(z, t) &= \frac{1}{a_{44}}\left(1 + \frac{a_{43}^2}{A}\right)h_x z - \frac{a_{43}}{A}m_x z, \\
\hat{v}_0(z, t) &= -\frac{1}{2A}(a_{44}m_x - a_{43}h_x)z^2,
\end{aligned}
\tag{11.5-5a-d}
$$

and consistent with (11.2-13), we have

$$m_x(t) = \pi R(R\mathcal{Q}_1 + \mathcal{Q}_2)T^*(1 - e^{-t/\tau})\cos\beta,$$
$$h_x(t) = -\pi R\mathcal{Q}_3 T^*(1 - e^{-t/\tau})\cos\beta. \qquad (11.5\text{-}6a,b)$$

Here $A \equiv a_{33}\, a_{44} - a_{34}^2$.

Note that, in this form, the quasi-static solution satisfies all the boundary conditions, including the nonhomogeneous ones at $z = L$, that is Eqs. (11.5-4).

A Special Case

For Eqs. (11.4-5)–(11.4-7) that correspond to thermally-induced bending vibration of unshearable booms, the quasi-static equation is

$$a_{33}\hat{v}_0'''' = 0, \qquad (11.5\text{-}7)$$

with the boundary conditions

at $z = 0$:

$$\hat{v}_0 = \hat{v}_0' = 0, \qquad (11.5\text{-}8a,b)$$

and at $z = L$:

$$a_{33}\hat{v}_0'' + m_x = 0 \quad \text{and} \quad a_{33}\hat{v}_3''' = 0. \qquad (11.5\text{-}9a,b)$$

The associated quasi-static solution is

$$\hat{v}_0(z, t) = -\frac{1}{2a_{33}}m_x z^2, \qquad (11.5\text{-}10)$$

where m_x is expressed by (11.5-6a); this coincides with that obtained by Thornton (1996).

11.5-3 Vibratory Solution

We use the representation:

$$[\overline{u}_0(z, t), \overline{v}_0(z, t), \overline{\theta}_y(z, t), \overline{\theta}_x(z, t)]$$
$$= \sum_{r=1}^{N}[U_r(z), V_r(z), Y_r(z), X_r(z)]q_r(t). \qquad (11.5\text{-}11)$$

Here $q_r(t)$ is the rth generalized modal coordinate, N is the mode number, while $U_r(z)$, $V_r(z)$, $X_r(z)$ and $Y_r(z)$ are trial functions. These are selected to be the eigenfunctions of the cantilevered beam in the absence of thermal effect. Treating the thermal terms as external loads, using the orthogonality of the eigenfunctions (see Librescu, 1975; Song et al., 1998; and Section 7.4), and including viscous damping one obtains the equation

$$\ddot{q}_r(t) + 2\zeta_r\omega_r\dot{q}_r(t) + \omega_r^2 q_r(t) = \frac{H_r(t)}{N_r} \qquad (r = 1, 2, .., N). \qquad (11.5\text{-}12)$$

Here ζ_r and ω_r denote the damping ratio and the undamped eigenfrequencies associated with the rth mode.

$$H_r(t) = -\int_0^L (h'_x U_r + m'_x X_r - h_x Y_r)dz, \tag{11.5-13a}$$

and

$$N_r = \int_0^L [b_1 U_r^2 + b_1 V_r^2 + (b_4 + \delta_n b_{14})X_r^2 \tag{11.5-13b}$$

$$+ (b_5 + \delta_n b_{15})Y_r^2]dz + \delta_m m_{tip}[U_r^2(L) + V_r^2(L)],$$

are the generalized loads and the norm, respectively.

$H_r(t)$ has to be determined in conjunction with (11.3-3) where $\hat{T}(z,t)$ is given by Eq. (11.2-12).

Special Case

For decoupled unshearable equation of transverse bending, the counterparts of (11.5-13) are

$$H_r(t) = \int_0^L m''_x V_r dz, \tag{11.5-14a}$$

and

$$N_r = \int_0^L b_1 V_r^2 dz + \delta_m m_{tip} V_r^2(L). \tag{11.5-14b}$$

From Eq. (11.5-12), the generalized coordinates $q_r(t)$ are determined as

$$q_r(t) = e^{-\zeta_r \omega_r t}[q_r(0)\cos\hat{\omega}_r t + \frac{\dot{q}_r(0) + \zeta_r \omega_r q_r(0)}{\hat{\omega}_r} \sin\hat{\omega}_r t]$$

$$+ \frac{1}{\hat{\omega}_r N_r}\int_0^t H_r(p)e^{-\zeta_r \omega_r (t-p)}\sin\hat{\omega}_r(t-p)dp, \tag{11.5-15}$$

where p is a dummy time variable, while $\hat{\omega}_r = (1 - \zeta_r^2)^{1/2}\omega_r$ denotes the rth frequency of damped free vibration.

Since the system is linear, the sum of the quasi-static and vibratory solutions according to (11.5-1), results in the solution of the thermally- induced vibrations.

11.6 NUMERICAL SIMULATIONS ON DYNAMIC RESPONSE

Determination of the thermally induced dynamic response of spacecraft boom systems is essential for their on-orbit retrieval and repair.

We consider a composite boom and a tip mass exposed to an incident heat flux applied instantaneously at $t = 0$; the data for the numerical simulations are supplied in Table 11.1. Note that the results on dynamic response can be used to determine the conditions for flutter instability; in the context of an aeroelastic system, this possibility was highlighted by Marzocca et al. (2002). Only the transverse deflection time-history will be displayed; the in-plane deflection exhibits a similar trend to the transverse one.

Figure 11.6-1 displays the time-history of transverse deflection of the boom tip. The boom carries a tip mass ($m_{tip} = 1kg$) built up of graphite-epoxy material layers in the sequence $[60°/ - 60°]$. Although the solar incidence angle is $\beta = 15$ deg., the oscillations about the quasi-static deflections decay rapidly due to the relatively high damping ratio $\zeta = 0.01$.

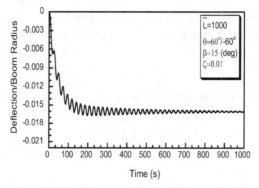

Figure 11.6-1: Transverse deflection time-history of the boom tip within the coupled thermal-structural analysis ($m_{tip} = 1$kg; $L = 10.92$m, $R = 1.092 \times 10^{-2}$m)

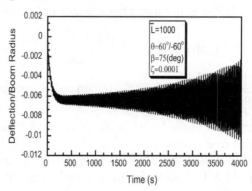

Figure 11.6-2: Unstable transverse deflection time-history of the boom tip in the coupled thermal-structural analysis ($m_{tip} = 1$kg; $L = 10.92$m, $R = 1.092 \times 10^{-2}m$)

For a relatively small damping ratio ($\zeta = 0.0001$) (see Fig. 11.6-2), even for a large solar incidence, $\beta = 75$ deg., the dynamic response is characterized by growing oscillations, leading to thermal flutter instability.

However, due to the large incident angle featured in the latter case, only a small component of the solar flux is absorbed by the boom, accounting for the relatively small magnitude of the total deflection response.

This is in sharp contrast to the former case that is characterized by a much smaller angle of incidence, and a much larger damping ratio.

This phenomenon is further illustrated in Fig. 11.6-3, which shows that for a relatively large damping parameter, $\zeta = 0.01$, the increase of the solar incidence leads to smaller total deflections of the stable boom-tip than for reduced solar incidence. In addition, as expected, for this level of damping, the increase of the incoming radiation angle is accompanied by a decrease of oscillation amplitudes.

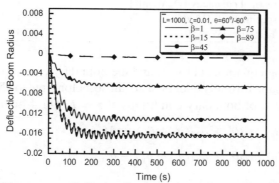

Figure 11.6-3: Stable transverse deflection time-history of the boom tip for selected incoming radiation angles β (deg.) as per the coupled thermal-structural analysis ($m_{tip} = 1\text{kg}$)

11.7 PIEZOELECTRIC VIBRATION CONTROL

11.7-1 Preliminaries

Spacecraft booms exposed to solar radiation exhibit thermally induced vibrations, and in some conditions can experience vibration amplitudes that can increase exponentially as time unfolds, this, yielding the thermal flutter instability. To eliminate this and reduce the vibration amplitude in the stable operational regime, an active feedback control should be implemented. The precise pointing operation of a flexible boom needs careful control, see Park (2002). We will describe a piezoelectric boundary moment control, as in Chapter 6, in which piezoactuators and sensors are spread over the entire beam span and activated out-of-phase (see Fig. 11.7-1).

Table 11.1: Boom properties for numerical simulations of dynamic response and instability

Data	Graphite-Epoxy
E_1 [N/m^2]	2.07×10^{11}
$E_2 = E_3$ [N/m^2]	5.17×10^9
G_{12} [N/m^2]	3.10×10^9
$G_{23} = G_{31}$ [N/m^2]	2.55×10^9
$\nu_{21} = \nu_{31}$	0.00625
ν_{32}	0.25
h (boom wall thickness) [m]	2.35×10^{-4}
ρ (boom mass density) [kg/m^3]	1.528×10^3
α (boom absorptivity)	0.92
$\alpha_1; \alpha_2$ (thermal expansion coefficients) [K^{-1}]	$1.1 \times 10^{-6}; 25.2 \times 10^{-6}$
σ (Stephen-Boltzmann constant) [W/m^2 K^4]	5.670×10^{-8}
k (boom thermal conductivity) [W/mK]	1.731
c (boom specific heat) [J/kgK]	10.44×10^2
S_0 (solar heat flux) [W/m^2]	1.350×10^3

The governing equations (11.4-2) and the boundary conditions (11.4-3) and (11.4-4a,b) remain formally identical to those belonging to the smart boom, with the exception of boundary conditions at $z = L$, (11.4-4c,d), that become

$$\delta\theta_y : a_{22}\theta'_y + a_{25}(v'_0 + \theta_x) = \tilde{M}_y,$$
$$\delta\theta_x : a_{33}\theta'_x + a_{34}(u'_0 + \theta_y) - m_x = \tilde{M}_x. \qquad (11.7\text{-}1a,b)$$

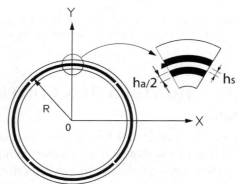

Figure 11.7-1: Configuration of the boom cross-section with embedded piezoactuators/sensors

Now, the stiffness quantities and inertia should include also the contributions from the piezoactuators/sensors. The control moment $\tilde{M}_x$ and $\tilde{M}_y$ are provided by Eqs. (6.6-19) or (6.6-20), depending on whether proportional or velocity feedback control is applied.

The control procedure is similar to that described in Chapters 6 and 7; as in Chapter 6, both proportional and velocity feedback controls will be applied.

A word of caution is in order. The piezoelectrically induced bending moment is based on the sensor signal. Due to the pyroelectric coupling, the temperature change produces an electric displacement, that in turn induces a piezoelectric voltage. In order to eliminate the inadvertent results generated by this effect, the voltage in the sensors caused by the pyroelectric effect may be estimated from the measured temperature, and removed from the sensor output voltage. This is achieved in the closed loop system, in the temperature compensator block, (see Inman et al., 1996).

The results supplied by Inman et al. (1996) and Friswell et al. (1997), reveal that in the absence of temperature compensation, more conservative results are obtained.

11.7-2 Open/Closed Loop Dynamic Response

Due to the inhomogeneous boundary conditions arising from the time-dependent thermal and piezoelectrically induced bending moments at the beam tip, the closed-loop boundary value problem is complex; we therefore use the Extended Galerkin Method (EGM).

The representation of the displacement quantities as given by (6.6-21) in the EGM gives an equation similar to (6.6-28). The quantities $\mathbf{q(t)}$, $\mathbf{M}$ and $\mathbf{K}$, as well as $\mathbf{P_p}$ and $\mathbf{P_v}$, are expressed by (6.6-24) and (6.6-26), respectively.

Now the load vector $\mathbf{Q(t)}$ is

$$\mathbf{Q}(t) = \left\{ \begin{array}{c} \displaystyle\int_0^L h_x(z,t)\mathbf{U}_0'dz \\[2mm] 0 \\[2mm] \displaystyle\int_0^L h_x(z,t)\mathbf{Y}_0'dz \\[2mm] \displaystyle\int_0^L m_x(z,t)\mathbf{X}_0'dz \end{array} \right\}. \tag{11.7-2}$$

For this case, h_x and m_x are determined from Eqs. (11.3-3a,d) in conjunction with (11.2-12).

11.8 RESULTS

11.8-1 Validation of the Solution Methodologies

In this chapter two methods have been applied: Boley-Barber-Thornton and Extended Galerkin Method.

We need to discuss to what extent they agree. We may ask various questions:

(i) Do predictions from the two solution methods agree? Since these methodologies are vastly different in principle, this issue is important,

(ii) Do the predictions provided by the EGM agree with those obtained by other methodologies?

To answer question (i) we consider the unactivated unshearable isotropic beam cantilever with boom properties provided in Table 1 by Johnston and Thornton (1998).

Figure 11.8-1 compares the transversal deflection time-history of the boom tip from *BBT* and *EGM*; they agree perfectly.

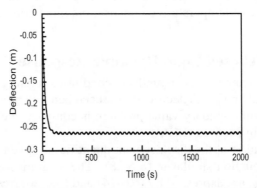

Figure 11.8-1: Transverse deflection time-history of the beam tip as predicted by BBT and EGM

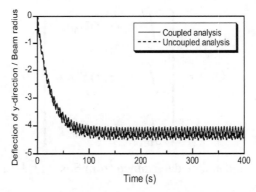

Figure 11.8-2: Dimensionless transverse deflection time-history of the beam tip obtained within the coupled and uncoupled thermal-structural analyses ($\beta = 80$ deg. and $\zeta = 0$)

For question (ii), Figures 11.8-2 and 11.8-3 compare the predictions obtained via the EGM with those in Figs. 7 and 8, respectively in Ko and Kim (2003), obtained via the finite element method (FEM); again there is perfect agreement.

In these graphs the boom material was considered to be isotropic; and the characteristics, following the data from Table 1 of Thornton and Kim (1993) are supplied in Table 2 of the paper by Ko and Kim (2003).

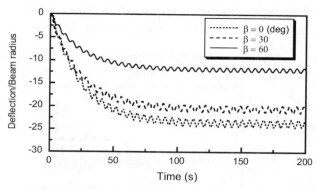

Figure 11.8-3: Dimensionless transverse deflection time-history of the beam tip for selected solar incident angles

11.8-2 Open/Closed Loop Response and Instability

The numerical simulations in Section 11.6 concern a boom manufactured of graphite-epoxy, with thermoelastic and geometric characteristics supplied in Table 11.1. The properties of piezoactuators/sensors are provided in Table 11.2. Figure 11.8-5 shows the open/closed loop dynamic response time-histories of transverse displacements of the boom tip; this plot shows that the uncontrolled boom features thermal flutter instability and the use of proportional feedback control stabilizes it.

Table 11.2: Properties of the piezoactuators/sensors

C_{11}^p [GPa]	(elastic coefficients)	138.99
C_{12}^p [GPa]		77.769
C_{13}^p [GPa]		74.294
C_{33}^p [GPa]		114.99
C_{44}^p [GPa]		25.598
h_p [mm]	(thickness of the piezoactuator/sensor)	0.046
ρ_p [kg/m^3]	(mass density)	7495
e_{31}^p [N/mV]	(piezoelectric constant)	5.201
α_1^p [K^{-1}]	(thermal expansion coefficient)	$1.2\ 10^{-6}$

Figure 11.8-5 shows the results for velocity feedback control. Although flutter is eliminated and the vibrations are contained, the former feedback control appears to be more efficient than velocity control.

However, stabilized, the steady amplitude response appears to be larger than in velocity feedback control. Figure 11.8-6 shows for the same parameters as in Figs. 11.8-4 and 11.8-5, the closed-loop dimensionless boom-tip deflection time histories for the thermally-compensated/non-compensated cases. These plots show that the thermally non-compensated case yields marginally more conservative deflection than the compensated one; as a result, at least for the present case, one can neglect thermal compensation

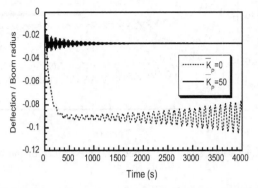

Figure 11.8-4: Open/closed loop dynamic response of the boom tip ($\theta = [60\,°/ - 60\,°]$, $\zeta = 0.0001$, $\beta = 89$ deg., $L = 5.91$m and $R = 1.092 \times 10^{-2}$m): dimensionless transverse displacement

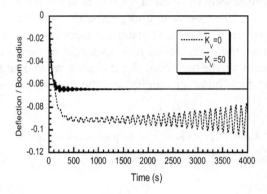

Figure 11.8-5: The counterpart of Fig. 11.8-4 for velocity feedback control

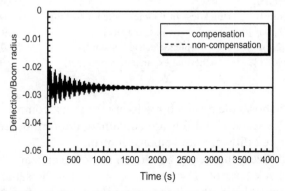

Figure 11.8-6: Closed-loop transverse deflection time-history. Thermal compensation/non-compensation cases ($\theta = [60°/ - 60°]$; $\zeta = 0.00001$, $\bar{K}_v = 0$, $\bar{K}_p = 50$)

11.9 CONCLUSIONS

In this chapter, we modeled spacecraft booms as a thin-walled composite beam of circular closed cross-sections, irradiated by a parallel flux of known intensity from a single distant source, at a given arbitrary angle with respect to the boom axis, and addressed related problems of thermally induced vibrations and instability.

The structural model includes damping ratio, lay-up of the composite structure, solar incidence-angle, heat flux level, boom relative length, tip mass, etc. As we showed, these factors affect vibration and instability.

In addition, it was put in evidence the beneficial implications of the feedback control on the dynamic response and the flutter instability. The excellent agreement with predictions obtained by other researchers was highlighted.

Note that the conditions for thermal flutter have been captured here from the dynamic response behavior, and not from an eigenvalue problem (see e.g. Murozono and Sumi, 1989, 1991). To get an eigenvalue problem, by using a brute method, the thermal term appearing in the right-hand side of the governing equation has to be ignored. This omission can yield inadvertent conclusions, e.g., that the boom is more prone to flutter at larger incident radiation angles than at smaller incident angles. This result stems from the inappropriate way of getting the eigenvalue problem. To capture in full thermal effects, the eigenvalue should be derived by starting with a geometrically non-linear boom model, and then carrying out the linearization about an equilibrium state. In various contexts this procedure was used by Xue and Mei (1993), Librescu et al. (1996a,b), Librescu and Lin (1997) and Stere and Librescu (2000).

We list a few references where the issue of the piezoelectric based control of structures exposed to temperature was addressed: Bruch et al. (1993) Chandrasekhara and Tenneti (1995), Sunar and Rao (1997), Lee and Saravanos (1997); Krommer and Irschik (1999), Oh et al. (2001), Tzou and Bao (1997), Chattopadhyay et al. (1999), Tauchert et al. (2000). Note the comprehensive survey papers by Saravanos and Heyliger (1999), Tauchert and Ashida (2003), and Altay and Dökmeci (2003).

REFERENCES

Altay, G. A. and Dökmeci, M. C. (2003) "Some Comments on the Higher Order Theories of Piezoelectric and Thermopiezoelectric Rods and Shells," *International Journal of Solids and Structures*, Vol. 40, pp. 4699–4706.

Beam, R. M. (1969) "On the Phenomenon of Thermoelastic Instability (Thermal Flutter) of Booms with Open Cross Section," NASA TN D-5222, June.

Boley, B. A. (1956) "Thermally Induced Vibrations of Beams," *Journal of Aeronautical Sciences*, Vol. 23, pp. 179–181.

Boley, B. A. and Barber A. D. (1957) "Dynamic Response of Beams and Plates to Rapid Heating," *Journal of Applied Mechanics*, Trans. ASME, September, pp. 413–416.

Bruch, J. C., Adali, S., Sadek, I. S. and Sloss, J. M. (1993) "Structural Control of Thermoelastic Beams for Vibration Suppression," *Journal of Thermal Stresses*, Vol. 16, pp. 249–263.

Chandrashekhara, K. and Tenneti, R. (1995) "Thermally Induced Vibration Suppression of Laminated Plates with Piezoelectric Sensors and Actuators," *Smart Materials and Structures*, Vol. 4, pp. 281–290.

Chattopadhyay, A., Li, I. and Gu, H. (1999) "Coupled Thermo-Piezoelectric-Mechanical Model for Smart Composite Laminates," *AIAA Journal*, Vol. 37, pp. 1633–1638.

Chelu, P. and Librescu, L. (2003) "Wavelet Technique Applied to Dynamic Response of Thin-Walled Composite Booms Exposed to Solar Radiation, Part I. Theoretical Formulation; Part II. Solution Methodology," in *Proceedings of the 5th International Congress on Thermal Stresses and Related Topics*, Blacksburg, VA, June 8–11, L. Librescu and P. Marzocca (Eds.), Vol. 2, TA-6-1-8.

Donohue, J. H. and Frish, H. P. (1969) " Thermoelastic Instability of Open Section Booms," *NASA-TN D-5310*, December.

Friswell, M. I., Inman, D. J. and Rietz, R. W. (1997) "Active Damping of Thermally Induced Vibrations," *Journal of Intelligent Material Systems and Structures*, Vol. 8, No. 8, pp. 678–685.

Gulick, D. W. (1994) "Thermally Induced Vibrations of an Axial Boom on a Spin-Stabilized Spacecraft," *AIAA-94-1556-CP*, AIAA Inc.

Inman, D. J., Rietz, R. W. and Wetherhold, R. C. (1996) "Control of Thermally Induced Vibrations Using Smart Structures," in *Dynamics and Control of Structures in Space*, C. K. Kirk and D. J. Inman (Eds.), Computational Mechanics Publications, Southampton, UK, pp. 3–16.

Johnston, J. D. and Thornton, E. A. (1998) "Thermally Induced Attitude Dynamics of Spacecraft with Flexible Appendage," *Journal of Guidance, Control and Dynamics*, Vol. 21, No. 4, pp. 581–587.

Ko, K-E. and Kim, J-H. (2003) "Thermally Induced Vibrations of Spinning Thin-Walled Composite Beams," *AIAA Journal*, Vol. 41, No. 2, pp. 296–303.

Krommer, M. and Irschik, H. (1999) "An Analogy Between Thermal and Piezoelectric Stresses Taking Into Account the Direct Piezoelectric and the Pyroelectric Effect," in *Thermal Stresses '99* (Proc. Third International Congress on Thermal Stresses), pp. 73–76.

Lee, H-J. and Saravanos, D. A. (1997) "Generalized Finite Element Formulation for Smart Multilayered Thermal Piezoelectric Composite Plates," *International Journal of Solids Structures*, Vol. 34, pp. 3355–3371.

Librescu, L. (1975) *Elastostatics and Kinetics of Anisotropic and Heterogeneous Shell-Type Structures*, Noordhoff International Publishers, Leyden, The Netherlands.

Librescu, L., Lin, W., Nemeth, M. P. and Starnes, Jr., J. H. (1996a) "Vibrational of Geometrically Imperfect Panels Subjected to Thermal and mechanical Loads," *Journal of Spacecraft and Rockets*, Vol. 33, No. 2, pp. 285–291.

Librescu, L., Lin, W., Nemeth, M. P. and Starnes, Jr., J. H. (1996b) "Frequency-Load Interaction of Geometrically Imperfect Curved Panels Subjected to Heating," *AIAA Journal*, Vol. 34, No. 1, pp. 166–177.

Librescu, L. and Lin, W. (1997) "Vibration of Thermomechanically Loaded Flat and Curved Panels Taking into Account Geometric Imperfections and Tangential Edge Restraints," *International Journal of Solids and Structures*, Vol. 34, No. 17, pp. 2161–2181.

Marzocca, P., Librescu, L. and Silva, W. A. (2002) "Unified Approach of Aeroelastic Response and Flutter of Swept Aircraft Wings in an Incompressible Flow," *AIAA Journal*, Vol. 40, No. 5, pp. 801–812.

Mindlin, R. D. and Goodman, L. E. (1950) "Beam Vibration with Time-Dependence Boundary Conditions," *Journal of Applied Mechanics*, Vol. 17, pp. 377–380.

Murozono, M. and Sumi, S. (1989) "Thermally Induced Bending Vibration of Thin-Walled Boom with Closed Section Caused by Radiant Heating," *Memoirs of the Faculty of Engineering*, Kyushu University, Vol. 49, No. 4, pp. 273–290.

Murozono, M. and Sumi, S. (1991) "Thermal Flutter of Thin-Walled Circular Section Beams Subjected to Radiant Heating," *Proceedings of the Fourth Conference of Asian-Pacific Congress on Strength Evaluation*, International Academic Publishers, pp. 676–681.

Oguamanam, D. C. D., Hansen, J. S. and Heppler, G. R. (2004) "Nonlinear Transient Response of Thermally Loaded Laminated Panels," *Journal of Applied Mechanics*, Trans. ASME, Vol. 71, January, pp. 49–56.

Oh, I-K., Han, J-H. and Lee, I. (2001) "Thermopiezoelectric Snapping of Piezolaminated Plates Using Layerwise Nonlinear Finite Elements," *AIAA Journal*, Vol. 39, pp. 1188–1197.

Park, S-Y. (2002) "Thermally Induced Attitude Disturbance Control for Spacecraft with a Flexible Boom," *Journal of Spacecraft and Rockets*, Vol. 39, No. 2, pp. 325–328.

Petrof, R. C. and Raynor, S. (1968) "The Temperature Distribution in Rotating Thick-Walled Cylinders Heated by Radiation," *International Journal of Heat and Mass Transfer*, Vol. 11, pp. 427–438.

Saravanos, D. A. and Heyliger, P. R. (1999) "Mechanics and Computational Models for Laminated Piezoelectric Beams," *Appl. Mech., Rev.*, Vol. 52, pp. 305–320.

Song, O. and Librescu, L. (1993) "Free Vibration of Anisotropic Composite Thin-Walled Beams of Closed Cross-Section Contour," *Journal of Sound and Vibration*, Vol. 167, No. 1, pp. 129–147.

Song, O., Ju, J. S. and Librescu, L. (1998) "Dynamic Response of Anisotropic Thin-Walled Beams to Blast and Harmonically Oscillating Loads," *International Journal of Impact Engineering*, Vol. 21, No. 8, pp. 663–682.

Song, O., Yoon, I. and Librescu, L. (2001) "New Contributions on Thermally Induced Bending Vibration and Thermal Flutter of Spacecraft Booms Subjected to Solar Heating," in *Proceedings of the Fourth International Congress on Thermal Stresses*, Osaka, Japan, June 8–11, pp. 497–500.

Song, O., Yoon, I. and Librescu, L. (2002) "Thermally Induced Bending Vibration of Composite Spacecraft Booms Subjected to Solar Heating," in *Proceedings of 43rd AIAA/ASME/ASCE/AHS/ASC Structures Structural Dynamics and Materials Conference, 10th AIAA/ASME/AHS Adaptive Structures Conference*, April 22–25, Denver, CO, Paper AIAA-2002-1250.

Song, O., Yoon, I. and Librescu, L. (2003a) "Thermally Induced Bending Vibration of Composite Spacecraft Booms Subjected to Solar Heating," *Journal of Thermal Stresses*, Vol. 41, pp. 2005–2022.

Song, O., Yoon, I. and Librescu, L. (2003b) "Thermally Induced Vibration Control of Composite Spacecraft Booms via Piezoelectric Strain Actuation," *Proceedings of the 5th International Congress on Thermal Stresses and Related Topics*, Blacksburg, VA, June 8–11, L. Librescu and P. Marzocca (Eds.), Vol. 1, MM-4-3-1-4.

Stere, A. and Librescu (2000) "Nonlinear Thermoaeroelastic Modeling of Advanced Aircraft Wings Made from Functionally Graded Materials," in *Proceedings of the 41st AIAA/ASME/ASCE/AHS/ACS Structures, Structural Dynamic Conference and Exhibition*, Paper AIAA#2000-1628, Atlanta, GA, April 3–6, pp. 72–82.

Sumi, S., Murozono, M., Imoto, T. (1988) "Thermally-Induced Bending Vibrations of Thin-Walled Boom with Tip Mass Caused by Radiant Heating," *Technology Reports of Kyushu University*, Vol. 61, No. 4, pp. 449–455.

Sumi, S., Murozono, M., Nakazato, S. and Tatsuyama, Y. (1990) "Thermally-Induced Bending Vibration of Thin-Walled Boom with Tip Mass," *Technology Reports of Kyushu University*, Vol. 63, No. 2, pp. 167–175.

Sunar, M. and Rao, S. S. (1997) "Thermopiezoelectric Control Design and Actuator Placement," *AIAA Journal*, Vol. 35, pp. 534–539.

Tauchert, T. R., Ashida, F., Noda, N., Adali, S. and Verijenko, V., (2000) "Developments in Thermopiezoelasticity with Relevance to Smart Composite Structures," *Composite Structures*, Vol. 48, pp. 31–38.

Tauchert, T. R. and Ashida, F. (2003) "Control of Transient Response in Intelligent Piezothermoelastic Structures," *Journal of Thermal Stresses*, Vol. 26, pp. 559–582.

Thornton, E. A. and Foster, R. S. (1991) "Dynamic Response of Rapid Heated Structures, in *Progress in Astronautics and Aeronautics*, S. N. Atluri (Ed.), AIAA, Washington, DC, pp. 451–476.

Thornton, E. A. and Kim, Y. A. (1993) "Thermally Induced Bending Vibrations of a Flexible Rolled Up Solar Array," *Journal of Spacecraft and Rockets*, Vol. 30, No. 4, pp. 438–448.

Thornton, E. A. (1996) "Thermal Structures for Aerospace Applications," *AIAA Education Series*, American Institute of Aeronautics and Astronautics, Inc., Reston, VA.

Tzou, H. S. and Bao, Y. (1997a) "Nonlinear Piezothermoelasticity and Multi-Field Actuations, Part I: Nonlinear Anisotropic Piezothermoelastic Shell Laminates, *Journal of Vibration and Acoustics*, Vol. 119, pp. 374–381.

Tzou, H. S. and Bao, Y. (1997b) "Nonlinear Piezothermoelasticity and Multi-Field Actuations, Part II: Control of Nonlinear Deflection, Buckling and Dynamics," *Journal of Vibration and Acoustics*, Vol. 119, pp. 382–389.

Xue, D. Y. and Mei, C. (1993) "Finite Element Nonlinear Panel Flutter with Arbitrary Temperatures in Supersonic Flow," *AIAA Journal*, Vol. 31, No. 1, pp. 154–162.

Chapter 12

AEROELASTICITY OF THIN-WALLED
AIRCRAFT WINGS

12.1 PRELIMINARIES

The use of laminated composite structural systems in aeronautical/aerospace vehicles was stimulated by the development of high modulus, high strength, low weight fiber-reinforced composite material systems, and their exotic characteristics such as directionality and ply-stacking sequence.

The ability to tailor their mechanical characteristics by orienting the fibers in preferred directions constitutes one of the overwhelming arguments for their use. The various elastic and structural couplings inherently featured by advanced structural composites can be exploited to enhance the response characteristics of flight vehicles. A most spectacular product of this technology was, among others, the possibility to eliminate, without weight penalties, the occurrence of the chronic aeroelastic divergence instability that has prevented for a long time the free use of the swept-forward wing aircraft.

This idea was prompted first time by Krone (1975) and followed, in the context of both static and dynamic aeroelasticity, by Sherrer et al. (1981), Hertz et al. (1982), Weisshaar (1978, 1979, 1980, 1981), Niblett (1980), Hollowell and Dugundji (1982), Lottati (1985, 1987), Oyibo and Berman (1985), Oyibo (1989), Lee and Miura (1987), Lee and Lee (1990), Lee et al. (1994), Librescu and Simovich (1988), Librescu and Khdeir (1988), Librescu and Song (1990, 1992), Librescu and Thangjitham (1991), Yamane and Friedmann, (1993), Karpouzian and Librescu (1994, 1995, 1996), Chattopadhyay et al. (1996) and Jha and Chattopadhyay (1997), Cesnik et al. (1996), Gern and Librescu (1998, 1999a,b, 2000 and 2001), Kim and Lee (2001), and most recently by Qin (2001), Qin et al. (2002, 2003), Qin and Librescu (2003 a,b) and Gasbarri et al. (2002). Finite element models have been developed by Kapania and Castel (1990) and Lee (1995).

The tailoring concept applied to composite lifting surfaces, in general, and to forward-swept wings, in particular, has been thoroughly discussed in survey papers by Weisshaar (1987) and Shirk et al. (1986). Comprehensive paradigms to optimize and actively control composite wings have been developed (see Livne et al., 1990, 1993).

Composites structures present inherent challenges: complexities arising from the anisotropic nature of fibrous composite materials, and the multitude of structural couplings.

The advantages can be reached, if, and only if the effects induced by these couplings are well understood; for an accurate prediction of flight vehicle response, adequate structural models must be developed. More powerful analytical tools for aeroelastic response and instability predictions are needed. Until now most studies have used a solid beam model.

In this chapter we use a refined structural model in the form of a thin-walled beam (see Librescu and Song, 1990, 1992; Librescu et al., 1996a,b; Qin, 2001; Qin et al., 2002, 2003; Qin and Librescu, 2003).

For aircraft wings, since that a most beneficial elastic coupling from the aeroelastic response and instability standpoints is the bending-twist one, proper ply-angle distributions that generate such a coupling should be implemented.

The governing equations as well as the boundary conditions for cantilevered beams featuring such an elastic coupling will be used in this chapter.

12.2 GOVERNING EQUATIONS

Consider a swept aircraft wings modeled as a single-cell thin-walled composite beam. Take the global coordinate system (x, y, z) with its origin at the wing root.

The coordinates x and z are referred to as the chordwise (positive rearward) and spanwise coordinates, respectively, while y is the coordinate normal to the plane (x, z), positive upward (see Figs. 12.2-1).

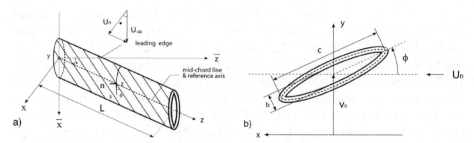

Figure 12.2-1: *(a)* Geometric configuration of the aircraft wing modeled as a thin-walled beam. *(b)* Geometry of the deformed configuration of wing normal cross-section

As before, we adopt the circumferentially asymmetric configuration (CAS). For cantilevered wings of non-uniform cross-section the related governing equations are

$$\delta v_0 : \ [a_{55}(v_0' + \theta_x)]' - \ (\underline{a_{56}\phi''})' \ + p_y(z, t) = b_1 \ddot{v}_0,$$

$$\delta \theta_x : \ (a_{33}\theta_x')' + (a_{37}\phi')' - a_{55}(v_0' + \theta_x) + \ \underline{a_{56}\phi''} \ = \underline{(b_4 + \delta_n b_{14})} \ \ddot{\theta}_x,$$

$$\delta \phi : \ - \ (\underline{a_{66}\phi''})'' \ + (a_{77}\phi')' + \ \underline{[a_{65}(v_0' + \theta_x)]''} \ + (a_{73}\theta_x')' + m_z(z, t)$$

$$= [(b_4 + b_5) + \delta_n(b_{14} + b_{15})]\ddot{\phi} - \underline{\underline{[(b_{10} + \delta_n b_{18})\ddot{\phi}']'}}.$$

$$(12.2\text{-}1a\text{-}c)$$

The associated homogeneous boundary conditions are as follows:

At $z = 0$:

$$v_0 = 0, \ \theta_x = 0, \ \phi = 0, \ \underline{\phi'} \ = 0, \qquad (12.2\text{-}2a\text{-}d)$$

and at $z = L$:

$$\delta v_0 : \ a_{55}(v_0' + \theta_x) - \ \underline{a_{56}\phi''} \ = 0,$$

$$\delta \theta_x : \ a_{33}\theta_x' + a_{37}\phi' = 0,$$

$$\delta \phi : \ - \ (\underline{a_{66}\phi''})' \ + a_{77}\phi' + \ \underline{[a_{65}(v_0' + \theta_x)]'}$$

$$+ a_{73}\theta_x' + \ \underline{\underline{(b_{10} + \delta_n b_{18})\ddot{\phi}'}} \ = 0,$$

$$(12.2\text{-}3a\text{-}d)$$

$$\delta \phi' : \ \underline{a_{65}(v_0' + \theta_x)} \ - \ \underline{a_{66}\phi''} \ = 0.$$

In these equations, $a_{37}(\equiv a_{73})$ and $a_{56}(\equiv a_{65})$ are stiffness quantities associated with the bending-twist and the twist- transverse shear coupling, respectively. In addition, $p_y(z, t)$ and $m_z(z, t)$ denote the aerodynamic lift (positive upwards) and twist moment about the reference axis (position nose-up), per unit span, respectively.

The unshearable counterpart of Eqs. (12.2-1)–(12.2-3) are provided next.

Governing equations:

$$\delta v_0 : \quad (a_{33}v_0'')'' - (a_{37}\phi')'' = p_y(z, t) - b_1\ddot{v}_0 + \underline{[(b_4 + \delta_n b_{14})\,\ddot{v}_0']'},$$

$$\delta\phi : \quad - \underline{(a_{66}\phi'')''} + (a_{77}\phi')' - (a_{73}v_0'')'$$

$$= -m_z(z, t) + [(b_4 + b_5) + \delta_n(b_{14} + b_{15})]\ddot{\phi} - \underline{\underline{[(b_{10} + \delta_n b_{18})\ddot{\phi}']'}}.$$

$$(12.2\text{-}4a,b)$$

The boundary conditions at $z = 0$ are:

$$v_0 = v_0' = \phi = \underline{\phi'} = 0, \tag{12.2-5a-d}$$

while the ones at $z = L$:

$$\delta v_0 : \quad (a_{33}v_0'')' - (a_{37}\phi')' = \underline{(b_4 + \delta_n b_{14})\ddot{v}_0'} \ ,$$

$$\delta v_0' : \quad a_{37}\phi' - a_{33}v_0'' = 0,$$

$$\delta\phi : \quad - \underline{(a_{66}\phi'')'} + a_{77}\phi' - a_{73}v_0'' = - \underline{\underline{(b_{10} + \delta_n b_{18})\ddot{\phi}'}},$$

$$(12.2\text{-}6a\text{-}d)$$

$$\delta\phi' : \quad \underline{\phi'' = 0} \ .$$

In these equations the terms underscored by a dashed line (- - - -); a curly (⌣) and two superposed dashed lines (====) identify the terms associated with the warping restraint, rotatory inertia and warping inertia, respectively.

The previously displayed governing equations are basic for the studies of static aeroelasticity (subcritical static response and divergence instability), and of the dynamic aeroelasticity (subcritical dynamic response and flutter instability). Moreover, depending upon the specific flight speed regime for which proper expressions of the aerodynamic load and moment, $p_y(z)$ and $m_z(z)$, respectively, have to be considered, these structural equations can be used in the study of the aeroelasticity of aircraft wings at any flight speed regime, i.e. incompressible or compressible. For the study of the divergence instability and of the static aeroelastic response, the inertia terms in Eqs. (12.2-1)–(12.2-3) or (12.2-4)–(12.2-6), have to be discarded. Several results on these issues will be supplied next.

12.3 STATIC RESPONSE AND DIVERGENCE INSTABILITY

12.3-1 Preliminaries

Consider a swept aircraft wing with spanwise uniformity. The angle of sweep, (considered positive for swept-back and negative for swept-forward wings), is measured in the $(x - z)$ plane of the wing from the direction normal to the airflow to the reference axis (that coincides with the z-axis) (see Fig. 12.2-1).

The wing is clamped normal to the reference axis, its effective length L being measured along this axis. All the parameters of the swept wing, such as the chord, locations of aerodynamic and centroidal centers are measured in the sections normal to the reference axis. For incompressible flight speed, strip-theory aerodynamics gives the lift force and the twist aerodynamic moment of the chordwise segment of unit width about the z- axis:

$$p_y(z) = q_n c a_0(\phi_0 + \phi - v_0' \tan \Lambda) - NW/2L,$$
$$m_z(z) = q_n c a_0 e(\phi_0 + \phi - v_0' \tan \Lambda) + q_n c^2 C_{MAC} - NWd/2L.$$
$$(12.3\text{-}1a,b)$$

In these equations

$$q_n = \frac{1}{2} \rho \, U_n^2 (= q \cos^2 \Lambda), \qquad (12.3\text{-}1c)$$

denotes the dynamic pressure normal to the leading edge of the swept wing; c the chord of the wing measured in the cross-section normal to the reference axis; Λ the angle of sweep; e the offset between the axis of aerodynamic centers and the reference axis; d denotes the distance between the line of the centers of mass and the reference axis; ϕ_0 the rigid angle of attack (measured in planes normal to the leading edge); C_{MAC} the wing section pitching moment coefficient; $W/2L$ the weight per unit wing span and N the load factor normal to the wing surface. Its expression is

$$N = \frac{2q_n c a_0}{W} \int_0^L (\phi_0 + \phi - v_0' \tan \Lambda) dz. \qquad (12.3\text{-}2)$$

In addition, the *corrected lift-curve slope* coefficient associated with the wing cross-sections normal to the spanwise coordinate, a_0, is expressed as

$$a_0 = \frac{2\pi \, AR}{AR + 4 \cos \Lambda}, \qquad (12.3\text{-}3)$$

where $AR (\equiv (2L)^2/S)$, is the wing aspect ratio and S its total area.

The static aeroelastic response is analyzed in the subcritical speed range, i.e., in the range defined by $q_n < (q_n)_D$, where $(q_n)_D$ denotes divergence dynamic pressure. Note that Eqs. (12.3-1) show that for $\Lambda < 0$, i.e., for swept-forward wings, the aeroelastic bending-twist coupling increases $p_y(z)$ and $m_x(z)$, which in turn reduces dramatically the divergence speed: for $\Lambda > 0$, the opposite trend takes place. These two phenomena are referred to as *wash-in* and *wash-out*, respectively. Whereas the goal of the subcritical aeroelastic analysis is the determination of the distribution of the effective angle of attack ϕ_{eff} and of the lift force as affected by the elastic deformations, the critical case involves the determination of divergence instability conditions. Clearly, the main target of tailoring applied to swept-forward wings is to yield a decrease of the effective angle of attack, and implicitly of the aeroelastic lift and thus, an increase of the critical divergence speed. The study of the subcritical static aeroelastic response requires the solution of an integral-differential system of equations obtained by inserting Eqs. (12.3-2) and (12.3-1) in (12.2-1) or (12.2-4); static instability requires the solution of an eigenvalue problem, where the divergence speed plays the role of eigenvalue.

For the static aeroelastic approach, the inertia terms in Eqs. (12.2-1), (12.2-3) and in their unshearable counterparts, Eqs. (12.2-4),(12.2-6), should be discarded, and the only loading terms to be retained are those associated with the aerodynamic lift and twist moment.

12.3-2 Special Cases

The equations show that for swept wing aircraft the equations of bending and twist are coupled. The source of the coupling is two fold, aerodynamical and structural; see Eqs. (12.3-1) and (12.2-1), (12.2-3). We can obtain closed form solutions only in special cases.

12.3-2a Pure Bending Divergence of Swept Wings

Consider an unshearable wing model. For large torsional stiffness (this yielding $\phi \to 0$), Eqs. (12.2-4) and (12.3-1) give their uncoupled bending counterpart:

$$(a_{33}v_0'')'' + q_n c a_0 v_0' \tan \Lambda = 0, \tag{12.3-4}$$

where v_0 must satisfy the boundary conditions

$$
\begin{array}{lll}
v_0 = 0, & v_0' = 0, & \text{at } z = 0, \\
a_{33}v_0'' = 0, & (a_{33}v_0'')' = 0, & \text{at } z = L.
\end{array} \tag{12.3-5a-d}
$$

Rayleigh's quotient in conjunction with the boundary conditions gives the divergence condition expressed in terms of the dynamic pressure as

$$(q_n)_D = -\frac{2 \int_0^L a_{33}(z)(v_0'')^2 dz}{\tan \Lambda \int_0^L a_0(z)c(z)(v_0^2)' dz}. \tag{12.3-6}$$

Ignoring the possibility that the lift curve slope a_0 is negative, which would correspond to a stalled wing, equation (12.3-6) shows that only swept- forward wings, $\Lambda \to -\Lambda$, can exhibit divergence instability in pure bending. Equation (12.3-6) extends results obtained by Librescu and Simovich (1988) to wings modeled as thin- walled beams. Equation (12.3-6), shows that in order to increase $(q_n)_D$ as much as possible, the bending stiffness a_{33} must be maximized. This can be achieved, without weight penalties (see Table 4.3-1) by designing the wing so that $K_{14} \to 0$. On the other hand, Eq. (12.3-1c), in conjunction with (12.3-6) show that q_D attains a minimum when $\Lambda = -45$ deg.

Remark 1

Replacing in (12.3-4) $v_0' \to V_3$ one obtains the equation $(a_{33} V_3')'' + \lambda V_3 = 0$, $(\lambda \equiv q_n c a_0 \tan \Lambda)$ that should be solved in conjunction with the boundary conditions

$$V_3 = 0 \text{ at } z = 0, \text{ and } V_3' = V_3'' = 0 \text{ at } z = L.$$

In spite of the fact that above equation is equivalent to (12.3-4), there is no assurance that a third order differential equation with real coefficients cannot have also complex eigenvalues. However, as it was shown, (see Flax, 1961), such imaginary eigenvalues cannot occur, and in fact this equation exhibits a denumerable infinite set of real eigenvalues.

Remark 2

One can readily be seen that $v_0(\eta) = A(6\eta^2 - 4\eta^3 + \eta^4)$ satisfies the boundary conditions (12.3-5) at $\eta = 0, 1$ where $\eta = z/L$. On its basis, one can get in conjunction with (12.3-6) a rather accurate solution of the divergence instability.

12.3-2b Divergence in Pure Torsion

If bending stiffness is infinite ($a_{33} \to \infty$), then Eqs. (12.2-4) yield $v_0 \to 0$. Applying the Rayleigh quotient procedure to the resulting decoupled twist equation, one obtains, in conjunction with the associated boundary conditions, the proper expression of the divergence speed

$$(q_n)_D \equiv q_D = \frac{\int_0^L [\, a_{66}(z)(\phi'')^2 + a_{77}(z)(\phi')^2] dz}{a_0 \int_0^L c(z)e(z)\phi^2 dz}. \qquad (12.3\text{-}7)$$

Its full determination has to be made in conjunction with the fulfillment of the boundary conditions:

$$\phi = 0, \quad \phi' = 0, \quad \text{at } z = 0, \qquad (12.3\text{-}8a,b)$$

and

$$\phi'' = 0, \quad a_{77}\phi' - (a_{66}\phi'')' = 0, \quad \text{at } z = L. \qquad (12.3\text{-}8c,d)$$

For wings modeled as solid beams (see Flax, 1961; Librescu and Simovich, 1988), Eq. (12.3-7) shows that pure torsion divergence can occur only for straight wings.

In addition, it shows that for straight wings the warping inhibition (identified by the term underscored by a dashed line) has a beneficial influence on the divergence instability.

A representation of $\phi(z)$ fulfilling all the boundary conditions can be obtained in a polynomial form. For the free warping case when the underscored terms in (12.3-7) and (12.3-8) vanish, we may take $\phi(z) = A \sin(\pi z / 2L)$.

12.3-3 Bending-Twist Divergence of Swept-Forward Wings

For unshearable beams, using the conditions corresponding to the decoupled divergence in pure bending and twist, Eqs. (12.3-6) and (12.3-7), respectively, we can derive an approximate expression for the coupled bending-twist divergence of *swept wings* (see Librescu et al., 1996b). This approximate expression is based on the extension of the ideas of Diederich and Budiansky (1948), further developed by Librescu and Simovich (1988): in its terms the divergence boundary can be approximated as a straight line that includes the two previously obtained decoupled expressions of the divergence instability as special cases.

We need several preliminary results. First convert, Eqs. (12.3-6) and (12.3-7) to the following convenient form:

$$(q_n)_D^{PB} = \frac{2(a_{33})_R \int_0^1 A_{33}(\tilde{v}_{0,\eta\eta})^2 d\eta}{(a_0)_R c_R L^3 \tan \Lambda \int_0^1 A_0 C(\tilde{v}_0^2)_{,\eta} d\eta} ; \quad (\Lambda \to -\Lambda) \qquad (12.3\text{-}9a)$$

$$(q_n)_D^{PT} = \frac{(a_{77})_R \int_0^1 \left[A_{66}\left(\frac{a_{66}}{a_{77}}\right)_R \frac{1}{L^2}(\phi_{,\eta\eta})^2 + A_{77}(\phi_{,\eta})^2 \right] d\eta}{(a_0)_R c_R e_R L^2 \int_0^1 A_0 C E \phi^2 d\eta} . \qquad (12.3\text{-}9b)$$

In these equations,

$$A_{ij}(z) \equiv a_{ij}(z)/(a_{ij})_R; \quad (A_0(z); \ C(z); \ E(z))$$
$$= (a_0(z)/(a_0)_R; \ c(z)/c_R; \ e(z)/e_R),$$
$$\tilde{v}_0 \qquad\qquad\qquad\qquad\qquad\qquad \equiv v_0/L, \qquad (12.3\text{-}9c\text{-}e)$$

where index R identifies the affiliation of a quantity to the wing root cross-section, $(\cdot)_{,\eta} \equiv d(\cdot)/d\eta$, while the superscripts PB and PT denote pure-bending and pure-twist, respectively.

Equations (12.3-9) suggest the incorporation of the following modified dynamic pressure coefficients in pure bending and pure torsion

$$(Q_n)^{PB} = \frac{q_n (a_0)_R \, c_R L^3 \tan \Lambda}{(a_{33})_R}, \tag{12.3-10a}$$

$$(Q_n)^{PT} = \frac{q_n (a_0)_R \, c_R e_R L^2}{(a_{77})_R}. \tag{12.3-10b}$$

Assimilating (12.3-10a,b) to the orthogonal Cartesian axes (x, y), one can define, in this coordinate system the line containing $(q_n)_D^{PB}$ and $(q_n)_D^{PT}$ as

$$(Q_n)^{PT} = (q_n)_D^{PT} + [(q_n)_D^{PT}/(q_n)_D^{PB}](Q_n)^{PB}. \tag{12.3-11}$$

This represents the approximate divergence boundary that corresponds to a swept wing. Equation (12.3-11) implies that for a wing rigid in bending, when $(Q_n)^{PB} = 0$, $(Q_n)^{PT} \to (q_n)_D^{PT}$; while for a torsionally rigid wing, when $(Q_n)^{PT} = 0$, $(Q_n)^{PB} \to -(q_n)_D^{PB}$. Replacing Eq. (12.3-10) in (12.3-11), representing for convenience $(q_n)_D^{PB}$ in Eq. (12.3-9a) as $(q_n)_D^{PB} = (\tilde{q}_n)_D^{PB}/\tan \Lambda$, and solving for $(q_n)_D$ one can derive the divergence speed of the swept wing in terms of the decoupled in bending and twist divergence speeds as

$$(q_n)_D = \frac{(a_{77})_R (q_n)_D^{PT}}{(a_0)_R c_R e_R L^2 \left[1 - \tan \Lambda \left(\dfrac{a_{77}}{a_{33}} \right)_R \dfrac{L}{e_R} \dfrac{(q_n)_D^{PT}}{(\tilde{q}_n)_D^{PB}} \right]}. \tag{12.3-12}$$

From Eq. (12.3-12) one can get two conclusions: i) for swept-forward wings, i.e. for the case $\Lambda \to -\Lambda$, a dramatic drop of the divergence speed is experienced, and ii) for swept-back wings one can find, by letting the denominator of (12.3-12) go zero, an optimal Λ_{cr} for which the condition of the *divergence-free* is obtained (see also Librescu and Simovich, 1988).

Full determination of the divergence speed has to be done in conjunction with the boundary conditions, (12.2-5) and (12.2-6) specialized for the static case. Equation (12.3-12) extends to thin-walled beams the divergence expression obtained by Librescu and Simovich (1988) for solid beams, as well as that by Diederich and Budiansky (1948).

12.4 NUMERICAL SIMULATIONS

12.4-1 Divergence Instability

The previous section has provided closed form solutions for special cases. Now, we consider the general case.

For numerical simulations the wing is modeled as a cantilevered single-layer thin-walled beam of biconvex cross-sections with dimensions, $L = 80$in.

(2.03m), $c = 10$in. (0.254m), $b = 2$in. (5.08cm), and $h = 0.4$in.(1.02cm). It is made of graphite-epoxy with elastic properties given in Section 6.4-1. For the numerical simulations related to the divergence instability and subcritical static aeroelastic response, both Laplace Transform Technique in the spatial domain and the Extended Galerkin Method have been used, and have provided an excellent agreement.

The effect of ply-angle considered in conjunction with that of the sweep angle on the divergence speed is illustrated in Fig. 12.4-1.

The figure shows plots of the normalized divergence speed $(q_n)_D/(q_n)_D^*$ versus the ply angle θ, where $(q_n)_D^* = 35.566$ psi. (245,261N/m^2) denotes the normalizing divergence speed corresponding to straight ($\Lambda = 0$) aircraft wings and zero ply-angle ($\theta = 0$). The results reveal that the range of ply-angles for which divergence instability is avoided, decreases with increasing forward sweep and increases with increasing sweep back angles. Note that, in the context of a solid beam wing model, Gern and Librescu (2000) found an identical trend to that in Fig. 12.4-1.

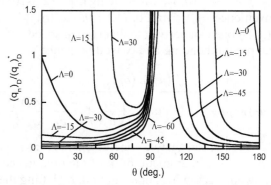

Figure 12.4-1: The normalized divergence speed versus the ply-angle for swept (Λ(deg.) $\lessgtr 0$) and straight ($\Lambda = 0$) wings

Figure 12.4-2a plots the divergence speed versus the ply-angle for two swept angles, $\Lambda = 0$ and $\Lambda = -60$ deg., and for the bending-twist-coupling stiffness $a_{37} = 0$ and $a_{37} \neq 0$: for $a_{37} = 0$ the divergence speed exhibits symmetry about $\theta = 90$ deg., whereas for $a_{37} \neq 0$ it is not symmetric. Moreover, for a large range of ply angles resulting in negative values of a_{37}, (see Fig. 12.4-2b), there are significant enhancements of the divergence instability behavior for swept wings, and especially for straight aircraft wings.

Although the bending-twist cross coupling is not the only factor influencing the aeroelastic static behavior, it is clear that its effect is dominant. By increasing its negative values as much as possible, we can achieve a reduction of the wash-in effect.

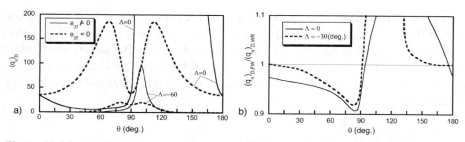

Figure 12.4-2: The divergence speed for unswept and swept composite wings. (a) With and without bending-twist coupling ($\Lambda = 0$, -60 deg.). (b) Effect of warping restraint ($\Lambda = 0$, -30 deg.)

Figure 12.4-2b presents a plot of the ratio $(q_n)_D^{FW}/(q_n)_D^{WR}$ as a function of the ply-angle θ for $\Lambda = 0$ and $\Lambda = -30$ deg., where $(q_n)_D^{FW}$ and $(q_n)_D^{WR}$ denote the divergence speeds corresponding to the free warping and the warping restraint wing models, respectively; in contrast to metallic wings (see Petre et al., 1961), the warping inhibition does not always yield an increase of the divergence speed. In other words, for composite aircraft wings, the free warping model does not yield always the most conservative results from the divergence instability point of view. Hence, for a rational aeroelastic design, the warping restraint effect must always be taken into consideration. Similar results for solid beam models have been reported by Librescu and Simovich (1988), Oyibo (1989), Librescu and Khdeir (1998), Librescu and Thangjitham (1991), and most recently by Hwu and Tsai (2002).

12.4-2 Subcritical Static Aeroelastic Response

The goal of the subcritical static aeroelastic response is the determination of the distribution of the effective angle of attack $\phi_{eff} (\equiv \phi_0 + \phi - v_0' \tan \Lambda)$, and, implicitly, of the aerodynamics lift as affected by the elastic deformations.

Determination of the response requires the solution of an integral-differential systems of equations obtained by replacing Eqs. (12.3-2) and (12.3-1) in Eq. (12.2-1) or (12.2-4).

To reduce the problem to a differential equation system, the load factor N is taken to be a constant which has to be evaluated iteratively; Extended Galerkin Method is used.

Figures 12.4-3a and 12.4-3b display the spanwise distribution of the normalized effective angle of attack, ϕ_{eff}/ϕ_0, for a swept and a straight wing with the ply angle θ and swept angle Λ acting as parameters, where the normalizing factor $\phi_0 = 5$ deg. is the rigid angle of attack. The plots provide a measure of the subcritical static aeroelastic response. The effective angle of attack constitutes a measure of the induced aeroelastic loads. For values of the ply-angle $\theta <$ 90 deg., the aeroelastic loads are amplified, whereas for $\theta > 90$ deg., for both forward and swept back wings, there is attenuation of the aeroelastic load. In all cases, the flight speed corresponds to a dynamic pressure $(q_n)_{Flight} = 3.446$ psi $(23,760 \ N/m^2)$. The basic conclusion is that tailoring can play a significant role not only in counteracting the detrimental wash-in effect, but also in containing the effect of elastic twist. This improvement of the subcritical static aeroelastic response of forward swept wing is basically due to the bending-twist coupling stiffness a_{37}.

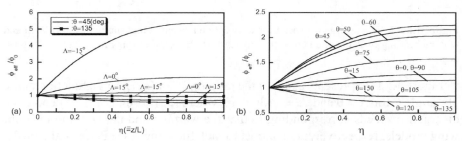

Figure 12.4-3: Spanwise distribution of the normalized effective angle of attack for two ply-angles θ (deg.). (a) Swept wing. (b) Straight wing

These results confirm that the coupling stiffness a_{37} plays a key role in controlling the subcritical static aeroelastic response and divergence instability. By increasing the negative value of a_{37} as much as possible for $\Lambda < 0$, the wash-in effect can be nullified and even converted into a wash-out effect.

Figure 12.4-4 displays the spanwise distribution of the subcritical aeroelastic response in terms of $\tilde{v}_0 (\equiv v_0/L)$, θ_x and ϕ for a straight wing with ply-angles θ (deg.) $= 45; 135$. As before, $(q_n)_{Flight} = 3.446$ psi. $(23,760 \ N/m^2)$ was prescribed. These plots again show that, for values of the ply-angle θ for which the cross-coupling stiffness a_{37} is negative, and in contrast to the case in which a_{37} is positive, smaller displacements are experienced.

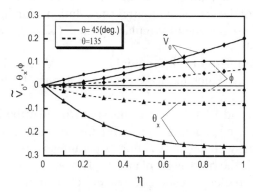

Figure 12.4-4: Subcritical normalized responses in bending ($\tilde{v}_0$), twist (ϕ), and rotation (θ_x) for a straight wing and for θ (deg.) $= 45$ and 135

The transverse shear flexibility generally has a detrimental effect on the subcritical aeroelastic response of swept-forward wings, in the sense that it exacerbates the wash-in effect. The results supporting this conclusion are not displayed here, but Karpouzian and Librescu (1994) found similar trends for a solid beam model. However, in contrast to other cases of ply angles, results for 55 deg. $< \theta <$ 90 deg. and for $\Lambda > 0$ reveal that the coupling stiffnesses, and especially a_{56}, plays a significant role also in this respect, and can render the effect of transverse shear flexibility either immaterial or slightly beneficial.

Table 12.4-1: Distribution along the wing span of ϕ_{eff}/ϕ_0 for $\Lambda = -60$ deg. for two selected values of the ply-angle and for shearable (w/TS) and unshearable (w/o TS) wing models

η	w/TS		w/o TS	
	$\theta = 75$ deg.	$\theta = 105$ deg.	$\theta = 75$ deg.	$\theta = 105$ deg.
0.0	1.0200	1.0100	1.0000	1.0000
0.1	1.2000	1.0200	1.1940	1.0160
0.2	1.4130	1.0290	1.4080	1.0260
0.3	1.5780	1.0380	1.5700	1.0350
0.4	1.7070	1.0430	1.7000	1.0400
0.5	1.8060	1.0470	1.8000	1.0450
0.6	1.8750	1.0514	1.8700	1.0490
0.7	1.9170	1.0534	1.9100	1.0510
0.8	1.9400	1.0533	1.9300	1.0520
0.9	1.9530	1.0545	1.9500	1.0530
1.0	1.9500	1.0524	1.9500	1.0530

An additional comment on the implication of transverse shear on the divergence instability and static response of swept aircraft wings is in order: Equations (12.3-1) imply that for swept-back wings ($\Lambda > 0$) the bending deformation tends to reduce the effective angle of attack, an opposite effect occurs for swept-forward wings. On the other hand, as is well-known, (see e.g. Librescu, 1975), transverse shear yields an increase of the bending deflection. This, in conjunction with (12.3-1), yields the conclusion that the wash-out and wash-in effects characterizing the swept-back and swept-forward wings, respectively, are further exacerbated by the inclusion of transverse shear. Tables 12.4-1 and 12.4-2 displaying ϕ_{eff}/ϕ_0 for Λ (deg.) $= -60$ and 30 underline these conclusions. Hwu and Tsai (2002) reported similar results for solid wings.

Table 12.4-2: Distribution along the wing span of ϕ_{eff}/ϕ_0 for $\Lambda = 30$ deg. for two selected values of the ply-angle and for shearable (w/TS) and unshearable (w/o TS) wing models

η	w/TS		w/o TS	
	$\theta = 75$ deg.	$\theta = 105$ deg.	$\theta = 75$ deg.	$\theta = 105$ deg.
0.0	0.9790	0.99300	1.0000	1.00000
0.1	1.1520	0.90000	1.1610	0.90200
0.2	1.3620	0.79500	1.3680	0.79600
0.3	1.5180	0.72000	1.5260	0.72200
0.4	1.6520	0.66400	1.6600	0.66500
0.5	1.7550	0.62500	1.7610	0.62500
0.6	1.8220	0.60200	1.8280	0.60200
0.7	1.8670	0.59000	1.8730	0.59100
0.8	1.8990	0.58500	1.9030	0.58500
0.9	1.9200	0.58400	1.9150	0.58400
1.0	1.9200	0.58600	1.9190	0.52500

12.5 DYNAMIC AEROELASTICITY OF AIRCRAFT WINGS

12.5-1 Preliminaries

In the previous sections we discussed the static aeroelastic instability and response of aircraft wings modeled as anisotropic thin-walled beams in an incompressible flow field.

The Extended Galerkin Method (EGM) that we used possesses enormous advantages over the Laplace Transformation Method (LTM), that is more intense than the EGM, but is an exact one. For an application of the LTM to the determination of divergence and flutter of aircraft wings modeled as solid beams, see Karpouzian and Librescu (1994, 1995, 1996).

Note that although approximate, within the EGM one can include the structural and aerodynamic variations along the wing span, enabling one to accomplish a 3-D analysis, while within the LTM, the structure and aerodynamics have to be considered uniform in the spanwise direction.

For spanwise uniform wings, the predictions obtained from EGM and LTM are in an excellent agreement. For aircraft wings modeled as solid beams, this perfect agreement was reported by Karpouzian and Librescu (1995). For these reasons, in discussing flutter instability and subcritical dynamic response we will use only the EGM.

At this point it should be recalled that the *modified strip analysis* applied by Yates (1958) to the flutter prediction of aircraft wings has many commonalities in its concept to the EGM, in the sense that EGM allows a modified strip analysis to be carried out.

Dynamic aeroelasticity problems are much more intricate than static ones: i) for the subcritical aeroelastic response the unsteady aerodynamic loads have to be expressed in the time-domain; flutter analysis, is carried out in the frequency domain and, ii) the unsteady aerodynamic loads that should be considered for various flight speed regimes and in the approach of the flutter and dynamic aeroelastic response are much more intricate than the ones used in the static aeroelasticity.

Since the unsteady aerodynamic equations and the dynamic aeroelastic instability and response require special developments, we will confine ourselves to a few results extracted from Qin (2001), Qin et al. (2002) and Qin and Librescu (2003), where the structural model of thin-walled beams was used. Additional references will also be provided.

12.5-2 General Considerations

We use the concept of aerodynamic indicial functions; these enable us to address both the aeroelastic response in the subcritical flight speed regime to arbitrary time-dependent external excitations, such as gusts, and explosive air blasts, and the flutter instability as well; for the former we work in the time domain, for the latter in the frequency domain.

When a feedback control capability is used, the time-domain representation of aerodynamic loads is needed for the determination of the closed-loop aeroelastic response. Moreover (see Marzocca et al., 2001, 2002), the time-domain approach enables one to capture the characteristics of the flutter boundary.

The indicial function method has a number of advantages: (1) an accurate description of aerodynamic characteristics, (2) the possibility of obtaining the unsteady airloads on an aircraft undergoing arbitrary small motions, (3) unified formulation of the aerodynamics in various flight speed regimes, and (4) the expressions of the indicial functions can be derived or approximated in various ways, via analytical, computational fluid dynamics, or experimental methods. Marzocca et al. (2003), showed that one can use indicial functions to carry-out nonlinear unsteady aerodynamic modeling of 2-D lifting surfaces.

For incompressible flows, the unsteady aerodynamic lift and twist moment are expressed in the time and frequency domains via the use of the Wagner and Theodorsen functions, respectively; these functions play the role of indicial aerodynamic functions. Since Wagner's and Theodorsen's functions are connected via a Laplace transform, one can conclude that in the incompressible flight speed regime, the aerodynamic lift and moment are obtainable in terms of a single indicial function, namely of the Wagner one (see e.g. Bisplinghoff et al., 1996; Fung, 1955).

By contrast in the compressible flight speed range encompassing the subsonic, transonic and the supersonic ranges, four specific indicial functions are needed for each of the various regimes.

For subsonic compressible, transonic (fully linearized) and supersonic flight speed regimes, appropriate analytical expressions for indicial functions were obtained by Bisplinghoff et al. (1996), Leishman (1988, 1993), Singh and Baeder (1997). Marzocca et al. (2002) carried out a study on the validation of 2-D indicial functions and showed that the supersonic indicial functions can accurately predict the aeroelastic behavior in the deep supersonic range. Note that in the compressible speed range, each indicial function is dependent on the flight Mach number.

The results that will be displayed next involve the aeroelastic response to gusts, and air blasts; the distinction between unsteady aerodynamic and gust loads, on one hand, and air blasts, on the other, was clearly established by Bisplinghoff et al. (1996) and this point of view was adopted here.

To evaluate the aeroelastic response to gusts of arbitrary shape in the incompressible flight we use the Küssner indicial function. In this context, sharp edged gusts, we have $W_F(\tau) = Y(\tau)V_G$ where V_G is a measure of the gust intensity, $Y(\tau)$ is the Heaviside function, while $\tau(\equiv 2U_\infty t/c)$ is the dimensionless time; we assume that V_G is uniformly distributed along the wing span. The associated lift and moment due to the penetration into a gust are obtained from Duhamel's integral; see Bisplinghoff et al. (1996), von Kármán and Sears (1938) and Dowell (1995). For compressible flows (subsonic and supersonic), proper indicial functions for determining the additional lift $L_g(\tau)$ and moment $M_g(\tau)$ due to the gust, per unit wing span, have to be used. These are provided in Bisplinghoff et al. (1996).

We assume that the time-dependent pressure pulses impacting the wing, reach their peak values so quickly that the wing structure can be considered to be loaded instantly and uniformly in both spanwise and chordwise directions. For this reason, the twist moment associated with these loads is immaterial; the global unsteady aerodynamic load and moment appearing in Eqs. (12.2-1) and their classical counterparts, Eqs. (12.2-4), can be cast most conveniently as:

$$
\begin{aligned}
p_y(z,t) &= L_a(z,t) + L_g(z,t) + L_b(t), \\
m_z(z,t) &= M_a(z,t) + M_g(z,t).
\end{aligned} \tag{12.5-1a,b}
$$

The expression of $L_b(t)$ for various blast pulses has already been recorded; the remaining expressions may be found in Qin et al. (2003) and Qin and Librescu (2003).

12.5-3 Selected Results

12.5-3a Validation of the Solution Methodology

Equations (12.2-4) and the associated boundary conditions (12.2-5) and (12.2-6) corresponding to a Bernoulli-Euler thin-walled beam are similar to those in classical aeroelasticity, where the concept of the solid beam model is used;

only the expressions for stiffness and inertia terms change. The exact results obtained by Goland and Luke (1948), form the benchmark for flutter predictions in the incompressible flight speed regime.

The results supplied in Table 12.5-1 reveal an excellent agreement. Moreover, the present predictions reveal better accuracy than a number of results scattered throughout the literature.

Table 12.5-1 shows that the EGM coupled with the Root Locus Method (referred by Rodden and Stahl (1969) to as the Transient Method), provides better flutter predictions than those based on the U-g method.

Table 12.5-1: Comparison of flutter results for the Goland's wing

Method	Flutter speed (km/hr)	Error[d] (%)	Flutter frequency (Hz)	Error[d] (%)
Exact (Goland & Luke, 1948)	494.1	—	11.25	—
EGM[a], N=9, *(Q. & L.) (V-g method)	493.2	-0.18	11.15	-0.89
EGM, N=7, *(Q. & L.) (V-g method)	493.1	-0.20	11.15	-0.89
EGM, N=9, *(Q. & L.) (Transient method)	494.5	0.08	11.04	-1.87
EGM, N=7 (Gern & Librescu, 1998)	493.6	-0.10	12.02	6.84
Steady-state solution (Patil el al., 2000)	488.3	-1.17	11.17	-0.71
COMBOF[b], 10 finite difference stations	484.2	-2.00	11.20	-0.44
COMBOF[b], 25 finite difference stations	493.1	-2.23	11.27	0.18
State vector: (Lehman (1982))	494.4	0.08	—	
LTM[c] (Karpouzian & Librescu, 1995)	495	0.18	11.20	-0.44

[a] EGM: Extended Galerkin's Method; [b] Housner & Stein (1974); [c] LTM: Laplace Transform Method; [d] Relative error $= \dfrac{[approx.] - [exact]}{[exact]}$; *Qin & Librescu (2003)

In the same context, the flutter characteristics obtained via a subcritical aeroelastic response analysis (see Marzocca et al., 2002), are $U_f = 490$ (km/h) and $\omega_F = 11$ Hz. This shows that the dynamic aeroelastic response can also provide accurate predictions of the flutter instability characteristics.

12.5-3b Results

We assume that the wing features a bi-convex cross-section, that the constituent material is graphite-epoxy, and the ply-angle configuration CAS, that induces bending-twist cross-coupling is applied; the wing geometric characteristics are supplied in Table 12.5-2.

Figure 12.5-1 shows the influence of the ply-angle on the dynamic aeroelastic response of the tip cross-section of a straight wing ($\Lambda = 0$), of $AR = 12$, flying at $M = 0.7$, and exposed to a sharp-edged gust of $V_G = 15$ m/s. For θ (deg.) = 120 and 135, as time unfolds, the oscillations damp out, while for $\theta = 105$ deg., the oscillations increase in time, corresponding to flutter. The main conclusion is that by tailoring, one can postpone the occurrence of the flutter instability at the expense of large dynamic response amplitudes.

Table 12.5-2: Geometric characteristics of the test wing

Parameters	Values
Chord ($2b^a$: m)	0.757
Hight ($2d^a$: m)	0.0997
Wall thickness (h : m)	0.0203
Layer thickness (: m)	0.0034
Constituent layers of the wall	6

^{a} the length is measured along the contour line.

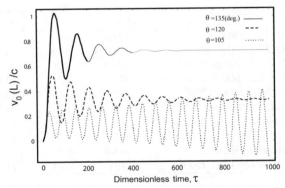

Figure 12.5-1: Effect of ply-angle on the dynamic response in plunging of the wing tip to a sharp-edged gust ($[\theta_6]$ lay-up, $AR = 12$, $\Lambda = 0$, $M_{Flight} = 0.7$)

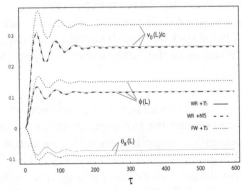

Figure 12.5-2: Effect of warping restraint and transverse shear on dynamic aeroelastic response in plunging, twist and rotation of the wing tip to a sharp-edged gust ($[75_6]$, $AR = 10$, $\Lambda = 30$ deg., $M_{Flight} = 0.7$)

For subsonic compressible flight, ($M = 0.7$), Fig. 12.5-2 shows the implications of the warping restraint (WR) and transverse shear (TS) on the dynamic
response of the tip cross- section of a swept wing ($\Lambda = 30$ deg.) of $AR = 10$.

The results reveal that even for large aspect ratio wings, the warping restraint
significantly improves the response, but transverse shear has little influence.
The results reported first by Librescu and Simovich (1988) for the aeroelastic
divergence of composite swept wings, and for flutter instability by Qin et al.
(2002), reveal that the warping restraint is not always beneficial: even for high
aspect ratio wings the free warping model does not always provide the most
critical case from the flutter instability point of view.

It is found that at special ply-angles the inherent increase in stiffness due
to the warping constraint, is outweighed by the decrease of the twist stiffness
induced at the respective stacking-sequence composite by the bending-twist
elastic coupling a_{37}. For a safe aeroelastic design, the warping constraint should
always be taken into consideration.

Figure 12.5-3 shows the influence of the warping restraint on the aeroelastic
response of the tip cross-section of an unswept wing of $AR = 6$, in supersonic
flight, $M = 1.5$, impacted by a sonic-boom pulse; the stacking sequence of the
wall is $[45_6]$.

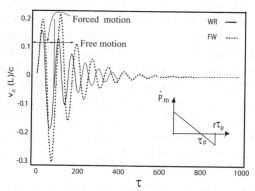

Figure 12.5-3: Effects of warping restraint on the dynamic response in plunging of the wing
tip in a supersonic flow ($M_{Flight} = 1.5$), and impacted by a sonic boom pulse, ($AR = 6$, $[45_6]$,
$\hat{P}_m (\equiv cP_m/b_1) = 10^{-3}$, $r = 2$, $\tau_p = 40$, $\Lambda = 0$)

The results show that the warping restraint model yields, specially in the
forced motion regime, response amplitudes lower than those predicted by the
free warping model. However, as Fig. 12.5-4 shows, the warping restraint
model can also have a response with larger amplitude than that corresponding
to the free warping model. This is due to the complex interaction between
warping restraint stiffening effect and the elastic coupling for the $[135_6]$ lay-up.
For the static aeroelastic response, a similar phenomenon was reported for a
solid-beam wing model by Librescu and Thangjitham (1991).

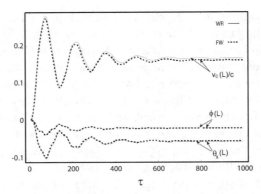

Figure 12.5-4: Warping restraint effect on the dynamic aeroelastic response of the wing to a sharp-edge gust ($M_{Flight} = 1.5$, $[135_6]$ lay-up, $AR = 6$, $\Lambda = 0$)

12.6 LIFTING SURFACE CONTROL USING SMART MATERIALS TECHNOLOGY

12.6-1 Preliminaries

Smart structures are likely to play an increasing role in the design of civil and military aircraft (see e.g. Kudva and Lockyear, 1999). Smart lifting surfaces will enhance aircraft performance, expand the flight envelope without weight penalties and without the danger of the occurrence of any aeroelastic instability, and improve ride quality (see also Lazarus et al., 1991; Heeg, 1991; Crawley, 1994; Nam and Kim, 1995; McGowan et al., 1996; Nam et al., 2001).

We present some results for the combined action of structural tailoring and of induced strain actuation provided by piezoelectric materials for the control of the static aeroelastic response and instability of swept/unswept aircraft wings in an incompressible flight speed (see Song et al., 1992; Weisshaar and Ehlers, 1992; Librescu et al., 1996a).

A swept forward wing aircraft has a very low divergence instability speed; any way to increase it without weight penalty would be important. We assume that the piezoactuators are distributed over the entire wing span and actuated out-of phase, while the host structure, of a bi-convex cross-section, is manufactured of graphite-epoxy material whose ply-angle fulfils the conditions $\theta(y) = -\theta(-y)$, for which the system of governing equations (12.2-1), with the boundary conditions (12.2-2) and (12.2-3) are applicable.

For the static problem, all the inertia terms in these equations should be discarded. For out-of-phase actuation, a boundary moment $\tilde{M}_x$ is piezoelectrically induced at the wing tip, $z = L$. The only equation that is affected, is the boundary condition (12.2-3b); it becomes

$$\delta\theta_x : \quad a_{33}\theta'_x + a_{37}\phi' = \tilde{M}_x. \tag{12.6-1}$$

Here $\tilde{M}_x$ is the piezoelectrically induced boundary moment given by Eq. (6.6-1a). We assume a simple control law requiring that the applied electric field $\mathcal{E}_3$ (on which $\tilde{M}_x$ depends), be proportional to the vertical deflection at the beam tip, $v_0(L)$:

$$\tilde{M}_x(L) = k_p v_0(L), \qquad (12.6\text{-}2)$$

where k_p is the feedback gain. Its dimensionless counterpart $K_p = k_p L^2 / [a_{33}|_{\theta=0}]$ will be used in the numerical simulations.

To these governing equations and the feedback control law (12.6-2), Eqs. (12.3-1) for the static aerodynamic loads should be added. Two different problems can be addressed: (i) one involving closed/open loop eigenvalue problem, enabling one to obtain the controlled/uncontrolled divergence speed, and (ii) related to the closed/open loop boundary value problem, yielding the controlled/uncontrolled static aeroelastic response.

12.6-2 Results

Figure 12.6-1 shows the variation of the divergence speed parameter $(q_n c a_0)_D$ as a function of the ply-angle, for selected values of the feedback gain K_p. Figures 12-6-1a and 12.6-1b, are related to a straight ($\Lambda = 0$) and swept-forward wing ($\Lambda = -30$ deg.), respectively. The combination of tailoring and induced strain actuation, leads to a significant increase of the divergence speed.

Figure 12.6-1 shows that in the region where the bending-twist cross coupling stiffness a_{37} attains a maximum value, i.e., around $\theta \cong 65$ deg., increase of the feedback gain greatly increases the divergence speed. Beyond that specific value of θ and irrespective of the feedback gain, the divergence speed follows closely the variation of the uncontrolled wing, where tailoring attains maximum efficiency. Figure 12.6-2a displays the normalized effective angle of attack ϕ_{eff}/ϕ_0 in conjunction with selected control gains; strong attenuations of aeroelastic loads are experienced for the controlled wing as compared to the uncontrolled one. Figure 12.6-2b reveals the effects of the ply and sweep angles.

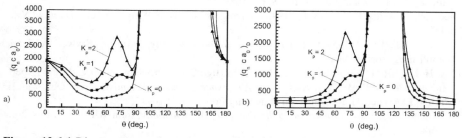

Figure 12.6-1 Divergence speed parameter against ply-angle for various feedback gains: (a) $\Lambda = 0$ and (b) $\Lambda = -30$ deg.

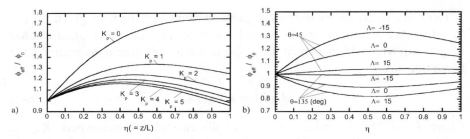

Figure 12.6-2 Spanwise distribution of the effective normalized angle of attack: (a) selected feedback gains ($\Lambda = -15$ deg. $\theta = 45$ deg., $q_{Flight} = 2$ psi, (13,792 N/m^2)), (b) selected ply-angles, θ (deg.), for three sweep angles ($K_p = 1$, $q_{Flight} = 2$ psi)

The results demonstrate once more that the application of both tailoring and piezoelectric control technology can yield dramatic improvements of the static aeroelasticity of aircraft wings.

12.6-3 Concluding Remarks

These results demonstrate the synergy achieved by the simultaneous application of tailoring and the piezoelectric control. This synergistic interaction was reported for flutter of an aircraft wing modeled as a solid beam by Gern and Librescu (1999b).

We hope that the advances in material science will be able to provide light-weight strain actuators characterized, by high actuation strains; this would make actuated aeroelastic control a truly successful technology.

REFERENCES

Bisplinghoff, R. L., Ashley, H. and Halfman, R. L. (1996) *Aeroelasticity*, Dover Publications Inc., New York.

Cesnik, C. E. S., Hodges, D. H. and Patil, M. J. (1996) "Aeroelastic Analysis of Composite Wings," *AIAA-96-1444-CP*.

Chattopadhyay, A., Zhang, S. and Jha, R. (1996) "Structural and Aeroelastic Analysis of Composite Wing Box Sections Using Higher-Order Laminate Theory," *AIAA-96-1567-CP*.

Crawley, E. F. (1994) "Intelligent Structures of Aerospace: A Technology Overview and Assessment," *AIAA Journal*, Vol. 31, No. 8, pp. 1689–1699.

Diederich, F. W. and Budiansky, C. (1948) "Divergence of Swept Wings," *NACA Technical Note*, 1680.

Dowell, E. H. (Ed.) (1995) *A Modern Course in Aeroelasticity*, 3rd Edn., Kluwer Academic Publishers, Boston, MA.

Flax, H. H. (1961) "Aeroelasticity and Flutter in High Speed Problems of Aircraft and Experimental Methods," Vol. VIII *High Speed Aerodynamic and Jet Propulsion*, eds. Donovan, H. F. and Lawrence, H. R., Princeton University Press, pp. 161–417.

Fung, T. C. (1955) *An Introduction to the Theory of Aeroelasticity*, John Wiley and Sons, New York.

Gasbarri, P., Mannini, A. and Barboni, R. (2002) "Dynamic Modeling of Swept Wing with PTFE Method," *Computers and Structures*, Vol. 80, pp. 1255–1260.

Gern, F. H. and Librescu, L. (1998) "Effects of Externally Mounted stores on Flutter Characteristics of Advanced Swept Cantilevered Aircraft Wings," *Aerospace Science and Technology*, Vol. 2, No. 5, pp. 321–333.

Gern, F. H. and Librescu, L. (1999a) "Aeroelastic Tailoring of Advanced Aircraft Wings Carrying External Stores," *Atti della Accademia delle Scienze di Torino, Classe di Scienze Fisiche, Mathematiche' e Naturali*, Quaderni 1, pp. 201–219 (Issue devoted to Placido Cicala).

Gern, F. H. and Librescu, L. (1999b) "Synergistic Interaction of Aeroelastic Tailoring and Boundary Moment Control on Aircraft Wing Flutter," in *Proceedings of the CEAS/AIAA/ICASE /NASA Langley International Forum on Aeroelasticity and Structural Dynamics*, June 22–25, Williamsburg, VA, pp. 719–733.

Gern, F. H. and Librescu, L. (2000) "Aeroelastic Tailoring of Composite Wings Exhibiting Nonclassical Effects and Carrying External Stores," *Journal of Aircraft*, Vol. 37, No. 6, pp. 1097–1004.

Gern, F. H. and Librescu, L. (2001) "Static and Dynamic Aeroelasticity of Advanced Aircraft Wings Carrying External Stores," *AIAA Journal*, Vol. 36, No. 7, pp. 1121–1129.

Goland, M. and Luke, Y. (1948) "The Flutter of a Uniform Cantilever Wing with Tip Weights," *Journal of Applied Mechanics*, Vol. 15, No. 1, pp. 13–20.

Heeg, J. (1991) "Flutter Suppression Via Piezoelectric Actuation," NASA TM-104120.

Hertz, T. J., Shirk, M. H., Ricketts, R. H. and Weisshaar, T. A., (1982) "On the Track of Practical Forward-Swept Wings," *Astronautics and Aeronautics*, Vol. 20, January, pp. 40–52.

Hollowell, S. J. and Dugundji, J. (1982) "Aeroelastic Flutter and Divergence of Stiffness Coupled, Graphite/Epoxy, Cantilevered Plates," *AIAA Paper 82-0722*, May.

Housner, J. M. and Stein, M. (1974) "Flutter Analysis of Swept-Wing Subsonic Aircraft With Parameter Studies of Composite Wings," *NASA TN-D7539*, September.

Hwu, C. and Tsai, Z. S. (2002) "Aeroelastic Divergence of Stiffened Composite Multicell Wings Structures," *Journal of Aircraft*, Vol. 38, No. 2, pp. 2242–2251.

Jha, R. and Chattopadhyay, A. (1997) "Development of a Comprehensive Aeroelastic Analysis Procedure for Composite Wings Using Laplace Domain Methodology," *AIAA 97-1026-CP*.

Kapania, K. K. and Castel, F. (1990) "A Simple Element for Aeroelastic Analysis of Undamaged and Damaged Wings," *AIAA Journal*, Vol. 28, No. 2, pp. 329–337.

Karpouzian, G. and Librescu, L. (1994) "A Comprehensive Model of Anisotropic Composite Aircraft Wings and Its Use in Aeroelastic Analyses," *Journal of Aircraft*, Vol. 31, No. 3, pp. 702–712.

Karpouzian, G. and Librescu, L. (1995) "Exact Flutter Solution of Advanced Composite Swept Wings in Various Flight Speed Regimes," in *Proceedings of the 36th AIAA/ASME/ASCE/AHS/ASC Structures, Structural Dynamics and Materials Conference*, New Orleans, LA, April 10–12, AIAA Paper 95-1382.

Karpouzian, G. and Librescu, L. (1996) "Non-Classical Effects on Divergence and Flutter of Anisotropic Swept Aircraft Wings," *AIAA Journal*, Vol. 34, No. 4, April, pp. 786–794.

Kim, D-H. and Lee I. (2001) "Transonic and Supersonic Flutter Characteristics of a Wing-Box Model with Tip Stores," *AIAA-2001-1464*.

Krone, N. J. Jr. (1975) "Divergence Elimination with Advanced Composites," AIAA Paper 75-1009, August.

Kudva, J. and Lockyear, A. J. (1999) "Exploiting Smart Technologies for Military Aircraft Applications – Perspectives on Development of a Smart Air Vehicle," in *40th AIAA/ASCE/AHS/ASC Structures, Structural Dynamics, and Materials Conference, AIAA/ ASME/AHS Adaptive Structures Forum, AIAA Forum on Non-Deterministic Approaches Conference and Exhibit*, April 12–15, St. Louis, Missouri, AIAA 99-1511.

Lazarus, K. B., Crawley, E. F. and Bohlmann, J. D. (1991) "Static Aeroelastic Control Using Strain Actuated Adaptive Structures," *Journal of Intelligent Material Systems and Structures*, Vol. 2, No. 3, July, pp. 386–410.

Lee, I. and Miura, H. (1987) "Static Aeroelastic Analysis for Generic Configuration Aircraft," *NASA TM 89423*, June, pp. 6–11.

Lee, I. and Lee, J. J. (1990) "Vibration Analysis of Composite Plate Wings," *Computers and Structures*, Vol. 37, No. 6, pp. 1077–1085.

Lee, I., Kim, S. H. and Miura, H. (1994) "Static Aeroelastic Characteristics of a Composite Wing," *Journal of Aircraft*, Vol. 31, No. 6, pp. 1413–1416.

Lee, U. (1995) "Equivalent Dynamic Beam-Rod Models of Aircraft Wing Structures," *Aeronautical Journal*, Vol. 99, December, pp. 450–457.

Lehman, L. L. (1982) "A Hybrid State Vector Approach to Aeroelastic Analysis," *AIAA Journal*, Vol. 20, No. 10, October, pp. 1442–1449.

Leishman, J. (1988) "Validation of Approximate Indicial Aerodynamic Functions for Two-Dimensional Subsonic Flow," *Journal of Aircraft*, Vol. 25, No. 10, pp. 914–922.

Leishman, J. (1993) "Indicial Lift Approximations for Two-Dimensional Subsonic Flow as Obtained from Oscillatory Measurements," *Journal of Aircraft*, Vol. 30, No. 3, pp. 340–351.

Librescu, L. (1975) *Elastostatics and Kinetics of Anisotropic and Heterogeneous Shell-Type Structures*, Noordhoff International Publishers, Leyden. The Netherlands.

Librescu, L. and Khdeir, A. A. (1988) "Aeroelastic Divergence of Swept-Forward Composite Wings Including Warping Restraint Effect," *AIAA Journal*, Vol. 26, No. 11, November, pp. 1373–1377.

Librescu, L. and Simovich, J. (1988) "General Formulation for the Aeroelastic Divergence of Composite Swept-Forward Wing Structures," *Journal of Aircraft*, Vol. 25, No. 4, April, pp. 364–371.

Librescu, L. and Song, O. (1990) "Static Aeroelastic Tailoring of Composite Aircraft Wings Modelled as Thin-Walled Beam Structures," in *Achievements in Composites in Japan and the United States*, Fifth Japan–U.S. Conference on Composite Materials, A. Kobayashi (Ed.), Kokon Shoin Co., Tokyo, pp. 141–149.

Librescu, L. and Song, O. (1992) "On the Static Aeroelastic Tailoring of Composite Aircraft Swept Wings Modelled as Thin-Walled Beam Structures," *Composites Engineering*, Vol. 2, Nos. 5–7 (Special Issue: *Use of Composites in Rotorcraft and Smart Structures*), pp. 497–512.

Librescu, L. and Thangjitham, S. (1991) "Analytical Studies on Static Aeroelastic Behavior of Forward Swept Composite Wing Structures," *Journal of Aircraft*, Vol. 28, No. 2, February, pp. 151–157.

Librescu, L., Meirovitch, L. and Song, O. (1996a) "Integrated Structural Tailoring and Control Using Adaptive Materials for Advanced Aircraft Wings," *Journal of Aircraft*, Vol. 30, No. 1, January–February, pp. 203–213.

Librescu, L., Meirovitch, L. and Song, O. (1996b) "Refined Structural Modeling for Enhancing Vibrations and Aeroelastic Characteristics of Composite Aircraft Wings," *La Recherche Aérospatiale*, No. 1, pp. 23–35.

Livne, E., Schmidt, L. A. and Friedmann, P. P. (1990) "Towards Integrated Multidisciplinary Synthesis of Actively Controlled Fiber Composite Wings," *Journal of Aircraft*, Vol. 27, No. 12, pp. 979–992.

Livne, E., Schmidt, L. A. and Friedmann, P. P. (1993) "Integrated Structure/Control/Aerodynamic Synthesis of Actively Controlled Composite Wings," *Journal of Aircraft*, Vol. 30, No. 3, May–June, pp. 387–394.

Lottati, I. (1985) "Flutter and Divergence Aeroelastic Characteristics for Composite Forward Swept Cantilevered Wing," *Journal of Aircraft*, Vol. 22, No. 11, pp. 1001–1007.

Lottati, I. (1987) "Aeroelastic Stability Characteristics of a Composite Swept Wings with Tip Weights For an Unrestrained Vehicle," *Journal of Aircraft*, Vol. 24, No. 11, pp. 793–812.

Marzocca, P., Librescu, L. and Chiocchia, G. (2001) "Aeroelastic Response of 2-D Lifting Surfaces to Gust and Arbitrary Explosive Loading Signatures," *International Journal of Impact Engineering*, Vol. 25, No. 1, pp. 41–65.

Marzocca, P., Librescu, L. and Chiocchia, G. (2002) "Aeroelasticity of Two-Dimensional Lifting Surfaces via Indicial Function Approach,"' *The Aeronautical Journal*, Vol. 39, No. 1057, pp. 147–153.

Marzocca, P., Librescu, L. and Silva, W. A. (2002) "Unified Approach of Aeroelastic Response and Flutter of Swept Aircraft Wings in an Incompressible Flow," *AIAA Journal*, Vol. 40, No. 5, pp. 801–812.

Marzocca, P., Librescu, L. and Chiocchia, G. (2002) "Aeroelastic Response of a 2-D Airfoil in a Compressible Flow Field and Exposed to Blast Loading," *Aerospace Science and Technology*, Vol. 6, No. 4, pp. 259–272.

Marzocca, P., Librescu, L. and Chiocchia, C. (2002) "Unsteady Aerodynamics in Various Flight Speed Regimes for Flutter/Dynamic Response Analyses," in *Proceeding of the 18th AIAA Applied Aerodynamic Conference*, Denver, CO, August 14–17, AIAA-2000-4229.

Marzocca, P., Librescu, L., Kim, D. and Lee, I. (2003) "Linear/Nonlinear Unsteady Aerodynamic Modeling of 2-D Lifting Surfaces Via a Combined CDF/Analytical Approach," *AIAA-2003-1925*, 7–10 April, Norfolk, VA.

McGowan, A.-M. R., Heeg, J. and Lake, R. C. (1996) "Results of Wing-Tunnel Testing from the Piezoelectric Aeroelastic Response Tailoring Investigation," *AIAA-96-1511-CP*.

Nam, C. and Kim, Y. (1995) "Optimal Design of Composite Lifting Surface for Flutter Suppressions with Piezoelectric Actuators," *AIAA Journal*, Vol. 33, No. 10, October.

Nam, C., Chattopadhyay, A. and Kim, Y. (2001) "Aeroelastic Control of Smart Composite Plate with Delaminations," *Journal of Intelligent Material Systems and Structures*, Vol. 11, pp. 868–876.

Niblett, L. T. (1980) "Divergence and Flutter of Swept-Forward Wings with Cross Flexibilities," RAE-TR-80047, April.

Oyibo, G. A. and Berman, J. H. (1985) "Anisotropic Wing Aeroelastic Theories with Warping Effects," *Second International Symposium on Aeroelasticity and Structural Dynamics*, Aachen, FRG, April.

Oyibo, G. A. (1989) "Some Implications of Warping Restraint on the Behavior of Composite Anisotropic Beams," *Journal of Aircraft*, February, Vol. 26, No. 2, pp. 187–189.

Patil, M. J., Hodges, D. H. and Cesnik, C. E. S. (2000) "Nonlinear Aeroelastic Analysis of Complete Aircraft in Subsonic Flow," *Journal of Aircraft*, Vol. 37, No. 5, pp. 753–760.

Petre, A., Stanescu, C. and Librescu, L. (1961) "Aeroelastic Divergence of Multicell Wings Taking Their Fixing Restraints into Account," *Revue de Mecanique Appliquee*, Vol. 19, No. 6, pp. 689–698.

Qin, Z. (2001) *Vibration and Aeroelasticity of Advanced Aircraft Wings Modeled as Thin-Walled Beams*, Ph.D. Thesis, Virginia Polytechnic Institute and State University, Blacksburg, Virginia, October, Department of Engineering Science and Mechanics.

Qin, Z., Marzocca, P. and Librescu, L. (2002) "Aeroelastic Instability and Response of Advanced Aircraft Wings at Subsonic Flight Speeds," *Aerospace Science and Technology*, Vol. 6, No. 3, pp. 195–208.

Qin, Z. and Librescu, L. (2003a) "Aeroelastic Instability of Aircraft Wings Modeled as Anisotropic Composite Thin-Walled Beams in the Compressible Flow," *Journal of Fluid and Structures*, Vol. 19, No. 1, pp. 43–61.

Qin, Z. and Librescu, L. (2003b) "Dynamic Aeroelastic Response of Aircraft Wings Modeled as Anisotropic Thin-Walled Beams," *Journal of Aircraft*, Vol. 40, No. 3, pp. 532–543.

Qin, Z., Librescu, L. and Marzocca, P. (2003) "Aeroelasticity of Composite Aerovehicle Wings in Supersonic Flows," *Journal of Spacecraft and Rockets*, Vol. 40, No. 2, pp. 162–173.

Rodden, W. P. and Stahl, B. (1969) "A Strip Method for Prediction of Damping in Subsonic Wind Tunnel and Flight Flutter Tests," *Journal of Aircraft*, Vol. 6, No. 1, pp. 9–17.

Sherrer, V. C., Hertz, T. J. and Shirk, M. H. (1981) "Wing Tunnel Demonstration of Aeroelastic Tailoring Applied to Forward Swept Wings," *Journal of Aircraft*, Vol. 19, Nov., pp. 976–903.

Shirk, M. H., Hertz, t. J. and Weisshaar, T. A. (1986) "Aeroelastic Tailoring – Theory, Practice and Promise," *Journal of Aircraft*, Vol. 23, No. 1, January, pp. 6–18.

Singh, R. and Baeder, J. D. (1997) "Direct Calculation of Three-Dimensional Indicial Lift Response Using Computational Fluid Dynamics," *Journal of Aircraft*, Vol. 34, No. 4, pp. 465–471.

Song, O., Librescu, L. and Rogers, C. A. (1992) "Application of Adaptive Technology to Static Aeroelastic Control of Wing Structures," *AIAA Journal*, Vol. 30, No. 12, pp. 2882–2889.

von Kármán, T. and Sears, W. R. (1938) "Airfoil Theory for Non-Uniform Motion," *Journal of the Aeronautical Sciences*, Vol. 5, No. 10, pp. 379–390.

Weisshaar, T. A. (1978) "Aeroelastic Stability and Performance Characteristics of Aircraft with Advanced Composite Swept Forward Wing Structures," AFFDL-TR-78-116, September.

Weisshaar, T. A. (1979) "Forward Swept Wing. Static Aeroelasticity," AFFDL-TR-3087, June.

Weisshaar, T. A. (1980) "Divergence of Forward Swept Composite Wings," *Journal of Aircraft*, Vol. 17, June, pp. 442–448.

Weisshaar, T. A. (1981) "Aeroelastic Tailoring of Forward Swept Composite Wings," *Journal of Aircraft*, Vol. 18, pp. 669–676.

Weisshaar, T. A. (1987) "Aeroelastic Tailoring-Creative Use of Unusual Materials," in *AIAA/ASME/ASCE/AHS 28th Structures, Structural Dynamics and Materials Conference*, Monterey, CA, April 9–10, AIAA Paper 87-0976.

Weisshaar, T. A. and Ehlers, S. M. (1992) "Adaptive Aeroelastic Composite Wings – Control and Optimization Issues," *Composites Engineering*, Vol. 2, Nos. 5–7, pp. 457–476.

Yates, E. C., Jr. (1958) "Calculation of Flutter Characteristics for Finite-Span Swept or Unswept Wings at Subsonic and Supersonic Speeds by a Modified Strip Analysis," NACA RM L57L10.

Yamane, T. and Friedmannn, P. P. (1993) "Aeroelatic Tailoring for Preliminary Design of Advanced Propellers with Composite Blades," *Journal of Aircraft*, Vol. 30, No. 1, pp. 119–126.

Chapter 13

OPEN-SECTION BEAMS

13.1 INTRODUCTION

The basic equations related to open-section beams (OSB) were supplied in Chapters 2, 3 and in the Appendix.

Due to their wide range of applications in most diverse engineering areas, such as space erectable booms installed on spacecraft; in aeronautical industry both as direct load-carrying and stiffener members; in offshore, marine and civil engineering, and, as I-section or cruciform-section beams in flex beams of bearingless helicopter rotor, there has been a great deal of research devoted to such structures.

We refer to the pioneering works by Vlasov (1940 – the original work in Russian, and 1961, its English translation), Goodier (1942), Djanelidze (1943), Hoff (1944), Timoshenko (1945), Reissner (1952, 1956) Nylander (1956) and Gere and Lin (1958), who addressed issues of modeling, buckling and vibration of metallic OSBs.

A great deal of work related to the thermal flutter instability of very long tubular booms known as STEMs (Storable Tubular Extendible Members) has been accomplished; see Frish (1967, 1970), Yu (1969), Beam (1969, 1973) and Murozono et al. (1985). These booms have been modeled as slender open-section cantilever metallic beams.

A variety of issues related to the modeling and static/dynamic behavior of metallic OSBs have been addressed by Tso (1965), Aggarwal and Cranch (1967), Barta (1967), Nemat-Nasser and Tsai (1969), Mei (1970), Yu (1972), Gay (1978), Hirashima and Iura (1977), Nishino and Hasegawa (1979), Pignataro et al. (1983), Kitipornchai and Wang (1986), Pan and Lu (1983), Muller (1983), Fouad (1984), Wekezer (1987, 1989), Roberts (1987), Senjanovic and Fan (1991a,b), Laudiero et al. (1991), Bažant and Cedolin (1991), McGee

(1992), Bank and Cofie (1992), Smith and Bank (1992), McGee et al. (1993), Barbero et al. (1993), Li et al. (1994), Ambrosini et al. (1995), Massa and Barbero (1968), and Musat and Epureanu (1999). With the increased use of advanced composites, linearized problems involving this type of structures have been investigated in a series of works by Bauld and Tzeng (1984), Chandra and Chopra (1991), Mottram (1992), Wu and Sun (1992), Badir et al. (1993), Craddock and Yen (1993), Sherbourne and Kabir (1995), Floros and Smith (1996), Pandey et al. (1995), Kaiser and Francescatti (1996), Loughlan and Atta (1997), Volovoi et al. (1999), Rand (1999), Song and Librescu (1995), Song et al. (2001), Maddur and Chaturvedi (1999, 2000), Lee and Kim (2001, 2002), Kollár (2001), and Qiao and Zou (2002), Jung et al. (1999, 2002). Nonlinear structural models of *metallic* OSBs have been developed by Ghobarach and Tso (1971), Barsoum (1970), Epstein and Murray (1976), Roberts and Azizian (1983), Attard (1986), Moore (1986), Ioannidis and Kounadis (1994), Trahair (1993), Polillo et al. (1998), Ronagh et al. (2000), Mohri et al. (2002), while for *composite* OSBs by Ascione and Feo (1995), Fraternali and Feo (2000) and Hodges et al. (1999). For the development of optimization design paradigms applied to open-section composite beams, the works by Savic et al. (2001) and Cardoso et al. (2002) are indicated.

Finally, the monographs by Gjelsvik (1981), Murray (1984) and Ojalvo (1990) provide a good deal of information related to the various aspects of the theory of metallic open-section beams.

The cross-sections in frequent use are with "I", "L", "U", "V" profiles. This chapter, will address a few issues related to the static response and free vibration of bisymmetric composite I-beams. The goal is to highlight a number of non-classical effects for composite beams that have been not sufficiently emphasized. Note that the two basic lamination schemes, CAS and CUS, are valid for both open and closed-cross section beams.

For an I-beam, these two configurations are illustrated in Figs. 13.1-1a,b.

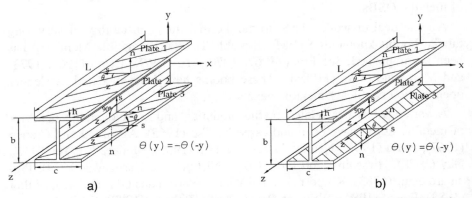

Figure 13.1-1: (a) Circumferentially asymmetric stiffness configuration (CAS). (b) Circumferentially uniform stiffness configuration (CUS)

13.2 BASIC EQUATIONS

13.2-1 Preliminaries

The points of the OSB are referred to a fixed Cartesian coordinate system, (x, y, z) with the z-axis parallel to the longitudinal axis of the beam. The contour of the beam, assumed to be uniform, is defined as the curved line, generated by crossing the middle surface of the beam by a plane transverse to its longitudinal axis.

We introduce a local coordinate system (n, s, z), where n and s are the normal and tangential coordinates to the contour, respectively (see Figs. 13.1-1). The circumferential length of the contour measured from an extremity of it (assumed to coincide with $s = 0$) is denoted by ℓ, while the span of the beam measured along the z-axis is denoted by L.

13.2-2 Kinematic Equations

Within the dynamic theory of TOBs the components of the 3-D displacement vector are expressed as

$$
\begin{aligned}
u(x, y, z, t) &= u_0(z, t) - y\phi(z, t), \\
v(x, y, z, t) &= v_0(z, t) + x\phi(z, t), \\
w(x, y, z, t) &= w_0(z, t) + \theta_x(z, t)\left[y(s) - n\frac{dx}{ds}\right] \\
&\quad + \theta_y(z, t)\left[x(s) + n\frac{dy}{ds}\right] - \phi'(z, t)[\bar{F}(s) - nr_t(s)],
\end{aligned}
\tag{13.2-1a-c}
$$

Here, see Eq. (2.1-29b),

$$
\bar{F}(s) = \int_0^s r_n(\bar{s})d\bar{s}.
\tag{13.2-2}
$$

Based on (13.2-1), the 1-D strain measures obtained as a special case of Eqs. (2.2-9)–(2.2-13) are

$$
\varepsilon_{zz} = \varepsilon_{zz}^{(0)} + n\varepsilon_{zz}^{(1)}; \quad \gamma_{sz} = \gamma_{sz}^{(0)} + n\gamma_{sz}^{(1)}; \quad \gamma_{nz} = \gamma_{nz}^{(0)},
\tag{13.2-3a-c}
$$

where

$$
\begin{aligned}
\varepsilon_{zz}^{(0)} &= w_0' + x(s)\theta_y' + y(s)\theta_x' - \phi''\bar{F}(s), \\
\varepsilon_{zz}^{(1)} &= \theta_y'\frac{dy}{ds} - \theta_x'\frac{dx}{ds} + \phi''r_t(s), \\
\gamma_{sz}^{(0)} &= (u_0' + \theta_y)\frac{dx}{ds} + (v_0' + \theta_x)\frac{dy}{ds}, \\
\gamma_{sz}^{(1)} &= 2\phi', \\
\gamma_{nz}^{(0)} &= (u_0' + \theta_y)\frac{dy}{ds} - (v_0' + \theta_x)\frac{dx}{ds}.
\end{aligned}
\tag{13.2-4a-e}
$$

13.2-3 Equations of Motion and Boundary Conditions

For a shearable beam loaded at its free end $z = L$, the equations of motion and the boundary conditions are obtained from Eqs. (3.2-13):

$$\delta u_0: \quad Q_x' = K_1,$$
$$\delta v_0: \quad Q_y' = K_2,$$
$$\delta w_0: \quad T_z' = K_3,$$
$$\delta \phi: \quad \delta_r B_w'' + M_z' = K_4 - K_9', \qquad (13.2\text{-}5\text{a-f})$$
$$\delta \theta_y: \quad M_y' - Q_x = K_7,$$
$$\delta \theta_x: \quad M_x' - Q_y = K_5,$$

where the inertia terms K_i are supplied in Table 3.2-1.

The related non-homogeneous boundary conditions are:

	Static BCs:	Geometric BCs:	
$\delta u_0:$	$Q_x = \underset{\sim}{Q}_x$	$u_0 = \underset{\sim}{u}_0$	
$\delta v_0:$	$Q_y = \underset{\sim}{Q}_y$	$v_0 = \underset{\sim}{v}_0$	
$\delta w_0:$	$T_z = \underset{\sim}{T}_z$	$w_0 = \underset{\sim}{w}_0$	
$\delta \theta_y:$	$M_y = \underset{\sim}{M}_y$	$\theta_y = \underset{\sim}{\theta}_y$	(13.2-6a-g)
$\delta \theta_x:$	$M_x = \underset{\sim}{M}_x$	$\theta_x = \underset{\sim}{\theta}_x$	
$\delta \phi:$	$\delta_r B_w' + M_z = \underset{\sim}{M}_z$	$\phi = \underset{\sim}{\phi}$	
$\underline{\delta \phi'}:$	$\delta_r B_w = \delta_r \underset{\sim}{B}_w$	$\delta_r \phi' = \delta_r \underset{\sim}{\phi}'.$	

13.2-4 CAS Configuration

13.2-4a Definition and Governing Equations

For this case, the ply lay-up in the flanges of the I-beam follows the rule

$$\theta(y) = -\theta(-y), \qquad (13.2\text{-}7)$$

whereas in the web, the ply-angle is considered invariably to be $\theta = 90$ deg. (see Fig. 13.1-1). We suppose that the flange and web elements are manufactured from the same material.

The governing equations for the shearable beam model split exactly into two featuring the following elastic couplings:

• Lag-Extension-Transverse Shear Coupling

$$\delta w_0: \quad a_{11} w_0'' + a_{14}(u_0'' + \theta_y') = b_1 \ddot{w}_0,$$
$$\delta u_0: \quad a_{14} w_0'' + a_{44}(u_0'' + \theta_y') = b_1 \ddot{u}_0, \qquad (13.2\text{-}8\text{a-c})$$
$$\delta \theta_y: \quad a_{22} \theta_y'' - a_{14} w_0' - a_{44}(u_0' + \theta_y) = (b_5 + \delta_n b_{15}) \ddot{\theta}_y.$$

Boundary Conditions for Clamped-Free Edges

at $z = 0$:

$$w_0 = u_0 = \theta_y = 0, \tag{13.2-9a-c}$$

at $z = L$:

$$\delta w_0 : a_{11}w_0' + a_{14}(u_0' + \theta_y) = 0,$$
$$\delta u_0 : a_{14}w_0' + a_{44}(u_0' + \theta_y) = \underset{\sim}{Q}_x, \tag{13.2-10a-c}$$
$$\delta \theta_y : a_{22}\theta_y' = 0.$$

We convert the governing equation systems to a dimensionless form by introducing the parameters in Table 13.2-1.

Table 13.2-1: Dimensionless parameters

Parameter	f_1	f_2	f_3	f_4	g_1	g_2	g_3	$\overline{f}_x$
Its expression	$\dfrac{a_{14}}{a_{11}}$	$\dfrac{a_{14}}{a_{44}}$	$\dfrac{a_{14}L^2}{a_{22}}$,	$\dfrac{a_{44}L^2}{a_{22}}$,	$\dfrac{b_1 L^2}{a_{11}}$,	$\dfrac{b_1 L^2}{a_{44}}$,	$\dfrac{(b_5 + \delta_n b_{15})L^2}{a_{22}}$	$\dfrac{\underset{\sim}{Q}_x}{a_{44}(0)}$

The dimensionless equations for harmonic oscillations are

$$\delta \overline{w}_0 : \overline{w}_0'' + f_1(\overline{u}_0'' + \theta_y') + g_1\omega^2 \overline{w}_0 = 0,$$
$$\delta \overline{u}_0 : \overline{u}_0'' + \theta_y' + f_2\overline{w}_0'' + g_2\omega^2 \overline{u}_0 = 0, \tag{13.2-11a-c}$$
$$\delta \theta_y : \theta_y'' - f_3\overline{w}_0' - f_4(\overline{u}_0' + \theta_y) + g_3\omega^2\theta_y = 0.$$

and for clamped-free BCs we have

at $\eta = 0$:

$$\overline{w}_0 = \overline{u}_0 = \theta_y = 0 \tag{13.2-12a-c}$$

and at $\eta = 1$:

$$\delta \overline{w}_0 : \overline{w}_0' + f_1(\overline{u}_0' + \theta_y) = 0,$$
$$\delta \overline{u}_0 : \overline{u}_0' + \theta_y + f_2\overline{w}_0' = \frac{\underset{\sim}{Q}_x}{a_{44}} \left(= \frac{\underset{\sim}{Q}_x}{a_{44}(0)} \frac{a_{44}(0)}{a_{44}(\theta)} \equiv \overline{f}_x \frac{a_{44}(0)}{a_{44}(\theta)} \right),$$
$$\delta \theta_y : \theta_y' = 0. \tag{13.2-13a-c}$$

Here, $a_{ij}(0) \equiv a_{ij}(\theta = 0)$.

COMMENT A

From Eqs. (13.2-8) and (13.2-12) it clearly appears that the lag-extension coupling is due to the cross-coupling stiffness a_{14}. For ply-angles $\theta = 0, 90$ deg. (see Fig. 13.2-2), a_{14} is zero, and as a result, the extension motion becomes decoupled from the lagging one. The same decoupling occurs in the case of non-shearable beams.

In this case Eqs. (13.2-8) through (13.2-10) split exactly into two independent groups associated with pure extension and pure lagging as:

Pure extension:

$$\delta w_0 : \ a_{11} w_0'' = b_1 \ddot{w}_0,$$

with the BCs:

$$w_0 = 0, \ \text{at } z = 0, \ \text{and } w_0' = 0, \ \text{at } z = L.$$

and

Pure lagging:

$$a_{22} u_0'''' + b_1 \ddot{u}_0 = (b_5 + \delta_n b_{15}) \ddot{u}_0''$$

with the BCs:

$$u_0 = u_0' = 0, \ \text{at } z = 0,$$

and

$$a_{22} u_0''' = (b_5 + \delta_n b_{15}) \ddot{u}_0' - \underset{\sim}{Q}_x \ \text{and} \ a_{22} u_0'' = 0, \ \text{at } z = L.$$

• Flap-Twist-Transverse Shear Coupling

Governing Equations

$$\delta v_0 : \ a_{55}(v_0'' + \theta_x') = b_1 \ddot{v}_0,$$

$$\delta \theta_x : \ a_{33}\theta_x'' + a_{37}\phi'' - a_{55}(v_0' + \theta_x) = (b_4 + \delta_n b_{14})\ddot{\theta}_x,$$

$$\delta \phi : \ -a_{66}\phi'''' + a_{37}\theta_x'' + a_{77}\phi'' = [b_4 + b_5 + \delta_n(b_{14} + b_{15})]\ddot{\phi}$$
$$- (b_{10} + \delta_n b_{18})\ddot{\phi}''. \tag{13.2-14a-c}$$

Clamped-Free Boundary Conditions

at $z = 0$:

$$v_0 = \theta_x = \phi = \phi' = 0, \tag{13.2-15a-d}$$

and at $z = L$:

$$\delta v_0 : \ a_{55}(v_0' + \theta_x) = \underset{\sim}{Q}_y,$$

$$\delta \theta_x : \ a_{33}\theta_x' + a_{37}\phi' = 0,$$

$$\delta \phi : \ -a_{66}\phi''' + a_{37}\theta_x' + a_{77}\phi' = \underset{\sim}{M}_z - (b_{10} + \delta_n b_{18})\ddot{\phi}',$$

$$\delta \phi' : \ a_{66}\phi'' = 0. \tag{13.2-16a-d}$$

In Eqs. (13.2-13) and (13.2-16) $\underset{\sim}{Q}_x$, $\underset{\sim}{Q}_y$ and $\underset{\sim}{M}_z$ denote the shear forces and the twist moment applied at the beam tip, $z = L$.

The stiffness quantities can be obtained from Table 4.3-1. For the I-beam, the a_{ij} and inertia coefficients are supplied in Tables 13.2-2 and 13.2-3, respectively.

Table 13.2-2: The non-zero 1-D stiffness quantities for the present bisymmetric cross-section I-beam

Quantity	a_{11}	a_{22}	a_{33}	a_{43}
Its expression	$2cK_{11}^{(1)} + bK_{11}^{(2)}$	$\dfrac{c^3}{6}K_{11}^{(1)} + bK_{44}^{(2)}$	$\dfrac{b^2 c}{2}K_{11}^{(1)} + \dfrac{b^3}{12}K_{11}^{(2)} + 2ck_{44}^{(1)}$	$-bcK_{12}^{(1)}$

Quantity	a_{44}	a_{55}	a_{66}	a_{77}
Its expression	$2cK_{22}^{(1)} + bA_{44}^{(2)}$	$bK_{22}^{(2)} + 2cA_{44}^{(1)}$	$\dfrac{b^2 c^3}{24}K_{11}^{(2)} + \dfrac{c^3}{6}K_{44}^{(1)} + \dfrac{b^3}{12}K_{44}^{(2)}$	$4cK_{53}^{(1)} + 2bK_{53}^{(2)}$

Table 13.2-3 The expression of the inertia coefficients

Quantity	b_1	b_4	b_5	b_{10}	b_{14}	b_{15}	b_{18}
Its expression	$\rho h(2c + b)$	$\rho h(\dfrac{b^2 c}{2} + \dfrac{b^3}{12})$	$\rho h\dfrac{c^3}{6}$	$\rho h\dfrac{b^2 c^3}{24}$	$\rho h^3\dfrac{c}{6}$	$\rho h^3\dfrac{b}{12}$	$\dfrac{\rho h^3}{12}(\dfrac{b^3}{12} + \dfrac{c^3}{6})$

Here superscripts (1) and (2) refer to the flange and web elements, respectively, while K_{ij} are provided by Eqs. (A.85).

We define the dimensionless spanwise coordinate $\eta(\equiv z/L)$, where $0 \leq \eta \leq 1$, and the dimensionless quantities displayed in Table 13.2-4.

Table 13.2-4: Dimensionless parameters

Quantity	v_0	f_5	f_6	f_7	f_8	g_4
Its expression	v_0/L	$\dfrac{a_{37}}{a_{33}}$	$\dfrac{a_{55}L^2}{a_{33}}$	$\dfrac{a_{37}L^2}{a_{66}}$	$\dfrac{a_{77}L^2}{a_{66}}$	$\dfrac{b_1 L^2}{a_{55}}$
Quantity	g_5	g_6	g_7	f_y	m_z	
Its expression	$\dfrac{(b_4 + \delta_n b_{14})L^2}{a_{33}}$	$[b_4 + b_5 + \delta_n(b_{14} + b_{15})]L^4/a_{66}$	$\dfrac{(b_{10} + \delta_n b_{18})L^2}{a_{66}}$	$\dfrac{\underset{\sim}{Q}_y}{a_{44}(0)}$	$\dfrac{\underset{\sim}{M}_z L^3}{a_{66}(0)}$	

Equations (13.2-14) become:

$$\delta\bar{v}_0 : \quad \bar{v}_0'' + \theta_x' + g_4\omega^2\bar{v}_0 = 0,$$
$$\delta\theta_x : \quad \theta_x'' + f_5\phi'' - f_6(\bar{v}_0' + \theta_x) + g_5\omega^2\theta_x = 0, \qquad (13.2\text{-}17\text{a-c})$$
$$\delta\phi : \quad \phi'''' - f_7\theta_x'' - f_8\phi'' - g_6\omega^2\phi + g_7\omega^2\phi'' = 0,$$

and the BCs become:

at $\eta = 0$

$$\bar{v}_0 = \theta_x = \phi = \phi' = 0, \qquad (13.2\text{-}18\text{a-d})$$

and at $\eta = 1$

$$\delta\bar{v}_0 : \quad \bar{v}'_0 + \theta_x - \bar{f}_y \frac{a_{44}(0)}{a_{55}(\theta)} = 0,$$

$$\delta\theta_x : \quad \theta'_x + f_5\phi' = 0, \qquad (13.2\text{-}19\text{a-d})$$

$$\delta\phi : \quad \phi''' - f_7\theta'_x - f_8\phi' + \bar{m}_z \frac{a_{66}(0)}{a_{66}(\theta)} + g_7\omega^2\phi' = 0,$$

$$\delta\phi' : \quad \phi'' = 0.$$

These equations considered in conjunction with the associated boundary conditions, have in special cases, exact solution. In general, the solution will be derived via the Extended Galerkin Method.

COMMENT B

Equations (13.2-14) and (13.2-16) reveal that the flapping-twist coupling is due to the stiffness a_{37}. For $\theta = 0$ and 90 deg., it vanishes (see Fig. 13.2-2), and as a result an exact decoupling of flapping and twist occurs. As it is shown next, this coupling subists also in the case of the non-shearable beams. Indeed, the unsherable counterpart of Eqs. (13.2-14) through (13.2-16) becomes:

$$\delta v_0 : a_{33}v_0'''' - a_{37}\phi''' + b_1\ddot{v}_0 = (b_4 + \delta_n b_{14})\ddot{v}_0'',$$

$$\delta\phi : a_{66}\phi'''' - a_{77}\phi'' + a_{37}v_0''' + [b_4 + b_5 + \delta_n(b_{14} + b_{15})]\ddot{\phi} = (b_{10} + \delta_n b_{18})\ddot{\phi}'',$$

with the BCs:

$$v_0 = v'_0 = \phi = \phi' = 0 \text{ at } z = 0,$$

and at $z = L$:

$$\delta v_0 : a_{33}v_0''' - a_{37}\phi'' = (b_4 + \delta_n b_{14})\ddot{v}'_0 - \underset{\sim}{Q}_y,$$

$$\delta v'_0 : a_{33}v_0'' - a_{37}\phi' = 0,$$

$$\delta\phi : a_{66}\phi''' - a_{77}\phi' + a_{37}v_0'' = (b_{10} + \delta_n b_{18})\ddot{\phi}' - \underset{\sim}{M}_z,$$

$$\delta\phi' : a_{66}\phi'' = 0.$$

These equations reveal that the stiffness a_{37} is responsible for the flapping-twist coupling in both shearable and unshearable cases. The decoupling takes place for only $\theta = 0$ and 90 deg.

Associated with the decoupled flapping-twist system of the I-beam, an interesting application was done by Herrmann and Nemat-Nasser (1967), that is also discussed by Païdousis (1998). It concerns a cantilevered I-beam with two pairs of flexible pipes attached to its web and conveying fluid.

In this context, the system becomes non-conservative, and as a result, both flutter and divergence instabilities are experienced.

Needless to say, for this case, the decoupled equations of flapping and twist as obtained here have to be adjusted as to include the effects of the pipes carrying fluid. In this sense, see above mentioned references.

13.2-4b Static Analytical Solution

• Lagging-Extension-Transverse Shear Coupling

For the static case, the exact solution of Eqs. (13.2-11) through (13.2-13) is

$$\bar{u}_0 = -\frac{C_2}{6}\eta^3 + \frac{C_2}{2}\eta^2 - \frac{C_1}{f_1}\eta; \quad \theta_y = \frac{C_2}{2}\eta^2 - C_2\eta; \quad \bar{w}_0 = C_1\eta.$$

$$(13.2\text{-}20\text{a-c})$$

where the coefficients C_i obtained from the boundary conditions at $\eta = 1$, are

$$C_1 = \frac{f_1\bar{f}_x}{f_1 f_2 - 1}\frac{a_{44}(0)}{a_{44}(\theta)}, \quad C_2 = C_1\left(f_3 - \frac{f_4}{f_1}\right). \qquad (13.2\text{-}21\text{a,b})$$

• Flap-Transverse Shear Elastic Coupling

This case is obtained from Eqs. (13.2-17), by taking $a_{37} = 0$. This implies, see Table 13.2-4, that for this case $f_5 = f_7 = 0$; the flap-transverse shear from the twist motions decouple.

For the flap-tranverse shear the exact static solution is

$$\bar{v}_0 = C_1\left(-\frac{\eta^3}{6} + \frac{\eta^2}{2} + \frac{\eta}{f_6}\right),$$

$$\theta_x = C_1\left(\frac{\eta^2}{2} - \eta\right),$$

where $\qquad\qquad\qquad\qquad\qquad\qquad\qquad\qquad (13.2\text{-}22\text{a-c})$

$$C_1 = f_6\bar{f}_y\frac{a_{44}(0)}{a_{55}(\theta)}.$$

• Torsional Response

For the twist equation one expresses the solution as

$$\phi(\eta) = \sum_{i=1}^{4} A_i e^{\lambda_i\eta}. \qquad (13.2\text{-}23)$$

The roots of the characteristic equations are $\lambda_{1,2,3,4} = 0, 0, \alpha, -\alpha$, where $\alpha = \sqrt{f_8}$, and the solution is:

$$\phi(\eta) = a + b\eta + ce^{\alpha\eta} + de^{-\alpha\eta}. \qquad (13.2\text{-}24)$$

The boundary conditions, yield the system

$$\mathbf{B}\mathbf{X} + \mathbf{F} = 0, \qquad (13.2\text{-}25)$$

where

$$\mathbf{B} = \begin{bmatrix} 1 & 0 & 1 & 1 \\ 0 & 1 & \alpha & -\alpha \\ 0 & 0 & \alpha^2 e^\alpha & \alpha^2 e^{-\alpha} \\ 0 & -\alpha^2 & 0 & 0 \end{bmatrix}, \quad \mathbf{X} = \{a, b, c, d\}^T, \quad \mathbf{F} = \{0, 0, 0, \bar{M}_z\}^T,$$

while

$$\bar{M}_z = \underset{\sim}{M}_z L^3 / a_{66}(\theta) (\equiv \bar{m}_z a_{66}(0) / a_{66}(\theta)).$$ (13.2-26a-d)

Equation (13.2-25) gives

$$\mathbf{X} = -\mathbf{B}^{-1} \mathbf{F},$$ (13.2-27)

from which, in conjunction with Eqs. (13.2-26) and (13.2-24) one obtains the solution of this boundary-value problem.

Alternatively, successive integrations and the fulfillment of boundary conditions gives the solution

$$\phi = C_1 \left[-\tan h\alpha + \alpha\eta + \frac{\sin h\alpha (1 - \eta)}{\cos h\alpha} \right],$$ (13.2-28a,b)

where

$$C_1 = \frac{\bar{m}_z}{\alpha^3} \frac{a_{66}(0)}{a_{66}(\theta)}.$$

• Other Special Cases
For the unshearable case, Eqs. (13.2-8) through (13.2-19) split exactly into two independent systems related to pure flapping and pure lagging.

• *Pure flapping*
The equations and static solutions are

$$a_{33} v_0'''' = 0,$$ (13.2-29)

with the boundary conditions:

At $\eta = 0$

$$v_0 = 0, \quad v_0' = 0,$$ (13.2-30a,b)

and at $\eta = 1$

$$a_{33} v_0'' = 0, \quad a_{33} v_0''' = -\underset{\sim}{Q}_y,$$ (13.2-30c,d)

so that

$$\bar{v}_0 = C_1 (\eta^3 - 3\eta^2),$$

where   (13.2-31a,b)

$$C_1 = -\frac{\bar{f}_y}{6} \frac{a_{44}(0) L^2}{a_{33}(\theta)}.$$

• *Pure lagging*

$$a_{22}u_0'''' = 0, \tag{13.2-32}$$

with the boundary conditions:

at $\eta = 0$:

$$u_0 = u_0' = 0, \tag{13.2-33a,b}$$

and at $\eta = 1$:

$$a_{22}u_0'' = 0, \quad a_{22}u_0''' = -\underset{\sim}{Q}_x. \tag{13.2-33c,d}$$

For this case,

$$\bar{u}_0 = C_1(\eta^3 - 3\eta^2),$$

where

$$\tag{13.2-34a,b}$$

$$C_1 = -\frac{\bar{f}_x}{6} \frac{a_{44}(0)L^2}{a_{22}(\theta)}.$$

• *Free-warping twist model*
The static solution is

$$\phi = C_1\eta,$$

where

$$\tag{13.2-35a,b}$$

$$C_1 = \bar{m}_z \frac{a_{66}(0)}{a_{77}(\theta)L^2}.$$

In these equations $\bar{f}_x$, $\bar{f}_y$ and $\bar{m}_z$ are provided in Tables 12.3-1 and 12.3-4, respectively.

13.2-5 CUS Configuration

We suppose that the flange and web elements are from graphite-epoxy; that in the web the ply-angle is $\theta = 90$ deg., and that the beam is loaded at the free-end cross section only.

The ply lay-up in the flanges of the I-beam follows the rule

$$\theta(y) = \theta(-y).$$

• **Flapping-lagging-transverse shear coupling**

Governing Equations:

$$\delta u_0 : \ a_{44}(u_0'' + \theta_y') + a_{43}\theta_x'' = b_1\ddot{u}_0,$$

$$\delta v_0 : \ a_{55}(v_0'' + \theta_x') = b_1\ddot{v}_0, \tag{13.2-36a-d}$$

$$\delta\theta_y : \ a_{22}\theta_y'' - a_{44}(u_0' + \theta_y) - a_{43}\theta_x' = (b_5 + \delta_n b_{15})\ddot{\theta}_y,$$

$$\delta\theta_x : \ a_{33}\theta_x'' + a_{34}(u_0'' + \theta_y') - a_{55}(v_0' + \theta_x) = (b_4 + \delta_n b_{14})\ddot{\theta}_x.$$

Clamped-Free BCs:

at $z = 0$:

$$u_0 = v_0 = \theta_y = \theta_x = 0, \qquad\qquad (13.2\text{-}37\text{a-d})$$

at $z = L$:

$$
\begin{aligned}
\delta u_0 &: a_{44}(u_0' + \theta_y) + a_{43}\theta_x' = \underset{\sim}{Q}_x, \\
\delta v_0 &: a_{55}(v_0' + \theta_x) = \underset{\sim}{Q}_y, \\
\delta\theta_y &: a_{22}\theta_y' = 0, \qquad\qquad\qquad (13.2\text{-}37\text{e-h}) \\
\delta\theta_x &: a_{33}\theta_x' + a_{34}(u_0' + \theta_y) = 0.
\end{aligned}
$$

The dimensionless coefficients that are used are supplied in Table 13.2-5.

Table 13.2-5: Dimensionless parameters used in the case of CUS beams

Quantity	$\hat{f}_1$	$\hat{f}_2$	$\hat{f}_3$	$\hat{f}_4$	$\hat{f}_5$	$\hat{f}_6$
Its expression	$\dfrac{a_{43}}{a_{44}L}$	$\dfrac{a_{44}L^2}{a_{22}}$	$\dfrac{a_{43}L}{a_{22}}$	$\dfrac{a_{43}L}{a_{33}}$	$\dfrac{a_{55}L^2}{a_{33}}$	$\dfrac{b_1L^2}{a_{44}}$

Quantity	$\hat{f}_7$	$\hat{f}_8$	$\hat{f}_9$	$\overline{f}_x$	$\overline{f}_y$
Its expression	$\dfrac{b_1L^2}{a_{55}}$	$\dfrac{(b_5 + \delta_n b_{15})L^2}{a_{22}}$	$\dfrac{(b_4 + \delta_n b_{14})L^2}{a_{33}}$	$\dfrac{\underset{\sim}{Q}_x}{a_{44}(0)}$	$\dfrac{\underset{\sim}{Q}_y}{a_{44}(0)}$

For harmonic motion the dimensionless equations are

$$
\begin{aligned}
\delta\overline{u}_0 &: (\overline{u}_0'' + \theta_y') + \hat{f}_1\theta_x'' + \hat{f}_6\omega^2\overline{u}_0 = 0, \\
\delta\overline{v}_0 &: \overline{v}_0'' + \theta_x' + \hat{f}_7\omega^2\overline{v}_0 = 0, \\
\delta\theta_y &: \theta_y'' - \hat{f}_2(\overline{u}_0' + \theta_y) - \hat{f}_3\theta_x' + \hat{f}_8\omega^2\theta_y = 0, \qquad (13.2\text{-}38\text{a-d}) \\
\delta\theta_x &: \theta_x'' + \hat{f}_4(\overline{u}_0'' + \theta_y') - \hat{f}_5(\overline{v}_0' + \theta_x) + \hat{f}_9\omega^2\theta_x = 0.
\end{aligned}
$$

Clamped-free BCs:

at $\eta = 0$:

$$\overline{u}_0 = \overline{v}_0 = \theta_y = \theta_x = 0, \qquad\qquad (13.2\text{-}39\text{a-d})$$

at $\eta = 1$:

$$\delta \bar{u}_0 : \; (\bar{u}_0' + \theta_y) + \hat{f}_1 \theta_x' = \frac{\underaccent{\approx}{Q}_x}{a_{44}} \left(= \frac{\underaccent{\approx}{Q}_x}{a_{44}(0)} \frac{a_{44}(0)}{a_{44}(\theta)} \equiv \bar{f}_x \frac{a_{44}(0)}{a_{44}(\theta)} \right),$$

$$\delta \bar{v}_0 : \; \bar{v}_0' + \theta_x = \frac{\underaccent{\approx}{Q}_y}{a_{55}} \left(= \frac{\underaccent{\approx}{Q}_y}{a_{44}(0)} \frac{a_{44}(0)}{a_{55}(\theta)} \equiv \bar{f}_y \frac{a_{44}(0)}{a_{55}(\theta)} \right),$$

$$\delta \theta_y : \; \theta_y' = 0, \tag{13.2-39e-h}$$

$$\delta \theta_x : \; \theta_x' + \hat{f}_4 (\bar{u}_0' + \theta_y) = 0.$$

There are a number of special cases where there are exact solutions:

• Flap and lag decoupled motion: Unshearable case
The dimensionless equations are

$$\bar{u}_0'''' - \hat{f}_2 \hat{f}_6 \omega^2 \bar{u}_0 + \hat{f}_8 \omega^2 \bar{u}_0'' = 0,$$

$$\bar{v}_0'''' - \hat{f}_5 \hat{f}_7 \omega^2 \bar{v}_0 + \hat{f}_9 \omega^2 \bar{v}_0'' = 0. \tag{13.2-40a,b}$$

For clamped-free BCs we have

at $\eta = 0$:

$$\bar{u}_0 = \bar{u}_0' = 0, \; \bar{v}_0 = \bar{v}_0' = 0, \tag{13.2-41a-d}$$

and at $\eta = 1$:

$$\bar{u}_0''' + \hat{f}_8 \omega^2 \bar{u}_0 + \bar{f}_x \frac{a_{44}(0)L^2}{a_{22}} = 0, \; \bar{u}_0'' = 0,$$

$$\bar{v}_0''' + \hat{f}_9 \omega^2 \bar{v}_0 + \bar{f}_y \frac{a_{44}(0)L^2}{a_{33}} = 0, \; \bar{v}_0'' = 0. \tag{13.2-41e-h}$$

The static solution is

$$\bar{u}_0 = \frac{C_1}{6} \eta^3 - \frac{C_1}{2} \eta^2, \; C_1 = -\bar{f}_x \frac{a_{44}(0)L^2}{a_{22}},$$

$$\bar{v}_0 = \frac{\hat{C}_1}{6} \eta^3 - \frac{\hat{C}_1}{2} \eta^2, \; \hat{C}_1 = -\bar{f}_y \frac{a_{44}(0)L^2}{a_{33}}. \tag{13.2-42a-d}$$

These results show that for unshearable I-beams, in the absence of the loads $\bar{f}_x$ and $\bar{f}_y$, both transversal and lateral deflections are zero.

With the CUS lay-up there is twist-extension coupling. Since $a_{17} = 0$, the twist becomes decoupled from the extension.

• Equation for the warping restrained twist

$$\delta \phi : \; -\phi'''' + \tilde{f}_1 \phi'' + \tilde{f}_2 \omega^2 \phi - \tilde{f}_3 \omega^2 \phi'' = 0. \tag{13.2-43}$$

Clamped-free BCs:

at $\eta = 0$:

$$\phi = \phi' = 0, \qquad \text{(13.2-44a,b)}$$

at $\eta = 1$:

$$\delta\phi : \ -\phi''' + \tilde{f}_1\phi' - \tilde{f}_3\omega^2\phi' - \overline{m}_z\frac{a_{66}(0)}{a_{66}} = 0,$$

$$\delta\phi' : \ \phi'' = 0, \qquad \text{(13.2-44c,d)}$$

where the dimensionless coefficients are

$$\tilde{f}_1 = \frac{a_{77}L^2}{a_{66}}, \ \tilde{f}_2 = \frac{[b_4 + b_5 + \delta_n(b_{14} + b_{15})]L^4}{a_{66}},$$

$$\tilde{f}_3 = \frac{(b_{10} + \delta_n b_{18})L^2}{a_{66}}, \ \overline{m}_z = \frac{\underline{M}_z L^3}{a_{66}(0)}. \qquad \text{(13.2-45a-d)}$$

Considering $\phi(\eta) = e^{\lambda\eta}$, from the characteristic equation $\lambda^4 - \tilde{f}_1\lambda^2 = 0$, one gets the roots $\lambda_{1,2,3,4} = 0, \ 0, \ \alpha, \ -\alpha$, where $\alpha = \sqrt{\tilde{f}_1}$, and

$$\phi(\eta) = C_1 + C_2\eta + \sum_{i=3}^{4} C_i e^{\lambda_i\eta}. \qquad \text{(13.2-46)}$$

The BCs in conjunction with (13.2-26) yield the matrix equation

$$\begin{bmatrix} 1 & 0 & 1 & 1 \\ 0 & 1 & \alpha & -\alpha \\ 0 & 0 & \alpha^2 e^\alpha & \alpha^2 e^{-\alpha} \\ 0 & -\alpha^2 & 0 & 0 \end{bmatrix} \begin{Bmatrix} C_1 \\ C_2 \\ C_3 \\ C_4 \end{Bmatrix} + \begin{Bmatrix} 0 \\ 0 \\ 0 \\ \overline{m}_z a_{66}(0)/a_{66} \end{Bmatrix} = \begin{Bmatrix} 0 \\ 0 \\ 0 \\ 0 \end{Bmatrix}, \qquad \text{(13.2-47)}$$

or in condensed form as

$$\mathbf{A}\mathbf{x} + \mathbf{f} = 0, \qquad \text{(13.2-48)}$$

where $\mathbf{x} = \{C_1, \ C_2, \ C_3, \ C_4\}^T$.

The solution of (13.2-48) is $\mathbf{x} = -\mathbf{A}^{-1}\mathbf{f}$ and

$$C_1 = \frac{C_2}{\alpha}\frac{(1 - e^{2\alpha})}{(1 + e^{2\alpha})}, \ C_2 = \frac{\overline{m}_z a_{66}(0)}{\alpha^2 a_{66}(\theta)},$$

$$C_3 = -\frac{C_2}{\alpha}\frac{1}{(1 + e^{2\alpha})}, \ C_4 = \frac{C_2}{\alpha}\frac{e^{2\alpha}}{(1 + e^{2\alpha})}. \qquad \text{(13.2-49a-d)}$$

Equations (13.2-49) in conjunction with (13.2-46) provide the solution for $\phi(\eta)$.

• Extension

Governing equation

$$\overline{w}_0'' + f_1 \omega^2 \overline{w}_0 = 0,$$

where (13.2-50)

$$f_1 = \frac{b_1 L^2}{a_{11}}.$$

Clamped-free BCs:

$$\overline{w}_0(0) = 0, \quad \overline{w}_0'(1) = \frac{T_z}{a_{11}} \left(= \frac{T_z}{a_{11}(0)} \frac{a_{11}(0)}{a_{11}(\theta)} \equiv \bar{t}_z \frac{a_{11}(0)}{a_{11}(\theta)} \right). \quad (13.2\text{-}51\text{a,b})$$

For the static case one gets the solution

$$\overline{w}_0(\eta) = \frac{T_z}{a_{11}(0)} \frac{a_{11}(0)}{a_{11}} \eta \equiv \bar{t}_z \frac{a_{11}(0)}{a_{11}} \eta. \quad (13.2\text{-}52)$$

For more intricate cases, either the Extended Galerkin Method or the Laplace Transform can be used.

Via the Extended Galerkin Method the static solution should be obtained by representing the kinematic variables as

$$\left(\bar{u}_0(\eta), \ \bar{v}_0(\eta), \ \theta_x(\eta), \ \theta_y(\eta) \right)$$

$$= \sum_{j=1}^{N} \left(a_j u_j(\eta), \ b_j v_j(\eta), \ c_j x_j(\eta), \ d_j y_j(\eta) \right), \quad (13.2\text{-}53)$$

where $u_j(\eta)$, $v_j(\eta)$, $x_j(\eta)$ and $y_j(\eta)$ are trial functions that have to fulfill all the kinematic boundary conditions. Inserting representation (13.2-53) into the energy functional one obtains the following matrix equation:

$$\mathbf{AC} + \mathbf{f} = 0, \quad (13.2\text{-}54)$$

where $\mathbf{A}$ is the stiffness matrix, $\mathbf{f}$ is the loading vector, while $\mathbf{C}$ is a constant vector defined as

$$\mathbf{C}^T = \{a_1, a_2, \ldots a_N, \ b_1, b_2, \ldots b_N, \ c_1, c_2, \ldots \ c_N, \ d_1, d_2, \ldots d_N\}. \quad (13.2\text{-}55)$$

From Eq. (13.2-54), one obtains:

$$\mathbf{C} = -\mathbf{A}^{-1}\mathbf{f}. \quad (13.2\text{-}56)$$

This equation considered in conjunction with (13.2-53), results in the solution of the boundary value problem.

In the next sections, results on statics and free vibration of clamped-free I-beams, obtained via the EGM will be presented.

13.2-6 Results

13.2-6a CAS Beam Configuration

• Static Response

In all numerical simulations the material of the beam is graphite-epoxy, whose elastic characteristics have been supplied in Table 6.4-2. For the dimensions of the flanges and web denoted in contracted form as $b \times c \times h \times L$, the following three scenarios labeled as (A), (B) and (C) are considered:

$$(A) \implies (10 \times 10 \times 0.4 \times 80)\text{in.}$$
$$(B) \implies (5 \times 5 \times 0.4 \times 80)\text{in.} \qquad (13.2\text{-}57\text{a-c})$$
$$(C) \implies (2 \times 2 \times 0.4 \times 80)\text{in.}$$

To get a better understanding of the effects of the ply-angle, we supply graphs depicting the variation with θ of the various stiffness quantities.

Figures 13.2-1 and 13.2-2 correspond to Case (A) and show the variations, as a function of θ, of the extensional, bending (lateral and transversal), and of the transverse shear stiffnesses, as well as of the coupling stiffnesses, while in Fig. 13.2-3, for scenarios (A) and (B), the variations vs. θ of both the warping and of the torsional stiffnesses. These plots help to optimize such structures by using the tailoring technique.

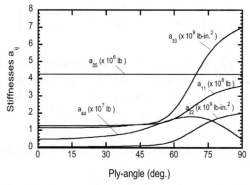

Figure 13.2-1: Variation with ply-angle of the extensional (a_{11}), bending (a_{22}, a_{33},) and shear (a_{44}, a_{55}) stiffnesses. Case (A)

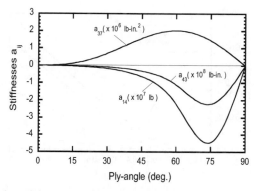

Figure 13.2-2: Variation of the coupling stiffnesses (a_{14}, a_{37} and a_{43}) with the ply-angle, Case (*A*)

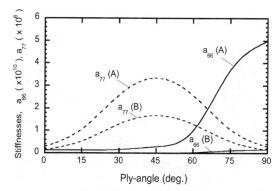

Figure 13.2-3: Variation of the warping (——) and twist (- - - - -) stiffnesses , a_{66} and a_{77}, respectively, with the ply-angle, for two cross-section scenarios, cases (*A*) and (*B*)

Figures 13.2-4 and 13.2-5 display the distributions of the lagging and flapping displacements along the beam span, as a function of the ply-angle, for the shearable and unshearable beam models (Case (*A*)). The transverse shear effect is more prominent in the flapping than in lagging degree of freedom, in the sense that in the former case the unshearable model strongly underestimates the deflection, whereas in the latter one this trend is greatly attenuated.

Figures 13.2-6 and 13.2-7 display the distributions along the beam span of rotations θ_x and θ_y. As θ increases, both θ_x and θ_y decrease, θ_y more than θ_x.

Finally, the distributions of the twist angle for the free warping and warping restraint torsional models are compared in Fig. 13.2-8; the warping inhibition reduces the twist angle.

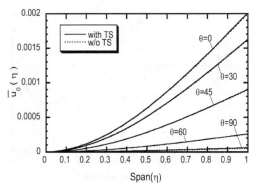

Figure 13.2-4: Distribution along the beam span of the lagging deflection as influenced by the incorporation/discard of transverse shear, and by the ply-angle θ (deg.), ($\bar{f}_x = 0.00001$)

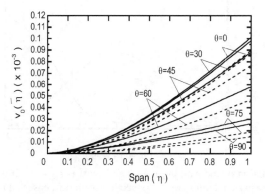

Figure 13.2-5: The counterpart of Fig. 13.2-4 for the flapping deflection ($\bar{f}_y = 0.00001$)

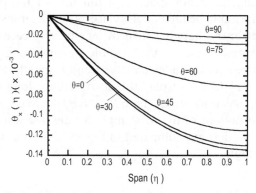

Figure 13.2-6: Distribution along the beam span of the rotation θ_x (rad.) as influenced by the ply-angle, θ (deg.), ($\bar{f}_y = 0.00001$)

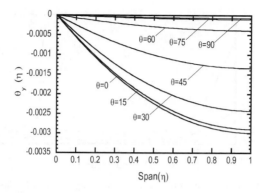

Figure 13.2-7: The counterpart of Fig. 13.2-6 for θ_y ($\bar{f}_x = 0.00001$)

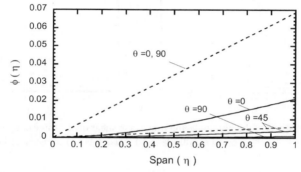

Figure 13.2-8: Distribution of the twist angle (rad.) along the beam span as influenced by the ply-angle, θ (deg.), determined as per the free (- - - -) and restrained (————) warping models ($\bar{m}_z = 0.1$)

Figure 13.2-3 shows that the twist stiffness is a maximum at $\theta = 45$ deg., and minimum at $\theta = 0$ and 90 deg.: corresponding to these ply-angles, minimum and maximum twist angles are experienced.

Figure 13.2-3 shows that the warping stiffness increases monotonously with the increase of the ply-angle from $\theta = 0$ till $\theta \cong 50$ deg., and increases steeply until $\theta = 90$ deg.: the twist angle will experience a continuous decay in the former ply-angle range and a rapid one in the latter ply-angle range.

Figure 13.2-9 shows the effects on the beam tip twist angle played by the tailoring, for the three cross-section scenarios, and of the inclusion/neglect of the warping restraint.

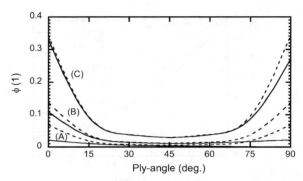

Figure 13.2-9: Effect of ply-angle on the beam tip twist angle (rad.) in Cases (A), (B) and (C), for free (- - - -) and restrained (———) warping models $(\overline{m}_z = 0.1)$

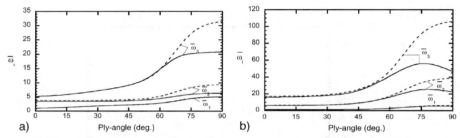

Figure 13.2-10: Effects of the ply-angle and transverse shear on the three coupled eigenfrequencies; shearable (———), nonshearable (- - - -), Case (A). a) transversal bending-twist; b) lateral bending-extension

• Free Vibration

Figure 13.2-10 shows the effect of transverse-shear on the first three natural frequencies of the transversal bending-twist coupled mode as a function of the variation of the ply-angle; i) the first natural frequency is marginally influenced by the transverse shear, and ii) the higher mode frequencies are insensitive to the transverse shear until a certain ply-angle, $\theta \approx 50$ deg. Beyond that value, with the increase of the ply-angle, a steep increase of frequencies is obtained if transverse shear is neglected. Their trend of variation is explainable on the basis of the variation with θ of the warping (a_{66}) and transverse shear (a_{55}) stiffnesses.

Figure 13.2-11 highlights effects of the warping restraint and of the ply-angle on the first three natural frequencies in transverse bending-twist coupled mode. The effect of the warping restraint can be explained on the basis of the variation with θ of the warping (a_{66}) and twist (a_{77}) stiffnesses (see Fig. 13.2-3).

The effect of the lateral bending-extension coupling stiffness, a_{14}, as a function of θ, is highlighted in Fig. 13.2-12; from a certain ply-angle ($\theta \approx 60$ deg.), the coupling reduces the frequencies in pure lateral bending. This trend can be explained by the variation of the coupling stiffness a_{14} vs. θ as depicted in Fig. 13.2-2.

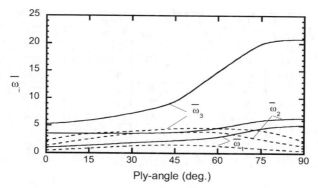

Figure 13.2-11: Effects of ply-angle and warping inhibition on the three coupled transversal bending-twist eigenfrequencies; warping restraint (———), free warping (- - - -), Case (A)

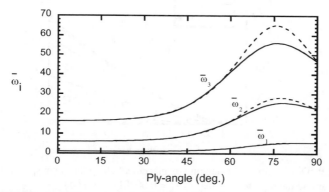

Figure 13.2-12: Effects of ply-angle and of the lateral bending-extension coupling stiffness a_{14} on the first three coupled eigenfrequencies. $a_{14} \neq 0$ (———); $a_{14} = 0$ (- - - -), Case (A)

Table 13.2-6 displays for shearable and unshearable beams, their variation for selected ply-angles, and for the three scenarios of the I-beam cross-section. Equations (13.2-8) show that for unshearable beams the lateral bending becomes decoupled from the extension. In this case, the natural frequencies that are supplied correspond to lateral bending.

This table shows that the frequencies in lateral bending are strongly influenced by transverse shear and for higher mode frequencies, especially in Case (A). The influence is less for Cases (B) and (C).

Table 13.2-6: Effect of transverse shear on the lateral bending-extension coupled eigenfrequencies for three different cross-section scenarios (the numbers in brackets denote percent difference), and for shearable and non-shearable models

$\bar{\omega}_i$	θ (deg.)	Case (A) Shearable	Case (A) N-Shearable	Case (B) Shearable	Case (B) N-shearable	Case (C) Shearable	Case (C) N-shearable
$\bar{\omega}_1$	0	1.0	1.004 (0.4)	0.525	0.526 (0.1)	0.266	0.266 (0)
	45	1.493	1.499 (0.4)	0.765	0.766 (0.1)	0.347	0.347 (0)
	90	5.48	6.25 (14.1)	3.022	3.13 (1.2)	1.257	1.265 (0.6)
$\bar{\omega}_2$	0	6.06	6.22 (2.6)	3.26	3.28 (0.6)	1.662	1.665 (0.18)
	45	9.025	9.28 (2.8)	4.75	4.79 (0.8)	2.17	2.173 (0.14)
	90	22.422	38.72 (72.7)	15.866	19.58 (23.4)	7.604	7.922 (4.2)
$\bar{\omega}_3$	0	16.171	17.08 (5.6)	9.0	9.15 (1.7)	4.638	4.657 (0.4)
	45	24.008	25.51 (6.3)	13.105	13.34 (1.8)	6.059	6.08 (0.3)
	90	46.704	106.41 (127.8)	36.873	54.57 (48.0)	20.212	22.165 (9.7)

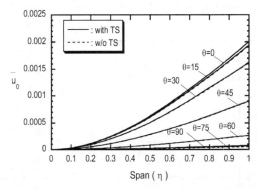

Figure 13.2-13: Distribution of the lagging displacement along the beam span as influenced by transverse shear and the ply-angle θ (deg.) ($\bar{f}_x = 0.00001$), Case (A)

13.2-6b CUS Beam Configuration

• Static Response

Figure 13.2-13 depicts the distribution of the lagging displacement along the beam span, as a function of the ply-angle, for the dimensionless transversal tip load $\bar{f}_x = 0.00001$. and for shearable and unshearable beams; this plot shows that the effect of transverse shear on the lagging displacement is small. The lagging displacement decreases with the increase of the ply-angle, i.e. with the increase of the lateral bending stiffness a_{22} (see Fig. 13.2-1).

Figure 13.2-14 reveals that for the same loading conditions, the flapping deflection decreases with the increase of the ply-angle, i.e. with the increase of the transversal bending stiffness a_{33}.

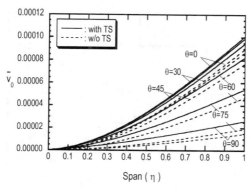

Figure 13.2-14: The counterpart of Fig. 13.2-13 for the flapping deflection ($\bar{f}_x = 0.00001$), Case (A)

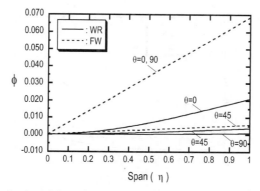

Figure 13.2-15: Distribution of the twist (rad.) along the beam span as influenced by the warping restraint and the ply-angle θ (deg.) ($\bar{m}_z = 0.1$), Case (A)

On the other hand, in the present case, and in contrast to the previous one, the transverse shear effect is very strong: the unshearable model underestimates the flapping deflection as compared to its shearable beam counterpart.

Figure 13.2-15, shows for an applied twist moment at the beam tip, the distribution of the twist angle along the beam span for selected ply-angles and for free and restrained warping beam models. As for CAS configuration, the increase of the ply-angle tends to decrease the twist angle. However, there are two main differences: i) in the CAS case, the warping restraint has a much stronger effect than in CUS; there is a larger diminution of the twist along the beam span. This is related to the increase of the ply-angle when comparing CAS with CUS.

Figure 13.2-16, depicts the variation of the extensional displacement for selected values of the ply-angle, and for a fixed value of the extensional load applied at the beam tip; the increase of the ply-angle increases the extensional stiffness a_{11}, and decreases the extensional displacement.

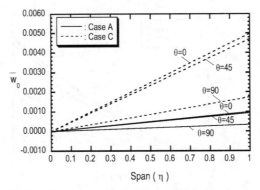

Figure 13.2-16: Distribution of extensional displacement along the beam span as influenced by the ply-angle θ (deg.); ($\bar{t}_z = 0.001$), Cases (A) and (C)

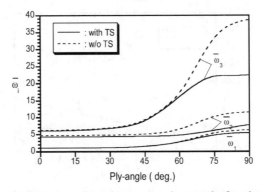

Figure 13.2-17: Effects of the ply-angle and transverse shear on the first three flap-lag-transverse shear coupled eigenfrequencies ($\bar{\omega}_i \equiv \omega_i/\omega_0$, $\omega_0 = 94.7126$ rad/s), Case (A)

Since there is a larger extensional stiffness in the Case (A) compared to (C), the effect of ply-angle is less strong in the former case than in the latter. These results show the advantages of composites in the static case.

• Free Vibration

Figure 13.2-17 shows the influence of the transverse shear and the ply-angle, on the first three natural frequencies in flap-lag motion; the first natural frequency is very little influenced by transverse shear. With the increase of the mode number, the frequencies are more and more influenced by this effect. The drop of natural frequencies due to the inclusion of transverse shear appears from a ply-angle, $\theta > 55$ deg.

Figure 13.2-18 displays the variation of the fundamental frequency in pure torsion, as a function of the ply-angle, for the two scenarios (A) and (C) of the I-beam, and for free and restrained warping beam models. In Case (C) the warping restraint has little effect, in Case (A) it is strong.

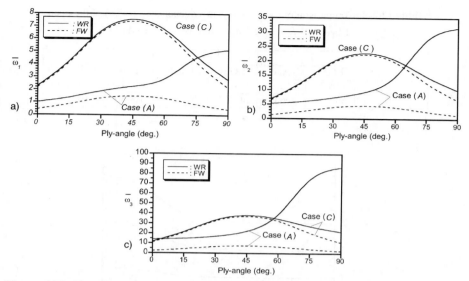

Figure 13.2-18: Variation of the dimensionless torsional frequencies $\overline{\omega}_i (\equiv \omega_i / \omega_0, \omega_0 = 116.3495$ rad/s), with the ply-angle for Cases (A) and (C). (a) variation of $\overline{\omega}_1$; (b) variation of $\overline{\omega}_2$; (c) variation of $\overline{\omega}_3$

However, for larger warping stiffnesses at higher ply-angles, the frequencies in Case (A) are larger than those in Case (C).

The variation with that of the ply-angle of the first three extensional natural frequencies is highlighted in Fig. 13.2-19; the trend of variation of the extensional stiffness a_{11} with θ (see Fig. 13.2-1), appears to be more prominent for the higher natural frequencies than for the fundamental one.

Tables 13.2-7 through 13.2-9 display the first three decoupled flapping and lagging natural frequencies for the I-beam predicted via the exact and the Extended Galerkin Method, and for various boundary conditions; the first three torsional natural frequencies, WR case, $C - F$ boundary conditions; and the first three extensional natural frequencies, $C - F$ boundary conditions, Case (A).

Table 13.2-7: Comparison of bending (flap-lag decoupled) natural frequencies ω_i (rad/s) of I-beams (Warping restraint included), Case (A) (C-clamped; S-simply support; F-free); (L, F) $\Rightarrow$ (lag, flap) modes

Ply-angle (deg.)	BCs	ω_1		ω_2		ω_3	
		EGM	Exact	EGM	Exact	EGM	Exact
0	C-F	94.7 (L)	94.7	408.9 (F)	408.9	574.0 (L)	574.0
	S-S	264.2 (L)	264.2	1020.9 (L)	1020.8	1053.2 (F)	1053.2
	C-C	580.0 (L)	580.0	1514.8 (L)	1514.2	1617.6 (F)	1617.6
90	C-F	518.6 (L)	518.6	740.3 (F)	740.3	2123.7 (L)	2123.7
	S-S	1299.0 (L)	1299.0	1655.9 (F)	1655.9	3528.8 (L)	3528.8
	C-C	1828.4 (L)	1828.4	1888.4 (F)	1888.4	3711.5 (L)	3711.5

Table 13.2-8: Comparison of torsional eigenfrequencies ω_i (rad/s) of the open I-beam, WR case, clamped-free BCs, Case (A)

Ply-angle (deg.)	ω_1		ω_2		ω_3	
	EGM	Exact	EGM	Exact	EGM	Exact
0	116.35	116.35	610.34	610.34	1624.98	1624.98
90	596.07	596.07	3670.6	3670.62	10080.0	10080.0

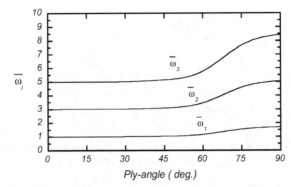

Figure 13.2-19: Variation of the three dimensionless extensional eigenfrequencies with the ply-angle; $\overline{\omega}_i (\equiv \omega_i/\omega_0, \omega_0 = 5320.55$ rad/s), Case (A)

The results reveal the excellent accuracy of predictions obtained via EGM. Note that for these ply-angles, there is the difference between CUS and CAS.

Table 13.2-9: Comparison of extensional eigenfrequencies ω_i (rad/s) of the open I-beam, clamped-free BCs, Case (A)

Ply-angle (deg.)	ω_1		ω_2		ω_3	
	EGM	Exact	EGM	Exact	EGM	Exact
0	5320.55	5320.55	15961.6	15961.6	26603.1	26602.7
90	8993.37	8993.37	26980.1	26980.1	44967.3	44966.9

13.3 CONCLUSIONS

The couplings induced by the directional feature of composites can be used as to improve the static and dynamic response of I-beams. Warping constraint and transverse shear have important effects.

All these results highlight the necessity of getting a good and complete as possible understanding of the implications of these non-classical effects. Only on this basis one can achieve a reliable design of these structures.

REFERENCES

Aggarwal, H. R.. and Cranch, E. T. (1967) "A Theory of Torsional and Coupled Bending Torsional Waves in Thin-Walled Open Section Beams," *Journal of Appl. Mech.*, Trans. ASME, Vol. 34, No. 2, pp. 337–343.

Ambrosini, R. D., Riera R. D. and Danesi, R. F. (1995) "Dynamic Analysis of Thin-Walled and Variable Open Section Beams with Shear Flexibility," *International Journal for Numerical Methods in Engineering*, Vol. 38, pp. 2867–2885.

Ascione, L. and Feo, L. (1995) "On the Mechanical Behavior of Thin-Walled Beams of Open Cross-Section: An Elastic Non-linear Theory," *International Journal of Engineering Anal. Design*, Vol. 2, pp. 14–34.

Attard, M. M. (1986) "Nonlinear Theory of Non-Uniform Torsion of Thin-Walled Open Beams," *Journal of Thin-Walled Structures*, Vol. 4, pp. 101–134.

Badir, A. M., Berdichevski, V. L. and Armanios, E. A. (1993) "Theory of Composite Thin-Walled Opened-Cross-Section Beams," in *Proceedings of the AIAA/ASME/ASCE/AHS/ASC*, 34th Structures, Structural Dynamics and Materials Conference, La Jolla, CA, April 22, pp. 2761–2770.

Bank, L. C. and Cofie, E. (1992) "A Modified Beam Theory for Bending and Twisting Open-Section Composite Beams-Numerical Verification," *Composite Structures*, Vol. 21, pp. 29–39.

Barbero, E. J., Lopez, R., Anido and Davalos, F. (1993) "On the Mechanics of Thin-Walled Laminated Composite Beams," *Journal of Composite Materials*, Vol. 27, No. 8, pp. 806–829.

Barsoum, R. S. (1970) "A Finite Element Formulation for the General Stability Analysis of Thin-Walled Members," Ph.D. Thesis, Cornell University, Ithaca, NY.

Barta, T. A. (1967) "On the Torsional-Flexural Buckling of Thin-Walled Elastic Bars with Mono-symmetric Open Cross-Section," in *Thin-Walled Structures*, A. H. Chilver (Ed.), pp. 60–86, John Wiley & Sons, New York.

Bauld, N. R. and Tzeng, L. S. (1984) "A Vlasov Theory for Fiber-Reinforced Beams with Thin-Walled Open Cross-Sections," *International Journal of Solid and Structures*, Vol. 20, No. 3, pp. 277–297.

Bažant, Z. P. and Cedolin, L. (1991) *Stability of Structures*, Chapter 6, Oxford University Press, New York, Oxford.

Beam, R. M. (1969) "On the Phenomenon of Thermoelastic Instability (Thermal Flutter) of Booms with Open Cross Section," NASA TN D-5222.

Beam, R. M. and Yagoda, H. P. (1973) "On the Torsional Static Stability and Response of Open Section Tubes Subjected to Thermal Radiation Loading," *International Journal of Solids and Structures*, Vol. 9, pp. 151–175.

Cardoso, J. B., Sousa, L. G., Castro, J. A. and Valido, A. J. (2002) "Optimal Design of Laminated Composite Beam Structures," *Structural Multidisciplinary Optimization*, Vol. 24, pp. 205–211.

Chandra, R. and Chopra, I. (1991) "Experimental and Theoretical Analysis of Composite I-Beams with Elastic Coupling," *AIAA Journal*, Vol. 29, No. 12, pp. 2197–2206.

Craddock, J. N., Yen, S. C. (1993) "The Bending Stiffness of Laminated Composite Material I-Beams," *Composites Engineering*, Vol. 3, No. 11, pp. 1025–1038.

Djanelidze, G. Iu (1943) "Variational Formulation of the V. Z. Vlasov's Theory of Thin-Walled Beams," *Prikladnaia Mathematika Mechanika, (Applied Mathematics and Mechanics)*, Vol. VII, No. 6 (in Russian).

Epstein, M. and Murray, D. W. (1976) "Three-Dimensional Large Deformatoin Analysis of Thin-Walled Beams," *International Journal of Solids and Structures*, Vol. 12, pp. 867–876.

Floros, M. W. and Smith, E. C. (1996) "Finite-Element Modeling of Open-Section Composite Beams with Warping Restraint Effects," *AIAA Journal*, Vol. 35, No. 8, pp. 1341–1347.

Fouad, A. M. (1984) "Complex Frequency Analysis of Damped Thin-Walled Beams with Open Cross Section," University Microfilms International, Ann Arbor, Michigan 48106, Ph.D. Dissertation, University of New Hampshire.

Fraternali, F. and Feo, L. (2000) "On a Moderate Rotation Theory of Thin-Walled Composite Beams," *Composites: Part B*, Vol. 31, pp. 141–158.

Frish, H. P. (1967) "Thermal Bending Plus Twist of Thin-Walled Cylinder of Open Section with Application to Gravity Gradient Booms," NASA TND-4069.

Frish, H. P. (1970) "Thermally Induced Vibration of Long Thin-Walled Cylinders of Open Section," *Journal of Spacecraft and Rockets*, Vol. 7, pp. 897–905.

Gay, D. (1978) "Influence of Secondary Effects on Free Torsional Oscillations of Thin-Walled Open Section Beams," *Journal of Applied Mechanics*, Trans. ASME, **45**, pp. 681–683.

Gere, J. M. and Lin, Y. K. (1958) "Coupled Vibration of Thin-Walled Beams of Open Cross Section," *Journal of Applied Mechanics*, Trans. ASME, Vol. 25, pp. 373–378.

Ghobarah, A. A. and Tso, W. K. (1971)" A Non-Linear Thin-Walled Beam Theory," *International Journal of Mechanical Science*, Vol.13, pp. 1025–1038.

Gjelsvik, A. (1981) *The Theory of Thin Walled Bars*, A Wiley-Interscience Publication, New York.

Goodier, J. N. (1942) "Torsional and Flexural Buckling of Bars of Thin-Walled Open Section Under Compressive and Buckling Loads," *Journal of Applied Mechanics*, Vol. 9, Trans. ASME, Vol. 64, pp. A-103-107.

Herrmann, G. and Newat-Nasser, S. (1967) "Instability Modes of Cantilevered Bars Induced by Fluid Flow Through Attached Pipes," *International Journal of Solids and Structures*, Vol. 3, pp. 39–52.

Hirashima, M. and Iura, M. (1977) "On the Derivation of Fundamental Equations of Curved and Twisted Thin-Walled Open Section Members," Waeda University, No. 79.

Hoff, N. J. (1944) "A Strain Energy Derivation of The Torsional Flexural Buckling Loads of Straight Columns of Thin-Walled Open Sections," *Quarterly of Applied Mathematics*, Vol. 1, pp. 341–345.

Hodges, D. H., Harursampath, D., Volovoi, V. V. and Cesnik, C. E. S. (1999) "Non-Classical Effects in Non-linear Analysis of Pretwisted Anisotropic Strips," *International Journal of Non-Linear Mechanics*, Vol. 34, pp. 259–277.

Ioannidis, G. I. and Kounadis, A. N. (1994) "Lateral Post-Buckling Analysis of Monosymmetric I-Beams Under Uniform Bending," *Journal of Constructional Steel Research*, Vol. 30, pp. 1–12.

Jung, S. N., Nagaraj, V. V. and Chopra, I. (1999) "Assessment of Composite Rotor Modeling Techniques," *Journal of the American Helicopter Society*, Vol. 44, No. 3, pp. 188–205.

Jung, S. N., Nagaraj, V. T. and Chopra, I. (2002) "Refined Structural Model for Thin and Thick-Walled Composite Rotor Blades," *AIAA Journal*, Vol. 40, No. 1, pp. 105–116.

Kaiser, C. and Francescatti, D. (1996) "Theoretical and Experimental Analysis of Composite Beams with Elastic Coupling," in *20th Congress of the International Council of the Aeronautical Sciences*, ICAS-96-5.5.4, Sorrento, Napoli, Italy, September 8–13.

Kitipornchai, S., Wang, C. M. (1986) "Buckling of Monosymmetric I-Beams under Moment Gradient," *Journal of Structural Engineering*, Vol. 112, No. 4, pp. 781–799.

Kollár, L. P. (2001) "Flexural-Torsional Vibration of Open Section Composite Beams with Shear Deformation," *International Journal of Solids and Structures*, Vol. 38, pp. 7543–7558.

Laudiero, F., Savoia, M. and Zaccaria, D. (1991) "The Influence of Shear Deformations on the Stability of Thin-Walled Beams under Non-Conservative Loading," *International Journal of Solids and Structures*, Vol. 27, No. 11, pp. 1351–1370.

Lee, J. and Kim, S. E. (2001) "Flexural-Torsional Buckling of Thin-Walled I-Section Composites," *Computers & Structures*, Vol. 79, pp. 987–995.

Lee, J. and Kim, S. E. (2002) "Free Vibration of Thin-Walled Composite Beams with I-Shaped Cross-Section," *Composite Structures*, Vol. 55, pp. 205–215.

Li, D-B., Chui, Y. H. and Smith, I. (1994) "Effect of Warping on Torsional Vibration of Members with Open Cross-Sections," *Journal of Sound and Vibration*, Vol. 170, No. 2, pp. 270–275.

Loughlan, J. and Atta, M. (1997) "The Behavior of Open and Closed Section Carbon Fibre Composite Beams Subjected to Constrained Torsion," *Composite Structures*, Vol. 38, No. 1–4, pp. 631–647.

Maddur, S. S. and Chaturvedi, S. K. (1999) "Laminated Composite Open Profile Sections: First Order Shear Deformation Theory," *Composite Structures*, Vol. 45, pp. 105–114.

Maddur, S. S. and Chaturvedi, S. K. (2000) "Laminated Composite Open Profile Sections: Nonuniform Torsion of I-Sections," *Composite Structures*, Vol. 50, pp. 159–169.

Massa, J. C. and Barbero, E. J. (1998) "A Strength of Materials Formulation for Thin Walled Composite Beams with Torsion," *Journal of Composite Materials*, Vol. 32, No. 17, pp. 1560–1594.

McGee, O. G. (1992) "Effect of Warping-Pretwist Torsionally Clamped-Pinned Thin-Walled Open Profile Bars," *International Journal for Numerical Methods in Engineering*, Vol. 35, pp. 325–349.

McGee, O. G., Owings, M. I. and Harris, J. W. (1993) "Torsional Vibrations of Pretwisted Thin-Walled Cantilevered I-Beams," *Computers & Structures*, Vol. 47, No. 1, pp. 47–56.

Mei, C. (1970) "Coupled Vibrations of Thin-Walled Beams of Open Section," *International Journal of Mechanical Science*, Vol. 12, pp. 883–891.

Mohri, F., Azrar, L. and Potier-Ferry, M. (2002) "Lateral Post-Buckling Analysis of Thin-Walled Open Section Beams," *Thin-Walled Structures*, Vol. 30, No. 12, pp. 1013–1036.

Moore, D. B. (1986) "A Non-Linear Theory for the Behavior of Thin-Walled Sections Subjected to Combined Bending and Torsion," *Journal of Thin-Walled Structures*, Vol. 4, pp. 101–134.

Mottram, J. T. (1992) "Lateral-Torsional Buckling of Thin-Walled Composite I-Beams by the Finite Difference Method," *Composites Engineering*, Vol. 2, No. 2, pp. 91–104.

Muller, P. (1983) "Torsional-Flexural Waves in Thin-Walled Open Beams," *Journal of Sound and Vibration*, Vol. 87, No. 1, pp. 113–141.

Murozono, M., Hashimoto, Y. and Sumi, S. (1985) "Thermally-Induced Vibration and Stability of Booms with Open Cross Section Caused by Unidirectional Radiant Heating," (in Japanese) *Journal of the Japan Society for Aeronautical and Space Sciences*, Vol. 33, No. 383, pp. 719–727.

Murray, N. W. (1984) *Introduction to the Theory of Thin-Walled Structures*, Clarendon Press, Oxford.

Musat, S. D. and Epureanu, B. I. (1999) "Study of Warping Torsion of Thin-Walled Beams with Open Cross-Section Using Macro-Elements," *International Journal of Numerical Methods in Engineering*, Vol. 44, pp. 853–868.

Nemat-Nasser, S. and Tsai, P. F. (1969) "Effect of Warping Rigidity on Stability of a Bar Under Eccentric Follower Force," *International Journal of Solids and Structures*, Vol. 5, pp. 271–279.

Nishino, F. and Hasegawa, A. (1979) "Thin-Walled Elastic Members," *Journal of the Faculty of Engineering*, The University of Tokyo (B), Vol. XXXV, (2), pp. 109–190.

Nylander, H. (1956) "Torsion, Bending, and Lateral Buckling of I-beams," *Trans. of the Royal Institute of Technology*, Stockholm, Sweden, No. 102.

Ojalvo, M. (1990) *Thin-Walled Bars with Open Profiles*, Olive Press.

Païdoussis, M. P. (1998) *Fluid-Structures Interactions. Slender Structures and Axial Flow*, Vol. 1, Academic Press, New York.

Pan, Y. H., Lu, S. Y. (1983) "Increase of Warping Rigidity in Open Section of a Containership by Stiffening Plates," *Journal of Ship Research*, Vol. 27, No. 4, pp. 265–270.

Pandey, M. D., Kabir, M. Z. and Sherbourne, A. N. (1995) "Flexural-Torsional Stability of Thin-Walled Composite I-Section Beams," *Composites Engineering*, Vol. 5, No. 3, pp. 321–342.

Pignataro, M., Rizzi, N. and Luongo, A. (1983) *Stabilita, Bifurcazione e Comportamento Post-critica Delle Strutture Elastiche* (in Italian) Sec. 6.1, ESA, Roma.

Polillo, V. R., Garcia, L. F. T. and Villac, S. F. (1998) "Discussion about Geometrically Nonlinear Formulations for Combined Flexure and Torsion of Thin-Walled Open Bars," *Journal of the Brazilian Society of Mechanical Sciences*, Vol. XX, No. 1, pp. 103–115.

Qiao, P. Z. and Zou, G. P. (2002) "Free Vibration Analysis of Fiber-Reinforced Plastic Composite Cantilever I-Beams," *Mechanics of Advanced Material and Structures*, Vol. 9, No. 4, pp. 359–373.

Rand, O. (1999) "Theoretical Model for Thin-Walled Composite Beams of Open Cross-Sectional Geometry," in *American Helicopter Society, 55th Annual Forum Proceedings*, May 25–27, Washington, DC, pp. 356–367.

Reissner, E. (1952) "On Nonuniform Torsion of Cylindrical Rods," *Journal of Mathematics and Physics*, Vol. 31, pp. 214–221.

Reissner, E. (1956) "Note on Torsion with Variable Twist," *Journal of Applied Mechanics*, Vol. 23, Trans. ASME, Vol. 78, pp. 315–316.

Roberts, T. M. and Azizian, Z. G. (1983) "Nonlinear Analysis of Thin-Walled Bars of Open Cross-Section," *International Journal of Mechanical Sciences*, Vol. 25, No. 8, pp. 565–577.

Roberts, T. M. (1987) "Natural Frequencies of Thin-Walled Bars of Open Crossection," *Journal of Engineering Mechanics*, Vol. 113, No. 10, pp. 1584–1593.

Ronagh, H. R., Bradford, M. A. and Attard, M. M. (2000) "Nonlinear Analysis of Thin-Walled Members of Variable Cross-Section, Part I. Theory," *Journal of Computers & Structures*, Vol. 77, pp. 285–299.

Savic, V. Tuttle, M. E. and Zabusky, Z. B. (2001) "Optimization of Composite I-Sections Using Fiber Angles as Design Variables," *Composite Structures*, Vol. 53, pp. 265–277.

Senjanovic, I. and Fan, Y., (1991a) "Pontoon Torsional Strength Related to Ships with Large Deck Opening," *Journal of Ship Research*, Vol. 35, No. 4, pp. 339–351.

Senjanovic, I. and Fan, Y. (1991b) "On Torsional and Warping Stiffness of Thin-Walled Girders," *Thin-Walled Structures*, Vol. 11, pp. 233–276.

Sherbourne, A. N. and Kabir, M. Z. (1995) "Shear Strain Effects in Lateral Stability of Thin-Walled Fibrous Composite Beams," *Journal of Engineering Mechanics*, May, pp. 640–647.

Smith, S. J. and Bank, L. C. (1992) "Modification to Beam Theory for Bending and Twisting of Open-Section Composite Beams-Experimental Verification," *Composites Structures*, Vol. 22, No. 3, pp. 169–177.

Song, O. and Librescu, L. (1995) "Dynamic Theory of Open Cross-Section Thin-Walled Beams Composed of Advanced Composite Material," *Journal of Thermoplastic Composite Materials*, Vol. 8, No. 2, pp. 225–238.

Song, O., Librescu, L. and Jeong, N-H. (2001) "Static Response of Thin-Walled Composite I-Beams Loaded at Their Free-End Cross Section: Analytical Solution," *Composite Structures*, Vol. 52, No. 1, pp. 55–65.

Timoshenko, S. P. (1945) "Theory of Bending, Torsion and Buckling of Thin-Walled Members of Open Cross-Section," *Journal of the Franklin Institute*, Vol. 239, No. 3, 4, 5, pp. 201–219, 249–268, 343–361.

Trahair, N. S. (1993) *Flexural-Torsional Buckling of Structures*, Chapman and Hall, London.

Tso, W. K. (1965) "Coupled Vibrations of Thin-Walled Elastic Beams," *Journal of the Engineering Mechanics Division*, Proceedings of ASCE, June, pp. 33–52.

Vlasov, V. Z. (1961) *Thin-Walled Elastic Beams*, 2nd Edition, Jerusalem, Israel Program for Scientific Translation. (First Edition: Stroizdat, Moscow, 1940.)

Volovoi, V. V., Hodges, D. H., Berdichevski, V. L. and Sutyrin, V. G. (1999) "Asymptotic Theory for Static Behavior of Elastic Anisotropic I-Beam," *International Journal of Solids and Structures*, Vol. 36, pp. 1017–1043.

Wekezer, J. W. (1987) "Free Vibrations of Thin-Walled Bars with Open Crossections," *Journal of Engineering Mechanics*, Vol. 113, No. 10, pp. 1441–1453.

Wekezer, J. W. (1989) "Vibrational Analysis of Thin-Walled Bars With Open Crossection," *Journal of Structural Engineering*, Vol. 115, No. 12, pp. 2965–2978.

Wu, X. X. and Sun, C. T. (1992) "Simplified Theory for Composite Thin-Walled Beams," *AIAA Journal*, Vol. 30, No. 12, pp. 2945–2951.

Yu, Y-Y. (1969) "Thermally Induced Vibration and Flutter of a Flexible Boom," *Journal of Spacecraft and Rockets*, Vol. 6, pp. 902–910.

Yu, Y-Y. (1972) "Variational Equation of Motion for Coupled Flexure and Torsion of Bars of Thin-Walled Open Section Including Thermal Effects," *Journal of Applied Mechanics*, Trans. ASME, Vol. 38, No. 2, June, pp. 502–506.

APPENDIX A:
THE CONSTITUTIVE EQUATIONS

A.1 INTRODUCTION

In order to provide a reasonably self-contained basis for the development of 1-D constitutive equations of thin-walled beam theory, in general, and of the theory presented in this work, in particular, and to avoid any interruption in the course of our exposition, several elements on the constitutive equations of the 3-D anisotropic elasticity theory will be summarized. For similar reasons, a number of basic elements which concern the 3-D constitutive equations of piezoelectric materials, strictly related to the developments in Chapter 3 will also be supplied.

A.2 LINEARLY ELASTIC 3-D ANISOTROPIC CONTINUUM

The kinematic equations and the equations of motion are valid for every continuous medium, irrespective of its physical properties.

For an elastic body, in addition to the mentioned equations, the ones defining its physical properties have to be considered. If the strains are small enough, the physical behavior of an elastic anisotropic material can be approximated by a linear relationship between stress and strain components; such a material is termed *Hookean*, (see Wempner, 1981).

Composite materials consist of a reinforcement, generally in the form of fibers of high thermal and mechanical performance, dispersed in a surrounding matrix in an appropriate pattern. Such materials exhibit a blend of properties superior to those of their individual constituents. The matrix bonds the reinforcement together and distributes loads among the fibers; the reinforcement supports the mechanical loads. Due to their high specific stiffness and strength, resistance to corrosion as well as due to their tailorability property, such structures are prime candidates in aeronautical, aerospace, naval, and civil applications; they can operate in severe and complex environmental conditions.

Changes in temperature are commonplace in composite materials, both during fabrication and use. There are two important effects emerging from the existence of the temperature field (see Daniel and Ishai, 1994; Vinson and Sierakowski, 2002). First, most materials expand when heated and contract when cooled, and secondly, the stiffness changes so that it becomes softer and weaker as it is heated. The combination of high temperature and high humidity has a detrimental effect on the structural performance of fiber-reinforced polymeric composites. To utilize in full the potential of composite materials, a method to predict their deformation and response characteristics due to changes in temperature and moisture absorption must be available (see Daniel and Ishai, 1994).

The ingestion of moisture leads to a linear variation of deformation with swelling, similar to the thermal loading. The constitutive equations incorporating changes in the temperature (T) and the specific moisture concentration (M) from some reference levels T_0 and M_0, respectively, at which the body is considered to be stress free, can be expressed in Cartesian tensor form as:

$$\sigma_{ij} = C_{ijmn}\epsilon_{mn} - \lambda_{ij}T - \mu_{ij}M \quad (i, j, m, n = 1, 2, 3), \quad \text{(A.1)}$$

where C_{ijmn}, λ_{ij} and μ_{ij} are the elasticity, the stress-temperature tensor and the stress-moisture tensor of the anisotropic body, respectively; these are fourth and second rank tensors, respectively.

This linear relationship characterizes the anisotropic Hookean solid and it represents an extension of the Duhamel-Neumann thermoelastic constitutive equations to the hygrothermoelastic case. We employ conventions of *free* (unrepeated) and *summation* (repeated) indices.

The components C_{ijmn}, λ_{ij} and μ_{ij}, satisfy the following symmetry conditions:

$$\begin{aligned}
C_{ijmn} &= C_{jimn}, \quad C_{ijmn} = C_{ijnm}, \quad C_{ijmn} = C_{mnij}, \\
\lambda_{ij} &= \lambda_{ji}, \\
\mu_{ij} &= \mu_{ji}.
\end{aligned} \quad \text{(A.2a-e)}$$

The first symmetry relation follows directly from the symmetry of the stress tensor $\sigma_{ij} = \sigma_{ji}$, while the second arises from the symmetry of the strain tensor ϵ_{ij}. The third symmetry relation is a consequence of the interchangeability rule

$$\frac{\partial^2 \mathcal{W}}{\partial \epsilon_{ij} \partial \epsilon_{mn}} = \frac{\partial^2 \mathcal{W}}{\partial \epsilon_{mn} \partial \epsilon_{ij}}, \quad \text{(A.3)}$$

where, for the present case

$$\mathcal{W} = \frac{1}{2} C_{ijmn} \epsilon_{ij} \epsilon_{mn} - \lambda_{ij} \epsilon_{ij} T - \mu_{ij} \epsilon_{ij} M. \quad \text{(A.4)}$$

This form of (A.4) implies the existence of a free-reference-state characterized by $\sigma_{ij} = 0$ and $\epsilon_{ij} = 0$ at $T = T_0$ and $M = M_0$.

By virtue of these symmetries, the elastic and thermal coefficients can be represented in matrix form as:

$$[C_{ijmn}] \equiv \begin{bmatrix} C_{1111} & C_{1122} & C_{1133} & C_{1123} & C_{1131} & C_{1112} \\ & C_{2222} & C_{2233} & C_{2223} & C_{2231} & C_{2212} \\ & & C_{3333} & C_{3323} & C_{3331} & C_{3312} \\ & & & C_{2323} & C_{2331} & C_{2312} \\ & Symm. & & & C_{3131} & C_{3112} \\ & & & & & C_{1212} \end{bmatrix}, \quad \text{(A.5a)}$$

$$[\lambda_{ij}] \equiv \begin{bmatrix} \lambda_{11} & \lambda_{12} & \lambda_{13} \\ & \lambda_{22} & \lambda_{23} \\ Symm. & & \lambda_{33} \end{bmatrix}, \quad \text{(A.5b)}$$

$$[\mu_{ij}] \equiv \begin{bmatrix} \mu_{11} & \mu_{12} & \mu_{13} \\ & \mu_{22} & \mu_{23} \\ Symm. & & \mu_{33} \end{bmatrix}. \quad \text{(A.5c)}$$

A body exhibiting such mechanical and thermal properties is termed *triclinic*.

For this general case, the matrices C_{ijmn}, λ_{ij}, and μ_{ij} are fully populated; there are 36 *nonzero* elastic coefficients, and 9 thermal and 9 hygrothermal coefficients. The symmetry properties in the general case of anisotropy, reduce the number of *independent* elastic coefficients to 21, and the thermal and hygrothermal ones to 6 each.

A.3 MATERIAL SYMMETRY

The various symmetries can reduce the number of independent coefficients C_{ijmn}, λ_{ij}, and μ_{ij}. We examine the effect on the strain energy density function of a rotation of axes.

A.3-1 One Surface of Symmetry (Monoclinic Hookean Material)

Suppose that through each point of the elastic body there is a surface $x_3 = 0$ with respect to which the energy functional remains invariant when x_3 is replaced by $-x_3$, then

$$C_{1123} = C_{1131} = C_{2223} = C_{2231} = C_{2312} = C_{3112} = 0,$$
$$C_{3323} = C_{3331},$$
$$\lambda_{13} = \lambda_{23} = 0, \quad \text{(A.6a-d)}$$
$$\mu_{13} = \mu_{23} = 0.$$

Within this kind of symmetry, the elastic coefficients containing the index 3 either once or thrice, and the hygrothermal coefficients containing the index 3 once should vanish. The matrices in Eqs. (A.5) reduce to

$$
[C_{ijmn}] \equiv
\begin{bmatrix}
C_{1111} & C_{1122} & C_{1133} & 0 & 0 & C_{1112} \\
 & C_{2222} & C_{2233} & 0 & 0 & C_{2212} \\
 & & C_{3333} & 0 & 0 & C_{3312} \\
 & & & C_{2323} & C_{2331} & 0 \\
 & Symm. & & & C_{3131} & 0 \\
 & & & & & C_{1212}
\end{bmatrix},
\qquad \text{(A.7a)}
$$

$$
[\lambda_{ij}] \equiv
\begin{bmatrix}
\lambda_{11} & \lambda_{12} & 0 \\
 & \lambda_{22} & 0 \\
Symm. & & \lambda_{33}
\end{bmatrix},
\qquad \text{(A.7b)}
$$

$$
[\mu_{ij}] \equiv
\begin{bmatrix}
\mu_{11} & \mu_{12} & 0 \\
 & \mu_{22} & 0 \\
Symm. & & \mu_{33}
\end{bmatrix}.
\qquad \text{(A.7c)}
$$

Such a material, referred to as *monoclinic* or *monotropic* (see Wempner, 1981), is characterized by 13 *independent* elastic coefficients, 4 thermal and 4 hygro coefficients. The axis x_3 normal to the surface of symmetry is referred to as the *principal material direction*.

A.3-2 Three Planes of Symmetry (Orthotropic Material)

If the material exhibits elastic and thermal symmetries at each point with respect to the orthogonal surfaces $x_3=0$ and $x_2=0$, then *additional* coefficients, besides the ones given by Eqs. (A.6), will vanish:

$$
\begin{aligned}
C_{1112} &= C_{3312} = C_{2331} = C_{2212} = 0, \\
\lambda_{12} &= 0, \quad \mu_{12} = 0.
\end{aligned}
\qquad \text{(A.8a-c)}
$$

For this material

(i) The elastic coefficients in Eqs. (A.6) and (A.8), containing the indices 1 and/or 3 once or thrice, and the thermal and the hygrothermal coefficients including the indices 1 and/or 3 once should vanish,

(ii) There are number of 9 *independent* elastic coefficients, as well as 3 thermal and 3 hygrothermal coefficients that characterize the orthotropic solid,

(iii) Being two orthogonal surfaces of material symmetry, the third mutually orthogonal surface will constitute another surface of symmetry, and,

(iv) The coordinate axes normal to the surfaces of elastic symmetry are referred to as *the principal material directions*.

For an orthotropic body, the thermoelastic matrices are

$$
[C_{ijmn}] \equiv
\begin{bmatrix}
C_{1111} & C_{1122} & C_{1133} & 0 & 0 & 0 \\
 & C_{2222} & C_{2233} & 0 & 0 & 0 \\
 & & C_{3333} & 0 & 0 & 0 \\
 & & & C_{2323} & 0 & 0 \\
 & Symm. & & & C_{3131} & 0 \\
 & & & & & C_{1212}
\end{bmatrix},
\qquad \text{(A.9a)}
$$

$$
[\lambda_{ij}] \equiv
\begin{bmatrix}
\lambda_{11} & 0 & 0 \\
 & \lambda_{22} & 0 \\
 Symm. & & \lambda_{33}
\end{bmatrix},
\qquad \text{(A.9b)}
$$

$$
[\mu_{ij}] \equiv
\begin{bmatrix}
\mu_{11} & 0 & 0 \\
 & \mu_{22} & 0 \\
 Symm. & & \mu_{33}
\end{bmatrix}.
\qquad \text{(A.9c)}
$$

A.3-3 Transverse Isotropy

A transversely isotropic material is one in which, at each point there is one principal material direction about which there is rotational symmetry. If, the principal direction is parallel at each point to the axis x_3, and there is rotational symmetry about the x_3 axis, then the surface x_1 - x_2 is the surface of isotropy; the matrices of elastic, thermal and hygrothermal coefficients become

$$
[C_{ijmn}] \equiv
\begin{bmatrix}
C_{1111} & C_{1122} & C_{1133} & 0 & 0 & 0 \\
 & C_{1111} & C_{1133} & 0 & 0 & 0 \\
 & & C_{3333} & 0 & 0 & 0 \\
 & & & C_{3131} & 0 & 0 \\
 & Symm. & & & C_{3131} & 0 \\
 & & & & & (C_{1111} - C_{1122})/2
\end{bmatrix},
\qquad \text{(A.10a)}
$$

$$
[\lambda_{ij}] \equiv
\begin{bmatrix}
\lambda_{11} & 0 & 0 \\
 & \lambda_{11} & 0 \\
 Symm. & & \lambda_{33}
\end{bmatrix},
\qquad \text{(A.10b)}
$$

$$
[\mu_{ij}] \equiv
\begin{bmatrix}
\mu_{11} & 0 & 0 \\
 & \mu_{11} & 0 \\
 Symm. & & \mu_{33}
\end{bmatrix}.
\qquad \text{(A.10c)}
$$

There are 5 *independent* elastic coefficients, 2 thermal hygrothermal coefficients. One of the materials exhibiting the transverse isotropy properties is the *pyrolytic graphite*, which due to its special thermo-mechanical properties is likely to be an excellent candidate for being used in the thermal protection of aerospace vehicles. Moreover, *unidirectional fiber reinforced composites* belong to the class of transverse isotropy; in such a case the isotropy plane is normal to the axis of the fibers. If the fibers are aligned in the x_1-direction, the x_2-x_3 plane is the plane of isotropy, and the indices 2 and 3 are interchangeable. For this case

$$C_{2222} = C_{3333}, \quad C_{1133} = C_{1122},$$
$$C_{3131} = C_{1212}, \quad C_{2323} = (C_{2222} - C_{2233})/2, \tag{A.11a-f}$$
$$\lambda_{22} = \lambda_{33}, \quad \mu_{22} = \mu_{33}.$$

The matrices of elastic and thermal coefficients become

$$[C_{ijmn}] \equiv \begin{bmatrix} C_{1111} & C_{1122} & C_{1122} & 0 & 0 & 0 \\ & C_{2222} & C_{2233} & 0 & 0 & 0 \\ & & C_{2222} & 0 & 0 & 0 \\ & & & (C_{2222} - C_{2233})/2 & 0 & 0 \\ Symm. & & & & C_{1212} & 0 \\ & & & & & C_{1212} \end{bmatrix},$$

$$\tag{A.12a}$$

$$[\lambda_{ij}] \equiv \begin{bmatrix} \lambda_{11} & 0 & 0 \\ & \lambda_{22} & 0 \\ Symm. & & \lambda_{22} \end{bmatrix}, \tag{A.12b}$$

$$[\mu_{ij}] \equiv \begin{bmatrix} \mu_{11} & 0 & 0 \\ & \mu_{22} & 0 \\ Symm. & & \mu_{22} \end{bmatrix}. \tag{A.12c}$$

A.3-4 Isotropic Hookean Material

If there are no preferred directions, the role of the indices 1,2 and 3 are fully interchangeable. In this case

$$C_{1111} = C_{2222} = C_{3333},$$
$$C_{1122} = C_{1133} = C_{2233},$$
$$C_{1212} = C_{3131} = C_{2323} \equiv (C_{1111} - C_{1122})/2, \tag{A.13a-e}$$
$$\lambda_{11} = \lambda_{22} = \lambda_{33},$$
$$\mu_{11} = \mu_{22} = \mu_{33}.$$

As a result, the matrices of elastic and hygrothermal coefficients become

$[C_{ijmn}]$

$$
\equiv \begin{bmatrix}
C_{1111} & C_{1122} & C_{1122} & 0 & 0 & 0 \\
 & C_{1111} & C_{1122} & 0 & 0 & 0 \\
 & & C_{1111} & 0 & 0 & 0 \\
 & & & (C_{1111} - C_{1122})/2 & 0 & 0 \\
 & & & & (C_{1111} - C_{1122})/2 & 0 \\
Symm. & & & & & (C_{1111} - C_{1122})/2
\end{bmatrix},
$$

(A.14a)

$$
[\lambda_{ij}] \equiv \begin{bmatrix}
\lambda_{11} & 0 & 0 \\
 & \lambda_{11} & 0 \\
Symm. & & \lambda_{11}
\end{bmatrix},
$$

(A.14b)

$$
[\mu_{ij}] \equiv \begin{bmatrix}
\mu_{11} & 0 & 0 \\
 & \mu_{11} & 0 \\
Symm. & & \mu_{11}
\end{bmatrix}.
$$

(A.14c)

The body has two independent elastic coefficients, and one thermal and one hygrothermal coefficient. An account of the elastic and hygrothermal coefficients for various types of elastic symmetries is presented in Table A.1.

Table A.1: Summary of 3-D elastic and hygrothermal coefficients

Type of material symmetry	Number of elastic independent coefficients	Number of thermal (hygro) independent coefficients	Number of nonzero elastic coefficients	Number of nonzero thermal (hygro) coefficients
Triclinic	21	6	36	9
Monoclinic	13	4	20	5
Orthotropic	9	3	12	3
Transversely Isotropic	5	2	12	3
Isotropic	2	1	12	3

For orthotropic, transversely isotropic, and isotropic materials the shear and normal components of stress and strain tensors are not coupled.

Triclinic and isotropic materials represent two extreme types of anisotropy: the former has one surface of symmetry, the latter an infinity. For a thorough discussion of the various types of anisotropy, see Bogdanovich and Pastore (1996).

A.4 ALTERNATIVE FORM OF THE 3-D CONSTITUTIVE EQUATIONS

Recall the general constitutive equations of elastic anisotropic Hookean solids:

$$\sigma_{ij} = C_{ijmn}\epsilon_{mn} - \lambda_{ij}T - \mu_{ij}M, \tag{A.15a}$$

$$\epsilon_{ij} = S_{ijmn}\sigma_{mn} + \alpha_{ij}T + \beta_{ij}M. \tag{A.15b}$$

Equation (A.15b) represents the inverted counterpart of Eq. (A.15a). Here S_{ijmn} denotes the compliance tensor, while α_{ij} and β_{ij} denote the thermal expansion and the moisture-swelling tensors, respectively. Their components follow the same pattern of *nonzero* and number of independent coefficients in the various symmetry cases as their inverted counterparts. In order to establish the relationship between (S_{ijmn}, C_{ijmn}), (λ_{ij}, μ_{ij}), and $(\alpha_{ij}, \beta_{ij})$ we will replace σ_{mn} in Eq (A.15b) by its expression in Eq. (A.15a). This yields (Librescu, 1990)

$$\epsilon_{ij} = S_{ijmn}\left[C_{mnpq}\epsilon_{pq} - \lambda_{mn}T - \mu_{mn}M\right] + \alpha_{ij}T + \beta_{ij}M. \tag{A.16}$$

Collection and identification of the coefficients associated with ϵ_{ij}, T and M results in :

$$\begin{aligned}
S_{ijmn}C_{mnpq} &= (\delta_{ip}\delta_{jq} + \delta_{iq}\delta_{jp})/2, \\
\alpha_{ij} - S_{ijmn}\lambda_{mn} &= 0, \\
\beta_{ij} - S_{ijmn}\mu_{mn} &= 0,
\end{aligned} \tag{A.17a-c}$$

where δ_{ij} denotes the Kronecker delta ($\delta_{ij} = 0$ if $i \neq j$, and $\delta_{ij} = 1$ if $i = j$). Equation (A.17a) expresses the fact that C_{ijmn} and S_{mnpq} are inverse tensors, whereas the remaining two equations establish a relationship between α_{ij} and λ_{ij}, and between β_{ij} and μ_{ij}, respectively. Multiplication of equations ($A.17b$) and ($A.17c$) by C_{mnpq} and employment of Eq. ($A.17a$) yields

$$\lambda_{pq} = \alpha_{ij}C_{ijpq}, \qquad \mu_{pq} = \beta_{ij}C_{ijpq}. \tag{A.18a,b}$$

In light of Eqs. (A.18) and ($A.17b$, c), the constitutive equations (A.15) become

$$\begin{aligned}
\sigma_{ij} &= C_{ijmn}\left[\epsilon_{mn} - \alpha_{mn}T - \beta_{mn}M\right], \\
\epsilon_{ij} &= S_{ijmn}\left[\sigma_{mn} + \lambda_{mn}T + \mu_{mn}M\right].
\end{aligned} \tag{A.19a,b}$$

This form of the constitutive equations will be used later.

A.5 TRANSFORMATION OF MATERIAL COEFFICIENTS

In Section A.3.2 the characteristics of an orthotropic material were described assuming that its axes of material symmetry coincide with the geometrical axes (i.e. with the global axes of convenience).

However, the principal directions of orthotropy of the constituent layer materials often do not coincide with the geometrical axes whose orientation is dictated by the character of the problem to be investigated or by the boundary conditions.

In addition, for a laminated composite structure the principal directions of orthotropy are different in each constituent layer, this being part of the implementation of the tailoring technique.

In such cases, the transformation of the constitutive relations with respect to the global coordinate system is required.

If the constitutive behavior in the principal material coordinate system $(x_1', x_2', x_3' (\equiv x_3))$ is known, the constitutive law in the axes of convenience, (x_1, x_2, x_3), as represented in Fig. A.1, is found by applying the tensor transformation rules.

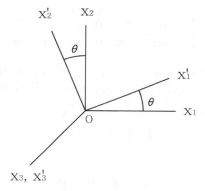

Figure A.1: Original and rotated material coordinate systems

The constitutive equations referred to the primed coordinate system, after an in-plane rotation of the x_1 - x_2 axes about the x_3-axis, assume the form:

$$\sigma_{i'j'} = C_{i'j'k'l'}(\epsilon_{k'l'} - \alpha_{k'l'}T - \beta_{k'l'}M), \tag{A.20}$$

where

$$\begin{aligned}
\sigma_{i'j'} &= a_{i'l}a_{j'm}\sigma_{lm}, \\
\epsilon_{k'l'} &= a_{k'p}a_{l'q}\epsilon_{pq}, \\
\alpha_{k'l'} &= a_{k'p}a_{l'q}\alpha_{pq}, \\
\beta_{k'l'} &= a_{k'p}a_{l'q}\beta_{pq}.
\end{aligned} \tag{A.21a-d}$$

In Eqs. (A.21), $a_{i'j}$ denotes the direction cosine of the angle between the $x_{i'}$ and x_j axes, so that

$$[a] \equiv a_{i'j} = \begin{bmatrix} m & n & 0 \\ -n & m & 0 \\ 0 & 0 & 1 \end{bmatrix} \tag{A.22}$$

where $m \equiv \cos\theta, n \equiv \sin\theta, \theta \in [0, 2\pi]$ being the counterclockwise rotation angle about the positive x_3-axis. Note that $[a]$ is an orthogonal matrix which satisfies the relationship $[a][a]^T = \mathbf{I} = [a]^T[a], ([a]^T \equiv [a]^{-1})$ where superscript T denotes the matrix transpose, while $\mathbf{I}$ is the identity matrix whose diagonal elements are the unites.

In matrix form Eq. $(A.21a)$ is given by

$$\{\sigma'\} = [T_3]\{\sigma\}. \tag{A.23}$$

Here

$$T_3(\theta) = \begin{bmatrix} m^2 & n^2 & 0 & 0 & 0 & 2mn \\ n^2 & m^2 & 0 & 0 & 0 & -2mn \\ 0 & 0 & 1 & 0 & 0 & 0 \\ 0 & 0 & 0 & m & -n & 0 \\ 0 & 0 & 0 & n & m & 0 \\ -mn & mn & 0 & 0 & 0 & m^2 - n^2 \end{bmatrix} \tag{A.24}$$

is the transformation matrix, while $\{\sigma'\}$ and $\{\sigma\}$ denote the stress tensors in the primed and unprimed coordinates, respectively. Similarly, the transformation of strains is given by

$$\{\gamma'\} = [\tilde{T}_3]\{\gamma\}, \tag{A.25}$$

where

$$\tilde{T}_3(\theta) = \begin{bmatrix} m^2 & n^2 & 0 & 0 & 0 & mn \\ n^2 & m^2 & 0 & 0 & 0 & -mn \\ 0 & 0 & 1 & 0 & 0 & 0 \\ 0 & 0 & 0 & m & -n & 0 \\ 0 & 0 & 0 & n & m & 0 \\ -2mn & 2mn & 0 & 0 & 0 & m^2 - n^2 \end{bmatrix}, \tag{A.26}$$

and $\gamma_{ij} \equiv \epsilon_{ij}$ for $i = j$ and $\gamma_{ij} \equiv 2\epsilon_{ij}$ for $i \neq j$. Here ϵ_{ij} denote the tensorial strain components. It is apparent that the following relationship between $\tilde{T}_3$ and T_3 holds valid

$$\tilde{T}_3(\theta) = [T_3^{-1}]^T = [T_3(-\theta)]^T. \tag{A.27}$$

Replacement in Eq. (A.20) of Eqs. (A.21) and (A.25) results, in view of Eqs. (A.24), (A.26) and (A.27), in the constitutive equations expressed in the global coordinates (x_1, x_2, x_3), where the elastic and hygrothermal properties in the principal material coordinates (x_1', x_2', x_3') are assumed to be known. In matrix form these are given as:

$$\{\sigma\} = [C](\{\gamma\} - \{\alpha\}T - \{\beta\}M), \tag{A.28}$$

where

$$\{\sigma\} = \{\sigma_{11}, \sigma_{22}, \sigma_{33}, \sigma_{23}, \sigma_{31}, \sigma_{12}\}^T,$$
$$\{\gamma\} = \{\epsilon_{11}, \epsilon_{22}, \epsilon_{33}, \gamma_{23}, \gamma_{31}, \gamma_{12}\}^T,$$
$$\{\alpha\} \equiv [T_3]^{-1} \{\alpha'\} = \{\alpha_{11}, \alpha_{22}, \alpha_{33}, 0, 0, \alpha_{12}\}^T, \quad \text{(A.29a-d)}$$
$$\{\beta\} \equiv [T_3]^{-1} \{\beta'\} = \{\beta_{11}, \beta_{22}, \beta_{33}, 0, 0, \beta_{12}\}^T,$$

while

$$[C] = [T_3]^{-1} [C'] [\tilde{T}_3], \quad \text{(A.30a)}$$

where

$$[C'] \equiv [C_{i'j'k'l'}] \text{ and } [T_3(\theta)]^{-1} = [T_3(-\theta)]. \quad \text{(A.30b,c)}$$

Expressed in full, Eq. (A.28) becomes

$$
\left\{
\begin{array}{c}
\sigma_{11} \\ \sigma_{22} \\ \sigma_{33} \\ \sigma_{23} \\ \sigma_{31} \\ \sigma_{12}
\end{array}
\right\}
=
\left[
\begin{array}{cccccc}
C_{1111} & C_{1122} & C_{1133} & 0 & 0 & C_{1112} \\
C_{2211} & C_{2222} & C_{2233} & 0 & 0 & C_{2212} \\
C_{3311} & C_{3322} & C_{3333} & 0 & 0 & C_{3312} \\
0 & 0 & 0 & C_{2323} & C_{2331} & 0 \\
0 & 0 & 0 & C_{3123} & C_{3131} & 0 \\
C_{1211} & C_{1222} & C_{1233} & 0 & 0 & C_{1212}
\end{array}
\right]
\left\{
\begin{array}{c}
\epsilon_{11} - \alpha_{11}T - \beta_{11}M \\
\epsilon_{22} - \alpha_{22}T - \beta_{22}M \\
\epsilon_{33} - \alpha_{33}T - \beta_{33}M \\
\gamma_{23} \\
\gamma_{31} \\
\gamma_{12} - \alpha_{12}T - \beta_{12}M
\end{array}
\right\}.
$$
$$\text{(A.31)}$$

where the symmetry of the stiffness matrix $[C]$ is implied. In Eqs. (A.30a) and (A.29c,d), $[C]$, $\{\alpha\}$ and $\{\beta\}$ denote, respectively, the matrices of transformed elastic and hygrothermal coefficients in the global coordinates.

We will use Voigt's contracted notation in place of the tensor indicial notation. This consists of replacing in the elastic and hygrothermal coefficients of a pair of indices by a single one, through the replacement rules: $11 \rightarrow 1; 22 \rightarrow 2; 33 \rightarrow 3; 23$ (or $32) \rightarrow 4; 31$ (or $13) \rightarrow 5; 12$ (or $21) \rightarrow 6$.

On the basis of Eqs. (A.30b) and (A.25), Eq. (*A.30a*) can be written in full as:

$$C_{11} = C_{1'1'}m^4 + 2(C_{1'2'} + 2C_{6'6'})m^2 n^2 + C_{2'2'}n^4,$$
$$C_{12} = (C_{1'1'} + C_{2'2'} - 4C_{6'6'})m^2 n^2 + C_{1'2'}(m^4 + n^4),$$
$$C_{13} = C_{1'3'}m^2 + C_{2'3'}n^2,$$
$$C_{16} = -C_{2'2'}mn^3 + C_{1'1'}m^3 n - (C_{1'2'} + 2C_{6'6'})mn(m^2 - n^2),$$
$$C_{22} = C_{2'2'}m^4 + 2(C_{1'2'} + 2C_{6'6'})m^2 n^2 + C_{1'1'}n^4,$$
$$C_{23} = C_{1'3'}n^2 + C_{2'3'}m^2,$$
$$C_{26} = C_{1'1'}mn^3 - C_{2'2'}m^3 n + (C_{1'2'} + 2C_{6'6'})mn(m^2 - n^2), \quad \text{(A.32a-m)}$$
$$C_{33} = C_{3'3'},$$

$$C_{36} = (C_{1'3'} - C_{2'3'})mn,$$
$$C_{44} = C_{4'4'}m^2 + C_{5'5'}n^2,$$
$$C_{45} = (C_{5'5'} - C_{4'4'})mn,$$
$$C_{55} = C_{4'4'}n^2 + C_{5'5'}m^2,$$
$$C_{66} = (C_{1'1'} + C_{2'2'} - 2C_{1'2'})m^2n^2 + C_{6'6'}(m^2 - n^2)^2.$$

In addition, the elements of matrices α and β are, respectively:

$$\alpha_1 = \alpha_{1'}m^2 + \alpha_{2'}n^2,$$
$$\alpha_2 = \alpha_{2'}m^2 + \alpha_{1'}n^2,$$
$$\alpha_3 = \alpha_{3'}, \hspace{3cm} \text{(A.33a-d)}$$
$$\alpha_6 = 2(\alpha_{1'} - \alpha_{2'})mn,$$

and

$$\beta_1 = \beta_{1'}m^2 + \beta_{2'}n^2,$$
$$\beta_2 = \beta_{2'}m^2 + \beta_{1'}n^2,$$
$$\beta_3 = \beta_{3'}, \hspace{3cm} \text{(A.34a-d)}$$
$$\beta_6 = 2(\beta_{1'} - \beta_{2'})mn.$$

In Eqs. (A.32) through (A.34), $C_{I'J'}$, $\alpha_{I'}$ and $\beta_{I'}$ denote the elastic and hygrothermal coefficients of an orthotropic material expressed in principal material directions. The relationships in Eqs. (A.31) through (A.34) reveal that in non-principal coordinates (or off-axis coordinates), an orthotropic body is characterized, likewise a monoclinic material, by 13 independent elastic constants and 4 independent thermal/hygrothermal coefficients. In this case, the coupling between normal and shear effects is present. However, C_{16}, C_{26}, C_{36}, α_6, and β_6 are not independent, but merely linear combinations of the remaining constants. Since the material, in its principal material directions is orthotropic, it is called a *generally orthotropic* material.

Equations (A.32) through (A.34) show that when the principal and the global (geometric) coordinates coincide (implying $\theta = 0$), C_{ij}, α_i, and β_i reduce to $C_{i'j'}$, $\alpha_{i'}$, and $\beta_{i'}$, respectively. In such a case the material behaves as a truly orthotropic one, and the equations (A.31) become:

$$\begin{Bmatrix} \sigma_1 \\ \sigma_2 \\ \sigma_3 \\ \sigma_4 \\ \sigma_5 \\ \sigma_6 \end{Bmatrix} = \begin{bmatrix} C_{11} & C_{12} & C_{13} & 0 & 0 & 0 \\ C_{12} & C_{22} & C_{23} & 0 & 0 & 0 \\ C_{13} & C_{23} & C_{33} & 0 & 0 & 0 \\ 0 & 0 & 0 & C_{44} & 0 & 0 \\ 0 & 0 & 0 & 0 & C_{55} & 0 \\ 0 & 0 & 0 & 0 & 0 & C_{66} \end{bmatrix} \begin{Bmatrix} \epsilon_1 - \alpha_1 T - \beta_1 M \\ \epsilon_2 - \alpha_2 T - \beta_2 M \\ \epsilon_3 - \alpha_3 T - \beta_3 M \\ \gamma_4 \\ \gamma_5 \\ \gamma_6 \end{Bmatrix}.$$

$$\text{(A.35)}$$

It should be noticed that when the coordinate transformations have to be implemented, the tensor indices must be employed.

A.6 ALTERNATIVE REPRESENTATIONS

Equations (A.32) through (A.34) show that the stiffness quantities of a generally orthotropic material depend on the angle of orientation of its principal material directions. The technology taking advantage of this property, in the sense of modifying the response of the structure according to predetermined goals by changing fiber orientation angle, is referred to as *structural tailoring*. It has been used successfully in the aeronautical industry and is likely to play a significant role in the design of future generations of aeronautical/aerospace vehicles, of helicopter and turbine blades, of space stations, satellites, etc. For practical purposes an alternative representation of equations (A.32) through (A.34) will be presented. In this representation a separation of the quantities that are invariant to the rotation of axis (and which represent the truly composite material properties) and of those depending on the rotation of the principal material directions is applied.

To this end, and following Tsai and Hahn (1980) we use the elementary trigonometric identities:

$$
\begin{aligned}
m^4 &= (3 + 4\cos 2\theta + \cos 4\theta)/8, \\
m^3 n &= (2\sin 2\theta + \sin 4\theta)/8, \\
mn^3 &= (2\sin 2\theta - \sin 4\theta)/8, \\
m^2 n^2 &= (1 - \cos 4\theta)/8, \\
n^4 &= (3 - 4\cos 2\theta + \cos 4\theta)/8.
\end{aligned}
\qquad \text{(A.36a-e)}
$$

Direct substitution of Eqs. (A.36) into Eqs. (A.32) through (A.34) yields:

$$
\begin{aligned}
C_{11} &= U_1' + U_2' \cos 2\theta + U_3' \cos 4\theta, \\
C_{22} &= U_1' - U_2' \cos 2\theta + U_3' \cos 4\theta, \\
C_{12} &= U_4' - U_3' \cos 4\theta, \\
C_{66} &= U_5' - U_3' \cos 4\theta, \\
C_{16} &= (U_2' \sin 2\theta)/2 + U_3' \sin 4\theta, \\
C_{26} &= (U_2' \sin 2\theta)/2 - U_3' \sin 4\theta,
\end{aligned}
\qquad \text{(A.37a-l)}
$$

$$
\begin{aligned}
C_{13} &= U_8' + U_9' \cos 2\theta, \\
C_{23} &= U_8' - U_9' \cos 2\theta, \\
C_{55} &= U_{10}' + U_{11}' \cos 2\theta,
\end{aligned}
$$

$$C_{45} = U'_{11} \sin 2\theta,$$
$$C_{44} = U'_{10} - U'_{11} \cos 2\theta,$$
$$C_{36} = U'_{9} \sin 2\theta,$$

and

$$\alpha_1 = V'_1 + V'_2 \cos 2\theta,$$
$$\alpha_2 = V'_1 - V'_2 \cos 2\theta,$$
$$\alpha_6 = V'_2 \sin 2\theta, \qquad\qquad\qquad \text{(A.38a-d)}$$
$$\alpha_3 = \alpha_{3'},$$

and

$$\beta_1 = P'_1 + P'_2 \cos 2\theta,$$
$$\beta_2 = P'_1 - P'_2 \cos 2\theta,$$
$$\beta_6 = P'_2 \sin 2\theta, \qquad\qquad\qquad \text{(A.39a-d)}$$
$$\beta_3 = \beta_{3'}.$$

The quantities U'_i, V'_i, and P'_i depend only on the properties of the orthotropic material and are independent of ply-angle θ).

Their expressions are:

$$U'_1 = (3C_{1'1'} + 3C_{2'2'} + 2C_{1'2'} + 4C_{6'6'})/8,$$
$$U'_2 = (C_{1'1'} - C_{2'2'})/2,$$
$$U'_3 = (C_{1'1'} + C_{2'2'} - 2C_{1'2'} - 4C_{6'6'})/8,$$
$$U'_4 = (C_{1'1'} + C_{2'2'} + 6C_{1'2'} - 4C_{6'6'})/8,$$
$$U'_5 = (C_{1'1'} + C_{2'2'} - 2C_{1'2'} + 4C_{6'6'})/8,$$
$$U'_8 = (C_{1'3'} + C_{2'3'})/2,$$
$$U'_9 = (C_{1'3'} - C_{2'3'})/2, \qquad\qquad\qquad \text{(A.40a-m)}$$
$$U'_{10} = (C_{5'5'} + C_{4'4'})/2,$$
$$U'_{11} = (C_{5'5'} - C_{4'4'})/2,$$
$$V'_1 = (\alpha_{1'} + \alpha_{2'})/2,$$
$$V'_2 = (\alpha_{1'} - \alpha_{2'})/2,$$
$$P'_1 = (\beta_{1'} + \beta_{2'})/2,$$
$$P'_2 = (\beta_{1'} - \beta_{2'})/2.$$

In Eqs. (A.37) through (A.39), the positive sign of θ is opposite to that considered in the tensorial transformation, i.e. it is positive when measured from the on-axis (i.e. the symmetry axes) of a ply to the laminate coordinate system. This sign reversal of the angle θ, enables one to emphasize the contribution of an off-axis ply-angle to the laminate. This representation is useful in the tailoring analysis of laminates. Each constituent lamina could have a different ply orientation.

The coordinate transformation (A.32) through (A.34) yield the following invariants

$$
\begin{aligned}
C_{11} &+ C_{22} + C_{33} + 2(C_{12} + C_{23} + C_{31}) \\
&= C_{1'1'} + C_{2'2'} + C_{3'3'} + 2(C_{1'2'} + C_{2'3'} + C_{3'1'}), \\
C_{11} &+ C_{22} + C_{33} + 2(C_{44} + C_{55} + C_{66}) \\
&= C_{1'1'} + C_{2'2'} + C_{3'3'} + 2(C_{4'4'} + C_{5'5'} + C_{6'6'}), \\
\alpha_1 &+ \alpha_2 + \alpha_3 = \alpha_{1'} + \alpha_{2'} + \alpha_{3'}, \\
\beta_1 &+ \beta_2 + \beta_3 = \beta_{1'} + \beta_{2'} + \beta_{3'}.
\end{aligned}
\tag{A.41a-d}
$$

A.7 ELASTIC COEFFICIENTS OF ORTHOTROPIC MATERIALS IN TERMS OF ENGINEERING CONSTANTS

In practice, the coefficients C_{ij} for the orthotropic material are defined in terms of the engineering constants. To this end, as a first step, from Eq. (A.17a) the relationships between C_{ij} and S_{ij} for an orthotropic body are derived. These are:

$$
\begin{aligned}
C_{11} &= (S_{22}S_{33} - S_{23}^2)/S, \\
C_{12} &= (S_{13}S_{23} - S_{12}S_{33})/S, \\
C_{22} &= (S_{33}S_{11} - S_{13}^2)/S, \\
C_{13} &= (S_{12}S_{23} - S_{13}S_{22})/S, \\
C_{33} &= (S_{11}S_{22} - S_{12}^2)/S, \\
C_{23} &= (S_{12}S_{13} - S_{23}S_{11})/S, \\
C_{44} &= 1/S_{44}, \\
C_{55} &= 1/S_{55}, \\
C_{66} &= 1/S_{66},
\end{aligned}
\tag{A.42a-i}
$$

where

$$
S = S_{11}S_{22}S_{33} - S_{11}S_{23}^2 - S_{22}S_{13}^2 - S_{33}S_{12}^2 + 2S_{12}S_{23}S_{13}.
\tag{A.43}
$$

The components of the compliance tensor are expressed in terms of the engineering constants as

$$[S_{ij}] = \begin{bmatrix} \dfrac{1}{E_1} & -\dfrac{\nu_{21}}{E_2} & -\dfrac{\nu_{31}}{E_3} & 0 & 0 & 0 \\[2mm] -\dfrac{\nu_{12}}{E_1} & \dfrac{1}{E_2} & -\dfrac{\nu_{32}}{E_3} & 0 & 0 & 0 \\[2mm] -\dfrac{\nu_{13}}{E_1} & -\dfrac{\nu_{23}}{E_2} & \dfrac{1}{E_3} & 0 & 0 & 0 \\[2mm] 0 & 0 & 0 & \dfrac{1}{G_{23}} & 0 & 0 \\[2mm] 0 & 0 & 0 & 0 & \dfrac{1}{G_{31}} & 0 \\[2mm] 0 & 0 & 0 & 0 & 0 & \dfrac{1}{G_{12}} \end{bmatrix} . \tag{A.44}$$

Substitution of the expressions of S_{ij} as given by (A.44) into the right-hand side members of Eqs. (A.42) and (A.43) gives the relationships between C_{ij} and the engineering coefficients as:

$$\begin{aligned}
C_{11} &= E_1(1 - \nu_{23}\nu_{32})/\Delta, \\
C_{22} &= E_2(1 - \nu_{31}\nu_{13})/\Delta, \\
C_{33} &= E_3(1 - \nu_{12}\nu_{21})/\Delta, \\
C_{12} &= E_1(\nu_{21} + \nu_{31}\nu_{23})/\Delta, \\
C_{13} &= E_1(\nu_{31} + \nu_{21}\nu_{32})/\Delta, \\
C_{23} &= E_2(\nu_{32} + \nu_{12}\nu_{31})/\Delta, \\
C_{44} &= G_{23}, \\
C_{55} &= G_{31}, \\
C_{66} &= G_{12}.
\end{aligned} \tag{A.45a-i}$$

Here E_1, E_2, E_3 denote the Young's moduli along the principal material directions; G_{12}, G_{23}, G_{31} denote the shear moduli, ν_{12}, ν_{23}, ν_{31} denote the Poisson's ratios while

$$\Delta = 1 - \nu_{12}\nu_{21} - \nu_{32}\nu_{23} - \nu_{13}\nu_{31} - 2\nu_{21}\nu_{13}\nu_{32}. \tag{A.46}$$

From symmetry considerations of the compliance matrix $[S_{ij}]$, the following reciprocal relationship hold

$$\frac{E_i}{\nu_{ij}} = \frac{E_j}{\nu_{ji}} \quad (i, j = 1, 2, 3) \tag{A.47}$$

For a *unidirectional composite reinforced in the x_1-direction* we have

$$E_2 = E_3 \equiv E, \quad E_1 \equiv E',$$
$$\nu_{13} = \nu_{12} \equiv \nu', \quad \nu_{23} = \nu_{32} \equiv \nu,$$
$$G_{23} \equiv G = E/2(1+\nu), \quad G_{12} = G_{13} \equiv G', \quad \text{(A.48a-g)}$$
$$\nu_{21} = \nu_{31} = \nu' E/E'.$$

For a transversely isotropic solid whose plane of isotropy is parallel at each point to the plane x_1-x_2, the elastic constants are

$$E_1 = E_2 \equiv E; \ E_3 = E'; \ G_{12} \equiv G(\equiv E/2(1+\nu)); \ G_{23} = G_{31} \equiv G',$$
$$\nu_{31} = \nu_{32} \equiv \nu'; \ \nu_{12} = \nu_{21} \equiv \nu; \ \nu_{13} = \nu_{23} \equiv \nu' E/E'. \quad \text{(A.49a-g)}$$

For these two cases, the associated stiffness coefficients C_{ij} are obtainable by replacing Eqs. (A.48) and (A.49) in (A.45).

For completeness, the elastic stiffness coefficients C_{ij} for this type of transverse isotropy are

$$C_{11} = \frac{E(E\nu'^2 - E')}{\Delta} = C_{22},$$
$$C_{12} = -\frac{E(E\nu'^2 + E'\nu)}{\Delta},$$
$$C_{13} = -\frac{EE'\nu'(1+\nu)}{\Delta} = C_{23}, \quad \text{(A.50a-e)}$$
$$C_{33} = -\frac{E'^2(1-\nu^2)}{\Delta},$$
$$C_{44} = C_{55} = G',$$

where $\Delta = (1+\nu)(2E\nu'^2 + E'\nu - E')$.

The pyrolytic graphite and its alloys feature this type of transverse isotropy. Due to its very low thermal conductivity coefficient in the thickness direction as compared to that in the tangential direction, it is used in the design of thermal protection of aerospace vehicle systems.

A.8 ANISOTROPIC THIN-WALLED BEAMS

A.8-1 Introduction

A thin-walled composite beam consists of a number of laminae bonded and/or cured together. The theory of thin-walled beams is essentially a 1-D approximation of the 3-D theory of elasticity. This should result in a system of ordinary differential governing equations expressed in terms of the 1-D displacements measures $u_0(x_2)$, $v_0(x_2)$, $w_0(x_2)$, the 1-D rotational measures $\theta_1(x_2)$, $\theta_2(x_2)$, $\phi(x_2)$ and the warping measure W_M. To obtain the system of 1-D governing equations, the constitutive equations have to feature the same 1-D character. An intermediate stage, consisting of the derivation of 2-D constitutive equations will be considered; the 2-D stress resultants and stress couples should relate the 2-D strain measures, where $x_1 (\equiv s)$ and $x_2 (\equiv z)$ are the independent coordinates involved in these equations.

A.8-2 3-D Equations for a Lamina

Consider a laminated composite thin-walled beam consisting of N orthotropic laminae referenced to a nonprincipal material coordinate system. For the k-th constituent lamina of the laminate, the constitutive equations are

$$
\begin{Bmatrix} \sigma_{ss} \\ \sigma_{zz} \\ \sigma_{nn} \\ \sigma_{zn} \\ \sigma_{sn} \\ \sigma_{sz} \end{Bmatrix}_{(k)} = [C]_{(k)} \begin{Bmatrix} \epsilon_{ss} - \alpha_s T - \beta_s M \\ \epsilon_{zz} - \alpha_z T - \beta_z M \\ \epsilon_{nn} - \alpha_n T - \beta_n M \\ \gamma_{zn} \\ \gamma_{sn} \\ \gamma_{sz} - \alpha_{sz} T - \beta_{sz} M \end{Bmatrix}_{(k)}, \qquad (A.51a)
$$

where

$$
[C]_{(k)} \equiv \begin{bmatrix} C_{11} & C_{12} & C_{13} & 0 & 0 & C_{16} \\ C_{12} & C_{22} & C_{23} & 0 & 0 & C_{26} \\ C_{13} & C_{23} & C_{33} & 0 & 0 & C_{36} \\ 0 & 0 & 0 & C_{44} & C_{45} & 0 \\ 0 & 0 & 0 & C_{45} & C_{55} & 0 \\ C_{16} & C_{26} & C_{36} & 0 & 0 & C_{66} \end{bmatrix}_{(k)}. \qquad (A.51b)
$$

Since the material properties in Eqs. (A.51) are affiliated to the k-th layer, for the sake of identification, the matrices are associated with the subscript (k). Unless otherwise specified, the indices s, z, and n will replace the former indices 1, 2, and 3, respectively.

As in the theory of plates and shells (see e.g. Librescu, 1975, 1990), we will invoke the fact that σ_{nn} is negligible when compared with the other stress components. The condition $\sigma_{nn} = 0$ leads to the following transverse normal strain

$$
\epsilon_{nn} = -\frac{C_{13}}{C_{33}}(\epsilon_{ss} - \alpha_s T - \beta_s M) - \frac{C_{23}}{C_{33}}(\epsilon_{zz} - \alpha_z T - \beta_z M)
$$

$$
-\frac{C_{36}}{C_{33}}(\gamma_{sz} - \alpha_{sz} T - \beta_{sz} M) + \alpha_n T + \beta_n M. \qquad (A.52)
$$

As a result, Eq. (A.51a) is recast in the form:

$$
\left\{
\begin{array}{c}
\sigma_{ss} \\
\sigma_{zz} \\
\sigma_{zn} \\
\sigma_{sn} \\
\sigma_{sz}
\end{array}
\right\}_{(k)}
= [\bar{Q}]_{(k)}
\left\{
\begin{array}{c}
\varepsilon_{ss} - \bar{\alpha}_s T - \bar{\beta}_s M \\
\varepsilon_{zz} - \bar{\alpha}_z T - \bar{\beta}_z M \\
\gamma_{zn} \\
\gamma_{sn} \\
\gamma_{sz} - \bar{\alpha}_{sz} T - \bar{\beta}_{sz} M
\end{array}
\right\}_{(k)}, \qquad (A.53)
$$

where $[\bar{Q}]$ is the matrix of the *reduced elastic coefficients* given by

$$
[\bar{Q}] =
\begin{bmatrix}
\bar{Q}_{11} & \bar{Q}_{12} & 0 & 0 & \bar{Q}_{16} \\
\bar{Q}_{21} & \bar{Q}_{22} & 0 & 0 & \bar{Q}_{26} \\
0 & 0 & \bar{Q}_{44} & \bar{Q}_{45} & 0 \\
0 & 0 & \bar{Q}_{54} & \bar{Q}_{55} & 0 \\
\bar{Q}_{61} & \bar{Q}_{62} & 0 & 0 & \bar{Q}_{66}
\end{bmatrix}, \qquad (A.54a)
$$

where

$$
\bar{Q}_{IJ} = C_{IJ} - \frac{C_{I3} C_{J3}}{C_{33}} = \bar{Q}_{JI}, \quad (I, J = 1, 2, 6), \qquad (A.54b)
$$

$$
\bar{Q}_{LM} \equiv C_{LM}, \quad (L, M = 4, 5) \qquad (A.54c)
$$

while

$$
\bar{\alpha}_I = \alpha_I - \frac{C_{I3}}{C_{33}} \alpha_3, \quad \bar{\beta}_I = \beta_I - \frac{C_{I3}}{C_{33}} \beta_3, \qquad (A.54d)
$$

are the *reduced* stress-temperature and stress-moisture coefficients, respectively.

For an orthotropic body:

$$
\bar{Q}_{11} = \frac{E_1}{1 - \nu_{12}\nu_{21}}; \quad \bar{Q}_{12} = \frac{E_1 \nu_{21}}{1 - \nu_{12}\nu_{21}} = \frac{E_2 \nu_{12}}{1 - \nu_{12}\nu_{21}}
$$

$$
\bar{Q}_{22} = \frac{E_2}{1 - \nu_{12}\nu_{21}}; \quad \bar{Q}_{66} = \bar{C}_{66} = G_{12}; \quad \bar{Q}_{44} = G_{23}; \quad \bar{Q}_{55} = G_{31}.
$$

$$
(A.55a\text{-}f)
$$

A.8-3 2-D Stress-Resultants and Stress-Couples

In order to obtain a 2-D variant of the constitutive equations, the 2-D stress resultants and stress couples have to be defined. As in the theory of plates and shells, these quantities are defined so that these 2-D quantities are statically equivalent to their 3-D counterparts. Suppose that the kth lamina occupies the domain $n_{(k-1)} < n < n_{(k)}, k = 1 \ldots N$, we define:

(a) *The membrane stress resultants*

$$\left\{ \begin{array}{c} N_{ss} \\ N_{zz} \\ N_{sz} \end{array} \right\} = \sum_{k=1}^{N} \int_{n_{(k-1)}}^{n_{(k)}} \left\{ \begin{array}{c} \sigma_{ss} \\ \sigma_{zz} \\ \sigma_{sz} \end{array} \right\}_{(k)} dn, \qquad (A.56)$$

(b) *The transverse shear stress resultants*

$$\left\{ \begin{array}{c} N_{zn} \\ N_{sn} \end{array} \right\} = \sum_{k=1}^{N} \int_{n_{(k-1)}}^{n_{(k)}} \left\{ \begin{array}{c} \sigma_{zn} \\ \sigma_{sn} \end{array} \right\}_{(k)} dn, \qquad (A.57)$$

and

(c) *The stress couples*

$$\left\{ \begin{array}{c} L_{zz} \\ L_{sz} \end{array} \right\} = \sum_{k=1}^{N} \int_{n_{(k-1)}}^{n_{(k)}} \left\{ \begin{array}{c} \sigma_{zz} \\ \sigma_{sz} \end{array} \right\}_{(k)} n\,dn. \qquad (A.58)$$

Here $n_{(k)}$ and $n_{(k-1)}$ denote the distances from the middle surface of the cross-section to the upper and lower surfaces of the k-th layer, respectively. The $n_{(k)}$ values are signed numbers varying from $n_{(0)} = -h/2$ to $n_{(N)} = h/2$ in a laminate of N layers and of total thickness h (see Fig. A.2). Equations (A.56) through (A.58) show that the stress resultants and stress couples do not depend on n, but are functions of (s, z) coordinates.

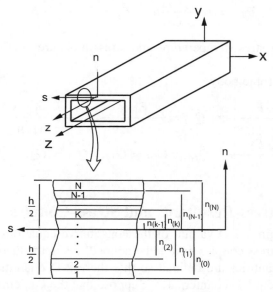

Figure A.2: Geometry of an N-ply laminate

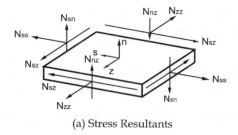

(a) Stress Resultants

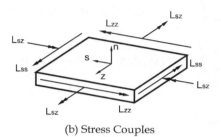

(b) Stress Couples

Figure A.3: Stress resultants and stress couples on a beam element

Equations (A.56) through (A.58) involve the assumption that the reference surface of the thin-walled beam is *shallow* in the sense of $h/R_1 << 1$, where R_1 denotes the radius of curvature of the mid-surface in the s-direction. The stress resultants have the units of force per unit length, while the stress couples have the units of moment per unit length. The positive sign convention for the stress resultants and stress couples is displayed in Fig. A.3.

The symmetry of the stress tensor and of the shallowness of beam middle surface, implies $N_{sz} = N_{zs}$ and $L_{sz} = L_{zs}$.

A.8-4 A First Step Toward Obtaining the Constitutive Equations of Thin-Walled Open Beams

Upon substituting Eqs. (A.53) into (A.56) through (A.58), we obtain the 2-D constitutive equations. As an example, consider the circumferential (hoop) membrane stress resultant N_{ss}. In conjunction with Eq. (A.53), (see Song, 1990)

$$
\begin{aligned}
N_{ss}(s, z) &= \sum_{k=1}^{N} \int_{n_{(k-1)}}^{n_{(k)}} \sigma_{ss}^{(k)} dn \\
&= \sum_{k=1}^{N} \int_{n_{(k-1)}}^{n_{(k)}} [\bar{Q}_{11}(\epsilon_{ss} - \bar{\alpha}_s T - \bar{\beta}_s M) + \bar{Q}_{12}(\epsilon_{zz} - \bar{\alpha}_z T - \bar{\beta}_z M) \\
&\quad + \bar{Q}_{16}(\gamma_{sz} - \bar{\alpha}_{sz} T - \bar{\beta}_{sz} M)]_{(k)} dn.
\end{aligned}
\tag{A.59}
$$

With the definitions of *stretching* and *bending-stretching coupling* stiffness quantities A_{ij} and B_{ij}, respectively

$$A_{ij} = \sum_{k=1}^{N} \int_{n(k-1)}^{n(k)} \bar{Q}_{ij}^{(k)} \, dn, \quad B_{ij} = \sum_{k=1}^{N} \int_{n(k-1)}^{n(k)} \bar{Q}_{ij}^{(k)} n \, dn, \qquad \text{(A.60a,b)}$$

as well as those of hygrothermal hoop forces (per unit length)

$$N_{ss}^{T}(s, z) = \sum_{k=1}^{N} \int_{n(k-1)}^{n(k)} T(s, z, n)[\bar{Q}_{11}\bar{\alpha}_s + \bar{Q}_{12}\bar{\alpha}_z + \bar{Q}_{16}\bar{\alpha}_{sz}]_{(k)} \, dn,$$

$$N_{ss}^{M}(s, z) = \sum_{k=1}^{N} \int_{n(k-1)}^{n(k)} M(s, z, n)[\bar{Q}_{11}\bar{\beta}_s + \bar{Q}_{12}\bar{\beta}_z + \bar{Q}_{16}\bar{\beta}_{sz}]_{(k)} \, dn,$$

$$\text{(A.61a,b)}$$

Eq. (A.59) becomes:

$$\begin{aligned}
N_{ss}(s, z) =\, & A_{11}\epsilon_{ss} + A_{12}\epsilon_{zz}^{(0)} + B_{12}\epsilon_{zz}^{(1)} \\
& + A_{16}\gamma_{sz}^{(0)} + 2B_{16}W_M - N_{ss}^{T} - N_{ss}^{M}.
\end{aligned} \qquad \text{(A.62)}$$

Proceeding thus for the other stress resultant quantities, we obtain

$$\begin{Bmatrix} N_{ss} \\ N_{zz} \\ N_{sz} \end{Bmatrix} = \begin{bmatrix} A_{11} & A_{12} & A_{16} \\ A_{12} & A_{22} & A_{26} \\ A_{16} & A_{26} & A_{66} \end{bmatrix} \begin{Bmatrix} \epsilon_{ss} \\ \epsilon_{zz}^{(0)} \\ \gamma_{sz}^{(0)} \end{Bmatrix} + \begin{bmatrix} B_{12} & B_{16} \\ B_{22} & B_{26} \\ B_{26} & B_{66} \end{bmatrix} \begin{Bmatrix} \epsilon_{zz}^{(1)} \\ 2W_M \end{Bmatrix}$$

$$- \begin{Bmatrix} N_{ss}^{T} \\ N_{zz}^{T} \\ N_{sz}^{T} \end{Bmatrix} - \begin{Bmatrix} N_{ss}^{M} \\ N_{zz}^{M} \\ N_{sz}^{M} \end{Bmatrix}, \qquad \text{(A.63)}$$

where, in addition to Eqs. (A.61), we have also

$$N_{zz}^{T}(s, z) = \sum_{k=1}^{N} \int_{n(k-1)}^{n(k)} T(s, z, n)[\bar{Q}_{12}\bar{\alpha}_s + \bar{Q}_{22}\bar{\alpha}_z + \bar{Q}_{26}\bar{\alpha}_{sz}]_{(k)} \, dn,$$

$$N_{sz}^{T}(s, z) = \sum_{k=1}^{N} \int_{n(k-1)}^{n(k)} T(s, z, n)[\bar{Q}_{16}\bar{\alpha}_s + \bar{Q}_{26}\bar{\alpha}_z + \bar{Q}_{66}\bar{\alpha}_{sz}]_{(k)} \, dn. \quad \text{(A.64a,b)}$$

N_{zz}^{M} and N_{sz}^{M}, are obtainable from Eq (A.64) by applying the substitutions

$$T \rightarrow M; \quad \bar{\alpha}_s \rightarrow \bar{\beta}_s, \quad \bar{\alpha}_z \rightarrow \bar{\beta}_z, \quad \bar{\alpha}_{sz} \rightarrow \bar{\beta}_{sz} \qquad \text{(A.65a-d)}$$

The transverse shear stress resultants, N_{zn} and N_{sn}, are

$$N_{zn}(s, z) = A_{44}\gamma_{zn} + A_{45}\gamma_{sn},$$
$$N_{sn}(s, z) = A_{45}\gamma_{zn} + A_{55}\gamma_{sn}. \tag{A.66a,b}$$

By virtue of in-plane cross-section non-deformability implying $\gamma_{sn} = 0$, Eqs. (A.66) reduce to

$$N_{zn}(s, z) = A_{44}\gamma_{zn},$$
$$N_{sn}(s, z) = A_{45}\gamma_{zn}. \tag{A.67a,b}$$

The transverse shear stiffness quantities A_{LM} are

$$A_{LM} = \sum_{k=1}^{N} k_{LM}^2 \, \bar{Q}_{LM}^{(k)}(n_{(k)} - n_{(k-1)}), \quad (L, M = 4, 5) \tag{A.68}$$

where k_{LM}^2 are transverse shear correction factors.

If we assume that transverse shear stresses have a parabolic distribution across the laminate thickness, and define a continuous weighting function (see Vinson and Sierakowski, 2002), we obtain

$$f(n) = \frac{5}{4}\left[1 - \left(\frac{n}{h/2}\right)^2\right], \tag{A.69}$$

an alternative expression for A_{LM}:

$$A_{LM} = \frac{5}{4} \sum_{k=1}^{N} \bar{Q}_{LM}^{(k)}\left[n_{(k)} - n_{(k-1)} - \frac{4}{3h^2}(n_{(k)}^3 - n_{(k-1)}^3)\right]. \tag{A.70}$$

For a single layered beam, $N = 1$, $n_{(k)} = h/2$, $n_{(k-1)} = -h/2$, and

$$A_{LM} = \frac{5}{6}\bar{C}_{LM}h \quad (L, M = 4, 5). \tag{A.71}$$

This form can also be obtained from Eq. (A.68), by prescribing Reissner's transverse shear correction factor $k_{LM}^2 \equiv k^2 = 5/6$.

Similarly, the stress couples are expressible in matrix form as:

$$\left\{\begin{array}{c} L_{zz} \\ L_{sz} \end{array}\right\} = \left[\begin{array}{ccc} B_{12} & B_{22} & B_{26} \\ B_{16} & B_{26} & B_{66} \end{array}\right] \left\{\begin{array}{c} \epsilon_{ss} \\ \epsilon_{zz}^{(0)} \\ \gamma_{sz}^{(0)} \end{array}\right\} + \left[\begin{array}{cc} D_{22} & D_{26} \\ D_{26} & D_{66} \end{array}\right] \left\{\begin{array}{c} \epsilon_{zz}^{(1)} \\ 2W_M \end{array}\right\}$$

$$- \left\{\begin{array}{c} L_{zz}^T \\ L_{sz}^T \end{array}\right\} - \left\{\begin{array}{c} L_{zz}^M \\ L_{sz}^M \end{array}\right\}. \tag{A.72}$$

In these relationships, both the stiffness quantities B_{ij} and D_{ij} occur. The latter as well as the *thermal and hygric moments* (per unit length) appearing in Eqs. (A.72) are defined as:

$$D_{ij} = \sum_{k=1}^{N} \int_{n_{(k-1)}}^{n_{(k)}} \bar{Q}_{ij}^{(k)} n^2 dn. \tag{A.73}$$

$$L_{zz}^{T}(s, z) = \sum_{k=1}^{N} \int_{n_{(k-1)}}^{n_{(k)}} T(s, z, n)(\bar{Q}_{12}\bar{\alpha}_s + \bar{Q}_{22}\bar{\alpha}_z + \bar{Q}_{26}\bar{\alpha}_{sz})_{(k)} n dn,$$

$$L_{sz}^{T}(s, z) = \sum_{k=1}^{N} \int_{n_{(k-1)}}^{n_{(k)}} T(s, z, n)(\bar{Q}_{16}\bar{\alpha}_s + \bar{Q}_{26}\bar{\alpha}_z + \bar{Q}_{66}\bar{\alpha}_{sz})_{(k)} n dn,$$

$$L_{zz}^{M}(s, z) = \sum_{k=1}^{N} \int_{n_{(k-1)}}^{n_{(k)}} M(s, z, n)(\bar{Q}_{12}\bar{\beta}_s + \bar{Q}_{22}\bar{\beta}_z + \bar{Q}_{26}\bar{\beta}_{sz})_{(k)} n dn,$$

$$\text{(A.74a-d)}$$

$$L_{sz}^{M}(s, z) = \sum_{k=1}^{N} \int_{n_{(k-1)}}^{n_{(k)}} M(s, z, n)(\bar{Q}_{16}\bar{\beta}_s + \bar{Q}_{26}\bar{\beta}_z + \bar{Q}_{66}\bar{\beta}_{sz})_{(k)} n dn.$$

A.8-5 Remarks on Stiffness Quantities

Equations (A.63), (A.67), and (A.72) are the 2-D thermoelastic constitutive equations for the laminated composite anisotropic thin-walled beams. In their expression, in contrast to their isotropic single-layer counterpart, a variety of coupling effects are present. These couplings constitute new degrees of freedom in the hands of the designer. Among them, the stiffness quantities B_{ij} induce a coupling between membrane stress resultants (see Eqs. (A.63)) and the extension strain $\epsilon_{zz}^{(1)}$ varying through the beam thickness, and the warping W_M, and between the stress couples (see Eqs. (A.72)) and the in-plane strain components.

The stiffness terms A_{16}, A_{26} introduce a coupling between the circumferential (hoop) and the axial membrane stress resultants and the membrane shear strain components. Since the material properties are piecewise constant through the thickness of the beam, the stiffness quantities can be expressed as:

$$A_{ij} = \sum_{k=1}^{N} \bar{Q}_{ij}^{(k)}[n_{(k)} - n_{(k-1)}], \quad B_{ij} = \frac{1}{2}\sum_{k=1}^{N} \bar{Q}_{ij}^{(k)}[n_{(k)}^2 - n_{(k-1)}^2],$$

$$D_{ij} = \frac{1}{3}\sum_{k=1}^{N} \bar{Q}_{ij}^{(k)}[n_{(k)}^3 - n_{(k-1)}^3]. \tag{A.75a-c}$$

For uniform thermal and moisture fields throughout the thickness [i.e, when $T \equiv T(s, z)$ and $M \equiv M(s, z)$], the expressions for the associated thermal and hygric forces and moments, become

$$N_{ss}^T(s, z) = T(s, z) \sum_{k=1}^{N} \{\bar{Q}_{11}\bar{\alpha}_s + \bar{Q}_{12}\bar{\alpha}_z + \bar{Q}_{16}\bar{\alpha}_{sz}\}_{(k)}[n_{(k)} - n_{(k-1)}],$$

$$N_{zz}^T(s, z) = T(s, z) \sum_{k=1}^{N} \{\bar{Q}_{12}\bar{\alpha}_s + \bar{Q}_{22}\bar{\alpha}_z + \bar{Q}_{26}\bar{\alpha}_{sz}\}_{(k)}[n_{(k)} - n_{(k-1)}],$$

$$N_{sz}^T(s, z) = T(s, z) \sum_{k=1}^{N} \{\bar{Q}_{16}\bar{\alpha}_s + \bar{Q}_{26}\bar{\alpha}_z + \bar{Q}_{66}\bar{\alpha}_{sz}\}_{(k)}[n_{(k)} - n_{(k-1)}],$$

$$L_{zz}^T(s, z) = \frac{1}{2}T(s, z) \sum_{k=1}^{N} \{\bar{Q}_{11}\bar{\alpha}_s + \bar{Q}_{12}\bar{\alpha}_z + \bar{Q}_{16}\bar{\alpha}_{sz}\}_{(k)}[n_{(k)}^2 - n_{(k-1)}^2],$$

$$L_{sz}^T(s, z) = \frac{1}{2}T(s, z) \sum_{k=1}^{N} \{\bar{Q}_{16}\bar{\alpha}_s + \bar{Q}_{26}\bar{\alpha}_z + \bar{Q}_{66}\bar{\alpha}_{sz}\}_{(k)}[n_{(k)}^2 - n_{(k-1)}^2].$$

(A.76a-e)

N_{ss}^M, N_{zz}^M, N_{sz}^N, L_{zz}^M and L_{sz}^M, are obtainable from their thermal counterparts, Eqs. (A.76), by using the substitutions, Eq. (A.65). The laminate stacking sequence code specifies the ply composition, the exact location and sequence of various plies.

For an unsymmetric laminate this prescription is done from the bottom of the laminate, to the top, while for a symmetric laminate, from the middle surface to the top surface. Several standard notations are given next: $[0_3^o/90_2^o/45^o/45^o/90_2^o/0_3^o]_T \equiv [0_3^o/90_2^o/45^o]_S$, where subscripts T and S stand for total number of plies and symmetric sequence respectively, $[\pm\theta^o]_n = [+\theta^o/-\theta^o/+\theta^o/-\theta^o \cdots]$; $[\pm 45^o]_n = [+45^o/-45^o/+45^o/-45^o]$, where, the subscript n indicates that the sub-laminate $[\pm\theta^o]$ is repeated n times.

There are a number of special laminate configurations that, in addition to their technological importance, lead to simplifications of the 2-D constitutive law. These simplifications depend basically on the *odd* and *even* character of stiffness $\bar{Q}_{ij}$ and of thermal $\bar{\alpha}_i$ and hygrothermal $\bar{\beta}_i$ expansion coefficients, as well as on the stacking sequence characteristics of the laminate. The odd and even features of $\bar{Q}_{ij}$, $\bar{\alpha}_i$, and $\bar{\beta}_i$ (i.e. the change and the invariance in sign of the respective quantities respectively, when ply-angle $\theta \rightarrow -\theta$) can easily be revealed when using Eqs. (A.32) through (A.34) and are summarized by

$$\bar{Q}_{ij}(\theta) = \begin{cases} -\bar{Q}_{ij}(-\theta) & \text{if } (ij) = (16), (26), (45), (36) \\ \bar{Q}_{ij}(-\theta) & \text{if } (ij) = (11), (12), (13), (23), (22), (44), (55), (66) \end{cases}$$

$$\bar{\alpha}(\theta) = \begin{cases} -\bar{\alpha}(-\theta) \text{ if } i = 6 \\ \bar{\alpha}(-\theta) \text{ if } i = 1, 2, 3 \end{cases} \tag{A.77a,b}$$

whereas for $\bar{\beta}_i(\theta)$ a similar behavior as for $\bar{\alpha}_i(\theta)$ is obtained. Based on the odd and even properties of a generally orthotropic material, special classes of laminate configurations can be distinguished.

A.8-6 Selected Classes of Laminate Configurations

(a) *Symmetric Laminates*

This case requires both geometric and material symmetry properties as well as symmetry of the hygrothermal distributions about the middle surface of the beam. For the conditions of symmetry stated above,

$$B_{ij} = 0. \tag{A.78}$$

For the same case, when the thermal and hygrothermal fields are uniform across the laminate thickness, Eqs. (A.74) imply

$$L_{zz}^T = L_{sz}^T = L_{zz}^M = L_{sz}^M = 0, \tag{A.79}$$

which means that there are no hygrothermal induced moments (per unit length).

(b) *Balanced Laminates*

To every lamina with ply-orientation $+\theta$, there exists a companion laminae of identical material properties and thickness of orientation $-\theta$ *somewhere* within the laminate. For this case

$$A_{16} = A_{26} = 0, \tag{A.80a}$$

and

$$N_{sz}^T = N_{sz}^M = 0. \tag{A.80b}$$

For a balanced and symmetric laminate, $B_{ij}=0$, $A_{16} = A_{26} = 0$, $N_{sz}^T = N_{sz}^M = 0$ and $L_{sz}^T = L_{sz}^M = 0$. It clearly appears that the use of *unbalanced* angle-ply laminates (i.e, of such laminates for which A_{16}, A_{26} and D_{26} are different from zero), is essential when the tailoring technique should be implemented. A balanced laminate can be symmetric, antisymmetric or asymmetric.

(c) *Antisymmetric Laminates*

For this case, to every lamina with orientation $+\theta$ and location n with respect to the middle surface of the beam, there is another lamina with orientation $-\theta$ at the location $-n$. This laminate consists of an even number of plies. This configuration yields the simplification

$$A_{16} = A_{26} = D_{16} = D_{26} = 0. \tag{A.81}$$

Moreover, if the antisymmetric laminate consist of pairs of $+\theta_i$ and $-\theta_i$ orientations, symmetrically located about the middle surface and having the same thickness and elastic properties, besides Eq. (A.81), we have

$$B_{11} = B_{22} = B_{12} = B_{66} = 0. \tag{A.82}$$

Additional simplifications can occur for cross-ply laminated composite thin-walled beam structures. For this case the reader is referred to the monographs by Daniel and Ishai (1994), Bogdanovich and Pastore (1996), Vinson and Sierakowski (2002) and Reddy (2004), where other ply sequence configurations of practical importance are also considered.

A.8-7 Equations for Open Cross-Section Beams

One of the assumptions in general use within the theory of thin-walled beams is that the stress component σ_{ss} is negligible compared with the remaining stress components. This means that N_{ss} can be assumed negligible in the constitutive equations. In light of this assumption and in conjunction with Eq. (A.62)

$$\epsilon_{ss} = \frac{1}{A_{11}}(-A_{12}\epsilon_{zz}^{(0)} - A_{16}\gamma_{sz}^{(0)} - 2B_{16}W_M - B_{12}\epsilon_{zz}^{(1)} + N_{ss}^T + N_{ss}^M). \tag{A.83}$$

Replacement of Eq.(A.83) into the constitutive equations (A.63b,c) yields the final form of 2-D constitutive equations for thin-walled open beams:

$$\left\{ \begin{array}{c} N_{zz} \\ N_{sz} \end{array} \right\} = \left[\begin{array}{cccc} K_{11} & K_{12} & K_{13} & K_{14} \\ K_{21} & K_{22} & K_{23} & K_{24} \end{array} \right] \left\{ \begin{array}{c} \epsilon_{zz}^{(0)} \\ \gamma_{sz}^{(0)} \\ W_M \\ \epsilon_{zz}^{(1)} \end{array} \right\} - \left\{ \begin{array}{c} \hat{N}_{zz}^T \\ \hat{N}_{sz}^T \end{array} \right\} - \left\{ \begin{array}{c} \hat{N}_{zz}^T \\ \hat{N}_{sz}^M \end{array} \right\},$$

$$\left\{ \begin{array}{c} L_{zz} \\ L_{sz} \end{array} \right\} = \left[\begin{array}{cccc} K_{41} & K_{42} & K_{43} & K_{44} \\ K_{51} & K_{52} & K_{53} & K_{54} \end{array} \right] \left\{ \begin{array}{c} \epsilon_{zz}^{(0)} \\ \gamma_{sz}^{(0)} \\ W_M \\ \epsilon_{zz}^{(1)} \end{array} \right\} - \left\{ \begin{array}{c} \hat{L}_{zz}^T \\ \hat{L}_{sz}^T \end{array} \right\} - \left\{ \begin{array}{c} \hat{L}_{zz}^M \\ \hat{L}_{sz}^M \end{array} \right\}.$$

$$\tag{A.84a,b}$$

The constitutive equations (A.67a,b) associated with the transverse shear stress resultants, remain unchanged. Within this variant of constitutive equations, the modified stiffness quantities K_{ij} are defined by:

$$K_{11} \equiv A_{22} - A_{12}^2/A_{11},$$
$$K_{12} \equiv A_{26} - A_{12}A_{16}/A_{11} = K_{21},$$
$$K_{13} \equiv 2(B_{26} - A_{12}B_{16}/A_{11}),$$
$$K_{14} \equiv B_{22} - A_{12}B_{12}/A_{11} = K_{41},$$
$$K_{22} \equiv A_{66} - A_{16}^2/A_{11},$$
$$K_{23} \equiv 2(B_{66} - A_{16}B_{16}/A_{11}),$$
$$K_{24} \equiv B_{26} - A_{16}B_{12}/A_{11} = K_{42}, \qquad \text{(A.85a-m)}$$
$$K_{43} \equiv 2(D_{26} - B_{12}B_{16}/A_{11}),$$
$$K_{44} \equiv D_{22} - B_{12}^2/A_{11},$$
$$K_{51} \equiv B_{26} - B_{16}A_{12}/A_{11},$$
$$K_{52} \equiv B_{66} - B_{16}A_{16}/A_{11},$$
$$K_{53} \equiv 2(D_{66} - B_{16}^2/A_{11}),$$
$$K_{54} \equiv D_{26} - B_{12}B_{16}/A_{11},$$

while the modified hygrothermal forces and moments are:

$$\hat{N}_{zz}^T \equiv N_{zz}^T - A_{12}N_{ss}^T/A_{11},$$
$$\hat{N}_{zz}^M \equiv N_{zz}^M - A_{12}N_{ss}^M/A_{11},$$
$$\hat{N}_{sz}^T \equiv N_{sz}^T - A_{16}N_{ss}^T/A_{11},$$
$$\hat{N}_{sz}^M \equiv N_{sz}^M - A_{16}N_{ss}^M/A_{11}, \qquad \text{(A.86a-h)}$$
$$\hat{L}_{zz}^T \equiv L_{zz}^T - B_{12}N_{ss}^T/A_{11},$$
$$\hat{L}_{zz}^M \equiv L_{zz}^M - B_{12}N_{ss}^M/A_{11},$$
$$\hat{L}_{sz}^T \equiv L_{sz}^T - B_{16}N_{ss}^T/A_{11},$$
$$\hat{L}_{sz}^M \equiv L_{sz}^M - B_{16}N_{ss}^M/A_{11}.$$

A.8-8 2-D Constitutive Equations for Closed Cross-Section Beams

The procedure of deriving the constitutive relationships for closed cross-section beams is similar to that used for open beams. For thin-walled beams of closed cross-section, the difference lies in the expression of the torsional function. In conjunction with Eq. (2.1-62) the expression of the shear strain component is

$$\gamma_{sz} = \gamma_{sz}^{(0)} + \gamma_{sz}^{(t)}, \qquad \text{(A.87a)}$$

where

$$\gamma_{sz}^{(0)}(s, z) = \gamma_{xz}(z)\frac{dx}{ds} + \gamma_{yz}(z)\frac{dy}{ds}, \qquad (A.87b)$$

and

$$\gamma_{sz}^{(t)} = \Psi W_M. \qquad (A.87c)$$

Herein

$$\Psi \equiv \psi + 2n; \quad W_M \equiv \phi', \qquad (A.88a,b)$$

where ψ is the torsional function that is expressed by Eq. (2.1-44c) considered in conjunction with (2.1-36c) and (2.1-39a), or in an inclusive form by Eq. (9.1-2).

Proceeding as for open cross-section beams, and assuming that h and G_{sz} are uniform along the beam circumference we obtain the following 2-D constitutive equations

$$\left\{ \begin{array}{c} N_{ss} \\ N_{zz} \\ N_{sz} \end{array} \right\} = \left[\begin{array}{ccc} A_{11} & A_{12} & A_{16} \\ A_{12} & A_{22} & A_{26} \\ A_{16} & A_{26} & A_{66} \end{array} \right] \left\{ \begin{array}{c} \epsilon_{ss} \\ \epsilon_{zz}^{(0)} \\ \gamma_{sz}^{(0)} \\ +2\frac{\Omega}{\beta}W_M \end{array} \right\}$$

$$+ \left\{ \begin{array}{c} B_{12} \\ B_{22} \\ B_{26} \end{array} \right\} \epsilon_{zz}^{(1)} + 2 \left\{ \begin{array}{c} B_{16} \\ B_{26} \\ B_{66} \end{array} \right\} W_M - \left\{ \begin{array}{c} N_{ss}^T \\ N_{zz}^T \\ N_{sz}^T \end{array} \right\} - \left\{ \begin{array}{c} N_{ss}^M \\ N_{zz}^M \\ N_{sz}^M \end{array} \right\},$$

$$N_{zn} = A_{44}\gamma_{zn},$$

$$N_{sn} = A_{45}\gamma_{zn}, \qquad (A.89a\text{-}d)$$

$$\left\{ \begin{array}{c} L_{zz} \\ L_{sz} \end{array} \right\} = \left[\begin{array}{ccc} B_{12} & B_{22} & B_{26} \\ B_{16} & B_{26} & B_{66} \end{array} \right] \left\{ \begin{array}{c} \epsilon_{ss} \\ \epsilon_{zz}^{(0)} \\ \gamma_{sz}^{(0)} + 2\frac{\Omega}{\beta}W_M \end{array} \right\} + \left\{ \begin{array}{c} D_{22} \\ D_{66} \end{array} \right\} \epsilon_{zz}^{(1)}$$

$$+ 2 \left\{ \begin{array}{c} D_{26} \\ D_{66} \end{array} \right\} W_M - \left\{ \begin{array}{c} L_{zz}^T \\ L_{sz}^T \end{array} \right\} - \left\{ \begin{array}{c} L_{zz}^M \\ L_{sz}^M \end{array} \right\}.$$

Here

$$\Omega \equiv \frac{1}{2}\oint r_n ds, \quad \beta \equiv \oint ds. \qquad (A.89e,f)$$

Note that the warping measure W_M strongly intervenes in the constitutive equations of both open and closed cross-section beams.

Remark:
One should observe that due to cross-section nondeformability implying that $\gamma_{sn} = 0$, the term $N_{sn}\,\gamma_{sn}$ in the strain energy becomes immaterial. For this reason, the stiffness A_{45} does not appear in the 1-D stiffnesses a_{ij}.

A.8-9 Final Form of 2-D Constitutive Equations

The smallness of the hoop stress resultant N_{ss} combined with Eq (A.89a) yields

$$\epsilon_{ss} = -\frac{1}{A_{11}}\left(A_{12}\epsilon_{zz}^{(0)} + A_{16}\gamma_{sz}^{(0)} + 2A_{16}\frac{\Omega}{\beta}W_M\right.$$

$$\left. + 2B_{16}W_M + B_{12}\epsilon_{zz}^{(1)} - N_{ss}^T - N_{ss}^M\right). \tag{A.90}$$

Replacing Eq. (A.90) in Eqs. (A.89a,d) leads to

$$\left\{\begin{array}{c} N_{zz} \\ N_{sz} \end{array}\right\} = \left[\begin{array}{cccc} K_{11} & K_{12} & K_{13} & K_{14} \\ K_{21} & K_{22} & K_{23} & K_{24} \end{array}\right] \left\{\begin{array}{c} \epsilon_{zz}^{(0)} \\ \gamma_{sz}^{(0)} \\ W_M \\ \epsilon_{zz}^{(1)} \end{array}\right\} - \left\{\begin{array}{c} \hat{N}_{zz}^T \\ \hat{N}_{sz}^T \end{array}\right\} - \left\{\begin{array}{c} \hat{N}_{zz}^M \\ \hat{N}_{sz}^M \end{array}\right\},$$

$$\left\{\begin{array}{c} L_{zz} \\ L_{sz} \end{array}\right\} = \left[\begin{array}{cccc} K_{41} & K_{42} & K_{43} & K_{44} \\ K_{51} & K_{52} & K_{53} & K_{54} \end{array}\right] \left\{\begin{array}{c} \epsilon_{zz}^{(0)} \\ \gamma_{sz}^{(0)} \\ W_M \\ \epsilon_{zz}^{(1)} \end{array}\right\} - \left\{\begin{array}{c} \hat{L}_{zz}^T \\ \hat{L}_{sz}^T \end{array}\right\} - \left\{\begin{array}{c} \hat{L}_{zz}^M \\ \hat{L}_{sz}^M \end{array}\right\}.$$

$$\tag{A.91a,b}$$

As expected, the constitutive equations (A.89b,c) remain unchanged. The stiffness quantities that are affected by this change are K_{13}, K_{23}, K_{43}, and K_{53}. Their expressions are

$$K_{13} = 2\left(A_{26} - \frac{A_{12}A_{16}}{A_{11}}\right)\frac{\Omega}{\beta} + 2\left(B_{26} - \frac{A_{12}B_{16}}{A_{11}}\right),$$

$$K_{23} = 2\left(A_{66} - \frac{A_{16}^2}{A_{11}}\right)\frac{\Omega}{\beta} + 2\left(B_{66} - \frac{A_{16}B_{16}}{A_{11}}\right),$$

$$K_{43} = 2\left(B_{26} - \frac{B_{12}A_{16}}{A_{11}}\right)\frac{\Omega}{\beta} + 2\left(D_{26} - \frac{B_{12}B_{16}}{A_{11}}\right), \tag{A.92a-d}$$

$$K_{53} = 2\left(B_{66} - \frac{B_{16}A_{16}}{A_{11}}\right)\frac{\Omega}{\beta} + 2\left(D_{66} - \frac{B_{16}^2}{A_{11}}\right).$$

The remaining stiffness quantities K_{ij} defined by Eq. (A.85) remain unaltered. Note that the constitutive equations for multicell closed section beams are similar to those for single cell counterparts. Thus, Eqs.(A.91) and (A.89b,c) can be applied to multicell beam structures as well.

A.8-10 Unified Form of 2-D Constitutive Equations

A unified expression of 2-D constitutive equations, applicable to both open and closed cross-section beams will be displayed next. This is given by

$$
\begin{Bmatrix} N_{zz} \\ N_{sz} \\ L_{zz} \\ L_{sz} \\ N_{zn} \\ N_{sn} \end{Bmatrix} = \begin{bmatrix} K_{11} & K_{12} & K_{13} & K_{14} & 0 \\ K_{21} & K_{22} & K_{23} & K_{24} & 0 \\ K_{41} & K_{42} & K_{43} & K_{44} & 0 \\ K_{51} & K_{52} & K_{53} & K_{54} & 0 \\ 0 & 0 & 0 & 0 & A_{44} \\ 0 & 0 & 0 & 0 & A_{45} \end{bmatrix} \begin{Bmatrix} \epsilon_{zz}^{(0)} \\ \gamma_{sz}^{(0)} \\ W_M \\ \epsilon_{zz}^{(1)} \\ \gamma_{zn} \end{Bmatrix}
$$

$$
- \begin{Bmatrix} \hat{N}_{zz}^T \\ \hat{N}_{sz}^T \\ \hat{L}_{zz}^T \\ \hat{L}_{sz}^T \\ 0 \\ 0 \end{Bmatrix} - \begin{Bmatrix} \hat{N}_{zz}^M \\ \hat{N}_{sz}^M \\ \hat{L}_{zz}^M \\ \hat{L}_{sz}^M \\ 0 \\ 0 \end{Bmatrix} . \tag{A.93}
$$

With the exception of K_{13}, K_{23}, K_{43}, and K_{53} whose expressions are different for open and closed cross-section beams, the remaining stiffness quantities are formally similar for both cases; the constitutive equations, Eqs. (A.93), can be used for both open and closed section beams.

The constitutive equations of thin-walled beam theory used in the literature are also displayed here. In present notations, these are

$$
\begin{Bmatrix} N_{zz} \\ N_{sz} \end{Bmatrix} = \begin{bmatrix} K_{11} & K_{12} \\ K_{21} & K_{22} \end{bmatrix} \begin{Bmatrix} \epsilon_{zz}^{(0)} \\ \gamma_{sz}^{(0)} \end{Bmatrix} . \tag{A.94}
$$

Clearly, Eqs. (A.93) are much more comprehensive than (A.94) that do not include bending and the thickness-wise contributions, and as such, are likely to describe more accurately the behavior of thin-walled beams.

A.8-11 Two Structural Coupling Configurations –
CUS and CAS

We consider two structural configurations which exhibit special structural couplings. Considered first by Rehfield and Atilgan (1989), these structural configurations are referred to as *circumferentially uniform stiffness* (CUS) and *circumferentially asymmetric stiffness* (CAS) configurations. For a thin-walled
beam of rectangular cross-section,(CUS) implies the ply-angle distribution
$\theta(y) = \theta(-y)$ of the top and bottom walls of the box beam (*flanges*) and
of $\theta(x) = \theta(-x)$ of the lateral walls (*webs*). The latter one, (CAS), implies
$\theta(y) = -\theta(-y)$, and $\theta(x) = -\theta(-x)$ (see Fig. A.4).

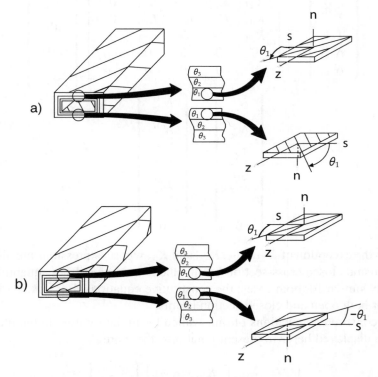

Figure A.4: a) Circumferential uniform stiffness (CUS) configuration. b) Circumferentially
asymmetric stiffness (CAS) configuration

For the CAS configuration (see Eq. A.77), the stiffness quantities $\bar{Q}_{16}$, $\bar{Q}_{26}$,
$\bar{Q}_{36}$, $\bar{Q}_{45}$ of each of the layers in the bottom flanges are the negative value of
the layers counterpart in the top flanges, that is:

$$\bar{Q}_{16}^{(U)} = -\bar{Q}_{16}^{(L)}; \quad \bar{Q}_{26}^{(U)} = -\bar{Q}_{26}^{(L)}; \quad \bar{Q}_{36}^{(U)} = -\bar{Q}_{36}^{(L)}; \quad \bar{Q}_{45}^{(U)} = -\bar{Q}_{45}^{(L)}.$$
$$\text{(A.95a-d)}$$

This implies that in this case the following relationships hold

$$A_{16}^{(U)} = -A_{16}^{(L)}; \ A_{26}^{(U)} = -A_{26}^{(L)}; \ A_{36}^{(U)} = -A_{36}^{(L)}; \ A_{45}^{(U)} = -A_{45}^{(L)};$$
$$D_{16}^{(U)} = -D_{16}^{(L)}; \ D_{26}^{(U)} = -D_{26}^{(L)} \ D_{36}^{(U)} = -D_{36}^{(L)}; \ D_{45}^{(U)} = -D_{45}^{(L)}.$$

$$(A.96a\text{-}h)$$

Similar types of relationships appear also in the opposite webs of the beam.

For CUS, the stiffness quantities $\bar{C}_{16}, \bar{C}_{26}, \bar{C}_{36}, \bar{C}_{45}$ in the opposite members exhibit the same sign, a trend which is valid also for the same components of the matrices $[A]$ and $[D]$. The CUS and CAS configurations are also referred to as the *antisymmetric* and *symmetric* configurations, respectively, see Smith and Chopra (1991).

Similar ply-angle configurations can be implemented in open beams, as well. As an example, in Figs. (13.1-1a) and (13.1-1b), the CAS and CUS are shown in a *I* beam.

A.8-12 Additional Remarks

The derivation of 2-D form of the constitutive equations is but an intermediate step. The goal is to convert these constitutive equations to a 1-D form, where only the single independent variable z (representing the longitudinal axis of the beam) will be involved. This step is carried out in Chapter 3.

A.9 PIEZOELECTRIC CONSTITUTIVE EQUATIONS

A.9-1 Preliminaries

As is well known, piezoelectric materials generate an electrical charge in response to a mechanical deformation or, conversely, provide mechanical strain when an electric field is applied across them. As a result, piezoelectric materials are excellent candidates for the role of sensors and actuators in the technology of smart structures (see e.g. Bailey and Hubbard, 1985; Tzou and Zhong, 1992; Tzou, 1993; Tzou and Anderson, 1992; Hanagud et al., 1992; Sunar and Rao, 1999; Chopra, 2002). In applications, piezoceramic materials are sandwiched between conductive surfaces (electrodes) and polarized in a suitable direction (see Fig. A.5). To model such structures, the constitutive equations of piezoelectric media have to be considered. In the sequel, several basic elements concerning these equations will be presented; for details, see e.g. Maugin (1988) and Eringen and Maugin (1990).

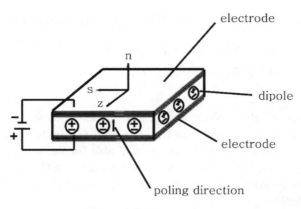

Figure A.5: Actuator layer

A.9-2 Piezoelectric Medium

In contracted indicial notations, the constitutive equations of a 3-D piezoelectric continuum are

$$\sigma_i = C_{ij}^{\mathcal{E}} \, S_j - e_{ri} \, \mathcal{E}_r,$$
$$D_r = e_{rj} \, S_j + \epsilon_{r\ell}^{S} \, \mathcal{E}_\ell, \qquad (r, \ell = 1, 2, 3) \qquad (A.97a,b)$$

In Eqs. (A.97) σ_i and $S_j(i, j = \overline{1, 6})$ denote the stress and strain components, respectively, where

$$S_j = \begin{cases} S_{pr} \text{ for } p = r, \, j = 1, 2, 3 \\ 2S_{pr} \text{ for } p \neq r, \, j = 4, 5, 6. \end{cases} \qquad (A.98)$$

$C_{ij}^{\mathcal{E}}$, e_{ri} and $\epsilon_{r\ell}^{S}$ denote the elastic coefficients (measured under constant electric field), the piezoelectric tensor and the dielectric tensor (measured under constant strain), while $\mathcal{E}_r$ and D_r denote the electric field intensity and electric displacement vector, respectively. In Eqs. (A.97) the summation convention over repeated indices is implied.

While Eq. (A.97a) describes the **converse** piezoelectric effect consisting of the generation of mechanical stress or strain when an electric field is applied, Eq. (A.97b) describes the **direct** piezoelectric effect consisting of generation of an electrical charge under a mechanical force. In a piezoelectric adaptive structure the direct effect is used for distributed sensing, while the converse effect is used for the active distributed control.

Equations (A.97) are valid for the most general case of anisotropy, i.e., for triclinic crystals. In this case, the piezoelectric continuum is characterized by 21 independent elastic constants, 18 independent piezoelectric constants, and 6 independent dielectric constants.

For a thickness-polarized continuum, the constitutive equations are the same as for a piezocrystal of the hexagonal $6mm$ symmetry class. This represents a transversely isotropic piezoelectric, the n-axis being a sixfold axis of symmetry.

In this case (see Eringen and Maugin, 1990), the piezoelectric continuum is characterized by five independent elastic coefficients $C_{11} = C_{22}$, $C_{13} = C_{23}$, C_{33}, $C_{44} = C_{55}$, $C_{66}(\equiv (C_{11} - C_{12})/2)$, three independent piezoelectric coefficients $e_{15} = e_{24}$, $e_{31} = e_{32}$ and e_{33} and two independent dielectric constants, $\epsilon_{11} = \epsilon_{22}$ and ϵ_{33}. We consider only this case.

The electric field vector $\mathcal{E}$ is represented in terms of the transverse normal component $\mathcal{E}_3$ (implying that $\mathcal{E}_1 = \mathcal{E}_2 = 0$).

Note due to the uniform voltage distribution, $\mathcal{E}_3$ is independent of space (but possibly dependent on time).

A.9-3 2-D Piezoelectric Constitutive Equations

We will assume that the master (or host) structure is composed of m layers, while the actuator is composed of ℓ piezoelectric layers. The actuators can be distributed over the entire span of the beam (see Figs. 6.6-1a and 9.2-1), or can be in the form of patches. Along the circumferential s, spanwise z and thickness n directions, they are distributed according to the law (see Fig. A.6):

$$
\begin{aligned}
R_{(k)}(n) &= Y\left(n - n_{(k-1)}\right) - Y\left(n - n_{(k)}\right), \\
R_{(k)}(s) &= Y\left(s - s_{k1}\right) - Y\left(s - s_{k2}\right), \qquad \text{(A.99a-c)} \\
R_{(k)}(z) &= Y\left(z - z_{k1}\right) - Y\left(z - z_{k2}\right),
\end{aligned}
$$

where $Y(\cdot)$ denotes Heaviside's distribution, $R(\cdot)$ is a spatial function describing the distribution of actuator patches, while $(n_{(k-1)}, n_{(k)})$, (s_{k1}, s_{k2}) and (z_{k1}, z_{k2}) denote the top and bottom heights of the piezopatch measured across the beam thickness, and its location along the beam circumference and span, respectively.

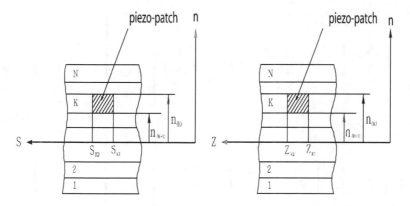

Figure A.6: Piezopatch location.

Proceeding as for Eq. (A.53), we find the 3-D constitutive equations for the transversely isotropic piezoelectric layers expressed in the coordinates (s, z, n):

$$
\left\{ \begin{array}{c} \sigma_{ss} \\ \sigma_{zz} \\ \sigma_{sz} \end{array} \right\}_{(k)} = \left[\begin{array}{ccc} \bar{Q}_{11} & \bar{Q}_{12} & 0 \\ \bar{Q}_{12} & \bar{Q}_{11} & 0 \\ 0 & 0 & \bar{Q}_{66} \end{array} \right]_{(k)} \left\{ \begin{array}{c} S_{ss} \\ S_{zz} \\ S_{sz} \end{array} \right\}_{(k)}
$$
$$
- \left\{ \begin{array}{c} \bar{e}_{31}^{(k)} \, \mathcal{E}_3^{(k)} \, R_{(k)}(n) \, R_{(k)}(s) R_{(k)}(z) \\ \bar{e}_{31}^{(k)} \, \mathcal{E}_3^{(k)} \, R_{(k)}(n) \, R_{(k)}(s) R_{(k)}(z) \\ 0 \end{array} \right\}, \tag{A.100a}
$$

$$
\sigma_{zn}^{(k)} = C_{44}^{(k)} \, S_{zn}^{(k)} . \tag{A.100b}
$$

where

$$
\bar{Q}_{11} = \bar{Q}_{22} = \frac{E}{1 - v^2}; \quad \bar{Q}_{12} = \frac{Ev}{1 - v^2},
$$
$$
\bar{Q}_{66} = G_{12} = \frac{E}{2(1 + v)}; \quad C_{44} = G_{23} \equiv G', \tag{A.101a-d}
$$
$$
\bar{e}_{31} = e_{31} - \frac{Ev'}{E'(1 - v)} e_{33}. \tag{A.101e}
$$

In Eq. (A.100a), the last terms identify the actuation stresses induced by the applied electric field.

The 2-D stress resultants and stress couples of the global structure can be obtained via the integration of the 3-D stress components through the total laminate thickness, i.e. through the thickness of the host structure and of piezoelectric layers. Assuming that the hoop stress is small, and replacing the notation for the strain components used in the piezoelectricity theory by the standard one in frequent usage in beam theory (see Eqs. (A.91)), we express the 2-D stress-resultants and stress couples as:

$$
\left\{ \begin{array}{c} N_{zz} \\ N_{sz} \end{array} \right\} = \left[\begin{array}{cccc} K_{11} & K_{12} & K_{13} & K_{14} \\ K_{21} & K_{22} & K_{23} & K_{24} \end{array} \right] \left\{ \begin{array}{c} \epsilon_{zz}^{(0)} \\ \gamma_{sz}^{(0)} \\ W_M \\ \epsilon_{zz}^{(1)} \end{array} \right\} - \left\{ \begin{array}{c} \tilde{N}_{zz} \\ 0 \end{array} \right\},
$$

$$\tag{A.102a,b}$$

$$
\left\{ \begin{array}{c} L_{zz} \\ L_{sz} \end{array} \right\} = \left[\begin{array}{cccc} K_{41} & K_{42} & K_{43} & K_{44} \\ K_{51} & K_{52} & K_{53} & K_{54} \end{array} \right] \left\{ \begin{array}{c} \epsilon_{zz}^{(0)} \\ \gamma_{sz}^{(0)} \\ W_M \\ \epsilon_{zz}^{(1)} \end{array} \right\} - \left\{ \begin{array}{c} \tilde{L}_{zz} \\ 0 \end{array} \right\},
$$

and

$$N_{zn} = A_{44}\gamma_{zn}, \quad N_{sn} = A_{45}\gamma_{zn}. \qquad \text{(A.102c,d)}$$

In these equations K_{ij} denote the modified local stiffness coefficients of the adaptive structure, while $\tilde{N}_{zz}$ and $\tilde{L}_{zz}$ denote the piezoelectrically induced stress resultant and stress couple, respectively. Their expressions are:

$$\tilde{N}_{zz}(s, z) = \mathcal{E}_3 \left(1 - \frac{A_{12}}{A_{11}}\right)(n_2 - n_1)\bar{e}_{31} R(s, z),$$

$$\tilde{L}_{zz}(s, z) = \mathcal{E}_3 \bar{e}_{31}(s, z)(n_2 - n_1)\left[\frac{1}{2}(n_1 + n_2) - \frac{B_{12}}{A_{11}}\right] R(s, z). \quad \text{(A.103a,b)}$$

Note that in Eqs. (A.102) and (A.103), the stiffness quantities A_{ij}, B_{ij} and D_{ij} as given by Eqs.(A.75) and K_{ij} by Eqs. (A.85) and (A.92),have to be based on the **total** number of constituent layers $N = m + \ell$, for the host (m) and the piezoelectric layers (ℓ). One additional remark is in order here: while the material of the host structure can exhibit any type of anisotropy, the piezoac-tuators/sensors as considered here feature transversely isotropic properties, the surface of isotropy being parallel at each point to the beam mid-surface.

In Eqs. (A.103) the following notation was used:

$$R_{(k)}(s, z) \equiv R_{(k)}(s) R_{(k)}(z). \qquad \text{(A.104)}$$

For $z_1 = 0$ and $z_2 = L$, the piezoactuator is spread over the entire beam span.

REFERENCES

Bailey, T. and Hubbard, J. E., Jr. (1985) "Distributed Piezoelectric-Polymer Active Vibration Control of a Cantilever Beam," *Journal of Guidance, Control and Dynamics*, Vol. 9, No. 5, pp. 605–611.

Bogdanovich, A. E. and Pastore, C. M. (1996) *Mechanics of Textile and Laminated Composites with Applications to Structural Analysis*, Chapman & Hall, London.

Chopra, I. (2002) "Review of State of Art of Smart Structures and Integrated Systems," *AIAA Journal*, Vol. 40, No. 11, pp. 2145–2187.

Daniel, I. M. and Ishai, O. (1994) *Engineering Mechanics of Composite Materials*, Oxford University Press, New York.

Eringen, A. C. and Maugin, G. A. (1990) *Electrodynamics of Continua I., Foundations and Solid Media*, Spring-Verlag, New York.

Hanagud, S., Obal, M. W. and Calise, A. J. (1992) "Optimal Vibration Control by the Use of Piezoelectric Sensors and Actuators," *Journal of Guidance, Control and Dynamics*, Vol. 15, No. 5, pp. 1199–1206.

Librescu, L. (1975) *Elastostatics and Kinetics of Anisotropic and Heterogeneous Shell-Type Structures*, Noordhoff International Publishing, Leyden, Netherlands.

Librescu, L. (1990) "Theory of Composite Structures-Lecture Notes," Virginia Polytechnic Institute and State University, Blacksburg, VA.

Maugin, G. A. (1988) *Continuum Mechanics and Electromagnetic Solids*, North-Holland, Amsterdam.

Reddy, J. N. (2004) *Mechanics of Laminated Composite Plates and Shells: Theory and Analysis*, 2nd Edition, CRC Press, Boca Raton.

Rehfield, L. W. and Atilgan, A. R. (1989) "Toward Understanding the Tailoring Mechanisms for Thin-Walled Composite Tubular Beams," in *Proceedings of the First USSR–U.S. Symposium on Mechanics of Composite Materials*, 23–26 May, Riga, Latvia SSR, S. W. Tsai, J. M., Whitney, T. W. Chou and R. M. Jones (Eds.), ASME Publ. House, pp. 187–196.

Smith, E. C. and Chopra, I. (1991) "Formulation and Evaluation of an Analytical Model for Composite Box-Beams," *Journal of the American Helicopter Society*, Vol. 36, No. 3, pp. 23–35.

Song, O. (1990) "Modeling and Response Analysis of Thin-Walled Beam Structures Constructed of Advanced Composite Material," Ph.D. Thesis Virginia Polytechnic Institute and State University, Blacksburg, VA.

Sunar, M. and Rao, S. S. (1999) "Recent Advances in Sensing and Control of Flexible Structures via Piezoelectric Material Technology," *Applied Mechanics Reviews*, Vol. 52, No. 1, pp. 1–16.

Tsai, W. and Hahn, M. T. (1980) *Introduction to Composite Materials*, Technomic Publishing Co., Westport, Connecticut.

Tzou, H. S. and Zhong, J. P. (1992) "Adaptive Piezoelectric Structures: Theory and Experiment," in *Active Materials and Adaptive Structures, Materials and Structures Series*, G. J. Knowles (Ed.), Institute of Physics Publ., pp. 219–224.

Tzou, H. S. and Anderson, G. L. (1992) *Intelligent Structural Systems*, Kluwer Academic Publisher, Norwell, MA.

Tzou, H. S. (1993) *Piezoelectric Shells, Distributed Sensing and Control of Continua*, Kluwer Academic Publishers, Dordrecht.

Vinson, J. R. and Sierakowski, R. L. (2002) *The Behavior of Structures Composed of Composite Materials*, Kluwer Academic Publishers, Dordrecht.

Wempner, G. (1981) *Mechanics of Solids with Applications to Thin Bodies*, Sijthoff & Noordhoff, Alphen aan den Rijn, The Netherlands; Rockville, Maryland, USA.

SUBJECT INDEX

A

Acceleration feedback control, 346, 351–353
Active distributed control, 590
Actuating voltage, 150
Adaptive
  beams, 219
  capability, 453
  materials technology, 3
  rotating thin-walled beams, 279
  structure technology, 144, 225
Admissible functions, 130, 142
Aerodynamic indicial functions, 513
Aeroelasticity
  dynamic, 502, 513
  static, 502, 503
Aeroelastic control, 518
Aeroelastic instability 3
  dynamic, 513, 514
  static, 499, 503, 504, 508
Aeroelastic response
  dynamic, 515
  static, 503, 509, 518
Aeroelastic tailoring, 2
Aircraft wing, 2, 122, 500
Angle of incidence, 475
Angle of sweep, 503
Angular velocity, 438
Anisotropic, 2, 557, 558
Anisotropy, 233, 243, 395, 559, 590
Antisymmetric laminates, 582

Augmented performance index, 196
Axisymmetric beam, 399

B

Balanced laminates, 582
Bang-bang control, 207
Base vectors
  contravariant, 272
  covariant, 271, 373
Beam taper ratio, 204
Bending-twist coupling, 131, 136, 504, 508, 510, 515, 519, 530
Bending-twist divergence, 506
Bernoulli-Euler beam model, 32, 169, 207, 376, 382
Betti's reciprocal theorem, 112, 116
Bi-convex cross-section, 131, 140, 183, 202, 507
Bifurcation of natural frequencies, 412, 428, 448, 457, 466
Bilinear form, 111
Bimoment, 58
Blast load, 165, 170, 203
Boley-Barber-Thornton (BBT) approach, 484, 491, 492
Boom, 475
Boundary moment control, 279, 453, 489, 518
Boundary-value problem, 113
Box beam, 154
Bredt-Batho equations, 23

Mechanics

SOLID MECHANICS AND ITS APPLICATIONS
Series Editor: G.M.L. Gladwell

Aims and Scope of the Series

The fundamental questions arising in mechanics are: *Why?*, *How?*, and *How much?* The aim of this series is to provide lucid accounts written by authoritative researchers giving vision and insight in answering these questions on the subject of mechanics as it relates to solids. The scope of the series covers the entire spectrum of solid mechanics. Thus it includes the foundation of mechanics; variational formulations; computational mechanics; statics, kinematics and dynamics of rigid and elastic bodies; vibrations of solids and structures; dynamical systems and chaos; the theories of elasticity, plasticity and viscoelasticity; composite materials; rods, beams, shells and membranes; structural control and stability; soils, rocks and geomechanics; fracture; tribology; experimental mechanics; biomechanics and machine design.

Mechanics

SOLID MECHANICS AND ITS APPLICATIONS
Series Editor: G.M.L. Gladwell

Mechanics

SOLID MECHANICS AND ITS APPLICATIONS
Series Editor: G.M.L. Gladwell

Mechanics

SOLID MECHANICS AND ITS APPLICATIONS
Series Editor: G.M.L. Gladwell

Mechanics

SOLID MECHANICS AND ITS APPLICATIONS

Series Editor: G.M.L. Gladwell

Mechanics

SOLID MECHANICS AND ITS APPLICATIONS
Series Editor: G.M.L. Gladwell

Mechanics

SOLID MECHANICS AND ITS APPLICATIONS
Series Editor: G.M.L. Gladwell

129. G.E.A. Meier and K.R. Sreenivasan (eds.): *IUTAM Symposium on One Hundred Years of Boundary Layer Research.* Proceedings of the IUTAM Symposium held at DLR-Göttingen, Germany, August 12–14, 2004. 2005   ISBN 1-4020-4149-7
130. H. Ulbrich and W. Günthner (eds.): *IUTAM Symposium on Vibration Control of Nonlinear Mechanisms and Structures.* 2005   ISBN 1-4020-4160-8
131. L. Librescu and O. Song: *Thin-Walled Composite Beams.* Theory and Application. 2006   ISBN 1-4020-3457-1
132. G. Ben-Dor, A. Dubinsky and T. Elperin: *Applied High-Speed Plate Penetration Dynamics.* 2006   ISBN 1-4020-3452-0